Roller Compacted Concrete Rapid Damming Technology

(Volume Ⅰ)

中文作者　田育功
译　　者　吴正桥　余　洋

Chinese Writer　Tian Yugong
Translators　Wu Zhengqiao　Yu Yang

黄河水利出版社
·郑州·

Abstract

This book describes the theory, method, experience and engineering practice of rapid damming technology with roller compacted concrete(RCC). Citing a large number of rich first-hand RCC test and research results, tender/bidding documentation and engineering examples of rapid damming, with full and accurate contents, this is a practical technology book with very high theoretical level and rich engineering practice.

The book includes:13 topics of development level of RCC damming, design and rapid construction of RCC dam, examples of raw materials and projects, mixing proportion design with examples, research and examples of RCC performance, admixture research and application, the role of rock powder in RCC, construction technology of grout enriched vibrated concrete(GEVR), key technology for rapid construction of RCC, temperature control and cracking prevention, quality control of RCC, core sample pressurizing water and in-situ shear resistance and RCC cofferdam work.

This book is a valuable reference for vast engineering and technical personnel engaged in the structure, design, research, construction and supervision in water conservancy and hydropower engineering sector, as well as teachers and students of related majors in colleges and universities.

图书在版编目(CIP)数据

碾压混凝土快速筑坝技术 ＝ Roller Compacted Concrete Rapid Damming Technology：全三册／吴正桥等译：田育功著. —郑州：黄河水利出版社，2021.8

ISBN 978－7－5509－3069－8

Ⅰ.①碾… Ⅱ.①吴… ②田… Ⅲ.①碾压土坝－混凝土坝 Ⅳ.①TV642.2

中国版本图书馆 CIP 数据核字(2021)第 165756 号

出 版 社:黄河水利出版社　　　　　　　　　　　网址:www.yrcp.com
　　地址:河南省郑州市顺河路黄委会综合楼14层　　邮政编码:450003
发行单位:黄河水利出版社
　　发行部电话:0371－66026940、66020550、66028024、66022620(传真)
　　E-mail:hhslcbs@126.com
承印单位:广东虎彩云印刷有限公司
开本:787 mm×1 092 mm　1/16
印张:42
字数:1 500 千字
版次:2021 年 8 月第 1 版　　　　　　　　　　印次:2021 年 8 月第 1 次印刷
定价(全三册):198.00 元

Translator's Preface

Roller compacted concrete (RCC) gravity dam has shown rapid construction speed, low cost and good engineering quality and other advantages. In the recent year, the rapid development of RCC dams in China, especially the completion of high DAMS over 100m, damming RCC materials and construction technology of China is indeed a kind of technical concept with unique characteristics of China. Carefully analyzing various technologies, their contents are very rich, and the connotation is also deep, and the application is very convenient and handy, to be easy to master. What more valuable is that, whenever a large RCC dam is completed, there are almost always innovations and improvements, so that the development of the technology is constantly rooted in the source of project construction. For technical progress in various aspects, many engineering technicians have written a large number of special treatises or monographs, well welcomed by industry peers. In order to better promote the design and construction concept of Chinese RCC dam in the world, Wu Zhengqiao and others translated the book with the permission of the author.

The translation results are divided into three volumes, division of labor is as follows:

Chief editor: Wu Zhengqiao.

Other editor:

(1) Volume Ⅰ: Wu Zhengqiao, Yu Yang.

(2) Volume Ⅱ: Yang Haiyan, Li Jianqiang, Yao Desheng, Zhao Chuntao, Zhou Hongmin, Bao Dixiao, Li min, Sun Shunan, Zhao Jianli.

(3) Volume Ⅲ: Zhang Baorui, He Qingbin, Peng Xiaochuan, Li Miao, Xu Lizhou, Pei Xianghui, Xu Ting, Liu Zhuo, Zhang Kai, Zhao Lin.

In this process of translating this book, Tian Yugong gave great support and help, thank you!

In addition, in the process of publishing this book, I would like to thank Wang Xiaohong and Yu Ronghai from the editorial department of China Water Resources Bei Fang investigation, Design & Research Co. LTD and the Yellow River Water Publishing House for their strong support.

Wu Zhengqiao

August 2021

Preface

In recent years, China has successively completed 200-metre roller compacted concrete(RCC) gravity dams and several 100-metre level thin double-curvature arch RCC dams. In construction process, all these dams have shown rapid construction speed, low cost and good engineering quality and other advantages. These high dam construction achievements signify that, China's RCC damming technology has made a breakthrough in overall development in a great number of engineering practices, through earnestly summing up experience, actively carrying out exploration and test study, continuous innovation and development. The technology level is getting higher and higher, getting more mature. These new technical achievements and engineering achievements have also brought many beneficial effects to international dam industry.

Looking back to early days, due to hazy knowledge, in RCC dam projects in China, "drying hard" concrete commonly applied in foreign countries was generally applied. As a result, the lack of bleeding at pouring face often caused poor inter-layer bonding, forming a "multi-layer steamed bread" phenomenon. This caused doubts and reservations about the quality of RCC mass of dam. In view of this condition, with a spirit of tenacity, many engineering technical personnel and researchers are continuing to explore, adjust and improve materials of concrete and construction technical actions, finally, a technology mode of "sub-plastic state" RCC with good bleeding performance, being easy to compact and being able to prevent vibration-press subsidence has been formed, widely used nowadays. It has better solved the problem of concrete, hot joint bonding, thoroughly improved impervious performance and shear resistance of course face, solved the safety and impervious problem for the construction of 200- meter high RCC gravity dam. Compared with some foreign RCC damming technologies, damming RCC materials and construction technology of China is indeed a kind of technical concept with unique characteristics of China. Carefully analyzing various technologies, their contents are very rich, and the connotation is also deep, and the application is very convenient and handy, to be easy to master. What more valuable is that, whenever a large RCC dam is completed, there are almost always innovations and improvements, so that the development of the technology is constantly rooted in the source of project construction. For technical progress in various aspects, many engineering technicians have written a large number of special treatises or monographs, well welcomed by industry peers.

The greatest advantage of RCC damming technology is its rapid speed, which has strong vitality. Generally, for concrete dam with a height of over 100 m, it can be completed for 2 – 3 years adopting RCC damming technology. Compared with normal concrete damming technology, the construction duration can be shortened by one-third and more. Engineering practice has proved that, RCC damming technology is credible and reliable in quality, of which the advantage is beyond any doubt.

The design is the key to technical innovation of rapid RCC damming. Technical innovation not only requires a solid and scientific technical foundation, but also allows failure and discussion, and also requires the courage to bear failure and responsibility, to achieve the objective of technical innovation through practice.

"Interlayer combination, temperature control and cracking prevention" is the core technology of rapid RCC damming. In recent years, with the increase in the height and volume of RCC dams, it has become common practice to continuously pour RCC in high-temperature seasons in order to crash the schedule or shorten the duration. As a result, temperature control measures for RCC have become more and more stringent,

which it has no different from normal concrete, to make temperature control measures of RCC show a trend more and more complicated. In this respect, RCC dams in Thailand, Cambodia, Laos, Myanmar and other countries located in subtropical region shall be learnt from. The RCC dams in these countries do not adopt cooling water pipe temperature control measures, and the design indicators adopt a single strength index and with a single aggregate gradation, the design concepts and temperature control measures of these dams are worth learning and thinking about.

The construction of RCC dam has the characteristics of one-time, and it is particularly important that construction quality is being under control all the time. Especially, for on-site RCC construction, dynamic control of VC value, timely rolling, mist spray moisturizing and covering for curing and other construction links concern directly the quality of interlayer bonding and the performance of temperature control and cracking prevention, which must be raised to the height of quality problem to understand their importance.

The Author of this book has collected a large number of RCC dam engineering practical data and multifaceted scientific test and research results in China, with full and accurate contents. While sorting out and analyzing these achievements, combined with personal exploration and study of RCC dam participated in the construction of RCC dams personally, the Author has carried out in-depth display and interpretation, and has more realization and understanding of "plastic RCC", to expand the discussion in the book. RCC damming is still a developing technology. For all concrete materials, construction process, dam engineering design, construction management, temperature control and cracking prevention, there is still room for improvement and development. It is beneficial for technical development to constantly summarize, analyze, study and exchange. After reading this book, I have benefited a lot and expanded my eyesight.

Former Director of the RCC Damming Professional Board of China Society for Hydropower Engineering

Wang Shengpei

Beijing, March 2010

Foreword

Roller compacted concrete(RCC) damming technology is a major technical innovation in the history of world damming. RCC damming technology is favored by global dam industry because of its rapid construction speed, short duration, low investment, safe and reliable quality, high degree of mechanization, simple construction, strong adaptability and green environmental protection, etc. In particular, the damming period of can be significantly shortened by one-third and more compared to similar normal concrete dams, to show strong advantage of vigorous development, and inject a fresh air and vitality into dam construction, which is in line with the development direction of good, rapid and economical.

The "rapid" is the greatest advantage of RCC damming technology, which is its strong vitality. Although the adoption of RCC damming technology has only been for 20 years more, the speed of damming and the large number of dams are unable to be matched by other damming technologies. As of the end of 2008, there were 180 RCC dams(including cofferdams) completed or under construction in China, ranking first globally. The greatest charm of RCC damming technology is its compatibility. RCC has the characteristics of concrete, conforming to the rules of water-cement ratio, and its cross-section design is the same as that of normal concrete dam, regardless of RCC gravity dam or arch dam; meanwhile, its construction has the characteristics of rapid construction of earth-rock dam. A large number of engineering practice has proved that, RCC dam has become one of the most competitive dam types.

Using RCC damming technology, the dam is bold and generous and real. The internal quality of the dam is good, and the appearance quality is beautiful, so its quality is not inferior to normal concrete dams. Although the RCC damming is bold, the construction is not rough at all. RCC damming technology is very delicate, to be a well-organized and standardized construction. The construction site is sparsely staffed, and the concrete roller compacting is carried out in an orderly manner from placing, spreading to rolling. On the contrary, for the pouring of normal concrete dam, due to that the normal concrete is restricted by pouring strength and temperature control and other factors, dam body is divided into blocks by transverse and longitudinal joints. For concrete construction, columnar pouring is adopted, which results in a large amount of formwork workload, and the amount of merging joint grouting is great and of long period. The dam construction often presents a large number of people on placement surface and appears busy, lacking of well-organized, unified and bold of RCC construction.

RCC damming is a systematic project integrating the research, design, construction, quality control and other aspects. What mostly reflects the characteristics of rapid RCC damming is whole-block placement and thin-lift concreting. Because of the change in damming technology, the design concept of complex layout and dam engineering structure has been changed. The complex layout design shall not only consider starting from meeting the requirements of rapid construction of RCC and simplifying the layout of the dam, but also require carrying out in-depth study on dam structure, temperature stress and overall performance. The adoption of RCC damming is a promotion to the design. The design concept must advance, the dam layout shall be different from normal concrete dam, to start from thin-lift concreting and whole-block placement, simple and rapid damming technology, combined with the characteristics of RCC itself, the simpler the dam's structural layout, the more obvious its advantages.

"Interlayer bonding, temperature control and cracking prevention" is the core technology of RCC rapid damming. Through a large number of tests and studies and engineering examples, for problems of great or low fluctuations in artificial sand and gravel powder content of RCC, the successful application admixing fly ash or admixing rock powder instead of sand, precise control of rock powder mixing, dynamic control of VC value and whole-block placement inclined layer rolling and other technologies have effectively improved the rollability, liquefaction and bleeding and interlayer bonding quality of RCC, promoting the development of RCC rapid damming technology. The "grout-mortar ratio" has become one of important parameters in the design of RCC mixing proportion, with the same important effect as three major parameters of water-cement ratio, sand ratio and water consumption. Rock powder has got more and more attention in RCC, become an indispensable part of RCC materials. Hydraulic RCC has been developed into non-slump semi-plastic concrete. The worrying problem of interlayer bonding quality has been well solved, significantly improved the ultimate tensile value and frost resistance.

Looking back at test history of drilling and coring of RCC and on-site water pressurizing tests, core samples at early stage with the longest length of 60 cm more have been developed into ones of ultra-long 16 m more, and core samples longer than 10 m are not uncommon. Drilling and coring have fully proved that the maturity of rapid RCC damming technology gets gradually maturing, reflecting the change in the properties of RCC from one side, which is indeed very beneficial to improve the quality of interlayer bonding and impervious performance of dam body.

RCC also has its dual nature. Because RCC adopts low cement content and high admixtures, it is required to adopt advanced design concepts to constantly innovate and deepen the study of its frost resistance, ultimate tensile value and carbonization and other properties(compared to normal concrete).

Over 20 years since 1988 when starting studying dam dam technology of left auxiliary dam RCC test at Longyangxia Dam, the Author has personally witnessed the development and growth of China's RCC damming technology, who is one of major participants, implementers, researchers and promoters of RCC dam technology of China, mastered a large number of first-hand test and research results, with rich engineering practical experience, having novel viewpoints in RCC damming technology. This book is achievement data of the test and research, construction technology, technical consultation, water storage acceptance and construction management in Longshou, Linhekou, Baise, Guangzhao, Jin'anqiao, Kalasuke and other RCC dams participated in and presided over by the Author in recent years, compiled on the basis of over 40 papers (thesis) published at home and abroad, in which many of valuable first-hand data are disclosed for the first time, being one of windows to understand and master RCC damming technology of China.

RCC damming technology of China ranks among global leading level, which is inseparable from continuous efforts and innovations of vast research, design, technical consultation, supervision, and construction personnel of a large number of water conservancy and hydropower projects in China, especially massive, scientific and hard works and strong supports of the RCC Damming Profession Board. Here, I would like to express my highest respect and heartfelt gratitude to elder Wang Shengpei, who has contributed his whole life and made outstanding contributions to China's RCC damming technology, and colleagues of the RCC Damming Profession Board! I would like to express my heartfelt gratitude to leaders and colleagues who have supported my technical work for long term!

In view of the characteristics of hydraulic RCC, this book makes a quite detailed analysis and research from the mechanism and application of RCC rapid damming. Through the discussion of rapid RCC damming technology and the analysis on typical engineering examples, the Author summarize in time, and think more calmly, to discuss and research from both the pros and cons, always insisting on putting the quality and safety of the dam in the important position of the first priority, so that the RCC damming technology is more viable,

safer and more perfect.

    This book is compiled in the intention of constructing best and first-class RCC dam in the world. Because there are many research topics on RCC damming technology, restricted by the Author's capability and limited time, it is inevitable to make mistakes and mistakes for sorting out, analyzing and studying and giving examples for so many topics, therefore, readers are expected to criticize and correct!

The Author

July 2010

Contents

Translator's Preface

Preface

Foreword

Chapter 1 General (1)
  1.1 Development Level of RCC Damming Technology in China (1)
  1.2 Development History of RCC Dam (10)
  1.3 Popularization and Application of RCC Damming Technology (12)
  1.4 Discussion on Key Rapid Damming Technologies of RCC (16)
  1.5 Innovation and Reflection on RCC Rapid Damming Technology (23)

Chapter 2 Design and Rapid Construction of RCC Dams (27)
  2.1 Overview (27)
  2.2 Design and Rapid Construction of RCC Dam (28)
  2.3 Design Index of RCC and Material Zoning of Dam (40)
  2.4 Introduction to RCC Dam Design (42)
  2.5 Conclusion (64)

Chapter 3 Raw Materials and Project Examples (66)
  3.1 General (66)
  3.2 Cement Properties and Project Cases (68)
  3.3 Performance and Quality Test of Fly Ash (78)
  3.4 Performance of Hydraulic Concrete Admixtures (88)
  3.5 Aggregate Properties and Project Cases (95)
  3.6 Conclusion (106)

Chapter 4 RCC Mix Proportion Design and Example (108)
  4.1 Overview (108)
  4.2 RCC Mix Proportion Parameter Selection (109)
  4.3 Design Basis and Content of Mix Proportion (133)
  4.4 RCC Mix Proportion Design Method (134)

Chapter 5 Research and Application of RCC Performance (141)
  5.1 Overview (141)
  5.2 Properties and Influencing Factors of RCC (142)
  5.3 Study on Relationship of Admixture, VC Value, Temperature with Setting Time (169)
  5.4 Autogenous Volume Deformation, Adiabatic Temperature Rise and Creep Tests (174)
  5.5 Test Study on Properties of RCC with Manufactured Sand and River Sand (186)

Chapter 6 Research and Application of RCC Admixture (193)
  6.1 Overview (193)
  6.2 Micro Analysis and Study of RCC Admixture (195)
  6.3 The Research and Application of SL Admixture in Gelantan Project. (232)

Chapter 7　Research and Utilization of Rock Powder in RCC (253)
7.1　Overview (253)
7.2　Limestone Rock Powder (255)
7.3　Influence of Rock Powder on Performance of RCC (259)
7.4　Project Examples of RCC with Rock Powder Replacing Sand (262)
7.5　Study on Utilization of Baise Diabase Manufactured Sand and Rock Powder in RCC (271)

Chapter 8　Construction Technology of GEVR (315)
8.1　Overview (315)
8.2　Mix Proportion Test of GEVR (318)
8.3　Construction Technology of GEVR (320)
8.4　Test Study on GEVR Mixed with Fiber in Impervious Area (323)
8.5　Application of GEVR in Baise RCC Main Dam (329)

Chapter 9　Key Technology for Rapid Construction of RCC (334)
9.1　Overview (334)
9.2　Key Technology for Rapid Construction of RCC (336)
9.3　Jin'anqiao Dam Roller Compacted Concrete Fast Construction Key Technique (374)
9.4　RCC Construction of Longshou Arch Dam in Cold and Dry Area (389)

Chapter 10　Temperature Control and Cracking Prevention of Roller Compacted Concrete (399)
10.1　Overview (399)
10.2　Basic Information and Standards of Temperature Control (401)
10.3　RCC Temperature Control Measures (413)
10.4　Technical Innovation and Discussion on Temperature Control and Anti-cracking (425)
10.5　The Temperature Control in RCC Gravity Dam in One Hydro-junction Project (430)

Chapter 11　RCC Quality Control and Project Cases (456)
11.1　Overview (456)
11.2　Quality Control and Evaluation Regulations (458)
11.3　Other Quality Control Measures and Discussions of RCC (469)
11.4　Application of Nuclear Densimeter in RCC (475)

Chapter 12　Core Drilling, Pump-in and In-situ Shear Tests (482)
12.1　Overview (482)
12.2　Core Drilling of Dam RCC (484)
12.3　Field Pump-in Test of RCC (493)
12.4　Performance Test of RCC Core Samples (500)
12.5　In-situ Shear Test of RCC Dams (504)
12.6　Core Drilling and Pump-in Test of Dam RCC (508)
12.7　Performance Test of RCC Core Sample of Jin'anqiao Dam (515)
12.8　On-site In-situ Shear Test of RCC Dams (519)

Chapter 13　RCC Cofferdam Construction and CSG Damming Technology (531)
13.1　RCC Cofferdam (531)
13.2　Cemented Sand and Gravel (CSG) Damming Technology (532)
13.3　Design and Rapid Construction of Longtan RCC Cofferdam (535)
13.4　CSG Mix Proportion Design and Application of Upstream Cofferdam of Gongguoqiao Dam (544)

Appendix A　National Method Hydraulic Concrete Mix Proportion Test Method ……………… (561)
　A.1　Foreword ……………………………………………………………………………… (561)
　A.2　Characteristics of Construction Method ……………………………………………… (562)
　A.3　Applicable Scope ……………………………………………………………………… (562)
　A.4　Technological Principle ……………………………………………………………… (562)
　A.5　Construction Process Flow and Key Points of Work ……………………………… (562)
　A.6　Material and Equipment ……………………………………………………………… (569)
　A.7　Quality Control ………………………………………………………………………… (572)
　A.8　Safety Measures ……………………………………………………………………… (572)
　A.9　Environmental Protection Measures ………………………………………………… (573)
　A.10　Benefits Analysis …………………………………………………………………… (573)

Appendix B　National Construction Method GEVR Construction Method in RCC Dam Construction ……………………………………………………………………… (575)
　B.1　Foreword ……………………………………………………………………………… (575)
　B.2　Characteristics of Construction Method ……………………………………………… (575)
　B.3　Applicable Scope ……………………………………………………………………… (576)
　B.4　Technological Principle ……………………………………………………………… (576)
　B.5　Construction Technology and Operation Points …………………………………… (576)
　B.6　Mechanical Equipment Configuration ……………………………………………… (578)
　B.7　Quality Control ………………………………………………………………………… (578)
　B.8　Safety Measures ……………………………………………………………………… (579)
　B.9　Environmental Protection Measures ………………………………………………… (579)
　B.10　Technical and Economic Analysis ………………………………………………… (579)
　B.11　Project Examples …………………………………………………………………… (579)

Appendix C　Example of Construction Methods: Construction Method of RCC for Dam of a Certain Project ……………………………………………………………… (581)
　C.1　General Principles …………………………………………………………………… (581)
　C.2　Normative References ……………………………………………………………… (581)
　C.3　Terminology …………………………………………………………………………… (582)
　C.4　Concrete Compaction Process Flow Chart ………………………………………… (585)
　C.5　Control and Management of Raw Materials ……………………………………… (586)
　C.6　Selection of Roller Compacted Concrete Mix Ratio and Issuing of Charger Sheet …………… (588)
　C.7　Inspection and Acceptance before Placement Construction of Roller Compacted Concrete … (588)
　C.8　Concrete Mixing and Management ………………………………………………… (591)
　C.9　Concrete Transport …………………………………………………………………… (593)
　C.10　Construction Management inside Block …………………………………………… (594)
　C.11　Construction under Special Climatic Conditions ………………………………… (604)
　C.12　Quality Control Management ……………………………………………………… (605)
　C.13　Management of Placement Surface after Rolling and Placement Finishing ……………… (609)

Main References ……………………………………………………………………………… (610)

Chapter 1

General

1.1 Development Level of RCC Damming Technology in China

1.1.1 Technical Advantages of RCC Damming

RCC(Roller Compacted Concrete) damming technology is a major technical innovation in the history of world damming. RCC damming technology is favored by global dam industry because of its rapid construction speed, short duration, low investment, safe and reliable quality, high degree of mechanization, simple construction, strong adaptability and green environmental protection, etc.

The "rapid" is the greatest advantage of RCC damming technology, which is its strong vitality. The greatest charm of RCC damming technology is its compatibility. RCC has the characteristics of concrete, conforming to the rules of water-cement ratio, and its cross-section design is the same as that of normal concrete dam, regardless of RCC gravity dam or arch dam; meanwhile, its construction has the characteristics of rapid construction of earth-rock dams, therefore, advantages of RCC damming technology are unmatched by other damming technologies. A lot of engineering practice has proved that, RCC dam has become one of the most competitive dam types, being in line with the trend of good and fast development.

Beginning in the 1960s, various countries began to test and research RCC, and since that Japan completed the world's first RCC dam, Shimajgawa Gravity Dam (dam height 89 m) in 1980, RCC dams have developed extremely rapidly. As of the end of 2018, there had been over 824 RCC dams with a height of 15 m all over the world.

The research and application of RCC damming technology in China has developed from the exploratory period in early 1980s, to transition period in early 1990s, to mature period in late 1990s, till current innovation and leading period. Over 30 years, RCC damming technology has been introduced, digested, absorbed, enhanced and continuously independently innovated and developed. The application of RCC damming technology in China is far ahead of the world in terms of the number of dams, dam type and the level of complete set of construction technologies, having made remarkable progress and achievements, and accumulated valuable experience, formed a complete set of complete theory and construction technology system of rapid RCC damming with characteristics of China and independent intellectual property rights.

China's RCC damming technology has low cement content, high admixture (flyash) content, medium cementious material content, high rock powder content, mixing admixture, low VC value, thin-layer spreading and entire cross-section rolling continuous rising construction and other characteristics, and the technology of constructing RCC dams in any area with high temperature, extreme cold, wet and rainy, dry and low rainfall and other areas have been fully mastered. Especially in the past 10 years, RCC damming technology has become more and more mature in China, ranking first in the world. It is recognized globally that, China is the leading country in RCC damming technology. At present, the dam of Longtan Hydropower Station Project with a height up to 192 m and the dam of Guangzhao Hydropower Station with a height up to 200.5 m have been completed, marking that China's RCC damming technology has stepped on 200-metre level. This fully shows strong vitality of rapid RCC damming technology. A large number of engineering practices have proved that, the quality of RCC damming technology is credible and reliable.

Since the completion of the first pit-mouth RCC gravity dam in 1986, during the period from the "Seventh Five-Year Plan" to the "Tenth Five-Year Plan", China has actively adopted "Four-New" technologies of new materials, new technologies, new processes and new equipment, and developed comprehensive and systematic research and tests on material properties, dam structure design, impervious system, temperature control, construction technology and related construction technology supporting equipment of RCC dams and so on, which has greatly promoted rapid development of RCC damming technology. As of the end of 2009, according to incomplete statistics, there are as many as 300-metre RCC dams (including RCC cofferdam, cementitious sand gravel(CSG) cofferdam, cementitious sand gravel(CSG) permanent dam, etc) completed. See Table 1.1 for above 100-metre height RCC dams completed and under construction in China. See Table 1.2 for RCC cofferdams, CSG cofferdams and CSG permanent dams completed in China.

1.1.2 Development Level of RCC Damming Technology

The superiority of RCC dam has been recognized by the dam industry all over the world. In any country, as long as a RCC dam is completed, its advantages of rapid, economic and safe superiority will immediately be highlighted, being quickly promoted. Both developed and developing countries have constructed RCC dams. In recent years, the number of Asian countries adopting RCC damming technology has been increasing dramatically, such as Vietnam, Cambodia, Thailand, Malaysia, Myanmar, Mongolia and other countries. At present, it has become an indisputable fact that, RCC dam has basically replaced traditional concrete gravity dams.

Relevant data show that, RCC dams higher than 200 m to be constructed overseas mainly include: tansang Dam in southern Myanmar is about to commence. The dam is 235 m high, with a total volume of 8.65 million m^3 and an installed capacity of 7 100 MW. It will be the largest RCC dam constructed so far. In addition, Nam Ngum Three-cascade Power Station in Laos is under design, with a planned height of 220 m; nan Theun 1 in Laos under construction has a dam height of 180 m. Basha dam in Pakistan is also under design, and the height may be close to 300 m. The height of RCC gravity dam is comparable to that of normal concrete dam, without mandatory limit.

1.1.2.1 RCC Gravity Dam

The RCC gravity dam in China started from the Kengkou. In the 1980s, the height of RCC dam was 60–80 m. Since the late 1990s, the number of RCC high dams above 100 m has increased year by year. At present, the main dams of Jin'anqiao (160 m), Guandi (168 m), Gongguoqiao (105 m), Guanyinyan (159 m), Longkaikou (119 m), Ludila (140 m), Ahai (138 m) and other large hydropower stations under construction are all RCC gravity dams. They are based on full-section RCC dam technology, and the concrete quantity is huge. Up to now, more than thirty 100 m RCC gravity dams have been completed or under construction, and gravity dams have become amajor dam type for RCC dams.

1.1.2.2 RCC Arch Dam

The arch dam of Puding Hydropower Station built in 1993 in China pioneered the damming technology of RCC arch dam in China. The highest dam height of the arch dam of Puding Hydropower Station is 75 m, which is the highest RCC arch dam in the world at that time, and it has established its leading position in the world from the beginning. The success of this technology has realized a new breakthrough in RCC damming technology and promoted the rapid development and improvement of RCC arch dam. Up to now, the 100 m high arch dams built or under construction include arch dams of Shapai (132 m), Shimenzi (109 m), Linhekou (100 m), Dahuashui (134.5 m), Bailianya (104.6 m), Yunlong Grade Ⅲ (135 m), Tianhuaban (110.5 m) and other hydropower stations, as well as ultra-thin arch dams of Longshou (thickness-height ratio of 0.170), Zhaolaihe (thickness-height ratio of 0.176), and other hydropower stations. At present, China is the leading country in the world in the construction of RCC arch dams with the largest number, the highest arch dams and the thinnest shape.

1.1.2.3 RCC Cofferdam

The RCC cofferdam is safer than the earth rock cofferdam, and the dam crest can overflow; with small cross section of RCC cofferdam, the area of the foundation pit can be reduced effectively, and the foundation pit construction can be quickly recovered; moreover, the diversion tunnel is significantly shortened and the diversion standard can be significantly reduced; at the same time, it has the function of temporary water retaining and power generation and bringing benefits into play in advance. For example, the Phase Ⅲ Upstream Cofferdam of the Three Gorges Project adopts concrete gravity cofferdam, with maximum height of 115 m. From December 16, 2002 to April 16, 2003, its placing was completed. It took only 4 months to complete the total volume of 1 100 000 m³ of RCC. The storage capacity reached 12.4 billion m³, to ensure that the Phase Ⅲ Cofferdam of the Three Gorges Project is put into water-retaining power generation on schedule. At present, the new technology of CSG dam has been applied in cofferdam engineering, and its damming technology has great potential in cofferdam engineering.

Table 1.1 Above 100 m height RCC dams completed and under construction in china

S/N	Project name	Location	River	Dam height (m)	Dam type	Total storage capacity ($\times 10^8 m^3$)	Installed capacity (MW)	Year of completed
1	Guxian	Yichuan, Shanxi; Jixian, Shanxi	Yellow River	215	weight	165.57	2 100	Under construction
2	Huangdeng	Lanping, Yunnan	Lancang River	203	weight	16.7	1 900	2018
3	Guangzhao	Qinglong, Guizhou	Beipan River	200.5	weight	32.45	1 040	2009
4	Longtan	Tian'e, Guangxi	Hongshui River	192	weight	188	4 900	2009
5	Guandi	Xichang, Sichuan	Yalong River	168	weight	7.6	2 400	2013
6	Wanjiakouzi	Xuanwei, yunnan; Liupanshui, Guizhou	Gexiang River	167.5	—	2.793	180	2016
7	Xiangjiaba	Yibin, Sichuan	Jinsha River	162	weight	51.63	7 750	2018
8	Jin'anqiao	Lijiang, Yunnan	Jinsha River	160	weight	8.9	2 400	2012
9	Guanyinyan	Huaping, Yunnan	Jinsha River	159	weight	20.72	3 000	2015
10	Tuoba	Weixi, Yunnan	Lancang River	158	weight	10.39	1 400	Under construction
11	Xinmiao	Yingjing, Sichuan	Yinghe River	147	weight	0.618 9	81	Under construction
12	Sanhekou	Fuping, Shaanxi	Maopingxi River	145	Double curvature arch dam	7.1	45	2020

Continued Table 1.1

S/N	Project name	Location	River	Dam height (m)	Dam type	Total storage capacity ($\times 10^8 \text{m}^3$)	Installed capacity (MW)	Year of completed
13	Xiangbiling	Huize, Yunnan	Niulanjiang River	141.5	Double curvature arch dam	2.484	240	2017
14	Sanliping	Fangxian, Hubei	Nanhe River	141	Double curvature arch dam	4.99	70	2013
15	Ludila	Yongsheng and Binchuan, Yuannan	Jinsha River	140	weight	17.18	2 160	2013
16	Wunonglong	Weixi, Yunnan	Lancang River	137.5	weight	2.72	990	2018
17	Yunlong River Grade III	Shien, Hubei	Qingjiang Tributary	135	Double curvature arch dam	0.43	40	2008
18	Dahuashui	Zunyi, Guizhou	Qingshui River	134.5	Double curvature arch dam	2.765	200	2008
19	Shapai	Wenshan, Sichuan	Caopo River	132	Double curvature arch dam	0.18	36	2002
20	Ahai	Ninglang, Yunnan	Jinsha River	132	weight	8.85	2 000	2013
21	Lizhou	Muli, Sichuan	Muli River	132	Double curvature arch dam	1.897	355	2016
22	Tuxikou	Xuanhan, Sichuan	Qianhe River	132	Double curvature arch dam	1.6	51	Under construction
23	Jiangya	Cili, Hunan	Loushui	131	weight	17.4	300	2000
24	Baise	Baise, Guangxi	Youjiang River	130	weight	56.6	540	2006
25	Hongkou	Ningde, Fujian	Huotong Creek	130	weight	4.497	200	2008
26	Shangbaishi	Fu'an, Fujian	Dongxi mainstream	129	weight	2.487	17	Under construction
27	Geliqiao	Pingba, Guizhou	Qingshui River	124	weight	0.774	150	2010
28	Yongding Bridge	Hanyuan, Sichuan	Shahe River	123	weight	0.165 9	—	2013
29	Huangzang Temple	Qilian, Qinghai	Heihe River	123	weight	4.03	49	Under construction
30	Kalasuke	Fuyun, Xinjiang	Irtysh	121.5	weight	24.19	140	2011
31	Wudu	Jiangyou, Sichuan	Fujiang River	121.3	weight	5.72	150	2012
32	Shannipo	Liupanshui, Guizhou	Beipan River	119.4	Double curvature arch dam	0.85	185.5	2015

Continued Table 1.1

S/N	Project name	Location	River	Dam height (m)	Dam type	Total storage capacity ($\times 10^8 \text{m}^3$)	Installed capacity (MW)	Year of completed
33	Yunkou	Lichuan, Hubei	Wuni River	119	Double curvature arch dam	0.35	300	2009
34	Guangyuan	Guilin, Guangxi	Wupai River	119	Double curvature arch dam	0.716 6	24	Under construction
35	Silin	Sinan, Guizhou	Wujiang River	117	weight	15.93	1 050	2009
36	Dagu	Sangri, Tibet	Yarlung Zangbo River	117	weight	0.552 8	660	Under construction
37	Longkaikou	Dali Yunnan	Jinsha River	116	weight	5.58	1 800	2013
38	Tingzikou	Cangxi, Sichuan	Jialing River	116	weight	40.67	1 100	2014
39	Suofengying	Xiuwen, Guizhou	Wujiang River	116	weight	2.012	600	2006
40	Three Gorges Upstream Cofferdam	Yichang, Hubei	Yangtze River	115	weight Demolished in 2006	120	9 800 Left bank unit	2003 Temporary
41	Pengshui	Pengshui, Chongqing	Wujiang River	113.5	weight	14.65	1 750	2007
42	Gelantan	Jiangcheng, Yunnan	Lixian River	113	weight	4.09	450	2009
43	Luopo	Lichuan, Hubei	Lengshui River	112	Double curvature arch dam	0.86	300	2009
44	Dachaoshan	Lincang, Yunnan	Lancang River	111	weight	9.4	1 350	2003
45	Mianhuatan	Yongding, Fujian	Dinjiang River	111	weight	20.35	600	2001
46	Yantan	Dahua, Guangxi	Hongshui River	110	weight	33.5	1 810	1995
47	Shimenzi	Manasi, Xinjiang	Manasi River	110	Double curvature arch dam	0.501	64	2000
48	Guanyinyan	Shuicheng, Guizhou	Moon River	109	Double curvature arch dam	0.261 7	3.2	2019
49	Mamaya	Guanling, Guizhou	Beipan River	109	weight	1.365	540	2016
50	Jinghong	Jinghong, Yunnan	Lancang River	108	weight	11.4	1 750	2009
51	Huanghuazhai	Changshun, Guizhou	Getu River	108	Double curvature arch dam	1.748	540	2010
52	Madushan	Gejiu, Yunnan	Hongjiang River	107.5	weight	5.51	300	2011

Continued Table 1.1

S/N	Project name	Location	River	Dam height (m)	Dam type	Total storage capacity ($\times 10^8 m^3$)	Installed capacity (MW)	Year of completed
53	Zhaolai River	Changyang, Hubei	Zhaolai River	107.5	Double curvature arch dam	0.703	36	2006
54	Tianhuaban	Zhaotong, Yunnan	Niulanjiang River	107	Double curvature arch dam	0.787	180	2011
55	Dahuaqiao	Lanping, Yunnan	Lancang River	106	weight	2.93	900	2018
56	Shaping	Xuan'en, Hubei	Baishui River	105.8	Double curvature arch dam	0.982	46	2010
57	Three Gorges Left Guide Wall	Yichang, Hubei	Yangtze River	100.5	weight	—	—	2000
58	Gongguoqiao	Yunlong, Yunnan	Lancang River	105	weight	3.16	9 000	2012
59	Wacun	Tianlin, Guangxi	Tuoniang River	105	weight	5.36	230	2019
60	Dahe	Douyun, Guizhou	Laidi River	105	Double curvature arch dam	0.437 6	—	2015
61	Jiexu	Sangri, Tibet	Yarlung Zangbo River	105	weight	0.362	560	Under construction
62	Bailianya	Huoshan, Anhui	Manshui River	104.6	Double curvature arch dam	4.6	50	2008
63	Linjiang	Linjiang, Jilin	Yalu River	104	weight	18.35	400	2016
64	Linhekou	Nangao, Shaanxi	Lanhe River	100	Double curvature arch dam	1.47	72	2003
65	Zhuchang River	Panxian, Guizhou	Zhuchang River	102	weight	0.442	4.75	2019
66	Mangshan	Zhangyi, Hunan	Changle River	101.3	weight	1.33	18	2019
67	Shuikou	Mingxi, Fujian	Minjiang River	101	NC weight	29.7	1 400	1996
68	hatuo	Yanhe, Guizhou	Wujiang River	101	weight	9.01	1 120	2013
69	Shankouyan	Pingxiang, Jiangxi	Yuanhe River	99.1	Double curvature arch dam	1.05	12	2014

Continued Table 1.1

S/N	Project name	Location	River	Dam height (m)	Dam type	Total storage capacity ($\times 10^8 m^3$)	Installed capacity (MW)	Year of completed
70	Madu River	Wufeng, Hubei	Siyang River	99	Double curvature arch dam	0.246 3	51	—
71	Lijia River	Lantian, Shanxi	Mangchuan River	98.5	Double curvature arch dam	0.569	—	2009 – 2014
72	Dayakou	Zhenkang, Yuannan	Nanpeng River	98	Double curvature arch dam	1.7	102	2012 – 2016
73	Longqiao	Lichuan, Hubei	Yujiang River	95	Double curvature arch dam	0.26	60	2007
74	Hongling	Qiongzhong, Hainan	Wanquan River	94.9	weight	6.62	62.4	2015
75	Fengman Rebuild	Jilin, Jilin	The second Songhua River	94.5	weight	103.77	1 480	2020

Table 1.2   RCC cofferdams, CSG cofferdams and CSG permanent dams completed in China

S/N	Name of cofferdam	Location	River	Weir height (m)	Type	Year of completion	Remarks
1	Yantan Upstream Cofferdam	Bama, Guangxi	Hongshui River	54.3	Arch	1988	It is a pilot works of the Three Gorges Project. After completion, the water flows through the cofferdam for many times, with discharge of 400 m^3/s
2	Yantan Downstream Cofferdam	Bama, Guangxi	Hongshui River	40.2	weight	1988	It is a pilot works of the Three Gorges Project. After completion, the water flows through the cofferdam for many times, with discharge of 400 m^3/s
3	Wan'an Upstream Cofferdam	Wan'an, Jiangxi	Ganjiang River	23.6	weight	1990	In 1992, the inflow is 12 900 m^3/s, the outflow is 300 m^3/s, and the operation is normal
4	Longitudinal cofferdam of water inlet	Minqing, Fujian	Minjiang River	48	weight	1990	The operation is normal after completion
5	Shanzai Upstream Cofferdam	Lianjiang, Fujian	Aojiang River	23	weight	1992	The operation is normal after completion
6	Shuidong Upstream Cofferdam	Youxi, Fujian	Youxi Creek	21.6	Arch	1992	There is no transverse joint, and the water retaining condition is good. In the dry season of the second year

Continued Table 1.2

S/N	Name of cofferdam	Location	River	Weir height (m)	Type	Year of completion	Remarks
7	Wuqiangxi Upstream Cofferdam	Yuanshui, Hunan	Yuanshui	40.8	weight	1992	The average daily increase is 0.6 m. the construction period is 60 d, and the cost is low
8	Jiangya Upstream Cofferdam	Cili, Hunan	Loushui	36	Arch	1995	The upstream face is vertical and the downstream face is 1:0.45. It is a concentric circular arch
9	Gaobazhou Longitudinal Cofferdam	Zhicheng, Hubei	Qingjiang River	27.5	weight	1997	The operation is normal after completion
10	Dachaoshan Upstream Cofferdam	Yunxian, Yunnan	Lancang River	54	weight	1998	It is a double curved arch, built in 88 d, without any longitudinal joint. In 1998, the water overflow was for 7 times and the operation was normal
11	Hongjiang Downstream Cofferdam	Gongjiang River, Hunan	Yuanshui	34.6	weight	1999	The construction period is 68 d, the upstream and downstream sections are made of grout enriched vibrated concrete (GEVR), and the operation is normal
12	Geheyan Upstream Cofferdam	Changyang, Hubei	Qingjiang River	37	Arch	1998	It has experienced 4 flood seasons and overflow for 16 times, with the maximum flow of 10 700 m^3/s
13	Longshou Downstream Cofferdam	Zhangye, Gansu	Heihe River	8	weight	2000	Construction in severe cold winter in Northwest China
14	Three Gorges RCC Left Guide Wall	Yichang, Hubei	Yangtze River	100.5	weight	2000	When pouring under high temperature, cracks appear in autumn
15	Letan Upstream Cofferdam	Xincheng, Guangxi	Hongshui River	31	—	2002	Upstream and downstream cofferdams and longitudinal cofferdams, January-May 2002
16	Baise Upstream Cofferdam	Baise, Guangxi	Youjiang River	—	weight	2002	The downstream face of the cofferdam is of precast concrete formworks
17	The Ⅲ phase cofferdam of Three Gorges	Yichang, Hubei	Yangtze River	115	weight	2003	Phase Ⅲ upstream cofferdam
18	Leidatan Upstream Cofferdam	Mile, Yunnan	Nanpan River	24	weight	2003	Downstream slope 1:0.6

Continued Table 1.2

S/N	Name of cofferdam	Location	River	Weir height (m)	Type	Year of completion	Remarks
19	Longtan Upstream Cofferdam	Tian'e, Guangxi	Hongshui River	73.7	weight	2004	It ensures the flood protection, explores the concrete raw materials and construction mix proportion suitable for Longtan Dam, and develops the construction technology
20	Longtan Downstream Cofferdam	Tian'e, Guangxi	Hongshui River		weight	2004	
21	Goupitan Upstream Cofferdam	Yuqing, Guizhou	Wujiang River	72.6	weight	2005	Upstream cofferdam (constructed by No. 8 Bureau)
22	Goupitan Downstream Cofferdam	Yuqing, Guizhou	Wujiang River	60.9	weight	2005	Downstream cofferdam (under hydropower construction by An'neng)
23	Pengshui Upstream Cofferdams	Chongqing	Wujiang River	40	weight		Upstream cofferdam (constructed by No. 8 Bureau)
24	Xiangjiaba Upstream Cofferdam	Yibin, Sichuan	Jinsha River	94	weight	2008	Upstream Phase II Longitudinal Cofferdam
25	Lizehangdian Cofferdam	Hechuan, Chongqing	Jialing River	28	weight	2020	RCC Guide wall
26	Jiemian Downstream Cofferdam	Youxi, Fujian	Junxi	13.54	Trapezoid	2004	First use of CSG technology
27	Daotang Upstream Cofferdam	Songtao, Guizhou	Pingnan River	4	Trapezoid	2005	CSG Cofferdam
28	Hongkou Upstream Cofferdam	Ningde, Fujian	Huotong Creek	35.5	weight		CSG cofferdam
29	Gongguoqiao Upstream Cofferdam	Dali, Yunnan	Lancang River	50	Trapezoid	2009	CSG Cofferdam
30	Shatuo Upstream Cofferdam	Yanhe, Guizhou	Wujiang River	24	Trapezoid	2010	CSG Cofferdam
31	Mamaya Downstream Cofferdam	Guanling, Guizhou	Beipan River	17.2	Trapezoid	2012	CSG Cofferdam
32	Dahuaqiao Upstream Cofferdam	Lanping, Yunnan	Lancang River	50	Trapezoid	2016	CSG Cofferdam
33	Shoukoupu reservoir	Yanggao, Shanxi	Heishui River	61.6	Trapezoid	2019	The permanent CSG dam
34	Jinjigou reservoir	Yingshan, Sichuan	Qujiang	33	Trapezoid	2019	Asymmetric CSG dam

Continued Table 1.2

S/N	Name of cofferdam	Location	River	Weir height (m)	Type	Year of completion	Remarks
35	Miaojiang reservoir	Miaojiang, Guizhou	Miaojiang River	50	Trapezoid	2020	Asymmetric
36	Maomao Rive mountain pond	Leishan, Guizhou	Maomao River	18.2	Trapezoid	2019	Asymmetric
37	Huayujing mountain pond	Xixiu, Guizhou	—	16	Trapezoid	2019	Asymmetric
38	Lianggan mountain pond	Fuquan, Guizhou	—	20.3	Trapezoid	2019	Asymmetric
39	Chonglaxi mountain pond	Tongren, Guizhou	—	14.8	Trapezoid	2019	Asymmetric
40	Shanyin reservoir	Putian, Fujian	Luxi	52	Trapezoid	Under construction	Asymmetric

Note: All No. 3 Bureau, No. 4 Bureau, No. 7 Bureau, No. 8 Bureau, No. 11 Bureau, Minjiang Bureau and Shaanxi Engineering Bureau are under SinoHydro; the Armed Police Detachment refers to units affiliated to the Armed Police Hydropower Headquarters; and Guiyang Institute, Northwest Research Institute, Kunming Institute, etc., refer to the affiliated units of HydroChina.

1.2 Development History of RCC Dam

RCC dams are produced in the fierce competition between normal concrete dams and earth-rock dams. The development of soil mechanics theory has relaxed the restrictions on building materials for earth-rock dams. The use of large-scale construction machinery has accelerated the construction of earth-rock dams, and reduced the cost of earth-rock dams, so that earth-rock dams are under vigorous development. However, while the cost is reduced, earth-rock dams also have some disadvantages, such as the complexity and interference of the construction, especially the safety of the imperviousness system is obviously lower than that of the concrete dam. The main reasons are as follows: firstly, considering the characteristics of damming materials, flood overtopping is not allowed, and dam body settlement is inevitable due to erosion of damming materials; secondly, due to the flatter slope of upstream slope and downstream slope and the huge volume of the dam body, the construction site is large and the diversion tunnel is long with high water-diversion standard; thirdly, the flood releasing structure cannot be combined with the dam body, so it is necessary to provide a specific and separated water releasing structure; fourthly, earth-rock dams need large-scale exploitation of local materials for damming, which will inevitably cause great damages to the surrounding environment and ecology.

Due to characteristics of concrete dams and earth-rock dams, people began to seek a new type of dam to combine the safety of concrete dams with the high efficiency construction of earth-rock dams. In 1960s, several projects were carried out based on this idea.

From 1961 to 1964, the Alpa Gera concrete gravity dam with the height of 172 m was completed in the Alps in Italy. Due to the limitation of mountain height and construction time, two inclined rail bucket trolleys were used to transport concrete, and then dump trucks were used for pouring. The dam is made of lean concrete. The pouring layer is 70 cm thick from one bank of the river valley to the other bank. The traditional block pouring method is adopted. The cooling water pipe in the dam is canceled. The bulldozer is used for spreading. The plug-in vibrator suspended at the back of the bulldozer is used for compaction. The concrete after vibration is cut by a joint cutter, and a transverse joint is formed at the specified position of the dam body. In order to ensure the imperviousness of the dam, steel sheets are laid on the upstream face of the dam in addition to the joint and water stop.

In 1965, two gravity wing walls with the height of 18 m were built at Manicougan I dam in Quebec, Canada. The interior was made of lean concrete, paved with bulldozers and vibrated by plug-in vibrators. The upstream face was made of rich mortar concrete, which was formed by vertical sliding formworks, and the downstream face was made of precast concrete blocks.

In 1970, the "concreterapid construction conference" was held in California, US. J. M. Raphael's paper *Optimal Gravity Dam* suggested using cement sand gravel materials for damming, and efficient earth rock transportation machinery and compaction machinery for construction. Due to the cementation and stabilization of cement, the shear strength of the material is improved and the cross section of the dam is reduced; the continuous pouring method similar to the earth-rock damming can not only shorten the construction period but also reduce the construction cost.

In 1972, the " Economic Construction Conference of Concrete Dam " was held in California, US. R. W. Cannon's paper *Construction of Concrete Dams with the Soil Compaction Method* further developed Raphael's idea. Cannon introduced the test results of using dump trucks for transportation, front loaders for paving, and vibrating rollers for compaction of lean concrete without slump. He suggested that the upstream and downstream dam surfaces shall be made of rich mortar concrete and formed by sliding formworks.

In 1973, at the 11th International Dam Conference, A. I. B. Moffat's paper *Study on Lean Concrete for Gravity Damming* recommended that the lean concrete used on the subgrade of England in the 1950s shall be used for the construction of concrete dams and compacted by road construction machinery. He predicted that the cost of a dam with a height of more than 40 m could be reduced by 15%.

The first large-scale application of RCC is the Tunnel of Tarbela Dam constructed in Pakistan in 1975. In 1974, the outlet of the spillway tunnel of the dam was destroyed by flood. The repair must be completed before the snow melt in spring, and the required construction must be extremely fast. Therefore, RCC was used for repair. 350 000 m^3 RCC was poured in 42 d. The average daily pouring strength was 8 400 m^3/d, and the maximum daily pouring strength was 18 500 m^3/d, which was the highest pouring strength of RCC at that time.

The Shimajigawa gravity dam built in Japan in 1980 is the first RCC dam in the world. The dam is 89 m high, and the volume of RCC reaches 165 000 m^3 (accounting for 52% of the total concrete volume of 317 000 m^3). The amount of cementing materials used in RCC dam body is 120 kg/m^3, including 30% of fly ash. 3 m thick normal concrete is poured on the upstream face of the dam to prevent seepage. The thickness of the compaction layer is 50 cm and 70 cm. Continuous pouring shall be carried out 1 – 3 d after each compaction layer is rolled. The transverse joint of the dam body is formed by a joint cutter.

In 1982, the United States built the world's first fully RCC gravity dam, Willow Creek Dam. The dam is with the height of 52 m and axis length of 543 m, without longitudinal and transverse joints. The amount of cementing materials used in the internal RCC is only 66 kg/m^3, including 47 kg/m^3 for cement, 19 kg/m^3 for fly ash. The thickness of the compaction layer is 30 cm and the volume of RCC poured is 337 000 m^3. RCC is placed in less than 5 months, which is 1 – 1.5 years shorter than conventional concrete dam. Due to the small amount of cementious materials, a large amount of water leakage occurred in case of partially impoundingof of reservoir, and then grouting was carried out. The cost of Willow Creek Dam is only about 40% of that of normal concrete gravity dams and 60% of that of rockfill dams, which fully shows the great advantages of RCC dams in terms of speed and cost. Its completion greatly promotes the rapid development of RCC dams in the United States and other countries in the world.

Compared with foreign countries, RCC dams in China started late, but developed rapidly. The research on RCC damming technology in China began in 1979. After a large number of indoor studies, field rolling compaction tests of large test blocks were carried out in the Concrete Pavement Project of Gongzui Hydropower Station in Sichuan Province in 1981 and in Xiamen Airport Construction Site in Fujian Province in 1983. In the last test, it was found that 50% fly ash was added into 152 kg/m^3 RCC cementing materials. As a result, the rollability, compactness and uniformity of RCC were significantly improved, and the compressive strength generally met the design requirements, showing the superiority of RCC with a large amount of fly ash. In 1984 and 1985, the test scale was further expanded, and RCC was applied to the non-main works or non-main parts of several hydropower projects. They include Tongjiezi Cement Tank Foundation, Niurixigou Auxiliary Dam, Shaxikou Longitudinal Cofferdam and Switch Station Retaining Wall, Gezhouba Shiplock Lower Guide Wall, etc. These semi-productive tests have carried out more comprehensive practical drills on the construction of mass RCC, trained the team, improved the construction technology and construction management level, and laid a foundation for the application of RCC damming technology.

In order to apply the technical achievements made to concrete dams, Fujian Kengkou Gravity Dam is selected as the industrial test project of RCC damming technology. According to the test research results of previous years and the practical experience of RCC dams built abroad at that time, Kengkou RCC dam is determined to be carried out according to the following principles: high content of fly ash, low cement content, no longitudinal and transverse joints in dam body, construction in low-temperature seasons, full-section layered rolling, continuous pouring rising and asphalt mortar imperviousness system. After rolling compaction for more than 6 months, the first RCC dam with the height of 56.8 m was built in May 1986. In the process of Kengkou RCC damming test, special research and field observation have been carried out on RCC workability control, compaction law, initial setting and interval time between layers, construction of heterogeneous concrete on placing surface, simplification of downstream surface formwork, variation characteristics of dam body temperature and construction organization and management, which has accumulated valuable experience for promoting and improving RCC damming technology.

Since the completion of Kengkou RCC dam, RCC dams have developed rapidly in China. Some projects that have been designed or even constructed have been subject to construction of RCC dam, and RCC dams are actively adopted. From then on, from the initial construction of RCC dams with the height of 50 – 60 m to the construction of RCC high dams with the height of more than 100 m, China's RCC damming technology has developed rapidly.

1.3 Popularization and Application of RCC Damming Technology

1.3.1 RCC Damming Board

The RCC Damming Board of China Society for Hydropower Engineering and the RCC Damming Board of Chinese Hydraulic Engineering Society (hereinafter referred to as the Board) aim at promoting the application of RCC damming technology and promoting the progress of water conservancy and hydropower technology. The rapid development and successful application of RCC damming technology in China is inseparable from a lot of hard work and lifelong efforts of experts of the Board for many years. The RCC Damming Board of China Society for Hydropower Engineering was established in Tongjiezi Project Site of Sichuan Province on June 11, 1990. The RCC Damming Board of Chinese Hydraulic Engineering Society was reorganized and established in

Datian County, Fujian Province in November 1996 by the RCC Group of Construction Board of Chinese Hydraulic Engineering Society. The RCC Damming Board of China Society for Hydropower Engineering is affiliated to the Hydropower Headquarters of the Chinese People's Armed Police Force; the RCC Damming Board of Chinese Hydraulic Engineering Society is affiliated to the construction and management station of the Ministry of Water Resources of the People's Republic of China.

In order to enable more units instructure, design, construction, management, teaching, etc. to understand and master the RCC damming technology, the Board began to develop group members after its establishment. Group members enjoy the right to participate in academic exchanges and other activities organized by the Board, and can obtain the technical data and scientific and technological information provided by the Board, as well as preferential technical consultation and services.

Both Boards have a group of senior experts with rich theoretical knowledge and practical experience, and technical backbones with certain expertise in materials, structure, design, construction, teaching, management, etc., so as to fully grasp the technical trend of RCC. The Boardprovides technical consulting services for relevant design, construction and construction units combined with the actual construction of water conservancy and hydropower projects. According to the different conditions and stages of each project, the Board organizes different types of experts to undertake the following tasks respectively: special research, formulation of method statement, preparation of construction organization design, scheme review, on-site technical guidance, etc. Up to now, the Board has provided technical consulting services for more than 100 RCC projects in China (such as Puding, Longshou, Baise, Longtan, Guangzhao, Kalasuke and Jin'anqiao) and foreign countries, and organized RCC damming technology training combined with the projects. Through engineering practice, a large number of experts and technical backbones in design, construction, scientific research, teaching, operation and management are cultivated.

Over the years, the two Boards have worked in parallel in organizing Chinese and foreign academic exchange activities, publishing academic journals, developing group members, or carrying out the work of other societies. The Board has paid close attention to the development trend of international RCC damming technology and learned the advanced international experience with an open mind. The Board has organized a group to visit Spain, France, etc., and held discussions with Western experts. At the national academic exchange conference held in China, relevant experts who have participated in international conferences and visited abroad are invited to introduce the experience and development trend of RCC technology abroad. At the same time, Chinese experts have also introduced the development of RCC technology in China at the international conference held in the United States, Spain and other countries. The Board has assisted Chinese National Board on Large Dams to hold international RCC dam technology seminar exchange conference in Beijing, Dujiangyan, Sichuan, and Guiyang, Guizhou in China.

1.3.2 Popularization of Early RCC Damming Technology

The popularization of early RCC damming technology in China mainly relies on the Department of Science and Technology of the Ministry of Water Resources of the People's Republic of China, the Leading Group of Popularization of RCC Damming Technology, the General Administration of Hydropower Construction, the Expert Group of RCC Damming Technology, the RCC Group of the Construction Board of Chinese Hydraulic Engineering Society and other organizations, as well as leaders, experts, scholars and technicians engaged in popularization of RCC technology application, to widely carry out scientific and technological research, test and application popularization activities of RCC. The Board and various organizations have played a very positive and effective role in the research, popularization and application of RCC damming technology, and made great achievements and contributions.

At the initial stage of popularization and application of RCC damming technology, there is an extremely lack of professionals, and a large number of talents who are familiar with RCC damming technology are needed. From March to April, 1986, the Department of Science and Technology of the Ministry of Water Resources of the People's Republic of China organized more than 200 people from units in design, construction, scientific research, colleges and universities of water conservancy and hydropower system in China to explain the RCC technology on the site of the Kengkou Dam in Fujian Province in three batches, which plays a catalyst role in popularization and improvement of RCC technology in China; in December 1988, a RCC technology training course was held in Fuzhou. Nearly 80 trainees from units in design, construction, scientific research and management from all over the country participated in the training. These trainees became the backbone and new force for popularization and application of RCC damming technology in the future.

In March 1987, the Department of Science and Technology of the Ministry of Water Resources of the People's Republic of China organized the publication of *Technical Trend of Rolled Compacted Concrete*. By June 1999, a total of 32 issues have been published. The editorial policy of *Technical Trend of Rolled Compacted Concrete* is as follows: exchange engineering design and construction technology of RCC in China, publicize and introduce the latest achievements and scientific research achievements of RCC technology in China, and introduce the development and trend of RCC technology in foreign countries. *The Technical Trend of Rolled Compacted Concrete* has effectively promoted the development, popularization and application of RCC damming technology in China.

1.3.3 RCC Damming Technology Exchange Conference

Before the establishment of RCC Board, a RCC exchange conference was held by the Department of Science and Technology of the Ministry of Water Resources of the People's Republic of China in Dahua, Guangxi. From the establishment in 1990 to 2008, the RCC Board organized, hosted and held national RCC damming technology exchange seminars, symposiums and international academic conferences of RCC dam in Tongjiezi (Sichuan), Sanya, Beijing, Longyou, Xiamen, Baise, Nanjing, Longtan, Guiyang, Pu'er, etc. for 21 times. See Table 1.3 for details of previous RCC damming technology exchange seminars. The national RCC damming technology exchange seminar has put forward many valuable opinions and suggestions on the development and research direction of RCC damming technology in China, which has effectively promoted the continuous development and technological innovation of RCC damming technology in China.

Among national RCC damming technology exchange seminars held from 1990 to 2008, the RCC Board has actively, conscientiously and responsibly collected and compiled the proceedings for the technology exchange conference. It has successively edited *RCC Damming-Design and Construction, Proceedings of 96' RCC Damming Technology Exchange Conference, Atlas of RCC Dams in China, Proceedings of International Symposium on RCC Dams, Roller Compacted Dams in China (Chinese and English version), Proceedings of* 2000 *National RCC Damming Technology Experience Exchange Conference, Proceedings of* 2001 *National RCC Technology Exchange Conference, Proceedings of* 2002 *National RCC Technology Exchange Conference, Proceedings of* 2003 *National RCC Technology Exchange Conference, Proceedings of* 2004 *RCC Damming Technology Symposium, Proceedings of* 2004 *National RCC Damming Technology Symposium, Proceedings of Special Conference* 2005 *on RCC Materials and Quality Inspection,* 2006 20 *Years of Roller Compacted Dams in China-from Kengkou Dam to Longtan Dam, Proceedings of the Fifth International Symposium on Roller Compacted Concrete (RCC) Dams in* 2007 *(Chinese and English version),* 2008 *RCC Damming Technology,* etc. The proceedings of RCC damming technology has greatly promoted the continuous improvement of design, research and construction damming technology, and formed a relatively complete theoretical system and a complete set of complete construction method for RCC damming technology.

Table 1.3   Historical RCC damming technology exchange seminars

S/N	Name and theme of the conference (or project referred to)	Time	Location	Number of people
1	RCC Damming Technology Exchange Seminar	1989-04	Dahua, Guangxi	92
2	RCC Damming Technology Exchange Seminar (Tongjiezi)	1990-11	Tongjiezi	87
3	RCC Damming Technology (Nondestructive Testing) Exchange Seminar	1990-12	Sanya, Hainan	56
4	International Symposium on RCC Damming	1991-11	Beijing	160
5	1992 National RCC Damming Technology Exchange Conference	1992-10	Kunming, Yunnan	125
6	1994 RCC Damming Material Seminar	1994-03	Beijing	55
7	1995 National RCC Damming Exchange Seminar	1995-04	Guilin, Guangxi	120
8	Celebration of the 10th Anniversary of Kengkou RCC Dams and Technology Exchange Conference	1996-11	Datian, Fujian	160
9	1997 RCC Admixture Seminar	1997-05	Longyou, Zhejiang	100
10	1998 National RCC Damming Technology Exchange Conference (Jiangya)	1998-11	Cili, Hunan	106
11	1999 National RCC Damming Technology Exchange Conference (Fenhe Reservoir II)	1999-05	Taiyuan, Shanxi	133
12	2000 National RCC Damming Technology Exchange Conference (Mianhuatan)	2000-11	Xiamen, Fujian	146
13	2001 National RCC Damming Technology Exchange Conference (Shapai)	2001-08	Chengdu, Sichuan	148
14	2002 National RCC Damming Technology Exchange Conference (Three Gorges Phase III Cofferdam)	2002-10	Yichang, Hubei	159
15	2003 National RCC Damming Technology Exchange Conference (Baise)	2003-11	Baise, Guangxi	215
16	Symposium on Damming Technology of RCC High Arch Dam in 2004 (Zhaolaihe)	2004-04	Yichang, Hubei	166
17	2004 National RCC Damming Technology Exchange Conference (Suofengying)	2004-10	Guiyang, Guizhou	205
18	2005 RCC Material and Quality Testing Conference	2005-11	Nanjing, Jiangsu	90
19	Celebration of the 20th Anniversary of Kengkou Dams and Technical Exchange Meeting of Longtan 200-metre RCC Dam	2006-06	Tian'e, Guangxi	248
20	The Fifth International Symposium on Roller Compacted Concrete (RCC) Dams	2007-11	Guiyang, China	386
21	2008 Construction Technology of RCC Dams in China (Gelantan)	2008-10	Pu'er, Yunnan	147

1.3.4 Prizes of RCC Damming Technologies

In terms of scientific research, Chinese scientific and technological workers have made a series of scientific and technological breakthroughs at different levels, such as damming materials, mix proportion design, dam structure design, imperviousness form, placing method, rolling compaction technology, interlayer bonding, temperature control, construction quality monitoring and control, safety monitoring and feedback analysis, etc. around two dam types (RCC gravity dams and RCC arc dams). The national key scientific and technological projects include the research on the construction technology of Puding RCC arch dams during the "Eighth Five-year Plan" period, the research on design of Longtan high RCC gravity dams during the "Ninth Five-year Plan" period, and the research on the rapid damming technology of high RCC arch dams during the "Ninth Five-year Plan" Period. Nearly 100 units have participated in the scientific tests, and the personnel involved are experts, professors and engineering technicians in design, scientific research, materials, construction, teaching, mechanical equipment, basic theory, etc. With the birth of a number of significant scientific research achievements, the thorough understanding of engineering technicians of RCC damming technology has been rapidly improved. A large part of scientific and research achievements obtained have been directly transferred to the design and construction of projects, achieving obvious technical and economic benefits, and laying a solid theoretical and scientific foundation for the popularization and application of RCC damming technology. Over the past 20 years, the RCC damming technology of China had been awarded more than 20 scientific and technological progress awards at national, provincial and ministerial levels in total, and the world-famous 200-metre RCC Dam of Longtan Hydropower Station was awarded the Milestone Award by the International Commission on Large Dams; kengkou, Puding and Shapai RCC High Arch Dams were awarded the National Prize I and Prize II for Scientific and Technological Progress respectively; jiangya, Dachaoshan, Longshou and other projects were awarded the Class I and Class II Scientific and Technological Progress Awards at Provincial and Ministerial Levels and the Flying Goddess Award; mianhuatan and Dachaoshan Hydropower Station Projects, based on superior RCC dam engineering quality, were awarded the Luban Award; 100 m Zhaolaihe RCC Dome Arch Dam was awarded the Annual Top Ten Innovation Awards of Chinese Enterprises; the Project on Research and Use of the Rock powder of Diabase Manufactured Sand in RCC of Baise Project was awarded the Class III Scientific and Technological Progress Award in China Electric Power; research on Key Construction Technologies of 200 m RCC Gravity Dam of Longtan Hydropower Station was awarded the Special Award for Scientific and Technological Progress of Sinohydro Corporation. Above awards show that RCC damming technology has a higher technological content.

1.4 Discussion on Key Rapid Damming Technologies of RCC

"Interlayer bonding, temperature control and cracking prevention" are taken as the core technologies of rapid RCC damming. The design, scientific research, construction and management shall always focus on above core technologies for comprehensive technical innovation.

1.4.1 Design of RCC Dam

1.4.1.1 Design and Rapid Construction of Dam Body

The design of RCC dam is particularly important in the application of rapid RCC damming technology. RCC dams require special design. The purpose of RCC damming technology is to simplify rather than hinder construction. The main difference between RCC and normal concrete is that the mix proportion and the construction technology of concrete materials have been changed, but the material performance of RCC is basically the same as that of normal concrete. In the design, the design standard of the dam is not reduced for selecting RCC, so the design section of RCC dam is the same as that of normal concrete dam. The dam is an important part of complex structures. During arrangement of RCC dam, the technical characteristics of rapid construction of RCC concreting in whole-block and thin-lifts shall be fully considered in the design, and various structures in the complex are reasonably arranged. The RCC dam shall be separated from power generation structures and outlet structures as far as possible. The RCC parts of the dam shall be relatively concentrated, so as to reduce the holes in the dam, simplify the dam structure, expand the scope of RCC as far as possible and minimize the interference to rapid RCC construction.

1.4.1.2 RCC Design Age

When the design age of RCC is determined, it is necessary to make full use of the later strength of RCC and simplify the temperature control measures in view of the characteristics of RCC such as small amount of cement, large admixture, slow temperature rise of hydration heat and low early strength. Therefore, the design age of RCC shall be 180 d or 360 d, and the indexes such as impermeability, frost resistance, tensile strength and ultimate tensile value shall be of the same design age as that of the compressive strength. The RCC later strength increase is related to the types and content of cement, admixtures, aggregates, additives and others. Massive engineering research results show that, generally, the increase rate of compressive strength of RCC at 28 d, 90 d and 180 d is roughly $1:(1.4-1.7):(1.7-2.0)$. At present, the design age of RCC is mostly 90 d, resulting in a failure to give full play to the advantage of later strength of RCC. The design age of 180 d or 360 d is worth studying and discussing.

1.4.1.3 Matching of Design Index of RCC

The mismatching of the design indexes of hydraulic concrete such as compressive strength, design age, frost resistance grade and ultimate tensile value lasts for a long time. Since the Three Gorges Phase II Project, the frost resistance grade has been taken as an important index to evaluate the concrete durability of the dam in the design. After that, China has listed the frost resistance grade as the main design index for concrete durability in the south and other mild regions. Especially the design index of frost resistance grade of the external and internal concrete of the dam has been higher and higher in recent years. The concrete design strength of the inner or upper part of RCC dam is generally 15 MPa, which does not match with the frost resistance grade and ultimate tensile value. Actually, by core drilling of the hardened concrete of the dam, the frost resistance grade and the ultimate tensile value of concrete core samples are lower than that of the concrete sampled at outlet. Taking the single frost resistance grade as an index to evaluate the concrete durability requires continuously in-depth study.

1.4.2 Key Technologies for Mix Proportion Design

1.4.2.1 RCC Definition

RCC refers to the concrete that is spread with the semi-plastic concrete mixture without slump in thin-lifts and rolled and compacted by vibrating roller, with bleeding performance. The definition of RCC in recent years is completely different from that of early hard RCC, which is also the inevitable result of the development process of hydraulic RCC rapid damming technology from practice to theory to practice. Since the full-section RCC damming technology was adopted in China from the late 1990s, the traditional "RCD" construction method and impervious structure have been changed. The dam is constructed by full-section RCC damming technology, and its imperviousness depends on itself, which requires that RCC shall be with a long enough initial setting time, and its surface is fully bleeding and elastic after being rolled and compacted by vibrating roller, so that the upper RCC aggregate can be embedded into the lower RCC that has been rolled, thoroughly changing the weak layer of RCC that is easy to form "layer-cake" seepage channel, with many layers. Therefore, RCC gradually transfers from hard concrete to semi-plastic concrete.

The construction of semi-plastic RCC without slump can prove: the interlayer bonding quality of RCC is good. From 2007 to the first half of 2009, China has taken the super-long core samples with length of greater than 15 m (15.03 m, 15.33 m, 15.85 m, 15.30 m, 16.49 m, 16.55 m, etc.) from Longtan, Guangzhao, Gelantan, Jinghong, Jin'anqiao, Kalasuke and other RCC dams respectively. The overall evaluation of most RCC dam core drilling and pump-in test shows that the permeability could not meet the design requirements, and the friction coefficient f' and the cohesive strength c' are greater than the designed control indexes. The appearance of core samples is smooth and compact with uniform aggregate distribution, no matter the hot joints caused by continuous spreading or cold joints, the interlayer bonding is good without obvious layer joint. The apparent density of RCC is consistent with that tested by nuclear densimeter on site, and the quality of RCC dam meets the design requirements.

1.4.2.2 Mixture Performance Test as a Key in Mix Proportion Design

RCC and normal concrete are common and certainly different in mix proportion design. The test shows that: the physical and mechanical properties such as strength, imperviousness, frost resistance and shear resistance of RCC body are not inferior to that of normal concrete. However, RCC construction is subject to whole-block placement and thin-lift, and imperviousness depends on the dam body itself. The mix proportion design shall meet the requirements of not only body strength, impermeability and other performances but also interlayer bonding of the dam. Therefore, the mixture performance test shall be the focus of the RCC mixture design test, and the standard of the fresh RCC shall be the fully bleeding. It is necessary to change the design concept that mix proportion design of normal concrete attaches importance to the performance of hardened concrete and despises the performance of mixture, which is the key of RCC mix proportion test.

1.4.2.3 Grout-mortar Ratio as a Key Parameter in Mix Proportion Design

After more than 20 years of extensive test research and damming practice, the mix proportion design of RCC has been mature, and a set of relatively complete theoretical system has been formed. The grout-mortar ratio, as the key technology for RCC mix proportion design, has become one of the important parameters of mix proportion design, and has the equally important effect with water-cement raido, sand ratio, unit water consumption and other parameters. In the mix proportion design of RCC, the (PV) value of grout-mortar ratio gets more and more attention. The grout-mortar ratio refers to the ratio of the volume of grout (including particle with particle size less than 0.08 mm) and mortar volume, i.e. grout-mortar volume ratio, referred to as "grout-mortar ratio". According to the practical experience of full-section RCC damming in recent years, when the content of rock powder in manufactured sand is controlled at about 18%, the grout-mortar ratio is generally not lower than 0.42. It can be seen that the grout-mortar ratio intuitively reflects a proportional relationship among RCC materials, and is an important index for evaluating RCC mixture performance.

1.4.3 Rapid Construction of Blinding RCC

For RCC dam, its bottom width is larger and foundation restraint range is higher because of no longitudinal joints. In order to prevent cracks in foundation concrete, the allowable temperature of foundation shall be controlled. The foundation of the dam is located on an uneven bedrock surface, therefore, it is necessary to pour the blinding concrete with a certain thickness for leveling and consolidation grouting before RCC pouring, and then RCC construction can be performed. Since the blinding is constructed with normal concrete, on the one hand, the concrete for blinding is of high strength, the cement consumption is large and the hydration heat is high, which is not conducive to temperature control; on the other hand, due to the small placing surface of the concrete for the blinding, large amount of formwork and low construction strength, the concreting for the blinding is one of key factors restricting the rapid construction of RCC dam.

The construction practice in recent years shows that RCC can achieve the same quality and performance as those of normal concrete. Therefore, the bedrock surface is immediately concreted with RCC synchronously after being leveled with low-slump normal concrete, which can obviously accelerate the blinding concrete construction, and give full play to the advantages of RCC such as low adiabatic temperature rise, small foundation temperature difference and beneficial crack resistance. For example, the blinding concrete in Baise

Project is constructed by spreading RCC immediately after the bedrock surface is leveled with normal concrete. Generally, blinding concrete pouring is performed in winter or at a low temperature, so RCC (instead of normal blinding concrete) is adopted to quickly construct the foundation cushion at a low temperature, which can effectively control the foundation temperature difference, speed up the consolidation grouting and prevent deep cracks in the dam foundation concrete.

1.4.4 Trucks + Full Pipe Chutes as the Fastest Transport Methods

Transport of RCC into placing surface is always one of the key factors restricting rapid construction. During transport of RCC into placing surface, by the application of a variety of transport and placement methods such as transport by trucks, conveying of belt conveyors, revolve of negative pressure chutes and aggregate bins and vertical transport of cable cranes or tower cranes, a large number of construction practices can prove that the direct transport of RCC by trucks into placing surface is the most effective way of rapid construction, which can greatly reduce the intermediate links and the concrete temperature rise. At present, the RCC dam is getting higher and higher. In the dam body at narrow river valley, the roads on the dam are of great height difference, so trucks cannot directly enter the placing surface. Vertical transport of RCC at intermediate links can be carried out by means of full pipe chutes, that is, combined transport of trucks outside placing surface + full pipe chutes + trucks on placing surface. Due to the increase of the size of the full pipe chutes, the traditional negative pressure chutes are completely replaced. At present, the section size of the full pipe chutes has reached 800 mm × 800 mm, and the downward inclination angle is generally 40° – 50°. Therefore, the aggregate bins on placing surface are removed, and the trucks outside placing surface unload the materials directly into the trucks on placing surface through the discharging bins and the full pipe chutes, so that the transport is very simple and fast. At present, the height difference of the full pipe chutes has exceeded 100 m, so the difficult problem of vertical transport is solved.

1.4.5 Key Technology to Ensure the Interlayer Bonding Quality

1.4.5.1 Liquefied Bleeding as an Important Criterion for the Evaluation of Rollability

The VC value has a significant impact on the performance of RCC. In recent years, a large number of construction practices prove that the key points for the control of on-site RCC are VC value of mixtures and setting time, and the dynamic control of VC value is the key to ensure the rollability and interlayer bonding of RCC.

Spraying is only to change the placing surface microclimate, so as to achieve moisture preservation and cooling. Liquefied bleeding of RCC is the mortar extracted from concrete liquefaction under the action of rolling by vibrating roller, thethin-lift of surface mortar is the key to ensuring the quality of interlayer bonding. Liquefied bleeding has become an important criterion for evaluating the rollability of RCC. Under the premise of satisfying the rollability, the main factors of the RCC mix proportion affecting the liquefied bleeding are whether it is rolled in time after pouring and gradual loss of VC value caused by temperature.

1.4.5.2 Timely Rolling as the Key of Ensuring the Interlayer Bonding Quality

Timely rolling is the key to ensuring the quality of interlayer bonding after paving RCC. RCC is a layered structure, and it is easy for its interlayer joint surface to become a leakage channel. The characteristics of RCC pouring are concreting in thin-lifts and whole-block and rolling. The quality of the interlayer bonding of RCC dams is closely related to the mix proportion design. The well-designed mix proportion plays an important role in the interlayer bonding. With the good construction mix proportion, after rolling by vibrating roller for 2 – 5 times, the layer surface can be fully bleeding. Timely rolling of RCC after pouring and paving is the key to ensuring the quality control of the interlayer bonding, which is directly related to the imperviousness and overall quality of the dam.

At present, the specification stipulates that the thickness of the interlayer after rolling is 30 cm. RCC dam with a height of 100 m has 300 layers of joints, so the imperviousness performance, interlayer shear resistance and overall performance of the dam with such many joints are very unfavorable. In the early days, some scholars and scientific researchers in China always had doubts and disputes about the interlayer bonding of RCC dams and the imperviousness of the dam. The quality of the interlayer bonding has been the main topic of research in the industry for many years. The tensile and shear strength of layer joint plays a leading role in the design of RCC dams, especially in areas with high earthquake occurrence. Therefore, the design has taken the shear resistance parameters f' and c' of the RCC dam layer (joint) surface as the design control index.

1.4.6 Technological Innovation for Increasing Rolling Layer Thickness and Reducing Joint Surface

RCC dams have many layer joints, and rolling is time-consuming and power-wasting, which seriously restricts the rapid construction of RCC. Therefore, it is very necessary to carry out technical innovations in the thickness of the rolling layer and the time of rolling, to study whether to improve the thickness of the rolling layer, reduce the rolling joint surface and increase the lifting layer height of RCC construction. For example, from the traditional thickness of 30 cm to 40 cm, 50 cm, 60 cm or even more thickness, it is generally possible to increase the 3 m lifting layer to 6 m or higher. Because of the imperviousness performance of the RCC body is not inferior to that of normal concrete, the increase in the thickness of the rolling layer can significantly accelerate the RCC damming, effectively reduce the layer joint in the dam body, and improve the impermeability. Exploratory test research has been carried out in related domestic projects for improving the test research on the thickness of the rolling layer.

The increase of the thickness of the rolling layer requires in-depth research on the matching of the rigidity and stability of the formwork, the quality of the vibrating roller and the excitation force. As the thickness of the RCC layer increases, GEVR jacking, grouting, and vibrating will become the construction difficulties. Based on the practical experience of the RCC dams of Baise, Guangzhao, Jin anqiao, etc., the use of machine-mixed GEVR can solve the construction problem of thick layer of GEVR. The pouring of machine-mixed GEVR is the same as that of RCC. After pouring, the concrete is vibrated with a vibrator, which not only simplifies the operation procedure of GEVR, but also guarantees the quality of large-volume GEVR.

1.4.7 Technological Innovation of GEVR

1.4.7.1 Technological Innovation of Hole Forming for GEVR

GEVR is a major technological innovation in China using the full-section RCC damming technology. GEVR is the rich mortar concrete formed by spreading grout in the RCC during the paving construction of RCC, which is vibrated and compacted by vibrator. The GEVR is mainly used on the surface of the dam impervious area, the periphery of the formwork, the bank slope, the gallery, the hole, the part with reinforcement, etc. The GEVR can significantly reduce the interference to the RCC construction, but the quality of GEVR construction is directly related to the imperviousness performance of the dam, so it gets a lot of attention.

The key to affecting the construction quality of GEVR is the quality of hole forming. The grouting method of the GEVR in RCC has experienced the top, layered, cutting and jack grouting methods. A lot of engineering practices show that the jack grouting method has become the mainstream method at present. Jack grouting method is to use the jack with 40 – 60 mm to drill holes on the surface of RCC paved. At present, all the hole forming is made by stepping jacks manufactured. Because manual hole forming is laborious and timeconsuming, with poor effect, it is difficult for the hole depth to reach the required depth, causing the grout cannot penetrate the bottom and the periphery in the hole, so the hole depth is the key to the quality of GEVR construction.

Based on years of construction experience, the mechanized jacks need to be studied by technical innovation. The mechanized jacks can be based on the principle of portable vibratory rammer, to transform the end of the vibratory rammer into a jack, and install a single-rod or multi-rod jack at the end of the rammer, which can effectively increase the depth of the hole and efficiency, reduce labor intensity, and significantly improve the construction quality of GEVR.

1.4.7.2 Technical Innovation of Grout Concentration Uniformity

After the hole forming meets the requirements of GEVR, grout uniformity and grouting are extremely critical construction techniques affecting the quality of GEVR. After the grout used for GEVR is prepared at the grouting station, the grout is transported to the grout truck on the placing surface through pipelines. Due to the characteristics of the grout, it is easy to produce precipitation, resulting in uneven grout concentration. Injecting grout with uneven concentration into GEVR will cause changes in water-cement raido and directly affect the quality of GEVR.

In order to prevent the precipitation of grout and ensure the uniformity of grout, it is necessary to conduct technical innovation for the traditional grout truck and study the feasibility of installing the mixer in the grout truck. The grout truck shall be technically modified according to the working principle of the mixer. Before each grouting, mix the grout in the grout truck with a mixer, so that the grout is uniform and reaches the designed density, and then perform grouting for the RCC with holes.

1.4.8 Technical Route of Temperature Control Measures for Simplifying

One of the advantages of RCC is to simplify or cancel temperature control. Early RCC dams have a low height, and make full use of low-temperature seasons and low-temperature periods for construction, and most of them do not take the temperature control measures. However, in recent years, due to the increase in the height and volume of RCC dams, in order to expedite or shorten the construction period, it has become a common practice to continuously pour RCC in high-temperature seasons and high-temperature periods, so that temperature control measures have become more and more stringent. The temperature control measures of RCC dams are the same as those of normal concrete dams, without difference. The temperature control measures of RCC dams are becoming more and more complicated, which will have certain negative impact on the simple and rapid construction of RCC.

The burying of cooling water pipes on the placing surface will greatly interfere with the rapid construction of RCC. The cancellation of the cooling water pipes of the dam will be a challenge to the temperature control of RCC. The cancellation of the cooling water pipe does not mean absence of control to the temperature of the RCC, but the technical route of the temperature control measures is different. The temperature control measures are mainly taken in the temperature control of the RCC before placing and during rolling, to strictly control the placing temperature to comply with the standard. The main technical route to cancel the cooling water pipe temperature control measures: firstly, do not pour RCC in high-temperature seasons and periods, try to use low-temperature seasons or periods to pour RCC, to effectively reduce the maximum temperature rise of dam concrete; secondly, strictly control the temperature at the outlet of the RCC. The control of temperature at the outlet is conducted with the conventional temperature control method, that is, control the temperature of cement and fly ash into the tank, pre-cool aggregates, and mix with cold water or ice. Although there is a certain cost to control the temperature of the RCC at the outlet, it can significantly reduce the interference to the rapid construction of the on-site RCC; thirdly, strictly control the temperature rise of RCC. The key to preventing the temperature from rising is timely rolling, spray moisturizing on the placing surface to change the microclimate, and cover insulation materials on the rolled concrete in time.

Temperature control has become one of the key factors to restrict the rapid RCC construction. The standards and technical routes of temperature control at different heights of RCC dams need to be carefully studied to break the single temperature control deadlock and passive situation. How to ensure that the temperature control standard and rapid construction reach an optimal combination point is a new problem to be solved.

1.4.9 Technical Innovation of Cracking Prevention for Dam Surface

Cracks are a common problem in concrete dams. The dam with cracks has long troubled people. The temperature and temperature stress of RCC are different from those of normal concrete dams. Although RCC dams have low hydration heat and fewer cracks on the dam surface in the early stages, the high temperature in the RCC dam lasts for a long time, so that the surface of the dam is weathered by low temperature, cold wave, exposure, drying and wetting, causing the temperature difference between the inside and outside of the dam, and easy appearance of surface cracks. The period of more surface cracks in RCC dams is usually after the dam is built. The RCC dam is a layer joint structure with multiple layers and low tensile strength between layers. In case of water storage or low water temperature of the RCC dams, and cold shock of water or temperature drop, the temperature difference between the inside and outside of the dam is large and easily causes layer or horizontal cracks, even vertical cracks.

Based on the long duration of high temperature in RCC dams and many layer joints, in order to prevent the disadvantages of large temperature difference between the inside and outside of the dam and easily causing surface cracks, it is necessary to carry out technical innovations in the protection of the dam surface, that is, to "dress the dam". The use of polymer cement flexible waterproof paint to protect the surface of the dam in an all-round way, which can give twice the result with half the effort.

Prevention of cracks on the surface of RCC dams shall be focused on, because most of cracks of RCC dams are surface cracks. Under certain conditions, they can develop into deep cracks that are often uneasily dealt with. Therefore, strengthening the surface protection of concrete dams is very important. Coating the surface of the dam with a polymer cement flexible waterproof material with a thickness of 3 – 5 mm can play a good role in surface protection, that is, it can prevent seepage and cracks, keep moisture and insulation, also can enhance the durability of concrete, with simple and fast construction and low cost. Compared with conventional XPS insulation plate for concrete surface protection, the polymer cement flexible waterproof material has great advantages.

1.4.10 Stepped Overflow Channel Flow Dissipation Technology

The RCC dam with a small variation in single-width discharge flow is based on the stepped spillway flow dissipation technology. The advantage is that it can increase the energy dissipation rate and reduce the size of the stilling basin. More importantly, it significantly simplifies the construction difficulty of the overflow surface and can accelerate the construction. For example, the flood discharge and energy dissipation of dams in Dachaoshan, Suofengying and other domestic hydropower stations is performed with a combined energy dissipation method of "X" wide-flange piers + stepped overflow dam surface + stilling basin, and the downstream non-overflow dam surface is also stepped.

The design scheme of the overflow surface of the overflow dam section is often a major factor affecting the construction period of RCC dams. Because most of the overflow surface design uses arc surface for flow discharge and energy dissipation, the construction of overflow energy dissipation arc surface is complex and difficult, and it is constructed in stages, that is, the construction of the overflow surface is carried out after the

RCC construction. Due to the different stresses between new and old concrete, it is easy for the interlayer bonding of abrasion and cavitation resistance concrete to be inconsistent, and it is difficult to guarantee the quality.

The spillway is subject to stepped flow dissipation. Firstly, it is simple in size, reducing construction interference and complexity; secondly, the high-grade stepped impact and abrasion-resistant concrete can rise synchronously with RCC, to effectively accelerate the construction; thirdly, the stepped discharge and energy dissipation effect is good, and the technical and economic benefits are significant. The disadvantage is that it is not very artistic and coordinated with the appearance of the dam. At present, the stepped discharge energy dissipation scheme is mostly used in foreign countries and less used in domestic projects. The use of stepped energy dissipation schemes in the overflow dam section of RCC dams needs to change the traditional concept, and make an analysis and comparison technically and economically, so that the advantages of stepped energy dissipation schemes can be promoted and applied, which has a very positive effect on shortening the construction period of the dam.

1.5 Innovation and Reflection on RCC Rapid Damming Technology

1.5.1 Innovation on RCC Rapid Damming Technology

1.5.1.1 Timely Revision of Technical Standards

Technical standards and regulations are relatively complete in China, but the revision of regulations and standards has not been able to keep pace with the times, and has not formed a good synergy at the same time. Separation is the main reason for the lag in the revision of technical specifications. It is impossible to make communication as the reform and opening up. Closing each other in a small circle seriously restricts the development of technical innovation. It is recommended that relevant national departments organize the experts, scientific research and engineering technicians in design, scientific research, construction, equipment, teaching and other related disciplines for joint formulation and revision of the standards. At the same time, it is necessary to ensure a certain amount of funds. Technical clauses with doubts and different views need to be demonstrated through tests to ensure the correctness and universality of the technical clauses. For example: RCC dam design, RCC definition, imperviousness method, mix proportion design, temperature control, etc. require a large number of personnel with practical experience and engineering adaptation to participate in the preparation of the standard, so that the standard of can keep pace with the times to guide the design, scientific research and construction.

1.5.1.2 Simplification of Design Innovation of RCC Dams

From a perspective of engineering science, the design direction of "full RCC" dams shall be more important, rather than the "compound RCC dams". For the upper part of the RCC low dam or the RCC high dam, the design shall be scientific and innovative based on the characteristics of its imperviousness standard and low strength grade index. If one grade and one strength grade are used for the design of RCC dams, it can greatly simplify the mutual interference of placing surface of RCC construction, and can give full play to the technical advantages of the RCC rapid damming. Because grout-mortar ratio adopted in current design of RCC mix proportion is relatively great, in the range of 30 m from the upper part of the dam, it can be considered to use III gradation or II gradation and one design grade of RCC for construction, and can significantly reduce the interference of the two types of concrete construction. For example, the successful full-section construction of the RCC dam of Yunnan Hongpo Reservoir with a single Ⅲ gradation RCC is a good example.

1.5.1.3 Difference Between Mix Proportion Design and Construction Mix Proportion of the Early Scientific Research

In the early stage of the project, the RCC mix proportion test research conducted by the scientific research unit is only a preliminary and fundamental report of results, and there are often certain differences with the construction mix proportion test results. Here are analysis reasons: firstly, the on-site construction of hydropower projects is affected by changes in hydrology, meteorology, topography, geological conditions and construction conditions, and there are big differences between the scientific research tests conducted under the standard conditions and the construction site; secondly, the scientific research test conditions and actual on site test conditions change. For example, the raw materials used in the scientific research test do not match the actual construction materials on site; thirdly, lack certain construction experience and master a few actual construction situations of the project. For example, the dynamic control of VC value, the precise control of rock powder content, the influence of interrelation of parameters (VC value and additive content, air content and setting time). The biggest difference in RCC mix proportion design between the scientific research test and on site construction is how to carry out mix proportion design by mixture performance and construction performance of the fresh RCC mixture.

1.5.1.4 In-depth Research on Cementing Grout Enriched Materials

The performance of RCC with rich grout and low VC value follows the traditional normal concrete performance. Therefore, the imperviousness system completely relies on the imperviousness of the dam, which significantly simplifies the design and layout of RCC dams. There is a difference between rich cementing grout materials and rich cementing materials. The rich cementing grout materials consider the content of rock powder in the sand or rock powder mixed, which effectively improves the grout-mortar ratio. In the future, the rich cementing grout materials will be subject to in-depth test research on RCC. The research directions are as follows: mainly the role of active and inert admixtures and micro rock powder in the cementing materials, the effect of rolling by vibrating roller on the liquefied bleeding and interlayer bonding of RCC with rich cementing grout, research on the impact of the amount of different cementing materials on impermeability, frost resistance, temperature stress and other durability and anti-cracking properties of RCC.

1.5.1.5 Discussion on Technical Innovation for Reducing the Maximum Aggregate Size

For the Ⅲ gradation RCC, the maximum aggregate size is 80 mm. In order to reduce the modulus of elasticity of diabase aggregates, quasi Ⅲ gradation RCC is used for the inside of the Baise Project, and the maximum aggregate size is 60 mm. The construction proved that the Baise Project adopts quasi Ⅲ gradation RCC, which has strong aggregate separation resistance, good rollability, fast liquefied bleeding, and tight interlayer bonding, to ensure the quality of RCC construction, reduce the modulus of elasticity, and improve the crack resistance. The experience of Baise Project also shows that the cement consumption of quasi Ⅲ gradation RCC is the same as that of Ⅲ gradation RCC, and the temperature control problem is not worrying. Therefore, the design shall carry out technical innovation, in-depth research, advance with the times, and study the feasibility of using one grade for RCC and the maximum aggregate size of 60 mm (in recent years, RCC of the single grade of 50 mm is mostly used abroad). In this way, from aggregate processing, mixing plant production, to RCC construction, it significantly accelerates RCC construction and gives twice the result with half the effort.

1.5.1.6 Reflection on the Change of Microclimate by Spray Moisturizing on the Placing Surface

The spray moisturizing on the placing surface of the RCC and the change of the microclimate of the placing surface are essentially a matter of investment. Meteorological data shows that the temperature difference between sunny days and cloudy and rainy days is generally about 10. If the dam placing surface is really formed into a misty and humid microclimate, changing the climatic conditions around the placing surface plays a very important role in improving the interlayer bonding and temperature control and cracking prevention measures of the RCC, but it is not easy. In fact, most of the spray moisturizing of the placing surface is only to spray or sprinkle the paving or rolling concrete on the placing surface. In order to reduce the cost of the undertaker, the nozzles of the spray gun are generally the nozzles of the cleaning machine. The author thinks that the spray moisturizing on the placing surface of RCC is the key to ensuring the quality of interlayer bonding, and it is necessary to carry out the technical innovation in nozzles of spray equipment and spray gun. The equipment is recommended to be purchased by the owner, as the main materials such as cement, steel, and fly ash used in the project are supplied by the owner. This can ensure the quality of the spray equipment, effectively improve the microclimate of RCC pouring, and improve the quality of interlayer bonding and reduce the temperature rise of concrete, so as to truly achieve the effect of temperature control and cracking prevention.

1.5.1.7 Reflection on Bid Winning at an Unreasonably Low Price

In recent years, due to the irregular bidding in the construction market, the unreasonable bid winning at a low price, the uneven quality of the undertaker, and the insufficient or low investment in scientific research, the lack of implementation, etc., it has seriously affected the play and quality of RCC rapid damming technology, which are worthy of our reflection on issues such as bidding and project approval and research of scientific research programs.

1.5.2 Reflection on RCC Rapid Damming

When discussing the advantages of RCC damming technology, it is necessary to analyze the best plan of RCC dams according to the value engineering method. From design to construction to project construction management, all factors must be considered at the same time. It is necessary to reflect, summarize and analyze the success experience and shortcomings of RCC damming technology. Based on years of practical experience in RCC test research and damming technology, the author deeply realizes that the rapid RCC damming technology still needs to be carefully summarized, and is subject to continuously technological innovations from complex layout, design indexes, mix proportion design, precise mixing, transportation plan, pad RCC, rolling layer thickness, GEVR, interlayer bonding and temperature control and cracking prevention, so as to achieve continuous improvement and development of RCC rapid damming technology to a higher level.

As the RCC damming technology has made great achievements, it has not concealed the existing problems. There are also many painful lessons in the development of RCC dams. For example, a RCC gravity dam in Brazil collapsed after impoundment, and RCC dams built in a number of countries have to be repaired as defective dams. Some RCC dams built in China also have quality problems, especially those with a height below 100 m built on some tributaries of a river. After the completion of some dams, the permeability is high, the recovery rate of core samples is low, and the shear resistance of interlayer bonding is poor. After impounding, the dam leaks seriously, so the dam must be treated by grouting and imperviousness, thus affecting the project quality and causing the loss of benefit. On the one hand, these defective RCC dams suffer management problems, but the defects are mainly caused by the construction contractor's little understanding of RCC damming technology.

The durability of a dam refers to its service life. When inspecting the construction of the Three Gorges Dam in 1999, Premier Zhu Rongji said that, "The construction of the Three Gorges Project is a permanent plan, which is related to the destiny of the country". As an academician of Chinese Academy of Sciences and Chinese Academy of Engineering, Pan Jiazheng said at the Quality Meeting of Three Gorges Project that, "The Three Gorges Project is a permanent project". Therefore, the safety and durability of dams are the key to the RCC dam research in the future. All participating parties must attach great importance to the safety and durability of the dams and keep a clear understanding of the characteristics of RCC damming technology, and strictly control the quality.

1.5.3 The Information-based Management is the Only Way to Realize the Innovation

In order to carry on the RCC damming technology, it is necessary to master the knowledge of RCC damming technology and realize the information-based scientific management according to the modern enterprise management mode. The information management is the only way to keep pace with the times and the development direction of rapid RCC damming technology innovation, which is also an inevitable trend.

It must break the traditional and old management mode in order to realize the scientific information management; the realization of information management is a concrete manifestation of scientificization, refinement and standardization, which can maximize the benefits; the information management can be applied in various aspects, such design, construction, technology, quality, schedule, safety and environmental protection.

For example, the famous Walmart supermarkets are distributed all over the world. Routine sales situations can be mastered at any time through information management. In terms of routine egg sales situation, the egg sales situation can be clearly found at any time through information management by computer.

The information management has been applied in Xiluodu, Dagangshan and other hydropower station projects. For example: such as the management of planned application amount, actual application amount, remaining amount and recovery amount of tie rod screws and nuts of frameworks on the dam concreting surface; the management of labor productivity, quota quantification, and the rate of excellent/good quality; the management of field operating efficiency, such as cable crane operation, lifting cycle time of concrete per tank, vibrating roller rolling times, and spreading and paving. The comprehensive information management of materials, technology, quality, schedule, completion rate and others is implemented.

The technical innovation can only bring into play the advantages of RCC damming technology from one aspect, while the management is another most important aspect for rapid RCC damming. In order to give full play to the rapid damming technology, it is necessary to implement information management.

Currently, the degree of hydropower development in China is less than 50%, and there is still a big gap with the developed countries in the world where the degree of hydropower development reaches over 90%. Therefore, we shall give full play to the mature advantages of rapid RCC damming technology. Under the permission of geological conditions and damming materials, the rapid RCC dam construction technology can effectively shorten the construction duration, save investment and get rid of the outdated concept of "long hydropower construction period", thus highlighting its advantages of price and technology.

Chapter 2

Design and Rapid Construction of RCC Dams

2.1 Overview

The design of RCC dams is particularly important in the application of rapid RCC damming technology. The design of RCC dams is closely related to its performance, which also directly affects its construction progress. Due to different concepts of imperviousness system, mix proportion design and construction technology of early RCC dams, RCC dams body are larger than those of normal concrete dams, and the concrete volume in the dam is larger. Therefore, the early RCC dams are mainly gravity dams. The "RCD" imperviousness system is mainly applied to RCC gravity dams. In other words, normal concrete is used outside, while super hard RCC is used inside. Therefore, the VC value of early RCC is very large, which is generally controlled within 20 – 30 s (15 – 25 s in China). In addition, its fluctuation range is large. Therefore, early RCC is defined as super hard or hard concrete. The excessive VC value can cause that RCC mixtures become loose and lose the stickiness. During mixing, transportation, discharging and paving, RCC mixtures separate each other easily, but the aggregates with large particle sizes congregate. Meanwhile, the cohesiveness, rollability, liquefied bleeding and interlayer bonding quality of RCC mixtures is poor. In this case, water seepage channels are easily formed on the layer (joint) surface, namely the so-called "layer-cake". In order to ensure the imperviousness performance of dams, normal concrete with a thickness of 1 – 3 m is usually used as the imperviousness materials outside the dams. Therefore, the designed section of early RCC gravity dams is larger than that of normal concrete gravity dams.

The biggest difference between RCC dams and normal concrete dams in design is the anti-sliding stability of layer joint surface of RCC dams. Since RCC dams adopt full-section damming technology, and RCC is poured in a whole-block andthin-lift by continuous rolling method, which is completely different from concrete columnar pouring mode of normal concrete dams, the anti-sliding stability and imperviousness performance of layer joint surface is the key to the design of RCC dams. Therefore, special design indexes of shear strength of layer joint surface of RCC dams are proposed. With the continuous innovation of RCC damming technology, the scientific development of materials and the change of imperviousness system, the design concept and design method of RCC dams have been completely changed.

The complex arrangement mode is of strategic significance to the rapid RCC dam construction. The arrangement mode of the complex of RCC dams and power generation structures has great effects on the rapid construction of RCC dams. The arrangement mode of underground powerhouses or headrace-type powerhouses is used for many RCC dams, which avoids the interference to rapid RCC dam construction. However, The arrangement mode of powerhouse at dam toe is used for a few RCC gravity dams, because headrace penstock and inlets fare arranged inside the dam, which seriously restricts the rapid construction of RCC dams. The outlet structures of RCC dam mainly adopt the surface outlet overflow mode. High dams are provided with middle outlets or low outlets for flood discharge, sediment discharge and downstream water supply.

When the height of RCC gravity dam is above 70 m, the slope ratio of downstream dam is general 0.75. When the height exceeds 100 m, the slope ratio at the lower part of the upstream dam is generally within 0.2 – 0.3. The thickness-height ratio of RCC arch dam is generally within 0.17 – 0.30. In recent years, transverse joints of RCC gravity dams are generally within 20 – 30 m wide. When the width of transverse joint is more than 25 m, a short transverse joint is set in the middle of dam section to prevent the vertical crack of dam surface. For RCC arch dams, joints are mainly transverse joints and induced joints.

There is an optimal combination point between RCC arch dam design and rapid RCC damming technology, especially dome dams. If we blindly pursuing small thickness-height ratio and overhang degree, the size of arch dam will be very small. Although concrete is saved, it is not good for rapid RCC construction. Up to now, the RCC volume of 100-metre RCC arch dam in China, such as Longshou, Shapai, Shimenzi, Linhekou, Dahuashui, Zhaolaihe and other RCC dome dams generally within 200 000 – 300 000 m^3. Generally speaking, the construction period of RCC arch dam is generally about 2 years, and the monthly maximum concreting volume is only about 30 000 m^3. The advantages of rapid RCC damming technology have not been brought into full play in the construction of arch dams. The shape of arch dam determines that the area of dam concreting surface is too narrow. It is impossible to give full play to the advantages of large machinery construction. Moreover, the circulation utilization rate of a large number of irregular formwork used during the construction of arch dams is very low, causing high formwork cost. Therefore, it is necessary to consider the characteristics of rapid mechanized construction in the design of RCC arch dams. The RCC gravity arch dam design scheme can give full play to the advantages of rapid RCC damming technology and improve the circulation utilization rate of formwork, and the damming speed and costs may be greater than those of ultra-thin RCC dome dams, so the comprehensive analysis is required to be performed in the design.

The high-level design is the premise and foundation to ensure the rapid construction of RCC dams. RCC damming technology is fundamentally different from damming technology, so the design shall not be limited to the design idea of normal concrete dams. Based on the characteristics of rapid construction of RCC dams, the design shall be elaborately optimized in the aspects, such as complex arrangement, dam body design, dam body structure, material subarea and temperature control, which can give twice the result with half the effort. Similarly, during construction, operating personnel shall be familiar with the detailed design characteristics of RCC dam projects, so as to design and control the construction organization reasonably regarding linear construction duration, key projects and construction difficulties.

2.2 Design and Rapid Construction of RCC Dam

2.2.1 Complex Arrangement and Rapid Construction

Complex arrangement mode is very important to the rapid construction of RCC dam. Compared with the quantity, cost and construction period, the complex arrangement mode of RCC dam is of strategic significance. When arranging RCC dams, the technical characteristics of rapid construction of RCC that is poured in a whole-block and thin-lift shall be taken into consideration in the design, and various structures in the complex are reasonably arranged. For RCC dams, the most ideal complex arrangement mode is that various structures are arranged outside the dam as far as possible based on the complex arrangement principle of earth-rock dam. The dam is the most important part of complex structures. The design specifications for RCC dams clearly points out that: when arranging RCC dams, the structures used for water release, power generation, irrigation, water supply, shipping and others shall be arranged scientifically and reasonably in combination with project tasks, and RCC dams shall arranged separately from power generation structures and outlet structures as far as possible.

No matter in the narrow valley or on the wide river, the powerhouse type of RCC gravity dam in China is mainly underground powerhouse. In general, RCC arch dams do not adopt the arrangement mode of powerhouse at dam toe, which is related to the topography of narrow "V-shaped" valley where the arch dam is located. The arrangement mode that the powerhouse is separated from the dam body is very beneficial to the rapid construction of RCC dam by large machinery. It can also balance the quantities of each structure in the complex, without taking up the linear construction duration. Meanwhile, the standard requirements for diversion and flood prevention can be greatly reduced by taking advantages of the discharge characteristics of RCC dam.

There are also a few RCC gravity dams that the power generation powerhouse is arranged at dam toe. Due to the topography, geology and hydrometeorology in the project area, construction duration or design and other factors, the power generation powerhouse is arranged at dam toe. For the dam with powerhouse at dam toe, headrace penstock, inlets, trash racks and other structures are mainly arranged in the dam body, which brings great difficulties and obstacles to rapid RCC construction. Construction practices prove that the arrangement mode of powerhouse at dam toe directly affects rapid RCC construction and has become the key to control the linear construction duration of RCC dam. It is very unfavorable to shorten the construction duration. In addition, the outlet structures of RCC dam are not molded during the construction period, and the powerhouse at dam toe cannot be used for the discharge of dam body, thus improving the design standards for diversion and flood prevention and increasing project investment.

In terms of the arrangement mode of outlet structures of RCC dam in China, open overflow outlets or overflow surface outlets are preferred, which is mainly to simplify the construction, reduce concrete subarea and facilitate RCC construction. For RCC dams with the height of over 100 m, gravity dams are generally arranged with 4 – 7 overflow outlets, while arch dams are generally arranged with 3 – 4 ones.

See Table 2.1 and Table 2.2 for complex arrangement cases of some RCC gravity dams and RCC arch dams in China.

Table 2.1   Complex arrangement cases of RCC gravity dams in China

S/N	Project title	Maximum dam height (m)	Dam volume ($\times 10^4 m^3$)		Installed capacity (MW)	Arrangement mode of outlet structures	Arrangement mode of headrace system and powerhouse	Year of dam completion
			Rolling	Total quantity				
1	Kengkou	56.8	4.5	6.04	1.5	3 overflow surface outlets	Headrace-type powerhouse at the right bank	1986
2	Tianshengqiao Grade II	60.7	13.1	49.4	1 320	9 riverbed overflow dams	Dam type headrace tunnels and powerhouse on the right bank	1993
3	Fenhe Reservoir II	88	36.2	44.5	96	3 overflow surface outlets and 4 flood discharge and sediment discharge bottom outlets	Headrace-type powerhouse at the right bank	1999
4	Jiangya	131	110	140	300	4 overflow surface outlets and 3 middle outlets	Bank-tower inlet and underground powerhouse on the right bank	1999
5	Dachaoshan	111	75.6	150	1 350	5 overflow surface outlets, 3 bottom outlets and 1 sediment discharge outlet	Inlet on the right bank and underground powerhouse	2001

Continued Table 2.1

S/N	Project title	Maximum dam height (m)	Dam volume (×10⁴ m³) Rolling	Dam volume (×10⁴ m³) Total quantity	Installed capacity (MW)	Arrangement mode of outlet structures	Arrangement mode of headrace system and powerhouse	Year of dam completion
6	Mianhuatan	113	50	61.5	600	3 overflow surface outlets and 1 flood discharge bottom outlet	Bank-tower inlet and underground powerhouse on the right bank	2001
7	Yantan	111	62.6	90.5	1 210	7 overflow surface outlets	Powerhouse at dam-toe	1992
8	Baise	130	212	258	540	4 overflow surface outlets and 3 middle outlets	Bank-tower inlet and underground powerhouse on the right bank	2005
9	Suofengying Hydropower Station	115.8	53	62	600	5 overflow surface outlets Steppe energy dissipation	Bank-tower inlet and underground powerhouse on the right bank	2005
10	Longtan	192	485	750	5 400	7 overflow surface outlets and 2 bottom outlets	Dam type inlet and underground powerhouse on the left bank	2007
11	Guangzhao	200.5	233	271	1 040	3 overflow surface outlets and 2 bottom outlets	Bank-tower inlet and underground powerhouse on the right bank	2008
12	Jin'anqiao	160	251.3	329.3	2 400	55 overflow surface outlets A bottom outlet on the left and a middle outlet on the right	Riverbed powerhouse at dam-toe	Under construction
13	Hongkou	130	62.7	72.7	200	4 overflow surface outlets	Headrace-type ground powerhouse on the left bank	2008
14	Pengshui	116.5	57	93.3	1 750	9 overflow surface outlets	Bank-tower inlet and underground powerhouse on the right bank	2008
15	Jufudu	95	48.0	80.0	285	5 overflow surface outlets and 1 sediment discharge bottom outlet	Dam type and headrace-type ground powerhouse	2007
16	Jinghong	108		339	1 750	7 overflow surface outlets and 2 sediment discharge bottom outlets	Powerhouse at dam-toe	2007
17	Leidatan Hydropower Station	83	20.4	34.0	108	5 overflow surface outlets and 2 bottom outlets	Dam type inlet with deep holes and underground powerhouse on the left bank	2006

Continued Table 2.1

S/N	Project title	Maximum dam height (m)	Dam volume ($\times 10^4 \text{m}^3$) Rolling	Dam volume ($\times 10^4 \text{m}^3$) Total quantity	Installed capacity (MW)	Arrangement mode of outlet structures	Arrangement mode of headrace system and powerhouse	Year of dam completion
18	Gelantan	113	94	139.9	450	5 overflow surface outlets, 2 middle outlets and 2 sediment discharge outlets	Headrace-type ground powerhouse on the left bank	2008
19	Silin	117	82.5	110	1 000	7 overflow surface outlets and 2 bottom outlets	Bank-tower inlet and underground powerhouse on the right bank	Under construction
20	Wudu Hydropower Station	120.3	138	161	150	7 overflow surface outlets and 2 bottom outlets	Powerhouse at dam-toe on the left bank	Under construction
21	Kalasuke Hydropower Station	121.5	235	267	140	4 overflow surface outlets, 1 middle outlet and 1 bottom outlet	Headrace-type ground powerhouse on the right bank	Under construction
22	Gongguoqiao	105	80.5	107	900	5 overflow surface outlets and 2 bottom outlets	Underground powerhouse on the right bank	Under construction
23	Longkaikou	119	229	330	1 800	5 overflow surface outlets, 4 middle outlets and 1 sediment discharge outlet	Powerhouse at dam-toe	Under construction
24	Ahai	138	160	300	2 000	5 overflow surface outlets and 3 flood discharge and sediment discharge bottom outlets	Powerhouse at dam-toe	Under construction
25	Guandi	168	253.5	297	2 400	5 overflow surface outlets and 2 bottom outlets	Bank-tower inlet and underground powerhouse on the right bank	Under construction
26	Ludila	140	154	197	2 100	25 overflow surface outlets and 2 bottom outlets	Bank-tower inlet and underground powerhouse on the right bank	Under construction
27	Guanyinyan	168		647.3	3 000	7 overflow surface outlets, 2 middle outlets and 2 sediment discharge outlets	Powerhouse at dam-toe	Under construction

Table 2.2　　　　Complex arrangement cases of RCC arch dams in China

S/N	Project title	Maximum dam height (m)	Dam volume ($\times 10^4 m^3$)		Installed capacity (MW)	Arrangement mode of outlet structures	Headrace system Arrangement mode of powerhouse	Year of dam completion
			Rolling	Total quantity				
1	Puding	75	10.3	13.7	75	4 open surface outlets	Headrace-type ground powerhouse on the right bank	1993
2	Longshou	75.5	19.3	21.7	59	2 overflow surface outlets, middle outlets and sediment discharge outlets	Gravity dam body inlet and headrace-type powerhouse on the bank	2001
3	Linhekou	100	23.1	29.5	72	5 flood discharge surface outlets and 1 flood discharge tunnel	Headrace system on the left bank and river-side powerhouse	2003
4	Shapai	130	36.4	38.3	36	2 flood discharge tunnels on the right bank	Headrace system on the right bank and ground powerhouse	2002
5	Shimenzi	109	18.9	21.5	6.4	3 flood discharge surface outlets and 1 flood discharge and sediment discharge outlet	Inlet on the left bank and ground powerhouse	2001
6	Zhaolaihe Hydropower Station	105	16.5	19.3	36	3 overflow surface outlets and an emptying tunnel reconstructed from the diversion tunnel	Bank-tower inlet and ground powerhouse on the right bank	2005
7	Dahuashui	134.5	54	64.7	180	3 overflow surface outlets and 2 middle outlets	Gravity dam inlet and headrace-type powerhouse on the left bank	2006
8	Saizhu	72	9.0	10.0	102	4 open surface outlets and 1 sediment discharge bottom outlet	Inlet with high water head on the right bank and headrace-type underground powerhouse	Design
9	Bailianya	104.6	54.1	62.9	50	3 flood discharge middle outlets and a flood discharge tunnel reconstructed from the diversion tunnel	Inlet at the upstream of right bank and underground powerhouse	2008
10	Longqiao	95	13.5	16.0	60	3 overflow surface outlets	Inlet at the upstream of right bank and downstream underground powerhouse	2007

Continued Table 2.2

S/N	Project title	Maximum dam height (m)	Dam volume ($\times 10^4 \text{m}^3$) Rolling	Dam volume ($\times 10^4 \text{m}^3$) Total quantity	Installed capacity (MW)	Arrangement mode of outlet structures	Headrace system Arrangement mode of powerhouse	Year of dam completion
11	Tianhuaban	113	18.2	36.0	180	3 overflow surface outlets, 2 middle outlets and 1 sediment discharge outlet	Bank-tower inlet on the right bank and headrace-type ground powerhouse	Under construction
12	Shannipo	119.4	21.5	26.7	180	3 overflow surface outlets and 2 middle outlets	Bank-tower inlet on the right bank and headrace-type powerhouse	Under construction
13	Yunlong River Grade III	135	17.5	18.3	40	3 overflow surface outlets and 2 middle outlets	Long inlet on the left bank and ground powerhouse	Under construction
14	Wanjiakouzi Hydropower Station	157.6	83.0	98.1	160	3 overflow surface outlets and 2 sediment discharge outlets	Headrace-type ground powerhouse on the right bank	Under construction
15	Luopoba Hydropower Station	114	18.2	20.7	25	3 overflow surface outlets	Inlet on the left bank and ground powerhouse	Under construction

2.2.2 Complex Arrangement and Rapid Construction Cases

2.2.2.1 Complex Arrangement and Rapid Construction Case of Baise Multipurpose Dam Project on Youjiang River, Guangxi

Baise Multipurpose Dam Project on Youjiang River, Guangxi is a large water conservancy project, which focuses on flood control and has comprehensive benefits such as power generation, irrigation, shipping and water supply. The project is located in the middle of Youjiang River at the upstream of the main stream of Yujiang River of Xijiang River system in Guangxi, and the dam is 22 km away from Baise City. Main structures of the complex are composed of main dam, underground power generation system, 2 auxiliary dams and navigation structure.

The main dam is a full-section RCC gravity dam, the axis of which is a broken line. The dam crest is 720 m long, which is divided into 27 dam sections. The crest elevation is 234 m. The maximum height of dam is 130 m. The total concrete volume in the dam is 2.58 million m³, including 2.12 million m³ of RCC. Power generation structures of complex arrangement are the arrangement mode of underground powerhouse on the left bank. The inlets are bank-tower inlets and arranged at the upstream of the left bank of the dam. 3 middle outlets and 4 surface outlets are arranged in the dam sections $4A^{\#} - 5^{\#}$. RCC construction is not preceded in high temperature season, but they are used for discharge in flood season. In this way, the diversion standard is low, and only one diversion tunnel is arranged, which effectively reduces project costs and balances the quantity of each structure in the complex. In particular, the arrangement mode of bank-tower inlets greatly simplifies the complexity of RCC dam body design, and gives full play to the advantages of rapid RCC damming technology.

The Baise Project adopts the arrangement mode that the main dam is separated from power generation powerhouse, which avoids the interference between RCC main dam and powerhouse, and meets the requirements of intensity and speed of RCC construction. The construction of RCC gravity dam is completed in only three dry seasons, which fully reflects the important role of design in the rapid construction of RCC dams.

2.2.2.2 Complex Arrangement and Rapid Construction Case of Longtan Hydropower Station Dam Project

Complex arrangement of Longtan Hydropower Station Dam Project. The design of Longtan Hydropower Station Dam Project has been studied for more than 40 years. In order to shorten the construction duration and save investment, a large number of optimization studies on the complex arrangement mode have been carried out. The arrangement scheme of underground powerhouse that is the most suitable for concrete gravity dam construction is adopted, with remarkable economic benefits. The complex of Longtan Hydropower Station Dam is mainly composed of main dam, flood discharge structures, navigation structures and headrace-type power generation powerhouse. The complex arrangement scheme approved through the preliminary design review is aimed at normal concrete gravity dams. In order to avoid the creep rock mass at the upstream of left bank, the arrangement scheme of powerhouse is "5 units at dam toe + 4 underground units". In the supplement stage of feasibility study, the complex arrangement scheme is studied in depth, focusing on two aspects: firstly, complex arrangement mode suitable for RCC dam construction; secondly, treatment method of creep rock mass at the left dam head.

During complex arrangement, in order to give full play to the advantages of rapid RCC construction, the RCC dam must be arranged separately from the inlets and diversion channels of power generation powerhouse. The parts of dam where the concrete can be rolled shall be relatively concentrated; the dam structure is simple, with little construction interference; the dam body is expanded to the greatest extent for RCC construction, so as to simplify construction technology, accelerate construction progress and reduce project investment. According to the research results, the final complex arrangement scheme is that 9 machines are arranged in the underground powerhouse on the left bank, which creates the good conditions for giving full play to the advantages of rapid RCC construction. This optimized arrangement scheme passes the review of China Renewable Energy Engineering Institute, and recognized by foreign experts from World Bank Special Advisory and HARZA in the United States and others. The advantages of the complex arrangement scheme adopted in the Longtan Hydropower Station Dam Project that the dam is arranged separately from power generation structures are as follows:

1. The dam construction is a key item in the linear construction duration, and the riverbed dam section is the key to restrict the power generation duration. Since the underground powerhouse is used in the complex arrangement scheme, there is no powerhouse behind the dam and no inlet and penstock in the dam. Therefore, the main ways for the project to bring benefits into full play in advance is to simplify the dam structure effectively, reduce the construction interference and accelerate the dam construction.
2. The flood season of Hongshui River is long and the flow is large, so there are certain risks in respect of construction, flood control and energy dissipation. The arrangement scheme of underground powerhouse can not only simplify diversion measures, but also make the arrangement of permanent outlet structures more flexible and the room for adjusting and optimizing the energy dissipation facilities larger.
3. There is no powerhouse behind the dam and no headrace penstock in the dam in this scheme, which enhances the rigidity of the upper dam, helpfully improves the stress of dam and enhances the seismic resistance of dam.
4. There is only one powerhouse in this scheme, which is convenient for operation, management and maintenance; when the normal pool level is raised subsequently, the dam heightening has few effects on the normal operation of hydropower station.

2.2.3 Design, Structure and Rapid Construction of Dam Body

2.2.3.1 Design and Rapid Construction of the Dam

The design of dam body conforms to the complex arrangement mode first, and the optimization and adjustment of specific structure positions is only local; then, when determining the position and structural type of each structure, it is necessary to consider to meet the requirements for the scale and functions, minimize the quantity and save investment; meanwhile, the RCC parts shall be relatively concentrated, so as to reduce the holes in the dam, simplify the dam structure, expand the scope of RCC as far as possible, minimize the interference to rapid RCC construction, give full play to the advantages of rapid RCC construction.

According to the design requirements of RCC gravity dam, the shape design of RCC gravity dam shall be simple and convenient for construction. The minimum width of dam crest shall not be less than 5 m. The slope of downstream dam shall be a plumb plane. The downstream slope ratio can be optimized according to the section of normal concrete gravity dam. The slope of upstream and downstream dam surface of RCC gravity dam is basically similar to that of normal concrete gravity dam, and the standard slope of gravity dam is 1:0.75. The upstream dam surface of middle dams and low dams, such as Kengkou dam, Gaobazhou dam, Mianhuatan dam and Jiangya dam, is usually a lead plane, and the downstream dam surface has the same slope value. However, for high dams, the upper part of the upstream dam surface is a vertical plane, and the lower part is a slope of 1:0.2 – 1:0.3. For example, the Dachaoshan Dam, Baise Dam, Longtan Dam, Guangzhao Dam, Jin'anqiao Dam and Guandi Dam, whose height is greater than 100 m.

Single-curvature dams or dome dams are adopted for most of the RCC arch dams, but their shape of arches are relatively simple compared with normal concrete arch dams, for the reason of accommodating the characteristics of RCC damming. The stress state of RCC arch dams is different from that of normal concrete arch dams. Because the RCC arch dam is rolled in full section in thin-lifts and whole-block, the arch has been formed after rolling. The temperature stress generated by the hydration heat temperature rise of the RCC during the construction would continuously affects the stress during operation as the dam temperature decreases. Whats more, because the arch of the RCC arch dam has been formed during construction, the dead weight stress distribution of the dam is also different from that of normal concrete arch dams.

Cases of dam body design and structure construction of some RCC gravity dams and arch dams completed or under construction in China are shown in Table 2.3 and Table 2.4.

Table 2.3　　Cases of dam body design and structure construction of some RCC gravity dams completed and under construction in China

S/N	Project title	Dam crest width (m)	Dam crest length (m)	Dam slope Upstream	Dam slope Downstream	Dam sections with joints	Imperviousness system
1	Kengkou	5	122.5	Vertical	1:0.75	No transverse joint	Centralized seepage control using asphalt mortar on the upstream face
2	Tianshengqiao Grade II	6.5	471.0	Lower part of the dam 1:0.4	1:0.54	26 dam sections	Normal concrete with RCD method
3	Fenhe Reservoir II	7.5	227.5	—	—	5 dam sections	II gradation RCC
4	Jiangya	12	369.8	Vertical	1:0.8	13 dam sections	II gradation RCC

Continued Table 2.3

S/N	Project title	Dam crest width (m)	Dam crest length (m)	Dam slope Upstream	Dam slope Downstream	Dam sections with joints	Imperviousness system
5	Dachaoshan	16	460.4	Vertical	1:0.75	23 dam sections	II gradation RCC
6	Mianhuatan	7	308.5	Vertical	1:0.75	7 dam sections	II gradation RCC with a thickness of 1/15 that of the head
7	Yantan	20	525.0	Vertical	1:0.65	Rolling 10 dam sections	Normal concrete with RCD method
8	Baise	10	720.0	Lower part of the dam 1:0.2	1:0.75	27 dam sections	II gradation RCC with imperviousness coating
9	Suofengying Hydropower Station	8	164.6	Lower part of the dam 1:0.25	1:0.70	9 dam sections	II gradation RCC
10	Longtan	18	761.26	Lower part of the dam 1:0.25	1:0.73	35 dam sections	II gradation RCC with imperviousness coating
11	Guangzhao	12	410.0	Lower part of the dam 1:0.25	1:0.75	20 dam sections	II gradation RCC with imperviousness coating
12	Jin'anqiao	16	640.0	Lower part of the dam 1:0.3	1:0.75	21 dam sections	II gradation RCC
13	Hongkou	6	374.0	Vertical	1:0.75	13 dam sections	II gradation RCC with imperviousness coating
14	Pengshui	20	325.53	Vertical	1:0.70	14 dam sections	II gradation RCC with waterproof coating
15	Jufudu	13	320.0	Lower part of the dam 1:0.2	1:0.8	13 dam sections	II gradation RCC
16	Jinghong	21	704.6	Vertical	1:0.8	27 dam sections	Abnormal II gradation RCC + permeable crystal
17	Leidatan Hydropower Station	8	201.5	Vertical	1:0.7	—	II gradation RCC
18	Gelantan	10	466	Lower part of the dam 1:0.2	1:0.75	16 dam sections	II gradation RCC with imperviousness coating
19	Silin	20	310.0	Vertical	1:0.70	16 dam sections	Abnormal II gradation RCC
20	Wudu Hydropower Station	10	727.0	Vertical	1:0.75	30 dam sections	Abnormal II gradation RCC

Chapter 2 Design and Rapid Construction of RCC Dams

Continued Table 2.3

S/N	Project title	Dam crest width (m)	Dam crest length (m)	Dam slope Upstream	Dam slope Downstream	Dam sections with joints	Imperviousness system
21	Kalasuke Hydropower Station	10	1 570	Lower part of the dam 1:0.15	1:0.75	83 dam sections	Abnormal Ⅱ gradation RCC
22	Gongguoqiao		356	Vertical	1:0.7	17 dam sections	Abnormal Ⅱ gradation RCC
23	Longkaikou	10	768	Lower part of the dam 1:0.20	1:0.75	30 dam sections	Abnormal Ⅱ gradation RCC
24	Ahai		482			19 dam sections	Abnormal Ⅱ gradation RCC
25	Guandi	14	516	Lower part of the dam 1:0.25	1:0.75	24 dam sections	Abnormal Ⅱ gradation RCC
26	Ludila	11	622	Lower part of the dam 1:0.20	1:0.75	29 dam sections	Abnormal Ⅱ gradation RCC
27	Guanyinyan	12	752	Lower part of the dam 1:0.1	1:0.75	30 dam sections	Abnormal Ⅱ gradation RCC

Table 2.4 Cases of dam body design and structure construction of some RCC arc dams completed and under construction in China

S/N	Project title	Dam crest width (m)	Arc length of dam crest (m)	Thickness-height ratio	Shape of arc	Parting of dam body	Imperviousness system
1	Puding	6.3	165.7	0.376	Asymmetric gravity arch dam	2 induced joints	Ⅱ gradation RCC with interlayer grouting
2	Longshou	5	140.84	0.17	Double curvature arch dam	2 induced joints	Ⅱ gradation RCC
3	Linhekou	6	311.0	0.282	Double curvature arch dam	5 induced joints 3 transverse joints	Ⅱ gradation RCC waterproof coating on lower part of the dam
4	Shapai	9.5	250.3	0.238	Three-center circular arc Single curvature arch dam	4 transverse joints	Ⅱ gradation RCC with waterproof coating
5	Shimenzi	5	176.5	0.285	Double curvature arch dam	Release joints on the abutment and short joint of the downstream crown	Ⅱ gradation RCC with imperviousness coating
6	Zhaolaihe Hydropower Station	6	198.05	0.17	Double curvature arch dam	3 induced joints 2 transverse joints	Ⅱ gradation RCC

Continued Table 2.4

S/N	Project title	Dam crest width (m)	Arc length of dam crest (m)	Thickness-height ratio	Shape of arc	Parting of dam body	Imperviousness system
7	Dahuashui	7	198.43	0.171	Double curvature arch dam	2 induced joints	Ⅱ gradation RCC
8	Saizhu	7	160.16	0.194	Double curvature arch dam	2 induced joints	Ⅱ gradation RCC
9	Bailianya	8	421.86	0.287	Double curvature arch dam	4 induced joints 5 transverse joints	Ⅱ gradation RCC with waterproof coating
10	Longqiao	6	155.64	0.23	Double curvature arch dam	4 transverse joints	Ⅱ gradation RCC
11	Tianhuaban	6	159.8	0.215	Double curvature arch dam	2 induced joints	Ⅱ gradation RCC
12	Shannipo	6	205.4	0.201	Double curvature arch dam	2 induced joints	Ⅱ gradation RCC
13	Yunlong River Grade Ⅲ	5.5	130.45	0.134	Double curvature arch dam	2 induced joints	Ⅱ gradation RCC
14	Wanjiakouzi Hydropower Station	9.0	378.3	0.210	Double curvature arch dam	6 induced joints 2 transverse joints	Ⅱ gradation RCC
15	Luopoba Hydropower Station	6	191.72	0.175	Double curvature arch dam	2 induced joints 2 transverse joints	Ⅱ gradation RCC

2.2.3.2 Structure of Dam and Rapid Construction

The structure of a RCC dam body is different from that of the normal concrete dam body. The construction method of RCC is fundamentally different from the column pouring of normal concrete; In order to accommodate the construction characteristics of continuous rolling compaction of large-area thin-layer pouring of the RCC, the RCC dam has no longitudinal joints. For example, the RCC gravity dams such as the Longtan and Guangzhao dams of 200 m high, and Baise, Jinanqiao and Guandi dams of vast area, such dams have no longitudinal joints. In recent years, the maximum transverse joint of RCC gravity dams completed is generally kept around 30 m. The transverse joint of dams such as the Guangguang, Longtan and Jinanqiao dams is mostly within 30 m. When the transverse joint exceeds 25 m, a short joint with a 3 m × 5 m depth is generally added in the middle of the upstream face of the dam block, which would effectively prevent cracks on the upstream face of the dam. As the improvement of the RCC construction technology and joint cutters, the joint forming has been very simple, and the dam joints are no longer a problem for whole-block construction of the RCC.

With the full-section RCC damming technology, Ⅱ RCC gradation with rich cementing materials and the GEVR on the dam surface are adopted by for seepage control of RCC dams. According to the data of projects completed or under construction, the ratio of the impervious structure thickness and the head is generally 1/15 – 1/12, and the thickness is 2 – 10 m based on the height of the dam. Based on the situation that some RCC dam placing surfaces are large and the construction is less susceptible, GEVR mixed by mixing plants can be adopted to ensure the quality and progress. For example, in the GEVR construction of projects such as the Baise, Guangzhao and Jin'anqiao projects, GEVR mixed by mixing plants is adopted by part of the construction based on practical construction conditions, which has achieved good results.

In recent years, in addition to the Ⅱ gradation RCC with richcementing materials on the upstream face and GEVR on the dam surface, the auxiliary waterproof materials on dam surface, generally, are also adopted for impervious structure of dams, i. e. , the polymer cement waterproof coating and imperviousness crystal are adopted for seepage control. Whats more, admixing imperviousness crystal directly into the GEVR grout is also adapted for seepage control in some projects, such as the Jinghong Project.

The drainage system of the dam is the key to the seepage control of the dam. A RCC dam must be equipped with a complete drainage system. As long as the drainage is arranged properly and prepared for unseen circumstances, the partial defects of the impervious structure will not cause sharp rise of the uplift pressure in the dam and layers. The drainage system of the dam is closely connected with the upstream impervious structure. The drainage system includes drainage galleries, vertical drainage pipes, etc. Methods such as drilling holes and embedded permeable pipes are often adopted by the vertical drainage holes of the dam body. Because the drilling holes are not easy to be blocked, and they have the best effect on drainage and the least interference in RCC construction. Embedded permeable pipes tend to interfere with the rolling of RCC construction and the rapid construction. Currently, the permeable pipes buried in dams are all finished permeable pipes made of nylon weaving, greatly improving the pipes quality. The early fragile non-fines drainage pipes are eliminated.

The arrangement of inside galleries of RCC dams is basically the same as that of normal concrete dams. In order to facilitate the construction of RCC, prefabricated galleries have been adopted in recent years. Its very efficient and quality-assured to splice and install the prefabricated galleries on the placing surface, which reduces the interference in the rapid RCC construction. However, too much or too complicated dam galleries will seriously affect the rapid RCC construction.

2. 2. 4 Discussion on Design of RCC Volumetric Weight and Dam Sections

Both the specification for load design of hydraulic structures and the design specifications for RCC dams have made requirements for the RCC volumetric weight. The dead weight standard of hydraulic structures can be calculated and determined according to the structure design dimensions and the materials volumetric weight. The volumetric weight of RCC for high dams should be determined by tests according to the material source, mix proportion, and construction conditions, and that for middle and low dams can be determined according to the parameters in similar projects. When no test data can be referenced, volumetric weight of 23. 5 – 24 kN/m^3 is feasible.

Volumetric weight is the most critical parameter for the design load of a gravity dam. The range of the volumetric weight determines the size of the dam section and its anti-sliding stability. A large quantity of project cases prove that the volumetric weight i. e. the apparent density of RCC is larger than that of the normal concrete, which mainly because the mix proportion and the construction method of RCC are quite different from these of the normal concrete. As the RCC mixture is in a state of semi-plastic concrete without slump, its cementing material consumption and the unit water consumption are less than these of the normal concrete, and the aggregate consumption is larger than that of the normal concrete. In addition, the RCC construction is performed by concreting in thin-lifts with vibrating mill for rolling, its compaction is significantly superior to that of the normal concrete.

The present RCC is semi-plastic concrete without slum, and the concerned anti-sliding stability of layer (joint) surface and weak surface of water seepage are solved. Therefore, when designing the section of a RCC dam, the volumetric weight i. e. the apparent density, of RCC should be carefully analyzed and studied. Although the section of a RCC dam is the same as that of the a normal concrete dam, its volumetric weight i. e. the apparent density, has been proved by large numbers of projects to be greater than that of normal concrete. In-depth research in design shall be performed to verify that whether the section of a RCC dam is more optimal than that of a normal concrete dam.

For example, manufactured diabase aggregate is adopted for the RCC of Baise Multipurpose Dam Project. The quasi Ⅲ gradation (maximum grain size of 60 mm) and the Ⅱ gradation respectively adopted into the interior and exterior of the dam with the RCC apparent density of 2 650 kg/m^3 and 2 600 kg/m^3 respectively, and the apparent density of normal concrete of 2 600 kg/m^3 and 2 560 kg/m^3 respectively. The results show that the RCC apparent density of the same gradation is respectively 50 kg/m^3 and 40 kg/m^3 heavier than that of the normal concrete. The apparent density is respectively converted into the volumetric weight of 26 kN/m^3 and 25.5 kN/m^3, both are greater than the design volumetric weight 24.5 kN/m^3.

Likewise, manufactured basalt aggregate is adopted for the RCC of Jin'anqiao and Guandi hydropower station projects. The Ⅲ gradation RCC apparent density of the dam interior is 2 630 kg/m^3 and 2 660 kg/m^3 respectively, and the Ⅲ gradation apparent density of normal concrete is 2 620 kg/m^3 and 2 620 kg/m^3 respectively. The RCC apparent density is respectively 10 – 40 kg/m^3 heavier than that of the normal concrete. The apparent density is respectively converted into the volumetric weight of 25.7 kN/m^3 and 26.1 kN/m^3, both are greater than the design volumetric weight 24.5 kN/m^3.

A large numbers of RCC projects show that the RCC apparent density is generally greater than that of the normal concrete, and it is necessary to analyze the same volumetric weight of RCC and normal concrete in the design. If the apparent density of Baise, Jinanqiao and Guandi dams are calculated according to the actual RCC volumetric weight value of 25.5 – 26.1 kN/m^3, it will be 6% – 8% larger than the design dead weight load of 24 kN/m^3. In this case, whether the dam section of a gravity dam as such can be optimized by about 6% is debatable.

The volumetric weight of concrete is closely related to the damming materials. When the aggregate density is small, the RCC dead weight load is smaller than the design volumetric weight value of 23.5 kN/m^3. The dam section cannot be reduced but need to be increased. For example, the RCC gravity dam of Hongkou Hydropower Station in Fujian has a low aggregate density hence it is difficult to reach the design value of RCC volumetric weight.

2.3 Design Index of RCC and Material Zoning of Dam

2.3.1 Analysis of RCC Design Indexes

The wide use of mineral admixtures in RCC has effectively delayed the hydration heat temperature rise and the development of earlier concrete strength. Therefore, it is necessary to keep the concrete design age to be 180 d or even 360 d. Due to the small cement amount and the high content of active admixtures content such as fly ash and others, the later RCC strength increases significantly. The later strength increase is related to the types and amount of cement, admixtures, additives, etc. A large number of project test results show that the growth rate of compressive strength of general RCC of 28 d, 90 d, and 180 d is roughly 1:(1.5 – 1.7):(1.7 – 2.0).

The mismatching of the design indexes of hydraulic concrete such as compressive strength, design age, frost resistance grade and ultimate tensile value lasts for a long time. The frost resistance grade has been taken as an important index to evaluate the concrete durability of dam in the Three Gorges Phase Ⅱ Project since 1997, then China has listed the frost resistance grade as the main design index for concrete durability in mild regions such as south China. Especially in recent years, the design indexes of frost resistance grade are becoming higher and higher in both external or internal concrete of the dam. Actually, according to the core samples drilled from hardened concrete of the dam, the concrete core samples have a lower frost resistance than that of the samples at the outlet. Taking the single frost resistance grade as an index to evaluate the concrete durability requires continuous in-depth study on the tests.

For example, for a RCC gravity dam in the mild southwestern region, the design indexes of the RCC inside the dam are: strength of 15 MPa, age of 90 d, and frost resistance of F100. The same frost resistance grade of F100 is adopted in the internal and external impervious areas of the RCC dam. Whether the reason is sufficient, and whether it is for the durability requirement or the crack resistance requirement is unclear. In order to meet the F100 frost resistance requirements of the internal RCC of the dam, a large amount of air entraining agent is needed. Due to the low design strength grade and large air content in internal RCC, the ultimate tensile value is reduced. Otherwise, a large amount of super RCC is needed to meet the F100 frost resistance requirements. A large number of test results show that the design indexes of concrete are not matched. There is a price to pay to meet one of the control indexes, i. e., using large amount of super concrete and more cement, and increasing temperature rise and stress. Practices indicate that large amount of super concrete is unfavorable for temperature control and crack resistance. Therefore, the designs shall follow the methodology of entirety theory. The design indexes of the dam RCC should be scientific and reasonable, realistic, and matching with each other, rather than mindlessly imitating and simply copying the other designs.

2.3.2 Discussion on RCC Design Age

When the design age of RCC is determined, it is necessary to make full use of the later strength of RCC and simplify the temperature control measures in view of the characteristics of RCC such as small amount of cement, large admixture, slow temperature rise of hydration heat and low early strength. The *Design Specifications for RCC Dams* SL 314—2004 specifies that: the compressive strength of RCC shall be tested with the strength of 180 d (or 90 d). The indexes such as impermeability, frost resistance, tensile strength and ultimate tensile value shall be tested with the concrete of same design age as that of the compressive strength test.

The specification for construction organization design in water conservancy and hydroelectric projects consists of specifications for water conservancy industry and power industry, i. e., "SL" refers to the water conservancy specification, and "DL (DL/T)" refers to the power specification. Throughout the design age of RCC, the compressive strength of concrete of 180 d is adopted by most of the RCC dams of water conservancy design, such as Miantiantan, Baise, Karasuke, and Wudu reservoirs. In the RCC dams of hydropower designs, by contrast, the compressive strength of concrete of 90 d is the mostly adopted while that of 180 d is the less adopted. For example, designs in projects such as Dazhaoshan, Lihekou, Shapai, Longtan, Guangzhao, and Jinanqiao projects. Whether the compressive strength of concrete of 90 d shall be adopted is questionable.

In early 1990s, the 180 d design age of concrete has been adopted for the high arch dam of Ertan Project in China. In recent years, the concrete compressive strength of 180 d design age has been adopted in the large super-high arch dams, such as the Laxiwa, Xiaowan, Xiluodu and Jinping Grade I projects with a dam height of 250 m, 292 m, 278 m and 305 m respectively. In recent years, although the construction of 100 m RCC dams has increased year by year, the overall height of the dam has not exceeded that of normal concrete dams. Currently, only Longtan and Guangguang dams are 200-metre RCC dams. Now, the gravity dam of the Jinsha River Xiangjiaba Hydropower Station has an installed capacity of 6 400 MW, a dam height of 162 m, and the concrete design age of 180 d (the lower part is optimized as RCC sections). Recently, the 180 d design age has been adopted in the RCC dams under construction in the hydropower designs such as Gongguoqiao, Guanyinyan, and Ludila projects. The design age and design indexes of RCC are worth pondering on, and there should be a breakthrough in the traditional concept on the design age.

2.3.3 Dam Material Zoning and Rapid Construction

The full-section damming technology is applied for RCC dams. Material zoning of the dam i. e. the concrete zoning has a great effect on rapid construction. Because different parts of the RCC dam have different requirements for concrete designs with different concrete design indexes and concrete gradations, interferences and confusions are caused in the construction. For example, the impervious area of upstream face of the dam is designed with Ⅱ gradation RCC with rich cementing materials; the inside of the dam is designed with Ⅲ gradation RCC; and the periphery of the formworks and galleries and dam slopes are constructed with GEVR. The RCC does not be applied to the same design index, which is very unfavorable for rapid RCC construction. According to relevant project experiences, the material zoning of RCC dams should be simplified as much as possible, and the scope of material zoning should be adjusted during construction due to structural constraints.

Compared with foreign RCC dams, the zoning materials of dams in domestic RCC dams is too complicated. For example, the height and type of some domestic RCC gravity dams are basically the same. Their height is far different from that in the Longtan and Guangzhao projects, and they are not 200 m high dams. However, the design indexes of lower, middle and upper parts are still applied into the materials zoning of dams. If the design route aims at increasing the concrete strength and ultimate tensile value to improve crack resistance, the route may be counterproductive. Because concrete strength grade and ultimate tensile value is improved by reducing the water-cement raido and increasing the amount of cement, it is unfavorable for temperature control and crack resistance. Therefore, there is an optimal combination point of the relationship between strength, ultimate tensile strength, temperature control and crack resistance. In the impervious area on the upstream face of the dam, the Ⅱ gradation enriched-grout RCC is used. Some dams also have different design indexes, which are very unfavorable for rapid RCC construction.

In recent years, the design has been studied, explored and applied in the zoning materials of RCC dams. For example, the full-section Ⅲ gradation RCC construction is adopted by the RCC gravity arch dam of Hongpo Reservoir in Kunming Province; and the Ⅱ gradation and Ⅲ gradation RCC of the Phase Ⅲ cofferdam of Three Gorges Project are applied to the design index $R_{90}15W8F50$; RCC with grade of $R_{180}15W4F25$ is adopted by the inner part of RCC main dam in Baise Project, and one gradation (quasi Ⅲ) gradation is adopted for the upper part. One design index and one gradation (the maximum grain size is 50 mm or 40 mm) are adopted generally by the foreign RCC dams in Myanmar of Burma, Thailand, Vietnam, etc. Due to the simplification of material zoning, the mutual construction interference is obviously reduced, and the rapid RCC construction is effectively accelerated. Regarding the material zoning of the dam, the design should draw on the material zoning experience of the dam in these projects, and continuously carry out technological innovation and in-depth research to provide technical support for the rapid RCC damming.

2.4 Introduction to RCC Dam Design

2.4.1 Introduction to Longtan RCC Dam Design

2.4.1.1 Project Overview

Longtan Hydropower Station is located at the upstream area of Hongshui River, and is 15 km from Tian'e County, Guangxi. The tributary area for the dam is 98 500 km², accounting for 71% tributary area of the Hongshui River. The power station is planned to be developed in two phases, i. e. the normal pool level is designed to be 400 m in long stage, and 375 m in the initial stage. The dam is a RCC gravity dam. In the initial stage, the normal pool level is 375 m, the maximum dam height is 192 m, the total storage capacity is 16.21 billion m^3, and the effective storage capacity is 11.15 billion m^3, which is an annual regulating reservoir with an installed capacity of 4.2 million kW and an average annual power generation of 15.67 billion kW · h, a guaranteed output of 1.234 million kW. In the later stage, when the normal pool level is 400 m, the maximum dam height is 216.5 m, the total storage capacity is 27.27 billion m^3, and the effective storage capacity is 20.53 billion m^3. It is an annual regulating reservoir with an installed capacity of 5.4 million kW, an average annual power generation of 18.71 billion kW · h, and a guaranteed output of 1.68 million kW.

2.4.1.2 Hydrometeorology and Engineering Geological Conditions

1. Hydrology and meteorology. Hongshui River is the middle and upper reaches of Xijiang River system in the Pearl River Basin. Hongshui River is 1 573 km long and its tributary area is 138 340 km^2. The basin features subtropical climate, and the runoff is mainly formed by rainfall. The annual runoff distribution is: the runoff in May-October accounts for 82.8% of the annual rainfall, and that in November-April of the following year accounts for 17.1% of the annual rainfall. The average annual discharge of the dam is 1 630 m^3/s, and the average annual runoff is 51.4 billion m^3. The sediment of the Hongshui River is mainly the suspended sediment. The average annual sediment transport rate of the dam is 1 660 kg/s, the average annual sediment runoff is 52.3 million t, and the average sediment concentration is 1.04 kg/m^3.

2. Project geology. The dam is located in a relatively stable block, which is a weak shock region with no regional active faults. There is no geological background for earthquakes. The regional earthquake risk is mainly affected by the peripheral earthquakes. The basic seismic intensity of the dam and the seismic intensity may induced by the reservoir is Level 7. The peripheral surface and groundwater divides of the reservoir are higher than the storage level of the reservoir. The reservoir basin is mainly composed of Triassic sandstone and shale. There is no leakage in the reservoir and the overall stability of reservoir bank is good.

   NSG materials are unavailable in local market thus concrete aggregates need to be produced and rolled by quarrying. The materials in the Dafaping and Macun quarries are all thick-heavy limestone and dolomitic limestone of the lower Permian system, with a very small amount of lime dolostone intercalation and flint nodules. The average saturated compressive strength of the materials is 81.3 MPa, the softening coefficient is 0.85 – 0.95, the saturated water absorption rate is 0.19%, and the density is 2.7 g/cm^3. The bedrocks of the both quarries are exposed, and there is basically no abandoned layer. The mining conditions are good. The reserves and qualities of both quarries can meet the design requirements.

2.4.1.3 Complex Arrangement of the Project

The main structures of multipurpose dam of Longtan Hydropower Station include the RCC gravity dam, flood discharge structures, navigation structures, and water conveyance and power generation system. The Longtan Hydropower Station is a Grade I project with a large (1) project scale. The main structures are designed according to Grade I, and the secondary structures are designed according to Grade III.

2.4.1.4 Main Structures

1. Water retaining structures and flood discharge structures. The dam of Longtan Hydropower Station is designed according to 500-year flood and checked according to 10 000-year flood.

   The RCC gravity dam is adopted by the water retaining structure. From right to left, it is the retaining dam section, the ship lift dam section, the riverbed retaining dam section, the overflow dam section and the elevator shaft dam section on the right bank, and the retaining dam section and the water inlet of the powerhouse dam section on the left bank; the total length of the dam axis is 761.26 m in initial stage and 849.44 m in later stage. The flood discharge structure is arranged in the riverbed dam section, with 7 surface outlets and 2 bottom outlets. The hole mouth size of the surface outlets is 15 m × 20 m (width × height, the same below). The bottom outlets are arranged on both sides of the surface outlets, the inlet elevation of bottom outlets is 290.00 m, and the controlled dimension is 5 m × 8 m. The function of the surface outlets flood is discharge and emptying reservoir, and that of the bottom outlets is later stage diversion, reservoir emptying and sand flushing etc. The flip buckets are used for energy dissipation by trajectory jet in narrow slit in the surface and bottom outlets.

Shape and structure of the dam. The dam shape is determined by the optimization method. The typical section shape parameters are shown in Table 1. After the optimization, the maximum bottom width of the overflow dam section is 168.58 m, and the ratio of the bottom width to the dam height(B/H) is 0.779, and the bottom width of the maximum water retaining dam section is 158.45 m with the B/H of 0.806, all of which are economical section shapes.

2. Water conveyance and power generation system. The water conveyance and power generation system of Longtan Hydropower Station is arranged on the left bank of the dam. Its main structures include water inlet, headrace tunnels, main powerhouse, bus tunnel, main transformer chamber, tailwater surge shaft, tailwater tunnel, tailwater outlet, switch station, outgoing line platform, and access tunnel. The water inlet 1, tailwater outlet, switch station, outgoing line platform and central controlstructure are arranged on the ground, the others are arranged underground on the left bank. The design return period of the flood of the underground powerhouse is 300 years, and the check return period of the flood is 1 000 years.

The final installation of the powerhouse is 9 units, which are arranged in the mountain behind the left bank dam. The water inlet is a dam type inlet with a single unit and a single tunnel headrace. There are 1 tailwater surge shaft and 1 tailwater tunnel for every three units. The headrace tunnel has a diameter of 10 m, and the tailwater tunnel has a diameter of 21 m.

The 9 units of the main powerhouse are arranged continuously, with a unit spacing of 32.5 m; the main and auxiliary erection bays are arranged at both ends of the powerhouse. The main erection bay is located at the right end of the main powerhouse with a length of 60.0 m. The auxiliary erection bay is 36.0 m long and is located at the left end of the main powerhouse; total length of the powerhouse 388.5 m; the net width of the main powerhouse is 28.5 m, and the span above the rock anchor beam is 30.3 m, and the erection bay has the same width as the main engine room. The total height of the powerhouse is 74.4 m.

The main transformer tunnel is located in the downstream of the powerhouse, with the axis parallel to the main powerhouse; the main transformer tunnel is arranged in two floors, the lower floor is the main transformer chamber, and the upper floor is the 500 kV cables and the neutral point equipment of main transformer; the total length of the main transformer tunnel is 405.5 m, with the width of 19.5 m. Height: 32.2 – 34.2 mm.

The GIS switch station and the outgoing line platform is centrally arranged on the hillside that is at about 500 m downstream of the left bank. The GIS switch station has an elevation of 335.00 m, the width of 17.5 m, the length of 217.0 m and the height of 20.0 m, and is arranged with the GIS erection bay, the relay protection room, the storage battery and the special room of various auxiliary equipment.

3. Navigation structure. The channels of the reach of the dam site are Grade channels (500 t). The total water head of the navigation structures in Longtan Hydropower Station can reach 181.00 (in later period), and the amplitude variation of upstream water level is 60.00 m. The navigation structure is arranged on the right bank with the Grade II vertical ship lift adopted, the maximum lifting height of the two grades ship lifts are 88.5 m and 92.5 m respectively, and the both lifts are connected by the passing channels. The axis of the ship lift is orthogonal to the dam axis, and the navigation structure is mainly composed of the upstream approach channel, the navigation dam section, the tower of the Grade ship lift, the middle passing channel, the tower of the Grade ship lift and the downstream approach channel, with a total length of 1 800.0 m.

2.4.1.5 Project Characteristics and New Technology Application

Longtan Project is a remarkable huge project, the creep high slope on the left bank, the high RCC dams, the huge underground powerhouse and the vertical ship lift with high lift are the highest world records, and the project characteristics are mainly shown in the following aspects:

1. The RCC damming technology is adopted to construct a 200 m high dam. The powerhouse is of full underground complex arrangement, which simplifies the arrangement of the flood discharge and energy dissipation structure, and under the condition that each structure meets the functional requirements and operates safely and conveniently, the RCC application range should be extended to the largest extent, so as to create a good condition for the RCC rapid construction and shortening the construction period.

   According to the development level of the RCC damming technology at home and abroad, the dam body concrete material parameters and the shape geometric parameters are selected reasonably, the dam body is determined by the optimum design method, and the systematic static and dynamic analysis of dam structure is analyzed, and under the condition of ensuring the safety operation of the dam, the dam should have the good stress conditions and the dam body consumes less concrete volume.

   According to the research results of the seepage control and temperature control analysis of the RCC dam body, the impermeability combination of the GEVR and the II gradation concrete is adopted as the impervious structure of the dam body in the 200-metre RCC dam, which simplifies the dam structure and is convenient for the RCC rapid construction.

   A large number of laboratory tests and field rolling tests have been conducted for the research of the RCC mix proportion, and the adequate research and demonstration of the RCC layer surface shear parameters have been conducted based on the characteristics of the dam height and dam body shape of Longtan Dam, which determine the mix proportion with good operating performance and physical and mechanical property that can meet the design requirements.

   A full set of design and research results, such as the RCC mix proportion, temperature control standards and measures, the RCC placing scheme and process, solve the technological problems of RCC continuous construction at high temperature and in rainy seasons, so as to create conditions of shortening the damming period.

2. The vertical ship lift with high lift is adopted to solve the navigation problem of the multipurpose dam. The vertical ship lift with high lift is adopted in Longtan Project to solve the navigation problems of the multipurpose dam, with the advantages of short passing time, large passing capacity, and no contradiction between navigation and power generation for water consumption. Both the total lift (181 m) of the vertical ship lift and the amplitude variation of upstream water level (60 m) of Longtan Project rank first in the world, the comprehensive innovation technology is adopted to solve a series of design problems that still have no mature experience to follow, such as the safe and stable operation of the vertical ship lift with high lift, the structural design of the head bay subjected to the two-way water load (the water head of 100 m along the direction of water flow and the water head of 70 m along the dam axis direction), the static force and the wind and seismic resistance design of the high-rise structure (110 m tower column) of the thin-wall, and the design of structures with large span middle channels.

3. The traditional method and the latest modern research means are comprehensively adopted to design the 400 m water inlet in the high slope on the left bank. The water inlet arrangement area of the underground powerhouse in the left bank is closely connected with the crooked, tilted and deformed natural slope, with the vertical depth of the toppling deformation about 30 – 76 m, and the volume about 12.88 million m³. The height of the slope excavated can reach 435 m, which is the typical high rocky slope with the anti-inclined layered structure. The slope foot is excavated with nine headrace tunnels with the diameter of 12 – 13.35

m, in order to save the excavated volume and the concrete volume, the water inlet dam section should be close to the slope foot. The water inlet dam section is closely connected with the rock slope, and the slope has the particularity with high use requirements and operation conditions which are different from these of the general slope. Through years of design and research and the scientific and technological research, the multiple mathematical and physical model analysis methods are adopted to study the dynamostatic steady deformation of the slope, steady deformation aging of slope and the interaction of the dam, slope and cave, so as to determine the design standard and the excavation and reinforce support scheme. The slope is excavated based on the principle of meeting the structural arrangement requirements of the power station water inlet and the basic self-stabilization of the slope; measures like drainage and the rock anchor reinforcement of the deep and shallow layers are taken for the comprehensive treatment of the slope, meanwhile, the basic treatment ideas of advance drainage and advance anchoring are emphasized, and strict requirements are put forward for the excavation and reinforcement procedures and the reinforce support procedures of the single-stage slope surface. During construction, the dynamic design idea should be adhered to, according to the exposed geological conditions and the monitoring data, the reinforcement measures and the construction procedures should be adjusted timely, and until October 2003, the treatment was completed. The prototype monitoring shows that the slope is stable.

4. The headrace system is composed of a dense cavern group. The headrace system is arranged in the underground of the left bank, it is a cavern group with large scale and dense caverns, the various methods like mathematical models and physical models are adopted to systematacially design and study the hydraulic properties of the cavern stability and the headrace system. Under the circumstances of adapting the actual geological conditions and ensuring the safety of the project, the good hydraulic characteristics and power quality are achieved by less quantity conducted in the arrangement of the headrace system. The excavation construction procedure and the support scheme of caverns are analyzed, compared and optimized in detail, the arrangement, excavation construction procedure and the support scheme of the construction adit are determined reasonably, which are applied and verified in practice during construction. A perfect monitoring system is designed, and the construction procedure, excavation scheme, support mode and support are monitored according to the monitoring data according to the idea of dynamic design in project construction.

2.4.2 Introduction to Guangzhao RCC Dam Design

2.4.2.1 Project Overview

Guangzhao Hydropower Station is located in the middle reach of Beipan River at the junction of Guanling County and Qinglong County of Guizhou Province, it is the leading power station in the main stream of the Beipan River, and the power station mainly generates electricity in combination with shipping and others. When the normal pool level is 745 m, the corresponding reservoir capacity is 3.135 billion m^3, and the regulated storage capacity is 2.037 billion m^3, it is an incomplete over-year regulation reservoir. The total installed capacity of the power station is 1 040 MW, the guaranteed output is 180.2 MW, and the average annual power generation is 2.754 billion kW·h. The controlled basin area of the dam is 13 548 km^2, the average annual discharge is 257 m^3/s, the average annual runoff is 8.11 billion m^3, and the average annual rainfall is 1 178.8 mm. The climate of the dam site is mild and moist, with the average annual temperature 18 ℃, the extreme maximum temperature 39.9 ℃ and the extreme minimum temperature −2.2 ℃. The power station in the power system mainly undertakes the peak regulation, frequency regulation, and emergency and load reserve.

2.4.2.2 Engineering Geological Condition

The dam area has flat bank slopes and complete land form, the both sides of the land form are relatively symmetric, it is the typical V-shaped transverse river valley, with huge mountain and narrow valley, and the

Chapter 2　Design and Rapid Construction of RCC Dams

upper land form is flat while the lower is steep with the slope angle about 45°. The middle and lower Triassic strata are exposed in the dam area, the lithology is composed of silty sand, mud shale, limestone and its transitional rocks, they are distributed in stratiform with obvious sedimentary rhythm. The strata exposed from the upstream to the downstream are Feixianguan Formation, Yongningzhen Formation and Guanling Formation in turn. The rock mass is hard and complete with high mechanical strength. The weathering degree of the rock mass is influenced by the lithology of the strata and the damage degree of the structure, the mud shale is strongly weathered, which is followed by transitional rocks, and then the limestone. The horizontal depth of weak weathering of the mud shale in both banks are 30 – 50 m, and the vertical weathering depth of the riverbed is about 10 m; the horizontal depth of the limestone weakly weathered in both banks are 15 – 25 m, and the vertical weathering depth of the riverbed is about 8 m, and these of the transitional rocks fall in between. The basic seismic intensity in the dam area is Level 6, while the seismic fortification of the dam is Level 7.

2.4.2.3　Layout of Complex

The multipurpose dam is mainly composed of the RCC gravity dam, the flood discharge system of dam body, the headrace system on the right bank, the ground powerhouse and the planned navigation structures on the left bank. Guangzhao Project is a Grade large (1) project, and the retaining dam, outlet structure, the headrace system and other permanent structures are Grade structures.

2.4.2.4　Barrage

The barrage is a RCC gravity dam with the dam crest total length of 410 m, the dam crest elevation of 750.50 m and the maximum dam height of 200.50 m. The dam crest width of the non-overflow dam section is 12 m, the maximum base width of the dam body is 159.05 m, and there are 20 dam sections in total. The non-overflow dam sections on the left and right banks are respectively 163 m and 156 m, and the length of the riverbed overflow dam and the bottom outlet dam section is 91 m. The flood discharge dam section is arranged in the middle of the main channel, including the surface outlet dam section with 3 outlets and 2 bottom outlet dam sections. The elevator shaft dam section is arranged in the right side of the bottom outlet dam section on the right bank, and the elevator shaft (staircase shaft) is connected with the gallery in the dam.

The concrete of the dam body include the GEVR, the normal concrete and the RCC. In addition of the parts of which the detailed structure and arrangement are required to use the normal concrete, and the parts, such as the dam downstream surface, the dam foundation surfaces of the both banks, the peripheries of the elevator shaft, the galleries and the peripheries of the orifice in the RCC zoning, and other parts that is not convenient for rolling construction, which are adopted with the GEVR, the RCC is adopted in the parts inside the dam body with the rolling condition.

The drainage system of the dam body is composed of the drainage gallery and the vertical drain hole curtain. The consolidation grouting with the hole depth of 8 – 12 m is arranged in the dam foundation.

2.4.2.5　Flood Release Structure

The outlet structure is composed of 3 surface outlets and 2 bottom holes in the dam body, the surface outlet discharges flood, and the bottom hole is responsible to empty the reservoir. The clear width of each surface outlet is 16 m, the weir crest elevation is 725 m, and each outlet is provided with the 16 m × 20 m (width × height) arc operating gate and the flat bulkhead gate. The energy dissipation by trajectory jet in narrow slit is adopted, and the outlet elevation of the flip bucket is 640 m, the shrinkage ratio of the narrow slit is 0.30, the anti-arc radius is 45 m, and the bucket angle is −10°. the design maximum discharge is 9 857 m/s. The two bottom holes are respectively located on the both sides of the overflow surface outlet, the floor elevation is 640 m, and the orifice size of the inlet section is 4 m × 6.5 m (width × height) with the bulkhead gate and the emergency gate set, and the outlet control dimension is 4 m × 6 m with the charging arc operating gate set. The disjunctive guide wall and inclined trajectory jet in narrow slit are adopted in the outlet, which make the water flow into the riverbed, the anti-arc radius of the inclined bucket is 45 m, with the maximum bucket angle 27° and the maximum discharge 1 597 m³/s. A 50 m long apron is set at dam toe.

2.4.2.6 Design of Dam Section

Main measures of basic shape design for the dam: RCC with rich cementing materials is adopted in the dam, and the dam is totally concreted by RCC. the GEVR and the II gradation RCC are adopted in the upstream face of the dam for imperviousness; the uplift pressure of the dam foundation surface is included into the pumping and pressure-relief effect; the economic section of dam is determined by the optimal method.

There is no deep slide instability along the dam foundation in Guangzhao Dam, the design of the dam economic section is controlled according to the conditions of dam stability, the stress of the dam foundation and the no tensile stress of the upstream dam surface, the stability and stress of the RCC layer surface at any elevation of the dam body can be satisfied by the material strength that meets the design requirements through the reasonable design of RCC mix proportion. Under the premise that the dam body section can meet the requirements of stability and stress, the minimal concrete volume of the whole dam body is the goal of dam section optimization. In order to make the arrangement of the dam body structure simple, only the basic profile of the maximum dam height is optimized in Guangzhao Hydropower Station, and the same upstream and downstream dam slopes are selected in different non-overflow dam sections. The basic cross section of the dam body is triangle, its vertex is set at the highest water level of the upstream, the optimization range of the upstream dam slope of the dam body is set within 0 – 0.30, and the optimization range of the downstream dam slope is within 0.5 – 0.9. The optimized parameters of the cross section include the elevation of the upstream and downstream dam slopes and the upstream break point. After optimization calculation, the optimized parameters of the basic section of the dam body proposed finally are: upstream dam slope 1:0.25, the elevation of the slope break point is 615.00 m; downstream dam slope 1:0.75, the elevation of the slope break point is 731.07 m.

2.4.2.7 Concrete Zoning and Impervious Design of the Dam

The dam body of Guangzhao Dam is designed based on the full-section RCC, the concrete materials of the dam body are divided into the normal concrete, the RCC and the GEVR. According to the operating conditions and the stress of the concrete in different parts of the dam body, the temperature environment during construction and other factors, the normal concrete of the dam body is divided into six zones of C I, C II, C III, C IV, C V and C VI, and the RCC is divided into five zones of R I, R II, R III, R IV and R V.

According to the material strength requirements, the RCC is divided into three zones of R I, R II and R V, which is bounded by the elevation of 600 m and 680 m. The combination of GEVR and II gradation RCC is adopted for imperviousness on the upstream face which is below the dam elevation of 710 m; the III gradation RCC combined by the GEVR and III gradation RCC is adopted for imperviousness on the upstream face which is above the dam elevation of 710 m. According to the acting heads of the dam, it is determined that the thickness of the GEVR above the elevation of 615 m is 0.80 m; the thickness of the GEVR under the elevation of 615 m is 1.20 m.

In order to strengthen the crack resistance of the upstream face under the effect of high water head, and limit the development of cracks, a layer of 25 @ 20 mm × 20 mm steel mesh is set on the upstream dam surface below the elevation of 640 m. The horizontal width of the upstream II gradation RCC is within 3 – 13 m according to the different acting heads, it is controlled by the distance between the downstream boundary and the drain hole curtain of the dam body which should be no less than 0.3 m, so that the drainage and pressure reduction of the hole curtain of the drainage hole can be guaranteed. In order to strengthen the reliability of the upstream impervious structure, a layer of polymer permeable capillary crystalline impervious materials are coated on the dam surface below the upstream dead water level.

2.4.2.8 Conclusion

After Guangzhao Dam has turned from a normal concrete gravity dam to a RCC gravity dam, its complex arrangement and the dam arrangement and feasibility study stage are basically unchanged, which have relatively less influence on the project construction. The maximum dam height is 200.5 m, which is the world class high RCC gravity dam. The dam arrangement, structure design, foundation treatment, the selection of the concrete damming materials and mix proportion, construction arrangement, the concrete system, the placing scheme, the determination of the temperature control standard and the temperature control and cracking prevention measures are very important. While drawing on the relevant project designs and construction experiences, according to the actual conditions of Guangzhao Project, and combined with the results of the relevant scientific research tests, detailed analyses and elaborate designs, the design of Guangzhao RCC gravity dam can be technically advanced, safe and reliable, and economical rational, which are beneficial to the rapid construction.

2.4.3 Introduction to Jin,anqiao RCC Dam Design

2.4.3.1 Project Overview

Jin'anqiao Hydropower Station is located at the mid reach of Jinsha River in Lijiang City, Yunnan Province, the left bank and the right bank respectively belongs to the Yongsheng County and the Gucheng Area in Lijiang City. The geographic location of the power station is proper, with the straight-line distance to Kunming, Panzhihua and Lijiang respectively 300 km, 130 km and 20 km.

The Jinsha River basin spans more than ten longitudes and latitudes, the landform is extremely complex, and the climatic characteristics vary greatly. The climate in the area varies greatly in horizontal and vertical directions, with obvious three-dimensional climate. The climate is mainly influenced by the west wind circulation of south branch of Qinghai-Tibet Plateau in winter half year, with sunny and dry weather and little rainfall; the climate is influenced by the strengthened southwest warm and moist air mass in summer half year, and the trace of the invasion is along the valley, then the rainfall is formed which is concentrated in flood season with large intensity. The average annual temperature of the dam area is 19.8 ℃, and the average annual rainfall is 1 078 m.

The project is mainly developed to generate electricity, with the comprehensive utilization of tourism development, shipping in the reservoir, the aquaculture and water and land conservation. The power station supplies power to the southern grid. The total installed capacity of the power station is 2 400 MW, the guaranteed output with operation (before the input of the upstream Hutiaoxia Reservoir) is 473.7 MW, with the annual power generation is 11.043 billion kW·h. After the input of the upstream Hutiaoxia Reservoir, the guaranteed output of the power station will increase to 1 351.3 MW, and the annual power generation will increase to 12.92 billion kW·h.

2.4.3.2 Engineering Geological Condition

The exposed strata in the dam area mainly is the upper part of the Upper Permian basalt formation, the lithology is the basalt, amygdaloidal basalt with the volcanic breccia lava and tuff, and there are 10 layers of tuff with good continuity distributed in the dam area. As influenced by the structure, the rocks of the basalt, amygdaloidal basalt and the volcanic breccia lava are hard, the basalt is relatively broken, and there are thicker fractured chloritized basalt rock mass, amygdaloidal basalt and volcanic breccia lava with relatively complete rock mass distributed in the riverbed; the bedding extrusion with argillic alteration often occurs in the tuff strata. The rock strata of the dam area is the uniclinal structure. The physico-geological function of the dam area is strong, which mainly are collapse, weathering of rock mass and unloading. The underground water is mainly bedrock fissure water, the water permeability of the rock mass is uneven, and the general trend is the gradual weakening from the surface to the deep.

The basalt quarry is located on the left bank of the Wulang River, the rock of the quarry are mainly the basalt and the amygdaloidal basalt, with volcanic breccia lava and tuff in several layers. In addition to the tuff and the full and strong weathering rock mass, the rocks are hard and relatively complete, with the saturated uniaxial compressive strength more than 60 MPa. The qualities and reserves of the quarries can meet the project requirements with good mining conditions and close haul distance.

2.4.3.3 Complex Work Arrangement

1. Project rank. The Jin'anqiao Hydropower Station is a large hydropower project focusing on power generation, with the check flood level of the reservoir 1 421.07 m, and the total capacity of reservoir is 913 million m^3; the normal water level is 1 418.00 m, the corresponding storage capacity is 847 million m^3, and the effective storage capacity is 346 million m^3. The maximum height of the RCC gravity dam is 160 m with the installed capacity of 2 400 MW. The project is a large(I) project, and the project grade is Grade I. The main structures are: the dam, the flood discharge structures and the headrace structures are the Grade structures. The secondary structures are Grade 3 structures.

2. Seismic fortification intensity. The seismic fortification intensity of the dam is Level 9, i.e., the dam will exceed the peak horizontal acceleration of the bedrock with a probability of 2% based on the reference period of 100 years. $a = 0.399g$ for the seismic design; the seismic fortification intensity of other structures are 8 magnitudes, i.e., the dam will exceed the peak horizontal acceleration $a = 0.246g$ of the bedrock with a probability of 5% based on the reference period of 50 years.

3. Complex work arrangement. The multipurpose dam of Jin'anqiao Hydropower Station Project is mainly composed of the barrage, powerhouse at dam toe of the riverbed, the overflow surface outlets in the dam body on the right bank, the discharging and sand flushing bottom outlets on the right bank, the stilling basin on the right bank, the sand flushing bottom outlets on the left bank and the access tunnels on the left bank.

The barrage is a RCC gravity dam, with the dam crest elevation of 1 424.00 m, the maximum dam height of 160 m and the dam crest length of 640 m. The barrage is composed of the non-overflow dam section, the sand flushing bottom outlets dam section on the left bank, the powerhouse dam section of the riverbed, the discharging and sand flushing bottom outlets on the right bank, the overflow surface outlets on the right bank and the non-overflow dam section on the right bank from left to right, there are 21 dam sections in total.

The non-overflow dam sections on the left bank are the $0^\#-5^\#$ dam sections, with the total length of 192 m and the dam crest height of 12 m. The sand flushing dam section on the left bank is the $6^\#$ dam section with the length of 30 m, and the sand flushing penstock extends from the underneath of the powerhouse erection bay to the downstream. The powerhouse dam sections at the dam toe in the middle of the river is the $7^\#-11^\#$ dam sections, with the total length of 156 m and the dam crest height of 26 m. The discharging and sand flushing bottom outlets on the right bank is the $12^\#$ dam section, which is arranged closely next to the right end wall of the powerhouse dam section, the length of dam section is 26 m, the dam crest height is 21 m and the discharging and sand flushing bottom outlet with 2 outlets is arranged. The overflow surface outlets dam sections on the right bank are the $13^\#-15^\#$ dam sections, the total length is 93 m, the dam crest height is 31 m, and five surface outlets are arranged with energy dissipation by hydraulic jump. The non-overflow dam sections on the right bank are the $16^\#-20^\#$ dam sections, with the total length of 183 m and the dam crest height of 12 m.

The single-unit and single-tunnel headrace mode is adopted for four units of the powerhouse at the dam toe with the pipe diameter of 10.5 m, it is the half-back penstock at the dam toe, with 2 m thick reinforced concrete wrapped. The water inlet of the power station is the vertical water inlet on the dam surface, with the elevation of 1 370.00 m. The dimensions of the main powerhouse is 213 m × 34 m × 79.2 m (length × width × height), with the 4 × 600 MW mixed-flow hydraulic turbine-generator unit inside. The erection bay and the main machine section are arranged from left to right. The installation elevation of the unit is 1 285.00 m, the elevation of the generator floor is 1 303.00 m, and the length of the single unit section is 34 m.

4. Main structures.

a) Roller compacted concrete dam.

Layout of dam body. The barrage is a concrete gravity dam, with the dam crest elevation of 1 424.00 m, the maximum dam height of 160 m and the dam crest length of 640 m. There are totally 21 dam sections. The downstream dam slope is 1:0.75, the beginning of the upstream dam slope is at the elevation of 1 330.00 m, the above is the vertical dam surface, and the dam slope below is 1:0.3.

There is no longitudinal joint in the dam, the arrangement of the transverse joints is based on the structural requirements of structures and the joints experiences of the RCC dam at home and aboard, on the basis of the normal concrete, the distance of the transverse joints are increased appropriately, and the maximum joint distance is controlled within about 30 – 34 m. In order to prevent the vertical crack on the upstream dam surface, when the dam section width is more than 30 m, the short joint with the depth of 3 m should be set on the II gradation RCC of the center line of the upstream dam section on the dam body. The dam sections of the riverbed are designed according to the joint stress of the powerhouse and the dam, the joint grouting is conducted in the joint of the powerhouse and the dam under the elevation of 1 278.5 m.

b) Dam body concrete zoning.

According to results of the static and dynamic stress analysis and the seepage analysis, upon the analogy with similar project experiences, the concrete zoning of the dam body structure should be designed. The following factors are mainly considered for the zoning design:

Strength requirements of the dam body: strength indexes of concrete in each subarea must meet the requirements of the specifications for the ultimate state of dam bearing capacity. According to the material mechanics method, plane finite element method (FEM) and 3D FEM, the stress of dam body is analyzed and reviewed. Material zoning of dam body is controlled according to stress results. Concrete of 90 d strength is adopted for the review under normal conditions and check conditions, and concrete of 180 d strength is adopted for the review under earthquake conditions.

Dam imperviousness requirements: the impervious layer on the upstream face of the dam is divided into 3 areas, with the elevation of 1 350.00 m as the boundary. The section below the boundary is concreted with $C_{90}20$ III gradation RCC, and the section above the boundary is concreted with $C_{90}15$ III gradation RCC, with $C_{90}20$ II gradation RCC adopted for imperviousness. According to different acting heads, impervious layers of different thickness are adopted, and the section below the upstream surface with an elevation of 1 335 m is concreted with 5 m thick II gradation impervious RCC; the section with an elevation of 1 335 – 1 398 m (dead water level) is concreted with 4 m thick II gradation impervious RCC; the section above elevation of 1 398 m is concreted with 3 m thick II gradation impervious RCC.

2.4.3.4 Main Design Features

1. Seismic steel bars on upstream and downstream dam surfaces. The seismic fortification intensity is Level 9. There are no successful examples of structure 100 m RCC dams in areas with such high seismic intensity at home and abroad. Therefore, the upstream and downstream dam surfaces are equipped with upstream and downstream seismic steel bars (deformed steel bar 28, spacing of 200 mm × 200 mm), which can reduce the sudden change of dam body shape, reduce the weight of dam crest, and use high-strength concrete in some weak parts, thus ensuring the seismic safety of the dam.

2. Fractured chloridized rock mass as dam foundation. The fractured chloritized rock mass is widely distributed

in the weakly weathered lower part of Jin'anqiao dam foundation to the slightly fresh area, which is "hard, brittle and broken". According to the special study, the fractured chloritized rock mass completely meets the requirements of dam stability and strength bearing capacity, and the deformation of dam body and dam foundation conforms to the general law, and the deformation value is not large.

3. Layout of powerhouse at dam toe. Jin'anqiao Hydropower Station has a powerhouse at dam toe. Due to the need of layout of water inlet, headrace penstock and gates, the use of RCC for damming is affected to some extent. Normal concrete is used for the parts affected by the gate piers, orifice and water inlet. The total volume of dam concrete is about 3.29 million m^3, including 2.51 million m^3 RCC and 780 000 m^3 normal concrete.

2.4.4 Introduction to Baise RCC Dam Design

2.4.4.1 Project Overview

Baise Multipurpose Dam Project on Youjiang River, Guangxi is a large water conservancy project, which focuses on flood control and has comprehensive benefits such as power generation, irrigation, shipping and water supply. The project is located in the middle of Youjiang River at the upstream of Yujiang River in Xijiang River system of Guangxi, and the dam is 22 km away from Baise City. The rainfall collection area is 19 600 km^2, the annual average discharge is 263 m^3/s and the annual runoff is 8.29 billion m^3. The average annual rainfall is 1 200 mm, the annual evaporation is 1 370 – 1 674 mm, the average annual temperature is 22.1 ℃, the normal pool level of the reservoir is 228 m, and the check flood level is 231.49 m; the total storage capacity is 5.66 billion m^3, the flood control capacity is 1.64 billion m^3, the regulated storage capacity is 2.62 billion m^3, which is an incomplete multi-year regulated reservoir. Combined with the application of floodwall in Nanning, the project can improve the flood control capacity of Nanning to the standard of 50 year flood prevention. The power station has the installed capacity of 540 MW and the annual power generation of 1.701 billion kW·h.

2.4.4.2 Engineering Geological Condition

The dam area is located in the West Guangxi depression of the Indosinian fold system in West Guangxi of paraplatform in South China, which is a relatively stable block between Yunnan-type structure and Guangxi-type structure. The basic seismic intensity of dam is Level 7.

Siliceous rocks, mudstone, argillaceous limestone and Variscan diabase of Devonian Liujiang Formation are exposed in the main dam area, with soft and hard rock layers. The diabase strip of the dam foundation is about 120 m thick, with hard lithology and weak water permeability. The interface between diabase and surrounding rocks is seriously altered, strongly weathered, and the rock mass is fractured, forming a weak zone with a certain scale.

The larger faults in dam foundation are F_6 and L_{46} faults. Fault F_6 is located on the right side of the riverbed, which breaks the diabase belt. The fracture zone width is 2 – 4 m in the upstream section and 0.6 – 1.4 m in the downstream section. The rock mass in the affected zone is fractured, with a width of 1 – 2 m and a local area of 24 m. F_{46} fault is located at the heel of No. 4 dam section of riverbed, which extends to downstream, narrows and then disappears, and is composed of altered diabase, filled with fully and strongly weathered cuttings and mud.

The foundation rock mass of the stilling basin is of various complex lithology, which is alternated with soft and hard rocks, with medium water permeability, strong bedding weathering, and the full and strong weathering depth is 20 – 60 m below the foundation surface. The structural fissures and interbedded crushed clay layers are well developed.

There is a lack of NSG materials in the dam area and limestone which can be used as manufactured aggregate within more than ten kilometers around the dam. The diabase on the right bank of the dam and the same belt of the dam foundation is creatively selected as the main dam concrete aggregate in engineering design, which is the first case in the world.

2.4.4.3 Layout of Complex and Main Engineering Structures

The project is classified as Class I project, and the water retaining structures are Grade I. The flood control standard is designed as per 500-year return period flood, and check flood as per 5 000-year return period flood, and design standard for flood discharge, energy dissipation and scour prevention structures is as per 100-year return period flood.

Major structures of hydraulic complex: main dam, underground power generation system, 2 auxiliary dams and navigation structures.

The main dam is located on the first diabase whose exposed thickness is only about 120 m in the dam area. In order to conform to the exposed shape of the diabase, the main dam is arranged along the exposed shape of diabase layer on the ground, and the dam axis is divided into three broken lines. It is a full-section RCC gravity dam, with crest length of 720 m, crest elevation of 234 m, width of 10 m and dam height of 130 m, and is divided into 27 dam blocks with length of 22 – 33 m. There are transverse joints between dam blocks but no longitudinal joints. A short joint with a depth of 3 m is added to the middle of upstream face of dam block with a length greater than 25 m.

Section of dam body: the section above the elevation of 146 m on the upstream face is a plumb plane, and the section below such elevation is 1:0.2. the downstream slope ratio of non-overflow dam section is 1:0.75 the downstream slope ratio of overflow dam is 1:0.7, the weir crest elevation is 210 m.

The section of foundation grouting drainage gallery is 3 m × 3.5 m with a shape of gate. The observation drainage gallery of 2.5 m × 3 m section with a shape of gate is located at elevations 155 m and 200 m on the upstream side of the dam, and the drainage gallery of 2.5 m × 3.5 m section with a shape of gate is located in the middle and downstream of the dam foundation.

II gradation RCC with enriched cementing materials is used to prevent seepage on the upstream face of the dam. 2 mm thick auxiliary imperviousness coating is set below the dead water level of elevation 203 m on the upstream dam surface.

$R_{180}150W2F50$ quasi III gradation RCC is used inside the dam. $R_{180}200W8F50$ II gradation RCC is used in the upstream impervious area of dam. The leveling layer of dam foundation is concreted with $R_{28}200W0F50$ quasi III normal gradation concrete. The overflow surface is concreted with $R_{90}400W8F100$ quasi III gradation impact and abrasion-resistant normal concrete. The outer surface of the dam, periphery of normal concrete inside the dam and the periphery of holes are concreted with GEVR. The all concrete is 42.5 moderate heat Portland cement.

Consolidation grouting is carried out within a certain range of dam foundation and upstream and downstream of the dam foundation, with the grouting depth of 12 – 15 m for the area above curtain in dam heel area, 8 m for dam area, 5 m for the rest parts, and 20 m for the upstream and downstream alteration zones outside dam foundation.

The impervious curtain is of double-row suspension type with a depth of 0.5 time the dam height.

Drainage of dam foundation: 3 rows of drainage holes are longitudinally arranged in the dam foundation. The main drainage hole is located at the downstream side of curtain in grouting drainage gallery, and the auxiliary drainage holes are arranged in two drainage galleries in the middle and downstream parts of dam foundation, and the holes are 20 m deep into rock.

Energy dissipation of outlet structures: combined energy dissipation of wide-flange piers at the surface outlet + drop flow at the middle outlet + underflow stilling basin. The overflow dam section is 88 m long and orthogonal to the river direction. It is provided with 4 surface outlets of 14 m × 18 m (width × height) and 3 middle outlets of 4 m × 6 m (width × height). The middle gate pier at the surface outlet is 8 m thick and 4 m thick for side gate pier.

The middle outlet is a pressure orifice, with an inlet bottom elevation of 167.5 m, which is arranged at the lower part of the middle pier at the surface outlet, and the outlet is a wide-flange pier without water.

The elevation of stilling basin bottom is 105 m. The elevation of the tailgate is 121 m, and the stilling basin is 82 m wide, 16 m deep and 124.617 m long. Measures such as thickening of concrete floor, consolidation grouting of foundation, and anchoring by anchor bars are taken.

2.4.4.4 Main Design Features

1. The RCC main dam area is separated with the powerhouse, which avoids the interference between them, and meets the requirements of intensity and speed of RCC construction.
2. Hard diabase is creatively used as manufactured aggregate for dam construction, which avoids many problems such as difficulty in mining, more waste materials, high cost, alkali active reaction of aggregate and so on caused by thin-lift, mud and flint in local limestone aggregate. Diabase is used as concrete aggregate for dam construction in this project, which is the first case in the world.
3. On the diabase wall with a thickness of 114 – 131 m and a riverbed exposure width of about 140 m, the dam is arranged in a broken line, and the adjustable range of the dam line has been controlled to centimeter level.
4. The main dam is constructed with full-section RCC, with II gradation RCC as the impervious zone for the dam surface, and the dam surface is coated with polymer cement waterproof coating as the auxiliary impervious layer. The use of moderate heat Portland cement with high content of fly ash, and high-density polyethylene plastic pipes embedded in RCC for forced cooling through natural river water effectively reduces the temperature rise of concrete, simplifies the temperature control measures of RCC construction, and is more economical, convenient and fast.
5. Quasi III gradation diabase aggregate (maximum particle size 60 mm) is used in the dam body, which obviously reduces the modulus of elasticity of the concrete, improves the seismic stress of the dam, and improves the temperature stress of the concrete.
6. The combined energy dissipation of wide-flange piers at the surface outlet + drop flow at the middle outlet + underflow stilling basin is adopted, which has achieved a breakthrough in the application of such energy dissipation in the 100 m high RCC dams.
7. $R_{90}400$ normal impact and abrasion-resistant concrete is used on the surface of overflow dam and stilling basin, and comprehensive measures are taken to improve the strength, crack resistance, impact and abrasion resistance of the concrete: diabase aggregate is used to strengthen the abrasion resistance of the concrete, moderate heat Portland cement is used to reduce the temperature rise of the concrete, low temperature pouring of concrete is controlled to reduce the maximum temperature of the concrete, and polypropylene staple fiber is mixed to improve the early crack resistance and impact resistance of the concrete.

2.4.5 Introduction to Guandi RCC Dam Design

2.4.5.1 Project Overview

Guandi Hydropower Station is located at the downstream of the main stream of Yalong River in Daluo Village at the junction of Xichang City and Yanyuan County in Liangshan Yi Autonomous Prefecture, Sichuan Province. It is the third cascade hydropower station in the five-stage hydropower development mode of Kala-Jiangkou reach of Yalong River. The upstream of the station is connected with the tail water of Jinping Cascade II Hydropower Station, and the reservoir area is about 58 km long. The downstream of the station is connected with Ertan Hydropower Station, which is about 145 km away from Ertan Hydropower Station. It is about 80 km away from

Chapter 2　Design and Rapid Construction of RCC Dams

Xichang Highway.

The main task of this Project is to generate electricity. Thenormal pool level of the reservoir is 1 330.00 m, the dead water level is 1 328.00 m, and the total storage capacity is 760 million m^3, and the daily storage is regulated.

This project is classified as Class Ⅰ project, with main hydraulic structures of Grade Ⅰ, secondary structures of Grade Ⅲ and temporary structures of Grade Ⅳ. The complex structures of the power station is mainly composed of left and right bank retaining dams, middle outlet dam sections and overflow dam sections (RCC gravity dam), stilling basin and water headrace and power generation system at the right bank. The underground powerhouse at the right bank is equipped with 4 × 600 MW units with a total installed capacity of 2 400 MW.

The manufactured aggregate of this project is the basalt yard on the upper side of the right dam shaller and the mountain on the upper edge of Zhuziba Gully.

2.4.5.2　Topographic and Geological Conditions

Guandi Hydropower Station is located in the downstream of the bend of Jinping River in Yalong River. The river flows generally from north to south, and three bends westward are formed in Guanfang-Daluo section. The dam of Guandi Hydropower Station is located on the bend of Daluo River in the most downstream, and Yalong River flows into Jinsha River in Panzhihua City after confluence with Anning River. The terrain of the region is high in northwest and low in southeast, of which the altitude is reduced from 5 000 – 4 000 m to about 2 000 m, and can be divided into three levels of planation surfaces at an altitude of 4 000 m, 3 000 m and 2 000 m above. Six levels of terraces can be seen scattered on both banks below 400 m the vertical height of the terrace above sea level.

The regional strata are bounded by Jinping Mountain-Xiaojinhe River, Songpan-Ganzi stratigraphic area in the west and Yangtze stratigraphic area in the east. The Yangtze stratigraphic area is divided into Yanyuan-Lijiang stratigraphic division and Kangdian stratigraphic division with Jinhe-Qinghe fracture as the boundary. The project area is located in Yanyuan-Lijiang stratigraphic division, and the main exposed strata are Proterozoic (Sinian system and Pre-sinian system): mainly rubble dolomite mixed with light gray dolomite angle, and the lower part is sandy shale mixed with marlstone and mudstone; upper Paleozoic (Devonian system-Permian system): the lower part is mainly limestone mixed with a small amount of sandstone, the upper part is basalt, and the top part is slate mixed with metamorphic sandstone and marl, etc.; mesozoic (Triassic system): sand shale and dolomitic limestone; cenozoic (Tertiary system): conglomerate mixed with sandstone; quaternary system: xigeda semi-consolidated siltstone, clay rock and various overburden layers.

2.4.5.3　Meteorological and Hydrological Conditions

Yalong River basin features Western Sichuan Plateau climate, which is mainly affected by the upper westerly circulation and southwest monsoon. The dam area has distinct dry and wet seasons. The dry season is from November to April with more sunshine, less humidity, large daily temperature difference and little rainfall, accounting for about 5% – 10% of the annual rainfall; the rainy season is from May to October, with humid climate and concentrated rainfall, accounting for more than 90% of the annual rainfall. The temperature in the basin increases from north to south, and the rainfall increases from north to south. According to the data of Xichang Meteorological Station, the average annual temperature is 17.0 ℃, the extreme maximum temperature is 39.4 ℃, the average annual relative humidity is 75%, and the maximum wind speed is 14.0 m/s; average annual rainfall is 1 142.3 mm.

The runoff in Yalong River basin mainly comes from atmospheric precipitation, and the flood is mainly caused by heavy rain. The flood season is from June to September, and the annual maximum flood occurs in July and August. The flood process is bimodal or multi-modal. Generally, the single-modal process lasts for

6 – 10 d and the bimodal process lasts for 12 – 17 d, which is generally characterized by relatively low flood peak, large flood volume and long duration.

2.4.5.4 Layout of Complex and Main Structures

Guandi Hydropower Station is a RCC gravity dam with a crest elevation of 1 334.00 m, a minimum foundation elevation of 1 166 m, a maximum dam height of 168.0 m, a maximum dam base width of 153.2 m and a crest axis length of 516 m; there are 24 dam sections in the whole dam, which are composed from left to right of left bank retaining dam sections $1^{\#} - 9^{\#}$, left middle outlet section $10^{\#}$, overflow dam sections $11^{\#} - 14^{\#}$, right middle outlet section $15^{\#}$ and right bank retaining dam sections $16^{\#} - 24^{\#}$; the overflow dam sections are provided with 5 overflow surface outlets, each with a clear width of 15 m and an overflow weir crest elevation of 1 311 m; the elevation of the bottom of the middle vent hole is 1 240.0 m, and the size of the vent hole is 5 m × 8 m; the lengths of left and right bank retaining dams along the dam axis are 190 m and 195 m respectively, and the lengths of overflow dam section and middle outlet dam section along the dam axis are 131 m, among which the lengths of left and right middle outlet dam sections are 22 m; the downstream of the overflow dam section is connected with a stilling basin. The side walls of the stilling basin are concrete sloping walls, with the top elevation of 1 224.0 m, the foundation elevation of 1 166.0 m and 1 180.0 m respectively, the floor elevation of 1 188.0 m and the stilling basin length of 145 m. The foundation cushion and dam crest of the retaining dam sections on both banks are constructed with normal concrete, galleries and periphery of holes are constructed with GEVR, and the rest parts are constructed with RCC; the overflow surface, gate piers and foundation cushion of the overflow dam sections are constructed with normal concrete, galleries and periphery of holes are constructed with GEVR, and the rest parts are constructed with RCC; the foundation cushion and dam crest of the middle outlet dam sections are constructed with normal concrete, the periphery of middle outlets is constructed with normal concrete, the upstream gate shaft is constructed with GEVR, and the rest parts are constructed with RCC; the upstream face of the dam is constructed with GEVR. The deep groove of stilling basin foundation is backfilled with RCC, and the rest parts are constructed with normal concrete. Dam foundation treatment includes consolidation grouting, contact grouting, curtain grouting and drainage holes. The depth of consolidation grouting holes in dam foundation is 6 – 18 m, the spacing between holes and rows is 3 m, with the layout of plum blossom; double rows of holes are applied for curtain grouting of dam foundation. The depth of main curtain holes is 0.5 – 0.7 time the dam height, and the depth of auxiliary curtain holes is 2/3 time the dam height. The curtain row spacing is 1.5 m, the hole spacing is 2.0 m, and the holes are staggered. A row of main drainage holes is set at the downstream of the impervious curtain, the depth of which is 0.5 time that of the main curtain hole, and the hole spacing is 3 m.

The excavation platform at the water inlet of the powerhouse is about 144.2 m long, with a width of 30 m, a front width of 40 – 90 m, and an excavation bottom elevation of 1 293 m. The maximum excavation height of the water inlet slope in this project is 41 m. The water inlet with an elevation 1 293.00 – 1 312.00 m is a vertical excavation slope, and the excavation slope ratio of elevation 1 312.00 – 1 334.00 m is 1∶0.3, and a 3.00 m wide berm is set at the elevation 1 312.00 m.

Geologically, $1^{\#}$ sliding mass on the right bank is called landslip on both sides of Zhuziba Gully at the right bank inlet. The landslip on the right side of Zhuziba Gully is located on the right side of the inlet, with and elevation of 1 293 – 1 320 m at the front edge and a steep wall at the rear edge with an elevation of 1 337 m, the average length is about 65 m along the gully. This landslip slope has been basically removed during the excavation of the inlet slope. After excavation, it is supported by spray anchors, wire meshes and anchor cables, and concrete frame beam is set.

2.4.6 Introduction to Longshou RCC Arch Dam Design

2.4.6.1 Project Overview

Longshou Hydropower Station is located at Yingluo Gorge at the outlet of main stream of Heihe River, which is about 30 km southwest of Zhangye City, Gansu Province. The design total installed capacity of the station is 59 MW (3×15 MW $+ 2 \times 7$ MW), the annual power generation is 183.6 million kW · h, the maximum dam height is 80 m, and the total storage capacity is 13.2 million m^3. It is a moderate-sized Grade II project. The dam, water headrace structures and powerhouse are Grade 3, and the seismic fortification intensity of the project area is Level 8. The main task of the power station is to relieve the shortage of local industrial and agricultural power supply and provide necessary electric energy, and to undertake peak regulation and phase regulation in Zhangye power grid.

2.4.6.2 Hydrometeorological Characteristics

Longshou Hydropower Station is located on the main stream of Heihe River in Zhangye City, in the hinterland of northwest China, with continental climate, hot summer, scarce rainfall, intense evaporation, severe winter and ice age as long as 4 months. The average annual precipitation in this area is 171.6 mm, the average annual evaporation is 1 378.7 mm, the average temperature is 8.5 ℃, the absolute maximum temperature is 37.2 ℃, the absolute minimum temperature is -33 ℃, and the maximum frozen soil depth is 1.5 m.

2.4.6.3 Geological Conditions of Dam Site

There is a fossil river terrace on the left bank of the dam, with the terrace ground elevation of about 1 722 m, bedrock weak weathering line elevation of about 1 697 m, bedrock roof exposed elevation of about 1 704 m, burden thickness of about 17 m, and steep bedrock on the left side of the terrace is exposed. The right side of the terrace is the main river channel, with the riverbed elevation of 1 680 m and bedrock weak weathering line elevation of about 1 652 m. The bank slopes below the elevation of 1 720 m on both sides of the main river channel are $50°-70°$ steep cliffs, and the bedrock is exposed. The overburden above the elevation of 1 720 m on the right bank is thick, the bedrock roof is flat, and the fault joints on the left bank of the main river channel are developed. The basic seismic intensity at the project site is 8, which is a typical area with high cold, high evaporation and high level of earthquake area.

2.4.6.4 General Project Layout

The multipurpose dam of Longshou Hydropower Station is composed of RCC arch dam, RCC gravity dam on the left bank, middle and surface outlet flood discharge structures, dam body emptying and sediment discharge outlet, dam intake structures and water headrace system, and ground powerhouse and switch yard on the left bank.

According to the topographic and geological conditions, combined with the layout characteristics of various structures in the complex, the barrage is arranged as a mixed dam on the plane, and the fault joints on the left bank of the main channel are developed. For safety reasons, it is proposed to build a RCC gravity dam, which will also serve as the abutment of arch dam, with a crest length of 47.16 m; RCC arch dam is set in the main channel, and the axis length of arch dam is 140.84 m; the bedrock roof above the elevation of 1 720 m on the right bank is flat, and a thrusted pier with a top length of 29.32 m is proposed, and the total length of the dam crest is 217.32 m.

Elevation of the arch dam crest is 1 751.50 m, the maximum center angle is 94.58, the minimum center angle is 54.79, the maximum radius of curvature is 54.5 m, and the minimum radius of curvature is 32.75 m, the maximum arc length of the dam crest is 140.84 m, the maximum dam height is 75.5 m, the dam crest thickness is 5.0 m, the dam base thickness is 13.5 m, the thickness-height ratio is 0.17, the maximum overhang degree of the crown cantilever is 1:0.08, the maximum overhang degree of the dam body is 1:0.189, and the volume of concrete in dam body is 68 300 m^3

The height of the gravity dam is 54.5 m, it is designed as an integral gravity dam regardless of dam section, its dead weight satisfies its own stability, and maintains the stability of the left arch end above elevation of 1 697.0 m. Its shape design shall meet the control indexes of stiffness, deformation and stress. The section of gravity dam is determined according to stability calculation and finite element stress analysis: the upstream is a plumb plane, with dam crest width of 30.0 m and dam base width of 65.43 m. The downstream dam slope is 1:0.65, the dam base elevation is 1 697.00 m and dam crest length is 47.16 m.

The thrusted pier is placed above the elevation of 1 720.0 m on the right bank, with a height of 31.50 m. Its shape is mainly controlled by stability. According to analysis, the dam crest width is 14.50 m, the dam base width is 30.25 m, the upstream face is vertical, the downstream face is a slope with a slope ratio of 1:0.5, and the dam crest length is 29.32 m.

The flood discharge structure is composed of middle outlets, surface outlets and a sediment discharge outlet. According to the characteristics of large amount of sediment in flood season, the water level of the reservoir can be lowered in flood season, and the sediment discharge outlet can be opened to discharge a large amount of sediment in the reservoir by using abundant inflow. The perennial flood and design flood are discharged by 3 middle outlets. The specific arrangement of surface outlets, middle outlets and sediment discharge outlet is as follows:

1. Surface outlet: two surface outlets are symmetrically arranged on both sides along the center line of arch dam, and are composed of WES overflow weir and anti-arc flip bucket. The clear width of the outlet orifice is 10.0 m, and the crest elevation is 1 741.00 m, energy dissipation by trajectory jet in narrow slit is adopted, with the flip bucket elevation of 1 735.206 m, the upstream cantilever length of 3.72 m, the outlet cantilever length of 2.99 m, and a total length of 14.22 m.
2. Mid-level outlet: three middle outlets are symmetrically arranged along the center line of arch dam, and the floor elevation is 1 710.00 m, the outlet orifice size is 5 m × 5.5 m, the inlet is provided with a 5 m × 6.6 m flat bulkhead gate, and the outlet is provided with a 5 m × 5.5 m flat operating gate. The floor of the outlet body is flat, and the roof is a 1:13.54 pressing slope. Mixed energy dissipation modes such as deflecting flow, narrow slit and drop flow are adopted.
3. Sediment flushing outlet: the sediment discharge outlet is arranged in the gravity dam on the left bank, with a floor elevation of 1 710.00 m, the inlet is provided with a 3 m × 7 m flat bulkhead gate, the floor is flat, and the roof is a 1:20 and 1:8 pressing slope, and the outlet is provided with a 3 m × 4 m arc operating gate.

The water headrace system is arranged on the left bank, and the water inlet is arranged in the gravity dam on the left bank. The total water headrace capacity is 109.3 m^3/s, and the water supply mode of 1 tunnel for 4 units is adopted. The water headrace system consists of water inlet, main pipe section, bifurcated pipe section and branch pipe section.

The powerhouse is arranged on the left bank of riverbed, about 110 m away from the barrage, with unit installation elevation of 1 684.4 m and unit spacing of 11 m. There are 3 × 15 MW and 2 × 7.0 MW units in the powerhouse, with the main powerhouse size of 64.04 m × 16.5 m × 31.17 m; the auxiliary powerhouse is arranged at the upstream of the main powerhouse, with a size of 47.9 m × 11 m × 20.8 m; two circuits of 110 kV outgoing poles are arranged in the switch yard, and a circuit of 110 kV outgoing line interval is reserved.

The switch yard is arranged on the fossil riverbed at the downstream of the powerhouse, with a size of 35 m × 42 m and an elevation of 1 724.0 m

2.4.6.5 Innovation of the Project

1. The civil tender of Longshou Hydropower Station Project was started in March 1999 and closed in October

1999. On May 30, 2001, the first unit was synchronized to generate electricity, and on July 9, all 4 units started to generate electricity, which was half a year ahead of the approved construction period after design optimization. During the operation of the power station, in the flood season of 2002, Longshou Dam suffered a 50-year return period flood; in September, 2003, the dam suffered an earthquake of Level 6.3 (less than 100 km from the epicenter), and the project operated safely and normally.

2. The engineering application of RCC in hyperbolic thin arch dams, special-shaped dams and complex arch dams with multiple openings has been solved. For seepage control, cracking prevention and frost resistance of arch dams, newer and more perfect research results are put forward.

3. NSG aggregate from Gobi Desert is adopted for Longshou Hydropower Station to take local materials. The concrete mix proportion of different parts and grades, including impervious concrete, crack-resistant concrete, dam main concrete and impact and abrasion-resistant concrete, have been studied.

4. In order to improve the crack resistance of concrete and simplify the temperature control measures, magnesium oxide and expansive agent are mixed in the concrete. In the application of Longshou Hydropower Station, despite the bad weather conditions, such mix proportion lead less cracking, and the concrete quality fully meets the design requirements.

5. In the research of construction technology, the technical problems such as concrete mixing, transport, placing and placing surface operation have been well solved based on the actual situation of the project. According to the characteristics of local projects, simple, practical and economical temperature control and cracking prevention measures have been adopted.

6. In the concrete rolling construction, there are breakthroughs in the construction in high temperature season, low temperature season and extreme air temperature conditions.

2.4.7 Introduction to Shapai RCC Arch Dam Design

2.4.7.1 Project Overview

Shapai Hydropower Station, located in Wenchuan County, Tibetan Qiang Autonomous Prefecture of Ngawa, Sichuan Province, is a cascade leading power station in the upstream of Caopo River, a tributary of Minjiang River. The hydropower station is developed by the combination of storage and headrace. The dam is located near Niuchanggou, Shapai Village, Caopo River. The powerhouse is located at Kechong platform about 5 km downstream, and the tail water of the power station is merged into the reservoir of Caopo Power Station. The dam is about 19 km away from the mouth of Caopo River, 47 km away from Wenchuan County and 136 km away from Chengdu.

The normal pool level of Shapai Reservoir is 1 866.0 m, the dead water level is 1 825.0 m, and the total storage capacity is 18 million m^3. The power station has the total installed capacity of 36 MW and the annual power generation of 179 million kW·h. The annual utilization hours are 4 791 h. This project is classified as Class Ⅲ project, with main structures of Grade Ⅲ.

The average annual discharge of the dam is 8.72 m^3/s, and the average annual temperature is 11.3 ℃.

The basic seismic intensity in the dam area is Level 7.

2.4.7.2 Engineering Geological Condition

The dam of Shapai Hydropower Station is located in a deep valley and the bedrock on both banks is exposed. The valley is V-shaped and basically symmetrical, and its width-height ratio is about 1.7, which is suitable for structure a concrete arch dam. The dam foundation is mainly massive granite (diorite) rocks, and Sc schist belts are distributed on the left bank of the dam foundation with an elevation of 1 800 m and the right bank with an elevation of 1 795 m. The dam foundation rock mass is not obviously relaxed, and the weathering is weak. The quality of rock mass is mainly Grade Ⅱ and locally Grade Ⅲ-1. After schist has been replaced by concrete, it can meet the requirements of arch dam foundation.

2.4.7.3 Layout of Complex and Main Engineering Structures

The complex work is mainly composed of RCC arch dam, two flood discharge tunnels on the right bank, headrace tunnel on the right bank, powerhouse and other structures. According to the natural conditions of the dam, the three-center circular single arch dam is adopted, the dam body is not provided with flood discharge structures, and two flood discharge tunnels are arranged on the right bank of the arch dam. The headrace system is arranged on the right bank of the arch dam, and the powerhouse is arranged on the Caopo River terrace 3.5 km downstream. During the construction period of the main works, cutoff cofferdam, tunnel diversion and damming all year round are adopted for diversion.

The RCC arch dam has a maximum dam height of 130 m, a top arch center line arc length of 250.25 m, a maximum center angle of 92.48, a thickness-height ratio of 0.238, and a dam volume of 383 000 m^3. 90 d-age 20 MPa RCC is adopted for the arch dam. II gradation RCC is adopted for dam body impervious layer, of which the thickness and indexes are as follows: the impervious layer with an elevation above 1 820 m is 3 m thick, and the impermeability grade is W6; the impervious layer with an elevation below 1 820 m is 6 – 8 m thick, and the impermeability grade is W8. LJP polymer waterproof coating is used as the auxiliary impervious layer for the upstream dam section with a dam surface elevation below 1 850 m.

Two flood discharge tunnels are arranged on the right bank of the arch dam: combined with the utilization of diversion tunnels, the No. 1 flood discharge tunnel is a vortex internal energy dissipation shaft tunnel, with the maximum discharge of 242 m^3/s, which is the first flood discharge tunnel with vortex internal energy dissipation in China; No. 2 flood discharge tunnel has a long steep slope with a gradient of 10%, with a maximum discharge of 211 m^3/s.

The headrace system is arranged on the right bank of arch dam, and the total length of the headrace tunnel is 3 500.92 m, the hole diameter is 3 m, and the diversion flow is 15.6 m^3/s. The surge shaft is of cylindrical impedance type, with a diameter of 4.5 m and a height of 99.36 m. The main powerhouse is 26 m long, 18.5 m wide and 31.55 m high, with 2 mixed flow units with single capacity of 18 MW installed.

During the construction period of the main works, cutoff cofferdam, tunnel diversion and damming all year round are adopted for diversion.

The foundation of the arch dam is mainly treated with impervious curtain, consolidation grouting, drainage system, chlorite schist dense zone replaced by concrete, prestressed anchor cable reinforcement of dam abutment, etc. The impervious curtain is mainly constructed in the grouting adit on both banks, and a small part is carried out in horizontal galleries of the dam body. Consolidation grouting is performed by uncovered heavy grouting and shallow pipe grouting, and modified concrete is adopted for replacement, which has little influence on the construction of RCC of the arch dam.

The arrangement features of the complex are separation of the powerhouse and dam, no flood discharge structures in the dam body, three-center circular single arch dam with simple structure and good stress and stability conditions. Such arrangement greatly simplifies the structure of RCC arch dam, creates extremely favorable conditions for the rapid construction of RCC, and easily ensures the construction quality.

2.4.7.4 Main Design Features of the Project

1. Complex arrangement of typical RCC arch dam: the arrangement of separation of the powerhouse and dam and no flood discharge structures in the dam body simplifies the structure of RCC arch dam, creates extremely favorable conditions for the rapid construction of RCC, and easily ensures the construction quality.
2. Design mode of a "full RCC dam": RCC is basically used in the arch dam except precast concrete or modified concrete around the hole structures inside the dam, foundation leveling layer, pump room and other

special parts.

3. Significant breakthroughs in the theory and technology of joints of RCC arch dam: the joint design theory and method of RCC arch dam are systematically established, and a reasonable joint design scheme of high arch RCC dam is put forward. The new technology of precast concrete gravity formwork joint formation is successfully adopted, and the joint repeated grouting technology suitable for RCC arch dam is successfully implemented.

4. An innovative breakthrough in RCC temperature control technology: it is the first time to realize the cooling technology of embedding high density polyethylene cooling pipes in RCC.

5. Important breakthroughs in high crack resistance rolling + concrete technology: low-brittleness delayed micro-expanding cement has been successfully developed, and RCC with high crack resistance has been optimally prepared by low-brittleness micro-expanding cement, manufactured granite aggregate with low ratio of elastic modulus to strength and fly ash. Crack resistance parameters were put forward for the first time to evaluate the crack resistance of RCC.

6. Development of micro-expanding RCC with MgO: the micro-expanding technology of adding MgO into RCC has been successfully adopted for arch dam cushion abutment (maximum length of 56 m, width of 44 m, and height of 12.5 m).

7. Successful application of impervious technology for RCC dam body: self-seepage control of II gradation RCC is the main imperviousness measure of the dam body. Considering the safety of the high arch dam, LJP synthetic polymer waterproof coating is also used as the auxiliary imperviousness measure.

8. Great progress for rapid construction technology of RCC: the construction process of high RCC arch dam has been dynamically simulated by computer. The new continuous forced mixing plant has been successfully applied, and the height of concrete conveyed by a vacuum chute reached 100 m. The adjustable full cantilever large formworks with alternating up and down and continuous rising has been developed and optimized to ensure the continuous construction of RCC. The application range of the modified concrete is expanded and the full RCC human placement is realized.

9. Systematical improvement of prototype observation technology of RCC arch dam: in this paper, "protective pipe erection method", "drilling method" and "pitting method" are put forward and used to bury and install the instruments in the dam. RcJ-1 buried joint tester suitable for RCC has been developed. The method, theory and analysis program for the analysis of prototype observation results and feedback analysis of high RCC arch dam have been established. The safety monitoring scheme of Shapai RCC Arch Dam is put forward.

2.4.8 Introduction to Linhekou RCC Arch Dam Design

2.4.8.1 Project Overview

Linhekou Hydropower Station is located on the main stream of Lanhe River in Langao County, Shaanxi Province. It is the third hydropower station in the cascade planning below Huali in the main stream of Lanhe River and the only control project in the whole basin of Lanhe River. The dam is located 1.0 km away from the upstream Linhe Township, about 7 km away from downstream Langao County and 80 km away from Ankang City. The powerhouse is located on the left bank of Lanhe River, with a highway mileage of about 5.5 km from the dam. Linhekou Hydropower Station is mainly used to generate electricity, which has comprehensive benefits such as aquaculture and tourism.

The main structures of the complex work are composed of RCC dome dam, surface flood discharge holes in the dam body, flood discharge tunnels, headrace tunnel, powerhouse and switch yard. The normal pool level is 512.0 m, the maximum dam height is 96.5 m, the total reservoir capacity is 147 million m^3, and the regulating capacity is 87.5 million m³, which is an incomplete annual regulating reservoir. Total installed capacity of the power station is 72 MW, guaranteed output is 11.7 MW, and the average annual power generation is 223 million kW · h.

2.4.8.2 Hydrology and Meteorology

The main stream of Lanhe River is 151.2 km, with a total head of 2 073 m, average gradient of the river channel of 13.07, and the tributary area of 2 128 km². Linhekou Hydropower Station is located on the main stream of the midstream of Lanhe River, and the controlled catchment area of the dam is 1 450 km².

According to the statistics of Langao Meteorological Station, the average annual temperature is about 15 ℃, the extreme maximum temperature is about 40.7 ℃, and the extreme minimum temperature is about −8.4 ℃, the average annual rainfall is about 1 000 mm, and the maximum wind speed is about 19.7 m/s over the years.

2.4.8.3 Geological Condition

The dam is located in Lanhe Gorge, with a narrow valley and steep bank slopes. The valley is V-shaped, and the both banks are basically symmetrical. The bedrock bank slope is below the elevation of 490 – 500 m, and the section above such elevation is covered by slope deposits. The height difference between the bank top and riverbed is 130 – 310 m, the river surface width is 23 – 40 m in dry season, and the valley width is 230 – 250 m at the dam top with an elevation of 515 m.

Major hydraulic structures are arranged on the batholith composed of Silurian metamorphic gravelly tuff, metamorphic tuff and slate, and the riverbed overburden is 5 – 8 m. The strata in the dam area are monoclinic, and the attitude of rock strata changes greatly. The overall strike is NW 290° – 310°, the trend is NE, the dip angle is 45° – 65°, the dry volumetric weight of dam foundation rock mass is 2.98 – 3.0 g/cm³, the dry compressive strength is 132 – 161 MPa, and the saturated compressive strength is 81 – 98 MPa. The longitudinal wave velocity of weakly and slightly weathered rock mass is 5 200 m/s, and that of strongly weathered relaxed rock mass is 3 200 – 3 400 m/s.

2.4.8.4 Layout of Complex

The complex of the power station is mainly composed of RCC dome dam, surface flood discharge holes in the dam body, flood discharge tunnels converted from diversion tunnels, pressure headrace tunnels, surge shafts, shore powerhouse and 110 kV switch yard. This project is classified as Class II large (2) project, with main structures of Grade II.

The dam is a dome dam with variable center and radius and uniform thickness, a crest elevation of 515 m and a maximum dam height of 96.5 m. The dam crest width is 6 m, the dam base width is 27.2 m, and the thickness-height ratio is 0.28. The surface flood discharge holes are arranged in the center of arch dam, with the hole mouth size of 9 m × 10.5 m and the crest elevation of 502 m. The headrace structures are arranged on the left bank, including the shore water inlet, pressurized headrace tunnel, surge shaft and pressure pipelines. The total length of headrace pipeline is 2 933 m. The powerhouse of the power station is a shore ground powerhouse, and the switch yard is arranged at the upstream of the powerhouse, which is of outdoor layout.

2.4.8.5 Main Structures of the Project

I Arch dam

1. Arch dam arrangement. The dam crest elevation of Linhekou Hydropower Station is 515 m, the foundation elevation is 418.5 m, the maximum dam height is 96.5 m, the arch radius of the upstream face is 172 m, the maximum center angle is 103.6, the arc length of upstream face dam crest is 311 m, and the width of dam base is 27.2 m. The reference plane orientation of the arch dam is NE67, which is consistent with the river flow direction in dam line area.

In order to meet the requirements of foundation grouting, drainage, traffic, observation and inspection, two layers of curtain grouting, drainage, traffic and observation galleries are set at the elevation of 438 m and 478 m in the dam, and one layer of drainage, traffic and observation gallery is set at the elevation of 493 m in the riverbed (mainly dam sections 4$^{\#}$ and 5$^{\#}$). Elevation of grouting curtains and draining adits on both banks is 438 m, 478 m and 515 m. The maximum overhang degree at the crown cantilever is 0.1; the maximum overhang degree on the right bank is 0.13; the overhang degree on the left bank is roughly controlled at 0.15, and the maximum overhang degree is 0.17.

2. Curtain grouting. There are 3 layers of grouting curtains and draining adits at the elevations of 515 m, 478 m and 438 m respectively in dam foundation on both banks. The adits on the left bank are 30 m, 35 m and 40 m long respectively, and those on the right bank are 30 m, 50 m and 40 m long respectively. There are two rows of curtain holes at the elevation of 515 m on the dam top, with a hole depth of 38 m; there are two rows of grouting tunnel curtain holes at the elevation of 478 m, with a hole depth of 42 m; there are two rows of foundation curtain holes at the elevation of 438 m, the first row of curtain holes is 40 m deep into the rock, and the second row of curtain holes is 25 m deep into the rock. Overlapping curtains are set in the curtain grouting tunnels at the elevations of 438 m and 478 m respectively.

3. Drainage design. The distance between the drainage holes is 3 m, and the drainage holes inside the grouting draining adit above elevation 478 m are drilled upward until 1 – 2 m into the dam foundation concrete, while the drainage holes outside grouting draining adit are drilled downward from the grouting draining adit with an elevation of 515 m to connect with the adit with an elevation of 478 m. The drainage holes with an elevation of 478 – 438 m are mainly drilled downward from the grouting draining adit with an elevation of 478 m to the grouting draining adit with an elevation of 438 m. The drainage holes below the elevation of 438 m are 20 m deep into the rock.

4. Impervious design of dam body. II gradation RCC with rich cementing materials and GEVR are used as the main impervious bodies for the dam, and LJP synthetic polymer waterproof coating is applied below the dead water level of the dam at elevation 485 m as the auxiliary impervious measures.

5. Design of dam joints and joint grouting system. There are five inducing joints in the dam (of which the No. 4 joint is partly a transverse joint) and three transverse joints. Inducing joints and transverse joints divide the dam body into nine dam sections from right to left, and the arc lengths of the crest on the upstream face are 22.28 m, 18.00 m, 34.00 m, 41.50 m, 49.33 m, 48.00 m, 36.00 m, 30.00 m and 31.90 m in turn.

The standard grouting area is formed by three sets of grouting pipelines and one set of exhaust pipelines. The height of the grouting area is 5.7 – 8.4 m. Repeated grouting can be performed in the dam joint grouting system, and multiple repeated grouting kits are installed on each joint grouting pipeline as grouting devices.

II Flood release structure

The outlet structure consists of surface holes on the dam body and the flood discharge tunnels reconstructed from the diversion tunnels.

5 flood discharge surface holes are arranged at the center of arch dam crest. The size of a single surface hole is 9 m × 10.5 m, the weir crest elevation is 502.0 m, and WES practical weir is adopted for weir surface. Differential flip bucket is used for energy dissipation. The check flood level is 512.5 m, the discharge volume in case that 5 holes are all opened is 3 080 m^3/s; when the design flood level is 511.10 m, the discharge volume in case that 5 holes are all opened is 2 488 m^3/s.

The flood discharge tunnel entrance is located on the left bank of the dam, which is reconstructed from the diversion tunnel. The elevation of flood discharge tunnel inlet roof is 464.0 m.

2.4.8.6 Main Design Features

The design feature of this power station is that the water retaining structure is a RCC dome dam with a dam height of 96.5 m, ranking first among similar dam types in China. The arc length of dam crest is 311 m, and the dam is the RCC arch dam with the longest arc length. The maximum overhang degree of the dam is 0.17, which is the largest in China.

Dam joints are formed by embedding induced plates, including transverse joints and induced joints. Repeated grouting system is embedded in the joints, which makes the construction simple and reliable. II gradation concrete with rich cementing materials is the main impervious material of the dam, GEVR is used to strengthen impervious layer, MP synthetic polymer material is used as the auxiliary impervious layer below elevation 485 m of the upstream face, which could ensure the reliable seepage control and convenient construction. The shape design of the dam is as simple as possible, and the layout of the gallery in the dam is optimized. The elevation of the foundation gallery is 19.5 m higher than that of the riverbed foundation surface, so the gravity drainage is adopted, and the dam foundation is not provided with a collecting well, which is convenient for operation and management, and gives full play to the characteristics of rapid rise of RCC.

2.5 Conclusion

1. The design of RCC dams is closely related to the performance of RCC. The development of materials science has completely changed the design concept and design method of RCC dams. In particular, RCC has been the semi-plastic concrete without slump because of its full-section damming technology, change of impervious shape, mix proportion of cement with rich cementing materials and low VC value. Its imperviousness, anti-freezing and ultimate tensile values are basically the same as those of normal concrete, and the concerned anti-sliding stability of layer (joint) surface and weak surface of water seepage are solved.

2. Complex arrangement mode is very important to the rapid construction of RCC dam. Compared with the quantity, cost and construction period, the complex arrangement mode of RCC dam is of strategic significance; the most ideal complex layout of RCC dam is to draw lessons from the design principle of complex layout of earth-rock dam, the other kinds of structures are arranged outside the dam to the greatest extent, and the RCC dam shall be separated from power generation structures and outlet structures as far as possible, and the bank-tower inlet shall be adopted, all such measures are the best technical scheme to ensure the rapid RCC construction.

3. The shape of RCC dam shall be simple, and the RCC parts shall be relatively concentrated, so as to reduce the holes in the dam, simplify the dam structure, expand the scope of RCC as far as possible, minimize the interference to rapid RCC construction, and give full play to the advantages of rapid RCC construction.

4. The structure of a RCC dam body is different from that of the normal concrete dam body. The construction method of RCC is fundamentally different from the column pouring of normal concrete; some machine-mixed GEVR is used in the construction according to the project practice, and the construction effect is good; the drainage holes of the dam body are drilled, and the galleries are prefabricated, which can effectively reduce the interference to the rapid RCC construction.

5. Volumetric weight is the most critical parameter for the design load of a gravity dam. The range of the volumetric weight determines the size of the dam section and its anti-sliding stability; a large quantity of project cases prove that the volumetric weight i.e., the apparent density of RCC is generally larger than that of the normal concrete; it is necessary to re-analyze the RCC with the same volumetric weight as normal concrete.

6. The problem of mismatch of design indexes of hydraulic concrete has a long history. Taking the single frost resistance grade as an index to evaluate the concrete durability requires continuous in-depth study on the tests; it is debatable that the design age of RCC shall be 90 d, and the 180 d (design age) compressive strength as that of high arch dam shall be adopted, so a breakthrough can be made in the design age.
7. In order to give full play to the advantages of RCC rapid damming, it is necessary to simplify or optimize the material zoning of RCC dam body as much as possible. In the material zoning of RCC dam body, technical innovation and deepening research shall be carried out scientifically according to the characteristics of RCC damming, so as to provide technical support for RCC rapid damming.

Chapter 3

Raw Materials and Project Examples

3.1 General

RCC is composed of cement, admixtures, sand, stone, additives, water, rock powder etc., which is a typical multiphase heterogeneous material. The stable quality of raw materials used in RCC is directly related to the reliability of concrete mix proportion test data, the consistency of RCC application and the stable control of mixture quality. Reasonable selection of raw materials and guaranteed of output are the key to ensure the quality and rapid construction of RCC, which must be paid great attention to.

The cementitious material of RCC is mainly composed of cement + admixture + micro-rock powder finer than 0.08 mm (determined according to actual project conditions), and the cementitious material is mixed with water to form cementing grout. In RCC, cementing grout wraps sand particles, fills the gaps between sands, and forms mortar together with sands. Mortar wraps stone particles and fills the gaps between stones. In RCC mixture, the cementing grout plays a role of "lubrication" between sand and gravel particles, which can adjust and improve the workability of the mixture. After the hydration reaction of the cementing grout, the hardened cementing grout will firmly cement the aggregates into a whole. Additive is aimed for reducing water, retarding, air entrainment and improving the performance in RCC, and is an important component of RCC. The rock powder has been an indispensable component of RCC.

Cement is the cementing material, which is widely used in water conservancy and hydropower projects, industrial and civil structures, roads and other construction projects. The main varieties of hydraulic concrete and hydraulic RCC commonly used are: portland cement, ordinary Portland cement and moderate heat Portland cement, low heat Portland cement, low heat Portland slag cement, etc. According to the latest national standard *Common Portland Cement* GB 175—2007(implemented on June 1, 2008), the strength grades 32.5 and 32.5 R of ordinary cement have been canceled. Because of the cancellation of low grade cement, *the application of admixture in concrete has made a greater breakthrough*. GB 200—2003 standard stipulates: moderate heat Portland cement is only of Grade 42.5. In recent years, stricter control requirements have been put forward regarding cement fineness (specific surface area), magnesium oxide (MgO), sulfur trioxide (SO_3), alkali content, hydration heat, compressive strength, breaking strength, tricalcium aluminate (C_3A), tetra calcium aluminoferrite (C_4AF) and cement temperature in the construction site for dam concrete, which effectively reduces the hydration heat temperature rise of concrete and controls the fluctuation of cement quality.

Admixtures are the main components of RCC cementing materials, and the admixture content in RCC in China generally accounts for 50% – 65% of cementing materials. Admixtures in RCC are generally active or inactive. Admixtures mainly include fly ash, granulated blast furnace slag, phosphorous slag, volcanic ash, tuff, limerock powder, etc. Admixtures can be mixed singly or in combination. For example, RCC with compound admixture of 50% phosphorous slag + 50% tuff, referred to as PT admixture, is adopted in Dachaoshan Project. The compound admixture of 50% manganese iron slag + 50% limerock powder, referred to as SL admixture, is adopted in Jinghong, Gelantan, Jufudu, Tuka River and other projects. As an admixture, fly ash always plays a dominant role in hydraulic concrete. The application research of fly ash in RCC is mature. Fly ash features large content and wide application, but also has the optimal performance among the admixtures.

Additives have been the fifth necessary material besides cement, admixture, coarse and fine aggregates and water. It is stipulated in the Specifications for Hydraulic Concrete Construction and Construction Specification for Hydraulic Roller Compacted Concrete that a proper amount of additive must be added to the hydraulic concrete. Because of the special construction method of RCC, in order to improve the layer bonding of RCC, setting time has been an extremely important performance of RCC mixture, which is directly related to the rollability, liquefied bleeding, interlayer bonding and construction quality of RCC. In recent years, frost resistance grade has been an important index for durability design of RCC in cold, mild or hot areas. In order to improve the frost resistance grade, air entraining agent is mixed in RCC. The additives used in RCC are mainly compound retarding high-range water reducing agent and air entraining agent. For example, retarding high-range water reducing agent and air entraining agent are mixed with RCC in Baise, Longtan, Jinghong, Jin'anqiao and other projects.

Aggregate refers to the sand and gravel used for mixing concrete, in which the fine aggregate is divided into natural sand and manufactured sand, and the coarse aggregate is divided into gravel and pebble. Sand and gravel aggregate is the main raw material of concrete, and aggregate generally accounts for about 80% – 85% of the volume of RCC, which has strict quality requirements. In the mix proportion design and mixing quality control of aggregate, the saturated and surface dry state is adopted for calculation, which is the biggest difference between hydraulic concrete and ordinary concrete in mix proportion design. The quality and quantity of aggregate determine the smooth construction and economy of the project. Therefore, the material yard must be selected correctly through strict exploration investigation, systematic physical and mechanical property test and economic comparison. The rapid construction of RCC, large and concentrated aggregate consumption, improper selection of aggregate or insufficient investigation will lead to the passive project construction. Therefore, special attention shall be paid to avoid any mistakes in aggregate selection. In the use of aggregate, it is also possible to use mixed aggregate, i. e. manufactured sand + river sand, pebble + gravel, according to the aggregate source and the actual production and processing of aggregate, which requires technical and economic analysis and comparison.

Any drinking water that meets the national standard can be used for mixing and curing concrete. Material content of concrete mixing water shall comply with requirements of *Specifications for Hydraulic Concrete Construction* DL/T 5144. Recent research data show that the air entraining agent has certain adaptability to the mixing water quality, and the construction water at the project site must be used in the initial mix proportion design.

Materials science is the foundation of all sciences. The quality of raw materials directly affects the construction quality of concrete, and is related to the strength, durability and overall performance of dams, which is the basis of project quality assurance. For this reason, special provisions are made for raw materials in standards such as Specifications for Hydraulic Concrete Construction, Construction Specification for Hydraulic Roller Compacted Concrete and Design Specifications for RCC Dams. Specific technical requirements for concrete raw materials are also put forward in the technical clauses of bidding documents for civil works of dams of water conservancy and hydropower projects.

This chapter mainly focuses on the quality requirements and performance of RCC raw materials, as well as project cases of raw material tests, and the following chapters describe the rock powder and admixtures.

3.2 Cement Poperties and Project Cases

3.2.1 Common Portland Cement

3.2.1.1 Definition of Common Portland Cement

According to the national cement standard GB 175—2007, the definition of common Portland cement is as follows: common Portland cement is the hydraulic cementing material made of Portland cement clinker, proper amount of gypsum and specified mixed materials.

3.2.1.2 Classification of Common Portland Cement

Common portland cement is divided according to the variety and content of mixed materials: portland cement and ordinary portland cement.

3.2.1.3 Strength Grade

1. The strength grade of portland cement is divided into: six grades 42.5, 42.5R, 52.5, 52.5R, 62.5 and 62.5R.
2. The strength grade of ordinary portland cement is divided into: four grades 42.5, 42.5R, 52.5 and 52.5R.
3. The strength grades of portland slag cement, pozzolanic Portland cement, fly ash portland cement and composite portland cement are divided into: 32.5, 32.5R, 42.5, 42.5R, 52.5 and 52.5R.

3.2.1.4 Technical Requirements for Common Portland Cement

The density of cement is a commonly used parameter in concrete mix proportion design, and the density of ordinary portland cement (ordinary cement for short) is generally $3.0 - 3.15$ g/cm^3.

Cement paste of standard consistency must be adopted when determining the setting time and soundness of cement. According to Chinese standards, the consistency of cement grout is determined by a consistometer, and the grout in which the test cone is sunk to a depth of 28 mm ± 2 mm is regarded as the grout of "standard consistency", and the water consumption at this time is the water consumption under standard consistency.

The main performance of ordinary portland cement depends on cement clinker, and its admixture is less and only plays an auxiliary role, so there is no fundamental difference between ordinary cement and portland cement in various properties. However, after all, ordinary cement is mixed with a small amount of admixture, and the overall performance is different from that of portland cement. Compared with portland cement, the early strength increase rate of ordinary Portland cement is lower, the frost resistance and abrasion resistance decrease slightly, but the sulfate resistance increases. Due to the limited content of admixture in ordinary cement, the performance of portland cement is not changed to a great extent, so this kind of cement has strong adaptability and is very popular with users, and can be widely used in various industries, civil structures and water conservancy and hydropower projects. See Table 3.1 for technical requirements of different varieties and strength grades of common portland cement.

Table 3.1　　　Technical indexes of common portland cement　　　Unit: %

Variety	Code	Loss on ignition (mass fraction)	Sulfur trioxide (mass fraction)	Magnesium oxide (mass fraction)	Specific surface area	Chloride ion (mass fraction)	Time of setting (min)	
							Initial setting	Final setting
Portland cement	P·I	3.0	≤3.5	≤5.0[a]	Not less than 300 m²/kg	0.06[c]	>45	<390
	P·II	≤3.5						
Common portland cement	P·O	≤5.0	≤3.5	≤5.0[a]			>45	<600
Portland-slag cement	P·S·A		≤4.0	≤6.0[b]	The residue larger than 80 μm on the sieve is not more than 10% or lager than 45 μm on the sieve is not more than 30%.			
	P·S·B		≤4.0	—				
Portland-pozzolana cement	P·P		≤3.5	≤6.0[b]				
Flyash-based Portland cement	P·F		≤3.5	≤6.0[b]				
Portland composite cement	P·C		≤3.5	≤6.0[b]				

Note: a. If the autoclave test of cement is qualified, the content (mass fraction) of magnesium oxide in cement can be relaxed to 6.0%.
　　b. If the content (mass fraction) of magnesium oxide in cement is greater than 6.0%, it is necessary to carry out the autoclave test of cement and pass the test.
　　c. When there are lower requirements, the index is determined by negotiation between the buyer and the seller.

3.2.1.5　Strength of Common Portland Cement

For different varieties and different strength grades of common portland cement, the strength at different ages shall be greater than the provisions in Table 3.2.

Table 3.2　　　Strength grade of common portland cement　　　Unit: MPa

Variety	Strength grade	Compressive strength		Bending strength	
		3 d	28 d	3 d	28 d
Portland cement	42.5	17.0	42.5	3.5	6.5
	42.5R	22.0		4.0	
	52.5	23.0	52.5	4.0	7.0
	52.5R	27.0		5.0	
	62.5	28.0	62.5	5.0	8.0
	62.5R	32.0		5.5	
Common portland cement	42.5	17.0	42.5	3.5	6.5
	42.5R	22.0		4.0	
	52.5	23.0	52.5	4.0	7.0
	52.5R	27.0		5.0	
Portland-slag cement Cinerite-based Portland cement Flyash-based Portland cement Portland composite cement	32.5	10.0	32.5	2.5	5.5
	32.5R	15.0		3.5	
	42.5	15.0	42.5	3.5	6.5
	42.5R	19.0		4.0	
	52.5	21.0	52.5	4.0	7.0
	52.5R	23.0		4.5	

3.2.2 Moderate Heat Portland Cement, Low Heat Portland Cement and Low Heat Portland Slag Cement

3.2.2.1 Definition of Moderate Heat Portland Cement, Low Heat Portland Cement and Low Heat Portland Slag Cement

According to the *Moderate Heat Portland Cement Low Heat Portland Cement Low Heat Portland Slag Cement* GB 200—2003, these three types of cement are defined as follows:

1. Moderate heat portland cement. The ground hydraulic cementing materials with medium hydration heat, which is made of portland cement clinker with appropriate composition with appropriate amount of gypsum, is called moderate heat portland cement (referred to as moderate heat cement), its strength grade is 42.5, and its code name is PMH.
2. Low heat portland cement. The ground hydraulic cementing materials with low hydration heat, which is made of portland cement clinker with appropriate composition with appropriate amount of gypsum, is called low heat Portland cement (referred to as low heat cement), its strength grade is 42.5, and its code name is PLH.
3. Low heat portland slag cement. The ground hydraulic cementing materials with low hydration heat, which is made of Portland cement clinker with granulated blast-furnace slag and appropriate composition with appropriate amount of gypsum, is called low heat Portland slag cement (referred to as low heat slag cement), its strength grade is 32.5, and its code name is PSLH.

3.2.2.2 Technical Indexes of Moderate Geat Portland Cement, Low Heat Portland Cement and Low Heat Portland Slag Cement

According to the standard requires of the *Moderate Heat Portland Cement Low Heat Portland Cement Low Heat Portland Slag Cement* GB 200—2003, the technical requirements of moderate heat portland cement, low heat portland cement and low heat portland slag cement are shown in Table 3.3, and the strength grade and hydration heat indexes are shown in Table 3.4.

The moderate heat cement is the main cement type used in hydraulic concrete, and the production process of moderate heat cement is basically the same as that of Portland cement. The main difference between both types of cement is that: according to the characteristics of hydraulic concrete, there are special requirements for certain components and mineral compositions of moderate heat cement clinker, and the clinker is not allowed to be mixed with mixed materials, which is a main difference compared to normal cement.

The main technical characteristics of moderate heat cement are as follows:

1. If the cement passes the autoclave test for soundness, the MgO in the clinker is allowed to be 6%.
2. The alkali content is negotiated by suppliers and users. If the cement may have a harmful reaction with the aggregate in the concrete, users can request low alkali content.
3. The reasonable control of moderate heat cement fineness is one of the keys to cement production. Generally, under the condition of sufficient strength and hydration heat meeting the standard, the specific surface area of cement is controlled within $280 - 350 \text{ m}^2/\text{kg}$.
4. The low hydration heat is one of the main characteristics of moderate heat cement. Generally, its heat release peak occurs at the stage of about 7 h hydration, but its heat release rate is only 60% of portland cement.
5. The setting time of moderate heat cement is normal, usually 2 – 4 h for the initial setting, and 3 – 6 h for the final setting.
6. The early strength of moderate heat cement is slightly lower than that of portland cement of the same grade.

On the one hand, in the application of moderate heat cement, we must pay attention to its alkali content to prevent the possible alkali-aggregate reaction from harming the concrete project. In addition, in order to further reduce the hydration heat and improve the corrosion resistance to reduce the influence of the alkali-aggregate reaction during the use of moderate heat cement, admixtures such as fly ash can be mixed into the concrete.

The chemical composition of the moderate heat Portland cement and low heat portland cement are not much different. However, due to the different firing schedules, the composition of the formed mineral is significantly different. The moderate heat portland cement clinker has a high C_3S content, while the low heat portland cement clinker has a high C_2S content. They are just the opposite, which determines that the performance of both types of cement is quite different.

Table 3.3　　Main technical characteristics of moderate heat cement and low heat cement

Type and standard	Limit of mineral composition of cement clinker (%)	Magnesium oxide (%)	Alkali content (%)	Sulfur trioxide (%)	Loss on ignition (%)	Specific surface area (m^2/kg)	Time of setting	
							Initial setting	Final setting
Moderate heat cement GB 200—2003	$C_3S \leqslant 55\%$; $C_3A \leqslant 6\%$; f-CaO $\leqslant 1\%$	<5.0	<0.6	<3.5	<3	>250	>60 min	<12 h
Low heat cement GB 200—2003	$C_2S \geqslant 40\%$	<5.0	<0.6	<3.5	<3	>250	>60 min	<12 h
Low heat portland slag cement GB 200—2003	$C_3A \leqslant 8\%$; f-CaO $\leqslant 1.2\%$; MgO $\leqslant 5\%$	<5.0	<1	<3.5	<3	>250	>60 min	<12 h

Note: 1. The MgO content in cement shall not exceed 5.0%. If the cement passes the autoclave test for soundness, the MgO content of the cement is allowed to be 6.0%.
　　　2. When the cement may have a harmful reaction with the aggregate in the concrete, and users request a low alkali, the alkali content of the cement shall not exceed 0.6%.

Table 3.4　　Strength grade and hydration heat of moderate heat cement and low heat cement

Type and standard	Compressive strength (MPa)			Bending strength (MPa)			Hydration heat (kJ/kg)	
	3 d	7 d	28 d	3 d	7 d	28 d	3 d	7 d
Moderate heat cement GB200—2003	12.0	22.0	42.5	3.0	4.5	6.5	251	293
Low heat cement GB200—2003	—	13.0	42.5	—	3.5	6.5	230	260
Low heat portland slag cement GB 200—2003	—	12.0	32.5	—	3.0	5.5	197	230

3.2.3　Mineral Composition of Cement Clinker

Calcium oxide, silicon oxide, aluminum oxide and ferric oxide in cement clinker exist as an aggregate of multiple minerals, rather than separate oxides, which is formed by the reaction of two or more oxides under the high-temperature calcination. The crystals are fine, and have a particle size is generally within 30 – 60 μm. Therefore, cement clinker is a type of artificial rock composed of multiple minerals, the crystals of which are fine, or a type of aggregate composed of multiple minerals.

3.2.3.1　Chemical Composition of Cement Clinker

Portland cement clinker is mainly composed of calcium oxide (CaO), silicon dioxide (SiO_2), aluminum oxide (Al_2O_3) and ferric oxide (Fe_2O_3), of which the proportion is generally within 62% – 67%, 20% – 24%, 4% – 7% and 2.5% – 6% respectively. These four oxides account for more than 95% of clinker. There are also few oxides with a proportion of less than 5%, such as magnesium oxide (MgO), sulfur anhydride (SO_3), titanium oxide (TiO_2), phosphorus oxide (P_2O_5) and alkali.

3.2.3.2 Mineral Composition of Cement Clinker

After high temperature calcination, CaO, SiO_2, Al_2O_3 and Fe_2O_3 in cement raw materials are combined into the main mineral compositions of clinker:

Tricalcium silicate $3CaO \cdot SiO_2$ (abbreviated as C_3S).

Dicalcium silicate $2CaO \cdot SiO_2$ (abbreviated as C_2S).

Tricalcium aluminate $3CaO \cdot Al_2O_3$ (abbreviated as C_3A).

Tetra calcium aluminoferrite $4CaO \cdot Al_2O_3 \cdot Fe_2O_3$ (abbreviated as C_4AF).

In addition, there are few oxides, such as free calcium oxides (fCaO), periclase (crystalline magnesium oxide), alkali-containing minerals and vitreous body.

In general, the content of C_3S and C_2S in clinker accounts for about 75%. Both of them are called silicate minerals; the content of C_3A and C_4AF accounts for about 22%. During calcination, the latter two minerals, magnesium oxide, alkali and others gradually melt into liquid at 1 200 – 1 280 ℃ to promote the smooth formation of C_3S, so they are called as fluxing minerals.

3.2.3.3 Hydration Reaction and Characteristics

The plastic mortar is formed by mixing Portland cement with an appropriate amount of water, which can cement the aggregates, such as sand and stone. After curing for a period of time, the aggregates will gradually become set cement with a certain mechanical strength. The hydration and hardening of cement is a complex process of physical, chemical and physicochemical changes. During this process, new hydration products are continuously generated and exothermic reactions occur, causing volume change and strength increase. In order to understand the whole extremely complicated hydration process of cement, it is necessary to understand the hydration characteristics of single minerals.

1. Tricalcium silicate (C_3S). During cement hydration, C_3S hydration speed is faster. It can rapidly make cement set, harden and form hydration products with a considerable strength. Therefore, the strength of C_3S develops relatively fast, with higher early strength and greater strength enhancement rate. The 28 d strength can reach 70% – 80% of the strength after a year. In terms of 28 d strength, the strength of C_3S among the four main minerals is the highest.

2. Dicalcium silicate (C_2S). When C_2S reacts with water, the early strength of hydration products is low due to slower hydration, but the later strength is high. It continues to hydrate even after several decades, and the strength still increases. Due to low hydration heat and good water resistance of C_2S, it is beneficial to properly increase the C_2S content of mass concrete or that of cement used in the aggressive projects.

    The content C_3S and C_2S of account for about 3/4 of cement clinker, so their hydration products have a great effect on the properties of set cement.

3. Tricalcium aluminate (C_3A). C_3A plays a role of flux during clinker calcination, which melts with C_4AF at 1 250 – 1 280 ℃ to form liquid, thus promoting the smooth formation of C_3S. The shape characteristics of C_3A crystals change with its cooling speed. Generally, they are drip shape when cooling quickly, and rectangular or columnar when cooling slowly.

    C_3A hydrates rapidly, with much heat release and short setting time. If there is no retarder such as gypsum in the cement, the concrete is easy to set rapidly. C_3A also hardens rapidly, and most of the strength is exerted within 3 d, so the early strength is exerted rapidly. However, the absolute value is not high, which is almost no longer increase, even shrink in the future.

    Duo to large shrinkage deformation and poor sulfate resistance of C_3A, during the production of sulphate resistant cement or the use of cement in the mass concrete projects, the C_3A content shall be controlled in the lower range.

During C_3A hydration, needle-like calcium sulphoaluminate crystals are generated first in the presence of gypsum, which is called as ettringite in nature. Meanwhile, C_3A volume expands during the generation of calcium sulphoaluminate crystals. Therefore, if hardened set cement is contact with sulfate solutions for a long time, there are generated calcium sulphoaluminate crystals in set cement, making set cement destroyed due to C_3A volume expansion.

4. C_4AF. C_4AF is also a mineral flux. It is easy to melt and can reduce the temperature at which the liquid phase appears and the viscosity of liquid phase, so it contributes to the formation of C_3S.

C_4AF hydration rate is between C_3A hydration rate and C_3S hydration rate in the early stage, but its subsequent development is not as good as that of C_3S. Its early strength is similar to that of C_3A, but the strength can continue to increase in the later period, which is similar to that of C_2S.

Its hydration products are affected by temperature and the concentration of calcium hydroxide in the solutions, which is closely related to the Al_2O_3/Fe_2O_3 content of this mineral. When the Al_2O_3 content of calcium ferroalumnates increases, the hydration of solid solutions will accelerate. If the Fe_2O_3 content of calcium ferroalumnates increases, the hydration reaction slows down.

C_4AF has better impact resistance and sulfate resistance, and its hydration heat is lower than that of C_3A. During the production of sulphate resistant cement or the use of cement in the mass hydraulic concrete projects, it is beneficial to increase the C_4AF content properly.

3.2.4 Brittleness Coefficient and Mineral Composition of Cement

3.2.4.1 Brittleness Coefficient and Mineral Composition of Cement

For mass concrete used for hydraulic structures and dams, as its interior is in the adiabatic state, the hydration heat of cement accumulates in the concrete, so that the temperature inside the dam can rise to 50 ℃ or above. The temperature difference with the concrete on the surface of the dam cooled more quickly can reach 10 ℃. Due to the expansion and contraction of objects, the great temperature difference between inside and outside the dam body will result in great tensile stress, which may cause concrete cracking, thus directly affecting project quality and dam safety. During mass hydraulic concrete construction, the most direct and effective technological approaches to reduce or eliminate the effects is to minimize the hydration heat of cement in addition to adopting reasonable construction technologies and temperature control measures. The factors that can make the hydration heat of cement and its heat release rate reduce include the mineral composition of clinker, the fineness of cement, admixtures, additives and others.

Moderate heat and low heat cement still belong to Portland cement series, and their clinker mineral compositions are still C_3S, C_2S, C_3A and C_4AF. Both the absolute value of hydration heat and relative heat release rate of C_3S, C_2S and C_3A from high to low are those of C_3A, C_3S and C_2S respectively. Obviously, the hydration heat of cement is reduced only by reducing the C_3A and C_3S content. The decrease of C_3S cement means the increase of C_2S cement. C_3S is the main strength component of silicate clinker. Although the hydration heat of C_2S is low, the early strength of C_2S is also slow. If the C_2S content is too much, the early strength of cement cannot be guaranteed. Therefore, the C_2S content shall not be excessively reduced. The design of clinker mineral composition shall mainly focuses on the decrease of C_3A proportion and the increase of C_4AF content correspondingly. In addition, the calorific value of free calcium oxide (fCaO) is also very high during digestion in water, which will increase the hydration heat of cement. Therefore, the content of fCaO shall be strictly controlled.

The brittleness coefficient of cement produced by different cement manufacturers is different as mineral content is different. The brittleness coefficient (K) of cement is the ratio of compressive strength to flexural strength of cement. If the brittleness coefficient is large and its ultimate tensile value is small, the crack resistance of cement is poor. Therefore, in order to improve the brittleness resistance of cement, the mineral component of C_2S and C_4AF in cement clinker shall increase as much as possible. Generally, C_4AF is more than 16%, which can significantly improve the flexural strength of cement and is very beneficial to improve the crack resistance of concrete. See Table 3.5 for comparison of brittleness coefficient and mineral component of cement in some domestic projects. According to statistical results, the brittleness coefficient of Baihua, Dongfeng, Changda ordinary cement, Yongbao and Western Yunnan moderate heat Portland cement is small, the content of magnesium oxide (MgO) is high, and the mineral component of low aluminum and high iron is obvious.

Table 3.5　　　　Comparison of brittleness coefficient and mineral component of cement in some domestic projects

Cement type	Brittleness coefficient ΔK	Mineral composition (%)				Magnesium oxide (%)	Engineering
		C_3S	C_2S	C_3A	C_4AF	MgO	
Baihua ordinary portland cement 425#	5.78	48.90	28.10	1.40	17.30	3.68	Shapai
Dongfeng ordinary portland cement 425#	5.41	47.60	26.10	2.70	16.80	3.80	
Guizhou portland cement 525#	6.20	50.40	21.40	7.50	14.20	2.30	Puding
Liuzhou portland cement 525#	6.91	59.00	14.60	8.20	15.60	1.20	Yantan
Jingmen moderate heat portland cement 525#	7.00	54.40	19.80	1.91	16.90	4.08	Three Gorges
Huaxin moderate heat portland cement 525#	7.30	52.50	23.90	2.20	16.30	4.39	
Shimen moderate heat portland cement 525#	6.49	50.40	20.30	3.20	20.10	4.39	
Sanjiang ordinary portland cement 425#	6.20	42.80	25.00	9.00	14.50	2.80	Shankou
Tiandong moderate heat portland cement 525#	6.62	51.40	22.80	2.91	17.40	2.08	Baise
Shimen moderate heat portland cement 525#	6.75	49.80	23.10	3.27	17.38	1.96	
Changda ordinary portland cement 42.5#	5.90	53.25	17.64	2.910	16.30	2.27	Guangzhao
Yongbao 42.5# moderate heat portland cement	5.95	52.40	20.13	3.65	17.83	4.16	Jin'anqiao
Western Yunnan 42.5# moderate heat portland cement	5.86	51.34	21.04	1.32	17.88	4.07	

3.2.4.2 Project Case of Low Brittleness Cement for Shapai Arch Dam

Shapai RCC arch dam is located in Wenchuan County, Aba Autonomous Prefecture, Sichuan Province. It is the leading hydropower station dam on Caopo River, a tributary of Minjiang River. The maximum dam height is 130 m. It was the highest full-section RCC arch dam in the world at that time (2001). It was listed as the supporting project of a national key scientific and technological research project for the "Ninth Five-year Plan" (*Research on RCC High Dam Construction Technology*).

The cement used in RCC of Shapai Project is "Xiongfeng" ordinary Portland 425# cement produced by dry process in Baihua Cement Plant of Aba Prefecture, Sichuan Province. The cement is one of the main materials for high crack resistance RCC in the research project for the "Ninth Five-year Plan". According to a large number of field test results of Shapai Project: the average content of C_3A and C_4AF in Baihua ordinary Portland cement 425# is 4.7% and 18.8% respectively. The formula of high iron and low aluminum in the production process of the cement is highlighted, and the brittleness coefficient of cement is less than 6. The RCC made of Baihua cement and gneiss granite with low modulus of elasticity is characterized by clearly low brittleness, large ultimate tensile value and high crack resistance. The content of MgO in cement is 3.68%, and its own MgO content is high, which is conducive to compensating the shrinkage of concrete. The physical, mechanical and thermal properties of cement meet the requirements of national standards. It reflects the high tensile and low modulus of elasticity characteristics of RCC for research.

3.2.5 Special Index Requirements for Cement in Large-scale Projects

3.2.5.1 Special Index Requirements for Cement

Medium hydration heat portland cement, low hydration heat portland cement or ordinary Portland cement are preferred in water conservancy and hydropower projects. In addition to complying with the national standards, some special requirements are put forward according to the specific use of projects.

For example, the fineness of cement, mainly expressed by specific surface area, it has a great influence on the hydration rate of cement. The particle size of cement clinker is fine, hydration is fast, and the hydration heat release is also accelerated. Therefore, the super fine cement will significantly increase the early hydration heat. Therefore, the specific requirements for the fineness of medium hydration heat portland cement, i.e. specific surface area, are put forward in mass concrete projects. According to project practice experience and test data in China, the fineness of cement has an important influence on the crack resistance of concrete. In order to obtain concrete with good crack resistance, the cement shall be slightly thicker, and the specific surface area of cement shall be generally controlled within 280 – 320 m^2/kg.

As for the mineral component in medium hydration heat portland cement, in order to reduce the hydration heat of cement, the content of tricalcium silicate (C_3S), tricalcium aluminate (C_3A), and tetracalcium aluminoferrite (C_4AF) shall be about 50%, less than 6% and more than 16% respectively. Moreover, because the content of C_3S and C_3A decreases, the hydration is relatively gentle, which is beneficial to crack healing. As for the alkali content of cement, in order to avoid alkali-aggregate reaction, the alkali content of cement clinker shall be controlled within 0.6%. With respect to the content of MgO, in order to make the volume of hardened concrete expand and compensate the shrinkage of concrete during cooling, the content of MgO in clinker of moderate heat Portland cement shall be controlled within 3.5% – 4.5%. At the same time, there are also requirements for the temperature of cement delivered to the construction site.

In the Three Gorges Project, in addition to the special index requirements for cement, the site supervisor is assigned to supervise the cement production. In recent years, large-scale water conservancy and hydropower projects have followed the above practice of the Three Gorges Project, such as Laxiwa, Xiaowan, Xiluodu, Jinping, Jin'anqiao and other projects, so as to ensure the quality of cement and fly ash at the time of delivery from the source.

3.2.5.2 Example of Special Index Requirements for Cement of Jin'anqiao Project

In Jinsha River Jin'anqiao Hydropower Station Project, the RCC gravity dam is adopted. The RCC is made of Lijiang Yongbao moderate hydration heat Portland cement. According to consultation opinions from experts, special internal control index requirements are put forward for moderate heat portland cement in the contract signed, and the internal control indexes of Jin'anqiao moderate heat Portland cement are shown in Table 3.6. The internal control indexes have more strict requirements on fineness, MgO content, hydration heat, 28 d compressive strength, flexural strength, etc. of moderate heat portland cement. At the same time, special requirements are proposed for the temperature of cement delivered to the construction site. The temperature of cement entering the mixer shall not be greater than 60 ℃, and the actual temperature of cement delivered to the site shall be controlled within 39 – 48 ℃, so as to effectively ensure the consistency and quality stability of RCC mix proportion of Jin'anqiao dam, thus playing an important role in temperature control and dam cracking prevention.

Table 3.6   Internal control indexes of moderate heat portland cement in Jin'anqiao project

S/N	Test items		Standard requirements for moderate heat portland cement GB 200—2003	Internal control index requirements of moderate heat portland cement in Jin'anqiao Project
1	Specific surface area(m^2/kg)		≥250	≤310
2	Content of magnesium oxide (%)		≤5.0	3.5 – 5.0
3	Alkali content (%)		≤0.6	≤0.6
4	SO_3 content (%)		≤3.5	≤3.0
5	Temperature control of cement delivered to construction site (℃)		—	≤60
6	Compressive strength(MPa)	3 d	≥12.0	≥12.0
		7 d	≥22.0	≥22.0
		28 d	≥42.5	47.5 ± 2.5
7	Bending strength(MPa)	3 d	≥3.0	≥3.0
		7 d	≥4.5	≥4.5
		28 d	≥6.5	≥8.0
8	Hydration heat(kJ/kg)	3 d	≤251	≤230
		7 d	≤293	≤281

3.2.5.3 Example of Contrast Test on Cement in Jin'anqiao Project

In order to further improve the quality of cement and fly ash in Jin'anqiao Project, Jin'anqiao Hydropower Station Co., Ltd. organized a contrast test on cement and fly ash in 2008. Participants of the contrast test include: site supervisor of China Structure Materials Academy, Owner's Central Laboratory, Contractor's Project Department Laboratory, Cement and Fly Ash Production Plant Laboratory, Standard Testing Laboratory, etc. Blind samples of moderate heat Portland cement are used for contrast test. See Table 3.7 for capacity verification and contrast test results of cement in Jin'anqiao Hydropower Station Project.

Contrast test results indicate: the main internal control indexes of cement, such as specific surface area, MgO, alkali content, compressive strength, flexural strength, hydration heat, have few errors with the standard results, which comply with national standards and requirements for internal control indexes specified in the contract. At the same time, it also indicates that the error of the test results of the construction site testing units is relatively large, which may be related to poor test conditions and personnel quality. Therefore, it is necessary to carry out contrast training for the testing laboratory according to the requirements of the *Accreditation and Evaluation Criteria for Laboratory Qualification* (GRSH [2006] No. 141), so as to improve the test level.

Through the development of contrast test on cement and fly ash, the cement quality is significantly improved and the cement quality is stable, which complies with not only specifications but also contract of internal control indexes of Jin'anqiao Project. Through the contrast test, the verification of the ability of personnel and facilities of testing units is greatly promoted, the test methods of each unit are further standardized, the testing level is improved, and the scientific basis for product quality evaluation is provided. At the same time, it also promotes site supervision, increases the monitoring of the production process of cement plant, strengthens the quality control of raw materials and fuel materials, stabilizes the clinker component, controls the specific surface area of grinding cement, and reduces the fluctuation of strength, which plays a positive role in ensuring the project quality, increasing mutual trust and improving the project completion data.

Chapter 3 Raw Materials and Project Examples

Table 3.7 Capacity verification and contrast test results of cement in Jin'anqiao Hydropower Station Project

Cement contrast test items		Standard unit	Contrast unit I		Contrast unit II		Contrast unit III		Contrast unit IV		Contrast unit V		Contrast unit VI	
		Standard results	Contrast results	Deviation from standard	Contrast results	Deviation from standard	Contrast results	Deviation from standard	Contrast results	Deviation from standard	Contrast results	Deviation from standard	Contrast results	Deviation from standard
Chemical analysis (%)	Loss on ignition	1.52	1.15	0.37	1.2	0.32	1.56	0.04	1.37	0.15	1.3	0.22	1.25	0.27
	Magnesium oxide	4.04	4.36	0.32	4.2	0.16	3.9	0.14	4.29	0.25	4.2	0.16	4.17	0.13
	Alkali content	0.54	0.53	0.01	0.52	0.02	0.47	0.07	0.49	0.05	0.45	0.09	0.5	0.04
	Sulfur trioxide	1.84	2.38	0.54	2.3	0.46	2.2	0.36	2.18	0.34	2.2	0.36	2.26	0.42
	Calcium oxide	59.52	61.12	1.6	59.91	0.39	60.57	1.05	60.33	0.81	59.67	0.15	59.98	0.46
Density (g/cm^3)		3.19	3.2	0.01	3.16	0.03	3.17	0.02	3.15	0.04	3.18	0.01	3.17	0.02
Specific surface area (m^2/kg)		319	308	11	322	3	326	7	296	21	326	7	334	25
Normal consistency (%)		25	24.2	0.8	24.4	0.6	24.8	0.2	24.2	0.8	24.5	0.5	24.2	0.8
Stability (pat test)		Qualified	Qualified	—	Qualified	—	Qualified	—	Qualified	—	Qualified	—	Qualified	—
Time of setting (min)	Initial setting	195	168	27	145	50	150	45	134	61	189	6	141	54
	Final setting	270	215	55	206	64	236	34	178	92	268	2	206	64
Compressive strength (MPa)	3 d	18.3	18.9	0.6	20.6	2.3	20.2	1.9	16.7	1.6	18.9	0.6	18.8	0.5
	7 d	28.1	28.8	0.7	29.4	1.3	30.3	2.2	25.9	2.2	28.8	0.7	28.3	0.2
	28 d	50.5	51.8	1.3	48.8	1.7	53.0	2.5	44.3	6.2	50.9	0.4	51.0	0.5
Bending strength (MPa)	3 d	3.7	4.2	0.5	4.7	1.0	4.3	0.6	4.0	0.3	4.4	0.2	4.0	0.2
	7 d	5.1	5.4	0.3	6.0	0.9	5.4	0.3	5.4	0.3	5.6	0.5	5.4	0.3
	28 d	7.9	8.3	0.4	8.3	0.4	8.0	0.1	7.7	0.2	8.1	0.2	8.2	0.3
Hydration heat (kJ/kg)	3 d	216	218	2	236	20	201	15	232	16	206	10	210	6
	7 d	267	279	12	257	10	261	6	282	15	262	5	260	7

3.3 Performance and Quality Test of Fly Ash

3.3.1 Technical Requirements of Fly Ash for Use in Hydraulic Concrete

Fly ash has been used in concrete for decades, and much successful experience has been obtained. Mixing fly ash into hydraulic concrete can significantly improve the performance of concrete mixture, and reduce the temperature rise of hydration heat, which is very beneficial to temperature control and cracking prevention. However, the effect of fly ash is mainly manifested in the later stage. In the case of a large amount of fly ash content, the early strength of concrete develops slowly. Although its early strength is low, its later strength and other properties increase significantly.

After the hydraulic concrete is poured, the process of putting it into use is long, so the biggest advantage of mixing fly ash into hydraulic concrete is to make use of the characteristics of the later strength of fly ash. In recent years, due to the continuous improvement of the quality of fly ash, especially the mass production of Grade I fly ash, fly ash has also changed from general mixture in the past to functional materials at present. Due to the difference between fly ash and other mixtures, fly ash particles are in the shape of micro beads. The higher the grade of fly ash, the finer the particles, and the more micro beads content, and the more obvious the improvement of concrete performance. For example, in the Three Gorges Project, high-quality Grade I fly ash (water demand ratio is no greater than 92%) is used as a functional material, to give full play to the morphological effect, micro-aggregate effect and pozzolanic effect of high content of spherical particles of Grade I fly ash, and play the role of solid water reducing agent.

According to *Fly Ash Used for Cement and Concrete* GB/T 1596—2005 and *Technical Standard of Flyash Concrete for Hydraulic Structures* DL/T 5055—2007, specific requirements for quality index and grade of fly ash are put forward. The technical requirements for fly ash used for hydraulic concrete are shown in Table 3.8.

Table 3.8　　Technical requirements for fly ash used for hydraulic concrete

Item		Technical requirements		
		Grade I	Grade II	Grade III
Fineness (residue on 45 μm square hole sieve)(%)		≤12.0	≤25.0	≤45.0
Water demand ratio(%)		≤95	≤105	≤115
Loss on ignition(%)		≤5.0	≤8.0	≤15.0
Moisture content(%)		≤5.0		
Sulfur trioxide (%)		≤1.0		
Free calcium oxide(%)	Class F fly ash	≤1.0		
	Class C fly ash	≤4.0		
Stability		Qualified		

When fly ash is used for reactive aggregate concrete, the alkali content of fly ash shall be limited, and its allowable value shall be determined by demonstration. The alkali content of fly ash is calculated by equivalent weight of sodium ($Na_2O + 0.658 K_2O$).

According to DL/T 5055—2007 standard, fly ash is divided into Class F and Class C. Class F fly ash is a kind of fly ash collected by burning anthracite or bitumite, and its fCaO is less than 1.0%; class C fly ash is a kind of fly ash collected by burning lignite or sub-bitumite, and its fCaO is less than 4.0% while the content of

CaO is generally greater than 10%. Class C fly ash contains high content of fCaO, so poor stability easily occurs. In order to ensure the project quality, the content of fCaO shall be controlled and the stability shall be conforming.

At the same time, the 2007 standard also puts forward specific technical requirements for mixing fly ash into hydraulic concrete. The strength design age of concrete mixed with fly ash shall make full use of the later performance of fly ash. Under the condition of ensuring the design requirements, the longer design age should be adopted as far as possible to obtain better technical and economic effect.

Although the maximum content of fly ash is specified in specifications for hydraulic concrete construction, the maximum amount of fly ash shall be determined after the test according to the concrete structure type, cement type and fly ash quality grade. According to the project experience in recent years, Grade II fly ash is mostly used in RCC dams, and the content is generally controlled ranging from 50% - 65%. This is related to the fact that thermal power plants in China can produce Grade II fly ash. Grade I fly ash is mainly used for large-scale water conservancy and hydropower projects and high-grade concrete, for example, Longtan RCC gravity dam at the level of 200 m. High-grade erosion resistance concrete and structural concrete use Grade I fly ash, which has played a good role in improving concrete construction performance, temperature control, cracking prevention and durability. At present, Grade III fly ash has been rarely used in permanent hydraulic structures.

As the most important admixture of RCC, fly ash is used in a large amount. In order to ensure the supply and quality stability of fly ash, 2 - 3 fly ash suppliers should be selected for use in projects. Fly ash of each manufacturer shall be stored and used in a fixed way to avoid affecting the quality and appearance color of concrete.

3.3.2 Physical and Chemical Properties of Fly Ash

3.3.2.1 Physical Properties of Fly Ash

Fly ash is the powder collected from flue gas of pulverized coal boiler in coal-fired power plant. Its physical properties are mainly density, bulk density, fineness, particle morphology, water demand, grain gradation (particle size distribution), etc. Physical properties of fly ash are closely related to chemical properties and mineral composition of fly ash. There are also many factors affecting the physical properties of fly ash. It is very important to know and understand the physical properties of fly ash for the application of fly ash in RCC. Although the research means of physical properties of fly ash can be relatively simple, there are not many systematic research results on the physical properties of fly ash at present.

1. Density. The density of fly ash ranges from 1.77 g/cm^3 to 2.43 g/cm^3. At present, the density of fly ash ranges mostly from $2.0 - 2.2$ g/cm^3. The density of fly ash is meaningful for the quality evaluation and control of fly ash. If the density changes, the quality fluctuation occurs to a certain extent. When fly ash is used as concrete mixture, the density of fly ash is usually one of the parameters in concrete mix proportion design. Generally, the most important factor affecting the density of fly ash is the content of CaO. According to the research results, the density of Class F low-calcium fly ash ($CaO \leqslant 10\%$) is generally low, with relatively large range of variation, while the density of Class C high-calcium fly ash ($CaO > 10\%$) is 19% higher than that of low-calcium fly ash.

2. Fineness. The utilization value of fly ash as a by-product is largely due to its fine particles and large specific surface area. Therefore, fineness is a very important performance index of fly ash. The particle size of fly ash is mainly distributed in the range of $0.5 - 300$ μm. The particle size of the fly ash of glass microsphere is in the range of $0.5 - 100$ μm, but the particle size of most of them is 45 μm. Modern thermal power plants usually collect dust with electrostatic precipitators, and the fineness of fly ash collected by different electric

fields varies greatly. Therefore, fineness is one of the main indexes to evaluate the grade of fly ash. Upon abundant test study and analysis, the fineness is greatly related to the content of CaO in fly ash. Class C high-calcium fly ash is usually fine, and most of the coarse fly ash has high quartz content.

Technical Standard of Flyash Concrete for Hydraulic Structures DL/T 5055 specifies the fineness of fly ash is expressed in the residue on sieve with pore size of 45 μm. According to a large number of test results at home and abroad, it is reasonable to use the standard sieve with pore size of 45 μm to determine the fineness of fly ash. Therefore, the current fly ash standard adopted by most countries in the world adopts percentage of residue on sieve with size of 45 μm as the fineness index of fly ash.

Relatively speaking, the specific surface area is more accurate to represent the fineness of fly ash, which can not only reflect the fineness of fly ash, but also reflect the particle shape of fly ash as a whole, and even reflect the amount of open space of fly ash particles. The specific surface area of fly ash is usually measured by Blaine method. In China, the specific surface area of fly ash varies from 800 cm^2/g to 5 500 cm^2/g, generally from 1 600 cm^2/g to 3 500 cm^2/g. Therefore, the effect of fly ash on improving the workability of concrete is very significant.

3. Color. The visible fly ash is gray powder. Generally, the lighter the color of undisturbed ash, the lower the loss on ignition of fly ash, and the lighter the color of fly ash of the same grade of coal, the finer the particle size of fly ash. Generally speaking, fine-grained pulverized coal is in relatively full combustion, so the loss on ignition of fly ash is relatively low. Since the color of fly ash can reflect the carbon content, as for the quality control and production control of fly ash, the color of fly ash is greatly affected by the combustion conditions of pulverized coal, and also related to the composition, water content, fineness and other factors of fly ash. Generally speaking, anthracite is darker than bitumite.

4. Particle morphology. It is necessary to observe the particle morphology of fly ash by means of microscope, scanning electron microscope (SEM) and other means. It is very common to observe the particle morphology of fly ash by SEM, because most of the particle size of fly ash can be observed by SEM, ranging from 1 μm to 400 μm. It can be observed by SEM that the fly ash with small particle size is spherical particle with smooth surface. The shape of fly ash with larger size (> 250 μm) is irregular. Some contain incomplete combustion substances, open spherical particles containing many fine and small fly ash particles, small fly ash particles adhering to the surface of large fly ash particles. Considerable non-spherical fly ash particles may be unburned carbon particles.

The content of CaO in fly ash has a great influence on the particle morphology of fly ash. CaO affects the particle size of fly ash because of reducing the degree of polymerization of aluminosilicate. The melts of Class C high-calcium minerals have low viscosity, so their melts can form smaller droplets and then cool to fly ash particles. Because the cooling rate of small droplets is faster, the content of vitreous body is higher. The viscosity also affects the particle shape of fly ash. It is easier to introduce air and form a hollow spherical structure because Class F low-calcium mineral has higher viscosity. This is also the main reason why the density of Class F low-calcium fly ash is lower than that of Class C high-calcium fly ash.

5. Water demand. Water demand is a very important physical property index for the application of fly ash. The water demand of fly ash can be defined as the water required when the mixture of fly ash and water reaches certain fluidity. The smaller the water demand of fly ash, the higher the project utilization value of fly ash. Main factors affecting the water demand of fly ash are fineness, particle morphology and grain gradation of fly ash. In addition, the water demand of fly ash is closely related to the density and loss on ignition of fly ash.

In engineering application, the water demand of fly ash is usually expressed in water demand ratio. If fly ash is used as cement concrete mixture, GB/T 1596 and DL/T 5055 specify that the water demand ratio of Grade Ⅰ, Grade Ⅱ and Grade Ⅲ fly ash is no more than 95%, 105%, and 115% respectively.

Because of many factors affecting the water demand of fly ash, the water demand of different fly ash is quite different. It is found that the water demand is directly proportional to the loss on ignition and fineness of fly ash.

6. Compressive strength ratio. As a concrete mixture, fly ash mainly uses its pozzolanic activity. The compressive strength ratio can accurately express this property of fly ash. GB/T 17671—1999 Method of Testing Cements-Determination of Strength specifies that the compressive strength ratio is the 28 d compressive strength ratio of cement mortar, and its value is the ratio of 28 d compressive strength of the test sample and 28 d compressive strength of the contrast sample.

7. Adsorption. Fly ash is a kind of particle composed of oxides of silicon, aluminum, calcium, iron and some trace elements oxides as well as unburned carbon. Some of these particles are spongy, porous, or hollow spherical particles. Because these particles are very fine, they have a very large specific surface area. In addition, fly ash is a kind of particle obtained by burning and cooling coal at a very high temperature. The surface of fly ash particle has certain activity. Therefore, fly ash has high adsorption. Therefore, when fly ash is used as concrete mixture, this adsorption performance directly affects the effect of concrete admixtures, especially the adsorption of air entraining agents is obvious. According to the research results of adsorption performance of fly ash, the adsorption capacity of fly ash is more than 75% of that of activated carbon powder.

3.3.2.2 Chemical Properties of Fly Ash

1. Chemical composition. The chemical composition of fly ash largely depends on the inorganic composition of raw coal and combustion conditions. According to the content difference, elements in fly ash can be divided into main elements and trace elements. In addition, another important chemical composition of fly ash is unburned carbon, which has a great influence on the application of fly ash. More than 70% of fly ash is usually composed of silicon oxide, aluminum oxide and iron oxide. Typical fly ash also contains oxides of calcium, magnesium, titanium, sulfur, potassium, nitrate and phosphorus. The chemical composition of fly ash varies greatly in different areas. Table 3.9 shows the statistical results of the content of main chemical elements in fly ash of 365 samples.

Table 3.9   Content of main chemical elements in fly ash   Unit: %

Element name	Content range	Mean value
O	—	47.83
Si	11.48 – 31.14	23.50
Al	6.40 – 22.91	15.26
Fe	1.90 – 18.51	3.84
Ca	0.30 – 25.10	2.31
K	0.22 – 3.10	1.04
Mg	0.05 – 1.92	0.52
Ti	0.40 – 1.80	0.71
S	0.03 – 4.75	0.32
Na	0.05 – 1.40	0.31
P	0 – 0.90	0.04
Cl	0 – 0.12	0.02
Others	0.50 – 29.12	4.30

The chemical composition of fly ash varies greatly in different areas. The systematic study on the chemical composition change of fly ash is of great significance to the utilization and recycling of fly ash. There are many research results on the chemical composition change of fly ash in the world. Statistical results of the content of main oxides of fly ash from 35 thermal power plants in some areas of China are listed in Table 3.10.

Table 3.10　　　　　　　　Chemical composition of fly ash in some areas of china　　　　　　Unit: %

Oxide	SiO_2	Al_2O_3	Fe_2O_3	CaO	MgO	NaO	K_2O	SO_3	Loos
Mean value	50.6	27.1	7.1	2.8	1.2	0.5	1.3	0.3	8.2
Range	33.9–59.7	16.5–35.1	1.5–19.7	0.8–10.4	0.7–1.9	0.7–1.9	0.2–1.1	9.0–1.1	1.2–23.6

2. Loss on ignition. A large number of test results show that, the larger the particle size, the higher the carbon content of fly ash, and the carbon content of fly ash particles with the size smaller than 45 μm is very low. Therefore, increasing the fineness of pulverized coal can greatly reduce the carbon content of fly ash. In other words, the larger the particle size of the fly ash, the higher the carbon content.

3. Pozzolanic activity. Siliceous or aluminosilicate materials have no or only weak gelling properties, but they will combine with CaO to form hydraulic solids in the presence of water, which is called pozzolanic property.

　　Fly ash, especially Class F low-calcium fly ash, is a typical pozzolanic material from the perspective of chemical composition. Such potential pozzolanic property of fly ash has been widely used in many engineering applications. Therefore, the pozzolanic property is the most basicproperty of fly ash.

　　Since portland cement can produce $Ca(OH)_2$ during hydration, the secondary reaction of fly ash with cement hydration product $Ca(OH)_2$ will occur in case that fly ash is mixed with portland cement. The reaction speed is faster than that of fly ash-lime mixtures. In view of this point, fly ash is used to prepare concrete or cement mortar instead of local cement, and thus the ratio of its compressive strength to that of reference concrete or cement mortar can reflect the pozzolanic activity of fly ash. The higher the ratio, the higher the pozzolanic activity of fly ash.

4. Reaction of fly ash with cement. The reaction of fly ash with cement is usually described as "the reaction of fly ash with cement hydration product $Ca(OH)_2$ to generate CSH gel", which is conventionally called "secondary reaction". The reaction of fly ash in the presence of cement is as follows: a layer of C-S-H gel shell is formed on the surface of fly ash particles (C-S-H gel is the hydration products of portland cement) at first. Then, the vitreous body on the surface of fly ash particles dissolves. This dissolution speed is usually affected by the solution with high concentration of alkaline hydration products in the pores of the basic system of cement. Finally, fly ash reacts with $Ca(OH)_2$ to form hydration products.

　　Generally speaking, the reaction of fly ash with cement will significantly affect the final properties of hardened cement mortar and concrete. If the CaO content of fly ash is different, there are many differences during the reaction of fly ash with cement. The main component of Class F low-calcium fly ash that can react with cement is vitreous body. Crystal phases in the fly ash particles, such as quartz, hematite and magnetite, are not reactive in the cement, and the vitreous body reacts slowly with cement at normal temperature, which is also the main factor of low early strength of concrete mixed with fly ash. However, the hydration reaction of fly ash with cement in the hydraulic mass concrete will be accelerated at the higher temperature of cement hydration heat and the degree will also be increased. The reaction speed of fly ash with cement is very fast under the steam curing or autoclaved conditions.

3.3.3 Fly Ash Test and Contrast Test

3.3.3.1 Fly Ash Quality Test in the Baise Project

Qujing Grade II fly ash, Panxian Grade II fly ash, Shimen Grade II fly ash and Xuanwei Grade II fly ash are used in the RCC main dam of the Baise Multipurpose Dam Project. The chemical components, quality and particle morphology characteristics of these four types of fly ash are tested.

1. See Table 3.11 for the chemical components of fly ash.

Table 3.11　　　　　　　　　　　Chemical components of fly ash　　　　　　　　　　　Unit: %

Sample name	Chemical composition									
	Loss	SiO_2	Al_2O_3	Fe_2O_3	CaO	MgO	SO_3	K_2O	Na_2O	R_2O
Qujing Grade II fly ash	3.00	54.45	25.31	8.21	3.22	0.88	0.43	0.83	0.25	0.80
Panxian Grade II fly ash	3.58	56.32	24.30	3.90	3.40	1.01	0.63	1.05	0.23	0.92
Shimen Grade II fly ash	1.40	52.02	34.79	3.78	3.68	0.70	0.25	1.06	0.68	1.38
Xuanwei Grade II fly ash	0.90	61.56	21.01	9.25	2.92	1.08	0.11	0.80	0.04	0.57

Note: Alkali content $R_2O = Na_2O + 0.658 K_2O$.

2. See Table 3.12 for the quality test results of fly ash.

Table 3.12　　　　　　　　　　　Quality test results of fly ash

Manufacturer (relevant standards)	Density (g/cm^3)	Fineness (%)	Water demand ratio (%)	Loss on ignition (%)	SO_3 (%)	Compressive strength ratio
Qujing Grade II fly ash	2.30	10.8	98	3.00	0.43	0.66
Panxian Grade II fly ash	2.37	11.6	102	3.58	0.63	0.66
Shimen Grade II fly ash	2.21	17.2	95	1.40	0.25	0.67
Xuanwei Grade II fly ash	2.33	11.6	100	0.90	0.11	0.60
DL/T 5055 Grade II fly ash	—	≤20	≤105	≤8	≤3	—

3. SEM pictures of particle morphology characteristics. See Figure 3.1 to Figure 3.8 for SEM pictures of particle morphology characteristics of four types of fly ash in the Baise Project. SEM pictures show that these four types of fly ash are all composed of glass microsphere and irregular vitreous body. Generally, the more the glass microsphere, the smaller the water demand ratio.

Figure 3.1　*SEM picture of qujing grade II fly ash(1)*

Figure 3.2　*SEM picture of qujing grade II fly ash(2)*

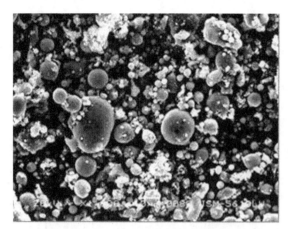

Figure 3.3 *SEM picture of panxian grade* Ⅱ *fly ash(1)(mixed with more irregular vitreous body)*

Figure 3.4 *SEM picture of panxian grade* Ⅱ *fly ash(2)(mixed with more irregular vitreous body)*

Figure 3.5 *SEM picture of shimen grade* Ⅱ *fly ash(1)*

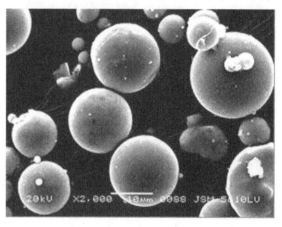

Figure 3.6 *SEM picture of shimen grade* Ⅱ *fly ash(2)(intact glass microsphere)*

Figure 3.7 *SEM picture of xuanwei grade* Ⅱ *fly ash(1)(irregular vitreous body is clearly visible)*

Figure 3.8 *SEM picture of xuanwei grade* Ⅱ *fly ash(2)*

4. X-ray diffraction analysis curve of fly ash. See Figure 3.9 for the X-ray diffraction analysis curve of four types of fly ash in the Baise Project, which are those of Qujing Grade Ⅱ fly ash, Panxian Grade Ⅱ fly ash, Shimen Grade Ⅱ fly ash and Xuanwei Grade Ⅱ fly ash from bottom to top. Mineral components of four types of fly ash are consistent, which are mainly amorphous vitreous body, α-quartz and mullite.

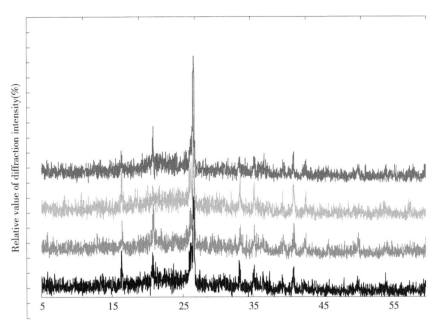

Figure 3.9 *X-ray diffraction diagram of fly ash*

3.3.3.2 Test of Grade I Fly Ash in the Longtan Project

The RCC gravity dam of Longtan Hydropower Station is 200.5 m high. Since the dam is very high and the reservoir is large, the quality level of fly ash must be high. The RCC of Longtan Hydropower Station is required to be mixed with Grade I fly ash. Grade I fly ash used in the Longtan Hydropower Station is mainly produced by Xuanwei, Luohuang, Xiangfan and Laibin B power plants. Physical and mechanical properties, chemical components and compressive strength rate of mortar with different fly ash content of four types of Grade I fly ash are tested respectively. See Table 3.13 to Table 3.15 for the test results respectively.

1. Test of physical and mechanical properties and chemical components of fly ash. Test results show that four types of fly ash meet the requirements of Grade I fly ash in *Technical Requirements of Fly Ash for Use in Hydraulic Concrete* DL/T 5055—1996.

Table 3.13　　　　Test results of physical and mechanical properties of fly ash

Fly ash type	Apparent density (g/cm^3)	Fineness (%)	Water demand ratio (%)	Loss on ignition (%)	Moisture content (%)	Compressive strength rate of 28 d mortar (%)
Xuanwei	2.93	6.5	95	0.96	0.12	74.6
Luohuang	2.53	6.7	92	2.78	0.07	86.1
Laibin B	2.33	4.7	95	3.33	0.12	77.4
Xiangfan	2.20	3.0	95	2.37	0.06	88.3
DL/T 5055—1996	—	≤12	≤95	≤5	≤1	—

Table 3.14　　　　Test results of chemical components of fly ash　　　　Unit: %

Fly ash type	CaO	SiO_2	Al_2O_3	Fe_2O_3	SO_3	MgO	K_2O	Na_2O	Alkali content
Xuanwei	1.84	62.4	20.4	9.9	0.41	1.93	0.67	0.08	0.52
Luohuang	3.94	44.45	28.5	15.98	1.80	1.81	1.01	0.71	1.37
Laibin B	3.43	60.65	25.04	6.39	0.52	1.87	1.75	0.21	1.36
Xiangfan	3.02	57.38	29.51	4.53	0.57	1.08	1.49	0.34	1.32

Table 3.15   Test results of compressive strength rate of fly ash mortar

Fly ash type	Apparent density (g/cm^3)	Fineness (%)	Water demand ratio (%)	Loss on ignition (%)	Moisture content (%)	Compressive strength rate of 28 d mortar (%)
Xuanwei	2.93	6.5	95	0.96	0.12	74.6
Luohuang	2.53	6.7	92	2.78	0.07	86.1
Laibin B	2.33	4.7	95	3.33	0.12	77.4
Xiangfan	2.20	3.0	95	2.37	0.06	88.3
DL/T 5055—1996	—	≤12	≤95	≤5	≤1	—

2. Strength test of fly ash mortar. Considering that the fly ash content in normal concrete is general 15% – 30%, and that of fly ash in RCC is 50% – 65%, the cement can be replaced by fly ash with a content equivalent to that of fly ash in normal concrete and RCC; according to the strength design age of dam concrete, the strength design age of fly ash mortar is the same as that of concrete, namely 7 d, 28 d and 90 d. See Table 3.16 for the strength test results of mortar with different content of fly ash. The test results show that:

a) The higher the content of fly ash, the lower the strength of mortar, showing a good linear relation.

b) The 28 d compressive strength rate of Xiangfan fly ash reaches 88.3%. The 28 d strength of cement mortar with 15% Xiangfan fly ash can reach that of cement mortar without fly ash, which is associated with the small fineness of Xiangfan fly ash (3.0%) that can give full play to micro-aggregate effect and pozzolanic effect and the smaller surface area as well as the rapid secondary reaction with $Ca(OH)_2$ precipitated from cement hydration.

c) When the same proportion of fly ash is added into the cement, the 28 d mortar compressive strength is successively that of cement mortar with Xuanwei fly ash, Laibin B fly ash, Luohuang fly ash and Xiangfan fly ash respectively from small to large, which is associated with the content of Al_2O_3 in the fly ash.

d) When the fly ash content is less than 25%, the 90 d mortar strength is basically equal to that of cement, and the strength difference of various fly ashes is relatively reduced. In particular, when the fly ash content is more than 50%, the later strength of Xuanwei fly ash with lower early strength is not much different from that of other fly ash.

Comprehensive analysis of fly ash test results: the fineness of Xuanwei fly ash is relatively large and the 28 d compressive strength rate is relatively low. Meanwhile, Xuanwei fly ash had the largest consumption in 2004, and the output supply is guaranteed. Therefore, Xuanwei fly ash is mainly used in the concrete mix proportion test.

Table 3.16   Strength test results of mortar with different content of fly ash

Fly ash type	F (%)	Compressive strength (MPa)			Bending strength (MPa)		
		7 d	28 d	90 d	7 d	28 d	90 d
Datum	0	37.9	57.4	68.6	7.1	8.7	9.8
Xuanwei fly ash	15	30.7	50.9	67.7	6.0	8.2	10.1
	25	26.8	44.9	65.4	5.2	7.7	9.9
	30	24.2	42.8	62.9	5.0	7.5	9.0
	50	14.8	28.4	48.5	3.3	5.8	8.5
	55	13.0	23.9	44.6	2.6	5.1	7.2
	60	10.8	20.3	39.0	2.5	4.4	7.3
	65	8.8	16.9	35.2	2.1	3.8	6.7

Chapter 3 Raw Materials and Project Examples

Continued Table 3.16

Fly ash type	F (%)	Compressive strength (MPa)			Bending strength (MPa)		
		7 d	28 d	90 d	7 d	28 d	90 d
Regression relation formular: $R7 = -0.45(F) + 37.71(R=0.997)$, $R28 = -0.65(F) + 59.89(R=0.995)$, $R90 = -0.60(F) + 74.96(R=0.960)$							
Laibin B fly ash	15	31.4	51.0	68.2	6.4	8.8	9.9
	25	26.6	47.0	65.2	5.7	8.5	10.1
	30	24.5	44.4	63.2	5.0	7.9	8.9
	50	14.2	29.0	48.3	3.3	5.7	7.6
	55	12.3	25.2	43.7	3.3	5.0	7.2
	60	10.1	21.2	38.0	2.6	4.8	6.9
	65	8.4	17.8	31.0	2.2	3.9	6.4
Regression relation formular: $R7 = -0.47(F) + 38.11(R=0.999)$, $R28 = -0.61(F) + 60.49(R=0.991)$, $R90 = -0.60(F) + 75.87(R=0.951)$							
Luohuang fly ash	15	33.4	52.3	69.2	6.2	8.5	10.1
	25	29.6	50.6	67.4	6.0	8.5	9.5
	30	28.9	49.4	65.5	5.6	8.1	9.0
	50	18.2	34.9	50.8	4.0	6.4	7.8
	55	16.6	31.6	46.3	3.8	6.1	7.5
	60	13.5	26.2	39.7	3.2	5.3	7.1
	65	12.0	22.7	34.4	3.0	4.7	6.4
Regression relation formular: $R7 = -0.42(F) + 39.42(R=0.991)$, $R28 = -0.55(F) + 61.38(R=0.975)$, $R90 = -0.57(F) + 74.46(R=0.938)$							
Xiangfan fly ash	15	34.3	57.6	72.1	6.3	8.8	10.9
	25	31.3	54.0	70.0	6.1	8.4	10.7
	30	29.0	50.7	68.6	5.8	8.7	10.6
	50	19.0	38.2	55.8	4.0	7.2	9.4
	55	16.3	34.2	53.4	3.7	6.7	8.8
	60	13.8	28.7	47.5	3.2	5.8	8.2
	65	11.6	24.6	40.4	2.7	4.9	8.1
Regression relation formular: $R7 = -0.43(F) + 40.22(R=0.991)$, $R28 = -0.55(F) + 63.68(R=0.962)$							

3.3.3.3 Contrast Test of Fly Ash in the Jin'anqiao Project

According to the test regulations in the Jin'anqiao Hydropower Station Project, the contrast test of cement and fly ash is carried out regularly. Through the contrast test, the test ability of laboratories of relevant units related to the project can be verified; the reliability of test results and equipment environment can be evaluated; the equipment can be verified and calibrated to eliminate system errors. Jin'anqiao Hydropower Station Co., Ltd. organized the contrast test of cement and fly ash in 2008. The blind sample of Grade Ⅱ fly ash was used for the contrast test. See Table 3.17 for the contrast test results of fly ash to verify the quality of fly ash used in the Jin'anqiao Hydropower Station Project.

Contrast test results indicate: there are small errors between the fineness of fly ash, water demand ratio, loss on ignition, SO_3 content, f-CaO content and CaO content and standard results, and the errors of the test results of site test units are relatively large, which may be related to the poor test conditions and personnel quality on site. The contrast test results show that the blind sample of fly ash meets Grade Ⅱ Class F fly ash index requirements.

Table 3.17　　　　Fly ash contrast test in the Jin'anqiao hydropower station project　　　　Unit: %

Contrast unit	Test results and errors	Fineness	Water demand ratio	Loss on ignition	Sulfur trioxide	Moisture content	Free calcium oxide	Calcium oxide
Standard laboratory	Result	22.3	98.0	2.85	0.18	0.2	0.13	3.56
Contrast unit I	Result	18.2	99.0	2.74	0.26	0.1	0.15	3.55
	Error	4.1	1.0	0.11	0.08	0.1	0.02	0.01
Contrast unit II	Result	19.7	94.0	2.7	0.3	0.1	0.1	2.68
	Error	2.6	4.0	0.15	0.12	0.1	0.03	0.88
Contrast unit III	Result	16.0	102	2.84	0.18	0.1	0.14	2.24
	Error	6.3	4.0	0.01	0	0.1	0.01	1.32
Contrast unit IV	Result	17.0	99.0	2.83	0.19	0.2	0.4	3.69
	Error	5.3	1.0	0.02	0.01	0	0.27	0.13
Contrast unit V	Result	20.83	98.0	3.0	0.2	0.1	0.2	3.6
	Error	1.47	0	0.15	0.02	0.1	0.07	0.04
Contrast unit VI	Result	21.0	97.0	2.89	0.3	0.09	0.06	2.91
	Error	1.3	1.0	0.04	0.12	0.11	0.07	0.65
Contrast unit VII	Result	21.0	99.0	2.82	0.5	0.1	0.06	2.91
	Error	1.3	1.0	0.03	0.32	0.1	0.07	0.65

3.4　Performance of Hydraulic Concrete Admixtures

3.4.1　Function of Admixtures in Hydraulic Concrete

The moderate admixtures in the concrete can improve the quality and performance, of concrete, reduce water consumption, save cement, reduce cost and speed up the construction progress. The admixtures have become the fifth necessary material in the hydraulic concrete besides cement, aggregate, mixtures and water. The addition of admixtures is an important measure to optimize the concrete mix proportion design.

　　The addition of admixtures is the most important technical measures to improve the performance of concrete. Currently, there are various concrete admixtures that can improve the performance of concrete and meet the requirements of construction, production, quality and performance. The most commonly used admixtures are concrete water reducing agent and concrete air entraining agent. The development of concrete water reducing agent can be divided into three stages. The first generation of ordinary water reducing agent mainly consists of calcium lignosulphonate, molasses and sugar calcium, but its water reducing rate is low. It is still used as the main raw material of naphthalene series composite admixtures now; the second generation of concrete water reducing agent is represented by ZB and JM naphthalene-based high-range water reducing agent, and replaces the first generation of water reducing agent with its good performance; the third generation of concrete water reducer is represented by polymer water reducing agent such as polycarboxylic acid and methacrylate graft copolymer, with high water reduction, high plastic retention, high enhancement. Its application effect in the self-compacting high performance concrete is remarkable.

　　There are many admixture manufacturers in China. The admixture to be used in any project shall be properly selected at the beginning of the project according to the requirements of the project design and construction technologies, and determined by strict adaptability test and demonstration according to the raw materials such as cement, mixtures and sand and aggregate to be used in the project. For the convenience of management, it is advisable to select 1 - 2 types of similar admixtures in a large and medium-sized project (including spare admixtures). In general, the type of admixtures shall not be replaced casually during construction. Compared with other raw materials, although the content of admixtures is small, it is very important to the quality of concrete. Therefore, it shall be strictly controlled after being demonstrated through the test. The quality and stability of admixtures shall be strictly tested during delivery and use according to relevant standards, and the admixtures shall be transported and stored in strict accordance with relevant standards.

High-range water reducing agent is mainly used as the admixtures of mass hydraulic concrete now. High-range water reducing agent has a strong dispersion effect on the cement, which can greatly improve the fluidity of cement mixtures. When the slump of concrete is the same, the high-range water reducing agent can greatly reduce the water consumption and significantly improve the strength of concrete at each age, thus reducing the consumption of cementing materials and saving costs. As the water demand for cement hardening is generally only about 20% of the mass of cement or cementing materials, the remaining concrete mixing water can easily form connected capillary channels during evaporation and loss, causing concrete defects. Water reducing agent used to reduce mixing water can significantly improve concrete microstructure, strength and impermeability, frost resistance and crack resistance and other performances.

In recent years, the design of water conservancy and hydropower projects, whether it is normal concrete or RCC, requires higher durability. Frost resistance grade is an extremely important index of durability, and concrete is designed with frost resistance grade whether it is in cold areas or southern mild and hot areas, or outside or inside dams. RCC construction features large placing surface, high pouring strength, and higher requirements for interlayer bonding and temperature control. In order to adapt to the construction characteristics of RCC, improve the performance of RCC mixtures, reduce the unit water consumption, cement consumption and hydration heat temperature rise, and improve RCC crack resistance and durability, admixtures must be mixed in RCC. Now set retarding high-range water reducing agent and air entraining agent are mainly adopted for RCC, which are required to have higher water reducing rate and retarding effect, and keep a certain air content, so that RCC mixtures can meet the quality requirements of construction, such as rollability, liquefied bleeding, interlayer bonding and durability. For this reason, the water reducing rate of admixtures in large projects has been put forward with specific index requirements.

The performance of RCC has been completely changed since the full-section damming technology has been adopted. The water reducing agent used in RCC in the past does not require high water reducing rate, while due to the change of RCC damming concept, with the improvement of hydraulic concrete quality requirements, the water reducing rate of water reducing agent and other quality requirements are getting higher and higher. For example, naphthalene-based high-range water reducing agent has been applied in Ertan, Three Gorges and other large water conservancy and hydropower projects with a water reducing rate of more than 18%. Diabase and basalt hard aggregates have been added into the RCC for Baise and Jin'anqiao projects, which leads to construction problems such as high unit water consumption of the concrete, short setting time, poor liquefied bleeding. It is necessary to greatly reduce unit water consumption of the concrete, especially when the water consumption is high, not only the water reducing rate of retarding high-range water reducing agent shall be higher, but also the admixture content shall be increased, so that the performance and construction performance of RCC with diabase and basalt aggregates can be obviously improved.

3.4.2 Technical Performance of Concrete Mixed with Admixtures

Admixtures commonly used in hydraulic concrete: air entraining agent, ordinary water reducing agent, hardening accelerating and water reducing agent, retarding and water reducing agent, air entraining and water reducing agent, high-range water reducing agent, retarding high-range water reducing agent, retarder and high-temperature retarder. According to the performance of RCC and the characteristics of rapid construction, high-range water reducing agent, retarder, retarding high-range water reducing agent and air entraining agent are commonly used for RCC in recent years. According to the quality requirements of *Technical Code for Hydraulic Concrete Admixtures* DL/T 5100—1999, the performance requirements of RCC concrete mixed with the four common admixtures are shown in Table 3.18, and the uniformity requirements of admixtures are shown in Table 3.19.

Table 3.18 Performance requirements of concrete mixed with four common admixtures

Test items		High-range water reducing agent	Retarder	Retarding high-range water reducing agent	Air entraining agent
Water reducing ratio(%)		≥15	—	≥15	≥6
Air content(%)		<3.0	<2.5	<3.0	4.5–5.5
Ratio of bleeding rate(%)		≤95	≤100	≤100	≤70
Time of setting (min)	Initial setting	−60 – +90	+210 – +480	−120 – +240	−90 – +120
	Final setting	−60 – +90	+210 – +720	−120 – +240	−90 – +120
Compressive strength ratio(%)	3 d	≥130	≥90	≥125	≥90
	7 d	≥125	≥95	≥125	≥90
	28 d	≥120	≥105	≥120	≥85
28 d ratio of shrinkage(%)		<125	<125	<125	<125
Frost resistance index		≥50	—	≥50	≥200
Effect on corrosion of steel bar		Indicate any corrosion on steel bars			
Influence on thermal performance		When used in mass concrete, the influence on 7 d hydration heat or 7 d adiabatic temperature rise of concrete should be specified			

Table 3.19 Uniformity requirements of admixtures

Test items	Indicator
Solid content or water content	(1) Within 3% of the specified value of the manufacturer for liquid admixtures (2) Within 5% of the specified value of the manufacturer for solid admixtures
Density	Within ±0.02 g/cm^2 of the specified value of the manufacturer for liquid admixtures
Chlorine ion content	Within 5% of the specified value of the manufacturer
Fluidity of cement paste	No less than 95% of the specified value of the manufacturer
Fineness	Percentage of residue on sieve 0.315 mm shall be less than 15%
pH	Within ±1 of the specified value of the manufacturer
Surface tension	Within ±1.5 of the specified value of the manufacturer
Reducing sugar content	Within ±3 of the specified value of the manufacturer
Total alkali content ($Na_2O + 0.658K_2O$)	Within 5% of the specified value of the manufacturer
Sodium sulfate	Within 5% of the specified value of the manufacturer
Ratio of bubble	Within 5% of the specified value of the manufacturer
Mortar fluidity	Within 5% of the specified value of the manufacturer
Insoluble substance content	Within 5% of the specified value of the manufacturer

The use of admixtures can also refer to the relevant regulations and requirements of the latest national standard *Concrete Admixtures* GB 8076—2008.

3.4.3 Quality and Performance of Common Admixtures

3.4.3.1 High-range Water Reducing Agent

High-range water reducing agent: admixtures which can reduce the mixing water consumption by more than 15% under the condition that the concrete slump is basically unchanged.

Water reducing agent is also called plasticizer or dispersant. Proper amount of water reducing agent in concrete during mixing can make cement particles disperse uniformly and release the water wrapped by cement particles, which can obviously reduce the water consumption of concrete. The water reducing agent is to improve the workability of concrete while keeping the mix proportion unchanged; or reduce water consumption and improve concrete strength under the condition of keeping the workability unchanged; or reduce the cement consumption while keeping the strength unchanged, saving the cement and reducing the cost. Meanwhile, the concrete is more uniform and dense after addition of the water reducing agent, which improves a series of physical and chemical properties of concrete, such as impermeability, frost resistance, erosion resistance, and also improves the durability of concrete. The water reducing agent plays a significant role in improving the setting time, liquefied bleeding, rollability, interlayer bonding and frost resistance of RCC. Admixtures play an essential role in concrete mix proportion test.

The performance of RCC has been completely changed since the full-section damming technology has been adopted. The water reducing agent used in RCC in the past does not require high water reducing rate, while due to the change of RCC damming concept, with the improvement of RCC quality requirements, the requirements for quality of water reducing rate are getting higher and higher. For example, diabase and basalt hard aggregates have been added into the RCC for Baise and Jin'anqiao projects, which leads to construction problems such as high water consumption for RCC, and short setting time, which requires a substantial reduction in the unit water consumption of concrete especially when the water consumption is high, the water reducing rate of naphthalene-based high-range water reducing agent is required to be more than 18%, and the content is increased to solve such construction problems.

3.4.3.2 Retarder

Coagulant: admixtures that can prolong the setting time of concrete.

The retarder can delay the setting and hardening time of concrete, keep the plasticity of concrete for a long time to facilitate the construction, slow down the hydration rate of concrete mortar, prolong the hydration heat release process, and be beneficial to the temperature control of mass concrete. The retarder can reduce the 1 d and 3 d strength of the concrete, but has no effect on the normal development of later strength.

The common retarder can prolong the initial setting time of concrete by 1 – 4 h. In order to meet the requirements of mass concrete construction in high temperature areas and seasons, the national "Eighth Five-Year Plan" Technological Research Project has developed a high temperature retarder, which can prolong the initial setting time of concrete to 6 – 8 h under the conditions of air temperature of (35 ± 2) ℃ and relative humidity of $(60 \pm 5)\%$, thus creating conditions for the smooth construction of mass hydraulic concrete in high temperature areas or seasons in China. For example, the retarding high-range water reducing agent widely used in the Three Gorges and Longtan projects can also make the initial setting time of concrete reach 6 – 8 h by appropriately increasing the content, which can meet the requirements of concrete construction in large placing surface in high temperature season around 35 ℃.

3.4.3.3 Retarding High-range Water Reducing Agent

Retarding high-range water reducing agent: admixtures with functions of retarder and high-range water reducing agent.

Retarding high-range water reducing agent is a composite admixture, which has the functions of retarding and water reduction. Retarding high-range water reducing agent is the leading product for RCC. Attention should be paid to that the air content of this admixture should not be too large during addition, because its own air bubbles are large and harmful, which is the main cause of slump or *VC* value loss, and also the symbol of unstable quality of retarding high-range water reducing agent.

Due to the requirements of RCC performance and construction characteristics, RCC retarding is the first priority. As there is a time process from RCC mixing to rolling, and the continuous rise of thin-lift rolling, most of the rolling layers are 3 m, which requires paving and rolling for 10 layers. In order to ensure good interlayer bonding quality between concrete layers, the interval time between construction layers must be controlled. The control standard of interlayer interval time is directly related to the quality of interlayer bonding. Generally, the allowable time for direct paving is less than the initial setting time. Climatic conditions, i. e., different seasons, temperature periods, sun exposure or cloudy days, have a great influence on the initial setting time of RCC. In the construction, the allowable time for direct paving is the initial setting time, so retarding is the first priority. It is required that the new RCC must have a long initial setting time to meet the construction requirements of interlayer bonding.

What's more, RCC requires high water reducing rate. Currently, RCC has high fly ash content, high rock powder content and low *VC* value, and because of the strong adsorption of fly ash to air and high water demand of high rock powder content, it is necessary to use retarding high-range water reducing agent and air entraining agent to ensure good workability of RCC, and meet the mechanical performance indexes such as impermeability and frost resistance required by design.

Construction joints and cold joints in RCC damming are weak links, which often form leakage channels and affect the stability against sliding, and must be dealt with seriously. The control standards and specific practices of various projects at home and abroad are different, but their essence is to limit the setting time. According to different construction seasons and temperature periods, many projects at home and abroad adopt different setting time standards and different contents, including winter and summer admixtures, which well meet the rapid construction requirements of RCC under various temperature conditions.

In recent years, Jiangsu Bote JM-Ⅱ, Zhejiang Longyou ZB-1RCC15 and Hebei Shijiazhuang DH-1 series of retarding high-range water reducing agents are mainly applied to hydropower projects, which significantly improves the working performance of RCC and the interlayer bonding quality.

3.4.3.4 Air Entraining Agent

Air entraining agent: it refers to an admixture that can introduce many independent and evenly distributed micro bubbles into concrete to improve the workability and durability of concrete.

Air entraining agent, as a kind of surface active material, is one of the admixtures commonly used in concrete. It enables a large number of evenly sealed small bubbles to be introduced into concrete from the air during mixing, so that the concrete contains a certain amount of air. A high-quality air entraining agent can introduce as many as one billion bubbles into concrete, with pore size of mostly 0.05 – 0.2 μm, and the bubbles are generally of discontinuously closed sphere with uniform distribution and sound stability. Because of bubbles, the volume of cement grout increases relatively, thereby improving the liquidity of concrete. Because of a large number of micro bubbles, the cohesiveness and water retention of concrete are also significantly improved. Because bubbles block the capillary channel in concrete and buffer the water pressure caused by water freezing in set cement, the impermeability and frost resistance of concrete are significantly improved. What's more, bubbles also enable reduction of modulus of elasticity of concrete, which is beneficial to improving the cracking resistance of concrete.

The relationship between the characteristics of RCC and the content of air entraining agent is described as follows. Due to the less dosage of cementing materials, large fly ash content, high content of rock powder in

RCC, construction method of vibratory compaction and other factors, the air entraining capacity of air entraining agent in RCC is much lower than that in normal concrete. In order to ensure that the frost resistance of RCC meets the frost resistance grade in design requirements, the content of air entraining agent shall be controlled according to the requirements for air content of concrete. By engineering practice, the content of air entraining agent in RCC is generally about several times that in normal concrete.

For example, Longshou RCC arch dam in Gansu, whose external and internal frost resistance grades of RCC are F300 and F200 respectively. According to the test study, mixing air entraining agent (up to 30 – 40 times that in normal concrete) into RCC enables that the air content of RCC is controlled within 5.0% – 5.5%, which meets the design requirements of up to F300 frost resistance grade.

In recent years, the air entraining agent has been changed from past colloidal shape to powdery shape, so it is easily prepared by cold water with good air entraining effect. At present, Jiangsu Bote JM-2000, Zhejiang Longyou ZB-1G and Hebei Shijiazhuang DH-9A air entraining agents are mainly used in projects.

3.4.4 Application Requirements of Admixtures in Projects

With the development of RCC damming technology, the present RCC is semi-plastic concrete without slump, and the VC value of outgoing concrete is generally 1 – 5 s. The VC value of on-siteplacing surface is controlled based on the principle of no sinking during rolling, so that the upper aggregate is embedded into the lower layer concrete after being rolled, and its surface is fully subject to bleeding, ensuring the interlayer bonding quality. Because of the gradual transition of RCC from past hard concrete to semi-plastic concrete, the performance of RCC is basically the same as that of normal concrete, the problem of difficult introduction of air content in RCC is solved, the content of air entraining agent significantly decreases with the decrease of VC value, and the frost resistance and ultimate tensile value of RCC are also improved significantly, so that the modulus of elasticity is reduced. So, high attention shall be paid to the use of admixtures. Compared with other raw materials, although the content of admixtures is small, it is very important to the quality of concrete. Therefore, it shall be strictly controlled after being demonstrated through the test.

With the improvement of design indexes (durability and frost resistance grade) of RCC, at present, the retarding high-range water reducing agent and air entraining agent can be compositely used in RCC. Especially, after air entraining agent is mixed into RCC, the rollability of fresh RCC is obviously improved, achieving great results, which not only meets the requirements of large surface pouring and paving, high-strength construction and continuously rising retardation of RCC, but also reduces water and improves durability.

With the continuous improvement of the full-section RCC damming technology, the performance requirements of RCC admixtures are higher and higher. It is required that the admixtures should have good compatibility with cement, and be easily prepared, without precipitation, but with good uniformity. At the same time, it is required that the admixtures should be with high water reduction, high retardation and stable air content. According to the specific situations of raw materials in specific projects, the admixture content still needs to be increased to meet the requirements of RCC construction.

There are many admixture manufacturers in China. The admixtures to be used in any project shall be properly selected before commencement of the project according to the requirements of the project design and construction technologies, and determined by strict adaptability test and demonstration according to the raw materials such as cement, and aggregate to be used in the project. Generally, the admixtures determined by optimization should not be replaced in engineering construction.

For example, on the basis of optimization of a large number of admixtures, Jiangsu Bote JM-II and Zhejiang Longyou ZB-1RCC15 retarding high-range water reducing agents as well as JM-2000 and ZB-1G air entraining agents are finally selected for Longtan Project. Selected admixtures are tested. The test results of uniformity of admixtures are shown in Table 3.20, and the performance test of concrete mixed with admixtures is shown in Table 3.21.

Table 3.20　　　　　　　　　　Test results of uniformity index of admixtures

Admixture varieties	Admixture name	Fineness (%)	pH value	Surface tension (MN/m)	Content of sodium sulfate (%)	Total alkali content (%)	Fluidity of cement paste (mm)
Retarding high-range water reducing agent	JM-Ⅱ	8.66	8.32	62.55	3.04	6.04	221.0
	ZB-1RCC15	0.87	11.62	65.65	2.50	7.42	195.0
Air entraining agent	JM-2000	—	8.47	39.40	—	—	97.0
	ZB-1G	—	9.06	36.95	—	14.56	95.0

Table 3.21　　　　　　　　　　Performance test results of concrete mixed with admixtures

Additive		Water reducing ratio (%)	Air content (%)	Ratio of bleedin grate (%)	Difference in setting time (min)		Compressive strength ratio (%)		
Variety	Mixing amount (%)				Initial setting	Final setting	3 d	7 d	28 d
JM-Ⅱ	0.5	21.2	1.4	59	+193	+193	158	168	149
ZB-1RCC15	0.5	20.5	2.3	48	+315	+305	142	143	135
JM-2000	0.006	6.7	4.8	61	+55	+112	97	98	98
ZB-1G	0.006	6.6	4.9	56	+50	+95	96	97	98
DL/T 5150—1999	Retarding high-range water reducing agent	≥15	<3.0	≤100	+120 – +240	+120 – +240	≥125	≥125	≥120
	Air entraining agent	>6.0	4.5 – 5.5	≤70	−90 – +120	−90 – +120	≥90	≥90	≥85

The result indicates: the cement grout mixed with two kinds of retarding high-range water reducing agents and air entraining agents optimized is with high fluidity, meeting the requirements of *Technical Code for Hydraulic Concrete Admixtures* DL/ 5100—1999, and achieving water reducing rate (internal control indexes of projects) of more than 18%. At the same time, the test results show that: When the content of retarding high-range water reducing agent is 0.5%, the water reducing rate of JM-Ⅱ and ZB-1RCC15 retarding high-range water reducing agents reaches 21.2% and 20.5% respectively, the air content is 1.4% and 2.3% respectively, the initial setting time is prolonged by +193 min and 315 min respectively, and the compressive strength rate of concrete mixed with admixtures for 3 d, 7 d and 28 d is more than 125%; the water reducing rate of JM-2000 and ZB-1G air entraining agents is 6.7% and 6.6% respectively, and the air content is in the range of 4.8% – 4.9%, with excellent performance, meeting the performance requirements for RCC in specifications and projects.

3.5 Aggregate Properties and Project Cases

3.5.1 Aggregate Quality Requirements

According to the provisions of *Specifications for Hydraulic Concrete Construction* DL/T 5144—2001 and *Construction Specification for Hydraulic Roller Compacted Concrete* DL/T 5112—2009, aggregates used in concrete are divided into fine aggregates and coarse aggregates by particle size of aggregates. The proportion of aggregates in RCC is significantly higher than that in normal concrete, which is mainly determined by the mix proportion and construction method of RCC and the different characteristics of normal concrete. As for RCC, its cementing material dosage and unit water consumption are less than that of normal concrete, and its compactness after rolling by vibrating roller in construction projects is better than that of normal concrete, so its density is higher than that of normal concrete. Its coarse aggregate is the same as that of normal concrete, but its maximum gradation and particle size are smaller than that of normal concrete. A large number of engineering practices prove that, inside the RCC the dam is generally Ⅲ gradation, and the maximum particle size of aggregate is 80 mm (The concrete in Baise Project is quasi Ⅲ gradation, and the maximum particle size is 60 mm). The RCC in imperviousness area outside the dam is generally Ⅱ gradation, and the maximum particle size of aggregate is 40 mm. The fineness modulus (FM) of sand for RCC and normal concrete is required to be basically the same, but the technical indexes of rock powder content and grain gradation of sand are quite different. In recent years, engineering practices prove that, the content of manufactured gravel powder is generally controlled at 16% – 22%. In some projects, the content of manufactured gravel powder has exceeded the limit of 22%, which is the biggest difference between sand index of RCC and that of normal concrete.

3.5.1.1 Quality and Performance of Fine Aggregate

1. FM and classification. Fine aggregate, also known as sand, is divided into manufactured sand and natural sand, with particle size of 0.16 – 5.0 mm. Fine aggregate shall be solid, clean and well graded. According to the requirements of sand for RCC, the FM of manufactured sand and natural sand should be respectively 2.2 – 2.9 and 2.0 – 3.0. The content of over-size particle shall be strictly controlled. The use of natural sand with FM less than 2.0 shall be demonstrated by test. The content of rock powder in manufactured sand (particle with size less than 0.16 mm in natural sand) should be controlled between 12% and 22%. The content of particle in rock powder (particle with size less than 0.08 mm in natural sand) should not be less than 5%. The optimal rock powder content shall be determined by test. The silt content of natural sand shall not be more than 5%. The water content of fine aggregate shall be stable, and that of dry saturated surface of manufactured sand should not exceed 6%. Other quality indexes shall meet the requirements of DL/T 5144. The quality requirements of sand for RCC and normal concrete are shown in Table 3.22.

    According to the FM of sand, the fine aggregate can be divided into coarse, medium and fine sand. The sand used in hydraulic concrete is required to be medium sand, that is, the FM of medium sand for RCC is 2.2 – 2.9, and that of natural sand should be 2.0 – 3.0; the FM of medium sand for normal concrete is 2.4 – 2.8, and that of natural sand should be 2.2 – 3.0. The use of the coarse sand or fine sand with FM exceeding foregoing range shall be demonstrated by test. The quality requirement index of sand shows that there are certain differences between sand for RCC and that for normal concrete in quality standards of FM and rock powder content, which should be noted during use.

2. Content of manufactured gravel powder. Based on a large number of projects and repeated tests, the manufactured sand for RCC contains higher rock powder content, which significantly improves the workability, liquefied bleeding, rollability, interlayer bonding, aggregate separation resistance, compaction

Table 3.22　　Quality requirements of fine aggregate

Item		Sand for RCC		Sand for normal concrete	
		Artificial sand	Natural sand	Artificial sand	Natural sand
Rock powder content(%)		12 - 22	—	6 - 18	—
Clay content (%)	≥C_{90}30, with frost resistance requirements	—	≤3	—	≤3
	< C_{90}30	—	≤5	—	≤5
Clay lump content		Not allowed	Not allowed	Not allowed	Not allowed
Solidity (%)	Concrete with frost resistance requirements	≤6	≤6	≤6	≤6
	Concrete without frost resistance requirements	≤10	≤10	≤10	≤10
Apparent density(kg/m³)		≥2 500	≥2 500	≥2 500	≥2 500
Content of sulfide and sulfate (converted to SO_3, mass %)		≤1	≤1	≤1	≤1
Organic content		Not allowed	Lighter than the standard color	Not allowed	Lighter than the standard color
Mica content (%)		≤2	≤2	≤2	≤2
Content of light substance (%)		—	≤1	—	≤1
Moisture content(%)		≤6	—	≤6	—

and other construction properties of RCC, as well as the impermeability, mechanics index and fracture toughness of hardened concrete; rock powder can be used as mixture to replace partial fly ash; proper increase of rock powder content facilitates increase of the output of manufactured sand, reduction of the cost and increase of technical economic benefits. Therefore, reasonable control of content of manufactured gravel powder is one of the important measures to improve the quality of RCC. In case of using the manufactured sand processed by limestone, the site construction of Linhekou Project can prove that various properties of concrete are better when the content of manufactured gravel powder processed by limestone in projects is controlled at 15% - 22% by rollability observation and compaction test; in case of using the manufactured sand processed by granite, the manufactured sand processed by aggregate in Mianhuatan Project is produced by dry process, and the rock powder content is about 17%, especially, fine particles of less than 0.08 mm in the manufactured sand account for about 30% of the rock powder content, which increases the function of inactive materials and improves the workability and impermeability of RCC, generating better effect; in case of using the manufactured sand processed by diabase or dolomite, the effect is better when the rock powder content is 20%, indicating that the rock powder of manufactured sand processed at different lithology is different. In recent years, the rock powder content of RCC is mostly controlled within the range of 16% - 22%, and the optimal rock powder content of the manufactured sand used in different projects shall be determined by test. By a large number of studies, the content of particle with size less than 0.08 mm in rock powder has a significant effect. According to the actual work conditions of Linhekou, Baise, Longtan and Guangzhao Projects, the content of particle with size less than 0.08 mm in rock powder can reach more than 15%.

3. Control of water content in sand. The water content of hydraulic concrete sand refers to the water content of dry saturated surface of sand. After sand absorbs water, a layer of water film is formed on the surface, causing sand volume expansion, which has been proved by test. For example, the manufactured sand

processed by limestone is used in Wanjiazhai Project. The test shows that the sand volume begins to expand rapidly when the water content of sand exceeds 6%. When the water content reaches 6% – 10%, the sand volume expands by about 15% – 30%, which has a great impact on the control of concrete mixing quality. So, great importance shall be attached to stability control of the water content of sand.

It is very important to control stability of the water content of produced sand, that is, to control the water content of the dry saturated surface of the manufactured sand not to exceed 6%. That is one of the main measures of controlling the water-cement ratio of concrete and the VC value or slump of the outgoing concrete, and is also the requirement to add ice when mixing the pre-cooling concrete. Generally, the amount of sand ranges mostly from 600 – 800 kg/m^3. As the water content of sand increases or decreases by 1%, the unit water consumption and the VC value of concrete shall increase or decrease correspondingly by 6 – 8 kg/m^3 and about 3 – 5 s respectively (The slump will increase or decrease correspondingly by 3 – 4 cm). The stability of the water content of sand has a great influence on the VC value of RCC, which is the key to ensure the stable quality of fresh concrete. For example, in the Jin'anqiao Project, it is difficult for the manufactured sand to dehydrate as the basalt is mixed with tuff, chlorite and other soft particles. The water content of sand often exceeds the control index requirement of 6%, and reaches about 10% in individual periods. In addition, the coarse aggregate also has certain water content. Without adding water to the mixing system, the outgoing RCC mixed has become normal concrete, which seriously affects the quality and construction of RCC. After this, addition of some dry sand in wet-process produced sand effectively reduces the water content and stability of the basalt manufactured sand, and ensures the normal construction of RCC.

The problem of instable water content of sand is highly valued in Ertan, Three Gorges, Longtan and other projects, and comprehensive measures such as vacuum dehydrator and dehydration screen are taken, with an obvious effect.

3.5.1.2 Quality and Performance of Coarse Aggregate

1. Maximum particle size and gradation for coarse aggregate. Coarse aggregate, also known as stone, is divided into gravel and pebbles, and its particle size ranges from 5.0 – 150 (120) mm. Aggregate shall be hard, rough, durable, clean and non-weathered. The particle shape shall be squoval as far as possible to avoid needle-plate like particles. Coarse aggregates are classified into four types: small-sized stones (5 – 20 mm), medium-sized stones (20 – 40 mm), large-sized stones (40 – 80 mm) and super-large-sized stones (80 – 150 mm) according to the principle of determining the particle size range.

    Coarse aggregates are divided into the following gradations according to the principle of determining the maximum particle size:

    a) I gradation: for 5 – 20 mm, the maximum particle size is 20 mm;

    b) II gradation: 5 – 20 mm and 20 – 40 mm with the maximum particle size of 40 mm;

    c) III gradation: 5 – 20 mm, 20 – 40 mm and 40 – 80 mm with the maximum particle size of 80 mm;

    d) IV gradation: 5 – 20 mm, 20 – 40 mm, 40 – 80 mm and 80 – 150 mm (or 120 mm) with the maximum particle size of 150 mm (or 120 mm).

    See Table 3.23 for quality requirements of coarse aggregates. When fine and coarse aggregates contain active aggregates, they must be demonstrated by special tests.

2. Coarse aggregate quality. Aggregate is an important condition for concrete. The aggregate quality is closely related to the concrete performance. The strength, pore structure, particle shape and size and modulus of elasticity of the aggregate directly affect the relevant properties of the concrete. Generally, the aggregate strength is higher than the design strength of concrete, because the aggregate mainly acts as a skeleton in the concrete, and the stress of the aggregate may greatly exceed the compressive strength of the concrete when

Table 3.23　　　　　　　　　　Quality requirements of coarse aggregate

Item		Coarse aggregate (5 – 150 mm)	Remarks
Clay content (%)	d_{20} and d_{40} particle size	≤1	
	d_{80} and d_{150} (d_{120}) particle size	≤0.5	
Clay lump content		Not allowed	
Solidity (%)	Concrete with frost resistance requirements	≤5	
	Concrete without frost resistance requirements	≤12	
Content of sulfide and sulfate (%)		≤0.5	Converted into SO_3 by mass
Organic content		Lighter than the standard color	If it is darker than the standard color, the contrast test for concrete strength shall be carried out, and the compressive strength ratio is no less than 95%.
Apparent density (kg/m³)		≥2 550	
Water absorption (%)		Lighter than the standard color	
Flat and Elongated Particle Content (%)		≤0.5	Converted into SO_3 by mass
Super-inferior size ratio (%)		Inspection with the circular sieve, its control standard: over size <5% and under size <10%	
Crushing index (%)		Choose according to different rock classes and concrete strength grades	

bearing load. The aggregate strength is not easily obtained by directly measuring the strength of individual aggregate, but is assessed by indirect methods. One method is to measure the crushing index for rock, and the other method is to sample and process it into a cube or cylinder sample on the rock as aggregate, to measure its compressive strength. The performance of rock is different in different water content states. In general, the strength decreases when the rock contains moisture, because the bonding force between rock particles is weakened by the infiltrated water film. If the rock contains some materials to be softened easily, the strength decrease is more obvious. Therefore, sometimes, the ratio of compressive strength under saturated state to that under dry state, namely, softening coefficient, is used to indicate the softening effect of the rock. The softening coefficient indicates strength reduction of the rock after being immersed in water.

Due to the wide distribution of water conservancy and hydropower projects, and various types of manufactured aggregate rocks, engineering practices show that manufactured aggregates of different rock types have different effects on the performance of RCC. Aggregate processing involves geological and petrological issues. Rock is a component of crust, and can be divided into three major categories according to the causes:

Magmatic rock: granite, basalt, andesite, diabase, syenite, diorite, pegmatite, etc.

Sedimentary rock: sandstone, mudstone, clay rock, limestone, dolomite, etc.

Metamorphic rock: slate, gneiss, marble, skarn, eclogite, etc.

In the use of aggregates, serious analysis and test research shall be conducted for the types of rocks used in specific projects. Aggregates have a great impact on the performance of RCC.

3.5.2 Various Rock Aggregates and Engineering Properties

3.5.2.1 Classification of Coarse Aggregate

It can be divided into three categories according to the rock components. Various rock aggregates and engineering properties are shown in Table 3.24.

Table 3.24   Various rock aggregates and engineering properties

Category	Rock name	Strength	Durability	Chemical stability	Apparent	Impurity	Shape of broken aggregates
Igneous rock	Granite and long rock	Good	Good	Good	Good	Yes	Good
	Dense long rock	Good	Good	Compared difference	Good	Yes	Good
	Basalt, diabase and gabbro	Good	Good	Good	Good	Rare	Good
	Peridotite	Good	Good	Compared difference	Good	Yes	Good
Sedimentary rock	Limestone and dolomite	Good	Good	Good	Good	Yes	Good
	Sandstone	Good	Good	Good	Good	Rare	Good
	Firestone	Good	Bad	Bad	Good	Relative large amount	Bad
	Conglomerate and breccia	Good	Good	Good	Good	Rare	Good
	Shale	Bad	Bad	—	Good	Yes	Good-bad
Metamorphic rock	Schist and gneiss	Good	Good	Good	Good	Rare	Better-bad
	Quartzite	Good	Good	Good	Good	Rare	Good
	Marble	Good	Good	Good	Good	Yes	Good
	Serpentinite	Good	Good	Good	Good-bad	Yes	Good
	Amphibolite	Good	Good	Good	Good	Rare	Good
	Slate	Good	Good	Good	Bad	Rare	Bad

3.5.2.2 Modulus of Elasticity and Apparent Density for Coarse Aggregate

Modulus of elasticity and apparent density for various rocks are shown in Table 3.25. In general, rocks with high modulus of elasticity also have high compressive strength. However, for the same kind of rock, its compressive strength is quite different due to the structure looseness or different compactness. For example, the ultimate compressive strength of limestone with cracks and the hardest limestone is 20 – 80 MPa and 180 – 200 MPa respectively. The modulus of elasticity has a good linear relation with apparent density for aggregate. Taking diabase in the Baise Project and basalt in the Jin'anqiao and the Guandi Project as an example, the modulus of elasticity of concrete is also high due to the high density of this kind of aggregate. For aggregates with high apparent density, in the Baise Project, quasi-Ⅲ gradation concrete is adopted and the maximum aggregate particle size is 60 mm to reduce the modulus of elasticity of RCC. At the same time, the content of manufactured gravel in diabase is more than 20%, and the low VC value of 1 – 5 s is adopted, effectively reducing the modulus of elasticity of RCC and improving the crack resistance of the dam.

Table 3.25   Modulus of elasticity and apparent density of rocks

Description	Modulus of elasticity ($\times 10^4$ MPa)	Apparent density (t/m^3)	Description	Modulus of elasticity ($\times 10^4$ MPa)	Apparent density (t/m^3)
Diorite	10.41	2.8 – 3.0	Serpentinite	7.24	—
Limestone	1.31 – 10.37	2.7 – 2.9	Marble	5.82 – 6.90	—
Granite	2.32 – 8.81	2.6 – 3.0	Dolomite	6.88	—
Basalt	7.67	2.7 – 3.3	Quartzite	6.05	—
Quartzite	7.74	—	Andesite	4.50 – 5.04	2.65 – 2.75
Taconite	5.62 – 7.58	2.4 – 2.6	Peridotite	3.40 – 4.44	3.0 – 3.5
Quartz trachyte	7.38	2.4 – 2.6	Slippery rock	3.06 – 4.46	—
Shale	7.34	2.7 – 3.0	Tuff	0.6 – 0.65	2.0 – 2.5

3.5.3 Three Main Methods of Manufactured Sand Producing

Hydropower projects are construction projects with very high technical content, and the requirements for manufactured gravel are very high. That is provided based on the strict requirements of safety, quality, construction period and other aspects. Therefore, the advanced manufactured gravel process of hydropower projects has always been the long-term goal of hydropower engineers. Key and difficult points are that it is necessary to meet the demand for quantity at the specified peaks of construction, and it must meet the quality requirements. Therefore, success of the sand and stone production line is mainly determined by the success of the sand production process. A lot of engineering practice shows that the quality and output of aggregate have a great effect on the rapid construction of RCC dams. Because the amount of manufactured sand accounts for more than 1/3 of the total aggregate, especially the content of rock powder in the manufactured sand is directly related to the rapid construction and cost-effectiveness of RCC, the reasonable selection of the processing and production methods of manufactured sand is the basic guarantee for the rapid construction of RCC during planning and design of the aggregate system.

Before the 1990s, the traditional sand production process is the manufactured sand production process for hydropower projects, where rod mills are adopted. However, because of low output, high operating cost, difficult sewage treatment and rock powder recovery in use, everyone is looking for more effective crushing equipment to replace rod mills for sand production.

The vertical shaft impact crusher was introduced to produce sand in China in 1990s. The main equipment, vertical shaft impact crusher, is produced by Barmac. The method of crushing aggregate by the vertical shaft crusher often directly determines the quality index for product particle shape and the particle size distribution. From the product's particle shape index, the stone-to-stone vertical shaft crusher is better than the stone-to-iron vertical shaft crusher, counterattack crusher, cone and gyrating crusher, and jaw crusher under normal conditions. At present, use of vertical shaft impact crushers and rod mills for producing sand is successful in the most domestic hydropower projects, which has been widely used in hydropower projects. However, because of the domain, use of concrete, source of aggregates and differences in rock types for hydropower projects, manufactured sand is still made with different production methods. Currently, the common sand production methods mainly include dry-process, wet-process and semi-dry sand production.

3.5.3.1 Dry-process Sand Production

Vertical shaft impact crushers with stone-to-iron crushing method are mainly adopted for dry-process sand production. A property of the dry-process sand production is that after rocks have been crushed, the yield of manufactured sand could be 40% – 50%, and the content of rock powder could be 20% – 25%, which are affected by the linear velocity of the vertical shaft impact crusher. The dry-process method could reduce water consumption and the water content of the produced sand, i. e. , the problem of low rock powder content is solved and the produced sand does not need to be dehydrated. However, the general grain gradation of the manufactured sand is obviously found to be "large at both ends but small in the middle". Although Baise, Mianhuatan, and Linhekou projects have greatly promoted the yield of manufactured sand and rock powder by dry-process sand production, but the dust problem is serious. Only by implementing enclosed production pattern can the dust problem be avoided.

The dry-process sand production has now combined the dust removal with the fines removal. For example, the fully dry-process sand production and the fine-removal equipment used in Ahai Project could not only reduce the dust for environment protection but also control the rock powder content of the produced sand by regulating the air volume of the dust remover. The fully-dry process provides the reliable technical support for the system to supply both normal sand and rolled sand. In addition, particles smaller than 0.08 m account for 88.6% of the materials removed by the dust remover, which plays an important role in ensuring the rock powder content in RCC and adjusting the grading of produced sand. The rock powder removed from the dry-process is easy to store and transport. By replacing part of fly ash in the concrete with the rock powder, the power could be reused, which transforms the waste into the resource, and greatly reduces the production and operation cost of manufactured sand.

The advantages and disadvantages of the dry-process sand production have been proved by various projects. This process features high yield, economy and considerable rock powder content. Nevertheless, the grain gradation continuity is poor and the dust problems shall be further studied. Successful cases can be learned from some domestic projects, such as Gelantan Project.

3.5.3.2 Wet-process Sand Production

The wet-process sand production is performed by the combination of vertical shaft impact crushers, rod mill and rock powder recycle device. The recycled rock powder will be mixed on the belt conveyor and placed into the finished products warehouse. In this process, the water content of the mixed sand can reach about 20%, and the sand is fluid. After placed into the warehouse, part of the fine sand and rock powder are separated from the coarse sand, thus the uniformity of the sand is poor, and the dehydration time is extended. Generally, the water content can be reduced to 6% below after dehydration for a week.

The recycle of rock powder of wet-process sand production is critical to the RCC sand. For instance, in the case of the wet-process sand production in Longtan Project, the rock powder recycle devices of the sand production system are basically operated normally, which can basically meet the requirements for the minimum rock powder content. This sand production practice shows that the content of rock powder yielded by rock powder recycle devices and sand scrapers is always about 17% (lower limit) and may be less than the lower limit. The recycle of rock powder, especially the recycle of the micro rock powder smaller than 0.08 mm, has long been a defect in the wet-process sand production, which requires further studies on the recycle conditions.

The wet-process sand production has a long dehydration cycle which could affect the yield of produced sand and requires larger warehouse capacity, and the micro rock powder loss can reach over 50% according to the data. The content of rock powder in the produced sand is low, which may cause heavy environment pollution. What's more, the conventional rod mill adopted for sand production requires higher steel consumption and production cost. In short, further studies are required on the recycle of rock powder, the uniformity of coarse and fine sand and the dehydration of the wet-process sand production.

3.5.3.3 Semi-dry Process Sand Production

Properties of semi-dry process sand production: semi-dry process sand production is a design concept that takes energy saving, environmental protection, intelligence and high quality into consideration in the whole production process, and can produce high-quality manufactured sand gravel without rod mills. The semi-dry process sand production replaces milling with crushing with the concept of crushing more and milling less. The semi-dry process sand production can meet or exceed the quality requirements of manufactured sand aggregate, and at the same time abandon or partially abandon some traditional sand production equipment with high energy consumption.

The in-depth study found that when using semi-dry process sand production to produce high-quality manufactured sand gravel, the advanced sand production process can also solve the problems of resource utilization, energy consumption, dust, noise, manufactured sand classification, fineness modulus and rock powder content, the recycling of silty sand and wastewater, the dehydration of silty sand, and the dust control in sand and gravel aggregate production. High quality and environmental protection overall embody the advancement of this process, which is energy-saving and environment-friendly.

Suofengying, Yunpeng, Shatuo, Sujiahe hydropower stations have adopted the semi-dry process sand production. According to various physical and mechanical properties of limestone and granite, with suitable production equipment selected and on the basis of summarizing sand production by other hard rocks in China, the whole-process semi-dry automatic technology is adopted to produce the gravel, which combines full recycling of sand and water, full greening of sites and factory management, and has been a modern and environment-friendly gravel processing system.

Improvement of the rock powder yield in wet-process joint sand production is a challenge. The dry-process or the semi-dry process with vertical shaft impact crushers is the trend in sand production. The key is the dust solution. The dry-process brings higher technical and economic benefits by lowering the water consumption of manufactured gravel system and the water content of the produced sand.

3.5.4 Examples of Projects with Aggregate Test

3.5.4.1 Examples of Projects with Silt Content Test of the Natural Sand

The sand for the RCC in China is mainly the manufactured sand, while some projects also apply natural sand. For example, projects in Longshou in Gansu, Kalasuke in Xinjiang, Xixi in Zhejiang and Gezhouba in Hubei. The natural sand is adopted in RCC in these projects. The natural sand is rather different with the manufactured sand regarding the application in RCC, and the main difference is that the natural sand consists no rock powder but has high silt content, which should be noted. Great controversies are found over the silt content test of the natural sand. According to the provisions of *Specifications for Hydraulic Concrete Construction* DL/T 5144—2001, the silt content of the natural sand refers to the amount of the particles smaller than 0.08 mm. The provisions do not clearly define whether the amount refers all silt content or includes a part of rock powder. Currently, according to the fact that there is few particles smaller than 0.16 mm in natural sand, the sand is generally replaced by additive rock powder or fly ash to effectively improve the grout-mortar ratio of RCC, and greatly make up the less cementing materials and insufficient mortar of RCC.

1. Natural sand for Longshou Project in Gansu Province. The main natural structure material for the Longshou Hydropower Station is sand and gravel, which is mainly produced in the sand and gravel yards Ⅲ 1 and Ⅲ 2 on the left bank of Heihe River, which are 1 - 2 km away from the downstream of dam axis. The natural gravel yards Ⅲ 1 and Ⅲ 2 are situated on the terraces Ⅲ and Ⅳ in the dam site, with flat terrain, 0 - 1.5 m thick silt loam on surface, and 6 m thick sand and gravel layer below the surface. These yards are rich in sand and gravel. See Table 3.26 for the quality of the natural sand and gravel.

    The Longshou Multipurpose Dam Project consists a RCC arch dam, a RCC gravity dam on the left bank, RCC trusted piers on the right bank, emptying, sediment discharge bottom outlets and the intake structures for the dam. The natural sand and gravel from Gobi Desert is adopted for RCC of Longshou Project. The aggregate are mined by rinsing the natural sand to remove the fine particles smaller than 0.08 mm. The

Chapter 3 Raw Materials and Project Examples

Table 3.26 Quality characteristics of natural structure sand and gravel for Longshou Project

Natural sand			Pebble		
Item	Test value (average)		Item	Test value (average)	
	Ⅲ1	Ⅲ2		Ⅲ1	Ⅲ2
Specific gravity (g/cm³)	2.70	2.68	Specific gravity (g/cm³)	2.74	2.72
Dry bulk density (g/cm³)	1.62	1.57	Dry bulk density (g/cm³)	1.83	1.82
Porosity (%)	39.6	40.0	Porosity (%)	33.45	31.63
Clay content(%)	5.61	6.73	Water absorption (%)	0.52	0.57
SO₃ content (%)	0.69	0.52	Needle slice content (%)	11.63	8.99
Content of the organic salt (%)	0.76	0.39	Soft particle content (%)	0.88	0
Organic content	Shallow	Shallow	Clay content(%)	0.22	0.3
Fineness module	2.35	2.06	SO₃ content (%)	0.22	0.32
Mean size	0.35	0.32	Organic content	Shallow	Shallow
—	—	—	Modulus grading	7.39	8.19

RCC production test found that the RCC has poor workability and rollability. After careful studies and analysis from all construction parties, it is generally believe that: firstly, the sand ratio of the RCC offered by the design institute is small. secondly, the content of the sand particles smaller than 0.16 mm is low. Most of these particles less than 0.16 mm are rinsed by water in the process of natural sand production, and only 3% – 5% of the particles are remained. Both factors are the main cause of poor workability and rollability of the RCC, resulting in the insufficient liquefied bleeding and exposed aggregate on compacted layers. In order to solve these problems, the sand gravel aggregate production mode is adjusted according to the features of Longshou Project: replacing the wet-process with the dry-process and increasing the content of the particles smaller than 0.16 mm, (especially the particles smaller than 0.08 mm, which are originally referred to the silt content). The grain gradation of sand has been improved, and the yield of the natural sand has been increased. The sand ratio in mix proportion of the RCC is properly increased. As a result, the service performance of the RCC is effectively improved.

2. Test of silt content by methylene blue solution of Xixi Project. The quality of manufactured sand produced for Zhejiang Xixi Reservoir Project is poor due to the processing system of rock and gravel. After many arguments and consulting with domestic famous experts, it is considered that adding proper amount of silty sand into the manufactured sand is beneficial to reduce the fineness modulus, increase the content of particles less than 0.08 mm in sand, and improve compaction and impermeability of RCC. The fineness modulus of the silty sand is low and usually less than 1. The content of particles smaller than 0.08 mm is 20% – 30%. According to the quality requirements of *Specifications for Hydraulic Concrete Construction* DL/T 5144—2001, the silt content in sand will be beyond the standard if the particles smaller than 0.08 mm are regarded as silt. Although the silty sand content is merely 10% – 20%, whether the silt content in the sand is beyond the standard would still be a concern. According to the tests of the content of gravel, silt, clay and fines specified in *Test Code for Aggregates of Hydraulic Concrete* DL/T 5151—2001, not all the particles smaller than 0.08 mm are the silt, but some of them are the sand fines. Further demonstration is needed on the silt content of the sand. The silt content in silty sand is searched according to the methylene blue solution test method specified in *Sand for Structure* GB/T 14684—2001, so as to provide the basis for the application of silty sand. The clay, natural silty sand, and the fresh rock powder from Baise Project in Guangxi Province and Xixi Project are adopted in the test, the clay is applied as pure silt and mixed with rock powder in different proportions, so as to test the silt content by methylene blue solution. The test

indicates that the usage of the methylene blue solution has a good linear relation with the silt content but no obvious relation with the type of rock powder. This has provided a reliable test method on the separation of the rock powder and silt smaller than 0.08 mm as well as the theoretical basis for the application of silty sand.

3.5.4.2 Examples of the Aggregate Application Projects

1. Gneiss aggregate applied in Shapai Arch Dam. The ratio of elastic modulus to strength of the gneiss granite aggregate applied in RCC of Shapai Arch Dam in Sichuan is low, and the formula of high iron and low aluminum is adopted in cement production, which makes the produced cement features low brittleness and high crack resistance. The ratio of elastic modulus to strength of the prepared RCC is about 40% lower than that of Puding and Yantan projects, 32% lower than that of Longtan Project and 27% lower than that of Dachaoshan Project; the ultimate tensile of the RCC $R_{90} 200^{\#}$ is about 85% higher than that of Puding Project, about 50% higher than that of Yantan Project, about 75% higher than that of Longtan Project and about 57% higher than that of Dachaoshan Project. The comparison indicates that the application of low elastic modulus aggregate and the cement of high iron and low aluminum have a significant effect on reducing the modulus of elasticity of RCC, which is extremely conducive to promoting the crack resistance of the concrete. The Shapai Project is a supporting project a national key scientific and technological research project for the "Ninth Five-year Plan" (*Research on RCC High Dam Construction Technology*). In this project, the optimization parameters and results of mix proportion in the "Ninth Five-year Plan" have been successfully applied. The research and successful application of the high crack resistance RCC materials has improved the ultimate tensile of the II gradation RCC for the dam to 1.36×10^{-4}, and the compressive modulus of elasticity to 16.65 GPa; the ultimate tensile of the III gradation RCC has been improved to 1.35×10^{-4}, and the compressive modulus of elasticity to 17.25 GPa. It reflects the high tensile and low modulus of elasticity characteristics of RCC for research. The ultimate tensile of the III gradation RCC core samples has been improved to 1.5×10^{-4}, and the compressive modulus of elasticity to 14.64 GPa. It reflects the high tensile and low modulus of elasticity characteristics of Shapai Dam RCC. Shapai RCC Dam has been put into operation for 10 years. Although the insulation and maintenance measures are not ideal, there is no crack in the dam, which effectively realizes the crack resistance of the RCC arch dam. Especially during the Wenchuan Earthquake in May 12, 2008, the Shapai Arch Dam was intact, despite being in the epicenter. The research and the successful application of the crack resistance materials used in Shapai Project has provided a useful lesson for promoting the crack resistance of RCC dams.

2. Diabase aggregate applied in Baise Project. The manufactured diabase aggregate applied in Baise Multipurpose Dam Project in Guangxi Province is the first case in China to apply such aggregate in mass hydraulic concrete. The manufactured diabase aggregate features a density of 3.0 g/cm³ above with good hardness and high modulus of elasticity. It is difficult to process such aggregate particularly because of the high rock powder content and poor size grading and water demand ratio much higher than that of the other manufactured aggregate in China. The apparent density of the RCC mixed can reach 2 650 kg/m³, which is 6%–10% higher than that of normal RCC. In order to reduce the modulus of elasticity of the concrete and improve the stress of dams, the maximum particle size of aggregate is reduced according to the characteristics of diabase aggregate and the advice of consulting experts at home and abroad, which can effectively reduce the elastic modulus of concrete. The concrete of quasi III gradation with 60 mm maximum particle size of the aggregate and quasi IV gradation with 100 mm maximum particle size of the aggregate has been adopted in Baise Project. Quasi III gradation and quasi IV gradation the first time in China to apply the concrete of these grades. The quasi III gradation RCC is used for Baise main dam. Due to the high rock powder content of 20%–24% in the manufactured diabase sand, the schemes with fly ash, high

content of additives and low *VC* value are replaced by rock powder, which has reduced the compressive elastic modulus of the hard diabase aggregate to 23.3 – 29.8 GPa from 41 GPa above. The reduction of aggregate particle size has an great effect on reducing the modulus of elasticity hence effectively improves the crack resistance of Baise Dam. The gallery of the dam is always dry during 5 years of operation, which reflects the excellent imperviousness performance of the dam.

3. Test of powder coating for coarse aggregate applied in Mianhuatan Project. In the production of coarse aggregate, powder coating may occur not only in dry-process production of manufactured aggregate, but also in semi-dry process production. In addition, the collision and falling of aggregate during transportation may also lead to powder coating of the aggregate. When the finished aggregate coated with powder is transported to the mixingstructure, the second sieving and washing are not adopted or the second sieving and washing are canceled, which weakens the bond stress between the rock powder coated on the surface of the coarse aggregate and the mortar, thus having a certain impact on the performance of RCC. For example, aggregate powder coating occurs in Mianhuatan, Longtan and Jin'anqiao projects. The influence of coarse aggregate coated with powder on the performance of RCC has been studied in Mianhuatan, Longtan and other projects.

The coarse and fine aggregates used in RCC of Mianhuatan Hydropower Station are manufactured gravels. In order to reduce the pollution of dust to the atmosphere, water injectors are installed in the coarse crusher and the middle crusher, and spraying devices are installed in the screening structure. The coarse aggregates produced by water injection and spraying are coated with a layer of rock powder which is difficult to remove. When mixing concrete, factors such as the loss of mixer blades are taken into account, and then the feeding sequence of sand → (cement + fly ash) → (water + additives) → coarse aggregates is adopted. Also, due to less water consumption in RCC mixing and limited mixing time, the rock powder cannot be fully separated from the surface of coarse aggregate during mixing, thus weakening the bond stress between aggregate and mortar, which will have a certain impact on concrete performance.

In order to research the influence of silty sand adhered to the surface of coarse aggregate on the concrete performance, it is necessary to find out the content and composition of silty sand adhered to the surface of coarse aggregate. Clean the large, medium and small aggregates separately, and dry and sieve the cleaned silty sand. The sieving results show that most of the coated silty sand is rock powder smaller than 0.16 mm, which is difficult to remove when coated on the surface of coarse aggregate. Because of the large surface area of small-sized stones, the content of adhered rock powder is also high, which has a great influence on the performance of concrete.

II gradation and III gradation RCC is adopted for the test. $R_{180}100^{\#}$, $R_{180}150^{\#}$ and $R_{180}200^{\#}$ RCC is adopted in Mianhuatan Project, with a longer design age and low compressive strength. The water-cement ratio of the three mix proportions is finally determined as 0.65, 0.60 and 0.55 respectively. The performance of RCC is compared between cleaned aggregate and aggregate coated with powder. The compression resistance, splitting tension, axial tension, impermeability and frost resistance of the concrete mixed with both aggregates are tested. The test results show that:

a) The washing of coarse aggregate has no obvious influence on the impermeability and frost resistance of the concrete. On the one hand, it is due to the long mixing age and small water-cement ratio, so the impermeability and frost resistance of RCC is higher. On the other hand, although the aggregate is coated with a layer of rock powder, it is tightly wrapped by mortar, the impermeability and frost resistance characteristics of the concrete under both conditions are not much different.

b) The compressive strength, splitting tensile strength and axial tensile strength of the concrete mixed with unwashed coarse aggregate are all reduced, and the average strength reduction at each age is as follows:

7.8% for compressive strength, 10.9% for splitting tensile strength, and 13.9% for axial tensile strength. The influence on tensile strength is significant, while normal concrete consumes a lot of water, which can reduce the influence of rock powder coated on the coarse aggregate on concrete performance to a certain extent, and its tensile strength is reduced by less than 5%.

c) With the increase of concrete age, the influence of rock powder coated on the coarse aggregate on the compressive strength of concrete is weakened, but the compressive strength, splitting tensile strength and axial tensile strength of RCC are reduced to varying degrees. Due to the constraint of durability, the water-cement ratio is lower and the strength surplus is more in the mix proportion design, so the concrete produced by coarse aggregate coated with rock powder can also meet the design requirements of impermeability, frost resistance, compressive strength and tensile strength.

A layer of rock powder is adhered to the surface of coarse aggregate produced by dry-process, and most of the power diameter is less than 0.16 mm, due to less water consumption of RCC and limited mixing time, the rock powder on the surface cannot be separated, which is bound to form a weak joint surface between cement mortar and coarse aggregate. This weak surface weakens the bonding force between aggregate and mortar in concrete, and reduces the compressive strength, axial tensile strength and splitting tensile strength to varying degrees. However, due to the constraint of durability, the designed water-cement ratio is small, and the strength margin of RCC is large. Therefore, it has no qualitative influence on the compressive strength, tensile strength, impermeability and frost resistance of the poured RCC. It is suggested that the influence of performance reduction factors should be taken into account in the design of similar engineering indexes.

3.6 Conclusion

1. Cement, admixtures and additives are important raw materials for concrete projects. Manufacturers and varieties should be determined through optimization test. The selected manufacturers should have sufficient production scale and successful experience in using their raw materials in various projects. The manufacturers shall be fixed on the basis of optimization. The raw materials provided shall have product certificates, and shall be reinspected as soon as possible after being transported to the construction site, so as to determine whether the raw materials are qualified and the quality changes. The raw materials at the construction site must be transported and stored according to different varieties, grades and factory numbers, and rain and moisture-proof measures should be taken for the storage site.

2. Medium hydration heat Portland cement, low hydration heat Portland cement or ordinary portland cement are preferred in water conservancy and hydropower projects. In addition to complying with the national standards, some special requirements are put forward according to the specific use of projects. It is generally required that the compressive strength and breaking strength of cement should be greater than 45 MPa and 8.0 MPa, the fineness should be slightly larger, the specific surface area should be controlled within 280 – 320 m^2/kg, and the MgO content in clinker should be controlled within 3.5% – 4.5%. In order to reduce the hydration heat of cement, it is required that the content of tricalcium aluminate (C_3A) and tetracalcium ferroaluminate (C_4AF) in the clinker should be less than 6% and 16% respectively, and the temperature of the cement mobilized in construction site shall follow the requirements.

3. The fineness, water demand ratio, loss on ignition, and sulfur trioxide content are the most important quality indexes of fly ash. Due to the continuous improvement of the quality of fly ash, especially the mass production of Grade I fly ash, fly ash has also changed from general mixture in the past to functional materials at present. The biggest advantage of mixing fly ash into hydraulic concrete is to make use of the characteristics of the later strength of fly ash. Due to the difference between fly ash and other mixtures, fly

ash particles are in the shape of micro beads. The higher the grade of fly ash, the finer the particles, and the more micro beads content, and the more obvious the improvement of concrete performance. Especially, the high-quality Grade I fly ash contains obvious spherical particles, and its shape effect, micro-aggregate effect and pozzolanic effect can play a good role as solid water reducing agent.

4. Additive has been the fifth necessary material besides cement, aggregate, admixture and water, which accelerates the realization of new RCC construction technology. Additive is a necessary means to improve the performance of RCC, so additive technology has been the key to the rapid development of RCC damming. Additive is an important technical measure to improve the optimization design of concrete mix proportion. In recent years, two kinds of additives, namely retarding high-range water reducing agent + air entraining agent, are mainly added into RCC, which not only meets the requirements of large RCC placing surface, high-strength construction and continuous rising retarding, but also achieves the purpose of water reduction and durability improvement. As the full-section damming technology is adopted for the RCC, the performance requirements of RCC admixtures are higher and higher. It is required that the admixtures shall have good compatibility with cement, and be easily prepared, without precipitation, but with good uniformity; at the same time, it is required that the admixtures should be with high water reduction, high retardation and stable air content; according to the actual situations of raw materials in specific projects, the admixture content still needs to be increased to meet the requirements of RCC construction.

5. Because the amount of manufactured sand accounts for more than 1/3 of total aggregate, especially the content of rock powder in the manufactured sand is directly related to the rapid construction and cost-effectiveness of RCC, the reasonable selection of the processing and production methods of manufactured sand is the basic guarantee for the rapid construction of RCC. Based on a large number of engineering and repeated tests, the manufactured sand for RCC contains higher rock powder content, which significantly improves the workability, liquefied bleeding, rollability, interlayer bonding, aggregate separation resistance, compactness and other construction properties of RCC, as well as the impermeability, mechanics index and fracture toughness of hardened concrete; rock powder can be used as mixture to replace partial fly ash; proper increase of rock powder content facilitates increase of the output of manufactured sand, reduction of the cost and increase of technical economic benefits. Therefore, reasonable control of content of manufactured gravel powder is one of the important measures to improve the quality of RCC. it is very important to control the water content of the dry saturated surface of the manufactured sand not to exceed 6%. That is the key to control the water-cement ratio of concrete and the *VC* value or outgoing RCC, and also the requirement to add ice for pre-cooling concrete.

6. Due to the wide distribution of water conservancy and hydropower projects, and various types of manufactured aggregate rocks, engineering practices show that manufactured aggregates of different rock types have different effects on the performance of RCC. In the production of coarse aggregate, powder coating may occur not only in dry-process production of manufactured aggregate, but also in semi-dry process production. In addition, the collision and falling of aggregate during transportation may also lead to powder coating of the aggregate. When the finished aggregate coated with powder is transported to the mixing structure, the secondary screening and washing are not adopted or the secondary screening and washing are canceled, which weakens the bond stress between the rock powder coated on the surface of the coarse aggregate and the mortar, thus having a certain impact on the performance of RCC.

Chapter 4

RCC Mix Proportion Design and Example

4.1 Overview

Material science is an important basis of modern science and technology. The quality of dam is ultimately reflected in the damming materials. The RCC mix proportion design is the best combination of damming materials. High-quality and scientific and reasonable mix proportion plays a decisive role in the rapid RCC damming, which is the basis of ensuring the quality of RCC dam and rapid construction. Mix proportion design is one of the key technologies for rapid RCC damming. It has a high technical content and is directly related to the success of rapid damming. It can achieve twice the result with half the effort and obtain significant technical and economic benefits.

Roller compacted concrete: It refers to the concrete that is spread with the semi-plastic concrete mixture without slump in thin-lifts and rolled and compacted by vibrating roller, with bleeding performance. The definition of RCC is completely different from that of early hard concrete. The development of things is a process from practice→theory→practice, which is also the inevitable result of the development process of RCC from "Roller Compacted Dam-Concrete (RCD)" damming technology to full-section damming technology. Over 20 years, RCC damming technology has developed from the early exploration period and transition period to the current mature period, and the RCC has also developed from ultra-hard concrete and hard concrete to semiplastic concrete without slump. Because RCC has become semi-plastic concrete without slump, its performance after hardening is basically the same as that of normal concrete, which conforms to the "watercement ratio" rule of concrete. Therefore, RCC design indexes and dam section design are the same as normal concrete dam.

Since the full-section RCC damming technology was adopted in China from 1990s, it has been emphasized that continuous interlayer paving and good interlayer bonding are required, RCC must meet the main performance requirements of imperviousness, frost resistance and ultimate tensile value, and the concrete inside the dam shall also meet the necessary temperature control and cracking prevention requirements. Therefore, the technical route of RCC mix proportion design in China is characterized by " two-low, two-high and dualadding", i. e. , low cement content and VC value, high admixtures and rock powder content, and mixing of retarding and water reducing agent and air entraining agent.

RCC and normal concrete are common and certainly different in mix proportion design. The main difference between RCC and normal concrete lies in the VC of rolling by vibrating roller, and the aggregate gradation and mortar content suitable for vibration rolling. Results of large amount of tests indicate: The physical and mechanical properties such as strength, imperviousness, frost resistance and shear resistance of RCC body are not inferior to that of normal concrete. However, RCC construction is subject to whole-block placement and thin-lift, and imperviousness depends on the dam body itself. The mix proportion design shall meet the requirements of not only body strength, impermeability and other performances but also joint surface combination of dam layers. Thus, it is required that RCC must have abundant cementing material mortar, proper aggregate separation resistance and long initial setting time. After being rolled and compacted by vibrating roller, the surface is fully bleeding and elastic, so that the upper RCC aggregate can be embedded into the lower RCC that has been rolled, which changes the weak layer of RCC that is easy to form "layer-cake" seepage channel, with many layers. At the same time, problems of frost resistance and low ultimate tensile value of RCC are also solved.

"Interlayer combination, temperature control and cracking prevention" is the core technology of rapid RCC damming. The mix proportion design of RCC must be carefully and scientifically designed around the core technology. Therefore, the mix proportion design of RCC shall focus on the performance test of mixture, so that the working performance of fresh RCC can meet aggregate separation resistance, rollability, liquefied bleeding and interlayer bonding according to construction requirements, rather than the performance obtained in the laboratory. It is necessary to change the design concept that mix proportion design of normal concrete attaches importance to the performance of hardened concrete and despises the performance of mixture.

The grout-mortar ratio *PV* has become the fourth major parameter in the mix proportion design of RCC and is an important index to evaluate the performance of RCC mixture. PV value plays the same important role as the three major parameters of water-cement ratio, sand ratio and unit water consumption, and is one of the important parameters in the mix proportion design of RCC.

In China, the design age of RCC is mostly 90 d or 180 d, so the design period of mix proportion is relatively long. Therefore, the mix proportion design test of RCC needs to be carried out in advance. In addition, it is required that the raw materials selected in the test shall be consistent with the actual raw materials used in the projects as far as possible, so as to avoid "two-skin" of raw materials, resulting in large difference between the test results and the actual construction.

4.2　RCC Mix Proportion Parameter Selection

The task of mix proportion design of RCC is essentially to meet the performance of strength, durability and crack resistance required by the design under the working conditions of mixture meeting the construction requirements, reducing the cement dosage as much as possible, and reasonably determining the dosage of water, cement, admixture, additive, sand and stone in 1 m^3 RCC.

With the development of full-section RCC damming technology, the design of RCC mix proportion in China is constantly innovated and perfected. The parameter selection of RCC mix proportion is a very important link in mix proportion design. On the premise of selection of three major parameters, the grout-mortar ratio *PV* has become a major parameter in the mix proportion design of RCC, and plays the same important role as the three major parameters of water-cement ratio, sand ratio and unit water consumption. The parameter selection of mix proportion of RCC is closely related to the properties of RCC such as admixture, additive, aggregate type, rock powder content, cementing material dosage and *VC* value. Especially the performance of fresh RCC is highly valued in the mix design.

In order to obtain the high-quality RCC mix proportion meeting the requirements of design index and construction rollability, in addition to certain mix proportion design and test experience, mix design and test personnel must also understand the structural type of RCC dam and construction characteristics of RCC, formulate a technical route for scientific and reasonable mix proportion design according to the RCC design index, design age, raw material characteristics, climate and terrain of dam site, construction conditions, etc. , thoroughly grasp the relationship between mix proportion parameters, so that the water-cement ratio, optimal sand ratio, unit water consumption, grout-mortar ratio, admixture, additive, *VC* value, rock powder content and other key parameters, and conduct careful mix proportion design always closely around the core technology of RCC "interlayer bonding, temperature control and cracking prevention".

According to the statistics of RCC mix proportion of some projects in China since 1990s, the mix proportion of Ⅲ gradation RCC in dams and Ⅱ gradation RCC in impervious area of main projects completed or under construction are listed in Table 4.1 and Table 4.2. The statistical analysis of more than 80 groups of RCC mix proportion parameters in the table is as follows.

4.2.1 Selection of Water Binder Ratio

water-cement ratio: It refers to the ratio of water per unit volume of concrete to cementing materials (The water is subject to the dry state of saturated aggregate surface), expressed in $W/(C+F)$.

The water-cement ratio is the key parameter and main factor to determine the strength and durability of concrete. Compared with normal concrete, RCC is variational in only mix proportion and construction process. Because RCC has become semi-plastic concrete without slump, its performance after hardening is basically the same as that of normal concrete, which conforms to the "water-cement ratio" rule of concrete. The compressive strength of RCC mainly depends on the water-cement ratio. There is a linear relationship between the reciprocal of the water-cement ratio (namely, binder-water ratio) and the compressive strength of RCC, namely, the famous Bolomey Formula. The water-cement ratio shall be selected through the test, and the relationship curve between the water-cement ratio and the compressive strength of RCC shall be established. According to test results, the water-cement ratio meeting the technical index requirements of design shall be selected.

Main factors affecting the water-cement ratio are closely related to the design index, design age, frost resistance grade, ultimate tensile value, aggregate performance, type and content of admixture and additive of RCC. The water-cement ratio is the most important parameter that affects and determines the durability and various properties of RCC. Therefore, great attention shall be paid to the selection of water-cement ratio in mix proportion design. In recent years, the frost resistance grade has become one of the main design indexes of RCC in the severe cold north or the mild south, and the frost resistance is a very important index for the durability of dam concrete.

At present, the design index of frost resistance grade of RCC is getting higher and higher. Because the design strength grade of RCC is generally low, a little cement and more admixture (fly ash) are used, so the design index of frost resistance grade and ultimate tensile value of RCC generally does not match with the strength grade, so the water-cement ratio shall be selected based on frost resistance grade and ultimate tensile value as control indexes. *Specifications for Hydraulic Concrete Construction* DL/T 5144 and *Code for Mix Design of Hydraulic Concrete* DL/T 5330 have specified the maximum allowable value of water-cement ratio. See Table 4.3 for the maximum allowable value of water-cement ratio of concrete.

According to the statistical analysis of mix proportion and water-cement ratio of RCC in some projects in Table 4.1 and Table 4.2, project cases show that: The water-cement ratio of Ⅲ gradation RCC in dams ($C_{90}15$ MPa and $C_{90}20$ MPa) is generally 0.50 – 0.60 and 0.45 – 0.50 respectively; the water-cement ratio of Ⅱ gradation RCC in impervious area ($C_{90}20$ MPa) is generally 0.45 – 0.50. Analysis results indicate that: The water-cement ratio of RCC selected in this range meets the design requirements.

4.2.2 Selection of Optimal Sand Ratio

Sand ratio: it refers to the percentage of sand in the volume ratio of sand to gravel in unit volume concrete. Because the density of sand and gravel is close, the mass ratio of sand to gravel is usually used instead of volume ratio to calculate the sand ratio. The sand ratio can be expressed in $W/(S+G)$.

Chapter 4 RCC Mix Proportion Design and Example

Table 4.1 Mix proportion of III gradation RCC for dams in some major projects in China

S/N	Project title	Design index of RCC	Mix proportion parameter						Material amount (kg/m³)					Cement type	Type of aggregate	Type of admixture
			water-cement ratio	Fly ash (%)	Sand ratio (%)	Water reducing agent (%)	Air entraining agent (1/10 000)	VC value (s)	Water content	Cement	Fly ash		Apparent density			
1	Kengkou	$R_{90}100S4$	0.70	57.1	36.8	0.2	—	15±5	98	60	80		2 374	Ordinary portland cement 425#	Manufactured tufflava rock powder, 10%–15%	Grade II fly ash
2			0.578	66.7	36	0.2	—	15±5	104	60	120		2 364	Ordinary portland cement 425#	River sand, fly ash and substituted sand, 20 kg/m³	Grade II fly ash
3	Yantan	$R_{90}150S4$	0.566	65.4	34	0.25	—	15±5	90	55	104		2 498	Ordinary portland cement 525#	Content of manufactured limestone aggregate and rock powder, 8%–18%	Grade II fly ash
4	Puding	$R_{90}150S4$	0.55	65	34	Triple compound 0.55	—	10±5	84	54	99		2 530	Ordinary portland cement 525#	Content of manufactured limestone and rock powder, 13%–17%	Grade II fly ash

Continued Table 4.1

S/N	Project title	Design index of RCC	Mix proportion parameter							Material amount (kg/m³)					Cement type	Type of aggregate	Type of admixture
			Water-cement ratio	Fly ash (%)	Sand ratio (%)	Water reducing agent (%)	Air entraining agent (1/10 000)	VC value (s)	Water content	Cement	Fly ash	Apparent density			
5	Jiangya	$C_{180}20W6F100$	0.58	60	33	0.4	—	7±4	93	64	96			Manufactured limestone	
6		$R_{90}C10D50$	0.60	62	34.5	0.6	6	2–6	90	57	93	2 498	Ordinary portland cement 425#	Content of manufactured limestone rock powder, 16%–20%	Grade I fly ash
7	Fenhe Reservoir II	$R_{90}C10D50$	0.56	63	33	0.6	6	2–6	90	60	101	2 450		Natural sand of limestone coarse aggregate	Grade I fly ash and substituted sand
8	Three Gorges Left Guide Wall	$R_{90}150\,D100S6$	0.50	58	34	0.5	5	3–7	87	73	101	2 450	Moderate heat cement 525#	Manufactured granite	Grade I fly ash
9		$R_{90}200\,D150S8$	0.50	51	34	0.5	5	3–7	89	87	91	2 450			
10	Hongpo (The dam is fully III gradation)	$R_{90}150$ Upstream	0.55	65	35	0.9	4	5–15	84	54	99+18	2 525	Ordinary portland cement 525R	Content of manufactured limestone rock powder, 13%–17%	Grade II fly ash and substituted sand
11		$R_{90}150$ Downstream	0.60	65	36	0.9	4	5–15	85	50	92+18	2 520			
12		$R_{180}100S4D25$	0.65	60	32	0.6	—	5–10	89	55	82	2 412	Ordinary portland cement 525#	Content of manufactured granite and rock powder, 17%–20%	Grade II fly ash
13	Mianhuatan	$R_{180}150S4D25$	0.56	60	32	0.6	—	5–10	89	64	95	2 412			
14		$R_{180}200S4D25$	0.50	60	32	0.6	—	5–10	89	72	106	2 412			

Chapter 4 RCC Mix Proportion Design and Example

Continued Table 4.1

S/N	Project title	Design index of RCC	Mix proportion parameter							Material amount (kg/m³)					Cement type	Type of aggregate	Type of admixture
			Water-cement ratio	Fly ash (%)	Sand ratio (%)	Water reducing agent (%)	Air entraining agent (1/10 000)	VC value (s)	Water content	Cement	Fly ash	Apparent density			
15	Shimenzi	$C_{90}15W6F100$	0.49	64	31	0.95	4	1–10	84	62	110	2 467	Ordinary portland cement 525R	Pebble and natural sand	Grade II fly ash
16	Gaobazhou	$R_{90}150$	0.52	50	31	0.4	1	5–8	91	88	88	2 438	Moderate heat cement 525#	Pebble and river sand	Grade II fly ash
17		$R_{90}200$	0.45	45	31	0.4	1	5–8	93	114	93	2 438			
18	Dachaoshan	$C_{90}15W4F25$	0.50	PT 60	34	0.75	—	2–5	87	67	107		Ordinary portland cement 525#	Manufactured basalt and rock powder, more than 15%	Phosphorous slag + tuff
19	Longshou	$C_{90}20W6F100$	0.48	65	30	0.9	8	0–5	85	62	115	2 450	Silicate 525#	Pebble Natural sand	Mixing MgO
20	Shapai	$R_{90}200$	0.50	50	33	0.75	1	2–8	93	93	93	2 480	Ordinary portland cement 425#	Content of manufactured granite and rock powder, 12%–19%	Gateway Grade II fly ash
21	Shankou Grade III	$R_{90}C10W6$	0.6	60 (65)	28	0.70	—	2–8	85.5	57 (49.5)	85.5 (93)	2 460	Ordinary portland cement 425#	Coarse and fine aggregate of river pebble	Grade III fly ash, and different temperature content

Continued Table 4.1

S/N	Project title	Design index of RCC	Mix proportion parameter							Material amount (kg/m³)					Cement type	Type of aggregate	Type of admixture
			Water-cement ratio	Fly ash (%)	Sand ratio (%)	Water reducing agent (%)	Air entraining agent (1/10 000)	VC value (s)	Water content	Cement	Fly ash	Apparent density			
22	Linhekou	R₉₀200S6D50	0.47	62	34	0.7	2	3–5	81	66	106	2 460	Moderate heat cement 525#	Content of manufactured limestone and rock powder, more than 18%	Quasi Grade I fly ash
23	Phase III cofferdam of the Three Gorges Project	R₉₀150W6F50	0.50	55	34	0.6	10	1–5	83	75	91		Moderate heat cement 525#	Content of manufactured granite and rock powder, 15%	Grade I fly ash
24	Baise (quasi Grade III)	R₁₈₀15S6D25	0.60	63	34	0.8	4	1–5	96	59	101	2 650	Moderate heat cement 525#	Manufactured diabase	Grade II fly ash
25	Suofengying Hydropower Station	C₉₀15W6F50	0.55	60	32	0.8	6	3–5	88	64	96+5	2 435	P·O 42.5	Manufactured limestone and rock powder, 17%–21%	Grade I fly ash mixed with MgO, 3%
26	Zhaolaihe Hydropower Station	C₉₀20W6F100	0.48	55	34	0.6	15	2–5	75	70	86	2 437	P·O 42.5	Content of manufactured limestone and rock powder, 18%–20%	Quasi Grade I fly ash

Chapter 4 RCC Mix Proportion Design and Example

Continued Table 4.1

S/N	Project title	Design index of RCC	Mix proportion parameter							Material amount (kg/m³)				Cement type	Type of aggregate	Type of admixture
			Water-cement ratio	Fly ash (%)	Sand ratio (%)	Water reducing agent (%)	Air entraining agent (1/10 000)	VC value (s)	Water content	Cement	Fly ash	Apparent density			
27	Pengshui	$C_{90}15W6F100$	0.50	60	35	0.6	6	3–8	83	66	100	2 528	P·MH 42.5	Content of manufactured limestone and rock powder, 14.5%	Grade I fly ash
28	Jinghong	$C_{90}15W6F50$	0.50	NH 60	33	0.5	2	3–5	80	64	96	2 441	P·MH 42.5	Natural sand	Double admixture of slag + rock powder
29		$C_{90}15W6F50$	0.50	50	35	0.5	2.5	3–5	80	80	80	2 440		Manufactured gravel	
30	Longtan	Lower part $C_{90}25W6F100$	0.41	55	33	0.6	2	2–7	79	85	108	2 465	P·MH 42.5	Content of manufactured limestone and rock powder, 16%–20%	Grade I fly ash
31		Central part $C_{90}20W6F100$	0.45	61	33	0.6	2	2–7	78	67	106	2 455			
32		Upper part $C_{90}15W6F50$	0.48	66	34	0.6	2	2–7	79	56	109	2 455			
33	Guangzhao	Lower part $C_{90}25W8F100$	0.45	50	34	0.7	4	3–5	78	83	83+14	2 483	P·O 42.5	Content of manufactured limestone and rock powder, 16%–20%	Grade II fly ash and substituted sand, 2%–3%
34		Central part $C_{90}20W6F100$	0.50	55	34	0.7	4	3–5	78	70	86+21	2 483			
35		Upper part $C_{90}15W6F50$	0.55	60	35	0.7	4	3–5	78	57	85+22	2 496			

Continued Table 4.1

S/N	Project title	Design index of RCC	Mix proportion parameter							Material amount (kg/m³)				Cement type	Type of aggregate	Type of admixture
			Water-cement ratio	Fly ash (%)	Sand ratio (%)	Water reducing agent (%)	Air entraining agent (1/10 000)	VC value (s)	Water content	Cement	Fly ash	Apparent density			
36	Silin	C₉₀15W6F50	0.54	60	35	0.6	13	2–6	80	59	89	2 470	P·MH 42.5	Content of manufactured aggregate and rock powder, 9%–15%	Grade II fly ash
37		C₉₀20W6F50	0.51	55	33	0.6	8	2–6	85	75	92	2 484	P·O 42.5		Double admixture of slag + rock powder
38	Gelantan	C₉₀15W4F50	0.50	SL 60	38	0.8	4	3–8	83	66	100	2 443	P·O 32.5	Content of manufactured limestone and rock powder, 16%–22%	Double admixture of slag + rock powder
39	Tuka River	C₁₈₀10W6F100	0.50	SL 55	38	0.60	12	3–8	95	85	105	2 440	P·O 42.5	Manufactured limestone Double-admixture and substituted sand, 5%	Double admixture of slag + rock powder
40	Jufudu	C₉₀15W4F50	0.55	SL 55	34	0.70	2	1–12	75	67	83	2 430	P·O 42.5	Tuffite	Double admixture of slag + rock powder

Continued Table 4.1

S/N	Project title	Design index of RCC	Mix proportion parameter							Material amount (kg/m³)					Cement type	Type of aggregate	Type of admixture
			Water-cement ratio	Fly ash (%)	Sand ratio (%)	Water reducing agent (%)	Air entraining agent (1/10 000)	VC value (s)	Water content	Cement	Fly ash	Apparent density			
41	WuduYinshui	Internal C_{180}20W8F50	0.55	60	35	0.7	10	3–5	84	61	92	2 472	P·MH 42.5	Content of manufactured aggregate and rock powder, 17.5%	Grade Ⅰ flyash
42		External R_{180}20W6F200	0.45	50	32	0.9	10	1–5	90	100	100	2 400			
43	Kalasuke Hydropower Station	Internal and lower part R_{180}20W4F50	0.53	62	30	0.9	6	1–5	90	65	105	2 400	P·O 42.5	Natural sand by water washing of granite gneiss gravel Limerock powder and substituted sand, 0–10%	Grade Ⅰ fly ash
44		Internal and upper part R_{180}15W4F50	0.56	62	30	0.9	6	1–5	90	61	100	2 400			

Continued Table 4.1

S/N	Project title	Design index of RCC	Mix proportion parameter							Material amount (kg/m³)					Cement type	Type of aggregate	Type of admixture
			Water-cement ratio	Fly ash (%)	Sand ratio (%)	Water reducing agent (%)	Air entraining agent (1/10 000)	VC value (s)	Water content	Cement	Fly ash	Apparent density			
45	Jin'anqiao	Lower part $C_{90}20W6F100$	0.47	60	33	1.2	25	1–5	90	76	115	2 630	P·MH 42.5	Manufactured basalt and rock powder, controlled as per 18%–19%, and rock powder and substituted sand in case of failing to meet requirements, 5%–8%	Grade II fly ash
46		Upper part $C_{90}15W6F100$	0.53	63	33	1.2	15	1–5	90	63	107	2 630			
47	Gongguoqiao	$C_{180}15W4F50$	0.50	55	36	0.8	5	3–7	90	81	99	2 440	P·MH 42.5	Manufactured sandy slate and rock powder >18%	Grade II fly ash
48	Guandi	Lower part $C_{90}25W6F100$	0.45	55	32	0.8	12	3–7	92	92	112	2 660	P·MH 42.5	Manufactured basalt aggregate and rock powder, controlled as per 16%–20%	Grade II fly ash
49		Central part $C_{90}20W6F100$	0.48	60	33	0.8	12	3–7	92	67	106	2 660			
50		Upper part $C_{90}15W6F100$	0.51	65	34	0.8	12	3–7	92	56	109	2 660			

Chapter 4　RCC Mix Proportion Design and Example

Table 4.2　Mix proportion of II gradation RCC for impervious areas of dams in some major projects in china

S/N	Project title	Design index of RCC	Mix proportion parameter					Material amount (kg/m³)					Cement type	Type of aggregate	Type of admixture
			Water-cement ratio	Fly ash (%)	Sand ratio (%)	Water reducing agent (%)	Air entraining agent (1/10 000)	VC value (s)	Water content	Cement	Fly ash	Apparent density			
1	Puding	$R_{90}200S6$	0.50	55	38	Triple compound 0.55	—	10±5	94	85	103	2 514	Ordinary portland cement 425#	Manufactured limestone	Grade III fly ash
2	Jiangya	$C_{90}20W12F100$	0.53	55	36	0.50	—	7±4	103	87	107			Manufactured limestone	
3	Fenhe Reservoir II	$R_{90}C20S8D150$	0.50	45	35.5	0.60	6	2–5	94	103	85	2 484	Ordinary portland cement 425#	Content of manufactured limestone rock powder, 16%–20%	Grade I fly ash
4		$R_{90}C20S8D150$	0.45	40	35	0.60	6	2–5	95	127	84	2 430	Ordinary portland cement 425#	Natural sand of limestone coarse aggregate	Grade I fly ash and substituted sand
5	Mianhuatan	$R_{180}150S4D25$	0.55	60	36	0.6	—	5–10	99	71	106	2 381	Ordinary portland cement 525#	Content of manufactured granite and rock powder, 17%–20%	Grade II fly ash
6		$R_{180}200S8D50$	0.50	60	36	0.6	—	5–10	99	80	118	2 381			
7	Shimenzi	$C_{90}25W8F300$	0.40	54	33	0.95	40	1–10	81	93	110	2 434	Ordinary portland cement 525R	Pebble and natural sand	Grade II fly ash
8	Gaobazhou	$R_{90}200$	0.48	45	35	0.4	1	5–8	109	114	93	2 407	Moderate heat cement 525#	Pebble and river sand	Grade II fly ash

Continued Table 4.2

S/N	Project title	Design index of RCC	Mix proportion parameter							Material amount (kg/m³)					Cement type	Type of aggregate	Type of admixture
			Water-cement ratio	Fly ash (%)	Sand ratio (%)	Water reducing agent (%)	Air entraining agent (1/10 000)	VC value (s)	Water content	Cement	Fly ash	Apparent density			
9	Dachaoshan	$C_{90}20W8F50$	0.50	PT 50	37	0.70	—	2–5	94	94	94		Ordinary portland cement 525#	Manufactured basalt	Phosphorous slag + tuff (PT)
10	Longshou	$C_{90}20W8F300$	0.43	53	32	0.7	40	0–5	91	100	112	2 420	Silicate 525#	Pebble and natural sand	Grade I fly ash mixed with MgO
11	Shapai	$R_{90}200\varepsilon_p 1.1$	0.53	40	37	0.75	2	2–8	102	115	77	2 482	Ordinary portland cement 425#	Content of manufactured granite and rock powder, 12%–19%	Gateway Grade II fly ash
12	Linhekou	$R_{90}200S8D100$	0.47	60	37	0.7	2	3–5	87	74	111	2 440	Moderate heat cement 525#	Content of manufactured limestone and rock powder, more than 18%	Quasi Grade I fly ash
13	Phase III cofferdam of the Three Gorges Project	$R_{90}150W6F50$	0.50	55	39	0.6	10	1–5	93	84	102		Moderate heat cement 525#	Content of manufactured granite and rock powder, 15%	Grade I fly ash

Continued Table 4.2

S/N	Project title	Design index of RCC	Mix proportion parameter							Material amount (kg/m³)					Cement type	Type of aggregate	Type of admixture
			Water-cement ratio	Fly ash (%)	Sand ratio (%)	Water reducing agent (%)	Air entraining agent (1/10 000)	VC value (s)	Water content	Cement	Fly ash	Apparent density			
14	Baise	$R_{180}20S10D50$	0.50	58	38	0.8	7	1–5	106	89	123	2 630	Moderate heat cement 525#	Manufactured diabase and rock powder, 20%–24%	Grade II fly ash
15	Suofengying Hydropower Station	$C_{90}20W8F100$	0.50	50	38	0.8	6	3–5	94	94	94	2 430	P·O 42.5	Manufactured limestone and rock powder, 17%–21%	Grade I fly ash mixed with MgO,3%
16	Zhaolaihe Hydropower Station	$C_{90}20W8F150$	0.48	50	37	0.6	15	3–5	85	88.5	88.5	2 403	P·O 42.5	Content of manufactured limestone and rock powder, 18%–20%	Quasi Grade I fly ash
17	Pengshui	$C_{90}20W10F150$	0.50	50	38	0.6	6	3–8	91	91	91	2 518	P·MH 42.5	Content of manufactured limestone and rock powder, 14.5%	Grade I fly ash
18	Jinghong	$C_{90}15W8F100$	0.45	NH 50	38	0.5	2	3–8	84	93	93	2 448	P·MH 42.5	Natural sand and gravel	Double admixture of slag + rock powder

Continued Table 4.2

S/N	Project title	Design index of RCC	Mix proportion parameter							Material amount (kg/m³)					Cement type	Type of aggregate	Type of admixture
			Water-cement ratio	Fly ash (%)	Sand ratio (%)	Water reducing agent (%)	Air entraining agent (1/10 000)	VC value (s)	Water content	Cement	Fly ash	Apparent density			
19	Longtan	$C_{90}25W12F150$	0.40	55	38	0.6	2	2–7	87	99	121		P·MH 42.5	Content of manufactured limestone and rock powder, 16%–20%	Grade I fly ash
20	Guangzhao Lower part	$C_{90}25W12F150$	0.45	50	38	0.7	6	3–5	83	92	92+23	2 450	P·O 42.5	Manufactured limestone and rock powder, 20%–24%	Grade II fly ash and substituted sand, 3%
21	Guangzhao Upper part	$C_{90}20W10F100$	0.50	55	39	0.7	6	3–5	86	77	95+23	2 455			
22	Silin	$C_{90}20W8F100$	0.48	50	38	0.6	13	2–6	91	95	95	2 454	P·MH 42.5	Content of manufactured aggregate and rock powder, 9%–15%	Grade II fly ash
23	Silin	$C_{90}20W8F100$	0.46	45	39	0.6	8	2–6	90	108	88	2 479	P·O 42.5		
24	Gelantan	$C_{90}20W8F100$	0.45	SL 55	34	0.8	5	3–8	93	93	114	2 402	P·O 32.5	Content of manufactured limestone and rock powder, 18%–20%	Double admixture of slag + rock powder

Continued Table 4.2

S/N	Project title	Design index of RCC	Mix proportion parameter						Material amount (kg/m³)				Cement type	Type of aggregate	Type of admixture
			Water-cement ratio	Fly ash (%)	Sand ratio (%)	Water reducing agent (%)	Air entraining agent (1/10 000)	VC value (s)	Water content	Cement	Fly ash	Apparent density			
25	Tuka River	$C_{180}7.5W4F50$	0.55	SL 60	34	0.60	12	3–8	88	64	96	2 460	P·O 42.5	Double-admixture and substituted sand, 5%	Double admixture of slag and rock powder and substituted sand
26	Jufudu	$C_{90}20W8F100$	0.44	SL 45	37	0.7	2	1–12	88	110	90	2 400	P·O 42.5	Tuffite	Double admixture of slag + rock powder
27	Wudu Yinshui	Impervious area $C_{180}20W8F50$	0.50	55	38	0.7	10	3–5	93	84	102	2 457	P·MH 42.5	Content of manufactured aggregate and rock powder, 17.5%	Grade I fly ash
28	Kalasuke Hydropower Station	Water level varying area $R_{180}20W10F300$	0.45	40	35	1.0	12	1–5	98	131	87	2 370	P·O 42.5	Natural sand by water washing of granite gneiss gravel	
29		Below dead water level $R_{180}20W10F300$	0.47	55	35.5	0.9	7	1–5	95	91	111	2 370		Limerock powder and substituted sand, 0–10%	Grade I fly ash

Continued Table 4.2

S/N	Project title	Design index of RCC	Mix proportion parameter						Material amount (kg/m³)				Cement type	Type of aggregate	Type of admixture
			Water-cement ratio	Fly ash (%)	Sand ratio (%)	Water reducing agent (%)	Air entraining agent (1/10 000)	VC value (s)	Water content	Cement	Fly ash	Apparent density			
30	Jin'anqiao	$C_{90}20W8F100$	0.47	55	37	1.0	20	1–5	100	96	117	2 600	P·MH42.5	Manufactured basalt, rock powder and substituted sand, 5%–8%	Grade II fly ash
31	Gongguoqiao	$C_{180}20W10F10$	0.46	50	38	0.8	5	3–7	100	109	109	2 420	P·MH 42.5	Manufactured sandy slate and rock powder >18%	Grade II fly ash
32	Guandi	Lower part $C_{90}25W10F100$	0.45	55	36	0.8	12	3–7	102	102	125	2 640	P·MH 42.5	Manufactured basalt aggregate and rock powder, controlled as per 16%–20%	Grade II fly ash
33		Upper part $C_{90}20W8F100$	0.48	55	37	0.8	12	3–7	102	102	96	2 640			

Note: 1. The mix proportion of RCC is construction mix proportion, and the VC value is the control value of outlet, under dynamic control.
2. In the table, F, PT and SL refer to fly ash, phosphorus slag and tuff, and iron ore slag and limerock powder respectively.

Table 4.3   Maximum allowable value of water-cement ratio of concrete

Location	Severe cold region	Cold region	Warm area
Above the upstream and downstream level (exterior of the dam)	0.50	0.55	0.60
Upstream and downstream level varying area (exterior of the dam)	0.45	0.50	0.55
Under the upstream and downstream level (exterior of the dam)	0.50	0.55	0.60
Foundation	0.50	0.55	0.60
Internal	0.60	0.65	0.65
Parts scoured by current	0.45	0.50	0.50

Note: Under the condition of environmental water erosion, the maximum allowable water-cement ratio of concrete outside the water level change area and underwater shall be reduced by 0.05.

The selection of the optimal sand ratio is to ensure that the fresh RCC mixture has good rollability, liquefied bleeding and interlayer bonding, and meets the VC value of construction requirements, and the sand ratio corresponding to the minimum unit water consumption is the optimal sand ratio.

The sand ratio has a great influence on the workability, working performance and water consumption of concrete. If the sand ratio is too high, the specific surface area of sand will increase, and the thickness of cement mortar layer playing the role of lubricating aggregate will be weakened. Therefore, the mixture will be dry and thick, with less fluidity; otherwise, under the condition of keeping the relative fluidity, it is necessary to increase the amount of cement mortar, namely, the amount of water and cementing materials, thus increasing the cost. If the sand ratio is too small, the amount of mortar in the voids of aggregate will be insufficient, resulting in poor fluidity of RCC, especially poor cohesiveness, water retention and rollability, difficult rolling and compaction, thus affecting the construction performance, strength, durability and some other performances of concrete. Therefore, the optimal sand ratio must be selected when the mix proportion parameter of RCC is determined.

The selection of optimal sand ratio of RCC is quite different from that of normal concrete. The selection of optimal sand ratio is more complicated, which is directly related to the type and content of admixture and additive, VC value and rock powder content. As for the outgoing fresh RCC, observe whether the mixture is muddy and uneasy to become loose by stepping and holding and squeezing; through the consistency test of mixture, i.e. VC value test, observe the speed of liquefied bleeding on the surface of VeBe consistometer. At the same time, invert the VC value sample after the test, and then observe the surface of the sample. If the mixture is smooth, dense and elastic, it can be preliminarily determined as the optimal sand ratio. The optimal sand ratio of RCC shall be selected with reference to the mix proportion experience of similar projects, and closely around the core technology of interlayer bonding, so that the RCC after pouring and paving must be fully bleeding, elastic and bright after rolling with vibrating roller, so as to ensure that the upper aggregate is embedded into the rolled and compacted concrete of the lower layer.

According to the statistical analysis of mix proportion and sand ratio of RCC in some projects in Table 4.1 and Table 4.2, project cases show that: Ⅲ gradation sand ratio of manufactured aggregate is generally 32% – 34% and that of Ⅱ gradation is 36% – 38%; the sand ratio of natural sand is generally 2% – 3% lower than that of manufactured sand. The main factors affecting the sand ratio are related to the type and shape of coarse aggregate, the grain gradation and fineness modulus of sand, especially the content of manufactured sand and rock powder. Based on a large number of engineering practices, the content of rock powder in sand has a significant impact on improving the workability of RCC.

4.2.3 Selection of Water Consumption Per Unit

Unit water consumption: it refers to the water consumption per cubic meter of concrete (subject to the dry state of saturated surface of sand and gravel aggregate), known as unit water consumption, referred to as water consumption. The unit water consumption is expressed in W.

The selection principle of unit water consumption is to minimize the unit water consumption on the premise of meeting the workability of fresh RCC, that is, meeting the VC value, rollability and liquefied bleeding requirements. Water consumption is directly related to the VC value of RCC, and there is a good correlation between water consumption and VC value. With the increase of water consumption, the VC value of RCC decreases. A large number of test results indicate that: When the VC value increases or decreases for 1 s, the water consumption decreases by about 1.5 kg/m^3.

The selection of unit water consumption is directly related to the rollability and cost-effectiveness of concrete. According to the statistical analysis of project cases in Table 4.1 and Table 4.2: the range of unit water consumption of RCC is large. Ⅲ gradation and Ⅱ gradation unit water consumption is 78 – 106 kg/m^3 and 83 – 110 kg/m^3 respectively. Main factors affecting unit water consumption are the type of aggregate, type and content of admixture and additive, content of rock powder, VC value, climate and other construction conditions.

For example, when the manufactured aggregate of natural aggregate and limestone is used, the water consumption of RCC is low; when the manufactured aggregate of diabase, basalt and other hard igneous rocks is used, the water consumption and admixture content of RCC increase significantly; when high-quality Grade Ⅰ fly ash is used, the water consumption of RCC can be significantly reduced under the water-reducing effect of small water demand ratio. High content of rock powder and low VC value can significantly improve the interlayer bonding quality of RCC, which is beneficial to the improvement of frost resistance and ultimate tensile value, but the demand for water consumption and admixture content is correspondingly increased.

4.2.4 Selection of Grout-mortar Ratio

Grout-mortar ratio: It refers to the ratio of the volume of RCC grout (cement, fly ash, water and micro rock powder with particle size less than 0.08 mm) and mortar volume, i.e. grout-mortar volume ratio, referred to as grout-mortar ratio (PV), expressed in PV value. Practice has proved that PV value is generally no less than 0.42, which increases not only the content of mortar in the mixture and fills voids of sand, but also the rollability of RCC mixture and the interlayer bonding strength.

The PV value of grout-mortar ratio is one of the important parameters of RCC mix proportion design. The grout-mortar ratio is directly related to the interlayer bonding, imperviousness performance and overall performance of RCC. The grout-mortar ratio is as important as water-cement ratio, sand ratio, unit water consumption and other parameters. In the mix proportion design of RCC, the PV value of grout-mortar ratio gets more and more attention. According to the practical experience of full-section RCC damming in recent years, when the content of rock powder in manufactured sand is controlled at about 18%, the grout-mortar ratio is generally not lower than 0.42. It can be seen that the grout-mortar ratio intuitively reflects a proportional relationship among RCC materials, and is an important parameter for evaluating the rollability, liquefied bleeding, interlayer bonding, and aggregate separation resistance of RCC for mixtures and construction performance.

Eigenvalues α and β are mainly adopted for early RCC mix proportion design. α is the grout filling coefficient, which is the ratio of grout volume to mortar pore volume, reflecting the situation that water, cement and admixture fill the sand voids; β is the mortar filling coefficient, which is the ratio of mortar volume to aggregate pore volume, reflecting the situation that water, cement, admixture and sand fill the coarse aggregate voids. According to construction experience, α is generally controlled as 1.1 – 1.3, and β is generally controlled as 1.2 – 1.5. As eigenvalues α and β are affected by sand grain gradation, rock powder content, aggregate size and other factors, the calculation conditions are complex, which leads to a large amplitude of eigenvalues α and β, and the evaluation is not intuitive. Therefore, in recent years, *PV* value of grout-mortar ratio has been the main parameter of mix proportion design and an important index to evaluate the performance of RCC.

It is different whether the air content is considered in the calculation of *PV* value of grout-mortar ratio. The *PV* value of grout-mortar ratio calculated by considering the air content is larger, while that calculated without considering air content is smaller. The author believes that the grout-mortar ratio can be calculated by absolute volume method without considering the air content. The reason is that RCC is semi-plastic concrete without slump. On one hand, the air content of the mixture is low, and the air is not easy to introduce; on the other hand, the air content of the mixture is extremely unstable and the air will be released quickly after vibration rolling. It is practical to calculate *PV* value of grout-mortar ratio by absolute volume method without considering air content.

Practices have proved that in recent years, the minimum *PV* value is generally not lower than 0.42, which can not only increase the grout content in the mixture to fill the voids of sand, but also effectively improve the rollability and interlayer bonding quality of the mixture. For example, the *PV* values of RCC applied in Baise, Longtan, Guangzhao, Gelantan, Jin'anqiao and Kalasuke are all higher than 0.42. In core drilling of RCC dam, the super-long core samples over 15 m or 16 m obtained from the above projects are the best evidence. On the other hand, due to the lack of deep understanding and research on the grout-mortar ratio, the interlayer bonding quality is very poor in the early "RCD" RCC damming technology. For example, in some early RCC dams, due to the poor research and understanding of admixtures and rock powder, the *PV* value of grout-mortar ratio is rather low. After the dam has been built, the interlayer leakage is serious, and the core samples taken out from boreholes are only 0.3 – 0.6 m long, and the recovery rate is low.

The main factor affecting *PV* value of grout-mortar ratio is the content of rock powder. According to the statistical analysis of Table 4.1 and Table 4.2: Generally, the dosage of RCC cementing material is 150 – 170 kg/m^3 inside the dam and 190 – 210 kg/m^3 outside the dam. If the rock powder content is not considered, the *PV* value of grout-mortar ratio is only 0.33 – 0.37 according to the calculation, which cannot guarantee the interlayer bonding quality of RCC. RCC dam is concreted by concreting in whole-block and thin-lifts without longitudinal joints, and the dam body has no longitudinal joints. Due to the requirements for temperature control and crack resistance, the role of rock powder in RCC is very important on the premise that it is impossible to increase the dosage of cementing materials, especially the micro-rock powder smaller than 0.08 mm, and the rock powder could increase cementing mortar to a certain extent. If the content of rock powder (particles smaller than 0.16 mm) in sand reaches about 18%, and the micro-rock powder smaller than 0.08 mm accounts for more than 30% of the rock powder, the *PV* value of grout-mortar ratio is generally not less than 0.42 as per calculation.

The research results for many years show that the mechanical index requirements of RCC still can be met if the content of rock powder is further increased to 22%. When the content of rock powder in sand is low and the *PV* value is lower than 0.42, the sand is generally replaced with rock powder or fly ash, so as to increase the content of rock powder or fine powder in sand to about 18%, which can significantly improve the mixture performance and construction performance of RCC.

4.2.5 Selection of Admixtures

Admixture: Admixtures refer to the active materials such as fly ash, double admixture (slag + limestone, phosphorous slag + tuff), volcanic ash, iron slag powder, phosphorous slag powder, silicon powder, etc., which are the main components of RCC cementing materials. Admixtures can be expressed by F, SL, PT, etc.

In the research and application of RCC admixtures, fly ash is always dominant as RCC admixture, so the research on the action mechanism of fly ash in RCC is deepened and the application is mature. Due to lack of fly ash or long transportation distance in a few projects, double admixture is used as RCC admixture according to technical analysis and comparison, which have achieved good results. For example, phosphorous slag + tuff mixed grinding admixture (PT) is adopted in Dachaoshan Project, while iron slag + rock powder admixture (SL) is adopted in Gelantan, Jinghong and Tuka River projects.

It can be seen from Table 4.1 and Table 4.2 that the RCC admixture is mainly II gradation fly ash, the fly ash content in III gradation RCC in the dam is generally 55% – 65%, and the fly ash content of Grade II RCC in the dam impervious area is generally 50% – 55%. The main factors affecting the fly ash content are RCC design indexes, age, and fly ash quality. Especially, the frost resistance grade, ultimate tensile value and strength grade do not match, and the design age is mostly 90 d (180 d in a few projects), which restricts the utilization of the later strength of fly ash concrete, increases the temperature control burden of the dam and limits the fly ash content. A large number of RCC dam practices have proved that the strength of RCC almost exceeds the design strength, and the surplus is large, which improves the elastic modulus of concrete, and increases the temperature stress of dam body, and is unfavorable to crack resistance. The core technology of rapid RCC damming is "interlayer bonding, temperature control and crack resistance", which is closely related to the quality and content of fly ash and other admixtures. Therefore, the selection of admixtures plays a very important role in the mix proportion design of RCC. It is one of the key technologies in the mix proportion design of RCC to conduct in-depth test and research on the variety, quality and content of admixtures.

4.2.6 Selection of Additives

Additive: additives refer to the inorganic materials, organic materials or inorganic & organic materials with no more than 5% of the mass of cementing materials added in concrete mixing, which aim at changing the workability of concrete, improving the strength and durability of the concrete.

Additive is one of the most important technical measures to improve the performance of concrete, which can effectively reduce the unit water consumption, reduce the amount of cementing materials, facilitate temperature control and improve the durability. In particular, additives have a remarkable effect on improving the performance, rollability, liquefied bleeding and interlayer bonding of RCC mixture. Additive has been an indispensable and important component of RCC.

It can be seen from Table 4.1 and Table 4.2 that RCC in China is mixed with additives, mainly naphthalene-based retarding high-range water reducing agent and air entraining agent. In recent years, the rapid RCC damming technology has developed vigorously in China, and RCC dam has been the mainstream dam type. The requirements for frost resistance of the dam have been continuously improved, and the factors such as low *VC* value and high rock powder content required for construction and interlayer bonding have made the additive content significantly higher than that of early RCC. The increase of additive content can effectively control the water consumption of RCC, reduce *VC* value, and improve the rollability, liquefied bleeding and interlayer bonding quality, which is one of the most important technical measures for RCC mix proportion design.

The frost resistance grade is an extremely important design index for the durability of RCC. Now the design index of frost resistance grade is becoming higher and higher. The RCC design indexes in Table 4.1 and Table 4.2 show that the frost resistance grade of Grade Ⅲ RCC inside the dam is generally F50 – F100, the frost resistance grade of Grade Ⅱ RCC outside the dam or in impervious area is generally F100, and the frost resistance grade outside the dam or in impervious area in severe cold areas has reached F300. For example, the frost resistance grade of RCC of external impervious areas in Longshou, Shimenzi, Kalasuke projects in severe cold areas is F300. In order to ensure the frost resistance of RCC, air entraining agent must be added in RCC to meet the frost resistance requirements. Because of the high rock powder content of RCC, air entrainment is more difficult than that for normal concrete. To reach the design air content, the content of air entraining agent is often several times that of normal concrete.

There are often some misunderstandings about the content of additives. The increase of the content of retarding high-range water reducing agent is considered to be the low water reducing rate; the content of air entraining agent is several times that of normal concrete, which is considered to have an impact on the strength and other properties of RCC, but actually there is no impact. The analysis of the main factors that increase the content of RCC additive compared with the past shows that:

First, the design indexes of RCC have been improved, for example, the frost resistance grade has been significantly improved compared with the past.

Second, the properties of raw materials such as RCC aggregate varieties have a significant impact on the additive content.

Third, it is required by the construction characteristics of RCC, such as high rock powder content, low VC value and retarding time.

The frost resistance is directly related to the air content of RCC. Air entraining agent must be added into RCC to meet the design frost resistance. Different frost resistance grade indexes have different requirements for air content. The frost resistance grade is F100, and the air content shall be controlled at 3.0% – 4.0%. The frost resistance grade is F150 – F200, and the air content shall be controlled at 4.0% – 5.0%. The frost resistance grade is F300, and the air content shall be controlled at 5.0% – 5.5%. Due to the high content of fly ash and rock powder in RCC, a large number of project tests show that fly ash and rock powder could strongly adsorb the air entraining agent. In order to meet the requirements of frost resistance grade, according to different aggregate varieties, the content of air entraining agent is generally 10/10 000 – 30/10 000, which can meet the air content control requirements of fresh concrete. The content of air entraining agent should be controlled according to the air content control requirements, and the outgoing air content of RCC should be controlled according to the upper limit.

Aggregate varieties have a great influence on additive content, for example, Diabase manufactured aggregate is used in Baise Project, while basalt manufactured aggregate is used in Jin'anqiao and Guandi projects. Diabase and basalt are igneous rocks, and the density of such hard rock aggregates is high, resulting in the apparent density of mixed RCC reaching 2 500 kg/m³, 2 630 kg/m³ and 2 660 kg/m³, which is rare in previous projects. Especially the technical problems of RCC caused by hard aggregates, such as high water consumption, poor rollability and liquefied bleeding, and low frost resistance and ultimate tensile value. A large number of tests show that increasing the additive content is the most effective technical measure to solve the problems above. The content of retarding high-range water reducing agent in RCC poured in Baise, Jin'anqiao and Guandi projects is 1.5%, 1.2% and 1.1% respectively, and the content of air entraining agent is

15/10 000 – 30/10 000 respectively. The unit water consumption of Grade Ⅲ and Grade Ⅱ CRR is 90 – 96 kg/m³ and 100 – 106 kg/m³ respectively, which is still relatively high.

However, when limestone aggregate RCC is used, the additive content and water consumption are obviously reduced. For example, limestone manufactured aggregate is used in RCC poured in Linhekou, Longtan and Guangzhao projects, with apparent density of RCC around 2 450 kg/m³, the content of retarding high-range water reducing agent is generally 0.6%, the content of air entraining agent is below 10/10 000, and the unit water consumption of Grade Ⅲ and Grade Ⅱ CRR is less than 80 kg/m³ and 85 kg/m³ respectively.

Construction practices have proved that dynamic control of additive content can effectively improve the interlayer bonding quality. Because the mix proportion design is carried out under the standard indoor temperature and meteorological conditions, there is a great difference with the field construction conditions. RCC construction is affected by the conditions of region, temperature, weather and different time periods, and the fluctuation of raw material quality. In order to make the RCC mix proportion meet the construction under various conditions, the mix proportion parameters are generally kept unchanged, and the performance of the mixture is controlled by adjusting the additive content. For example, during the high temperature period, additive content is increased in the daytime with sunshine and dry air, which can effectively reduce the *VC* value, prolong the retarding time, improve the rollability, improve the interlayer bonding quality, and keep the water-cement ratio unchanged due to the double action of water reducing and retarding by the additive.

4.2.7 Amount of Cementing Materials

Cementing material: Cementing materials refer to the material composed of cement and admixtures in RCC or mortar per unit volume, which can be expressed by C + F, C + PT, C + MH, etc.

The amount of cementing materials is directly related to various properties of RCC. Cement accounts for a small proportion in the cementing materials of RCC, while admixture accounts for a large proportion. It can be seen from Table 4.1 and Table 4.2 that the RCC in the dam body is of Grade Ⅲ, the amount of cementing materials with a strength grade of $C_{90}15$ MPa is generally 150 – 170 kg/m³, of which the cement consumption is 55 – 65 kg/m³; the amount of cementing materials with a strength grade of $C_{90}20$ MPa is generally 180 – 190 kg/m³, of which the cement consumption is 70 – 90 kg/m³; the amount of cementing materials in Grade Ⅱ concrete in impervious area is generally 190 – 220 kg/m³, of which the cement consumption is 85 – 100 kg/m³; the main factors affecting the amount of cementing materials are related to the design indexes of RCC (especially frost resistance grade and ultimate tensile value), aggregate variety, admixture quality and construction conditions.

With regard to the amount of cementing materials, *Construction Specification for Hydraulic Roller Compacted Concrete* DL/T 5112—2009 stipulates that the amount of cementing materials for RCC of permanent structures shall not be lower than 130 kg/m³. The amount lower than 130 kg/m³ should be demonstrated by a special test. There are always some one-sided understandings and misunderstandings about the amount of cementing materials in RCC. The meanings of RCC with rich cementing materials and RCC with rich cementing grout materials in RCC are different. Cementing materials mainly refer to the effective cementing materials composed of cement + active admixture, and the rich cementing mortar materials mainly include the micro rock powder below 0.08 mm in the cementing materials, but it is not considered in the calculation of water-cement ratio. A large number of project practices and test research results show that:

1. The permeability of RCC with rich cementing materials is very close to that of normal concrete, and their discreteness is basically the same.

2. When the amount of cementing materials reaches a certain value, the permeability of Grade Ⅲ RCC and Grade Ⅱ RCC is basically equal.
3. With the increase of RCC cementing mortar materials, the permeability of RCC gradually decreases, which not only shows the decrease of permeability, but also shows the decrease of dispersion coefficient.
4. When the amount of cementing materials (170 – 190 kg/m^3) and cementing mortar materials reach a certain value (PV value is above 0.42), the results of in-situ shear test of RCC show that the friction coefficient f' and cohesion c' are high, and the discreteness is very small.
5. The greatest effect of rock powder is to increase the PV value of RCC grout-mortar ratio, increase the content and volume of cementing mortar and improve the interlayer bonding quality.
6. The application of cementing mortar material with high rock powder content and RCC with low VC value eliminates the aggregate separation, which leads great changes in vertical transportation of RCC. Large full pipe chutes have replaced negative pressure chutes with low transportation strength and complex operation.
7. The RCC permeability cumulative curve in the pump-in test shows that the construction quality inspection standard is reasonable when 0.5 Lu is used as the design index of permeability of impervious area for RCC with the amount of cementing materials above 170 kg/m^3.

For example, in the mix proportion design of RCC in the early stage of Jin'anqiao Project, the content of rock powder in manufactured sand used in the scientific research and tests is low, and the VC value is controlled within 5 – 12 s. Because of the low content of rock powder and the large VC value, the elastic modulus of RCC is high, and the frost resistance and ultimate tensile cannot meet the design requirements. However, in the field RCC construction mix proportion test, the content of power stone in manufactured sand is controlled within 18% – 20% in the mix proportion design. In view of the low content of rock powder in basalt manufactured sand, the sand is replaced with the additive rock powder, and the VC value in outlet is controlled within 1 – 3 s. The PV value of grout-mortar ratio is over 0.44 as per calculation. The test results show that the slump-free semi-plastic RCC with rich cementing mortar materials significantly improves the construction performance of RCC, and also, the frost resistance and ultimate tensile are significantly improved, meeting the design requirements.

RCC with rich cementing mortar materials does not mobilize more cement but increases the content of rock powder and improve the grout-mortar ratio. For example, many dams such as Linhekou, Baise, Longtan, Guangzhao, Kalasuke, Jin'anqiao dams, do not mobilize high amounts of RCC cement and effective cementing materials. Therefore, we should have a comprehensive and correct understanding of RCC with rich cementing materials and RCC with cementing mortar materials.

4.2.8 Selection of VC Value

VC value: VC value refers to the pouring performance of RCC mixture, and the time measured by a VeBe consistometer in seconds (s).

The pouring performance of RCC, namely VC value, is one of the most important parameters of RCC mixture performance and mix proportion design. The VC value has a significant impact on the performance of RCC. The key points for the control of on-siteplacing surface are VC value and initial setting time, and VC value is the key to the rollability and interlayer bonding of RCC. The VC value in outlet should be adjusted in time according to the change of temperature and construction conditions, and should be dynamically controlled. In recent years, a large number of project practices show that it is more appropriate to control the on-site VC value within 2 – 8 s. The VC value in outlet should be changed according to the climatic conditions in dynamic selection and control within 2 – 5 s.

For example, in Fenhe River II Reservoir, the VC value is 2 – 4 s when the temperature exceeds 25 ℃ in summer; the VC value of Longshou Project is 0 – 5 s according to the characteristics of dry climate and large evaporation in Hexi Corridor; for Jiangya, Mianhuatan, Linhekou, Baise, Longtan, Guangzhao, Jin'anqiao and other projects, the VC value is mostly 1 – 5 s when the temperature exceeds 25 ℃. Because of the small VC value, RCC has good rollability from pouring to rolling, and the lower surface of the RCC can keep good plasticity before the upper RCC is poured. The control of VC value is based on the full bleeding and "elasticity", and the upper aggregate is embedded into the lower concrete after rolling.

4.2.9 Determination of Rock Powder Content

Rock powder: Rock powder refers to the particles smaller than 0.16 mm (0.075 mm abroad) in manufactured sand, or the processed particles smaller than 0.08 mm.

The latest *Construction Specification for Hydraulic Roller Compacted Concrete* DL/T 5112—2009 stipulates that the content of rock powder in manufactured sand should be controlled within 12% – 22%, and the optimum content of rock powder should be determined by tests. In recent years, a large number of project practices have proved that the optimum rock powder content of RCC is mostly controlled at about 18%. The optimum rock powder content needs to be determined by the mixture performance test and on-site rollability test.

The content of rock powder has a significant influence on the performance of RCC, and the greatest effect of rock powder is to increase the PV value of RCC grout-mortar ratio, and effectively improve the rollability and interlayer bonding quality; rock powder can play the role of micro-aggregate, improve the filling and compacting, and increase the volume of cementing mortar materials; when PV value is lower than 0.42, the sand is generally replaced with rock powder, so as to increase the content of rock powder in sand to about 18%, which can significantly improve the construction performance of RCC.

With the deepening research on the function of rock powder, more and more attention has been paid to rock powder in the mix proportion design of RCC. In the mix proportion design, the accurate method is adopted to control the calculation of the optimum rock powder content, which obviously improves the accuracy of the mix proportion design parameters of RCC and the coincidence of the test results.

Why should the content of rock powder be controlled by accurate method in mix proportion design? A large number of test results show that the water consumption of RCC increases by about 2 kg/m^3 for every 1% increase of rock powder content, and the fluctuation of rock powder content has a great influence on the water consumption of RCC; if the water consumption is constant, the VC value will fluctuate with the change of rock powder content, which will seriously affect the rollability; if the VC value is controlled according to the construction requirements, it will lead to the increase or decrease of water consumption, the change of water-cement ratio and the change of concrete strength, which is not allowed in quality control. Therefore, a large number of test results show that the determination of the optimum rock powder content is the key to ensure the stability of RCC mix proportion design parameters.

The test results are different with different test methods of rock powder content, and there is a big difference between dry method of rock powder test and water washing method of rock powder test. For example, the diabase manufactured sand is used for RCC of Baise Project. According to *Test Code for Aggregates of Hydraulic Concrete* DL/T 5151, dry-process sieve analysis measurement is adopted for diabase manufactured sand. The content of rock powder measured by drying is 16% – 20%, with FM = 2.7 – 3.0; the wet-process sieve analysis measurement is adopted, that is, the dried and weighed 500 g diabase manufactured sand is firstly sieved in water with a 0.16 mm sieve, and then dried for measurement. The content of rock powder content is 20% – 24%, with FM = 2.6 – 2.9. As different dry process and wet process are used to measure the content of rock powder in manufactured sand, test results are quite different, which easily causes that the quality of manufactured sand cannot meet the standard requirements. Therefore, in view of the specific aggregate type, the measurement methods of the content of rock powder in manufactured sand need to be further studied, and the practical measurement methods and standards that can reflect the performance of RCC shall be formulated.

4.3 Design Basis and Content of Mix Proportion

The mix proportion design of RCC shall follow *Code for Mix Design of Hydraulic Concrete* DL/T 5330—2015, *Test Code for Hydraulic Concrete* (*6 Roller Compacted Concrete*) SL 352—2006 and *Construction Specification for Hydraulic Roller Compacted Concrete* DL/T 5112—2009. The mix proportion design of RCC is mainly based on tender documents of projects, and the mix proportion design must be in accordance with technical terms of tender documents.

China's project construction, regardless of the size of projects, is subject to the bidding and tendering system. The tender documents for technical terms of the general contract for water conservancy and hydropower projects clearly specify field tests: the contractor shall, in accordance with the provisions of the "technical terms of the contract", build its own on-site material laboratory and provide sufficient qualified personnel and equipment; carry out sampling test for materials used in projects, such as cement, aggregate, admixture, additive and steel, as well as other materials specified by the supervisor; conduct field special tests (not limited to) : concrete construction mix proportion test, RCC field technology test, etc. ; the concrete mix proportion of various types of structures must be determined by test.

Therefore, the mix proportion design test of RCC must be carried out in accordance with technical terms of tender documents. The raw materials (cement, admixture, aggregate, additive, etc.) used in the concrete mix proportion test, the maximum water-cement ratio of concrete in the mix proportion design, and the content of admixture shall not only meet the regulations and specifications, but also meet the relevant requirements of technical terms of tender documents.

According to the requirements of technical terms of tender documents, the contractor shall submit a RCC mix proportion test plan with at least different ages (according to the age required by technical terms of different projects) for 28 d before the concrete mix design test. Test reports of at least 7 d, 28 d, 90 d and 180 d shall be provided. The curve of the relationship between water-cement ratio and compressive strength at different ages (7 d, 28 d, 90 d and 180 d) shall be provided. Each curve shall have at least four test points, and the data of each test point shall be obtained from at least 3 groups of test results. For the RCC mixed with admixtures of various percentages, the corresponding relationship shall be listed.

The RCC mix proportion test report shall at least include (but not limited to) the following contents:
1. All materials used and detailed description of the test data.
2. Detailed description of the test method, procedure and equipment.
3. Component proportion, batching, mixing, test, moulding and curing.
4. Qualification of materials and equipment during tests.
5. Detailed description of test results.
6. Conclusions.

The content of RCC mix proportion test shall also be tested according to the requirements of Table 4.4, and the corresponding test data shall be submitted.

Table 4.4   Content of RCC mix proportion test

Test features required		Maximum aggregate size, 80 mm	Maximum aggregate size, 40 mm	Concrete age (d)					
		After wet sieve	After wet sieve	Newly mixing	7	28	90	180	360
VC value (working degree)		√	√	√					
Temperature		√	√	√					
Air content		√	√	√					
Apparent density		√	√	√		√			
Water bleeding				√					
Time of setting	Initial setting	√	√	√					
	Final setting	√	√	√					
Compressive strength		√	√		√	√	√	√	√
Splitting tensile strength		√	√		√	√	√	√	√
Shear strength						√	√	√	√
Grade of impermeability		√	√				√	√	
Frost resistance grade		√	√				√	√	
Axial tensile strength		√	√		√	√	√	√	√
Ultimate tensile value		√	√		√	√	√	√	√
Modulus of elasticity under axial tension		√	√			√	√	√	√
Modulus of elasticity under static pressure		√	√			√	√	√	√
Creep test				√	√	√	√	√	√
Adiabatic temperature rise				√	√	√			

Note: "√" indicates that the test data shall be provided.

4.4 RCC Mix Proportion Design Method

4.4.1 Steps of Mix Proportion Design

The general steps of mix proportion design of RCC are as follows: preparation of test plan (according to design indexes of tender documents) →raw material test→mix proportion parameter selection test→relationship test of water-cement ratio and compressive strength → mix proportion design (preparation of strength and mix proportion design parameters) →mix proportion test (mixture performance, hardened concrete performance, etc.) →GEVR test→field rolling verification test of mix proportion.

What needs to be illustrated in the mix proportion design of hydraulic concrete is that aggregates are calculated according to the saturated and surface dry state. This is the biggest difference between the hydraulic concrete and the ordinary concrete in the mix proportion design, which needs to be emphasized in the mix proportion test. Under the condition of certain raw materials, the five main steps of selecting mix proportion design of RCC are as follows:

Step Ⅰ: preliminary selection of mix proportion design parameters. Preliminarily select mix proportion design parameters according to design indexes, type and content of admixtures, aggregate size, optimal gradation, VC value, etc., of RCC with reference to similar projects. It is necessary to adjust and select mix proportion parameters preliminarily designed through a large number of repeated mixture performance tests, such as water-cement ratio, water consumption, sand ratio, grout-mortar ratio, admixture content and rock powder content;

Step Ⅱ: test of relationship between water-cement ratio and compressive strength. Mix proportion parameters selected after trial mixing are subject to the test of relationship between water-cement ratio and compressive strength. At least three pieces of water-cement ratio test data and different admixtures are selected for the test to conduct the test of the curve of the relationship between water-cement ratio and compressive strength at different ages of 7 d, 28 d, 90 d and 180 d. The data of each test site shall be obtained from at least 3 groups of test results. At the same time, the development coefficient of the relationship between water-cement ratio of different admixtures and compressive strength and age is calculated;

Step Ⅲ: selection of mix proportion design parameters. Select mix proportion design parameters of RCC according to the curve of the relationship between water-cement ratio and compressive strength, and development coefficient of the relationship between strength and age, and with reference to design indexes (such as preparation strength, VC value, frost resistance grade and ultimate tensile value) and construction requirements;

Step Ⅳ: RCC mix proportion test determine the test mix proportion according to mix proportion design parameters, carry out the performance test of RCC mixtures, the strength, impermeability, frost resistance, ultimate tensile strength, modulus of elasticity, adiabatic temperature rise andothers of hardened concrete.

Step Ⅴ: site compaction verification test of mix proportion Determine the mix proportion of RCC according to the mixture performance and hardened concrete test results. Before the formal construction of dam RCC, the on-site raw materials and mixing system shall be used to carry out on-site productive RCC process test, and verify and adjust the construction mix proportion submitted, so that the determined mix proportion of RCC can meet the design and construction requirements.

4.4.2 Mix Proportion Design Method

The calculation method of mix proportion design of RCC is the same as that of normal concrete. Based on the saturated and surface dry state, the amount of each component of 1 m³ RCC is calculated by "absolute volume method" or "assumed apparent density method (hereinafter referred to as apparent density method, also known as mass method)".

4.4.2.1 Absolute Volume Method

Ⅰ Basic principles

1. The volume of 1 m³ fresh concrete mixture is equal to the sum of absolute volume of each component and air volume, there is:

$$1 = V_W + V_C + V_P + V_S + V_G + V_\alpha \tag{4-1}$$

That is: $1 = m_W/\rho_w + m_C/\rho_C + m_P/\rho_P + m_S/\rho_S + m_G/\rho_G + 0.01\alpha$ (4-2)

$$V_W = m_W/\rho,\ V_C = m_C/\rho_C,\ V_P = m_P/\rho_P,\ V_S = m_S/\rho_S,\ V_G = m_G/\rho_G,\ V_\alpha = 0.01\alpha$$

Where V_W, V_C, V_P, V_S, V_G and V_α = volume of water, cement, admixture, sand, stone and air in 1 m³ concrete, m³

m_W, m_C, m_P, m_S, m_G = mass of water, cement, admixture, sand and gravel in 1 m³ concrete, kg

ρ_w = density of water, kg/m³

ρ_C = density of cement, kg/m³

ρ_P = density of admixture, kg/m³

ρ_S = dry apparent density of dry saturated surface of sand, kg/m³

ρ_G = dry apparent density of saturated surface of stone, kg/m³

α = Air content percentage of concrete

2. Calculate the amount of cement and admixture according to the water-cement ratio and unit water consumption, namely:

Cement content: $\quad m_C = m_W / [m_W / (m_C + m_P)]$ (4-3)

Amount of fly ash: $\quad m_P = m_W / [m_W / (m_C + m_P)] \times$ Mixing amount(%) (4-4)

Where: $\quad m_W / (m_C + m_P) =$ water-cement ratio;

3. Calculate the mass of gravel according to the sand ratio, there is:

Sand mass: $\quad m_S = (V_S + V_G) \times (m_S / m_s + m_G) \times \rho_S$ (4-5)

Gravel mass: $\quad m_G = (V_S + V_G) \times [1 - (m_S / m_s + m_G)] \times \rho_G$ (4-6)

Where $V_S + V_G$ = volume of sand and gravel in 1 m³ concrete, m³

$m_S / (m_s + m_G)$ = Sand ratio, %

After the density, sand ratio and air content required for frost resistance of each material are known, the mass of sand and gravel can be calculated according to formulas (4-1) or (4-2).

Ⅱ Calculation steps of mix proportion

Step Ⅰ: calculate the amount of cementing materials. Calculate the amount of cementing materials of RCC according to the water-cement ratio and unit water consumption selected by the mix proportion design parameters, and respectively calculate the amount of cement and mixture according to the percentage of admixture;

Step Ⅱ: respectively calculate the volume of water, cement, mixture and air content according to the density of each material (The volume can be ignored due to the small content of additive);

Step Ⅲ: calculate the volume of aggregates according to the water consumption volume cementing material volume, air content volume and sand ratio, and then multiply it by the dry apparent density of dry saturated surface to calculate the aggregate mass.

Ⅲ Example of calculation of mix proportion by absolute volume method

The design index of Ⅲ gradation RCC for dams in a project is $C_{90}15W6F100$. Test condition: 42.5 moderate heat cement, Grade Ⅱ fly ash, limestone artificial gravel aggregate, 18% rock powder content, 0.6% of retarding high-range water reducing agent and 0.08% of air entraining agent; the density of water, cement, fly ash, sand and stone is 1 000 kg/m³, 3 200 kg/m³, 2 200 kg/m³, 2 600 kg/m³ and 2 700 kg/m³ respectively; mix proportion design parameters: The water-cement ratio, sand ratio, water consumption, fly ash content, air content and VC value is 0.50, 33%, 83 kg/m³, 60%, 3% (controlled by 3% -4%) and 3 -5 s respectively; aggregate grading: small-sized stones: medium-sized stone: large-sized stone = 30: 40: 30. The mix proportion is calculated by absolute volume method:

Cement content $\quad m_C = m_W / [m_W / (m_C + m_P)] \times (1 - 60\%) = 83/0.50 \times 40\% = 66.4 (\text{kg/m}^3)$

Amount of fly ash: $\quad m_P = m_W / [m_W / (m_C + m_P)] \times 60\% = 83/0.50 \times 60\% = 99.6 (\text{kg/m}^3)$

Water volume: $\quad V_W = m_W / \rho_w = 83/1\,000 = 0.083 (\text{m}^3)$

Volume of cement: $\quad V_C = [(m_C + m_P) \times (1 - 60\%)] / \rho_C = 66.4/3\,200 = 0.020\,8 (\text{m}^3)$

Volume of fly ash: $\quad V_P = [(m_C + m_P) \times 60\%] / \rho_P = 99.6/2\,200 = 0.045\,3 (\text{m}^3)$

Volume of air content: $\quad V_\alpha = 1\,000 \times \alpha = 1\,000 \times 3\% = 0.030 (\text{m}^3)$

Volume of aggregate: $\quad V_S + V_G = 1 - (V_W + V_C + V_P + V_\alpha)$

$\qquad = 1 - 0.083 - 0.020\,8 - 0.045\,3 - 0.030$

$$= 0.821 \, (\text{m}^3)$$

Calculate the mass of sand according to the sand ratio: $m_S = (V_S + V_G) \times m_S/(m_s + m_G) \times \rho_S = 0.821 \times 33\% \times 2\,600 = 704 (\text{kg/m}^3)$

Calculate the mass of stone according to the sand ratio: $m_G = (V_S + V_G) \times [1 - m_S/(m_s + m_G)] \times \rho_G = 0.821 \times 67\% \times 2\,700 = 1\,485 (\text{kg/m}^3)$

Conduct aggregate gradation according to the mass of stone: Small-sized stone: medium-sized stone: large-sized stone = 30 : 40 : 30, calculate the amount of aggregate with each particle size: small-sized stone = $1\,485 \times 30\% = 446(\text{kg/m}^3)$; medium-sized stone = $1\,485 \times 40\% = 594(\text{kg/m}^3)$; large-sized stone = $1\,485 \times 30\% = 446(\text{kg/m}^3)$.

4.4.2.2 Assumed Apparent Density Method

I Basic principles

The mass of 1 m³ fresh concrete mixture is equal to the sum of mass of each component, namely:

$$1 \text{ m}^3 \text{ apparent density} = m_W + m_C + m_p + m_S + m_G \tag{4-7}$$

Where m_W, m_C, m_P, m_S and m_g = mass of water, cement, admixture, sand and gravel in 1 m³ concrete, kg

II Calculation steps of mix proportion

Step I: assume the apparent density of 1 m³ mixture according to the experience of similar projects or calculate the apparent density of 1 m³ concrete by absolute volume method with the density of each known component. The assumed apparent density is decimal.

Step II: calculate the amount of gravel. Calculate the amount of cementing materials according to the water-cement ratio and water consumption, and then calculate the mass of gravel according to the assumed apparent density. The amount of gravel = 1 m³ apparent density - (the amount of cementing materials + water consumption). The mass can be ignored due to the small content of additive);

Gravel mass: $\qquad m_S + m_G = 1 \text{ m}^3 \text{ Apparent density} - (m_W + m_C + m_p) \tag{4-8}$

Sand mass: $\qquad m_S = (m_S + m_G) \times m_S/(m_s + m_G) \tag{4-9}$

Gravel mass: $\qquad m_G = (m_S + m_G) \times [1 - m_S/(m_s + m_G)] \tag{4-10}$

Where $m_S + m_G$ = dry saturated surface mass of gravel, kg/m³

$m_S/(m_s + m_G)$ = Mass sand ratio, %

Step III: adjust and modify the apparent density of unit volume concrete according to the calculated apparent density of 1 m³ concrete through the apparent density test of concrete mix proportion, to determine the amount of aggregate for construction mix proportion.

The mix proportion of RCC calculated with the absolute volume method is shown in Table 4.5.

III Example of calculation of mix proportion with the assumed apparent density method

The design index of III gradation RCC for dams in a project is $C_{90}15W6F100$. Test condition: 42.5 moderate heat cement, Grade II fly ash, limestone artificial gravel aggregate, 18% rock powder content, 0.6% of retarding high-range water reducing agent and 0.08% of air entraining agent; mix proportion design parameters: the water-cement ratio, sand ratio, water consumption, fly ash content, air content and VC value is 0.50, 33%, 83 kg/m³, 60%, 3% (controlled by 3% - 4%) and 3 - 5 s respectively; aggregate grading: small-sized stones: medium-sized stone: large-sized stone = 30 : 40 : 30. The mix proportion is calculated with the apparent density method. According to the apparent density of raw materials, the apparent density of III gradation limestone aggregate RCC is initially selected as 2 450 kg/m³ upon calculation and reference to the apparent density of similar projects. The apparent density method is used to calculate the following items:

Cement content: $\quad m_C = m_W/[m_W/(m_C + m_P)] \times (1 - 60\%) = 83/0.50 \times 40\% = 66.4 (\text{kg/m}^3)$

Amount of fly ash: $\quad m_P = m_W/[m_W/(m_C + m_P)] \times 60\% = 83/0.50 \times 60\% = 99.6 (\text{kg/m}^3)$

Gravel consumption: $m_S + m_G = 1 \text{ m}^3$ Apparent density $- (m_W + m_C + m_p)$
$$= 2\,450 - (83 - 66.4 - 99.6) = 2\,201(\text{kg/m}^3)$$

Sand mass: $m_S = (m_S + m_G) \times m_S/(m_s + m_G) = 2\,201 \times 33\% = 726(\text{kg/m}^3)$

Gravel mass: $m_G = (m_S + m_G) \times [1 - m_S/(m_s + m_G)] = 2\,201 \times (1 - 33\%) = 1\,475(\text{kg/m}^3)$

Conduct aggregate gradation according to the mass of stone: small-sized stone: medium-sized stone: large-sized stone = 30:40:30, calculate the amount of aggregate with each particle size: small-sized stone = $1\,475 \times 30\% = 443 \text{ kg/m}^3$; medium-sized stone = $1\,475 \times 40\% = 590 \text{ kg/m}^3$; large-sized stone = $1\,475 \times 30\% = 443 \text{ kg/m}^3$.

The mix proportion of RCC calculated with the assumed apparent density method is shown in Table 4.5.

Table 4.5　　　　　　　　　Mix proportion of Ⅲ gradation $C_{90}15W6F100$ RCC

Design methods	Mix proportion parameter					Material amount (kg/m³)							Apparent density (kg/m³)
	water-cement ratio	Sand ratio (%)	Fly ash (%)	Water reducing agent (%)	Air entraining agent (%)	Water	Cement	Fly ash	Sand	Stone	Water reducing agent	Air entraining agent	
Absolute volume method	0.50	33	60	0.6	0.08	83	66.4	99.6	704	1 485	0.996	0.133	2 439
Assumed density method	0.50	33	60	0.6	0.08	83	66.4	99.6	726	1 475	0.996	0.133	2 450

4.4.2.3 Determination of Apparent Density of Concrete

According to Table 4.5, different calculation methods are adopted for mix proportion design. The mix proportion parameters, and the consumption of water and cementing materials are the same for the absolute volume method and the assumed apparent density method. Due to the different apparent density of concrete, there are small differences in the mass of aggregates. The two calculation methods of mix proportion show that the apparent density condition of raw materials must be met when the absolute volume method is used to calculate the mix proportion. Meanwhile, the absolute volume method is more complex than the assumed apparent density method during the calculation of mix proportion.

The actual design and calculation of mix proportion are carried out according to three steps of absolute volume method, apparent density method and apparent density test of fresh concrete, which are interrelated. Steps for determination of apparent density of general construction mix proportion concrete are as follows:

Step Ⅰ: firstly calculate the apparent density of 1 m³ concrete by absolute volume method according to the apparent density of raw materials and the air content required for frost resistance. It shall be noted that the determination of air content of concrete has a great influence on the determination of apparent density. The volume of concrete with 1% air content is 0.01 m³, and the mass of concrete is about 24 – 25 kg/m³. Secondly, the influence of fluctuation of apparent density of raw materials is also one of the factors affecting the apparent density of concrete.

Step Ⅱ: the initially selected apparent density of concrete must be verified by the apparent density test of fresh concrete, to verify the accuracy of the designed apparent density. When the apparent density obtained from the test is greatly different from the designed apparent density, it is necessary to carry out the correction calculation. Correction coefficient = assumed apparent density/tested apparent density; corrected apparent density = assumed apparent density + (assumed apparent density × correction coefficient). See details in *Code for Mix Design of Hydraulic Concrete* DL/T 5330—2005.

Step Ⅲ: when the apparent density of concrete is determined through design and calculation of the mix proportion with the absolute volume method, the apparent density obtained is often not an integer. Because concrete is heterogeneous material, test results of a plate of fresh concrete are different when the apparent density test is carried out. Due to the loss of slump or *VC* value caused by gradual loss of fresh concrete, the air content decreases correspondingly, and the fluctuation of air content directly affects the apparent density. Therefore, when the apparent density of concrete is selected, the construction mix proportion is often an integer. Firstly, Test Code for Aggregates of Hydraulic Concrete requires the apparent density of aggregates to be 10 kg/m^3; secondly, the apparent density of concrete under mass control is rounded, which is convenient for the test personnel to remember and calculate; finally, the allowable deviation of concrete mass control for water, cementing materials and additives is ±1%, and that of aggregates is ±2%. Considering the above factors, in the construction mix proportion design, the apparent density of concrete construction mix proportion is generally an integer to 10 kg/m^3 and the larger value.

4.4.3 Trial Mixing Adjustment of Mix Proportion

The concrete mix proportion submitted should be adjusted in time according to the change of construction site conditions and the fluctuation of raw materials. However, key parameters such as water-cement ratio, unit water consumption and admixture are generally not allowed to be adjusted; generally, the sand ratio, rock powder content, gradation, additive content and other parameters are adjusted according to the law of the relation between mix proportion parameters according to the FM, rock powder content, aggregate under-size, temperature and air content change.

When the hydraulic concrete is mixed, aggregates shall be calculated according to the saturated and surface dry state. As aggregates are often not in saturated and surface dry state in actual condition, when the water content of aggregates exceeds the water absorption rate of the dry saturated surface or the absolute dry condition of aggregates cannot reach the saturated and surface dry state, in the concrete mixing, the unit water consumption of concrete shall be calculated according to the actual aggregate water content, deducting the aggregate surface water content or supplementary water absorption amount of dry saturated surface.

During the trial mixing adjustment, the mixer must be used, and the minimum mixing amount should not be less than 1/3 of the rated mixing amount of the mixer; at the same time, the minimum mixing amount shall be selected according to the maximum aggregate size. The general aggregate size is no less than 80 mm, and the minimum mixing amount shall not be less than 40 L.

4.4.4 Discussion on Mix Proportion Design

The mix proportion design of RCC is characterized by high technical content, long test cycle, high labor intensity, rich experience, etc. As raw materials and test conditions are largely different from raw materials used on site and construction conditions in the early stage of mix proportion design of RCC, the field application is a dynamic control and adjustment process after the determination of RCC mix proportion. According to the requirements of specifications and tender documents, the undertaker organizes experts to review RCC mix proportion submitted, and verify the conformity, rollability, interlayer bonding and other properties of RCC mix proportion through productive process test.

For example, a scientific research unit undertakes the RCC mix proportion test in cold and dry areas with large evaporation capacity in Northwestern China. Raw materials used for the mix proportion are transported to the place with high humidity in the south for test. When the mix proportion submitted is rechecked at the

Northwest project site, there is a big difference, which cannot meet work requirements of RCC mixtures. The main reason is different climatic conditions, causing that the water consumption, VC value and setting time of RCC are quite different from the actual situation in the field, which has been proved for many times by projects.

As for mix proportion design of RCC, preliminarily determine the water-cement ratio, sand ratio, water consumption, admixture content and other parameters according to the design indexes, raw materials and construction methods with reference to the mix proportion of similar projects. The key is that the water consumption shall meet the VC value of construction performance requirements of different grades and rock powder content. Due to the influence of raw materials, especially aggregate type, the water consumption of some projects is very high in order to meet the workability requirements of fresh RCC. Therefore, the water consumption and the amount of cementing materials must be controlled within the design range, temperature control and cracking prevention shall be carried out from the source, and the cement consumption shall be reduced, which is the key to the mix proportion design of RCC. Among measures such as controlling water consumption, reducing the amount of cementing materials and improving the construction performance of RCC, increasing the content of rock powder and additive is the most effective technical measure upon demonstration in a large number of project cases of RCC mix proportion.

Chapter 5

Research and Application of RCC Performance

5.1　Overview

RCC materials conform to the water-cement ratio rule of the concrete, and have the same performance as the normal concrete in strength, durability, density, etc. ; applying RCC rapid damming technology only changes the mix proportion and construction methods. In order to study the performance of hydraulic RCC, it is necessary to study and discuss the core technology of "interlayer bonding, temperature control and cracking prevention" based on the technical characteristics of full-section RCC rapid damming technology.

Hydraulic RCC is defined as semi-plastic concrete without slump, and its connotation is completely different from the definition of RCC in the early stage as super hard or hard concrete. There are different views and misunderstandings in the industry. The main reason is that the understanding of RCC rapid damming technology is still in the early concept. In the early stage, the RCC damming technology in China mainly absorbs foreign experience, and the impervious system is subject to the design concept of "RCD". Therefore, the hydraulic RCC in the early stage, similar to the RCC of highway and airport, is provided with large VC value, its strength is the main control index, and its performance is mainly used to bear strength load, without impervious function. Different functions and purposes of RCC decide its different definition. Therefore, RCC in the early stage is defined as super hard or hard concrete.

In 1993, the full-section RCC damming technology was firstly innovated in Puding Arch Dam in China, changing the impervious system of "RCD"; the dam completely depended on the RCC for impervious. Since then, the full-section damming technology is adopted for RCC dams. Due to the layer (joint) structure of RCC dams, the anti-sliding stability, impervious performance and interlayer bonding quality between layers (joints) have always been concerned by people. Therefore, RCC is required to be with good impervious performance, and to meet the requirements of anti-sliding stability and impervious performance between layers (joints). The interlayer bonding quality has become the key core technology of RCC rapid damming. The changes of RCC performance promote significant improvement of the interlayer bonding quality of RCC, and enable that the VC value of fresh RCC is getting smaller and smaller, which requires that the surface of fresh RCC must be fully bleeding and elastic from mixing, transportation, placing, paving to complete rolling, so as to ensure that the upper aggregates are embedded into the rolled and compacted RCC of the lower layer, which is an inevitable result required for RCC performance by the impervious performance of dams.

The performance of RCC is basically the same as that of normal concrete, but there are some differences. Compared with normal concrete, the performance of RCC: RCC is with a little cement and more admixtures and rock powder content, and its early strength is lower than that of normal concrete, but its later (long-age) strength increases significantly; the early deformation performance and durability of RCC are also inferior to that of normal concrete, which is mainly shown in the fact that the test results of ultimate tensile value and frost resistance of RCC are not as good as those of normal concrete; the adiabatic temperature rise of

RCC is obviously lower than that of normal concrete. The adiabatic temperature rise of RCC inside the dams is generally 15 – 18 ℃, which is very beneficial to the temperature control and cracking prevention of the dams.

As for the low strength and inferior ultimate tensile value and frost resistance of RCC in the early stage to that of normal concrete, by analysis, it is believed that: on the one hand, the long-age design index equal to the strength is not used in the design, and the characteristic of significant increase of strength of the RCC with more admixtures in the later stage is not fully utilized; on the other hand, it is about the impact of material composition and mix proportion, and in-depth study on mix proportion design and performance of mixtures is insufficient. At present, with the change of RCC definition, the ultimate tensile value, frost resistance grade and others of RCC have been greatly improved, and the difference between RCC and normal concrete has been significantly reduced, and the strength, durability and others of RCC are as good as that of normal concrete. Therefore, in the design, the concrete design index of dams is not reduced for selecting RCC. In recent years, RCC dams have become the mainstream dams and their height has reached 200 m, which is closely related to the improvement of RCC performance.

RCC performance mainly includes mixture performance, mechanical property, deformation, durability, thermal property, etc. These five properties are both related and mutually constrained, and closely related to design and construction. Factors affecting the performance of RCC are more complex, but they are mainly related to cement type, quality and content of admixtures, aggregate type, mix proportion design, construction process, etc., especially, RCC design age, rock powder content, *VC* value, additive quality and content and others have a great impact on the performance of RCC. The performance of fresh RCC mixtures is the most important performance of hydraulic RCC. Based on the premise that the performance of mixtures meets the construction requirements of rollability, liquefied bleeding and interlayer bonding quality, the hardened RCC is required to be with higher ultimate tensile value, lower modulus of elasticity, and good impermeability and frost resistance. In the performance test of RCC, except that the performance of mixtures is quite different from that of normal concrete, the hardened concrete test is the same as that of normal concrete, which can be carried out according to the *Test Code for Hydraulic Concrete* SL 352—2006. Main properties of hydraulic RCC include:

1. Mixture properties. *VC* value, air content, setting time, apparent density, etc.
2. Mechanical property. Compressive strength, splitting tensile strength, shear strength(calculated with c), etc.
3. Deformation properties. Ultimate tensile value (tensile strength), modulus of elasticity, creep, autogenous volume deformation, dry shrinking (wet bulging), etc.
4. Endurance quality. Impermeability, frost resistance, etc.
5. Thermal properties. Adiabatic temperature rise, temperature conductivity, thermal conductivity, specific heat coefficient, etc.

The chapter focuses on the perfect combination of RCC construction performance and hardened concrete performance through the analysis on performance research and influencing factors of RCC based on project cases.

5.2 Properties and Influencing Factors of RCC

5.2.1 Properties and Influencing Factors of RCC Mixtures

5.2.1.1 Properties of RCC Mixture

Properties of hydraulic RCC mixture are a basic guarantee for physical mechanics and durability of RCC. The properties of the mixture are directly related to rapid construction and interlayer bonding quality of RCC. It must have good workability and be convenient for construction, so as to ensure the good interlayer bonding quality. At the same time, it shall have the good physical mechanics and durability to ensure that the structure can safely bear the design load after the mixture is hardened.

Compared with the normal concrete mixture, the RCC mixture contains more aggregates and less cement. Although a large amount of fly ash and other admixtures are mixed, the amount of cementing material is still less than that of normal concrete. The mixture has no fluidity, poor cohesiveness, and no bleeding. To ensure that the RCC mixture has fluidity, it is necessary to overcome a larger yield stress than the normal concrete mixture. Therefore, the vibrating roller must be adopted for construction. In addition, if the mix proportion design is improper, the rock powder content is small and the VC value is large, it is easy to cause aggregate separation of the fresh RCC during discharging in the mixing structure and on site.

All modern RCC dams are based on the full-section RCC damming technology. "Interlayer bonding" has become the key core technology for rapid RCC damming. It is emphasized that the rolled and compacted layer surface must be fully bleeding and elastic, so as to ensure that the rolled and compacted aggregates of the upper layer are embedded into the lower concrete. It is required that the mixture shall bear the load of the vibrating roller during rolling compaction and traveling, without sinking. Therefore, the RCC mixture with good working performance shall have the VC value compatible with the construction equipment and environmental conditions (such as temperature and relative humidity); during pouring and paving, the mixture is uneasily separated and the retardation time is long; it can produce appropriate plastic deformation under the effect of the vibrating roller, and has good liquefied bleeding and rollability; at the same time, it is easy to compact under the effect of mechanical compaction of vibrating roller.

Main properties for hydraulic RCC mixture include: VC value, air content, apparent density, setting time, etc.

Ⅰ VC value

VC value of RCC is an extremely important index to measure the work and construction performance of the RCC mixture. The work is expressed by the VC value with s as the unit of measurement. RCC mixture is a kind of semi-plastic concrete without slump. The slump is zero, and it can be kneaded into a silt lump. The traditional slump test method cannot determine the VC value, so the VC value test must rely on auxiliary forces to complete the liquefied bleeding process. For many years, VeBe consistometer is still mainly used for measurement in the VC value test of RCC. At present, there is no method to evaluate all the characteristics of the RCC mixture.

Method of Measurement of Concrete Mixture with VeBe Consistometer ISO 4110—1979 is a standard accepted by the International Organization for Standardization. The equipment has the vibration frequency of 50 ± 3 Hz, no-load amplitude of 0.5 ± 0.1 mm, and nominal acceleration of $5g$. Measure the VC value of RCC with the VeBe consistometer. According to the *Test Code for Hydraulic Concrete* SL 352—2006, the RCC mixture is sieved through the 40 mm wet sieve, and then loaded in the measuring cylinder in two layers. The time (in s) required for the mixture from vibration to bleeding under the conditions of constant vibration frequency, amplitude and pressure (pressure of 4 900 Pa and gross mass of 17.75 kg). The VC value represents the energy required for the mixture vibration. The shorter the time, the easier it is to be vibrated. The VC value of fresh RCC is mainly evaluated by observing the speed of liquefied bleeding, the smoothness of the specimen appearance and the uniformity of the specimen section during the test. At present, the VC value is still mainly determined by experience to some extent.

Construction Specification for Hydraulic Roller Compacted Concrete DL/T 5112—2009 implemented from December 1, 2009 stipulates: "The VC value of RCC mixture shall be 2 – 12 s on site. The VC value in outlet should be changed according to the climatic conditions in dynamic selection and control within 2 – 8 s.

Ⅱ Air content

1. Air content effect. The purposeful entraining of 3.0% – 5.5% air into concrete can significantly improve

the durability and other properties of concrete. The most fundamental way to improve the frost resistance of RCC is to mix with the air entraining agent, so as to produce a large number of disconnected tiny bubbles in the concrete, thus improving the pore structure of concrete. The air entraining agent, a kind of surface active material, can entrain a large number of discontinuous tiny bubbles during concrete mixing, with the pore size of mostly 50 – 200 μm. The distribution is even and the concrete contains certain air. When the air content of RCC reaches 3.0% – 5.0%, it can effectively improve the working performance, frost resistance and impermeability of RCC. At the same time, it can also improve the toughness and deformation of concrete, and is useful for improving thermal performance.

Since 1990s, the design has taken the frost resistance grade as an important index for evaluating the durability of concrete in order to improve the durability of dam concrete. In cold regions in the north or subtropical mild regions in the south, the frost resistance grade has become an indispensable index for hydraulic concrete design. It is difficult to entrain air due to mixing of the large proportion of admixtures and characteristics of RCC mixtures. A large number of tests and studies have shown that the air entraining agent content shall increase in multiples (often more than ten or even dozens of times the content of normal concrete), so that RCC mixture has the same air content as normal concrete. Such as Longshou, Kalasuk, Jin'anqiao RCC dams, the content of air entraining agent is generally 0.10% – 0.30% of the cementing materials (0.007% – 0.012% for normal concrete content), so that the air content can reach 3.5% – 5.5%, thus ensuring that the frost resistance performance of RCC can meet F100 – F300 requirements.

The air content of RCC has the same performance as normal concrete. The content of air entraining agent in RCC is mainly based on the frost resistance grade and the maximum aggregate size, and the air content is determined through test. *Specifications for Hydraulic Concrete Construction* DL/T 5144—2001; when the frost resistance grade is no less than F200, the maximum aggregate size is 20 mm, 40 mm, 80 mm and 150 mm (120 mm) respectively, and the air content for reference is 5.5%, 5.0%, 4.5% and 4.0% respectively; when the frost resistance grade is no more than F150, the maximum aggregate size is 20 mm, 40 mm, 80 mm and 150 mm (120 mm) respectively, and the air content for reference is 4.5%, 4.0%, 3.5% and 3.0% respectively.

2. Air content effect. The air content has a great influence on the performance of RCC, especially in the mixture performance. With effect of tiny bubbles and certain volume, the air entraining agent not only reduces the water consumption and sand ratio, but also significantly improves the performance of the RCC mixture.

Air content has certain effect on dry shrinking that is mainly controlled by unit water consumption. The amount of cement in the mixture has a few impacts on dry shrinking except for the impact on water demand. The dry shrinking increases with the increase of water consumption, and the dry shrinking increases with the increase of the air entraining content. But mixing of the air entraining agent reduces the water consumption, without reducing the VC value, so the net shrinking shall not increase significantly.

If the air content of concrete is constant, the decrease in strength shall vary with the change of the maximum aggregate size, and the decrease in strength of concrete with large particle size shall be less. It shows that the lower grade RCC with more mortar has high air content, which has a greater impact on its strength.

The air content also has certain influence on the thermal properties. In ordinary mixtures, changes in air content, cement and water content have no significant effect on thermal properties. However, the thermal conductivity coefficient of hardened concrete is in inverse proportion to the air content and direct proportion to the amount of water and cement.

The air content has an effect on apparent density. The decrease in the apparent density of concrete is in direct proportion to the air content. If the air content increases, the apparent density decreases. Therefore, the optimal air content is interrelated with the strength, apparent density, frost resistance grade and performance of mixtures.

Ⅲ Time of setting

The setting time is an extremely important index for the performance of the RCC mixture, and directly related to the rapid construction, rollability and interlayer bonding quality of RCC. The setting time is divided into initial setting time and final setting time. The initial setting time indicates that RCC begins to harden, and is the limit to determine the construction time of RCC from concrete mixing to placing surface rolling compaction; the final setting time indicates that the concrete has hardened, and the mechanical strength of hardened concrete starts to develop rapidly.

There is a big difference in the setting time between RCC and normal concrete. The setting time of RCC is often shorter than that of normal concrete due to characteristics of RCC. With construction characteristics of the full surface pouring in a whole-block and thin-lift and continuous rolling compaction of RCC, the setting time of RCC is required to be long to ensure construction. Especially in high temperature periods or seasons, it is required to prolong the setting time. Generally, the initial setting time of fresh RCC is no less than 12 h to ensure rolling compaction of RCC from mixing to initial setting and subsequently continuous pouring and rolling compaction of the upper layer of RCC. Otherwise, the subsequent RCC can be constructed after it is treated as construction joints.

There are many factors that affect the setting time of RCC, including the compatibility of cement and additives, aggregate type, VC value, and construction environment temperature, wind speed, and sunshine and other factors. The loss of the setting time of fresh RCC is very fast under the high temperature period or the sun exposure. Especially for the hard aggregates with large density, although the RCC density is great, the setting time of the mixture is severely shortened. For example, diabase and the basalt aggregates are adopted for the RCC of Baise Dam and the Jin'anqiao Dam, the RCC apparent density respectively reach 2 650 kg/m^3 and 2 630 kg/m^3, while the setting time is severely shortened. By admixture optimization and increase of admixture content, the setting time of the mixtures is effectively prolonged, so as to ensure the rapid construction of RCC and the interlayer bonding quality.

Ⅳ Apparent density

The mass per unit volume of the RCC mixture is called the apparent density (kg/m^3). The apparent density is the basis for the material consumption designed and calculated in RCC mix proportion, the key index to evaluate the relative compaction of RCC on site, and a means to test the quality and the air content of mixtures. The RCC apparent density is also the important parameter for the design of concrete dam, and is very important for the concrete gravity dam which keeps stable by the volumetric weight, its apparent density is the value range of the volumetric weight, which determines the section size and the anti-sliding stability of the dam (see Section 2.2.4, Chapter 2 for details).

For the aggregates with large density, the good gradation and increased particle size can effectively increase the apparent density of the concrete. In case that the volume of the specimen cannot be calculated accurately, the drainage method can be used to quickly calculate the apparent density of the hardened concrete. The apparent density of the RCC mixture is related to the aggregate density, the gradation and the air content, and also related to the compaction degree of on-site construction. A large quantity of engineering projects proves that the apparent density of RCC is larger than that of the normal concrete, which mainly because the mix proportion and the construction method of RCC are quite different from these of the normal concrete. As the RCC mixture is in a state of semi-plastic concrete without slump, its cementing material consumption and the unit water consumption are less than these of the normal concrete, and the aggregate consumption is more than that of the normal concrete. In addition, the RCC construction is performed by full surface pouring in a thin-lift and whole-block with vibrating mill for rolling; its compaction is significantly superior to that of the normal concrete.

Due to the different aggregate varieties adopted, the RCC apparent density varies significantly. In general, the apparent density of RCC is 2 400 – 2 450 kg/m^3, while the apparent density of RCC with diabase and basalt can reach to 2 600 – 2 660 kg/m^3, and its density is very large.

The determination of the reference apparent density of RCC is the important basis to test the compactness and the relative compaction on site.

Reference apparent density: It is the average value of the large apparent density values obtained from the RCC with the given mix proportion in the laboratory test.

Degree of compaction: It refers to the ratio of the measured apparent density of the construction placing surface and the average reference apparent density obtained from the RCC laboratory test.

Relative compaction: It is an index to evaluate the compaction quality of RCC. For the external concrete of structure, the relative compaction should be no less than 98%; for the external concrete, the relative compaction shall be no less than 97%.

Determination methods of the reference apparent density: first, it is calculated by the absolute volume method, according to the design parameter of the RCC mix proportion and the density and air content of the raw materials, the material consumption of per cubic meter concrete can be calculated as the reference for trail mix; second, according to the reference mix proportion, the apparent density of mixtures can be verified by the mixing test; third, according to the control requirement and the scientific notation of mixtures, the reference apparent density of the concrete shall be kept in integer and take 10 kg/m^3 as the criterion.

The apparent density has a certain impact on the compressive strength of the hardened concrete. For example, the RCC main dam of the Baise Multipurpose Dam Project applies the diabase aggregate, and during the test of the RCC compressive strength, the relationship between the volumetric weight (the apparent density) and the compressive strength of the diabase aggregate concrete specimen (150 mm cube) has been tested, the result is shown in Table 5.1. The result shows that the volumetric weight and the strength of the hardened concrete specimen with the design indexes, ages, gradations from the same kind of the concrete have a certain relationship, that is, generally, the specimen with large volumetric weight will have high strength. Numbers of Specimens in the table with Nos. T-2228, T-2229 and TX-841 are $R_{28}15$ quasi-III gradation normal concrete, the 28 d compressive strength are respectively 16.5 MPa, 16.7 MPa and 17.2 MPa, and the corresponding measured volumetric weight are 2 581 kg/m^3, 2 587 kg/m^3 and 2 613 kg/m^3. The data shows that the magnitude of the apparent density has a certain impact on the concrete strength. The apparent density of the hardened $R_{180}15$ III gradation RCC specimen is basically in the range of (9 ± 0.1) kg/m^3. The test result also verifies again that the size of the water-cement ratio and the amount of the fly ash content are main factors to influence the concrete strength.

5.2.1.2 Influence of *VC* Value on RCC Performance

I *VC* value change process

A quantity of engineering practice proves that the *VC* value has a significant impact on the RCC performance. Over twenty years, the *VC* value of RCC has developed from the exploration stage that referred to and copied the value overseas in the early stage to the current stage of innovation and maturity, the control of *VC* value is gradually from large extent to small extent, and the RCC mixture gradually transits from the hard concrete to the semi-plastic concrete without slump, which changes the traditional "RCD" construction method and the impervious structure, so as to bring great changes to the RCC rapid damming technology.

Chapter 5 Research and Application of RCC Performance

Table 5.1 Test results of relationship between volumetric weight and compressive strength of concrete specimen with diabase aggregate (The specimen is 150 mm cube)

Test piece No.	Strength grade	Age (d)	Gradation	Single block mass (kg)	Single block load (kN)	Single block strength (MPa)	Average strength (MPa)	Measured volumetric weight (kg/m^3)	Designed volumetric weight (kg/m^3)
R-2555	$R_{28}20$ Rolling	28	II	8.865	314	14.0	13.8	2 726	2 600
				8.790	292	13.0			
				8.955	322	14.3			
R-2556	$R_{180}15$ Rolling	28	Quasi III	8.920	196	8.7	8.6	2 632	2 650
				8.875	183	8.1			
				8.850	200	8.9			
T-2228	$R_{28}15$ Normal state	28	Quasi III	8.655	376	16.7	16.5	2 581	2 560
				8.795	378	16.8			
				8.680	362	16.1			
T-2229	$R_{28}15$ Normal state	28	Quasi III	8.775	382	17.0	16.7	2 587	2 560
				8.650	378	16.8			
				8.770	370	16.4			
TX-841	$R_{28}15$ Normal state	28	Quasi III	8.875	440	17.8	17.2	2 613	2 560
				8.760	382	17.0			
				8.700	378	16.8			
R-2041	$R_{180}15$ Rolling	180	Quasi III	9.005	582	25.9	25.8	2 685	2 650
				9.005	574	25.5			
				9.180	584	26.0			
R-2043	$R_{180}15$ Rolling	180	Quasi III	9.060	578	25.7	25.7	2 686	2 650
				9.110	562	25.0			
				9.025	594	26.4			
R-2044	$R_{180}15$ Rolling	180	Quasi III	9.000	516	22.9	23.4	2 681	2 650
				9.015	522	23.2			
				9.130	540	24.0			
B-150	$R_{180}15$ Rolling	180	Quasi III	8.725	500	22.2	22.8	2 593	2 650
				8.730	508	22.6			
				8.800	528	23.5			
B-151	$R_{180}20$ Rolling	180	II	8.980	580	25.8	25.9	2 640	2 600
				8.875	586	26.0			
				8.875	584	26.0			
R-2049	$R_{180}15$ Rolling	180	Quasi III	9.290	542	24.1	22.6	2 711	2 650
				9.100	502	22.3			
				9.060	480	21.3			
R-2048	$R_{180}15$ Rolling	180	Quasi III	9.105	544	24.2	23.8	2 667	2 650
				8.950	520	23.1			
				8.950	540	24.0			
Spot inspection R-218	$R_{180}15$ Rolling	180	Quasi III	9.175	566	25.2	23.7	2 705	2 650
				9.110	496	22.0			
				9.115	540	24.0			

Construction Specification for Hydraulic RCC DL/T 5112—2009 (hereinafter referred to as "Version 2009 Specification") is implemented on December 1, 2009, and this is the fourth revision of the Construction Specification for Hydraulic RCC. In order to meet the needs of the RCC rapid damming technology, Specification in 2009 Version focuses on the quality and safety of hydraulic RCC construction in the revision content, so as to reflect the new level of the hydraulic RCC construction. The revision combines the practice of the hydraulic RCC construction in recent years, fully considers the suggestions and opinions in all aspects, and revises most chapters, especially makes major revision on the VC value of RCC mixtures pouring performance. Version 2009 Specification stipulates: "The VC value of RCC mixture shall be 2 – 12 s on site. The VC value in outlet should be changed according to the climatic conditions in dynamic selection and control within 2 – 8 s.

When the RCC construction technology was introduced in 1980s, the concept of high VC value overseas was introduced simultaneously, i. e., (20 ± 5) s. More than 10 RCC dams that represented by the Japanese Shimajigawa Project, the VC value of its RCC pouring performance stipulates that within the range of (20 ± 10) s, the RCC mix proportion and VC value of some Japanese projects are shown in Table 5.2.

Table 5.2   RCC mix proportion and VC value of some japanese projects

Project title	Maximum particle size (mm)	Water-cement ratio	Fly ash (%)	Sand ratio (%)	VC value (s)	Material amount (kg/m³)					Type of aggregate
						W	C	F	S	G	Additive	
Upper Part of Shimajigawa Dam	80	0.81	30	34	20 ± 10	105	91	39	749	1 476	0.325	Manual
Lower Part of Shimajigawa Dam	40	0.88	30	34	20 ± 10	105	84	36	752	1 482	0.30	Manual
Tamagawa Dam	150	0.73	30	30	20 ± 10	95	91	39	657	1 544	0.325	Manual
Okawa Dam	80	0.85	20	32	20 ± 10	102	96	24	686	1 500	0.3	Manual
Shin-nakano Dam	80	0.79	30	34	20 ± 10	95	84	36	723	1 415	0.3	Manual
Meili River	80	0.75	30	30	20 ± 10	90	84	36	668	1 588	0.3	Natural
Zhenye Dam	80	0.85	20	32	20 ± 10	102	96	24	726	1 552	0.3	Manual
Sun Ogawa	80	0.85	20	32	20 ± 10	102	96	24	706	1 500	0.3	Manual
Jingchuan	80	0.88	30	34	20 ± 10	105	84	36	752	1 582		Manual

The RCC in China has developed since early 1980s, it is still in the early exploration stage with the VC value basically referring to the regulations of Japan, America and other countries. The VC value specified in *Interim Regulations on Construction of Hydraulic RCC* SDJ—86 issued in 1986 is (20 ± 5) s; with the development of the RCC damming technology, the *Construction Specification for Hydraulic RCC* SL—94 revised in 1994 stipulates: The VC value in outlet shall be selected within 5 – 15 s; with the continuous development and extensive use of the RCC full-section damming technology, the new *Construction Specification for Hydraulic RCC* DL/T 5112—2000 (hereinafter referred to as "Version 2000 Specification") was reissued in 2000, and the Version 2000 Specification stipulates: The design pouring performance (VC value) of RCC mixture can be selected in 5 – 12 s, and the VC value in outlet shall be changed according to the climatic conditions in dynamic selection and control with the outlet value within 5 – 12 s.

The national *Construction Specification for Hydraulic RCC* has been revised for four times, which directly reflects the change process of the RCC VC value. Therefore, RCC in the early stage is defined as super hard or hard concrete due to the large VC value. Too large VC value leads to loose and non-adhesive RCC mixtures, during the mixing, discharging and paving process of RCC, the aggregates are easily to be separated and large aggregates may be concentrated. Meanwhile, the RCC mixture is poor in the cohesiveness, liquefied bleeding, rollability and interlayer bonding, which causes the water seepage channels between the layers, that is, the "layered-cake". Therefore, the imperviousness of RCC has always been taken into consideration, the imperviousness structure of RCC in early stage is mainly the outsourced normal concrete with higher grade than that of RCC, that is the so-called "RCD" impervious structure.

The Version 2009 Specification does not just narrow the control range of the VC value, its importance is the dynamic control of the VC value which is more flexible and suitable to actual circumstances. This is a major liberation for the constraints to the development and innovation of RCC damming technology.

Ⅱ Dynamic control of VC value

The modification of VC value of the RCC mixtures in Version 2009 Specification intensively reflects the construction practice results over the years, also reflects the deepening understanding of the hydraulic RCC performance, stipulates that the VC value in outlet shall be changed according to the change of the on-site construction conditions in dynamic selection and control, and under the condition of satisfying the on-site normal rolling, the low value can be adopted for on-site VC value.

The author counted the RCC mix proportion and the VC values of major projects since late 1990s, and the details are shown in Table 4.1 and Table 4.2 in Chapter 4. As can be seen from the Tables, the VC value in outlet is significantly lower than the 5 – 12 s specified in the Specification Version 2000, the VC value is significantly reduced, while the control range of each project is not exactly the same. The analysis shows that: China features vast territory and diverse climate, the environment and the climate conditions at the project site vary significantly, causing the diverse climatic conditions such as high temperature, high cold, humid and rainy and dry with little rain in the area of the project. The air temperature, sunlight and wind speed during construction have great influences on the VC value of RCC. The implemented VC value must be dynamically controlled during the RCC construction according to the geographic positions, climatic factors, raw materials (such as the mixtures, the aggregate lithology and the rock powder content) and other actual situations in the area of the project.

The key point of RCC on-site control is the VC value and initial setting time of the mixture, the control of VC value is the key of RCC rollability and interlayer bonding, and the VC value in outlet shall be adjusted according to the changes of air temperature and other conditions. For example, in Fenhe River Ⅱ Reservoir, the VC value is 2 – 4 s when the temperature exceeds 25 ℃ in summer; the VC value of Longshou Project is 1 – 5 s according to the characteristics of dry climate and large evaporation in Hexi Corridor; For Jiangya, Mianhuatan, Linhekou, Baise, Longtan, Guangzhao, Jin'anqiao and other projects, the VC value is mostly 1 – 5 s when the temperature exceeds 25 ℃. Because of the small VC value, RCC has good rollability from pouring to rolling, and the lower surface of the RCC can keep good plasticity before the upper RCC is poured. The control of VC value is based on the full bleeding and "elasticity", and the upper aggregate is embedded into the lower concrete after rolling.

In early stage, when the smaller VC value is adopted in the RCC for rolling, the "spring concrete" will appear, which is regarded as an influence on the concrete compactness, strength and other performances, this is a one-sided misunderstanding. Because after all the RCC is a kind of concrete, it conforms to the water-cement ratio rule of the concrete, and the RCC will not cause voids and pits due to the short of vibration like the normal concrete does.

Ⅲ Factors influencing VC value

1. Water content. The water consumption has a great impact on the VC value, according to test results of projects, such as Dachaoshan, Jiangya, Mianhuatan, Linhekou and Baise, the VC value increases or decreases by 1 s, the water consumption will be increased or decreased by 1.5 kg/m^3 accordingly. In case of adjusting the VC value simply by increasing the water consumption, it will have an impact on the RCC water-cement

ratio and strength, if the mix proportion is unreasonable, the placing surface bleeding will be easily caused especially under the circumstance of low content of the rock powder in the fine aggregate.

2. Additive. Firstly, admixture that is suitable for the performance of RCC in the project shall be selected, by adjusting the content of the admixture, the VC value can be changed to improve the performance of the RCC mixtures, so as to satisfy the RCC construction under the conditions of different climate and temperature. For example: in Guangdong Shankou, Mianhuatan, Linhekou, Baise and other projects, parameters of the RCC mix proportion remain unchanged under the high temperature periods, the aim to change the VC value can be reached by adjusting the admixture content.

3. Grout-mortar ratio PV. The grout-mortar ratio indicates the ratio of volume of the grout (water + cement + mixtures + 0.08 mm rock powder) and the volume of mortar in RCC. The grout-mortar ratio is a critical parameter in the design of the RCC mix proportion, which has a great impact on the possibility of the RCC, and it is no less than 0.42 in general. In Baise Project, due to the diabase property, the rock powder content of manufactured sand is as large as 20% – 24%, in which the content of the micro rock powder under 0.08 mm is up to 40% – 60%. Such a content has obvious impact on improving the grout-mortar ratio of RCC, and the actual grout-mortar ratio is 0.45 – 0.47 by calculation. As quasi-Ⅲ gradation RCC is adopted for the main dam of Baise, the maximum particle size of the aggregate is 60 mm with large rock powder content of manufactured sand, small VC value of the RCC mixtures, large grout-mortar ratio, good cohesiveness, equally distributed aggregate, rapid liquefied bleeding, and good rollability without bleeding, that is what people always said, entering the pouring unit with rain shoes, it shows the sufficient bleeding of the placing surface, which reflects the good interlayer bonding quality of the RCC.

Ⅳ Measures to adjust VC value

Methods to adjust VC value: first, adding water to the mixing plant directly; Second, spraying water directly on the placing surface; third, adjusting the content of the admixtures. The first two methods can easily change the water-cement ratio of RCC, in which the mixtures is prone to be segregated and bleeding, and the interlayer bonding is unsatisfied.

As the RCC mix proportion is conducted in the standard condition of temperature and humidity as required by the procedures, the conditions of construction site vary greatly, in order to ensure the construction of the RCC under the adverse natural climatic conditions like high temperature or large drying evaporation, the dynamic control and adjustment of the mix proportion must be conducted. Under the high temperature, sunlight and large evaporation, there are two technical measures adopted in general: first, mix proportion parameters shall be kept constant, and the content of the retarding high-range water reducing agent shall be adjusted appropriately to delay the initial setting time and lower the VC value; second, measures like mist spray and spraying water by a grinding roller shall be taken to change the microclimate of the placing surface, so as to reduce the temperature, keep the humidity of the RCC surface and reduce the loss of the VC value. These can effectively ensure the cohesiveness, liquefied bleeding and rollability of the RCC mixtures, so as to improve the interlayer bonding, imperviousness performance and overall performance of RCC.

5.2.2 Strength Performance and Influence Factors of RCC

Strength is the most important mechanical index of the hardened RCC, and also the important evaluation index of RCC quality control. RCC strength performance mainly includes the compressive strength, splitting tensile strength, shear strength and others. The compressive strength is its most important performance, and the main index and basis of the design of a RCC dam. The main factor influencing RCC strength is related to the water-cement ratio, the mixture content and the compactness, it is closely related to the quality of the cement and the mixture, while the mixture is related to the design age of the compressive strength. The molding method

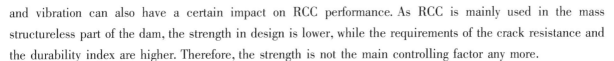

and vibration can also have a certain impact on RCC performance. As RCC is mainly used in the mass structureless part of the dam, the strength in design is lower, while the requirements of the crack resistance and the durability index are higher. Therefore, the strength is not the main controlling factor any more.

5.2.2.1 RCC Strength Performance

Ⅰ Compressive strength

The compressive strength of RCC is directly related to its compaction. Many researches regard the RCC mixture as the material like gravel soil, and study the relationship between the unit water consumption and the concrete apparent density of RCC; the relationship between the unit water consumption and the concrete compressive strength; The result shows that there is optimum water consumption. This indicates the relationship between the compressive strength and the compactness of RCC.

Ⅱ Splitting tensile strength

The splitting tensile strength (splitting tensile) of RCC is the same as that of the normal concrete, it decreases with the increase of the water-cement ratio, and increases with the increase of the compressive strength. Due to the complex tensile strength test, the splitting tensile strength is adopted to replace the tensile strength, with the value totally replacing the tensile strength, and the value tends to be conservative.

The concrete can bear the compressive stress with the allowable tensile strength rarely exceeding 10% of the compressive strength. In case the concrete being constrained, the tensile stress generated by the excessive restraint stress is bound to cause cracks.

The tensile strength is one of the most important factors influencing the concrete crack resistance, it is mainly composed of the tensile capacity of the cement mortar, the interface cementing property of the cement mortar and the aggregate, and the tensile capacity of the aggregate itself. The higher the concrete tensile strength is, the stronger its crack resistance is.

The ratio of the concrete tensile strength and the compressive strength is called the ratio of tension and compression. The ratio of tension and compression of the normal concrete is about 7% - 12% in general. The ratio of tension and compression of RCC (90 d), and RCC with high fly ash content is about 7% - 9%.

Ⅲ Shear strength

The shear strength is related to the anti-sliding stability of the dam body, and also related to the size of the dam section. In a sense, the shear strength of the dam body concrete is more important than the compressive strength. The shear strength of the RCC of the dam body has a variety of cases. The indoor and outdoor tests show that there is no big difference between the shear strength of RCC body and RCC inside the construction layer and that of the normal concrete. The concrete shear strength of the construction layer with continuous pouring (the layer poured continuously by RCC within the initial setting time) is equal to or inferior to that of the concrete inside the layer. If the layer has not been treated within the concrete initial setting time or there is the overhead aggregate position in the layer, the shear strength of the concrete is obviously lower than that of the concrete body. As the construction layer treated by cold joints (with the interval time between layers exceeding the concrete initial setting time, and mortar or grout is applied), the shear strength is lower than that of the concrete body, with the reduction degree related to the treatment measures of the layers.

The layer shear strength of RCC is a parameter concerned by the dam builders, especially in the high RCC dam. In order to improve the interlayer bonding quality of RCC construction, the improvement of the concrete shear strength of the construction layer is a key topic of the interlayer bonding quality research.

5.2.2.2 Principal Influence Factors of Strength

There are many factors influencing the RCC strength, including raw materials, water-cement ratio, aggregate performance, the variety, content and design age of different mixtures, molding condition, curing condition and other factors.

The uniformity of the RCC strength is the sign of successful on-site quality control. In case that the quality is not fully controlled during RCC production, the RCC strength will be changed greatly, so that the additional cement is required to ensure the concrete quality to comply with the design requirements. In addition, the more frequently the fineness (specific area) of the concrete changes, the more influences will be on the concrete compressive strength, and the uniformity controlled by the concrete will be also influenced.

Ⅰ water-cement ratio

For a long time, thousands of tests and in-depth research verify indisputably that the most primary factor influencing the concrete strength is the water-cement ratio. Under the condition of the maximum aggregate particle size with the constant air content, the concrete strength will directly change with the change of the water-cement ratio. A number of tests of the relationship between RCC compressive strength and water-cement ratio show that no matter the lean RCC or the rich gelling RCC with high fly ash, they are the same as the normal concrete that fully complies with the "water-cement ratio rule", i. e. , with the increase of the water-cement ratio, the compressive strength will decrease. Within a certain range, the compressive strength basically shows linear relation with the binder-water ratio $[(C+F)/W]$.

Ⅱ Aggregate impact

1. Sand ratio. When the sand fineness, content of the cementing material and the water-cement ratio are fixed, with the decrease of the sand ratio, the RCC compressive strength will increase gradually. While after the sand ratio decrease to a certain extent, the RCC strength will decrease. I. e. , there is a sand ratio that makes the RCC compressive strength maximum. This sand ratio is called the optimal sand ratio. When the sand ratio is large, the ratio of the grout volume of the cementing material and the sand void volume will be small, and when the sand ratio is small, the ratio of the mortar volume and the void volume of the coarse aggregate will be small, these all make the concrete compressive strength decrease.

2. Rock powder content in sand. When the sand ratio, the cementing material content and the water-cement ratio are fixed, with the increase of the rock powder content in sand, the RCC compressive strength will increase. For the rock powder contained in the manufactured sand, the rock powder content increase to 20%, the test result shows that it can still improve the compressive strength. The rock powder in the sand improve the construction performance of the concrete mixtures, the concrete mixtures are easy to be vibrated and compacted. The rock powder plays a role of the micro-aggregate in the hardened concrete, which improves the RCC strength.

3. Maximum particle size of coarse aggregate. Compared with the normal concrete, the gradation of coarse aggregates has a greater influence on the compaction effect of RCC. The apparent density of RCC increases with the increase of the maximum particle size of aggregates, but the increase of the maximum particle size easily causes the separation of coarse aggregates and affects the quality of rolling compaction. It must be pointed out that, the discontinuous gradation is not suitable for RCC because it has been proved that the aggregate separation of RCC subject to discontinuous gradation is serious. In recent years, the understanding of the maximum aggregate size of RCC is different from that of normal concrete, and the aggregate separation of RCC is closely related to the maximum particle size. For example, the quasi Ⅲ gradation with the maximum aggregate size of 60 mm is adopted in Baise Project, while the maximum aggregate size of 40 mm or 50 mm is used for RCC of Vietnam, Myanmar and Laos Projects, which significantly improves the aggregate separation of RCC.

Ⅲ Influence of formation vibration conditions on mechanical properties

Based on a large number of tests and studies and the author's practical experience in mix proportion design test for many years, the formation vibration time of RCC shall be based on the principle of full bleeding on the surface of the specimen, and shall not be controlled according to the time 2 – 3 times the *VC* value specified in the standard. The reason is that the test is a kind of research technical measure to simulate the actual construction, which must be consistent with the actual results, so as to provide a scientific basis for the quality and construction of the project.

RCC is based on the full-section damming technology. After the control over raw materials meets the requirements, RCC rolled and compacted is evaluated from the following aspects: mixing, transport, placing, paving, rolling compaction, spraying and curing. After rolling compaction, the RCC layer surface must be fully bleeding, with a thin-lift of grout. People feel elastic when walking on it, so as to ensure that the upper RCC aggregates are embedded in the lower concrete rolled and compacted, thus ensuring the interlayer bonding of RCC and improving the impermeability. Therefore, it is not suitable for spending the time 3 times the *VC* value on vibration formation of RCC specimens regardless of the strength, impermeability, frost resistance, ultimate tensile strength, modulus of elasticity, air content, apparent density, etc.

So it is questionable that, when RCC is subject to *VC* value test, RCC is fully bleeding within a very short period. Why is the time 2 – 3 times of *VC* value used in formation unable to meet the requirements of full bleeding?

Upon cause analysis: the VeBe consistometer and the vibrating table are fixed together to form a whole in the *VC* value test. During formation of RCC specimens, although the counterweight is added to the specimen mould, the vibration frequency and amplitude are completely different from that of *VC* value test when the VeBe consistometer is fixed with the shaking table due to separation of the specimen mould from the vibrating table. A large number of tests have proved that, the surface of RCC mixture is unable to be full bleeding when the time 2 – 3 times the *VC* value is used for control during specimen formation, which is inconsistent with the actual situation of full bleeding on construction site.

At the same time, due to the influence of boundary environmental conditions (such as water loss on the surface of the mixture and gradual loss of the *VC* value) during formation of RCC mixtures, vibration within the time 2 – 3 times the *VC* value is also one of the reasons that the surface of RCC cannot be fully bleeding. Therefore, the vibration control time of RCC shall be based on the standard of fully bleeding on the surface of the specimen.

1. Formation vibration time. According to test results, the compressive strength of concrete increases with the increase of formation vibration time. However, the strength of concrete will decrease if the vibration time is too long. This may be due to the excessive grout separated caused by excessive vibration, resulting in the reduction of internal grout and the stratification of aggregate, thus affecting the compressive strength.
2. Formation pressure. When the vibration time is constant, the compressive strength of concrete increases with the increase of specimen surface pressure. But when the pressure increases to certain extent, the compressive strength increases slightly. If the pressure increases too much, the block will jump during vibration formation, and the mixture cannot be compacted by vibration, resulting in decrease of compressive strength of concrete.
3. Vibration characteristics of formation vibration machinery. When the shaking table is used during formation of RCC, the vibration frequency and amplitude have an influence on the strength of RCC. When higher vibration is adopted for vibration, cement and admixture particles may be in larger relative movement, and the agglomeration structure of cementing materials will be disintegrated and liquefied. This is beneficial to improving the compaction of RCC. The test results of RCC specimens moulded by shaking table with the same frequency and different amplitudes show that the strength of RCC increases if the amplitude increases, because the liquefaction of the mixture increases with the increase of amplitude. But too large amplitude will

make the mixture jump, entraining air, which is not conducive to compaction by vibration.

Ⅳ Influence of curing on concrete performance

If concrete specimens are exposed to dry air without curing, their strength will stop increasing at the early age stage. After concreting, the strength of concrete exposed in dry air for 6 months is only half of that at the same age after wet curing for 14 d and exposure to dry air.

The curing temperature has an obvious influence on the strength growth. According to test results, the curing time at low temperature is longer than that at high temperature to reach the required strength. During continuous curing at high temperature for 28 d, the increase of strength changes with the change of temperature. During the age of 28 d, the higher the curing temperature, the higher the strength. But 28 d later, the trend is reversed.

Ⅴ Strength and age development coefficient

The strength is no longer an important index affecting the quality of RCC, because a large amount of admixtures (fly ash) are used in RCC, and its later strength increases significantly. On the other hand, the strength of RCC is controlled by other indexes. In order to meet the design index requirements of ultimate tensile value and frost resistance grade, RCC at the age of 90 d or 180 d needs to increase its strength appropriately to meet the requirements of crack resistance and durability index. Therefore, this is the main factor leading to the super strength of RCC.

The early compressive strength of RCC is lower and its increase is slower due to the use of more admixtures. After 28 d or 90 d, active silicon oxide, etc. in the admixture (for example, fly ash) and $Ca(OH)_2$ produced in the cement hydration process are in the secondary hydration reaction, and then the calcium silicate hydrate cement and other hydration products are generated, so that the hardened cementing material mortar is continuously compacted and the strength is continuously improved. Therefore, the strength growth rate of RCC at the age of 90 d and 180 d is much higher than that of RCC at the age of 28 d. See Table 5.3 for the average development coefficient of the compressive strength of RCC in some domestic projects. According to the results, compared with the compressive strength of RCC at the age of 28 d, the growth rate of that of 90 d is about 150% – 170%; the growth rate of that of 180 d is 180% – 220% compared with 28 d.

With the increase of fly ash content, the strength growth rate of RCC is remarkable in the later stage, showing that under the premise of meeting the design and construction requirements, the addition of fly ash significantly reduces the content of cement in concrete, which is very beneficial to the temperature control and crack resistance of RCC. At the same time, the results also show that the early strength of Jinghong and Gelantan RCC with double admixture NH (slag + rock powder) is high, but the strength development coefficient is obviously lower in the later period, which indicates that the strength growth of RCC with double admixture of slag and rock powder is significantly lower than that of fly ash RCC.

5.2.3 Deformation Properties and Influencing Factors of RCC

Deformation properties of RCC mainly include ultimate tensile value, modulus of elasticity, creep, dry shrinking, self volume deformation, Poisson's ratio, etc. The ultimate tensile value is an important index for crack resistance design of RCC.

5.2.3.1 Ultimate Tensile Deformation

The ultimate tensile strength of concrete refers to the maximum tensile deformation of concrete under tensile load. It is a very important factor affecting the crack resistance of concrete. The greater the ultimate tensile value, the higher the crack resistance of concrete. The ultimate tensile value and tensile strength are the main indexes to evaluate the crack resistance of concrete. Increasing the ultimate tensile value and tensile strength and reducing modulus of elasticity of concrete is an important measure to prevent dam cracking.

Chapter 5 Research and Application of RCC Performance

Table 5.3   Statistics of average development coefficient of compressive strength of RCC in some projects in China

Engineering	Design index of RCC	Gradation	Fly ash content (%)	Compressive strength of each age and development coefficient at the age of 28 d (%)				Remarks
				7 d	28 d	90 d	180 d	
Longshou	$C_{90}20W8F300$	II	53	72	100	143	—	Pebble and natural sand
	$C_{90}20W6F100$	III	65	75	100	156	—	
Linhekou	$R_{90}200D50S8$	II	63	68	100	152	—	Manufactured limestone Coarse and fine aggregate
	$R_{90}200D50S6$	III	65	59	100	160	—	
Mianhuatan	$R_{180}150S4D25$	II	60	71	100	147	205	Content of manufactured granite and rock powder, 17% – 20%
	$R_{180}200S8D50$	II	60	61	100	138	181	
	$R_{180}100S4D25$	III	60	62	100	174	211	
	$R_{180}150S4D25$	III	60	65	100	171	212	
	$R_{180}200S4D25$	III	60	66	100	153	197	
Baise	$R_{180}15S2D50$	Quasi III gradation	63	55	100	166	238	Manufactured diabase Coarse and fine aggregate
	$R_{180}20S10D50$	II	58	61	100	152	211	
Jinghong	$C_{90}15W6F50$	III	NH 60	76	100	118	120	NH double admixture (slag + rock powder) Natural sand and gravel
	$C_{90}15W6F50$	III	50	51	100	168	212	
	$C_{90}15W8F100$	II	NH 50	66	100	127	—	
Guangzhao	$C_{90}25W12F150$	II	50	57	100	136	—	Manufactured limestone Coarse and fine aggregate
	$C_{90}25W8F100$	III	50	55	100	138	—	
	$C_{90}20W6F100$	III	55	55	100	150	—	
	$C_{90}20W8F100$	II	55	60	100	153	—	
	$C_{90}15W6F50$	III	60	61	100	158	—	
Longtan	$C_{90}25W6F100$	III	55	50	100	157	194	Manufactured limestone Coarse and fine aggregate
	$C_{90}20W6F100$	III	60	42	100	165	195	
	$C_{90}15W4F50$	III	65	31	100	169	238	
	$C_{90}25W12F150$	II	55	52	100	147	166	
Gelantan	$C_{90}15W4F50$	III	SL 60	69	100	129	—	Manufactured limestone Coarse and fine aggregate SL (slag + rock powder)
	$C_{90}20W8F100$	II	SL 55	66	100	125	—	
Kalasuke Hydropower Station	$R_{180}150W4F50$	III	65	52	100	167	229	Coarse aggregates of artificial gneiss granite and water washed natural sand
	$R_{180}200W4F50$	III	60	64	100	161	207	
	$R_{180}200W6F200$	III	50	62	100	131	170	
	$R_{180}200W10F100$	II	50	64	100	150	184	
	$R_{180}200W10F300$	II	40	74	100	128	151	
Jin'anqiao	$C_{90}20W6F100$	III	60	50	100	171	203	Coarse and fine aggregates of artificial basalt
	$C_{90}20W8F100$	II	55	56	100	167	189	
	$C_{90}15W6F100$	III	63	42	100	189	218	

There are many factors affecting the ultimate tensile value of RCC, which are closely related to the amount of cementing materials, aggregate type, air content, design age, VC value, etc., especially the content of RCC cementing grout has a greater impact on the ultimate tensile value. Therefore, increasing the grout-mortar ratio, namely, increasing the content of rock powder and low VC value, is an effective technical measure to increase the ultimate tensile value.

Early RCC is defined as super hard or hard concrete. Its *VC* value is large, and ultimate tensile value is significantly lower than normal concrete. Since the full-section RCC damming technology was adopted, RCC has gradually transited to the present semi-plastic concrete without slump. With the deepening of the understanding and research level of rock powder content and grout-mortar ratio, the ultimate tensile value of RCC has been greatly improved, and the ultimate tensile test value of RCC is even greater than that of normal concrete in some projects (for example, Shapai Project). However, the ultimate tensile strength of RCC at the age of 90 d is still between $(0.7 - 0.8) \times 10^{-4}$, which is still lower than that of normal concrete of the same raw materials. The analysis results show that it is mainly related to the small amount of cement, the large amount of admixtures and the small amount of cementing materials.

In recent years, a large number of test results have shown that the ultimate tensile value of RCC increases and the modulus of elasticity deceases by using rich cementing mortar materials of high rock powder content and low *VC* value, which effectively improves the crack resistance of RCC dams. What needs to be explained here is that rich cementing materials and rich cementing mortar materials are two different concepts. Rich cementing mortar contains the micro rock powder with size of 0.08 mm, which effectively increases the grout content and improves the grout-mortar volume ratio.

For example, Baise, Guangguang and Jin'anqiao Projects adopt large grout-mortar ratio and low *VC* value, and the ultimate tensile value and modulus of elasticity are better than the test results of previous scientific research institutions, and meet the design requirements.

At present, in order to improve the crack resistance, the main way to improve the crack resistance is to increase the ultimate tensile value during design. There is an optimal bonding point between the index of ultimate tensile value and the design index, age, strength, etc., of RCC. If the design age of RCC is 90 d, to increase the ultimate tensile value means to reduce the water-cement ratio and increase the amount of cementing materials. However, the improvement effect of ultimate tensile value is not obvious, but the strength and modulus of elasticity increase, the temperature rise and stress of hydration heat increase, which is unfavorable to temperature control and crack resistance.

The ultimate tensile value is the measured value when the concrete axial tensile specimen reaches the failure point. The tensile value is a function of elasticity, creep and tensile strength, and its size depends on both concrete properties and speed of applying tensile load. It is generally expected that RCC has high ultimate tensile value and low modulus of elasticity, because the concrete can better withstand the change of temperature stress and improve the crack resistance.

The use of rich cementing mortar materials and low *VC* value can effectively improve the grout-mortar ratio and the ultimate tensile value of RCC. There are many factors that affect the ultimate tensile strength of RCC. For example, Shapai Project uses cement with low aluminum, high iron and low brittleness, and gneiss granite aggregates with low modulus of elasticity significantly improves the ultimate tensile value. The ultimate tensile value is more than 1.35×10^{-4}, which is rare in China. Baise and Jin'anqiao Projects adopt hard diabase and basalt aggregates. The test results of early scientific research institutions show that the ultimate tensile value is very low mainly due to the large *VC* value and insufficient understanding of rock powder content, resulting in less cementing mortar of RCC, that is, the *PV* value of grout-mortar ratio is small, generally less than 0.35, causing low ultimate tensile value and high modulus of elasticity. The results of construction mix proportion test show that the ultimate tensile value increases and the modulus of elasticity decreases as the mixture performance changes by taking technical measures of appropriate water-cement ratio, increasing the content of fly ash and admixture, replacing sand with rock powder and low *VC* value. It is the same as the reason why the frost resistance and ultimate tensile value of normal concrete are higher than those of

RCC. Therefore, the change of the performance of RCC mixtures has a great impact on the rapid RCC damming, which has a very important practical significance.

Due to the characteristics of RCC, its tensile strength and ultimate tensile value are generally lower than that of normal concrete. In recent years, technical innovations have been carried out to improve the tensile strength and ultimate tensile value of RCC. One is to mix with silicon powder into RCC and the other is to mix with fiber, which can effectively improve the crack resistance of RCC.

5.2.3.2 Modulus of Elasticity

The modulus of elasticity of concrete refers to the stress required for concrete to produce unit strain, which depends on the modulus of elasticity of aggregate itself and the grout ratio of concrete. RCC with high modulus of elasticity of aggregates has high modulus of elasticity, and RCC with high grout ratio can reduce modulus of elasticity. The higher the modulus of elasticity, the worse the temperature stress and crack resistance of concrete.

Concrete is not a real elastic material. In case of continuously increasing load, the stress-strain diagram of concrete can be represented by a curve. If a moderate load is applied in advance, the strain relation of fully hardened concrete is a straight line with certain slope from the practical point of view. The ratio of stress to strain obtained from the straight line section of the stress-strain curve is called "modulus of elasticity". When the load exceeds the working stress range, the stress-strain relationship will deviate from the straight line more and more, which means that the stress-strain relationship is no longer linear. However, when the stress is higher than 75% of the failure strength at the age of 28 d, the stress-strain ratio is quite consistent. Although the modulus of elasticity is not directly linear to the strength, generally speaking, the modulus of elasticity of high-strength concrete is also relatively high. The modulus of elasticity of ordinary concrete at the age of 28 d is in the range of 20 – 40 GPa.

As for most of materials, the modulus of elasticity does not change with the age, and the elastic recovery after unloading is equal to the elastic deformation under loading, which is independent of loading time. However, the modulus of elasticity of concrete generally increases with age, and this is particularly obvious for RCC, which is related to the significant strength growth of RCC. As the modulus of elasticity of concrete increases with age, concrete can expand and shrink freely when a large amount of hydration heat is released in the early stage, while the shrinkage of cooling in the later stage is restrained, resulting in large tensile stress.

In addition to the static method, measurement of the modulus of elasticity with the dynamic method, or the natural vibration frequency of specimens, or measurement of the velocity of acoustic waves passing through the specimen can be used to calculate the stress-strain relationship (The strain value corresponding to the test loading stress is directly measured). Generally, the dynamic method is mostly used to measure the damage of concrete specimens after freeze-thaw test or alkali aggregate reaction. The dynamic method is a fast and simple method to measure the modulus of elasticity without damaging the specimens in the test. The natural frequency or wave velocity ratio measured is lower, indicating that the modulus of elasticity has decreased and the quality of concrete has deteriorated.

A large number of RCC damming technology practices have proved that the strength of RCC almost exceeds the design strength, and the surplus is large, which improves the modulus of elasticity of concrete, and increases the temperature stress of dam body, and is unfavorable to crack resistance. For example, the average compressive strength of Longtan dam RCC is 36 – 37 MPa, but the modulus of elasticity reaches 70 GPa, which fully shows that both the strength and the modulus of elasticity are high.

Therefore, the strength of RCC is no longer the main factor to control technical indexes of RCC. The performance of RCC mixtures, interlayer bonding, temperature control and crack resistance are the core technologies of rapid construction of RCC, and these properties are directly related to the grout-mortar ratio of RCC. Therefore, it is necessary to pay more attention to the grout-mortar ratio in the mix proportion design.

5.2.3.3 Autogenous Volume Deformation

Under the condition of constant temperature and constant humidity, the volume deformation of concrete caused by the hydration of cementing materials is called autogenous volume deformation (autogenous deformation for short). The volume change of concrete in the process of hardening is mainly due to the different density of reactants and products before and after hydration reaction of cementing materials and water. The density of the product is less than that of the solid reactant. Although the volume of solid phase after hydration is larger than that before hydration, the total volume of cementing materials and water system is smaller (except for expansive cement). This chemical shrinkage reduction is the essence of hydration reaction of cementing materials.

The autogenous volume deformation of concrete includes both expansion and shrinkage. When the autogenous deformation is expansion, it can compensate the shrinkage caused by temperature drop, which is beneficial to the crack resistance of concrete. When the autogenous deformation is shrinkage, it is unfavorable to crack resistance of concrete. Therefore, the influence of autogenous deformation on crack resistance of concrete cannot be ignored. The autogenous volume deformation may show expansion occasionally, and shrinkage mostly, which deserves attention. It is completely the result of the chemical reaction inside the concrete and is related to the age. In addition, the autogenous volume deformation has nothing to do with the volume change caused by dry shrinkage or other external influences. The amplitude of variation of self-shrinkage is very large, which is the negligible 10×10^{-6} observed so far to more than 150×10^{-6}.

Excessive change of autogenous volume is harmful to concrete. Before the early tensile strength of concrete is fully formed, shrinkage occurs due to temperature drop, dry shrinkage, etc., and the restraint hardened concrete will produce cracks. Crack is not only a weakness that affects the design load capacity of concrete, but also seriously damages the durability and appearance of concrete. Water intrusion into cracks will damage the durability of concrete, and accelerate the leaching and corrosion of reinforcement. Concrete with cracks will be further damaged when exposed to freeze-thaw environment. When the concrete contains alkali active aggregate and cement with high alkali content (more than 0.6%) or is affected by water containing soluble sulfate, it will also produce disintegration. The uneven stress of concrete caused by the difference of volume change characteristics of various components will destroy its internal structure and affect the cementation between set cement and aggregate particles, especially after repeated expansion and shrinkage, it will also cause disintegration. In the case of restraints, the expansion of concrete will produce too high compressive stress and peel off at the joint.

The autogenous volume shrinkage is different from the dry shrinkage, which is not related to water consumption, but mainly depends on the characteristics and total amount of cementing materials; the shrinkage of concrete with rich cementing materials is larger than that of concrete with poor cementing materials. After concreting, the shrinkage is the most significant at the age of 60–90 d. Whether the volume change can cause cracks largely depends on resistance of internal force and external force to shrinkage. Mass concrete block is an example of external cracking caused by internal restraining force. Its surface is dried or cooled, but the internal is not affected by this effect.

Test results of autogenous volume deformation of RCC show that the autogenous volume deformation of RCC is smaller than that of normal concrete, which is about 50% of that of normal concrete. The amount of cement in RCC is small, and the main reaction of fly ash mixed with RCC occurs at the later stage. Therefore, it is inevitable that the autogenous volume deformation is small.

5.2.3.4 Dry Shrinkage

Dry shrinkage of concrete refers to the volume shrinkage deformation of concrete caused by water loss in unsaturated air. Dry shrinkage deformation is mainly due to the evaporation of pore water and capillary pore water in the drying process of concrete. The evaporation of pore water does not cause shrinkage of concrete. The evaporation of capillary pore water makes the water surface in pores retreat and the curvature of meniscus increase. Under the action of surface tension, the internal pressure of water is smaller than the external pressure. With the decrease of air humidity, the negative pressure in the capillary pores increases gradually, which produces shrinkage force and makes concrete shrink. When the water in the capillary pores is evaporated, if the drying continues, the absorbed water of cementing particles will also be partially evaporated. Due to the effect of molecular attraction, cementing particles losing the water film shorten the distance between particles and cause shrinkage.

Main factors affecting dry shrinkage of concrete include cement type, admixture type and content, aggregate type and content, admixture type and content, concrete mix proportion, medium temperature and relative humidity, curing conditions, concrete age, structural characteristics, carbonation effect, etc. Aggregate type has a great influence on dry shrinkage of concrete. According to relevant references, the dry shrinkage of concrete of sandstone aggregates is the largest, the dry shrinkage of concrete of limestone and quartzite aggregates is small, and the dry shrinkage of granite and basalt aggregates is medium. In terms of dam RCC, the dry shrinkage deformation at the age of 90 d is $(250-350) \times 10^{-6}$, and the dry shrinkage ratio of RCC with diabase manufactured sand with high content of rock powder at the age of 90 d can reach more than 600×10^{-6}, which is much larger than the temperature deformation $(150-200) \times 10^{-6}$ caused by hydration heat temperature rise of dam RCC. Therefore, if RCC is not properly cured, it is easy to produce dry shrinkage cracks on the surface.

Drying shrinkage is affected by many factors. In terms of importance, they include: unit water consumption, aggregate composition, rock powder content and initial curing duration. The total water demand of mixtures is the main factor affecting dry shrinkage. Therefore, it is necessary to reduce the unit water consumption of concrete as far as possible during the mix proportion design. RCC is mixed with a large number of admixtures (fly ashes). The water demand ratio of fly ash at different grades is different. The fly ash with high water demand ratio will increase dry shrinkage, while the high-quality fly ash with small water demand ratio will reduce the dry shrinkage. This effect is in direct proportion to the water demand ratio of fly ash. Therefore, high-quality fly ash with small water demand ratio is used for RCC, and the dry shrinkage ratio is also small.

The dry shrinkage ratio of RCC is different from that of normal concrete. Because RCC needs a higher content of rock powder to improve the grout-mortar ratio and rollability, the high content of rock powder is conducive to liquefied bleeding, interlayer bonding and compaction improvement, and it is also beneficial to the improvement of impermeability, frost resistance and ultimate tensile value, as well as the reduction of modulus of elasticity. However, too high content of rock powder is unfavorable to the dry shrinkage of hardened RCC. For example, diabase aggregate RCC is adopted in the Baise Project, and the content of sand powder of diabase manufactured sand is very high, and its dry shrinkage ratio is obviously higher than that of RCC with appropriate content of rock powder.

5.2.3.5 Creep

Creep refers to the case that concrete deformation increases with time under continuous load. The creep deformation is 1-3 times larger than the instantaneous elastic deformation. The creep deformation under the unit stress is called the creep degree (unit: $10^{-6}/MPa$). The creep of concrete has a great influence on the temperature stress of concrete. For mass concrete, the greater the creep, the greater the stress relaxation, and the more favorable it is to crack resistance of concrete.

The deformation of concrete under constantly continuous load can be divided into two parts: elastic deformation and creep deformation. Elastic deformation occurs immediately after loading and disappears immediately after unloading; Creep deformation develops continuously with time. In most concrete structures, the static load acts continuously and is the main part of the total load. Therefore, both instantaneous deformation and continuous plastic deformation must be considered when the deformation of such members is calculated. The progressive plastic deformation has a great influence on the stress development caused by slow temperature change and dry shrinkage. It is often called creep to distinguish it from a different plastic action. The plastic action of concrete, like the plastic flow of metal, is irrecoverable and can be regarded as a form of initial failure; However, creep is at least partially recoverable and can occur even at very low stresses.

The creep of concrete can be indefinitely continuous under continuous load. Two specimens subject to long-term test are still deformed after 20 years of continuous loading. But the rate of creep gradually decreases. According to the creep parameters obtained in the laboratory, the exact relationship between creep variables is obtained by computer program. The value of creep function (K) is relatively large when the concrete begins to bear load at early age, and smaller at later age. Concrete continues deformation at a decreasing rate with increase of time, but there is no obvious limit. Although a large number of test results can completely verify that the creep of concrete is unlimited, it is usually assumed that there is an upper limit of creep deformation.

The creep growth rate is directly affected by the change of water-cement ratio and loading strength. The creep increases with the increase of water-cement ratio, and the creep is approximately in direct proportion to the load. Many factors that increase the strength and modulus of elasticity will reduce the creep. Generally speaking, concrete made of loose grained aggregates (for example, some sandstone) has larger creep than concrete made of denser aggregates (for example, quartz or limestone).

In design, the numerical method of reduced modulus of elasticity is often used to consider creep approximately. When it is necessary to calculate the stress with a more precise relationship, for example, strain observation results of mass concrete, the mathematical analysis and prediction of creep shall be conducted with reference to the following characteristics:

1. Creep is a kind of hysteretic elastic deformation, which does not involve the failure or sliding of crystals, so it is not a plastic flow of viscous solids.
2. In the working stress region, creep is directly proportional to the stress. But when the stress is close to the ultimate strength of concrete, the creep growth rate is much faster than the stress growth rate.
3. If the influence of age on the change of concrete properties is considered, all creep is recoverable.
4. The creep of concrete has no plus or minus, and its ratio is equal, regardless of normal stress or negative stress.
5. The superposition principle is applicable to the creep.
6. The Poisson's ratio of creep strain is the same as that of elastic strain.

The research on the temperature creep stress of RCC dam is to simulate and calculate the temperature creep stress of RCC dam during construction and operation according to the proposed schemes regarding different dam gap spacing and RCC placing temperature based on the project experience under the safe condition of temperature creep stress during the construction and operation of dam, and summarize the variation law of dam temperature and stress distribution, so as to find an economical and reasonable scheme regarding dam gap spacing and RCC placing temperature and provide the design basis for the temperature control during dam concrete construction.

There are many factors affecting the creep. The creep of RCC is similar to that of normal concrete. It is also affected by the following factors. These include mainly:

1. Cement nature: when the crystal formation is slow and there are less crystals, the creep is greater;

2. Mineral components and gradation of aggregates: when the structure of aggregates is loose, with small density, poor gradation and many voids, the creep is greater;
3. Concrete mix proportion, especially water-cement ratio and coarse aggregate consumption: when the consumption of cementing materials and the water-cement ratio in the mixtures mixed based on the mix proportion is relatively large, and the consumption of coarse aggregate is small, the creep is greater;
4. Loading age and holding time: when loading, the age of concrete is short, and the strength of concrete is low, the creep is greater. The longer the holding time, the greater the creep;
5. Loading stress: the greater the loading stress, the greater the creep;
6. Structure dimension: the smaller the structure size, the greater the creep.

There are many factors affecting the value of concrete creep, mainly including: loading age, holding time, stress, load nature, humidity, aggregate content and elasticity, cement type, mix proportion and cementing material consumption. The creep is an important material nature affecting the temperature stress, which can make the temperature stress decrease. The greater the creep, the smaller the temperature stress. The creep of concrete is mainly related to the cementing material consumption. Compared with normal concrete, RCC contains less cementing materials, so the creep of RCC is generally smaller than that of normal concrete.

5.2.4 Durability and Influencing Factors of RCC

The greatest advantage of concrete is its durability, which is unmatched by other materials. The durable concrete shall enable to withstand the effects of various bad operating conditions, such as permeability, frost, erosion, weathering, chemical action and abrasion. The durability of hydraulic RCC is directly related to the service life and safe operation of the dam. The requirements for the durability of hydraulic RCC and normal concrete have been consistent each other in recent years, and the frost resistance grade of hydraulic RCC has been raised to a new level. The durability of hydraulic concrete is mainly measured and evaluated based on the frost resistance grade, which is one of the most important control indexes of the durability of hydraulic concrete. No matter in the south, north or hot and cold regions, the designed frost resistance grade of RCC reaches or exceeds F 50, F 100 and F 200. The frost resistance grade in the cold areas reaches F 300 or even F 400. The requirements will be higher. The air content of RCC is closely related to its durability, but there is a great difference between the air content of outgoing fresh concrete and that of RCC after actual pouring, reflecting that the air content of hardened concrete cannot meet the design requirements, thus causing that the core samples of hardened concrete cannot reach the designed frost resistance grade.

5.2.4.1 Impermeability

I Grade of impermeability

The imperviousness is the top priority of RCC dam as a water retaining structure, so the imperviousness performance is its most important performance. The impermeability of RCC is not inferior to that of normal concrete, but the RCC is spread and rolled in thin-lifts. The weak link of RCC dam lies in the imperviousness performance of layer joint surface. Therefore, the interlayer bonding quality is directly related to the imperviousness performance of dam, and the interlayer bonding technology is the core of rapid RCC damming technology.

There are special design requirements for the impermeability of hydraulic concrete. The impermeability grade is taken as the evaluation standard of permeability. The impermeability grade of concrete is divided into six grades (W2, W4, W6, W8, W10 and W12). The conversion relation between the impermeability grade and permeability coefficient of concrete is shown in Table 5.4. Due to high fly ash content of RCC, the impermeability is basically the same as the strength performance. The 90 d RCC is generally used for the impermeability test.

Table 5.4 Conversion relation between the impermeability grade and permeability coefficient of concrete

Grade of impermeability	Permeability coefficient (cm/h)
W2	7.05×10^{-5}
W3	4.93×10^{-5}
W4	2.81×10^{-5}
W5	2.16×10^{-5}
W6	1.50×10^{-5}
W7	1.14×10^{-5}
W8	9.40×10^{-6}
W9	6.89×10^{-6}
W10	6.20×10^{-6}
W11	5.42×10^{-6}
W12	4.64×10^{-6}

The impermeability of RCC not only related involves the water retaining function of concrete, but also directly affects the frost resistance and erosion resistance of concrete. The impermeability of RCC mainly depends on the mix proportion and compactness of concrete. The compaction of lean RCC is poor due to its less cementing materials consumption and higher water-cement ratio. RCC is defined as semi-plastic concrete without slump, with high fly ash content, high rock powder content and low VC value. The cementing material mortar in the RCC increases significantly, making it easy to roll and compact RCC during construction. The impermeability grade of 90 d concrete can generally reach above W8. The imperviousness effect of RCC dam is equivalent to or stronger than that of normal concrete dam.

Due to small water-cement ratio and the compaction during construction, there are less primary pores in the RCC with high fly ash content. With the increase of the age, the fly ash is gradually increase. Hydration products continuously fill primary pores, making coarse pores fine, fine pores blocked, and some connected pores closed. Therefore, the impermeability is obviously improved with the increase of the age. This is an obvious feature of the RCC with high fly ash content and high rock powder content.

The impermeability of RCC projects in practice is mainly controlled by the impermeability of horizontal construction joint surfaces. The intermittent time of horizontal construction joint surface and the methods and quality of cold joint treatment have a great effect on their impermeability. After proper treatment, the impermeability of horizontal construction surfaces can be obviously improved.

Ⅱ Factors affecting the impermeability

If hardened RCC is composed of very dense materials, it may be completely impermeable. However, it is actually impossible to completely fill the voids in the concrete aggregates with dense cement mortar. The actual water consumption is greater than that required for cement hydration in order to make the mixtures easy for construction. The excess water causes voids and holes, which may connect with each other to form a continuous channel. In addition, the absolute volume of hydration products is less than the total of original absolute volumes of cement and water. Therefore, it is impossible for hardened set cement cannot to occupy the same space as fresh cement grout. Meanwhile, there are also additional voids in the hardened set cement. Purposefully introduced air and entrained air can also cause voids in the concrete. However, as explained below, the air addition is to improve the impermeability of concrete rather than the permeability.

Hardened concrete has an inherent permeability. The water may penetrate into concrete through capillary pores or by means of pressure. Nevertheless, the permeability of RCC can be controlled so that the imperviousness performance of dam retaining structures can meet the design requirements without becoming a serious problem.

The inherent permeability of concrete can be ascertained by analyzing the internal structure of plastic concrete. After concreting, all solids, including cement particles and hydration products, are in an unstable equilibrium state. The sinking of solids forces water upward. In this case, it is easy to form a series of water seepage channels. Some water seepage channels may extend to the surface of concrete. Some larger aggregate particles gradually settle to a stable position where they form a skeleton structure by point contact or other contact means and continue to settle inside the skeleton. The sinking of mortar forces some water upward, making it gather at the bottom of large aggregate particles. Finally, the cement will separates out from cement grout in the sand. There are residual water pores in the cement grout after precipitation. For mixed concrete at the end of this stage, the original water (the main reason for harmful pore formation) is no longer evenly distributed in the cement grout, but fills in the larger voids under the aggregate particles. The reason why the internal pore structure of air-entraining concrete is different from the above structure is that the unconnected and isolated spherical bubbles greatly reduce bleeding phenomenon and water seepage channels.

Ⅲ　Analysis of impermeability test method

The Author thinks that, it is unscientific to carry out RCC impermeability test by the method of normal concrete, not meeting the actual situation of dam imperrviousness. According to the *Test Code for Hydraulic Concrete* SL 352—2006, the specimens for the impermeability test are made easily, but the seepage pressure on the concrete layer surface is different from the seepage characteristics of RCC dam. Relevant test results show that the anisotropy ratio of tangent and normal permeability coefficients of RCC layer surface reaches 3 – 4 orders of magnitude.

As long as RCC is vibrated and compacted according to the specifications, there is no impermeability problem. The weak link of RCC lies in interlayer bonding. For interlayer bonding impermeability test, it is necessary to change the traditional tangential impermeability test method into the normal impermeability test method, and deeply study RCC interlayer bonding impermeability test method.

5.2.4.2　Frost Resistance

Ⅰ　Freeze-thaw dmage mechanism

There are many explanations for the mechanism of concrete freeze-thaw damage. When water freezes, its volume increases by 9%, thus generating expansive force. There are usually two explanations for the mechanism of expansion force generation:

1. Hydrostatic pressure. When water freezes in the capillary pores, the excess water is forced to flow outward due to its volume expansion, and the outflow speed of squeezed water increases with the advancing of freezing face. However, the flow is resisted by the inner walls of capillary pore, thus generating hydrostatic pressure. The hydrostatic pressure depends on the resistance of water flow, the permeability of concrete and the length of channel (the shortest flow distance before hydrostatic pressure release).
2. Expansion force is derived from seepage pressure. When some water in the capillary pores freezes, frozen pure water separates out from the solutions containing alkali and other materials, making the concentration of local solutions increase to form a concentration difference, promoting the diffusion of cementing materials and water to the capillary pores and generating osmotic pressure.

Ⅱ　Frost resistance grade and air content

Frost resistance grade is one of the most important indexes to evaluate the durability of concrete. With the continuous development of full-section RCC damming technology, it is not new that the requirements for frost resistance grade of RCC dam are higher and higher in the cold northern region or the mild southern region. In the early stage, RCC frost resistance test is carried out based on the 28 d age. In addition, the research on RCC performance, especially the performance of mixtures, is relatively shallow. Therefore, the poor frost resistance of RCC with high fly ash content is inevitable. Its early frost resistance is worse than that of normal concrete. With the change of RCC definition in recent years, the research on RCC performance has been deepened. In particular, the roles of rich cementing mortar, low VC value and air entraining agent for RCC are more and more obvious. The frost resistance grade is tested based on the 90 d age (RCC strength is constant). The frost resistance of RCC has been the same as that of normal concrete. RCC prepared as per the designed mix proportion has met the design requirements of F300 frost resistance grade.

The most fundamental way to improve the frost resistance of RCC is to mix with the air entraining agent, so as to produce a large number of disconnected tiny bubbles in the concrete, thus improving the pore structure of concrete. For RCC exposed to the atmosphere, purposefully introducing a certain amount of air into RCC can greatly improve the freeze-thaw capacity of RCC. The air bubbles in the concrete can eliminate the pressure caused by the expansion of free water when it freezes. The air content of RCC shall be determined according to the frost resistance grade. Taking the Kalasuke RCC Dam in Xinjiang as an example. The frost resistance grade of RCC is designed according to the positions: F300 in the water level varying area of dam; F200 in the upper part of the downstream water level varying area; F100 below the dead water level; F50 inside the dam. For RCC at different frost resistance grades, the air content is controlled according to 5.0% – 6.0%, 4.5% – 5.5% and 4.0% – 5.0%, 3.0% – 4.0% respectively. The test results show that RCC frost-resistance can meet the different design requirements.

5.2.4.3 Abrasion Resistance

Whether RCC can be used on the surface of overflow dam and in the spillway or apron position, and withstand the action of water with high flow rate is closely related to the performances of concrete such as impact resistance, abrasion resistance and cavitation erosion resistance (abrasion resistance for short) in addition to the smooth of concrete surface. The main reasons for concrete surface abrasion are: cavitation erosion, abrasive material impact during water flow and others.

Cavitation erosion is one of the most destructive reasons. No matter how good the quality of concrete or other structure materials, their cavitation erosion resistance is poor. If there is an obstacle or abrupt change on the surface of concrete where the water with high flow rate flows through, a strong negative pressure zone will be formed near downstream positions. The negative pressure zone is quickly filled with the turbulence containing a small fast moving vapor cavitation. Vapor cavitations are formed upstream the negative pressure zone, flow through the zone and break at a point near downstream positions due to the increase of pressure in the water flow. When a cavitation breaks, the surrounding water flows to its center at high flow rate, thus accumulating great energy. The whole process of the formation, movement and break of cavitation is called cavitation erosion.

A small vapor cavitation can generate a serious impact. After gathering, they can not only damage concrete, but also make the hardest metal produce pits. Under the repeated action of high-energy impacts, pits or holes are formed eventually. Such phenomenon is called cavitation erosion. If there is a sudden jet or decompression when the water with high flow rate flows through the surface of discharge chute or surface of spillway, the cavitation will occur. Cavitation erosion may occur on the horizontal, inclined or vertical surfaces where water flows. In terms of concrete cavitation erosion on the stilling pool surface of spillway and near stilling piers, the water with high flow rate often causes negative pressure on the concrete surface during flood, thus causing cavitation erosion damage here. Sometimes metal or other materials with stronger cavitation erosion resistance than that of concrete to protected the surface of concrete in the places where low pressure cannot be avoided and in the dangerous areas. It is very effective to reduce the occurrence of cavitation and alleviate its effect on some structures by increasing the air content in the water flow at a suitable position upstream.

Concrete abrasion caused by abrasive materials in the water is as serious as cavitation damage, but it usually does cause catastrophic accidents like cavitation erosion damage. Turbulent flow is produced in the hydraulic jump area of the spillway and stilling basin of the sluice, which may especially wear and tear the dam. The action of water flow in these areas will take the pebbles, gravels and sand of downstream riverbed back to the stilling basin of concrete lining, and the action of water flow in the basin is equivalent to a ball mill. Even the best concrete can't bear such severe wear and tear. By contrast, cavitation erosion causes little or no abrasion to aggregate particles. Although the most serious wear and tear occurs in the places just mentioned above, similar damage may occur in diversion tunnels, channels and sewage pipes.

The abrasion resistance of concrete is related to the strength, composition materials and mix proportion of concrete. Prepare high-grade concrete with less cement grout by hard and wear-resistant aggregate and high-grade Portland cement, make the concrete surface level and smooth after vibrating and compacting, then the anti-abrasion performance will be high. This requirement for abrasion resistance of normal concrete is also applicable to RCC.

Currently, there is little research on the abrasion resistance of RCC at home and abroad. Relevant data show that under the condition that the compressive strength is basically the same, the abrasion resistance of RCC is better than that of normal concrete.

5.2.4.4 Chemical Corrosion Resistance

Like normal concrete, the chemical corrosion resistance of RCC mainly refers to the resistance to dissolution erosion, carbonic acid erosion, acid erosion, sulfate erosion and magnesium salt erosion. Now there is a lack of data on chemical erosion resistance of RCC. However, according to the erosion principle, RCC has strong resistance to dissolution erosion, magnesium salt erosion and sulfate erosion. Because RCC is mixed with a large amount of fly ash, the clinker is relatively reduced, the content of C_3S and C_4A is relatively reduced, the amount of $Ca(OH)_2$ precipitated by hydration is also small, which reacts with fly ash to generate stable calcium silicate hydrate and hydrated calcium aluminate in the secondary hydration reaction, so it has strong resistance to dissolution erosion. Because of the reduction of $Ca(OH)_2$, the resistance to sulfate erosion is also improved. In addition, because the content of hydrated calcium aluminate which is vulnerable to sulfate erosion is relatively reduced, the ability to resist sulfate erosion is also improved. As there is less $Ca(OH)_2$ in RCC, its ability to resist acid erosion is also strong.

One of the compounds generated after cement is combined with water is hydrated lime, which is quickly dissolved by water (more aggressive in case of dissolved carbon dioxide in water). Water seeps through cracks or along defective construction joints or through connected pores. The leaching of solid matters may seriously damage the concrete. The common white leachates or frost spots on the concrete surface are caused by leaching and accompanying carbonization and evaporation.

5.2.5 Thermal Properties of RCC

The thermal properties of RCC are the main basis for analyzing the temperature, temperature stress and temperature deformation in dam concrete, which mainly refer to various performances of concrete that generate or dissipate heat. RCC is mixed with a large amount of fly ash, and the total amount of cementing materials is less, so various thermal properties are different from those of normal concrete. Thermal properties of RCC include: Hydration heat of cementing materials, adiabatic temperature rise, specific heat, thermal conductivity, thermal diffusivity and coefficient of linear expansion, among which adiabatic temperature rise is the main index of temperature control calculation.

5.2.5.1 Hydration Heat of Cementing Materials

The heat released during cement hydration is called cement hydration heat. RCC is mixed with fly ash and other admixtures, and the hydration reaction of the admixtures also generates heat, so the heat released from the hydration process of cement with fly ash (or other admixtures) is called hydration heat of cementing materials.

The temperature rise and the temperature deformation is large of concrete hydration heat, the temperature stress is also large, and the crack resistance of concrete is poor. The main factors affecting the temperature rise of concrete hydration heat are the mineral composition of cement, the quality and content of admixtures (or addition materials), water consumption of concrete and cement consumption, etc.

The hydration heat of cementing materials is mainly related to cement varieties, content of cement mineral components, quality and content of admixtures and other factors. The hydration heat (in the case of specified water addition and curing temperature) of portland cement (pure clinker cement) is mainly related to the mineral composition of clinker. For the main minerals of cement, the heat generated by full hydration of C_3A is the largest, followed by C_3S, C_4AF and C_2S.

In addition to clinker, there are some mixed materials and a large number of admixtures in cementing materials. According to the test results of composite materials (including fly ash, iron slag and rock powder, phosphorous slag and tuff, volcanic ash, etc.) at home and abroad, the calorific value of cementing materials mixed with composite materials is only about 50% of that of cement clinker.

A large number of test results show that the hydration heat of cementing materials is reduced by adding fly ash. With the increase of fly ash content, the calorific value of fly ash decreases. This is because the hydration reaction of fly ash needs the hydration product $Ca(OH)_2$ of cement. The more fly ash is added, the less fly ash takes part in the reaction at the same age. Therefore, it seems that the hydration heat of cementing materials with different fly ash contents cannot be calculated by the same calorific value percentage. The hydration heat of cementing materials is measured by hydration heat test. From the obtained temperature process curve, it can be seen that the temperature peak of the temperature process curve decreases and the temperature peak is extended due to the addition of fly ash, and the temperature drops slightly slowly and the heat release is slightly larger at the later stage, and the total heat decreases.

The volume change caused by temperature involves both aggregate and volume change of cement grout. The volume change caused by dryness changes is generally considered to be mainly related to cement grout. However, the volume change caused by temperature and humidity will also produce similar disintegration. Chemical reactions between active aggregate and alkali (Na_2O and K_2O) in cement, and chemical reactions between soluble sulfate free in soil or groundwater and tricalcium aluminate (C_3A) in cement after contacting concrete structure will cause volume change and lead to damage.

5.2.5.2 Adiabatic Temperature Rise of RCC

Adiabatic temperature rise of concrete is the temperature rise of concrete under adiabatic conditions. Because of the poor thermal conductivity of concrete, the internal temperature rise of continuously poured mass concrete is close to the adiabatic temperature rise of concrete. The adiabatic temperature rise of laboratory concrete is measured under the condition that the concrete specimen neither loses heat nor absorbs heat from outside. However, due to the limitation of equipment and boundary conditions, it is difficult to directly measure the final temperature rise of concrete.

The calorific value of concrete is mainly caused by the hydration heat of cement. Although fly ash will also generate heat in hydration process, its calorific value is very small. The cement consumption of RCC is much less than that of normal concrete, so the adiabatic temperature rise is also very low. Meanwhile, there is a large content of fly ash in RCC, which has the characteristic of delayed heating. Therefore, the hydration heat of RCC has a slow temperature rise and a large temperature rise in the later period. The observation data of RCC dams built in the late 1980s and 1990s show that it usually takes decades for RCC dams to reach stable temperature.

The thermal properties of RCC in large and medium projects are determined by tests, and those in general projects can be determined with reference to similar engineering data. As the thermal performance of concrete depends on the thermal performance of water, cement, admixture and aggregate, the total heat per unit concrete can be calculated according to the amount of various materials in the concrete mix proportion.

The adiabatic temperature rise of concrete is the basis of temperature control and crack resistance calculation and temperature control design. The adiabatic temperature rise of concrete is measured by the laboratory under the boundary adiabatic conditions. Now the laboratory tests can only be done for 28 d age. After 28 d, the tester personnel will fit a calculation formula according to experience. This test method is mainly limited by instruments, equipment and boundary conditions, and cannot reflect the long-term heat release process of RCC, especially the slow temperature development process of RCC mixed with a large amount of fly ash, and there is a big error between the laboratory results and the temperature results measured in actual projects. Therefore, the adiabatic temperature rise measured by indoor tests is only for calculation and comparison, and there is still a big difference between it and the actual adiabatic temperature rise of the dam.

The development process of adiabatic temperature rise of RCC is related to the initial test temperature of concrete. In practical projects, the temperature rise of dam concrete is not equal to the adiabatic temperature rise obtained by indoor tests, and the highest temperature rise is not equal to the final adiabatic temperature rise of the concrete. In the process of RCC construction, due to continuous pouring and rolling, the boundary heat dissipation conditions of RCC are mainly affected by its own heat, especially the difficulty of internal concrete dissipating heat to the outside world. Therefore, the measured maximum temperature rise of RCC dam body is higher than the final adiabatic temperature rise of concrete.

Characteristics of adiabatic temperature rise of RCC: due to the small amount of cement and the large amount of admixtures in RCC, the adiabatic temperature rise is very low, obviously lower than that of normal concrete, which is beneficial to temperature control. However, the RCC dam has no longitudinal joints, and RCC is spread in whole-block and thin-lift and continuous rolling compaction. The heat dissipation conditions are completely different from those of normal concrete column pouring, and the heat dissipation effect in the dam is poor. Especially, the calorific value in the later period of RCC lasts for a long time, which requires the internal temperature of the dam to be stable, and it will take several decades. This is also the characteristic and disadvantage of RCC temperature control.

The main factors affecting the thermal performance of RCC depend on the concrete raw materials and mix proportion design, and have little to do with the construction method. The research on thermal properties of RCC must also focus on the characteristics of high admixture content, low initial strength and long hydration heat process.

5.2.5.3 Hermal Diffusivity, Heat Conductivity Coefficient, Specific Heat and Linear Expansion Coefficient

I Efficient of thermal conductivity

The thermal diffusivity of the concrete refers to the rate at which the same temperature reaches at various points during the cooling or heating of the concrete. It is an aggregative indicator reflecting heat diffusion of the concrete, expressed as "α", and in the unit of m^2/h. The greater the thermal diffusivity of the concrete, the quicker the same temperature is reached at various points. The thermal diffusivity of the concrete varies with the types of the concrete aggregate, consumption amount of aggregate, the apparent density of the concrete (or air content). Generally, the thermal diffusivity of the concrete decreases with the decrease of the apparent density of the concrete, the rise of the temperature, and the increase of the water content. The influence on the thermal diffusivity of the concrete by using different aggregates decrease in the following order: quartzite, dolostone, limestone, granite, rhyolite and basalt. The thermal diffusivity of the common concrete (also called as "heat diffusion coefficient") is from 0.002 – 0.006 m^2/h. There is no significant difference between thermal diffusivities of RCC and that of the normal concrete.

II Thermal conductivity factor

The heat conductivity coefficient of the concrete is the capacity that the concrete conducts the heat. It indicates the heat that passes through the surface of the concrete slab with an area of 1 m^2 and a thickness of 1 m within 1 hour when the temperature difference between two sides of the slab is 1 ℃. The heat conductivity coefficient is expressed as "λ", in the unit of kJ/(m · h · ℃). The heat conductivity coefficient of the concrete varies with the apparent density, temperature and moisture condition, it is also related with the amount of aggregate and the heat conductivity coefficient of the aggregate. The heat conductivity coefficient of the concrete generally increases with the increase of the apparent density of the concrete, the rise of the temperature and the increase of the water content. The heat conductivity coefficient of the concrete generally is $\lambda = 8 - 13$ kJ/(m · h · ℃). There is no significant difference between the heat conductivity coefficient of RCC and that of the normal concrete.

III Specific heat

The specific heat of the concrete indicates the heat that the unit mass of the concrete absorbs when the temperature rise (or fall) by 1 ℃, which is expressed by "c", in the unit of kJ/(kg · ℃). It increases with the increase of concrete water content and the decrease of aggregate consumption. The specific heat of the concrete generally is from 0.8 – 1.2 kJ/(kg · ℃). There is no significant difference between the specific heat of RCC and that of the normal concrete.

The relationship among above three coefficients of thermal performance is as follows:

$$\alpha = \lambda / (c\gamma)$$

Where α = thermal diffusivity of the concrete, m^2/h

λ = the heat conductivity coefficient of the concrete, kJ/(m · h · ℃)

c = specific heat of the concrete, kJ/(kg · ℃)

γ = apparent density of the concrete, kg/m^3

IV Linear expansion coefficient

Similar to most of engineering materials, the linear expansion occurs when the temperature of the concrete rises, the size of the expansion amount is expressed by the linear expansion coefficient "α", in the unit of 1/℃. The linear expansion coefficient indicates that the elongation in unit length when the temperature of the concrete increases by 1 ℃. It has something to do with the mix proportion of concrete and the nature of the aggregate. The linear expansion coefficient will decrease with the increase of the amount of aggregate used in the concrete. The linear expansion coefficient of the concrete with the quartzite used as the aggregate is the greatest, followed by the sandstone, granite, basalt and limestone decreases in turn. The linear expansion coefficient of the concrete generally is from $6 \times 10^{-6} - 12 \times 10^{-6}$/℃. There is no significant difference between the linear expansion coefficient of RCC and that of the normal concrete.

The linear expansion coefficient of the concrete refers to the deformation in the direction of length of the concrete caused by the change of unit temperature, in the unit of 10^{-6}/℃. The linear expansion coefficient (α) of the concrete is mainly dependent on the linear expansion coefficient of the aggregate, the relevant literatures show that the α-value of the quartzite aggregate concrete is the biggest one, the α-value of sandstone aggregate concrete is $(11-12) \times 10^{-6}/℃$, the α-value of the granite aggregate concrete is $(8-9) \times 10^{-6}/℃$, the α-value of limestone aggregate concrete is the smallest, which is just $(5-6) \times 10^{-6}/℃$. The smaller the α-value of the concrete, the smaller the temperature deformation (ΔT), the smaller the temperature stresses produced, therefore, its cracking resistance is stronger.

Ⅴ Interrelation of thermal properties

In the mass concrete, to allow the concrete to dissipate the excess heat and minimize the uneven volume change of the dam, the thermal properties become very important. The heat conductivity coefficient refers to the speed at which the heat is conducted through the unit area of an object when there is unit temperature difference on both sides of the unit thickness object. If the heat conductivity coefficient is divided by the specific heat multiplied by the apparent density $[\alpha = \lambda/(c\gamma)]$, the so called "thermal diffusivity (or thermal diffusivity)" can be obtained. The thermal diffusivity is the characteristic index that indicates that the concrete carries out heat exchange. The major factor that affects the thermal properties of the concrete is the mineral composition of the aggregate, which is the factor that is impossible to be clearly defined in the specification. The specification allows the adjustment to the requirement of the factors such as the ratio of the cement, admixture and sand, and even the water consumption, but they have slight impact on the thermal properties. Introducing certain amount of air into the concrete is an important factor, the introduction of the air content not only improves the performance of the fresh mixed concrete and the durability, but also forms good thermal barrier due to the air contained in the concrete.

5.3 Study on Relationship of Admixture, *VC* Value, Temperature with Setting Time

5.3.1 Summary

Baise Multipurpose Dam Project is located in subtropical region, the high temperature period is long, it is inevitable for the RCC to be constructed under high temperature condition, for the purpose of solving the adverse effect of serious shortening of setting time of RCC by diabase aggregate and hot climate, in September of 2002, 8 kinds of superior quality admixtures were selected for Baise Multipurpose Dam Project to widen the research and test.

The test results show that the mechanism for the setting time of the diabase aggregate RCC is totally different from that of the normal concrete, therefore, if the diabase aggregate RCC is mixed with different kinds of the admixtures, its setting time, water consumption and adaptability are also different. By comprehensive comparison from the viewpoint of meeting the performances of diabase aggregate RCC under high temperature climate such as the setting time, water consumption, workability, etc., the selected retardation type superplasticizer ZB-1RCC15 can meet the requirements of RCC construction. To determine the construction of diabase aggregate RCC under high temperature conditions, the test study on the relationship of mixing amount for the admixture ZB-1RCC15 with *VC* value, temperature and setting time is carried out.

5.3.2 Relationship of Mixing Amount of Admixture with Setting Time

The diabase sand powder is prepared by 20%, the setting time test is carried out under indoor condition of 18-22 ℃ and outdoor natural condition of 21-31 ℃, see Table 5.5 for the test result of relationship of the mixing amount of admixture ZB-1RCC15 with setting time, the result shows that:

Table 5.5 Test result of relationship of mixing amount of admixture ZB-1RCC15 with setting time

Test No.	Water-cement ratio	Gradation	Additive		VC value (s)	Initial setting time (h: min)		Cement type
			Variety	Mixing amount (%)		Indoor 18–22℃	Natural 21–31℃	
NR11-6-2	0.5	II	ZB-1RCC15	0.8	6	6:40	5:56	Tiandong moderate heat portland cement 525#
NR11-7-2	0.5	II	ZB-1RCC15	1.0	6	8:11	6:50	
NR11-1-2	0.6	Quasi III	ZB-1RCC15	0.8	6	5:50	4:45	
NR11-2-2	0.6	Quasi III	ZB-1RCC15	1.0	6	7:10	6:33	
NR11-5-2	0.6	Quasi III	ZB-1RCC15	1.2	5	8:34	7:58	

When the mixing amount of ZB-1RCC15 is 0.8%, the initial setting time for II gradation and quasi III gradation of RCC is 6 h 40 min and 5 h 50 min respectively indoor, and 5 h 56 min and 4 h 45 min respectively outdoor.

When the mixing amount of ZB-1RCC15 is 1.0%, the initial setting time for II gradation and quasi III gradation of RCC is 8 h 11 min and 7 h 10 min respectively indoor, and 6 h 50 min and 6 h 03 min respectively outdoor.

When the mixing amount of ZB-1RCC15 is 1.2%, the indoor and outdoor initial setting time for quasi III gradation RCC is 8 h 34 min and 7 h 58 min respectively.

The data analysis shows that:

When the mixing amount of the admixture to II gradation RCC increases from 0.8% to 1.0%: the initial setting time extends about 90 min under the condition that indoor temperature is from 18 ℃ to 22 ℃; the initial setting time extends about 120 min under the condition that natural temperature is from 21 ℃ to 31 ℃.

When the mixing amount of the admixture to quasi III gradation RCC increases from 0.8% to 1.0% and 1.2%: the initial setting time extends about 80 min and 160 min respectively under the condition that indoor temperature is from 18 ℃ to 22 ℃; the initial setting time extends about 35 min and 120 min under the condition that natural temperature is from 21 ℃ to 31 ℃.

Analysis of above test results show that: Under the circumstance that the mixing amount of the admixture is same, the initial setting time of II gradation RCC is 50 min more or less extended over that of quasi III gradation RCC. The setting time prolongs correspondingly with the increase of mixing amount of the admixture, that is the initial setting time of the RCC will extend 30 min more or less for each 0.1% increase of the admixture.

5.3.3 Relationship of *VC* Value and Setting Time

Under the condition that the mixing amount of the admixture is not changed and the temperature is the same, adopting smaller *VC* value can extend the initial setting time. See Table 5.6 for the test result for the relationship of *VC* value with setting time, the results show that, when the room temperature is from 19 ℃ to 20 ℃, if the *VC* value is 5.0 s, 6.0 s and 7.0 s, the corresponding initial setting time is 5 h 10 min, 4 h 52 min and 4 h 29 min respectively, if the *VC* value is increased by 2 s, the initial setting time will be shortened for 40 min more or less. The result shows that: Under the condition that the air temperature is same, decreasing the *VC* value of the RCC can extend the initial setting time correspondingly, that is the initial setting time will be extended for about 20 min for each 1 s decrease in the *VC* value.

Table 5.6   Test result of relationship of VC value with setting time (ZB-1RCC15 0.8%)

Test No.	Rock powder content (%)	ZB-1 RCC15 (%)	DH$_9$ (%)	Water content (kg/m^3)	VC value (s)	Temperature (℃)	Concrete temperature (℃)	Condition	Time of setting (h: min) Temperature (℃)	Initial setting	Final setting
KF1-7	24	0.8	0.015	106	5.0	23	24	Indoor	19-20	5:10	18:08
								Natural	17-30	5:13	15:27
KF1-8	24	0.8	0.015	103	6.0	22	23	Indoor	19-20	4:52	16:54
								Natural	20-33	4:35	10:25
KF1-9	24	0.8	0.015	100	7.0	23	24	Indoor	19-20	4:29	11:10
								Natural	26-33	4:15	8:30

5.3.4 Relationship of Temperature and Mixing Amount of Admixture with Setting Time

The RCC of Baise Multipurpose Dam Project was the largest RCC dam under construction at that time, the construction intensity was high, as the project is located in subtropical region, the period of high temperature is long, the difference in temperature between day and night is great and the quantity of evaporation is large, the RCC was inevitably constructed under high temperature conditions. The RCC is the semi plastic concrete, it was observed in the site operation that: there is very great influence on the setting time of the RCC by the temperature, quantity of evaporation, sunlight and diabase aggregate. To ensure the high intensity construction of diabase RCC under high temperature condition (25-35 ℃), it becomes necessary to study the relationship of high temperature climate at site with the setting time of the RCC and mixing amount of the admixture. See Table 5.7 for the test result on the relationship of temperature and mixing amount of admixture with setting time, the results show that:

For Nos. 1-1 to 1-8, when the mixing amount of ZB-1RCC15 (Test specimen 3) is 0.8%, the air temperature is 21 ℃, 25 ℃, 29 ℃, 33 ℃ and 37 ℃ respectively, the initial setting time of the RCC is 5 h 30 min, 4 h 25 min, 3 h 25 min, 3 h 15 min and 2 h 30 min correspondingly, that is if the air temperature raises by 16 ℃, the initial setting time will be shortened for about 3 h 00 min.

For Nos. 2-1 to 2-5, when the mixing amount of ZB-1RCC15 (Test specimen 3) is 1.0%, the air temperature is 21 ℃, 24 ℃, 27 ℃, 29 ℃ and 33 ℃ respectively, the initial setting time of the RCC corresponds to 8 h 15 min, 6 h 45 min, 6 h 05 min, 5 h 40 min and 5 h 10 min, that is if the air temperature raises by 12 ℃, the initial setting time will be shortened about 3 h05 min.

For Nos. 3-1 to 3-7, when the mixing amount of ZB-1RCC15 (Test specimen 3) is 1.5%, the air temperature is 19 ℃, 21 ℃, 24 ℃, 27 ℃, 29 ℃, 30 ℃ and 35 ℃ respectively, the initial setting time of the RCC is 14 h 55 min, 9 h 30 min, 7 h 53 min, 7 h 30 min, 6 h 25 min, 6 h and 5 h 48 min correspondingly, that is if the air temperature raises by 16 ℃, the initial setting time will be shortened for about 9 h.

Above test results show that, the temperature change has great influence on the setting time of the RCC, the relationship of the temperature, mixing amount of the admixture with the setting time shows that:

When the mixing amount of ZB-1RCC15 (Test specimen 3) is 0.8%, under the circumstance that the air temperature is less than 25 ℃, the initial setting time will be shortened for about 12 min for every 1 ℃ of temperature rise.

When the mixing amount of ZB-1RCC15 (Test specimen 3) is 1.0%, under the circumstance that the air

Table 5.7　Test results on the relationship of temperature and mixing amount of admixture with setting time

S/N	Test time	ZB-1RCC15 (Sample 3) (%)	DH_9 (%)	VC value (s)	Temperature (℃)	Average temperature (℃)	Time of setting(h: min) Initial setting	Final setting
1-1	14:00	0.8	0.015	3	23 – 20	21	5:30	16:32
1-2	9:00	0.8	0.015	3	19 – 23	21	5:00	12:25
1-3	10:00	0.8	0.015	3	22 – 28	25	4:25	10:20
1-4	15:00	0.8	0.015	3	20 – 28	25	4:42	11:42
1-5	9:00	0.8	0.015	8	27 – 31	28	4:04	11:10
1-6	9:00	0.8	0.015	7	25 – 32	29	3:25	9:40
1-7	15:00	0.8	0.015	3	33 – 32	33	3:15	8:20
1-8	16:00	0.8	0.015	7	37 – 36	37	2:30	6:46
2-1	8:00	1.0	0.015	3	19 – 23	21	8:15	16:50
2-2	8:30	1.0	0.015	3	21 – 28	24	6:45	12:00
2-3	14:00	1.0	0.015	3	28 – 24	27	6:05	16:10
2-4	8:00	1.0	0.015	3	25 – 33	29	5:40	11:48
2-5	9:00	1.0	0.015	4	27 – 37	33	5:10	10:30
3-1	14:00	1.5	0.015	3	23 – 17	19	14:55	30:25
3-2	8:00	1.5	0.015	3	19 – 23	21	9:30	18:30
3-3	8:00	1.5	0.015	3	21 – 28	24	7:53	15:53
3-4	14:00	1.5	0.015	3	28 – 24	27	7:30	17:30
3-5	8:00	1.5	0.015	3	25 – 33	29	6:25	15:46
3-6	14:00	1.5	0.015	3	34 – 26	30	6:00	15:52
3-7	14:00	1.5	0.015	4	37 – 30	35	5:48	19:08

temperature is less than 28 ℃, the initial setting time will be shortened for about 12 min for every 1 ℃ of temperature rise.

When the mixing amount of ZB-1RCC15 (Test specimen 3) is 1.5%, under the circumstance of sun exposure that the air temperature is more than 28 ℃, the initial setting time will be shortened for about 34 min for every 1 ℃ of temperature rise. It sufficiently shows that the setting time of RCC is shortened rapidly in the condition of sun exposure under high temperature climate.

The above results further demonstrate that, under the circumstance that the VC value is kept basically same, in case that the mixing amount of the admixture is different, the setting time of the RCC will be shortened correspondingly with the rise of the temperature. In the daytime, the air temperature increases gradually until the sunset in the evening, the initial setting time of RCC shortens rapidly under the condition of sun exposure; The air temperature decreases in the evening as the sun sets, the initial setting time of the RCC will extend gradually; It also shows that, when the air temperature is more than 25 ℃, the mixing amount of ZB-1RCC15 (Test specimen 3) needs to be increased from 0.8% to 1.0%; in the condition that is under the sun exposure when the air temperature is higher than 28 ℃, the temperature rises rapidly, the mixing amount of ZB-1RCC15 (Test specimen 3) needs to be further increased to 1.5%, which can meet the construction requirements of diabase RCC in high temperature climate.

Chapter 5 Research and Application of RCC Performance

5.3.5 Conclusion

1. The mechanism for the setting time of the diabase aggregate RCC is totally different from that of the normal concrete, therefore, the adaptability of different kinds admixtures to diabase aggregate RCC is also different. By comparison, the new product ZB-1RCC15 (Test specimen 3) manufactured by Zhejiang Longyou Admixture Factory with new material of macromolecular organic matters added can meet the construction requirements of RCC.

2. Under the circumstance of certain temperature and same mixing amount of the admixture, the initial setting time of II gradation RCC extends compared with that of quasi III gradation RCC. With the increase of the mixing amount of the admixture, the initial setting time of diabase RCC will be extended correspondingly; The initial setting time will extend 30 min more or less for every 0.1% increase in the mixing amount of the admixture.

3. Under the condition that the air temperature is same, decreasing the VC value of the RCC can extend the setting time, that is the initial setting time will be extended for about 20 min correspondingly for each 1 s decreased in the VC value.

4. According to the construction condition, when the air temperature is less than 25 ℃, the mixing amount of ZB-1RCC15 (Test specimen 3) is 0.8%; when the air temperature is more than 25 ℃, the mixing amount of ZB-1RCC15 (Test specimen 3) needs to be 1.0%; in the condition that the air temperature is more than is 28 ℃ and under the sun exposure, the mixing amount of ZB-1RCC15 (Test specimen 3) needs to be 1.5%, which can meet the setting time requirements of RCC in high temperature climate.

5. During the high temperature periods, by keeping the mix proportion parameter of the RCC unchanged, selecting different mixing amount of the admixture according to different periods of the air temperature, under the condition that the superposition action of VC value is minimized by increasing the mixing amount of the admixture, the setting time of the RCC is extended, this method is simple but easy to implement. The research result shows that the scheme that increases the mixing amount of the admixture is an effective technical approach to solve the setting time of diabase RCC in subtropical high temperature climate.

See Table 5.8 for the mix proportion of diabase aggregate RCC construction in high-temperature climate.

Table 5.8　　Mix proportion of of diabase aggregate RCC construction

S/N	Design requirements	Gradation	Water-cement ratio	Sand ratio (%)	Fly ash (%)	ZB-1 (%)	DH$_9$ (%)	VC value (s)	Material amount (kg/m³)			
									Water	Total cementing material	Cement	Coal ash	Unit weight
1-1	R$_{180}$15 S2D50	Quasi III	0.60	34	63	0.8	0.015	3–5	96	160	59	101	2 650
1-2						1.0	0.015	3–5					
1-3						1.5	0.015	3–5					
2-1	R$_{180}$20 S10D50	II	0.50	38	58	0.8	0.015	3–5	106	212	89	123	2 600
2-2						1.0	0.015	3–5					
2-3						1.2	0.015	3–5					

Note: 1. Raw material: medium-heat portland cement, class II fly ash, diabase coarse and fine aggregate, sand FM = 2.8 ± 0.2, rock powder content 20 ± 2%.

2. Conditions for the mixing amount of the admixture: in low temperature season and night, when the mixing amount of ZB-1RCC15 is 0.8%; when the air temperature is more than 25 ℃, the mixing amount of ZB-1RCC15 is 1.0%; when the air temperature is more than 28 ℃ and under sun exposure, the mixing amount of ZB-1RCC15 is 1.5%.

5.4 Autogenous Volume Deformation, Adiabatic Temperature Rise and Creep Tests

The autogenous volume deformation, adiabatic temperature rise and creep tests for the normal and RCC in dam of Jin'anqiao Hydropower Station is carried out in accordance with the requirements of technical terms in the bidding document and in compliance with *Test code for hydraulic concrete* SL 352—2006.

5.4.1 Raw Material

5.4.1.1 Cement

The cement is of Yongbao 42.5 medium-heat Portland cement, see Table 5.9 for results of physical and mechanical properties test of cement. The test results show that: The physical and mechanical properties of the cement are in compliance with the requirements in GB 200—2003 standard *Moderate Heat Portland Cement, Low Heat Portland Cement, Low Heat Portland Slag Cement*.

Table 5.9   Test result for physical and mechanical properties of the cement

Cement type	Physical properties					Indexes of mechanical properties						Hydration heat (kJ/kg)	
	Specific surface area (m^2/kg)	Standard consistency (%)	Stability	Time of setting (h: min)		Compressive strength (MPa)			Bending strength (MPa)				
				Initial setting	Final setting	3 d	7 d	28 d	3 d	7 d	28 d	3 d	7 d
42.5 moderate heat Portland cement	310	25.2	Qualified	2:31	3:24	19.8	32.1	50.2	4.2	5.5	8.0	212	260
GB 200—2003	≥250	—	Qualified	≥1:00	≤12:00	≥12.0	≥22.0	≥42.5	≥3.0	≥4.5	≥6.5	≤251	≤293

5.4.1.2 Fly Ash

The fly ash used is the Class II fly ash made by Panzhihua Liyuan Fly Ash Product Co., Ltd., see Table 5.10 for the result of quality test; the abrasion-resistance concrete used is Qujing Class I fly ash, see Table 5.10 for the result of quality test. The test results show that: The fly ash complies with the requirements of technical indexes in *Technical Specification of Fly Ash for Use in Hydraulic Concrete* DL/T 5055—2007.

Table 5.10   Fly ash quality test result

Fly ash type		Moisture content (%)	Fineness (%)	Water demand ratio (%)	Loss on ignition (%)	SO_3 (%)
Panzhihua Grade II		0.1	17.0	100	3.52	0.20
Qujing Grade I		0.1	7.8	95	2.93	0.49
DL/T 5055—2007	Grade I fly ash	≤1	≤25	≤105	≤8	≤3
	Grade II fly ash		≤12	≤95	≤5	

5.4.1.3 Rock Powder

The rock powder used is the limestone rock powder processed by Yongbao Cement Co., Ltd., see Table 5.11

for the results of its performance test. The fineness refers to the sieve margin that passes through 0.08 mm square mesh sieve.

Table 5.11　　　　　　　　　　　　　Mechanical property test result

Type of rock powder	Moisture content(%)	Fineness(%)
Limestone rock powder	0.1	16.6

5.4.1.4 Aggregate

The aggregate used for concrete mix proportion is the crushed basalt aggregate produced by the screening system of Sinohydro Bureau 8. See Table 5.12 for the result of manufactured basalt sand quality test.

Table 5.12　　　　　　　　　　　　Result of manufactured basalt sand test

Variety	Fineness module	Rock powder content(%)
Artificial sand	2.78	12.8
DL/T 5144—2001	2.2 - 2.9	10 - 22

5.4.1.5 Additive

The admixtures used is the JM-II retardation type superplasticizer manufactured by Jiangsu Sobute New Materials Co., Ltd. and the ZB-1G air entraining agent produced by Zhejiang Longyou Wuqiang Concrete Admixture Co., Ltd. The performance test of admixture-mixed-concrete is carried out for the admixture in compliance with *Technical Standard for Hydraulic Concrete Admixtures* DL/T 5100—1999 and *Concrete Admixtures* GB 8076—1997. See Table 5.13 for test results, results show that: the performance of concrete with water reducing agent and air entraining agent can meet the requirements of the standards.

Table 5.13　　　　　　　　Performance test results of concrete mixed with admixtures

Description	Additive		Water reducing ratio (%)	Air content (%)	Ratio of bleeding rate (%)	Difference in setting time (min)		Compressive strength ratio (%)		
	Variety	Mixing amount (%)				Initial setting	Final setting	3 d	7 d	28 d
Water reducing agent	JM-II	0.6	21.8	2.4	32.6	+310	+290	150	144	125
Air entraining agent	ZB-1G	0.005	7.3	5.4	11.5	+20	+17	98	96	93
GB 8076—1997	Retarding high-range water reducing agent		≥12	<4.5	≤100	> +90	—	≥125	≥125	≥120
	Air entraining agent		≥6	>3.0	≤70	-90 - +120	-90 - +120	≥95	≥95	≥90

5.4.2 Test Parameters for Mix Proportion of Dam Concrete

The items of autogenous volume deformation, adiabatic temperature rise and creep tests will be formed according to the mix proportion of RCC of Jin'anqiao Dam, see Table 5.14 for the test parameters of mix proportion of dam normal concrete, see Table 5.15 for the detail of the test parameters of mix proportion of dam RCC.

Table 5.14　　Test parameters for mix proportion of dam normal concrete

No.	Design ndex	Gradation	Water-cement ratio	Sand ratio (%)	Fly ash (%)	Water consumption (kg/m^3)	JM-II (%)	ZB-1G_9 (%)	Slump (cm)	Density (kg/m^3)
JC-1	C_{90}20W8F100	III	0.55	31	35	120	0.8	0.015	3–5	2 620
JC-2		IV	0.55	25	35	100	0.8	0.015	3–5	2 660

Table 5.15　　Test parameters for mix proportion of dam RCC

No.	Design index	Gradation	Water-cement ratio	Sand ratio (%)	Fly ash (%)	Water consumption (kg/m^3)	ZB-1 RCC15 (%)	ZB-1G (%)	VC value (s)	Density (kg/m^3)
JR-1	C_{90}15W6F100	III	0.53	33	63	90	1.2	0.15	1–3	2 630
JR-2	C_{90}20W6F100	III	0.47	33	60	90	1.0	0.30	3–5	2 630
JR-3	C_{90}20W8F100	II	0.47	37	55	100	1.0	0.30	3–5	2 600

5.4.3　Concrete Autogenous Volume Deformation Test

When the concrete is under the condition of constant temperature and moisture isolation, the volume deformation of concrete is caused by the hydration of cementing materials. See Table 5.16 for test results of autogenous volume deformation of dam normal and RCC, see Figure 5.1 for the relation curve of autogenous volume deformation value and age. It can be known from the test results and autogenous volume deformation curve that:

1. The autogenous volume deformation of normal concrete C_{90}20W8F100 and C_{90}50W8F150 are shrinkage type, its overall development trend is that the shrinkage of autogenous volume deformation will increase with the increase of the age. The trend of C_{90}20W8F100 concrete become slow after about 130 d, in two kinds of graded concrete, the shrinkage value of autogenous volume deformation for III gradation is higher than that for IV gradation.

2. See Table 5.16 for results of autogenous volume deformation test for the RCC, the results show that the type of autogenous volume deformation for the RCC is shrinkage:

　　For C_{90}15W6F100 III gradation of 1 d, 3 d, 7 d, 28 d, 90 d, 180 d and 360 d, the values of autogenous volume deformation are -2.2×10^{-6}, -4.5×10^{-6}, -8.3×10^{-6}, -22.2×10^{-6}, -30.5×10^{-6}, -32.4×10^{-6}, -34.0×10^{-6}.

　　For C_{90}20W6F100 III gradation of 1 d, 3 d, 7 d, 28 d, 90 d, 180 d and 360 d, the values of autogenous volume deformation are -3.5×10^{-6}, -5.5×10^{-6}, -9.2×10^{-6}, -24.5×10^{-6}, -35.0×10^{-6}, -36.0×10^{-6}, -37.0×10^{-6}.

　　For C_{90}20W8F100 II gradation of 1 d, 3 d, 7 d, 28 d, 90 d, 180 d and 360 d, the values of autogenous volume deformation are -3.5×10^{-6}, -7.5×10^{-6}, -14.5×10^{-6}, -23.5×10^{-6}, -37.8×10^{-6}, -39.5×10^{-6}, -40.3×10^{-6} respectively.

Table 5.16   Results of autogenous volume deformation test of dam normal and RCC

JC-1 normal state		JC-2 normal state		JR-1 rolling		JR-2 rolling		JR-3 rolling	
Age (d)	Deformation ($\times 10^{-6}$)	Age (d)	Deformation ($\times 10^{-6}$)	Age (d)	Deformation ($\times 10^{-6}$)	Age (d)	Deformation ($\times 10^{-6}$)	Age (d)	Deformation ($\times 10^{-6}$)
1	-3.1	1	-2.1	1	-2.2	1	-3.5	1	-3.5
2	-5.2	2	-3.1	2	-3.8	2	-4.0	2	-4.0
3	-6.8	3	-3.8	3	-4.5	3	-5.5	3	-7.5
5	-9.8	5	-4.8	4	-5.0	4	-6.0	4	-8.5
13	-18.5	13	-12.5	7	-8.3	7	-9.2	7	-14.5
28	-28.2	28	-20.5	14	-12.5	14	-15.5	14	-18.5
40	-37.5	40	-28.2	28	-22.2	28	-24.5	28	-23.5
50	-40.5	50	-30.5	40	-25.2	40	-23.8	40	-26.5
65	-42.2	65	-35.2	55	-27.2	55	-25.2	55	-29.0
78	-44.5	78	-40.5	75	-30.8	75	-34.5	75	-33.0
92	-46.5	92	-41.0	90	-30.5	90	-35.0	90	-37.8
105	-47.0	105	-41.5	105	-31.8	105	-35.5	105	-38.6
120	-48.5	120	-42.0	120	-31.6	120	-35.0	120	-39.0
130	-50.5	130	-43.5	130	-32.0	130	-35.5	130	-39.2
145	-51.0	145	-43.8	150	-32.0	150	-35.5	150	-39.3
150	-51.5	150	-44.0	160	-31.5	160	-36.0	160	-38.5
167	-51.5	167	-44.0	180	-32.4	180	-36.0	180	-39.5
180	-51.5	180	-44.1	210	-33.0	210	-36.5	210	-39.5
210	-52.0	210	-43.2	240	-33.5	240	-36.5	240	-39.0
240	-52.0	240	-42.5	270	-33.5	270	-37.0	270	-39.2
270	-53.0	270	-43.5	300	-34.0	300	-37.0	300	-39.3
300	-52.3	300	-44.0	330	-34.0	330	-37.0	330	-40.0
330	-53.5	330	-44.5	360	-34.0	360	-37.0	360	-40.3
360	-54.0	360	-45.0						

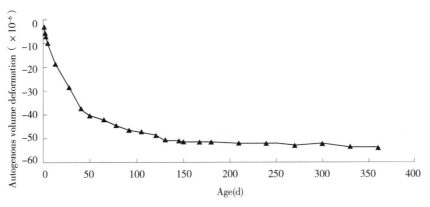

(a) No.: JC-1 normal concrete autogenous volume deformation curve with time

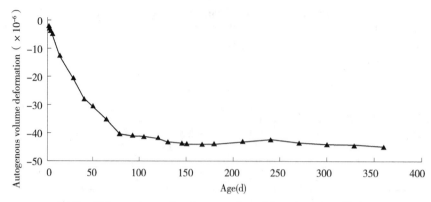

(b) No.: JC-2 normal concrete autogenous volume deformation curve with time

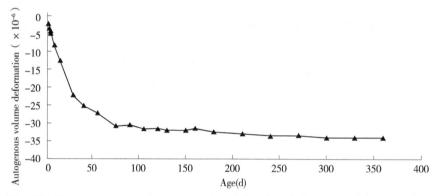

(c) No.: JR-1 Change process curve of autogenous volume deformation of roller compacted concrete with time

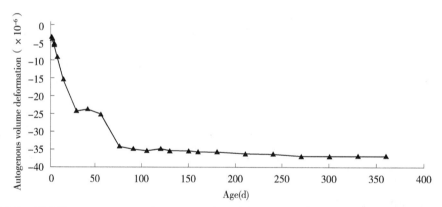

(d) No.: JR-2 Change process curve of autogenous volume deformation of roller compacted concrete with time

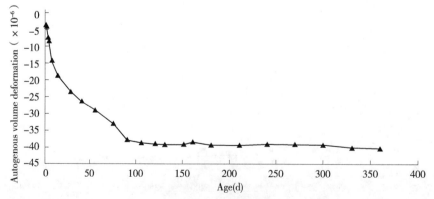

(e) No.: JR-3 Change process curve of autogenous volume deformation of roller compacted concrete with time

Figure 5.1 *Relation curve of autogenous volume deformation value and age change process*

5.4.4 Dam Concrete Adiabatic Temperature Rise Test

The temperature of the concrete is raised by the heat generated due to the hydration of the cement. The adiabatic temperature rise test of concrete is conducted under the adiabatic condition without heat exchange, and the temperature rise of concrete due to cement hydration is measured directly. Therefore, the lower the cement content per unit of concrete is, the smaller the adiabatic temperature rise. Full-grade is used in adiabatic temperature rise test. The adiabatic temperature rise test results of normal and RCC of dam for mix proportion are shown in Table 5.17; the adiabatic temperature rise duration curve is shown in Figure 5.2; the fitting equation of adiabatic temperature rise is shown in Table 5.18.

Table 5.17　　Adiabatic temperature rise test results of normal concrete and RCC of dam

Age	JC-1 normal state	JC-2 normal state	JR-1 rolling	JR-2 rolling	JR-3 rolling
Initial temperature of concrete	21.2	20.5	19.7	20.0	20.5
Age (d)	Temperature rise (℃)	Temperature rise (℃)	Temperature rise (℃)	Temperature rise (℃)	Temperature rise (℃)
0.13	0.4	0.4	0.3	0.3	0.3
0.25	0.8	0.8	0.6	0.6	0.7
0.38	1.8	1.6	1	1	1.1
0.5	2.9	2.6	1.3	1.4	1.8
0.75	6.2	5.8	2.1	2.2	2.6
1	8.0	7.9	2.8	3.1	3.4
1.25	9.9	8.8	3.4	4.1	4.6
1.5	11.0	10.4	4.0	4.5	5.1
2	12.5	12.3	5.1	5.8	6.6
3	15.5	14.5	6.8	7.6	8.4
4	17.4	16.2	8.2	8.8	9.6
5	18.9	17.5	9.4	10.0	10.8
6	19.9	18.7	10.3	10.8	11.7
7	20.9	19.3	11.1	11.6	12.6
8	21.4	19.7	11.6	12.3	13.4
9	21.8	20.1	12.1	12.8	14.0
10	22.1	20.3	12.6	13.3	14.3
11	22.4	20.5	13.0	13.7	14.5
12	22.6	20.7	13.4	14.0	14.8
13	22.7	20.9	13.7	14.2	15.0
14	22.8	21.0	13.9	14.5	15.3
15	23.1	21.1	14.2	14.8	15.5
16	23.5	21.2	14.4	15.0	15.8
17	23.8	21.3	14.6	15.3	16.1
18	24.1	21.5	14.8	15.5	16.5
19	24.2	21.7	15.0	15.7	17.0
20	24.4	21.7	15.1	15.9	17.2
21	24.5	21.8	15.2	16.0	17.4
22	24.6	21.8	15.3	16.2	17.6
23	24.7	21.9	15.4	16.3	17.8

Continued Table 5.17

Age	JC-1 normal state	JC-2 normal state	JR-1 rolling	JR-2 rolling	JR-3 rolling
24	24.9	22.0	15.5	16.4	18.0
25	25.0	22.0	15.6	16.5	18.1
26	25.1	22.1	15.7	16.6	18.3
27	25.1	22.2	15.8	16.7	18.4
28	25.1	22.2	15.9	16.8	18.6

Table 5.18 Fitting equation of adiabatic temperature rise of dam normal concrete and RCC

S/N	No. and type	Design index	Gradation	28 d adiabatic temperature rise value (℃)	Fitting equation	Final adiabatic temperature rise (℃)
1	JC-1 Normal concrete	$C_{90}20W8F100$	III	25.1	$T = 27.2t/(2.32 + t), t > 0.5$ d, $R = 0.9998$	27.2
2	JC-2 Normal concrete	$C_{90}20W8F100$	IV	22.2	$T = 25.4t/(3.32 + t), t > 0.75$ d, $R = 0.9990$	25.4
3	JR-1 Roller compacted concrete	$C_{90}15W6F100$	III	15.9	$T = 19.0t/(5.28 + t), t > 1$ d, $R = 0.9996$	19.0
4	JR-2 Roller compacted concrete	$C_{90}20W6F100$	III	16.8	$T = 19.7t/(4.90 + t), t > 1.25$ d, $R = 0.9999$	19.7
5	JR-3 Roller compacted concrete	$C_{90}20W8F100$	II	18.6	$T = 21.6t/(5.08 + t), t > 1.25$ d, $R = 0.9999$	21.6

The result indicates:

Adiabatic temperature rise of normal concrete. The adiabatic temperature rise of $C_{90}20W8F100$ III gradation concrete is 25.1 ℃ in 28 d and IV gradation concrete is 22.2 ℃ in 28 d. Under same design parameters, the adiabatic temperature rise of the IV gradation concrete is smaller than that of the III gradation, the main reason is that the unit cement content of the IV gradation concrete is less than that of the III gradation concrete.

Adiabatic temperature rise of RCC. The adiabatic temperature rise of $C_{90}15W6F100$ III gradation concrete is 15.9 ℃ in 28 d; The adiabatic temperature rise of $C_{90}20W6F100$ III gradation concrete is 16.8 ℃ in 28 d; The adiabatic temperature rise of $C_{90}20W8F100$ II gradation concrete is 18.6 ℃ in 28 d. The adiabatic temperature rise of RCC is directly related to the cement content and cementing material of unit concrete.

5.4.5 Creep Test of RCC

Under the action of long-term constant load, the concrete will be deformed as creep deformation along with time. The creep degree is the creep deformation under the action of unit stress and it is a function of loading age and time. When the concrete reaches the loading age, the cube compressive strength and static compressive

Chapter 5 Research and Application of RCC Performance

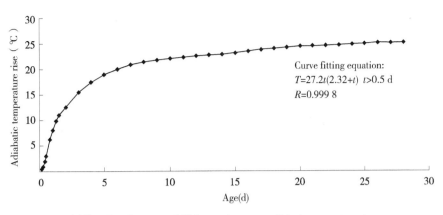

(a) No. : Duration curve of JC-1 normal concrete adiabatic temperature rise

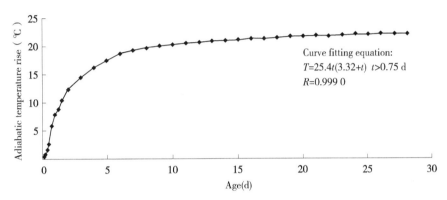

(b) No. : Duration curve of JC-2 normal concrete adiabatic temperature rise

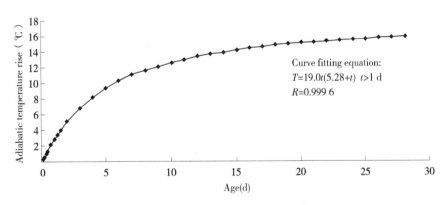

(c) No. : JR-1 Adiabatic temperature rise duration curve of roller compacted concrete

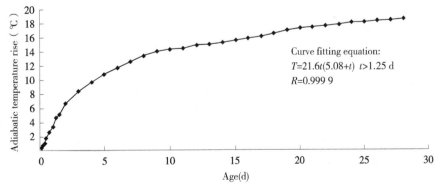

(d) No. : JR-3 Adiabatic temperature rise duration curve of roller compacted concrete

Figure 5.2 Duration curve of adiabatic temperature rise of dam concrete

elasticity modulus of each group of specimens are measured. Test results are shown in Table 5.19 below.

Table 5.19 Mechanical test results of concrete creep test

No. of specimen group	Age of the test specimen (d)	Compressive strength of cube specimen (MPa)	Static compressive elasticity modulus (GPa)
$C_{90}15W6F100$	7	8.8	10.5
	28	14.6	16.4
	90	21.2	22.3
	180	27.0	27.8
	365	32.2	33.4
$C_{90}20W6F100$	7	12.0	13.1
	28	20.3	20.8
	90	27.8	28.4
	180	33.6	34.2
	365	38.8	39.2

The compressive creep degree is calculated as follows:

$$C_c = \varepsilon_c \times \frac{1}{\sigma_c} \times \alpha$$

Where C_c = the creep degree of prototype concrete at a certain age, 10^{-6}/MPa

σ_c = the compressive stress of concrete specimen, MPa

α = mortar ratio (Calculated as 0.84)

The statistics of test results of C_{90} 15W6F100 concrete and C_{90} 20W6F100 concrete are shown in Table 5.20 and Table 5.21. The creep degree curve is shown in Figure 5.3 and Figure 5.4. According to the creep test results, the creep of early age loading is large and the creep degree decreases when the age increases; And, the creep degree increases with the increase of load holding time. The creep degree increases faster at the early stage of loading and the creep degree increases slowly after 90 d and tends to be stable.

Table 5.20 Creep test results of C_{90}15W6F100 concrete at 7 d, 28 d, 90 d, 180 d and 365 d

Creep results at 7 d		Creep results at 28 d		Creep results at 90 d		Creep results at 180 d		Creep results at 365 d	
Load holding time	Creepage degree ($\times 10^{-6}$/MPa)	Load holding time	Creepage degree ($\times 10^{-6}$/MPa)	Load holding time	Creepage degree ($\times 10^{-6}$/MPa)	Load holding time	Creepage degree ($\times 10^{-6}$/MPa)	Load holding time	Creepage degree ($\times 10^{-6}$/MPa)
2 h	0.9	2 h	0.3	2 h	0.7	2 h	0.1	2 h	0.1
6 h	5.9	6 h	0.5	6 h	2.0	6 h	2.2	6 h	0.2
1 d	26.9	1 d	1.0	1 d	4.1	1 d	4.8	1 d	0.5
2 d	37.4	2 d	8.7	2 d	6.7	2 d	6.1	2 d	1.8
3 d	43.5	3 d	13.3	3 d	8.1	3 d	6.8	3 d	2.5
4 d	47.9	4 d	16.5	4 d	9.3	4 d	7.3	4 d	3.1
5 d	51.2	5 d	19.0	5 d	10.4	5 d	7.8	5 d	3.5
6 d	54.0	6 d	21.0	6 d	11.4	6 d	8.1	6 d	3.8
7 d	56.3	7 d	22.8	7 d	12.3	7 d	8.4	7 d	4.1
8 d	58.4	8 d	24.3	8 d	13.2	8 d	8.6	8 d	4.3
9 d	60.1	9 d	25.6	9 d	14.0	9 d	8.9	9 d	4.6

Continued Table 5.20

Creep results at 7 d		Creep results at 28 d		Creep results at 90 d		Creep results at 180 d		Creep results at 365 d	
Load holding time	Creepage degree ($\times 10^{-6}$/MPa)	Load holding time	Creepage degree ($\times 10^{-6}$/MPa)	Load holding time	Creepage degree ($\times 10^{-6}$/MPa)	Load holding time	Creepage degree ($\times 10^{-6}$/MPa)	Load holding time	Creepage degree ($\times 10^{-6}$/MPa)
10 d	61.7	10 d	26.8	10 d	14.8	10 d	9.1	10 d	4.7
11 d	63.2	11 d	27.8	11 d	15.5	11 d	9.3	11 d	4.9
12 d	64.5	12 d	28.8	12 d	16.2	12 d	9.6	12 d	5.1
13 d	65.7	13 d	30.7	13 d	16.8	13 d	9.8	13 d	5.2
14 d	66.8	14 d	31.5	14 d	17.3	14 d	9.9	14 d	5.4
21 d	75.0	21 d	36.1	21 d	19.3	21 d	10.8	21 d	6.1
28 d	79.3	28 d	39.3	28 d	20.7	28 d	11.4	28 d	6.6
35 d	82.7	35 d	41.8	35 d	21.7	35 d	11.9	35 d	7.1
42 d	85.5	42 d	43.8	42 d	22.5	42 d	12.3	42 d	7.4
49 d	87.8	49 d	45.5	49 d	23.2	49 d	12.6	49 d	7.7
56 d	89.8	56 d	47.0	56 d	23.8	56 d	12.9	56 d	7.9
63 d	91.6	63 d	48.4	63 d	24.4	63 d	13.2	63 d	8.1
70 d	93.7	70 d	49.5	70 d	25.0	70 d	13.5	70 d	8.3
77 d	95.6	77 d	50.6	77 d	25.4	77 d	13.8	77 d	8.5
84 d	97.0	84 d	51.6	84 d	25.9	84 d	14.0	84 d	8.7
91 d	98.2	91 d	52.5	91 d	26.3	91 d	14.2	91 d	8.8
98 d	99.3	98 d	53.3	98 d	26.7	98 d	14.4	98 d	8.9
105 d	100.7	105 d	53.6	105 d	27.0	105 d	14.6	105 d	9.1
112 d	101.3	112 d	54.8	112 d	27.4	112 d	14.7	112 d	9.2
119 d	102.0	119 d	55.5	119 d	27.6	119 d	14.8	119 d	9.3
126 d	102.6	126 d	56.1	126 d	27.8	126 d	15.0	126 d	9.4
133 d	103.1	133 d	56.7	133 d	28.1	133 d	15.2	133 d	9.5
140 d	103.6	140 d	57.3	140 d	28.2	140 d	15.3	140 d	9.6
154 d	104.0	154 d	57.8	154 d	28.5	154 d	15.4	154 d	9.8
161 d	104.4	161 d	58.4	161 d	28.8	161 d	15.5	161 d	9.9
168 d	104.8	168 d	58.9	168 d	29.0	168 d	15.6	168 d	9.9
175 d	105.5	175 d	59.8	175 d	29.1	175 d	15.7	175 d	10.0
182 d	105.7	182 d	60.2	182 d	29.3	182 d	15.8	182 d	10.1
212 d	105.9	212 d	60.3	212 d	29.6	212 d	16.1	212 d	10.4
242 d	106.0	242 d	60.5	242 d	29.7	242 d	16.3	242 d	10.6
272 d	106.0	272 d	60.7	272 d	29.7	272 d	16.4	272 d	10.8
302 d	106.1	302 d	60.8	302 d	29.8	302 d	16.5	302 d	11.0
332 d	106.1	332 d	60.8	332 d	29.9	332 d	16.6	332 d	11.2
362 d	106.2	362 d	60.9	362 d	30.0	362 d	16.7	362 d	11.3

Table 5.21   Creep test results of $C_{90}20W6F100$ concrete at 7 d, 28 d, 90 d, 180 d and 365 d

Creep results at 7 d		Creep results at 28 d		Creep results at 90 d		Creep results at 180 d		Creep results at 365 d	
Load holding time	Creepage degree ($\times 10^{-6}$/MPa)	Load holding time	Creepage degree ($\times 10^{-6}$/MPa)	Load holding time	Creepage degree ($\times 10^{-6}$/MPa)	Load holding time	Creepage degree ($\times 10^{-6}$/MPa)	Load holding time	Creepage degree ($\times 10^{-6}$/MPa)
2 h	0.9	2 h	0.3	2 h	0.7	2 h	0.1	2 h	0.1
6 h	5.9	6 h	0.5	6 h	2.0	6 h	2.2	6 h	0.2
1 d	26.9	1 d	1.0	1 d	4.1	1 d	4.8	1 d	0.5
2 d	37.4	2 d	8.7	2 d	6.7	2 d	6.0	2 d	1.8
3 d	43.5	3 d	13.2	3 d	8.1	3 d	6.8	3 d	2.5
4 d	47.8	4 d	16.5	4 d	9.1	4 d	7.3	4 d	3.0
5 d	51.2	5 d	19.0	5 d	10.1	5 d	7.7	5 d	3.4
6 d	54.0	6 d	21.0	6 d	11.0	6 d	8.0	6 d	3.7
7 d	56.3	7 d	22.7	7 d	11.8	7 d	8.3	7 d	4.0
8 d	58.3	8 d	24.2	8 d	12.5	8 d	8.5	8 d	4.3
9 d	60.1	9 d	25.5	9 d	13.1	9 d	8.8	9 d	4.5
10 d	61.7	10 d	26.7	10 d	13.7	10 d	8.9	10 d	4.6
11 d	63.1	11 d	27.7	11 d	14.1	11 d	9.1	11 d	4.8
12 d	64.4	12 d	28.7	12 d	14.6	12 d	9.3	12 d	5.0
13 d	65.6	13 d	30.6	13 d	15.0	13 d	9.4	13 d	5.1
14 d	66.8	14 d	31.4	14 d	15.3	14 d	9.5	14 d	5.2
21 d	74.9	21 d	35.9	21 d	16.8	21 d	10.2	21 d	5.9
28 d	79.1	28 d	39.1	28 d	17.9	28 d	10.7	28 d	6.3
35 d	82.5	35 d	41.5	35 d	18.9	35 d	11.0	35 d	6.7
42 d	85.3	42 d	43.5	42 d	19.7	42 d	11.3	42 d	6.9
49 d	87.5	49 d	45.1	49 d	20.4	49 d	11.5	49 d	7.1
56 d	89.5	56 d	46.5	56 d	20.9	56 d	11.6	56 d	7.3
63 d	91.3	63 d	47.9	63 d	21.4	63 d	11.8	63 d	7.4
70 d	93.3	70 d	48.9	70 d	21.7	70 d	11.9	70 d	7.6
77 d	95.2	77 d	50.0	77 d	22.1	77 d	12.1	77 d	7.7
84 d	96.5	84 d	50.9	84 d	22.3	84 d	12.2	84 d	7.7
91 d	97.7	91 d	51.7	91 d	22.5	91 d	12.3	91 d	7.8
98 d	98.8	98 d	52.5	98 d	22.8	98 d	12.5	98 d	7.9
105 d	100.1	105 d	52.7	105 d	22.9	105 d	12.6	105 d	7.9
112 d	100.7	112 d	53.9	112 d	23.0	112 d	12.8	112 d	8.0
119 d	101.3	119 d	54.5	119 d	23.1	119 d	12.9	119 d	8.0
126 d	101.9	126 d	55.1	126 d	23.3	126 d	13.0	126 d	8.1
133 d	102.4	133 d	55.6	133 d	23.5	133 d	13.1	133 d	8.2

Continued Table 5.21

Creep results at 7 d		Creep results at 28 d		Creep results at 90 d		Creep results at 180 d		Creep results at 365 d	
Load holding time	Creepage degree ($\times 10^{-6}$/MPa)	Load holding time	Creepage degree ($\times 10^{-6}$/MPa)	Load holding time	Creepage degree ($\times 10^{-6}$/MPa)	Load holding time	Creepage degree ($\times 10^{-6}$/MPa)	Load holding time	Creepage degree ($\times 10^{-6}$/MPa)
140 d	102.8	140 d	56.1	140 d	23.6	140 d	13.2	140 d	8.2
154 d	103.2	154 d	56.6	154 d	23.7	154 d	13.4	154 d	8.3
161 d	103.5	161 d	57.1	161 d	23.9	161 d	13.6	161 d	8.3
168 d	103.9	168 d	57.6	168 d	24.0	168 d	13.7	168 d	8.4
175 d	104.4	175 d	58.3	175 d	24.3	175 d	13.8	175 d	8.5
182 d	104.5	182 d	58.4	182 d	24.4	182 d	13.9	182 d	8.6
212 d	104.5	212 d	58.5	212 d	24.6	212 d	14.0	212 d	8.8
242 d	104.5	242 d	58.5	242 d	24.7	242 d	14.1	242 d	8.9
272 d	104.6	272 d	58.6	272 d	24.8	272 d	14.2	272 d	9.0
302 d	104.6	302 d	58.6	302 d	25.1	302 d	14.4	302 d	9.1
332 d	104.7	332 d	58.7	332 d	25.3	332 d	14.5	332 d	9.2
362 d	104.7	362 d	58.8	362 d	25.4	362 d	14.6	362 d	9.2

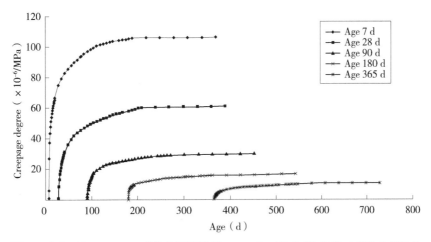

Figure 5.3 *Creep curve of* $C_{90}15W6F100$ *concrete at* 7 d, 28 d, 90 d, 180 d, 365 d

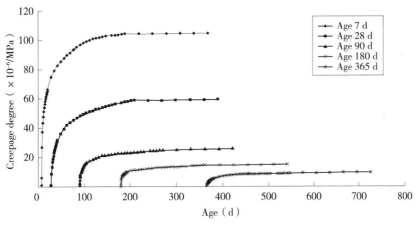

Figure 5.4 *Creep curve of* C9020W6F100 *concrete at* 7 d, 28 d, 90 d, 180 d, 365 d

5.5 Test Study on Properties of RCC with Manufactured Sand and River Sand

5.5.1 Foreword

Sand in concrete mainly plays the role of filling and rolling. The sand coarseness and gradation have great influence on the technical performance of concrete and project cost. Sand fineness refers to the average fineness of sand with different particle sizes mixed together. At the same content, if the sand is too coarse, the cohesion of concrete is poor, easy to separate and bleed water; The total specific surface area of fine sand increases and the sand has good cohesiveness; the mortar demand increases. When the mortar is insufficient, the strength and other properties decrease significantly. Therefore, medium sand with FM = 2.3 - 3.0 is better for concrete. However, the fineness modulus cannot fully reflect the gradation. Gradation refers to the composition of different grain sizes in sand. When there are more coarse particles in the sand, if the gap is filled by appropriate medium particles and a small amount of fine particles, and the porosity and specific surface area are small, the gradation is better.

Diabase aggregate is used in RCC main dam of Baise Multipurpose Dam Project. The diabase manufactured sand (hereinafter referred to as manufactured sand) lacks intermediate size grains due to the lithology of diabase. Most of grains are bigger than 2.5 mm (more than 30%). The content of rock powder is high (grains with size less than 0.16 mm account for 20% - 24%). There are many micro grains in the rock powder (The grains with size less than 0.08 mm account for 40% - 60% of the rock powder), so the gradation is poor. This kind of manufactured sand in RCC of Baise main dam project has influences on the water consumption, setting time, working performance and strength of RCC. According to the construction experience, the poorly graded sand, too coarse sand or too fine sand can be mixed together for improving sand gradation and concrete performance. The gradation improvement of river sand mixed with manufactured sand (mixed sand) is studied and the performance of RCC (RCC) with manufactured sand mixed with river sand is studied by tests according to the characteristics of Baise diabase manufactured sand and river sand through mixing the manufactured sand with river sand at ratio of 0, 20%, 30%, 40% and 100%.

5.5.2 Performance Test of Manufactured Sand Mixed with River Sand

Mixed sand is prepared according to mix proportion of manufactured sand and river sand at 100:0, 80:20, 70:30, 60:40, 0:100. The test results of physical properties of mixed sand are shown in Table 5.22. The gradation of mixed sand is shown in Table 5.23 and Figure 5.5. The test results show that:

1. The fineness modulus of mixed sand is 2.6 - 2.7 which meets the requirements of 2.2 - 2.9.
2. The content of rock powder is 23.71%, 20.62%, 17.11%, 15.15% which meet the standard requirements. The mud content in the manufactured sand and river sand is 3.17% which exceeds slightly the standard.
3. Grading curve: It is found from Figure 5.5 that the manufactured sand has more coarse and fine grains, less intermediate grains, poor gradation. The grain of river sand is small; The gradation curve of the mixed sand with the manufactured sand and river sand ratio of 80:20, 70:30, 60:40 is basically in the medium sand area and the gradation is good.

Chapter 5 Research and Application of RCC Performance

Table 5.22　　　　Test results of physical properties of mixed sand

Manufactured sand: river sand	Fineness module FM	Apparent density (kg/m³)	Bulk density (kg/m³)	Saturated dry water absorption rate of sand(%)	Rock powder content (%)	Voidage (%)
100:0	2.7	2 804	1 902	2.9	23.71	32.2
80:20	2.6	2 817	1 800	2.4	20.62	36.1
70:30	2.7	2 791	1 785	2.1	17.11	36.0
60:40	2.7	2 727	1 745	2.2	15.15	36.0
0:100	2.4	2 632	1 628	2.0	Mud content 3.17	38.1

Note: The porosity in the table is the calculation result after neglecting porosity.

Table 5.23　　　　Test results of grain gradation of mixed sand

Artificial sand: river sand	Screen residual at all sieve meshes (%)							Fineness module	Rock powder content (%)
	10.0 (mm)	5.0 (mm)	2.5 (mm)	1.25 (mm)	0.63 (mm)	0.315 (mm)	0.16 (mm)		
100:0	0	6.70	26.64	12.41	13.80	10.62	6.12	2.7	23.71
80:20	0	5.85	22.99	11.78	14.16	13.02	11.59	2.6	20.62
70:30	0	6.30	24.50	11.65	14.30	14.22	11.92	2.7	17.11
60:40	0	6.32	25.06	10.92	14.24	14.98	13.33	2.7	15.15
0:100	0	3.94	16.41	8.60	14.03	13.73	40.12	2.4	Mud content 3.17

5.5.3 Performance of RCC of Manufactured Sand Mixed with River Sand

The mixed sand is used to test the performance of RCC with the test condition as follows: Tiandong No. 525 moderate heat Portland cement, Shimen No. 42.5 low heat Portland cement, Qujing Grade II fly ash are used. The mixed sand is prepared according to the ratio of manufactured sand and river sand at 100:0, 80:20, 70:30, 60:40, 0:100. ZB-1RCC15 is tested at 0.8% content. The mix proportion of II gradation and quasi-III gradation RCC is adopted in the test. See Table 5.24 for mix proportion test parameters of RCC mixed with manufactured sand and river sand.

See Table 5.25 and Table 5.26 for the properties of RCC mixed with manufactured sand and river sand. Results show that:

1. Liquefaction and bleeding: due to the rough surface of diabase aggregate, the sand gradation is not good; The content of rock powder is high; The rolling effect is poor, therefore, the mixture of diabase aggregate RCC is slow in liquefaction and bleeding; Due to the good gradation, smooth surface and small friction of river sand, the mixture of diabase manufactured sand and river sand can improve the performance of RCC; the liquefying and bleeding are faster.

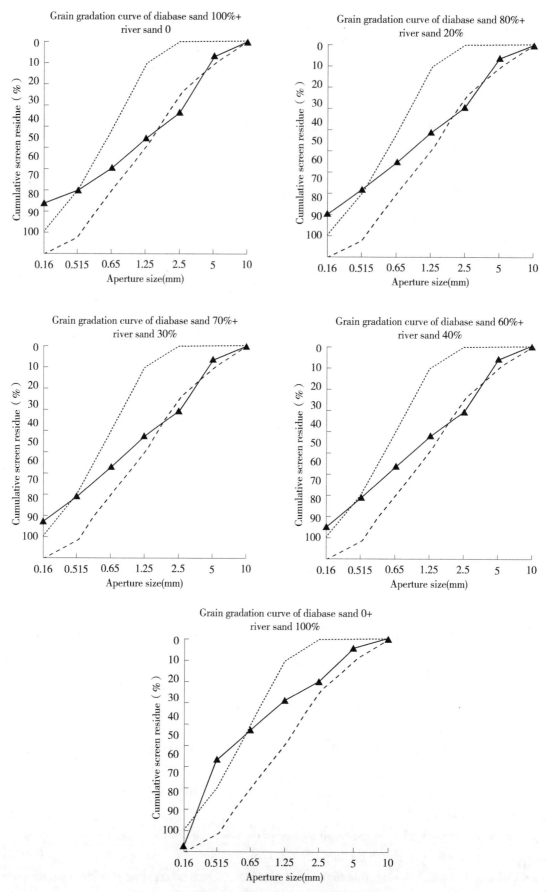

Figure 5.5 *Gradation curve of mixed sand*

Table 5.24 Mix proportion test parameters of RCC mixed with manufactured sand and river sand

Test No.	Level	Water-cement ratio	Manufactured sand + river sand (%)	Fly ash (%)	ZB-1 (%)	DH$_9$ (%)	Water content (kg/m^3)	(Small gravel: Middle gravel: Quasi large gravel:	Unit weight (kg/m^3)	VC value (s)
HS1-1 – 5	R$_{180}$15 S2D50	0.60	34	63	0.8	0.015	Controlled as per VC value	30:30:40	2 650	3 – 5
HS2-1 – 5	R$_{180}$20 S10D50	0.50	38	58	0.8	0.015	Controlled as per VC value	45:55	2 600	3 – 5

2. Water consumption, VC value: Due to the different properties of manufactured sand and river sand, when the VC value is basically the same, the water consumption of river sand RCC is 8 kg/m^3 less than that of manufactured sand RCC. The water consumption of RCC is gradually reduced when river sand content increases.

3. Time of setting: the initial setting time of RCC mixed with river sand is long; when the amount of river sand in the manufactured sand increases, the initial setting time of RCC is gradually extended. When the content of river sand is increased to 30%, the initial setting time is prolonged by about 1.00 h and the magnitude is small. It further reflects that diabase is the main cause to affect the setting time of RCC.

4. Air content: the air content of RCC mixed with river sand is the highest and the air content of RCC mixed with manufactured sand is the lowest. The result fully shows that: due to the characteristics of diabase aggregate, the air content of RCC is seriously reduced and low air content will affect the setting time and mixture performance of RCC.

Table 5.25 Performance of RCC mixed with manufactured sand + river sand (mixed sand)

Test No.	Gradation	Artificial sand: river sand	Water content (kg/m^3)	Mixture properties							
				VC value (s)	Air content (%)	Unit weight (kg/m^3)	Liquefaction and bleeding	Temperature (°C)	Initial setting	Final setting	Difference between initial setting times
HS1-1		100:0	102	5	1.6	2 668	General	17 – 20	5:18	13:32	0:00
HS1-2		0:100	94	6	3.0	2 740	Good	17 – 20	7:15	20:00	+1:57
HS1-3	Quasi III	80:20	97	5	1.8	2 692	Quite good	17 – 20	5:45	14:05	+0:27
HS1-4		70:30	96	6	1.9	2 723	Good	17 – 20	5:55	14:15	+0:37
HS1-5		60:40	95	5	2.0	2 728	Good	17 – 20	6:45	14:38	+1:27
HS2-1		100:0	112	6	1.7	—	General	25 – 34	7:25	15:24	0:00
HS2-2		0:100	104	5	3.5	—	Good	25 – 34	13:18	22:58	+5:53
HS2-3	II	80:20	107	5.5	1.9	—	Quite good	25 – 34	7:32	17:45	+0:07
HS2-4		70:30	106	5	2.0	—	Good	25 – 34	8:22	17:12	+0:57
HS2-5		60:40	105	5	2.1	—	Good	25 – 34	10:00	18:45	+2:35

Note: The raw material is Tiandong No. 525 moderate heat Portland cement; Grade II fly ash; Diabase manufactured aggregate, manufactured sand FM = 2.75, rock powder 22.8, 23.5%; River sand FM = 2.4, mud content 3.0%.

Table 5.26 Test results of RCC of manufactured sand and river sand (mixed sand)

Test No.	Cement type	Artificial sand: river sand	Fineness module FM	Rock powder (%)	Water content (kg/m³)	VC value (s)	RCC Liquefaction and bleeding	RCC temperature (℃)	Time of setting (h:min) Initial setting	Time of setting (h:min) Final setting	Difference between initial setting times (h:min)
NH4-1		100:0	2.75	22.8	102	5	General	17–20	5:45	14:25	0:00
NH4-2	R_{180}15S2D50 Rock cross-cut 42.5 Low heat	0:100	2.4	Mud 3.0	94	5	Good	17–20	7:40	26:55	+1:55
NH4-3		80:20	2.61	21.0	97	5	Quite good	17–20	6:00	14:55	+0:15
NH4-4		70:30	2.62	20.0	96	5	Good	17–20	6:55	15:15	+1:10
NH4-5		60:40	2.59	18.4	95	4	Good	17–20	7:05	15:30	+1:20

Note: 1. Raw material: Tiandong moderate heat portland and Shimen low heat Portland cement, Qujing Grade Ⅱ fly ash, grey manufactured sand FM = 2.60, rock powder 24.0%. FM of Youjiang river sand is 2.4; the mud content is 3%; the admixture is ZB-1RCC15.
2. The difference of initial setting time " + " indicates extending and " - " indicates shortening.

5.5.4 Properties of Hardened RCC with Manufactured Sand and River Sand

5.5.4.1 Mechanical Property

The test results of mechanical properties of RCC with manufactured sand and river sand are shownin table 5.27 and the compressive strength curve is shown. The result indicates: Compressive strengths of quasi-Ⅲ gradation RCC with 28 d, 90 d and 180 d are higher and the compressive strength of mixed sand RCC is lower; The regularity of compressive strength of Ⅱ gradation is not strong and is considered that it is caused by the test error. However, with the increase of river sand content, the compressive strength and splitting strength increase accordingly.

Table 5.27 Test results of mechanical properties of RCC with manufactured sand and river sand

Test No.	Concrete design requirements and location in the project	Gradation	Artificial sand: River sand	Water content (kg/m³)	Compressive/splitting strength (MPa) 7 d	28 d	28 d splitting	90 d	180 d
HS1-1			100:0	102	3.4	9.4	0.72	16.0	26.4
HS1-2	R_{180}15S2D50 Internal RCC	Quasi Ⅲ	0:100	94	6.4	12.1	1.07	19.5	29.6
HS1-3			80:20	97	3.4	9.8	0.86	18.2	28.3
HS1-4			70:30	96	3.4	10.3	0.79	17.4	28.3
HS1-5			60:40	95	3.2	11.0	0.78	17.2	28.8
HS2-1			100:0	112	11.5	17.9	1.45	31.4	37.2
HS2-2	R_{180}20S10D50 Upstream surface RCC	Ⅱ	0:100	104	10.1	17.2	1.05	26.4	38.6
HS2-3			80:20	107	9.3	13.2	1.38	27.8	37.4
HS2-4			70:30	106	10.4	17.7	1.28	24.4	37.5
HS2-5			60:40	105	10.9	16.5	1.26	30.7	43.6

5.5.4.2 Impermeability and Frost Resistance

The frost resistance and impermeability test results of RCC with manufactured sand + river sand (mixed sand) are shown in Table 5.28. The test results show that: Both frost resistance and impermeability of RCC with manufactured sand + river sand can meet the design requirements.

Chapter 5 Research and Application of RCC Performance

Table 5.28 Test results of frost resistance, impermeability, extreme tension and elastic modulus of RCC with manufactured sand and river sand (mixed sand)

Test No.	Artificial sand: river sand	Freezing resistance	Fenetration resistance	Ultimate tensile ε_p				Modulus of elasticity			
				Ultimate tension ε_p ($\times 10^{-6}$)		Axial tensile strength (MPa)		Compression modulus (GPa)		Axial compressive strength (MPa)	
				28 d	180 d	28 d	180 d	28 d	180 d	28 d	180 d
HS1-1	100:0	≥D50	≥S2	55	90	1.2	2.4	27.7	41.2	5.1	19.5
HS1-2	0:100	≥D50	≥S2	62	75	1.1	2.4	39.8	47.6	7.1	26.1
HS1-3	80:20	≥D50	≥S2	61	89	1.0	2.2	20.2	42.9	5.4	19.9
HS1-4	70:30	≥D50	≥S2	53	112	1.0	2.4	28.7	45.0	8.1	23.2
HS1-5	60:40	≥D50	≥S2	62	84	1.2	2.1	36.4	46.8	3.8	27.1
HS2-1	100:0	≥D50	≥S10	70	92	1.0	2.6	31.7	42.1	12.4	21.8
HS2-2	0:100	≥D50	≥S10	54	81	1.6	2.8	27.6	48.6	11.3	28.6
HS2-3	80:20	≥D50	≥S10	70	90	1.5	2.3	27.1	44.5	8.2	23.9
HS2-4	70:30	≥D50	≥S10	67	84	1.3	2.6	28.8	45.0	11.3	24.0
HS2-5	60:40	≥D50	≥S10	90	85	2.2	2.6	23.1	46.2	15.6	28.2

5.5.4.3 Ultimate Tensile and Elastic Modulus

The ultimate tensile and elastic modulus test results of RCC with manufactured sand, river sand and some river sand (20%, 30%, 40%) are shown in Table 5.28. The test results show that:

Ultimate tensile: the 28 d ultimate tensile of quasi-Ⅲ gradation RCC is higher than that of RCC mixed with river sand. The 180 d ultimate tensile is higher when 30% river sand is added; the 28 d ultimate tensile of Ⅱ gradation RCC mixed with 40% river sand is high and the 180 d ultimate tensile is high, the result shows that: when the manufactured sand is mixed with 30% – 40% river sand, the ultimate tensile of RCC is high.

Modulus of elasticity: generally, when the compressive strength and ultimate tensile value are high, the elastic modulus is also relatively high. The compression elastic modulus of diabase manufactured sand mixed with river sand also conforms to this law. The elastic modulus of quasi-Ⅲ gradation RCC with river sand at 28 d and 180 d is higher; The 28 d elastic modulus of Ⅱ gradation RCC with manufactured sand is high and the 180 d elastic modulus of Ⅱ gradation RCC with river sand is high.

5.5.4.4 Dry Shrinkage Property

Dry shrinkage is the long-term deformation performance of concrete, as the same as the wet expansion of concrete, it is caused by the change of moisture in concrete. When the concrete is hardened in water for a long time, the concrete will be slightly expanded; when the concrete hardens in the air, due to evaporation of water, the gel in concrete is gradually dried and contracted. The deformation is called as dry shrinkage. Dry shrinkage deformation can cause tensile stress on the surface of concrete, cause concrete cracking. The frost resistance, impermeability and corrosion resistance of concrete are reduced accordingly. It is very harmful to concrete.

The manufactured sand, river sand, mixed sand with 20%, 30% and 40% river sand are used as fine aggregate of RCC. The drying shrinkage properties are compared to study the dry shrinkage performance of RCC with manufactured sand, river sand and mixed sand. See Table 5.29 for dry shrinkage test results of RCC mixed with river sand and the relationship curve is shown. The test results show that:

The dry shrinkage rate of RCC mixed with river sand is basically the same in 3 d and 7 d. The dry shrinkage rate of 28 d, 90 d and 180 d is small.

The early dry shrinkage rate of RCC mixed with manufactured sand increases rapidly and the drying shrinkage of 180 d is over 600×10^{-6}.

Table 5.29   Mix proportion test of RCC with manufactured sand + river sand (mixed sand)

Test No.	Gradation	Artificial sand: river sand	Water content (kg/m³)	Drying shrinkage(×10⁻⁶)						
				3 d	7 d	14 d	28 d	60 d	90 d	180 d
HS1-1	Quasi Ⅲ	100:0	102	79	186	305	404	543	576	609
HS1-2		0:100	94	40	40	59	126	133	147	173
HS1-3		80:20	97	67	127	213	294	347	387	413
HS1-4		70:30	96	59	112	191	264	310	350	389
HS1-5		60:40	95	40	92	158	218	290	330	363
HS2-1	Ⅱ	100:0	112	99	198	344	469	542	595	641
HS2-2		0:100	104	40	40	59	138	151	178	198
HS2-3		80:20	107	60	158	244	336	375	421	461
HS2-4		70:30	106	59	145	224	303	350	396	429
HS2-5		60:40	105	40	119	185	258	324	363	403

Dry shrinkage rate of RCC with diabase manufactured sand mixed with river sand is smaller than the dry shrinkage rate of RCC with manufactured sand. When the amount of river sand increases, the dry shrinkage rate of RCC gradually decreases.

5.5.5 Conclusion

1. By mixing some river sand with diabase manufactured sand, the content of rock powder is reduced and the sand gradation is improved.
2. With the increase of the amount of river sand, the water consumption and gel material consumption of RCC are reduced accordingly. At the same time, the liquefaction and bleeding of RCC are improved, and the setting time of RCC is prolonged.
3. With the increase of river sand content, the compressive strength and splitting tensile strength of RCC are increased accordingly and the ultimate tensile and elastic modulus are increased accordingly.
4. With the increase of the amount of river sand in the manufactured sand, the dry shrinkage rate of RCC is obviously reduced. This is very beneficial to improve the crack resistance of RCC.
5. According to the research results, when the manufactured sand is mixed with river sand by 30%, the performance of RCC is obviously improved. The mix proportion of RCC with manufactured sand and river sand is shown in Table 5.30.

Table 5.30   Mix proportion of RCC with manufactured sand mixed and river sand

Test No.	Design requirements and location in the project	Gradation	Artificial sand: river sand	Water-cement ratio	Sand ratio (%)	Fly ash (%)	ZB-1 (%)	DH₉ (%)	Water content (kg/m³)
HS1-3	R_{180}15S2D50 Inside the dam	Quasi Ⅲ	80:20	0.60	34	63	0.8	0.015	95
HS1-4			70:30	0.60	34	63	0.8	0.015	94
HS1-5			60:40	0.60	34	63	0.8	0.015	93
HS2-3	R_{180}20S10D50 Upstream surface of dam	Ⅱ	80:20	0.50	38	58	0.8	0.015	105
HS2-4			70:30	0.50	38	58	0.8	0.015	104
HS2-5			60:40	0.50	38	58	0.8	0.015	103

Note: Moderate heat portland cement, Grade Ⅱ fly ash; diabase coarse and fine aggregate, sand FM = 2.75, rock powder 20 ± 2%; Youjiang river sand FM = 2.4, mud content ≤ 3.0%.

Roller Compacted Concrete Rapid Damming Technology

(Volume Ⅱ)

中文作者　田育功
译　　者　杨海燕　李建强　等

Chinese Writer　Tian Yugong
　　Translators　Yang Haiyan　Li Jianqiang et al

黄河水利出版社
·郑　州·

Abstract

This book describes the theory, method, experience and engineering practice of rapid damming technology with roller compacted concrete(RCC). Citing a large number of rich first-hand RCC test and research results, tender/bidding documentation and engineering examples of rapid damming, with full and accurate contents, this is a practical technology book with very high theoretical level and rich engineering practice.

The book includes:13 topics of development level of RCC damming, design and rapid construction of RCC dam, examples of raw materials and projects, mixing proportion design with examples, research and examples of RCC performance, admixture research and application, the role of rock powder in RCC, construction technology of grout enriched vibrated concrete(GEVR), key technology for rapid construction of RCC, temperature control and cracking prevention, quality control of RCC, core sample pressurizing water and in-situ shear resistance and RCC cofferdam work.

This book is a valuable reference for vast engineering and technical personnel engaged in the structure, design, research, construction and supervision in water conservancy and hydropower engineering sector, as well as teachers and students of related majors in colleges and universities.

图书在版编目(CIP)数据

碾压混凝土快速筑坝技术 = Roller Compacted Concrete Rapid Damming Technology：全三册／吴正桥等译：田育功著. —郑州：黄河水利出版社，2021.8

ISBN 978-7-5509-3069-8

Ⅰ.①碾… Ⅱ.①吴… ②田… Ⅲ.①碾压土坝-混凝土坝 Ⅳ.①TV642.2

中国版本图书馆 CIP 数据核字(2021)第 165756 号

出 版 社：黄河水利出版社　　　　　　　　　　　　　网址：www.yrcp.com
　　　　　地址：河南省郑州市顺河路黄委会综合楼 14 层　　邮政编码：450003
发行单位：黄河水利出版社
　　　　　发行部电话：0371-66026940、66020550、66028024、66022620(传真)
　　　　　E-mail：hhslcbs@126.com
承印单位：广东虎彩云印刷有限公司
开本：787 mm×1 092 mm　1/16
印张：42
字数：1 500 千字
版次：2021 年 8 月第 1 版　　　　　　　　　　　　　印次：2021 年 8 月第 1 次印刷
定价(全三册)：198.00 元

Translator's Preface

Roller compacted concrete (RCC) gravity dam has shown rapid construction speed, low cost and good engineering quality and other advantages. In the recent year, the rapid development of RCC dams in China, especially the completion of high DAMS over 100m, damming RCC materials and construction technology of China is indeed a kind of technical concept with unique characteristics of China. Carefully analyzing various technologies, their contents are very rich, and the connotation is also deep, and the application is very convenient and handy, to be easy to master. What more valuable is that, whenever a large RCC dam is completed, there are almost always innovations and improvements, so that the development of the technology is constantly rooted in the source of project construction. For technical progress in various aspects, many engineering technicians have written a large number of special treatises or monographs, well welcomed by industry peers. In order to better promote the design and construction concept of Chinese RCC dam in the world, Wu Zhengqiao and others translated the book with the permission of the author.

The translation results are divided into three volumes, division of labor is as follows:

Chief editor: Wu Zhengqiao.

Other editor:

(1) Volume Ⅰ: Wu Zhengqiao, Yu Yang.

(2) Volume Ⅱ: Yang Haiyan, Li Jianqiang, Yao Desheng, Zhao Chuntao, Zhou Hongmin, Bao Dixiao, Li min, Sun Shunan, Zhao Jianli.

(3) Volume Ⅲ: Zhang Baorui, He Qingbin, Peng Xiaochuan, Li Miao, Xu Lizhou, Pei Xianghui, Xu Ting, Liu Zhuo, Zhang Kai, Zhao Lin.

In this process of translating this book, Tian Yugong gave great support and help, thank you!

In addition, in the process of publishing this book, I would like to thank Wang Xiaohong and Yu Ronghai from the editorial department of China Water Resources Bei Fang investigation, Design & Research Co. LTD and the Yellow River Water Publishing House for their strong support.

Wu Zhengqiao

August 2021

Preface

In recent years, China has successively completed 200-metre roller compacted concrete(RCC) gravity dams and several 100-metre level thin double-curvature arch RCC dams. In construction process, all these dams have shown rapid construction speed, low cost and good engineering quality and other advantages. These high dam construction achievements signify that, China's RCC damming technology has made a breakthrough in overall development in a great number of engineering practices, through earnestly summing up experience, actively carrying out exploration and test study, continuous innovation and development. The technology level is getting higher and higher, getting more mature. These new technical achievements and engineering achievements have also brought many beneficial effects to international dam industry.

Looking back to early days, due to hazy knowledge, in RCC dam projects in China, "drying hard" concrete commonly applied in foreign countries was generally applied. As a result, the lack of bleeding at pouring face often caused poor inter-layer bonding, forming a "multi-layer steamed bread" phenomenon. This caused doubts and reservations about the quality of RCC mass of dam. In view of this condition, with a spirit of tenacity, many engineering technical personnel and researchers are continuing to explore, adjust and improve materials of concrete and construction technical actions, finally, a technology mode of "sub-plastic state" RCC with good bleeding performance, being easy to compact and being able to prevent vibration-press subsidence has been formed, widely used nowadays. It has better solved the problem of concrete, hot joint bonding, thoroughly improved impervious performance and shear resistance of course face, solved the safety and impervious problem for the construction of 200- meter high RCC gravity dam. Compared with some foreign RCC damming technologies, damming RCC materials and construction technology of China is indeed a kind of technical concept with unique characteristics of China. Carefully analyzing various technologies, their contents are very rich, and the connotation is also deep, and the application is very convenient and handy, to be easy to master. What more valuable is that, whenever a large RCC dam is completed, there are almost always innovations and improvements, so that the development of the technology is constantly rooted in the source of project construction. For technical progress in various aspects, many engineering technicians have written a large number of special treatises or monographs, well welcomed by industry peers.

The greatest advantage of RCC damming technology is its rapid speed, which has strong vitality. Generally, for concrete dam with a height of over 100 m, it can be completed for 2 – 3 years adopting RCC damming technology. Compared with normal concrete damming technology, the construction duration can be shortened by one-third and more. Engineering practice has proved that, RCC damming technology is credible and reliable in quality, of which the advantage is beyond any doubt.

The design is the key to technical innovation of rapid RCC damming. Technical innovation not only requires a solid and scientific technical foundation, but also allows failure and discussion, and also requires the courage to bear failure and responsibility, to achieve the objective of technical innovation through practice.

"Interlayer combination, temperature control and cracking prevention" is the core technology of rapid RCC damming. In recent years, with the increase in the height and volume of RCC dams, it has become common practice to continuously pour RCC in high-temperature seasons in order to crash the schedule or shorten the duration. As a result, temperature control measures for RCC have become more and more stringent,

which it has no different from normal concrete, to make temperature control measures of RCC show a trend more and more complicated. In this respect, RCC dams in Thailand, Cambodia, Laos, Myanmar and other countries located in subtropical region shall be learnt from. The RCC dams in these countries do not adopt cooling water pipe temperature control measures, and the design indicators adopt a single strength index and with a single aggregate gradation, the design concepts and temperature control measures of these dams are worth learning and thinking about.

The construction of RCC dam has the characteristics of one-time, and it is particularly important that construction quality is being under control all the time. Especially, for on-site RCC construction, dynamic control of VC value, timely rolling, mist spray moisturizing and covering for curing and other construction links concern directly the quality of interlayer bonding and the performance of temperature control and cracking prevention, which must be raised to the height of quality problem to understand their importance.

The Author of this book has collected a large number of RCC dam engineering practical data and multifaceted scientific test and research results in China, with full and accurate contents. While sorting out and analyzing these achievements, combined with personal exploration and study of RCC dam participated in the construction of RCC dams personally, the Author has carried out in-depth display and interpretation, and has more realization and understanding of "plastic RCC", to expand the discussion in the book. RCC damming is still a developing technology. For all concrete materials, construction process, dam engineering design, construction management, temperature control and cracking prevention, there is still room for improvement and development. It is beneficial for technical development to constantly summarize, analyze, study and exchange. After reading this book, I have benefited a lot and expanded my eyesight.

Former Director of the RCC Damming Professional Board of China Society for Hydropower Engineering

Wang Shengpei

Beijing, March 2010

Foreword

Roller compacted concrete(RCC) damming technology is a major technical innovation in the history of world damming. RCC damming technology is favored by global dam industry because of its rapid construction speed, short duration, low investment, safe and reliable quality, high degree of mechanization, simple construction, strong adaptability and green environmental protection, etc. In particular, the damming period of can be significantly shortened by one-third and more compared to similar normal concrete dams, to show strong advantage of vigorous development, and inject a fresh air and vitality into dam construction, which is in line with the development direction of good, rapid and economical.

The "rapid" is the greatest advantage of RCC damming technology, which is its strong vitality. Although the adoption of RCC damming technology has only been for 20 years more, the speed of damming and the large number of dams are unable to be matched by other damming technologies. As of the end of 2008, there were 180 RCC dams(including cofferdams) completed or under construction in China, ranking first globally. The greatest charm of RCC damming technology is its compatibility. RCC has the characteristics of concrete, conforming to the rules of water-cement ratio, and its cross-section design is the same as that of normal concrete dam, regardless of RCC gravity dam or arch dam; meanwhile, its construction has the characteristics of rapid construction of earth-rock dam. A large number of engineering practice has proved that, RCC dam has become one of the most competitive dam types.

Using RCC damming technology, the dam is bold and generous and real. The internal quality of the dam is good, and the appearance quality is beautiful, so its quality is not inferior to normal concrete dams. Although the RCC damming is bold, the construction is not rough at all. RCC damming technology is very delicate, to be a well-organized and standardized construction. The construction site is sparsely staffed, and the concrete roller compacting is carried out in an orderly manner from placing, spreading to rolling. On the contrary, for the pouring of normal concrete dam, due to that the normal concrete is restricted by pouring strength and temperature control and other factors, dam body is divided into blocks by transverse and longitudinal joints. For concrete construction, columnar pouring is adopted, which results in a large amount of formwork workload, and the amount of merging joint grouting is great and of long period. The dam construction often presents a large number of people on placement surface and appears busy, lacking of well-organized, unified and bold of RCC construction.

RCC damming is a systematic project integrating the research, design, construction, quality control and other aspects. What mostly reflects the characteristics of rapid RCC damming is whole-block placement and thin-lift concreting. Because of the change in damming technology, the design concept of complex layout and dam engineering structure has been changed. The complex layout design shall not only consider starting from meeting the requirements of rapid construction of RCC and simplifying the layout of the dam, but also require carrying out in-depth study on dam structure, temperature stress and overall performance. The adoption of RCC damming is a promotion to the design. The design concept must advance, the dam layout shall be different from normal concrete dam, to start from thin-lift concreting and whole-block placement, simple and rapid damming technology, combined with the characteristics of RCC itself, the simpler the dam's structural layout, the more obvious its advantages.

"Interlayer bonding, temperature control and cracking prevention" is the core technology of RCC rapid damming. Through a large number of tests and studies and engineering examples, for problems of great or low fluctuations in artificial sand and gravel powder content of RCC, the successful application admixing fly ash or admixing rock powder instead of sand, precise control of rock powder mixing, dynamic control of *VC* value and whole-block placement inclined layer rolling and other technologies have effectively improved the rollability, liquefaction and bleeding and interlayer bonding quality of RCC, promoting the development of RCC rapid damming technology. The "grout-mortar ratio" has become one of important parameters in the design of RCC mixing proportion, with the same important effect as three major parameters of water-cement ratio, sand ratio and water consumption. Rock powder has got more and more attention in RCC, become an indispensable part of RCC materials. Hydraulic RCC has been developed into non-slump semi-plastic concrete. The worrying problem of interlayer bonding quality has been well solved, significantly improved the ultimate tensile value and frost resistance.

Looking back at test history of drilling and coring of RCC and on-site water pressurizing tests, core samples at early stage with the longest length of 60 cm more have been developed into ones of ultra-long 16 m more, and core samples longer than 10 m are not uncommon. Drilling and coring have fully proved that the maturity of rapid RCC damming technology gets gradually maturing, reflecting the change in the properties of RCC from one side, which is indeed very beneficial to improve the quality of interlayer bonding and impervious performance of dam body.

RCC also has its dual nature. Because RCC adopts low cement content and high admixtures, it is required to adopt advanced design concepts to constantly innovate and deepen the study of its frost resistance, ultimate tensile value and carbonization and other properties(compared to normal concrete).

Over 20 years since 1988 when starting studying dam dam technology of left auxiliary dam RCC test at Longyangxia Dam, the Author has personally witnessed the development and growth of China's RCC damming technology, who is one of major participants, implementers, researchers and promoters of RCC dam technology of China, mastered a large number of first-hand test and research results, with rich engineering practical experience, having novel viewpoints in RCC damming technology. This book is achievement data of the test and research, construction technology, technical consultation, water storage acceptance and construction management in Longshou, Linhekou, Baise, Guangzhao, Jin'anqiao, Kalasuke and other RCC dams participated in and presided over by the Author in recent years, compiled on the basis of over 40 papers (thesis) published at home and abroad, in which many of valuable first-hand data are disclosed for the first time, being one of windows to understand and master RCC damming technology of China.

RCC damming technology of China ranks among global leading level, which is inseparable from continuous efforts and innovations of vast research, design, technical consultation, supervision, and construction personnel of a large number of water conservancy and hydropower projects in China, especially massive, scientific and hard works and strong supports of the RCC Damming Profession Board. Here, I would like to express my highest respect and heartfelt gratitude to elder Wang Shengpei, who has contributed his whole life and made outstanding contributions to China's RCC damming technology, and colleagues of the RCC Damming Profession Board! I would like to express my heartfelt gratitude to leaders and colleagues who have supported my technical work for long term!

In view of the characteristics of hydraulic RCC, this book makes a quite detailed analysis and research from the mechanism and application of RCC rapid damming. Through the discussion of rapid RCC damming technology and the analysis on typical engineering examples, the Author summarize in time, and think more calmly, to discuss and research from both the pros and cons, always insisting on putting the quality and safety of the dam in the important position of the first priority, so that the RCC damming technology is more viable,

safer and more perfect.

This book is compiled in the intention of constructing best and first-class RCC dam in the world. Because there are many research topics on RCC damming technology, restricted by the Author's capability and limited time, it is inevitable to make mistakes and mistakes for sorting out, analyzing and studying and giving examples for so many topics, therefore, readers are expected to criticize and correct!

The Author
July 2010

Contents

Translator's Preface
Preface
Foreword

Chapter 1　General .... (1)
　1.1　Development Level of RCC Damming Technology in China .... (1)
　1.2　Development History of RCC Dam .... (10)
　1.3　Popularization and Application of RCC Damming Technology .... (12)
　1.4　Discussion on Key Rapid Damming Technologies of RCC .... (16)
　1.5　Innovation and Reflection on RCC Rapid Damming Technology .... (23)

Chapter 2　Design and Rapid Construction of RCC Dams .... (27)
　2.1　Overview .... (27)
　2.2　Design and Rapid Construction of RCC Dam .... (28)
　2.3　Design Index of RCC and Material Zoning of Dam .... (40)
　2.4　Introduction to RCC Dam Design .... (42)
　2.5　Conclusion .... (64)

Chapter 3　Raw Materials and Project Examples .... (66)
　3.1　General .... (66)
　3.2　Cement Properties and Project Cases .... (68)
　3.3　Performance and Quality Test of Fly Ash .... (78)
　3.4　Performance of Hydraulic Concrete Admixtures .... (88)
　3.5　Aggregate Properties and Project Cases .... (95)
　3.6　Conclusion .... (106)

Chapter 4　RCC Mix Proportion Design and Example .... (108)
　4.1　Overview .... (108)
　4.2　RCC Mix Proportion Parameter Selection .... (109)
　4.3　Design Basis and Content of Mix Proportion .... (133)
　4.4　RCC Mix Proportion Design Method .... (134)

Chapter 5　Research and Application of RCC Performance .... (141)
　5.1　Overview .... (141)
　5.2　Properties and Influencing Factors of RCC .... (142)
　5.3　Study on Relationship of Admixture, VC Value, Temperature with Setting Time ...... (169)
　5.4　Autogenous Volume Deformation, Adiabatic Temperature Rise and Creep Tests ...... (174)
　5.5　Test Study on Properties of RCC with Manufactured Sand and River Sand ...... (186)

Chapter 6　Research and Application of RCC Admixture .... (193)
　6.1　Overview .... (193)
　6.2　Micro Analysis and Study of RCC Admixture .... (195)
　6.3　The Research and Application of SL Admixture in Gelantan Project. ...... (232)

Chapter 7　Research and Utilization of Rock Powder in RCC (253)
　7.1　Overview (253)
　7.2　Limestone Rock Powder (255)
　7.3　Influence of Rock Powder on Performance of RCC (259)
　7.4　Project Examples of RCC with Rock Powder Replacing Sand (262)
　7.5　Study on Utilization of Baise Diabase Manufactured Sand and Rock Powder in RCC (271)

Chapter 8　Construction Technology of GEVR (315)
　8.1　Overview (315)
　8.2　Mix Proportion Test of GEVR (318)
　8.3　Construction Technology of GEVR (320)
　8.4　Test Study on GEVR Mixed with Fiber in Impervious Area (323)
　8.5　Application of GEVR in Baise RCC Main Dam (329)

Chapter 9　Key Technology for Rapid Construction of RCC (334)
　9.1　Overview (334)
　9.2　Key Technology for Rapid Construction of RCC (336)
　9.3　Jin'anqiao Dam Roller Compacted Concrete Fast Construction Key Technique (374)
　9.4　RCC Construction of Longshou Arch Dam in Cold and Dry Area (389)

Chapter 10　Temperature Control and Cracking Prevention of Roller Compacted Concrete (399)
　10.1　Overview (399)
　10.2　Basic Information and Standards of Temperature Control (401)
　10.3　RCC Temperature Control Measures (413)
　10.4　Technical Innovation and Discussion on Temperature Control and Anti-cracking (425)
　10.5　The Temperature Control in RCC Gravity Dam in One Hydro-junction Project (430)

Chapter 11　RCC Quality Control and Project Cases (456)
　11.1　Overview (456)
　11.2　Quality Control and Evaluation Regulations (458)
　11.3　Other Quality Control Measures and Discussions of RCC (469)
　11.4　Application of Nuclear Densimeter in RCC (475)

Chapter 12　Core Drilling, Pump-in and In-situ Shear Tests (482)
　12.1　Overview (482)
　12.2　Core Drilling of Dam RCC (484)
　12.3　Field Pump-in Test of RCC (493)
　12.4　Performance Test of RCC Core Samples (500)
　12.5　In-situ Shear Test of RCC Dams (504)
　12.6　Core Drilling and Pump-in Test of Dam RCC (508)
　12.7　Performance Test of RCC Core Sample of Jin'anqiao Dam (515)
　12.8　On-site In-situ Shear Test of RCC Dams (519)

Chapter 13　RCC Cofferdam Construction and CSG Damming Technology (531)
　13.1　RCC Cofferdam (531)
　13.2　Cemented Sand and Gravel (CSG) Damming Technology (532)
　13.3　Design and Rapid Construction of Longtan RCC Cofferdam (535)
　13.4　CSG Mix Proportion Design and Application of Upstream Cofferdam of Gongguoqiao Dam (544)

Appendix A National Method Hydraulic Concrete Mix Proportion Test Method (561)
- A.1 Foreword (561)
- A.2 Characteristics of Construction Method (562)
- A.3 Applicable Scope (562)
- A.4 Technological Principle (562)
- A.5 Construction Process Flow and Key Points of Work (562)
- A.6 Material and Equipment (569)
- A.7 Quality Control (572)
- A.8 Safety Measures (572)
- A.9 Environmental Protection Measures (573)
- A.10 Benefits Analysis (573)

Appendix B National Construction Method GEVR Construction Method in RCC Dam Construction (575)
- B.1 Foreword (575)
- B.2 Characteristics of Construction Method (575)
- B.3 Applicable Scope (576)
- B.4 Technological Principle (576)
- B.5 Construction Technology and Operation Points (576)
- B.6 Mechanical Equipment Configuration (578)
- B.7 Quality Control (578)
- B.8 Safety Measures (579)
- B.9 Environmental Protection Measures (579)
- B.10 Technical and Economic Analysis (579)
- B.11 Project Examples (579)

Appendix C Example of Construction Methods: Construction Method of RCC for Dam of a Certain Project (581)
- C.1 General Principles (581)
- C.2 Normative References (581)
- C.3 Terminology (582)
- C.4 Concrete Compaction Process Flow Chart (585)
- C.5 Control and Management of Raw Materials (586)
- C.6 Selection of Roller Compacted Concrete Mix Ratio and Issuing of Charger Sheet (588)
- C.7 Inspection and Acceptance before Placement Construction of Roller Compacted Concrete (588)
- C.8 Concrete Mixing and Management (591)
- C.9 Concrete Transport (593)
- C.10 Construction Management inside Block (594)
- C.11 Construction under Special Climatic Conditions (604)
- C.12 Quality Control Management (605)
- C.13 Management of Placement Surface after Rolling and Placement Finishing (609)

Main References (610)

Chapter 6

Research and Application of RCC Admixture

6.1 Overview

RCC is characterized by large amount of admixtures and less amount of cement. When a large amount of admixtures are used in RCC, one advantage is that local material can be used, turning waste into treasure and reducing cost; green environmental protection, no thermal effect on the environment; secondly, it greatly improves the performance of RCC. The grind ability and interlayer bonding quality are improved. The temperature rise and temperature stress of concrete are effectively reduced, providing a strong guarantee for the rapid damming of whole-block and thin-layer pouring RCC, simplifying the construction technology, speeding up the construction progress.

Admixtures are the main components of RCC cementing materials, and the admixture content in RCC in China generally accounts for 50% – 65% of cementing materials. Admixtures in RCC are generally active or inactive. The so-called active and inactive admixture term is an artificial boundary to distinguish the activity of admixtures. In fact, as long as the mineral materials are ground to enough specific surface area, they will have certain chemical reaction activity with some hydration products of cement. The difference is only that the size of the activity and the time of its exertion and some components of some minerals may have adverse effects on the improvement of the properties of cement, stone and concrete.

The main admixtures are fly ash, iron ore slag, phosphorus slag, pozzolan, tuff, limerock powder, silicon powder, etc. The admixtures can be mixed alone, or can also be mixed together.

The admixture, fly ash has always been dominant in hydraulic concrete and the application of fly ash in RCC is mature. Fly ash is not only used in a large amount, but also widely used. Its performance is also the best among the admixtures. China is rich in fly ash resources, but because of its vast territory, the distribution of fly ash resources are extremely unbalanced. For example, the RCC dam projects of Dachaoshan, Jinghong, Jupudu, Gelantan, Dengke, Lazhai etc. in Yunnan, because the required fly ash is far away from the producing area, it's not economic to use this fly ash. In addition, a large number of hydropower projects have been started in recent years, as a result, the fly ash is in short supply. Therefore, based on the principle of local materials, for example, in Dachaoshan project, the scientific research and construction units have carried out a lot of tests and studies since the early 1990s, a new type of admixture (PT admixture) has been developed by mixing phosphorus slag (P) with local tuff (T). Its performance and content are similar to that of grade II fly ash, but the effect is similar, This pioneering work broadens the material source of RCC admixture and further understanding of admixtures. In Jinghong, Gelantan, Jufudu, Tuka River and other projects, 50% manganese iron slag and limerock powder are used at present, SL admixture for short; Pozzolanic ash is used as the admixture in the projects of Dengke and Lazhai. It is a kind of admixture made by mixing and grinding phosphorus slag and tuff (PT), iron ore slag and limestone (SL), fly ash and phosphorus slag (FP) and pozzolanic ash only. The effect is good.

Research and application of RCC admixture need to understand and master the characteristics of RCC. Its own characteristics are as follows:

1. The cement consumption is low, Generally, it is 55 – 90 kg/m^3.
2. The amount of admixture is large and accounts for 50% – 65% of the gel materials.
3. The material division of dam body is simple; the heat of hydration is low; the temperature control measures are simple; and whole-block and thin-lift continuous pouring can be carried out.
4. The dam body is not divided into longitudinal joints; full-section whole-block placement rolling can be conducted; The construction is simple and fast, that greatly reduces the formwork workload.
5. The RCC dam allows for crest overflow. It is safe and low in investment, having significant technical economic benefits.
6. It adapts to the main technical requirements and characteristics for the design of dams, especially good for lowering temperature stress and simplifying temperature control measures of dams.

Dams are the most important water control structures. The amount of concrete for dams varies from hundreds of thousands of cubic meters to millions of cubic meters, or even more than ten million cubic meters. If the dams are constructed with normal concrete, the demand of cement is in great. However, if the RCC damming is adopted, the demand for cement is greatly reduced, only about a half of that of normal concrete. Therefore, RCC damming technology is not only fast in construction, but also environmentally friendly. It is a kind of green concrete in modern society.

Due to plenty of admixtures used in the RCC, the cement content is effectively reduced, directly decreasing the emission of carbon and the generation of thermal effect. The RCC is consistent with the green concrete development direction. The late Academician Wu Zhongwei believed that, the green high-performance concrete (GHPC) is the development trend of concrete, and all concretes in the future shall be GHPC. The green concrete must meet the following conditions:

1. Saving resources and energies.
2. Not destroying environment, but more beneficial to environment.
3. Allowing for sustainable development, no harming to the development of later generations while meeting the demands of contemporary one.
4. Reducing the consumption of non-renewable mineral resources and the emission of pollutant, and fully utilizing wastes.

So, RCC itself is exactly the green concrete.

Admixtures have been indispensable for RCC cementitious materials for a moment. Its influences on the performance of RCC are mainly manifested in the following aspects:

1. With the micro-aggregate effect, the admixtures could improve the workability, increase the cohesion and reduce the segregation of mixture.
2. The admixtures could delay the occurrence of cement hydration heat peak, reduce the hydration heat, decrease the temperature rise of mass concrete, and reduce the temperature cracking due to its consistency with the RCC strength development law.
3. It is beneficial to delay the initial setting time of RCC mixture and very good for improving its crushability, liquefaction bleeding and inter-layer bonding.

This Chapter gives a full and systematic introduction of the quality, characteristics, amount and application of the admixtures by RCC performance study and application with different admixtures, offering a scientific basis and broader prospect for further expanding the selection and application of admixtures in RCC. The application, mechanism and construction method of admixtures for RCC is greatly different from that of normal concrete. It is necessary to study the admixture's adaptability to RCC and its long durability in a deep manner. The quality and safety of dams shall be always considered at the first place. It is necessary to keep technical innovation, make timely summary and calm consideration, and hold a discussion and research on the application of admixtures positively and negatively, so that the advantages and features of rapid RCC damming technology can be given full play to.

6.2 Micro Analysis and Study of RCC Admixture

6.2.1 Physical Indicators and Chemical Compositions of Admixtures

RCC admixtures mainly include: Basalt powder, limerock powder, phosphorous slag powder, tuff powder, iron-ore slag powder, pozzolan powder and fly ash. For all admixtures, physical indicators, chemical composition, microscopic analysis of granule morphology and mineral composition, particle distribution, mineral admixture's effect on cement mortar strength, hydration products and mechanism are tested and analyzed.

6.2.1.1 Physical Index

The tested density, water demand ratio, fineness (Specific Surface Area Method), saturated-surface-dried moisture retention, setting time of cement in admixture, cement mortar strength ratio (same fluidity, same water-cement ratio, 5 ages) as well as the hydration heat of admixtures are listed in Table 6.1 and Table 6.2 respectively. The tested density, fineness, stability, hydration heat, setting time and mortar strength of cement are listed in Table 6.3.

Table 6.1　　　　　Setting time and compressive strength rate of cement mortar in admixtures

Type of admixture	Density (g/cm^3)	Water demand ratio (%)	Specific surface area (m^2/kg)	Water absorption (%)	Setting time of cement with 30% admixture (h: min)		Compressive strength ratio (%)				
					Initial setting	Final setting	3 d	7 d	28 d	60 d	90 d
Cement	3.2		320	0.2	5:36	8:23	—	—	—	—	—
Basalt powder	2.9	110.5	270	0.1	7:33	10:23	43.8	44.0	44.7	44.9	52.9
Limerock powder	2.8	100.8	460	0.3	5:17	7:49	61.3	61.3	59.9	52.2	57.5
Phosphorous slag powder	2.8	100.8	570	0.3	4:56	7:35	66.9	68.3	58.3	51.0	69.1
Tuff powder	2.8	103.8	870	0.4	4:43	7:39	53.1	54.6	56.5	52.9	61.0
Iron-ore slag powder	2.9	98.5	300	0.5	4:34	7:57	48.8	50.4	54.0	67.7	64.1
Pozzolan powder	2.7	98.5	450	0.3	4:38	7:37	53.8	54.2	58.1	67.7	61.3
Fly ash	2.3	101.5	440	0.2	6:30	9:15	53.8	50.0	52.8	66.3	68.4

Table 6.2　　　　　　　　Test results of hydration heat of admixtures

Type of admixture		Basalt powder	Limerock powder	Phosphorous slag powder	Tuff powder	Iron-ore slag powder	Pozzolan powder	Fly ash
Hydration heat (kJ/kg)	1 d	71.3	96.7	94.9	82.8	109.8	97.8	108.6
	3 d	120.0	138.2	156.8	135.9	153.5	148.2	170.5
	5 d	150.4	167.1	192.3	168.1	193.6	180.2	207.0
	7 d	173.6	184.5	217.0	191.0	215.3	199.2	230.7

Table 6.3　　Test results of physical and mechanics performance of cement

Density (g/cm³)	Specific surface area (m²/kg)	Stability	Time of setting (h: min)		Hydration heat (kJ/kg)				Compressive strength (MPa)		Bending strength (MPa)	
			Initial setting	Final setting	1 d	3 d	5 d	7 d	7 d	28 d	7 d	28 d
3.2	320	Qualified	5:36	8:23	124.2	192.1	235.0	260.7	27.3	44.3	6.6	8.9

6.2.1.2 Chemical Composition

The chemical compositions of admixtures are analyzed in Table 6.4

Table 6.4　　Test Results of chemical compositions of admixtures　　Unit: %

Admixture	Basalt powder	Limerock powder	Phosphorous slag powder	Tuff powder	Iron-ore slag powder	Pozzolan powder	Fly ash
SiO_2	45.09	4.38	23.46	55.89	35.02	56.97	52.25
Al_2O_3	12.39	1.56	2.61	16.57	10.50	16.67	26.39
TFe_2O_3	13.83	0.76	1.04	7.30	1.94	7.09	5.91
MgO	4.05	0.49	1.56	4.35	8.20	3.95	3.15
CaO	7.97	48.00	47.92	5.30	37.27	5.38	3.24
Na_2O	2.41	0.03	0.25	3.45	0.16	3.37	0.23
K_2O	1.80	0.14	0.88	3.17	0.32	3.31	3.40
TiO_2	2.52	0.08	0.12	1.14	0.66	1.10	1.52
P_2O_5	0.55	0.03	1.05	0.44	0.03	0.46	0.21
MnO	0.30	0.02	0.04	0.10	1.79	0.10	0.05
SO_3	0.69	0.39	0.83	0.41	1.19	0.71	0.41
Loss on ignition	8.40	44.17	20.24	2.32	2.92	1.01	3.34

6.2.2 Microscopic Analysis of Granule Morphology and Mineral Composition of Admixture

6.2.2.1 Granule Morphology of Admixture (SEM Photo)

SEM analysis is conducted for mineral powders such as Basalt powder, limerock powder, phosphorous slag powder, tuff powder, iron-ore slag powder, pozzolan powder and fly ash, with the photos as follows Figure 6.1 to Figure 6.7.

6.2.2.2 Mineral Composition of Admixture

Mineral powders including Basalt powder, limerock powder, phosphorous slag powder, tuff powder, iron-ore slag powder, pozzolan powder and fly ash are analyzed with the X-ray diffractometer, as shown in Figure 6.8 to Figure6.14. See Table 6.5 for the main mineral compositions of admixtures.

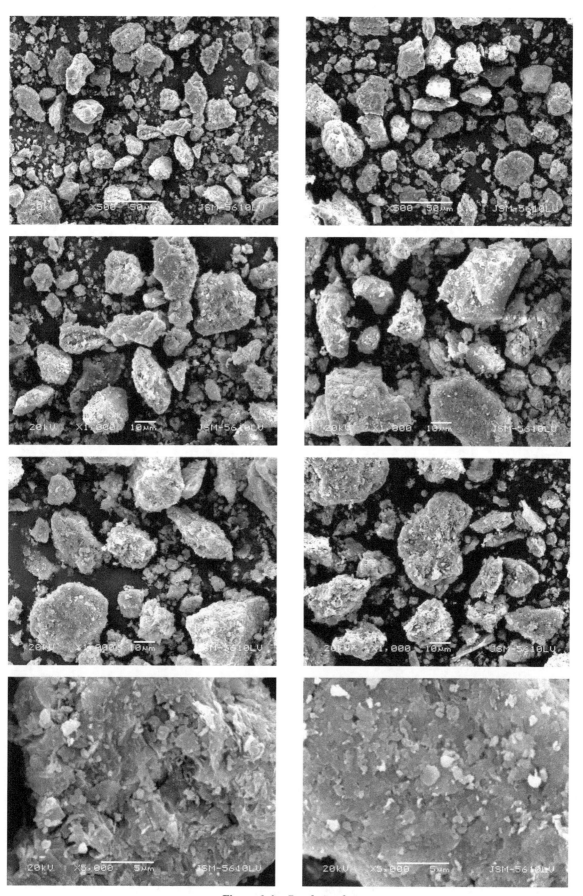

Figure 6.1 *Basalt powder*

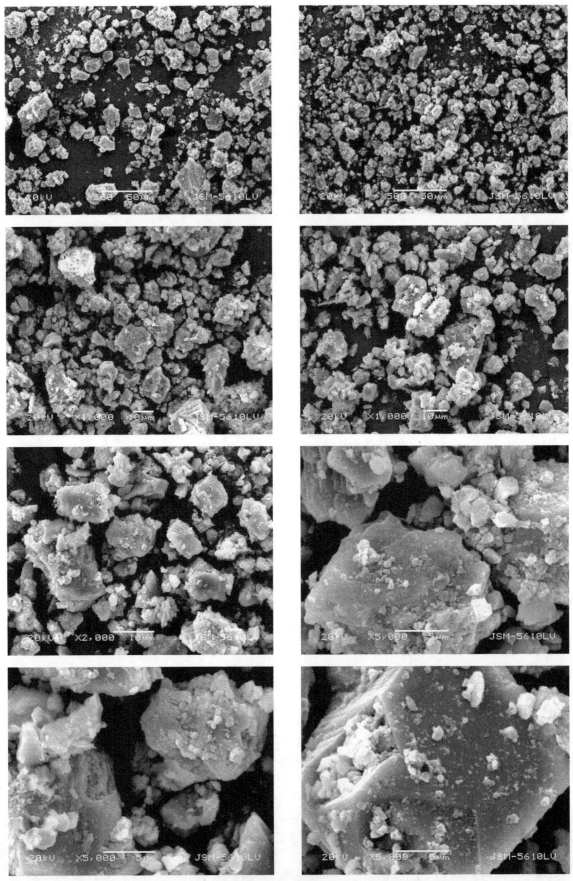

Figure 6.2　*Limerock powder*

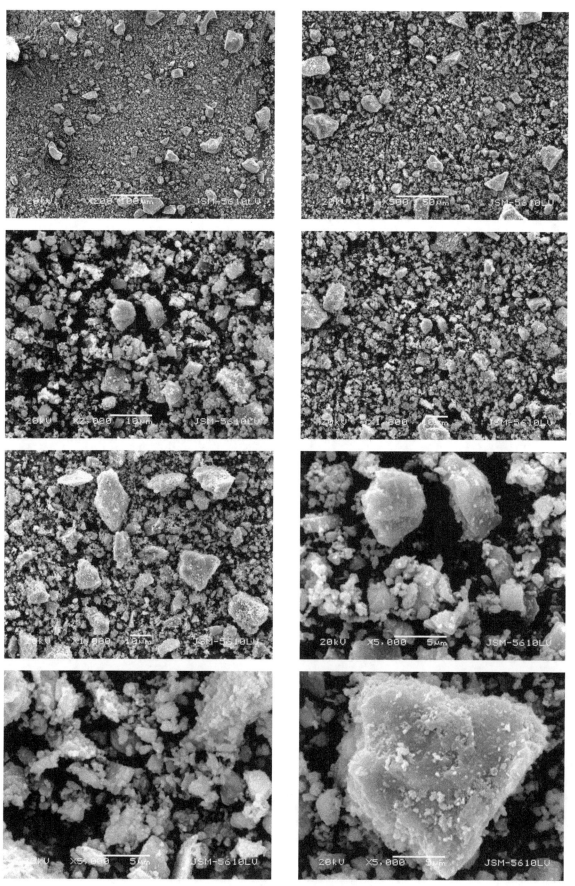

Figure 6.3 *Phosphorous slag powder*

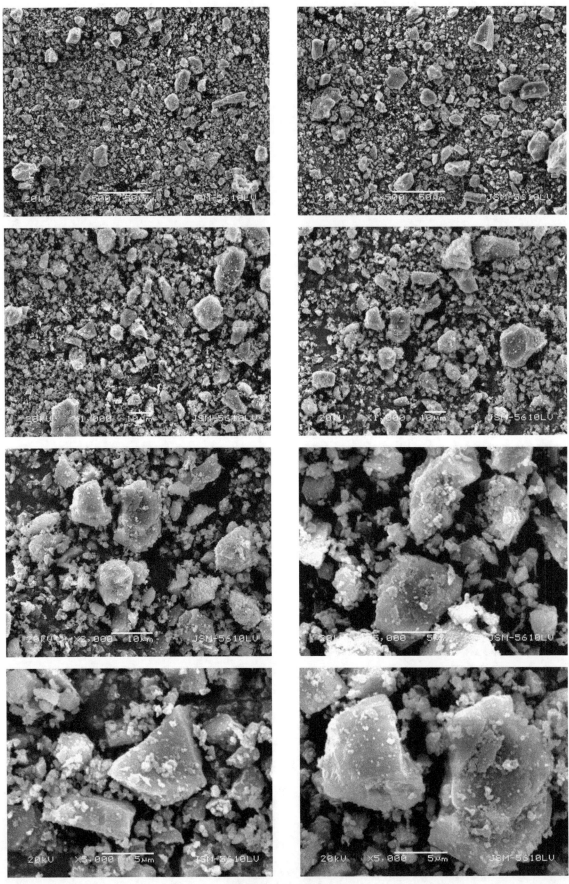

Figure 6.4 *Tuff powder*

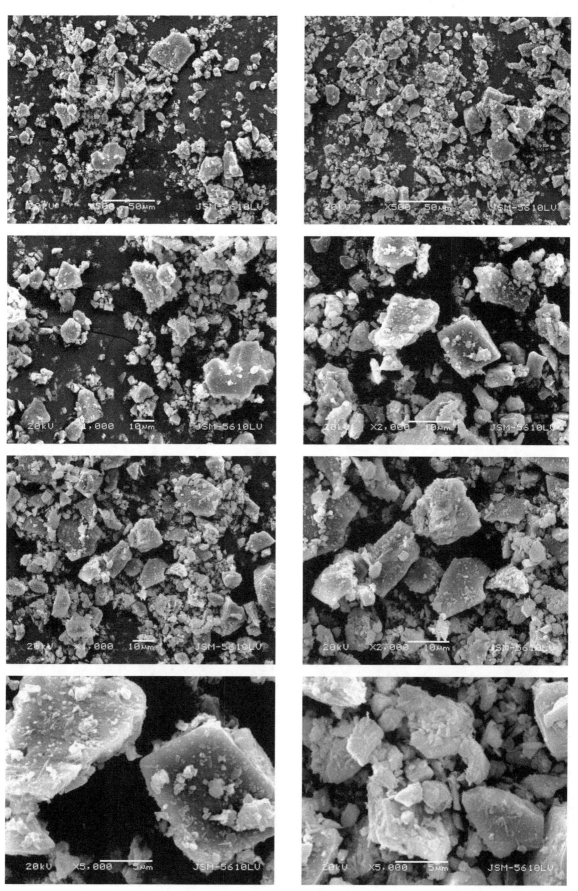

Figure 6.5 *Iron-ore slag powder*

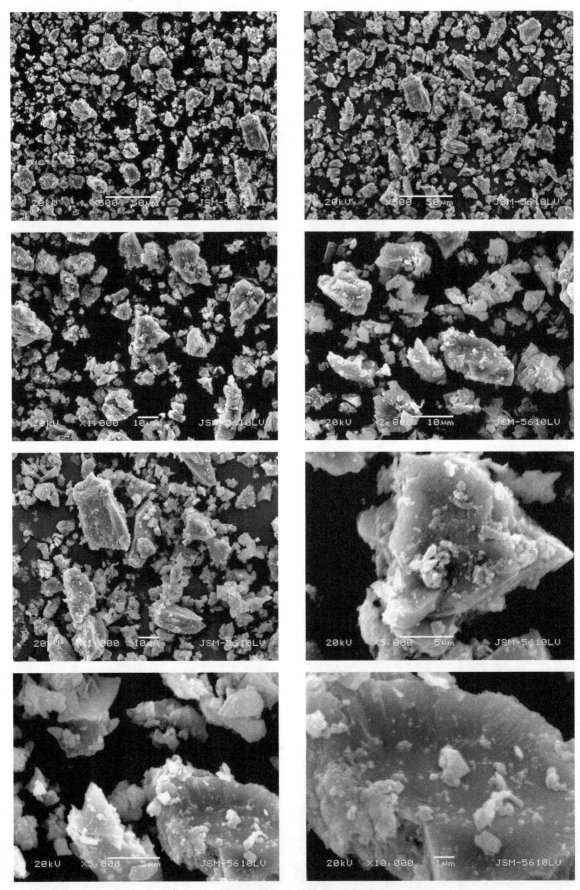

Figure 6.6　*Pozzolan powder*

Chapter 6 Research and Application of RCC Admixture

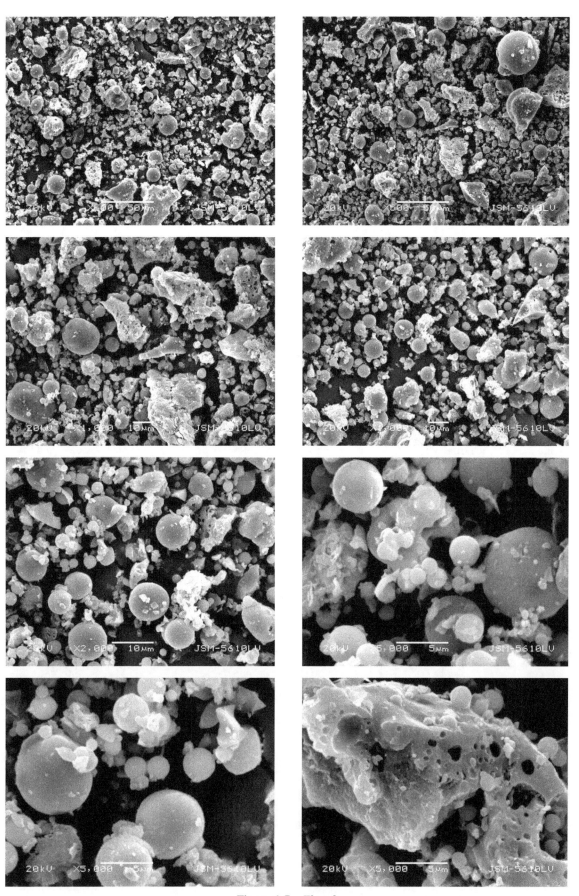

Figure 6.7 *Fly ash*

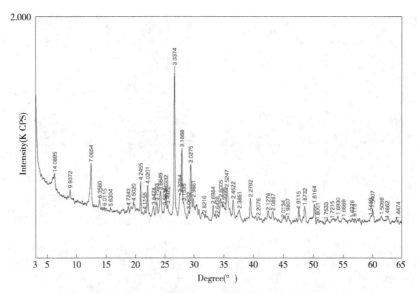

Figure 6.8　*X-ray diffraction spectrum of basalt powder*

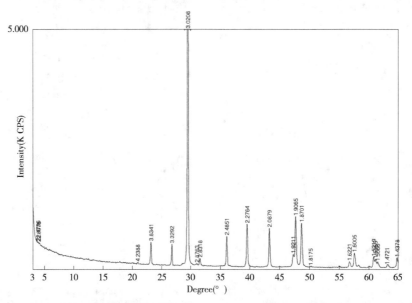

Figure 6.9　*X-ray diffraction spectrum of limerock powder*

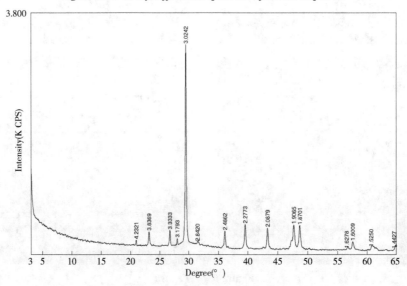

Figure 6.10　*X-ray diffraction spectrum of phosphorous slag powder*

Chapter 6 Research and Application of RCC Admixture

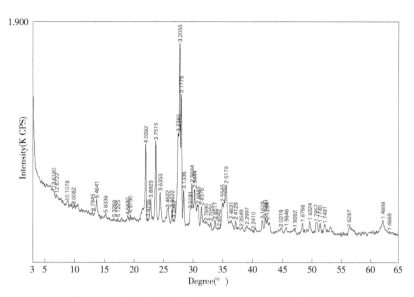

Figure 6.11 *X-ray diffraction spectrum of tuff powder*

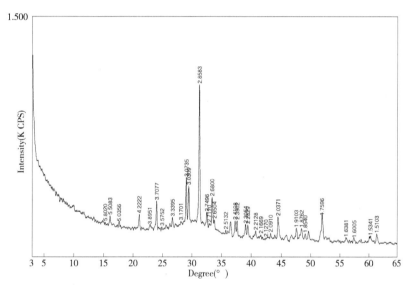

Figure 6.12 *X-ray diffraction Spectrum of Iron ore slag powder*

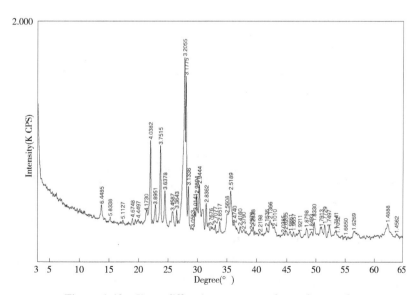

Figure 6.13 *X-ray diffraction spectrum of pozzolan powder*

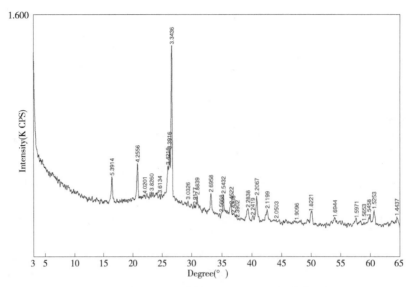

Figure 6.14 *X-ray diffraction spectrum of fly ash*

Table 6.5　　　　　　　　　　　Main mineral compositions of admixtures

Admixture	Main mineral compositions
Tuff powder	Feldspar (albite and anorthite predominantly)
Limerock powder	Calcite, quartz
Iron-ore slag powder	Melilite
Phosphorous slag powder	Calcite, quartz
Fly ash	Mullite, quartz
Basalt powder	Chlorite, mica, quartz, feldspar, calcite
Pozzolan powder	Feldspar (albite and anorthite predominantly)

6.2.3 Particle Distribution

6.2.3.1 Particle Distribution by Sieve Analysis

All kinds of mineral powders are conducted with sieve analysis, producing the particle distribution as shown in Table 6.6.

Table 6.6　　　　Test results of particle size of mineral admixtures　　　　Unit: %

Grain size (μm)	Basalt powder	Limerock powder	Phosphorous slag powder	Tuff powder	Iron-ore slag powder	Pozzolan powder	Fly ash
80	19.34	9.86	7.62	0.50	7.62	1.27	1.05
45	24.42	9.92	13.26	2.04	13.26	4.05	6.98
38	7.94	5.66	12.22	2.58	12.22	7.68	9.93
<38	48.30	74.56	66.9	94.89	66.9	87.01	82.04

6.2.3.2 Particle Distribution Analysis by Laser Particle Analyzer

All kinds of mineral powders are analyzed by JL-1155 laser particle analyzer, with the results listed as Table 6.7. Based on the analysis results in Table 6.7, Table 6.8 Cumulative Percent of Mass of Particles Passed and the accumulative curves from Figure 6.15 to Figure 6.22 are obtained.

Chapter 6   Research and Application of RCC Admixture

Table 6.7   Particle mass frequency distribution   Unit: %

Admixture (μm)	Tuff powder	Limerock powder	Iron-ore slag powder	Phosphorous slag powder	Fly ash	Basalt powder	Pozzolan powder
250 – 500	0	0	0	0	0	2.00	0.33
74 – 250	0.33	13.00	10.33	5.33	2.33	20.33	1.00
74 – 54	0	0	0	0	0	0	0
54 – 40	0	0	0	0	0	0	0
40 – 30	0	0	1.17	0	1.36	0.71	0.41
30 – 20	3.84	3.72	7.86	4.44	9.79	6.21	7.04
20 – 15	8.47	8.31	12.82	8.58	16.65	10.49	12.83
15 – 10	6.95	7.22	10.37	6.90	13.50	7.93	10.95
10 – 5	20.28	20.71	21.63	18.91	23.66	16.13	23.82
< 5	60.13	47.04	35.82	55.84	32.71	36.19	43.62

Table 6.8   Cumulative percent of mass of particles passed   Unit: %

Admixture (μm)	Tuff powder	Limerock powder	Iron-ore slag powder	Phosphorous slag powder	Fly ash	Basalt powder	Pozzolan powder
500	100.00	100.00	100.00	100.00	100.00	100.00	100.00
250	99.90	100.00	100.00	100.00	100.00	97.99	99.67
74	99.57	87.00	89.67	94.67	97.67	77.66	98.67
54	99.57	87.00	89.76	94.67	97.67	77.66	98.67
40	99.57	87.00	89.76	94.67	97.67	77.66	98.67
30	99.57	87.00	88.50	94.67	96.31	76.95	98.26
20	95.73	83.28	80.64	90.23	86.52	70.74	91.22
15	87.26	74.97	67.82	81.65	69.87	60.25	78.39
10	80.31	67.75	57.45	74.75	56.37	52.32	67.44
5	60.13	47.04	35.82	55.84	32.71	36.19	43.62

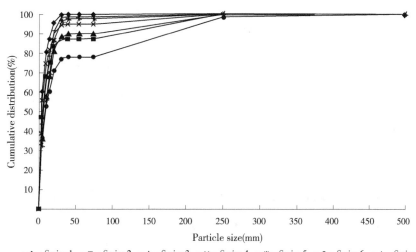

Note: Series 1: Tuff powder; Series 2: Limerock powder; Series 3: Iron-ore slag powder; Series 4: Phosphorous slag powder; Series 5: Fly ash; Series 6: Basalt powder; Series 7: Pozzolan powder

Figure 6.15   *Admixture particle distribution cumulative curve*

As shown in Table 6.8 and Figure 6.15, in several admixtures, the particle of tuff powder is finest, while that of Basalt powder is coarsest. According to the proportions of particles less than 30 μm, the particles from the finest to the coarsest one are as follows: tuff powder (99.57%), pozzolan powder (98.26%), fly ash (96.31%), phosphorus slag powder (94.64%), iron-ore slag powder (88.50%), limerock powder (87.00%), basalt powder (76.95%).

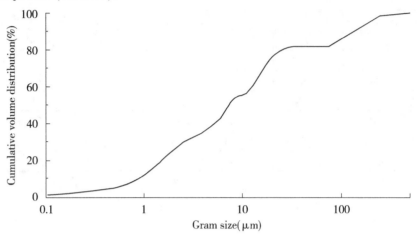

Figure 6.16 *Cumulative volume distribution curve of basalt powder*

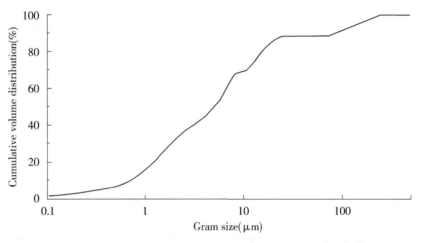

Figure 6.17 *Cumulative volume distribution curve of limerock powder*

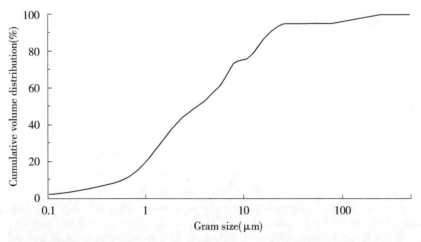

Figure 6.18 *Cumulative volume distribution curve of phosphorus slag powder*

Chapter 6 Research and Application of RCC Admixture

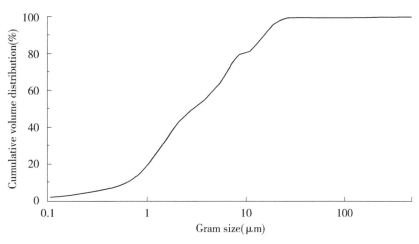

Figure 6.19 *Cumulative volume distribution curve of tuff powder*

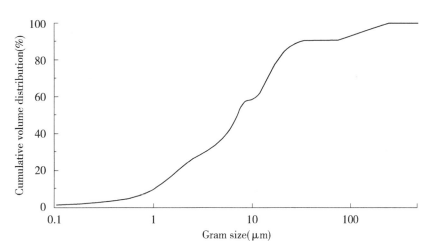

Figure 6.20 *Cumulative volume distribution curve of iron core slag powder*

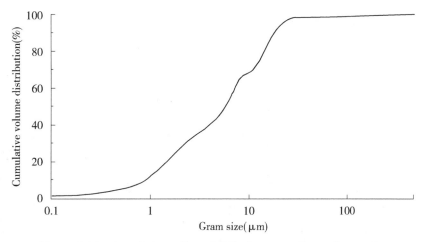

Figure 6.21 *Cumulative volume distribution curve of pozzolan powder*

6.2.4 Mineral Admixture's Effect on the Strength of Cement Mortar

6.2.4.1 Mineral Admixture's Effect on the Strength of Cement Mortar at a Constant Fluidity

Ⅰ Test methods

On the basis of 125 – 135 mm of cement fluidity jumping table, the cement is replaced with 30% mineral admixture at the same quantity to form a cement specimen. The mix proportion for test is shown in Table 6.9.

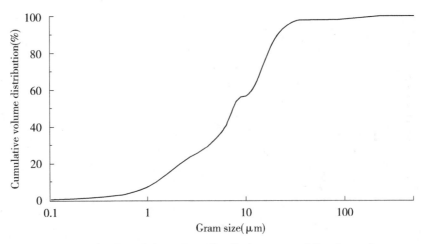

Figure 6.22 *Cumulative volume distribution curve of fly ash powder*

The strength test shall be carried out in accordance with *Methods of Testing Cements-Determination of Strength* (*ISO Cement Test Methods*) GB/T 17671—1999.

Table 6.9   **Mix proportion for testing**

No.	Type of admixture	Cement (g)	Admixture (g)	Standard sand (g)	W/B
J-XW	Basalt Powder	378	162	1 350	0.49
J-SH	Limestone rock powder	378	162	1 350	0.45
J-LZ	Phosphorous slag powder	378	162	1 350	0.45
J-NH	Tuff powder	378	162	1 350	0.46
J-KZ	Iron-ore slag powder	378	162	1 350	0.44
J-HS	Pozzolan powder	378	162	1 350	0.44
J-FA	Fly ash	378	162	1 350	0.45

Ⅱ Test Results & Analysis

Test results are summarized in Table 6.10.

Table 6.10   **Mineral admixture's effect on the strength of cement mortar at a constant fluidity**

No.		J-XW	J-SH	J-LZ	J-NH	J-KZ	J-HS	J-FA
Breaking strength ratio (%)	3 d	53.5	65.1	69.8	60.5	55.8	55.8	55.8
	7 d	57.6	69.7	72.7	65.2	66.7	65.2	60.6
	28 d	63.5	78.1	80.2	74.0	74.0	72.9	64.6
	60 d	69.8	78.9	77.7	73.8	81.6	74.7	79.8
	90 d	78.2	81.1	91.8	77.6	86.9	84.6	92.7
Compressive strength ratio (%)	3 d	43.8	61.3	66.9	53.1	48.8	53.8	53.8
	7 d	44.0	61.3	68.3	54.6	50.4	54.2	50.0
	28 d	44.7	59.9	58.3	56.5	54.0	58.1	52.8
	60 d	44.9	52.2	51.0	52.9	67.7	67.7	66.3
	90 d	52.9	57.5	69.1	61.0	64.1	61.3	68.4

Chapter 6 Research and Application of RCC Admixture

Continued Table 6.10

No.		J-XW	J-SH	J-LZ	J-NH	J-KZ	J-HS	J-FA
Brittleness coefficient	3 d	3.04	3.50	3.57	3.27	3.25	3.58	3.58
	7 d	3.29	3.78	4.04	3.60	3.25	3.58	3.55
	28 d	3.62	3.95	3.74	3.93	3.76	4.10	4.21
	60 d	3.35	3.44	3.41	3.72	4.31	4.71	4.32
	90 d	4.41	4.61	4.90	5.12	4.80	4.72	4.81

Test results show that, at a constant fluidity, these mineral admixtures will reduce the strength of cement. It is indicated that, if 30% cement is replaced with the mineral admixture at the same quantity, the strength of cement will decrease in 90 day age. In general tendency, the strength ratio will increase with the extension of age. That is to say, the mineral admixture will work efficiently with the extension of age. In addition, the mineral admixture will reduce the brittleness of cement. It suggests that, at the same strength, the mineral admixture will enhance the toughness of cement or concrete, beneficial for crack resistance.

6.2.4.2 Mineral Admixture's Effect on the Strength of Cement Mortar at a Constant Water-cement Ratio

Ⅰ Test methods

The cement is replaced with 30% mineral admixture at the same quantity to form a cement specimen. The mix proportion for test is shown in Table 6.11. The strength test shall be carried out in accordance with *Methods of Testing Cements-Determination of Strength* (*ISO Cement Test Methods*) GB/T 17671—1999.

Table 6.11 Mix proportion for testing

No.	Type of admixture	Cement(g)	Admixture(g)	Standard sand(g)	W/B
JAQ-C	Straight cement	450	0	1 350	0.50
JAQ-XW	Basalt Powder	315	135	1 350	0.50
JAQ-SH	Limestone rock powder	315	135	1 350	0.50
JAQ-LZ	Phosphorous slag powder	315	135	1 350	0.50
JAQ-NH	Tuff powder	315	135	1 350	0.50
JAQ-KZ	Iron-ore slag powder	315	135	1 350	0.50
JAQ-HS	Pozzolan powder	315	135	1 350	0.50
JAQ-FA	Fly ash	315	135	1 350	0.50

Ⅱ Test results & analysis

Test results are summarized in Table 6.12.

Table 6.12 Mineral admixture's effect on the strength of cement mortar at a constant water-cement ratio

No.		JAQ-C	JAQ-XW	JAQ-SH	JAQ-LZ	JAQ-NH	JAQ-KZ	JAQ-HS	JAQ-FA
Flexural strength(MPa) Flexural strength ratio (%)	3 d	4.4/100	2.6/59.1	2.4/54.5	2.4/54.5	2.1/47.7	1.9/43.2	1.9/43.2	2.3/52.3
	7 d	6.6/100	4.3/65.2	4.2/63.6	4.5/68.2	4.4/66.7	4.0/60.6	4.0/60.6	4.5/68.2
	28 d	8.9/100	6.5/73.0	6.2/69.7	6.7/75.3	6.4/71.9	6.6/74.2	6.6/74.2	7.0/78.7
	60 d	9.7/100	7.4/76.5	7.6/78.2	7.6/78.7	7.5/77.8	8.0/82.3	7.5/77.2	8.3/85.6
	90 d	10.2/100	7.3/71.9	7.33/72.0	8.8/86.4	7.8/76.3	8.5/83.7	8.0/78.4	9.4/92.3

Continued Table 6.12

No.		JAQ-C	JAQ-XW	JAQ-SH	JAQ-LZ	JAQ-NH	JAQ-KZ	JAQ-HS	JAQ-FA
Compressive strength(MPa) Compressive strength ratio (%)	3 d	16.5/100	8.4/50.9	8.5/51.5	8.1/49.1	8.0/48.5	6.0/36.4	6.0/36.4	8.0/48.5
	7 d	27.3/100	16.5/60.4	15.1/55.3	16.3/59.7	15.5/56.8	13.1/48.0	14.3/52.4	16.5/60.4
	28 d	44.3/100	26.0/58.7	24.3/54.9	26.7/60.3	27.0/60.9	26.6/60.0	25.8/58.2	27.7/62.5
	60 d	50.6/100	28.2/55.8	30.1/59.5	31.0/61.4	32.9/65.0	33.3/65.9	30.8/60.8	35.1/69.4
	90 d	51.8/100	31.9/61.7	33.0/63.7	38.0/73.3	37.1/71.7	39.4/76.1	35.6/68.7	44.8/86.6
Brittleness coefficient	3 d	3.75	3.23	3.54	3.38	3.81	3.16	3.16	3.48
	7 d	4.14	3.84	3.6	3.62	3.52	3.28	3.58	3.67
	28 d	4.98	4	3.92	3.99	4.22	4.03	3.91	3.96
	60 d	5.22	3.81	3.98	4.08	4.37	4.18	4.11	4.24
	90 d	5.09	4.37	4.5	4.31	4.78	4.62	4.46	4.77

The test results show that, at a constant water-cement ratio, these mineral admixtures will reduce the strength of cement. It is indicated that, if 30% cement is replaced with the mineral admixture at the same quantity, the strength of cement will decrease in 90 day age. In general tendency, the strength ratio will increase with the extension of age. That is to say, the mineral admixture will work efficiently with the extension of age. In addition, similar to that at a constant fluidity, the mineral admixture will reduce the brittleness of cement. It suggests that, at the same strength, the mineral admixture will enhance the toughness of cement or concrete, good to improve the crack resistance of cement or concrete.

6.2.5 Hydration Products and Hydration Mechanism

By testing nine cementitous materials shown in Table 6.13 under different hydration ages (14 d, 28 d, 60 d, 90 d and 120 d) with XRD, DSC/TG, SEM and other micro testing methods, the types, quantity, distribution, morphology and its evolution of hydration products are tested for conducting a comprehensive study of hydration mechanism.

Table 6.13   Mix proportion for testing

No.	C (%)	CH (%)	XW (%)	SH (%)	LZ (%)	NH (%)	KZ (%)	HS (%)	FA (%)	W/B
1	0	25	75	0	0	0	0	0	0	0.4
2	0	25	0	0	75	0	0	0	0	0.4
3	0	25	0	0	0	75	0	0	0	0.4
4	0	25	0	0	0	0	75	0	0	0.4
5	0	25	0	0	0	0	0	75	0	0.4
6	0	25	0	0	0	0	0	0	75	0.4
7	100	0	0	0	0	0	0	0	0	0.4
8	70	0	30	0	0	0	0	0	0	0.4
9*	70	0	0	30	0	0	0	0	0	0.4

Note: * Cements used are straight aluminous cement, primarily composed of CA, accounting for 90% or above.

6.2.5.1 Hydration Products

Ⅰ SEM analysis

The hydration products of 9 specimens at different ages listed in Table 6.13 are analyzed for morphology with SEM, with the analysis results shown in Figure 6.23 to Figure 6.31.

Hydration of (basalt powder 75% + calcium hydroxide 25%) at 14 d

Hydration of (basalt powder 75% + calcium hydroxide 25%) at 60 d

Hydration of (basalt powder 75% + calcium hydroxide 25%) at 120 d

Figure 6.23 *SEM photo of No. 1 specimen*

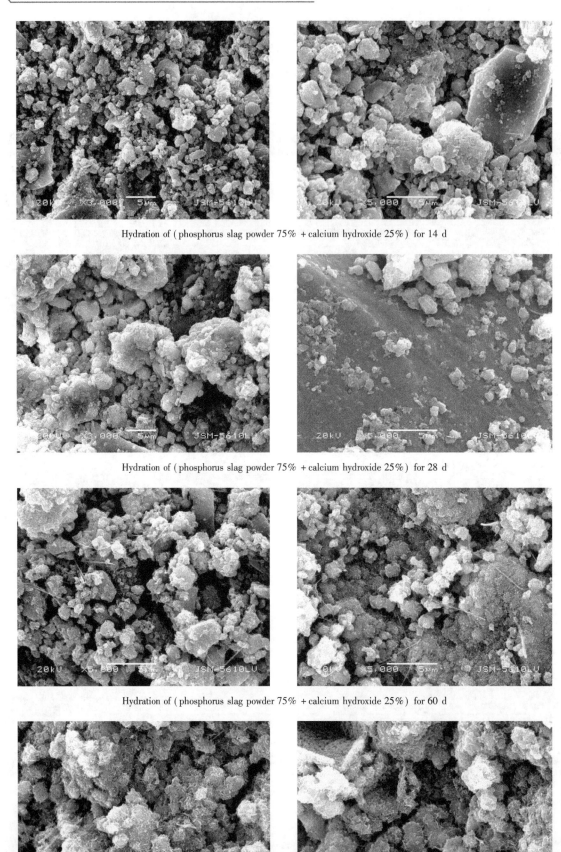

Hydration of (phosphorus slag powder 75% + calcium hydroxide 25%) for 14 d

Hydration of (phosphorus slag powder 75% + calcium hydroxide 25%) for 28 d

Hydration of (phosphorus slag powder 75% + calcium hydroxide 25%) for 60 d

Hydration of (phosphorus slag powder 75% + calcium hydroxide 25%) for 90 d

Hydration of (phosphorus slag powder 75% + calcium hydroxide 25%) for 120 d

Figure 6.24 *SEM photo of No. 2 specimen*

Hydration of (tuff powder 75% + calcium hydroxide 25%) for 14 d

Hydration of (tuff powder 75% + calcium hydroxide 25%) for 28 d

Hydration of (tuff powder 75% + calcium hydroxide 25%) for 60 d

Hydration of (tuff powder 75% + calcium hydroxide 25%) for 90 d

Hydration of (tuff powder 75% + calcium hydroxide 25%) for 120 d

Figure 6.25 *SEM photo of No.3 specimen*

Hydration of (iron-ore slag powder 75% + calcium hydroxide 25%) for 14 d

Hydration of (iron-ore slag powder 75% + calcium hydroxide 25%) for 28 d

Chapter 6 Research and Application of RCC Admixture

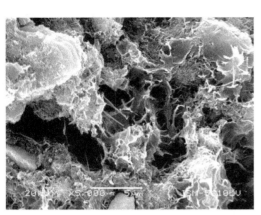 

Hydration of (iron-ore slag powder 75% + calcium hydroxide 25%) for 60 d

Hydration of (iron-ore slag powder 75% + calcium hydroxide 25%) for 90 d

Hydration of (iron-ore slag powder 75% + calcium hydroxide 25%) for 120 d

Figure 6.26 *SEM photo of No. 4 specimen*

Hydration of (pozzolan powder 75% + calcium hydroxide 25%) for 14 d

· 217 ·

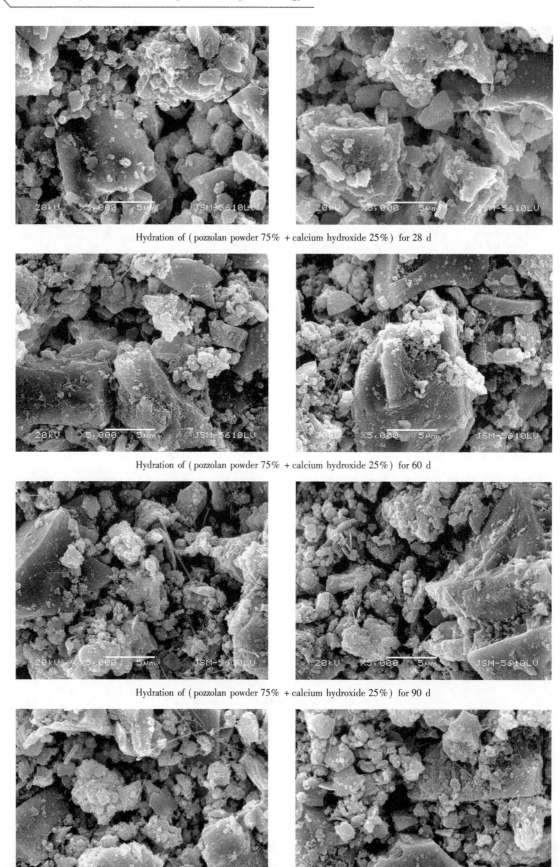

Figure 6.27 *SEM photo of No. 5 specimen*

Chapter 6 Research and Application of RCC Admixture

Hydration of (fly ash 75% + calcium hydroxide 25%) at for 14 d

Hydration of (fly ash 75% + calcium hydroxide 25%) at for 28 d

Hydration of (fly ash 75% + calcium hydroxide 25%) at for 60 d

Hydration of (fly ash 75% + calcium hydroxide 25%) at for 90 d

 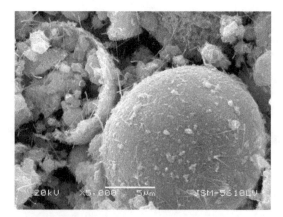

Hydration of (fly ash 75% + calcium hydroxide 25%) at for 120 d

Figure 6.28 *SEM photo of No.6 specimen*

Hydration of (cement 100%) for 14 d

Hydration of (cement 100%) for 28 d

Hydration of (cement 100%) for 60 d

Chapter 6 Research and Application of RCC Admixture

Hydration of (cement 100%) for 90 d

 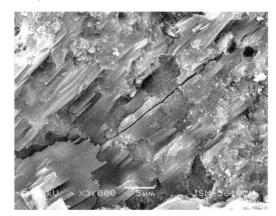

Hydration of (cement 100%) for 120 d

Figure 6.29 SEM photo of No. 7 specimen

Hydration of (cement 70% + basalt powder 30%) for 14 d

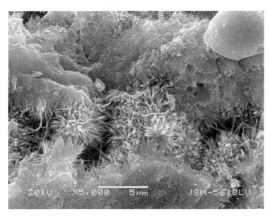 

Hydration of (cement 70% + basalt powder 30%) for 28 d

Hydration of (cement 70% + basalt powder 30%) for 60 d

Hydration of (cement 70% + basalt powder 30%) for 90 d

Hydration of (cement 70% + basalt powder 30%) for 120 d

Figure 6.30 *SEM photo of No. 8 specimen*

Hydration of (aluminous cement 70% + limerock powder 30%) for 14 d

Chapter 6 Research and Application of RCC Admixture

Hydration of (aluminous cement 70% + limerock powder 30%) for 28 d

Hydration of (aluminous cement 70% + limerock powder 30%) for 60 d

Hydration of (aluminous cement 70% + limerock powder 30%) for 90 d

Hydration of (aluminous cement 70% + limerock powder 30%) for 120 d

Figure 6.31 *SEM photo of No. 9 specimen*

As seen from Figure 6.23 to Figure 6.30 that, six admixtures (No. 1 – 6) have hydration reaction with Ca(OH)$_2$ at least at 14 d age, with the hydration products similar to that of cement hydration. More products will generate with the extension of hydration age. It suggests that, if such 6 mineral powders are added into the cement concrete, they will have some chemical reactivity, and will have secondary reaction with Ca(OH)$_2$, the hydration product of cement, contributing to the strength and other properties of concrete. A difference among them lies in the hydration speed and degree. Among them, the hydration of iron-ore slag powder is relatively quickest and most sufficient, followed by phosphorus slag powder, fly ash and tuff powder. The pozzolan powder and basalt powder have slower and inadequate reaction. This may be a sign of activity difference among them. As seen from Figure 6.31, the limerock powder can react with aluminous cement to generate hydration products. If the limerock powder is mixed into cement concrete, the limerock powder will have a reaction with the hydrated calcium aluminate, contributing to the strength and other properties of concrete.

II XRD analysis

With XRD, the hydration products of No. 1 – 6 and No. 9 specimen listed in Table 6.13 at different ages are analyzed, as shown in Figure 6.32 to Figure 6.43.

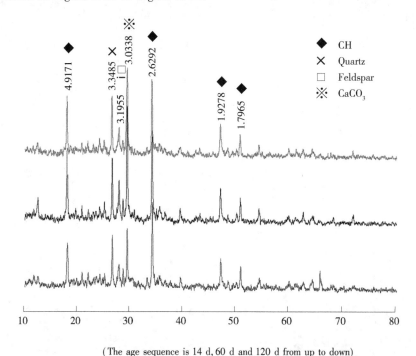

(The age sequence is 14 d, 60 d and 120 d from up to down)

Figure 6.32 *XRD diagram of No. 1 specimen (basalt powder 75% + calcium hydroxide 25%) at different ages*

It can be seen from Figure 6.32 to Figure 6.37 that, the amount of calcium hydroxide in No. 1 – 6 specimen gradually decreases with the extension of hydration age. It suggests that, with the extension of hydration age, the mineral powder will have more sufficient reaction with the calcium hydroxide, so the amount of calcium hydroxide decreases. This is same as observed by SEM. It show that, if these six mineral powders are mixed into the cement concreted, they will have certain chemical reaction activity and will have secondary hydration reaction with Ca(OH)$_2$, contributing to the strength and other properties of concrete.

From Figure 6.39 to Figure 6.43, the difference among No. 1 – 6 specimen can be roughly seen, that is, the amount of Ca(OH)$_2$ consumed in the hydration and the degree of hydration are different. Among them, the hydration of iron ore slag powder is the fastest and most sufficient. At 14 d, C-S-H can be seen in the hydration products. It can be seen from Figure 6.38 that limerock powder can react with the hydrated calcium aluminate

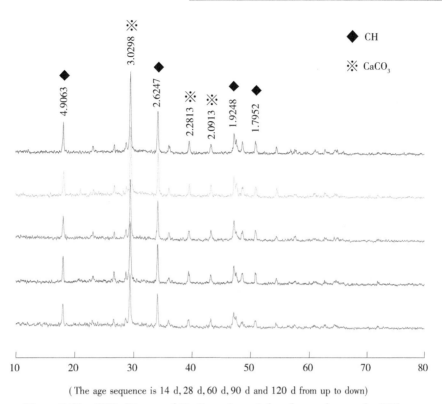

(The age sequence is 14 d, 28 d, 60 d, 90 d and 120 d from up to down)

Figure 6.33 *XRD diagram of No. 2 specimen (phosphorus slag powder 75% + calcium hydroxide 25%) at different ages*

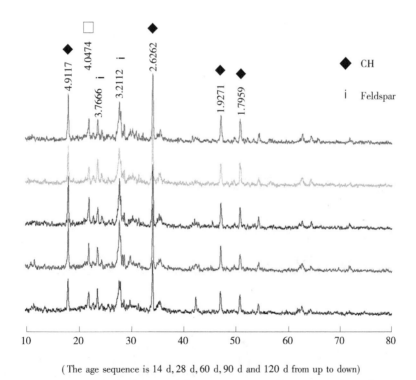

(The age sequence is 14 d, 28 d, 60 d, 90 d and 120 d from up to down)

Figure 6.34 *XRD diagram of No. 3 specimen (tuff powder 75% + calcium hydroxide 25%) at different ages*

to generate the hydrated calcium almuninate ($Ca_4Al_2CO_9 \cdot 11H_2O$ or $3CaO \cdot Al_2O_3 \cdot CaCO_3 \cdot 11H_2O$).

6.2.5.2 Study on Hydration Mechanism of Mineral Admixtures

By testing nine cementitous materials shown in Table 6.9 under different hydration ages (14 d, 28 d, 60 d, 90

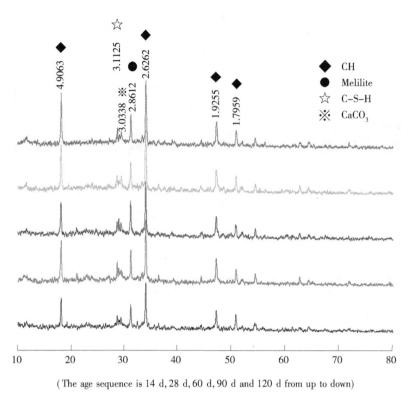

(The age sequence is 14 d, 28 d, 60 d, 90 d and 120 d from up to down)

Figure 6.35 *XRD diagram of No. 4 specimen (iron-ore slag powder 75% + calcium hydroxide 25%) at different ages*

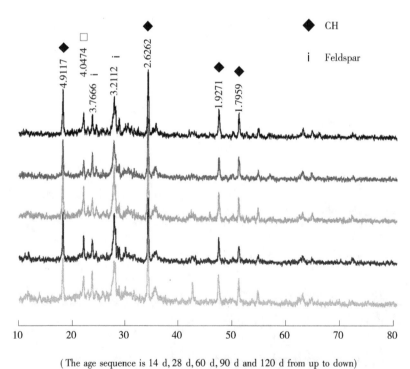

(The age sequence is 14 d, 28 d, 60 d, 90 d and 120 d from up to down)

Figure 6.36 *XRD diagram of No. 5 specimen (pozzolan powder 75% + calcium hydroxide 25%) at different ages*

d and 120 d) with XRD, DSC/TG, SEM and other micro testing methods, the types, quantity, distribution, morphology and its evolution of hydration products are tested, with the results in previous section. The influence of mineral admixtures on cement pore structure is studied by water absorption kinetics method to

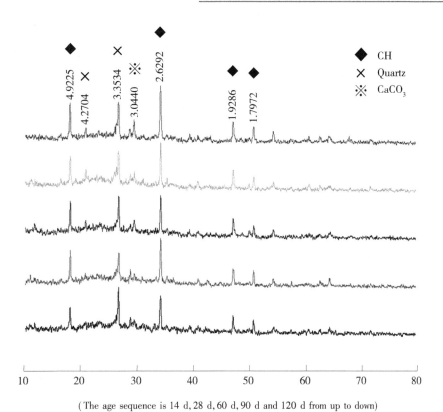

(The age sequence is 14 d, 28 d, 60 d, 90 d and 120 d from up to down)

Figure 6.37 *XRD diagram of No.6 specimen (fly ash 75% + calcium hydroxide 25%) at different ages*

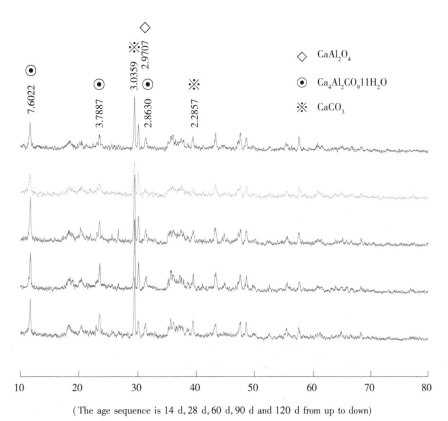

(The age sequence is 14 d, 28 d, 60 d, 90 d and 120 d from up to down)

Figure 6.38 *XRD diagram of No.9 specimen (aluminous cement 70% + limerock powder 30%) at different ages*

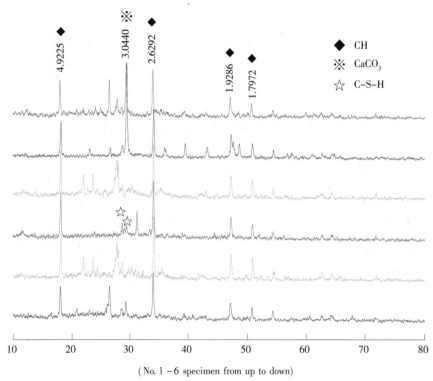

(No. 1 – 6 specimen from up to down)

Figure 6.39 *Specimen hydration XRD diagram at* 14 d

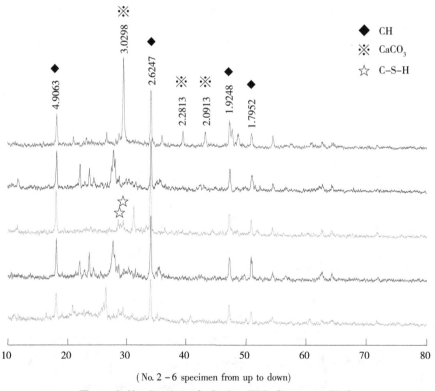

(No. 2 – 6 specimen from up to down)

Figure 6.40 *Specimen hydration XRD diagram at* 28 d

comprehensively study the hydration mechanism of mineral admixtures in cement.

Ⅰ Mineral admixtures' effects on cement pore structure-water absorption kinetics method

It is a common method to measure the parameters of porous materials such as concrete by soaking capillary porous materials in liquid. With this method, the apparent porosity, average pore size and pore uniformity of

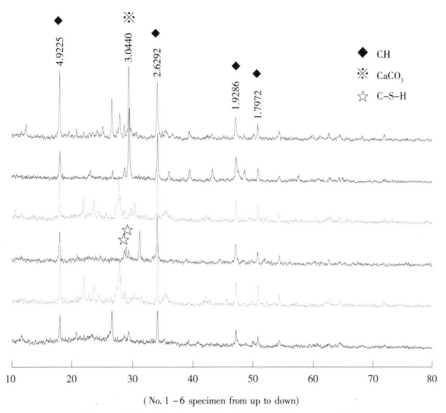

(No. 1 – 6 specimen from up to down)

Figure 6.41 *Specimen hydration XRD diagram at* 60 d

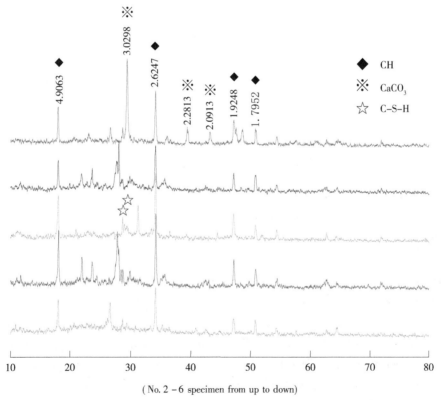

(No. 2 – 6 specimen from up to down)

Figure 6.42 *Specimen hydration XRD diagram at* 90 d

concrete structure can be measured.

The water absorption kinetics method is based on the water absorption curve of cement paste, mortar and

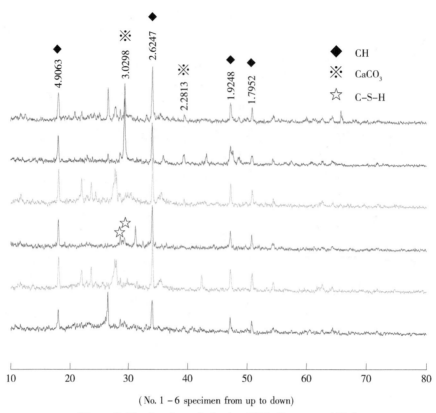

(No. 1 - 6 specimen from up to down)

Figure 6.43 *Specimen hydration XRD diagram at* 120 d

concrete with the characteristics of stable exponential function, approximating the exponential function of three parameters.

$$W_\tau W_{max} 1 - e^{\bar{\lambda}_\varepsilon \tau^\alpha}$$

Where W_τ and W_{max} = the elapsed water absorption time τ and the water absorption rate by weight of the porous material respectively

$\bar{\lambda}$ = an average pore diameter of the capillary pore in the material. Large, $\bar{\lambda}$ indicates that the average pore size of the material is large

α = the uniformity of capillary pore size, falling in the range of $0 \leqslant \alpha \leqslant 1$. For the single pore material, $\alpha = 1$. The smaller α is, the more inhomogeneous the pore size is

The test results are shown in Table 6.14.

The test results in Table 6.14 show that, the addition of admixtures could increase the absorption rate by quality of cementing materials at 7 d. At 28 d age, the cementing materials with limerock powder and phosphorus slag powder have the absorption rate by quality lower than that of cement paste. To 90 d, in addition to that with limerock powder and phosphorus slag powder, the cementing materials with iron ore slag powder and pozzolan powder have the absorption rate by quality lower that of cement paste. At the 120 d age, except that the cementing materials with tuff powder and fly ash have the absorption rate by quality higher than that of cement paste, those cementing materials with basalt powder, limerock powder, phosphorous slag powder, iron ore slag powder and pozzolan powder have the absorption rate lower than that of cement paste. Among them, the cementitous materials with limerock powder, phosphorus slag powder and iron ore slag powder have the absorption rate by quality lower than 50% of that of cement paste. Besides, both the uniformity of pore size and the average pore size are better than that of cement paste.

Table 6.14　　　　　　　　　Mineral admixtures' effects on cement pore structure

	No.	J-C	J-XW	J-SH	J-LZ	J-NH	J-KZ	J-HS	J-FA
7 d	Absorption rate by mass (%)	25.62	28.33	28.62	26.86	28.58	30.48	27.77	27.4
	α	0.77	0.95	0.89	0.91	0.84	0.54	0.57	0.53
	$\bar{\lambda}_1$	2.19	3.32	3.21	2.41	2.66	3.44	2.85	2.51
	$\bar{\lambda}_2$	2.77	3.54	3.71	2.63	3.22	9.85	6.28	5.6
14 d	Absorption rate by mass (%)	23.84	26.58	26.93	25.07	27.12	28.32	26.11	25.83
	α	0.84	0.98	0.94	0.72	0.9	0.92	0.96	0.86
	$\bar{\lambda}_1$	1.69	1.95	1.63	0.81	1.33	1.18	2.77	1.8
	$\bar{\lambda}_2$	1.88	1.98	1.68	0.74	1.38	1.2	2.9	1.97
28 d	Absorption rate by mass (%)	22.97	25.09	21.74	19.57	26.05	25.38	25	24.99
	α	0.66	0.75	0.47	0.44	0.66	0.44	0.56	0.5
	$\bar{\lambda}_1$	0.85	0.63	0.35	0.25	0.58	0.26	0.61	0.53
	$\bar{\lambda}_2$	0.78	0.55	0.11	0.04	0.44	0.05	0.41	0.28
90 d	Absorption rate by mass (%)	16.43	18.63	8.91	8.45	24.02	10.33	15.15	22.76
	α	0.65	0.58	0.31	0.37	0.6	0.35	0.41	0.63
	$\bar{\lambda}_1$	0.5	0.33	0.29	0.26	0.36	0.26	0.28	0.41
	$\bar{\lambda}_2$	0.34	0.15	0.02	0.03	0.18	0.02	0.05	0.25
120 d	Absorption rate by mass (%)	15.92	13.88	7.11	7.53	23.76	7.79	13.9	22.98
	α	0.64	0.48	0.36	0.48	0.74	0.37	0.51	0.89
	$\bar{\lambda}_1$	0.41	0.34	0.26	0.26	0.33	0.27	0.23	0.38
	$\bar{\lambda}_2$	0.25	0.11	0.02	0.06	0.22	0.03	0.05	0.34

Ⅱ　Hydration mechanism of admixtures in cement concrete

As seen from SEM, XRD, DSC, TG and DTG analysis and test results, it is found that these admixtures studied could have hydration reaction with $Ca(OH)_2$ or calcium aluminate hydrate at least from the 14 d, which will generate the product similar to the cement hydration product or the hydrated calcium aluminate ($Ca_4Al_2CO_9 \cdot 11H_2O$ or $3CaO \cdot Al_2O_3 \cdot CaCO_3 \cdot 11H_2O$). Some of these hydration products (such as calcium aluminate hydrate) can react with gypsum in cement to form hydrated calcium sulphoaluminate. More products will generate with the extension of hydration age. It suggests that, when these mineral powders are used as admixtures of cement concrete, they all have certain chemical reaction activity, contributing to the strength and other properties of concrete. A difference among them lies in the hydration speed and degree. Among them, the hydration of iron-ore slag powder is relatively quickest and most sufficient, followed by phosphorus slag powder, fly ash and tuff powder. The pozzolan powder and basalt powder have slower and inadequate reaction. As the admixtures have hydration reaction, the porosity of cementing materials decrease, the pore structure is optimized and the strength is improved.

The admixtures could improve the properties of concrete mixture and hardened concrete from the following aspects:

1. Form good gradation with cement particles, make the initial structure of cementing material more dense so as to reduce water and fill the gap between cement particles;
2. the active components in the admixture will have a secondary hydration reaction with some hydration products of cement, generating the hydration products with cementation performance, contributing to the improvement of the performance of cement paste and hardened concrete;
3. the core of admixture not involved in hydration serves as the micro aggregate in the cement paste and hardened concrete, contributing to the improvement of cement paste and hardened concrete.

6.2.6　Conclusion

Many mineral powders can be used as admixtures for concrete. After being ground to have a certain specific surface area, these mineral powders are mixed into the concrete to reduce water. Their roles in improving the properties of concrete mixture and hardened concrete include:
1. Form good gradation with cement particles, make the initial structure of cementing material more dense so as to reduce water and fill the gap between cement particles.
2. Active ingredients in admixture act secondary hydration with some hydrated products in cement, to generate hydration products with cementing property, contributing to the improvement of the performance of set cement and hardened concrete.
3. Admixture cores which are not involved into hydration can act as micro aggregates inside set cement and hardened concrete, to contribute to the improvement of some performance of the set cement and hardened concrete.

The so-called active and inactive admixture term is an artificial boundary to distinguish the activity of admixtures. In fact, as long as the mineral materials are ground to enough specific surface area, they will have certain chemical reaction activity with some hydration products of cement. The difference is only that the size of the activity and the time of its exertion and some components of some minerals may have adverse effects on the improvement of the properties of cement, stone and concrete.

6.3　The Research and Application of SL Admixture in Gelantan Project

6.3.1　Foreword

6.3.1.1　Project Profile

In order to adapt to continuous and rapid rolling and compacting construction of RCC and to reduce adiabatic temperature rise of mass concrete as much as possible, it is required to add admixture in large proportion inside RCC to reach the purpose of reducing cement consumption. However, with economic development of Yunnan Province and surrounding regions, there is rising shortage of market supply of flyash in Yunnan Province, within construction period of this Project (as well as several other cascade hydro-power stations developed by the Owner of this Project along main stream of Lixian River), the flyash of all manufacturers have been totally ordered. Therefore, the Owner made use of blast furnace slag produced by iron and steel enterprises in Yuxi Region and limestone mine produced by Jinggu Cement Plant, and these two materials were processed into blast furnace slag powder (hereinafter called slag for short, indicated by letter "S") and limerock powder (hereinafter called rock powder for short, indicated by letter "L"), after test and study and comparative analysis, slag S and rock powder L were selected and ground in proportionally-mixed by the ratio of 5∶5 and were processed into new type of admixture, called "SL" admixture for short. "SL" admixture not only can be used in Gelantan Project, but also can be used as concrete admixtures in other cascade hydro-power stations (such as Tuka River and Jufudu) developed on main stream of Lixian River

Lixian River originated from Jian Township, Yunnan Province, which is Level 1 tributary of Red River System. Gelantan Hydropower Station is located in downstream of Lixian River, which is 40 km far from China-Vietnam border and 50 km far from Jiangcheng County, Simao Prefecture, Yunnan Province. The drainage area above dam site is 17 170 km^2. Electricity generation is main task of Gelantan Hydropower Station, with total installed gross capacity of 450 MW. Gelantan Hydropower Station Project is classified as Grade Ⅱ, and project scale is of large-scale (2) type. Main structures include RCC gravity retaining dam, headrace tunnel, sand flushing tunnel and above-ground bank powerhouse. The maximum height of the dam is 113 m and dam crest is 466 m long. There is about 1.5 million m^3 of concrete in main works, including 900 thousand m^3 of RCC.

6.3.1.2 Blast Furnace Slag Micro Powder

Steel plant can produce waste slag in large quantity in process of steel production, which can turn into granulated blast-furnace slag with water quenching treatment after discharged in high temperature, granulated blast-furnace slag is one kind of potentially active cementing material. granulated blast-furnace slag must be ground and mixed with cement in use, it can be prepared into high performance concrete with various kinds, there are relatively mature research and applications on granulated blast-furnace slag acting as cement admixture or as high performance concrete admixture. However, there are less research and applications on slag powder (iron ore slag, ferromanganese slag etc) in RCC. There is shortage of flyash resource and transportation is difficult and costly in southwest region, admixture in large quantity is inevitably needed in hydro-power construction if economic and rapid RCC construction technique was adapted in dam construction. There is remarkable technical economy and environment protection significance for study on slag powder acting as admixture in RCC, which will develop much broader prospect for RCC dam construction technique.

Granulated blast furnace slag powder possesses potential hydraulicity and is high quality mixing material for cement and concrete. With rising of grinding technology and premixed concrete, fine-grinding granulated blast furnace slag (powder) is widely used. Since 1980s, some countries, including UK, USA, Canada, Japan, France and Australia etc, have enacted national standards in succession, which promotes remarkable development of slag powder application.

Slag powder acting as concrete admixture not only can replace cement in same quantity, but also can significantly improve many concrete properties, for example, new mixed slag powder concrete possesses feature of less bleeding and better plasticity. Hydration heat in slow way can prevent cracking caused by temperature rising inside mass concrete. It is possible to produce more ettringite crystallite to compensate shrinkage caused by overmuch fine powder in concrete. Hardened concrete possesses good feature of anti-sulfate and durability. Moreover, fine granulated slag not only can improve performance of cement and concrete, but also can increase added value of slag itself, and higher economic benefit can be gained.

Since 1990s, blast furnace slag powder was brought into use in some projects in some cities in our country, including Beijing, Shanghai, Changsha and Zhuhai etc. blast furnace slag powder was used in some projects, such as parking structure of Beijing Capital International Airport, overpass between new terminal and parking structure, Beijing Tongchan Mansion and complex structure on Shatanhou Street, and blast furnace slag powder has replaced 30% - 50% in cement consumption, it is proved that using is in good effect. blast furnace slag powder produced by some manufacturers has been even sold to Southeast Asia.

New enacted *Granulated Blast Furnace Slag Powder for Use in Cement and Concrete* GB/T 18046—2000 is an equivalent standard of *Standard Specification for Slag Cement for Use in Concrete and Mortars* ASTMC 989—94 and *Ground Granulated Blast-furnace Slag Used for Concrete* JISA 6206—1995, meanwhile, combining with actual condition of production and application situation of granulated blast furnace slag micro powder in China, granulated blast-furnace slag micro powder was graded as three levels according to 7 d and 28 d activity index and specific surface area.

The grading of slag micro powder: the specific surface area of Grade I slag micro powder is determined as 550 – 650 m²/kg, that of Grade II micro slag powder 450 – 550 m²/kg, and that of Grade III micro slag powder 350 – 450 m²/kg. But activity indexes of 7 d and 28 d must reach specified targets.

6.3.2 Blast Furnace Slag Powder and Limerock Powder

6.3.2.1 The Action Mechanism of Blast Furnace Slag

The blast furnace slag is generated by the combination of impurities (SiO_2, Al_2O_3 etc.) in ore and solvent (lime etc.) under high temperature condition in the process of iron smelting. Its main chemical compositions are CaO, SiO_2 and Al_2O_3. The amount of CaO, SiO_2 and Al_2O_3 can account for more than 90% in common slag. The hot melting slag can be formed into granulated blast-furnace slag with main vitreous structure and increases activity after fast cooling with water, compressed air or steam.

The slag with greater CaO content (percentage content) possesses higher activity. However, if CaO content was too high (exceeding 51%), the slag will have higher crystallization capacity and is easy to crystallize due to declining of meltdown slag viscosity. The vitreous structure in slag reduces now, and activity declines. The slag with greater content of Al_2O_3 will have higher activity, SiO_2 in slag can help the formation of vitreous structure, but if the SiO_2 content was too much, slag activity will be poor because there is no enough CaO and MgO to combine SiO_2. Therefore, the slag will possess the best activity under the condition that CaO and Al_2O_3 contents are higher and SiO_2 content is lower.

6.3.2.2 Common Methods to Assess Slag Activity

I Chemical analysis method

Chemical composition can be regarded as one aspect to assess slag activity. In accordance with the regulations in GB/T 203—1994, K value is adopted in assessment for slag quality factor.

$$K = \frac{CaO + MgO + Al_2O_3}{MnO + TiO_2 + SiO_2}$$

Where: CaO, MgO, Al_2O_3, SiO_2, MnO and TiO_2 = percentage by weight.

The higher quality factor, the higher activity of slag. Generally, granulated blast furnace slag has quality factor shall not be less than 1.2, however, it is obviously incomplete to assess slag activity just by chemical composition, because slag activity also relates to its internal structural condition. After all, chemical composition of slag can reflect its nature to some extent, therefore, it is still one of main methods to assess slag quality.

II Strength test method

Mix slag and silicate cement clinker & gypsum evenly into slag cement, and to keep mixing amount and fineness unchanged. Determine the strength of slag cement and pure clinker cement in 7 d and 28 d with standard test method respectively. Calculate the strength ratio in accordance with following formula:

$$R_{Ratio} = \frac{R_{28}}{R'_{28}(1 - X)}$$

Where R_{28} = strength of slag cement

R'_{28} = strength of pure cement clinker

X = slag mixing amount, %

When the strength ratio is equal to 1, it suggests that the slag has no activity. The greater strength ratio, the higher activity of the slag.

Grounded fine slag can react in hydration reaction to improve the activity only when its fineness (specific surface area) reached required degree. The slag powder fineness can directly affect its reinforcement performance, in principle, the larger specific surface area of slag powder, the better effect. However, if fineness requirement was too high, it is more difficult to grind, and the cost will increase sharply. Taking all factors into consideration, appropriate scope for specific surface area should be 400 – 600 m^2/kg. The slag powder with different specific surface areas has different influence on colloidal mortar strength, referring to Table 6.15. The result shows that the slag presents the best colloidal mortar strength when its specific surface area ranges from 400 – 600 m^2/kg, this also indicates that 50% slag powder can be mixed to replace equivalent cement, and colloidal mortar strength is high in late period, which indicates that slag powder has high activity.

Table 6.15 The influence of mixing slag powder with different specific surface area on colloidal mortar strength

Cement (%)	Slag (%)	Specific surface area (m^2/kg)	Compressive strength (MPa)			Compressive strength ratio (%)		
			7 d	28 d	90 d	7 d	28 d	90 d
100	0	—	38.5	51.6	64.3	100.0	100.0	100.0
50	50	400	24.1	59.5	77.9	62.6	115.3	121.2
50	50	405	34.5	69.1	91.1	89.6	133.9	141.7
50	50	485	41.1	71.5	91.6	106.8	138.6	142.5
50	50	538	43.2	72.4	92.2	112.2	140.3	143.4
50	50	578	32.1	71.1	90.2	74.3	137.8	140.3
50	50	647	60.2	74.5	87.4	156.4	144.4	135.9

Ⅲ The admixture of blast furnace slag powder with limerock powder

In actual application, it is more difficult to grind slag fine, considering the efficiency of grinding machine, it is already better if the slag grinding fineness can reach 400 – 500 m^2/kg. From enacted GB/T 18046—2000, as long as specific surface area of slag can be controlled from 420 – 450 m^2/kg, the requirements in the standard can be reached. In addition, it is not enough to only adopt specific surface area as quality index of slag powder, because fine ground slag powder from different grinding machines not always have same activity index (especially that value in 7 d age) even if they have same specific surface area.

The blast furnace slag powder and limerock powder are used as admixture in test, the blast furnace slag powder and limerock powder (hereinafter slag and rock powder admixture is abbreviated as SL admixture) are produced in Jinggu Cement Plant. In accordance with test methods in *Ground Granulated Blast Furnace Slag Used for Cement and Concrete* GB/T 18046—2000, to carry out test for blast furnace slag powder, its test results of physical performance refer to Table 6.16. For there is no relevant standard for limerock powder, to follow test methods of cement and flyash to carry out density test, fineness test and specific surface area test for limerock powder, its test results refer to Table 6.17. In accordance with *Methods for Chemical Analysis of Cement* GB/T 176—1996, to carry out chemical composition analysis for blast furnace slag powder and limerock powder, and 0.045 mm sieving test is adopted in fineness test. test result refers to Table 6.18.

From Table 6.16 to Table 6.18, it shows that various indicators of the slag can meet technical requirements in S75 level stated in *Ground Granulated Blast Furnace Slag Used for Cement and Concrete* GB/T 18046—2000.

Table 6.16　　Physical properties of slag powder

Item		Density (g/cm³)	Fineness (%)	Specific surface area (kg/m³)	SO₃ (%)	Moisture content (%)	Loss on ignition (%)	Ratio of mobility (%)	Strength ratio (%)	
									7 d	28 d
Inspection result		2.96	8.0	451	1.1	0.1	3.0	98	65	77
GB/T 18046—2000	S105	≥2.8	—	≥350	≤4.0	≤1.0	≤3.0	≥85	≥95	≥105
	S95	≥2.8	—	≥350	≤4.0	≤1.0	≤3.0	≥90	≥75	≥95
	S75	≥2.8	—	≥350	≤4.0	≤1.0	≤3.0	≥95	≥55	≥75

Table 6.17　　The test result of physical properties on rock powder

Item	Density (g/cm³)	Fineness(%)		Specific surface area (kg/m³)
		0.045 mm	0.08 mm	
Inspection result	2.76	60.8	29.2	—

Table 6.18　　The chemical composition of slag and rock powder

Type of admixture	Chemical composition(%)										
	SiO₂	Al₂O₃	Fe₂O₃	CaO	MgO	Na₂O	K₂O	SO₃	f(CaO)	TiO₂	Loss on ignition
Slag powder	34.18	9.30	3.52	36.93	9.62	0.31	0.60	1.13	0.06	0.69	3.00
Rock powder	1.81	1.74	0.25	54.64	0.44	0.04	0.65	—	0.37	0.04	40.03

6.3.3　The Mix Design for RCC with SL Admixture

6.3.3.1　Raw Materials in Test

1. Cement. The silicate cement used in the test are common silicate 32.5 cement and common silicate 42.5 cement (hereinafter called Jinggu P·O 32.5 and Jinggu P·O 42.5 for short respectively) in Taiyu manufactured by Jinggu Cement Plant, another kind of cement is common silicate 42.5 cement (hereinafter called Jianfeng P·O 42.5 for short) in Yexiang manufactured by Jianfeng Cement Plant.

2. Admixture. The blast furnace slag powder and limerock powder are used as admixture in the test, the slag and limerock powder are mixed into SL admixture with proportion of 50∶50, and then to carry out mix design test for RCC.

3. Aggregate. The aggregates used in the test mainly are limestone aggregates produced by Baishiyan material plant which is located at right side of dam, and secondly andesite aggregates are used in the test. For physical properties test result of two kinds of fine aggregates and coarse limestone & andesite aggregates, please refer to Table 6.19, Table 6.20 and Table 6.21 respectively. Results indicate that manufactured limestone sand and manufactured andesite sand have good gradation, the sand is medium type, therein, rock powder less than 0.16 mm account for 11% content. The aggregate can meet relevant regulations stated in *Specifications for Hydraulic Concrete Construction* DL/T 5144—2001.

4. Additive. Water-reducing agent uses HA-JC set retarding superplasticizer produced by Xiashankou Structure Material Factory in Pingxiang City Jiangxi Province and JM-Ⅱ (C) concrete set retarding superplasticizer developed by Jiangsu Structure Science Institute. These two kinds of water-reducing agents are powder-like

in tawny, with functions of high efficiency water reducing, delayed setting, strengthening and plastic retaining. The air entraining admixture is new type JM-2000 concrete air entraining admixture developed by Jiangsu Research Institute of Structure Science Co., Ltd. This air entraining admixture is a liquid in hazel.

Table 6.19　　　　　　　　　　Physical properties of two kinds of fine aggregates

Artificial sand		Fineness module	Apparent density of dry sand in saturated surface (kg/m^3)	Water absorption of dry sand in saturated surface (%)	Rock powder content (%)	Bulk accumulation density (kg/m^3)	Organic content	SO_3 content (%)
Measured results	Limestone	2.79	2 640	1.1	11.0	1 485	Lighter than the standard color	0.1
	Andesite	2.76	2 670	1.0	11.0	1 500	Lighter than the standard color	0.1
DL/T 5144—2001		—	≥2 500	—	6-18	—	Lighter than the standard color	<1.0

Table 6.20　　　　　　　　　Physical properties of coarse limestone aggregates

	Test Items	Gravel size (mm)			DL/T 5144—2001
		5-20	20-40	40-80	
1	Apparent density(kg/m^3)	2 700	2 680	2 680	≥2 550
2	Dry apparent density of saturated surface(kg/m^3)	2 660	2 650	2 660	≥2 550
3	Dry water absorption of saturated surface(%)	0.90	0.56	0.44	≤2.5
4	Close packing density (kg/m^3)	1 600	1 460	1 500	—
5	Voidage(%)	41	45	44	—
6	Powder content(%)	0.5	0.8	0.1	≤1.0(5-40 mm) ≤0.5(40-150 mm)
7	Organic content	Lighter than the standard color	Lighter than the standard color	Lighter than the standard color	Lighter than the standard color
8	Needle slice content(%)	7.1	1.4	1.4	≤15
9	Solidity(%)	2	4	2	≤5
10	Crushing index(%)	12.3	—	—	≤20

Table 6.21　　　　　　　　　Physical properties of coarse andesite aggregate

	Test Items	Gravel size (mm)			DL/T 5144—2001
		5-20	20-40	40-80	
1	Apparent density(kg/m^3)	2 830	2 850	2 870	≥2 550
2	Dry apparent density of saturated surface(kg/m^3)	2 800	2 830	2 860	≥2 550
3	Dry water absorption of saturated surface(%)	0.61	0.35	0.23	≤2.5
4	Close packing density (kg/m^3)	1 720	1 590	1 580	—

Continued Table 6.21

	Test Items	Gravel size (mm)			DL/T 5144—2001
		5 – 20	20 – 40	40 – 80	
5	Voidage (%)	40	44	45	—
6	Powder content(%)	1.0	0	0	≤1.0(5 – 40 mm) ≤0.5(40 – 150 mm)
7	Organic content	Lighter than the standard color	Lighter than the standard color	Lighter than the standard color	Lighter than the standard color
8	Needle slice content(%)	2	0	0	≤15
9	Solidity(%)	2.3	0.63	0.24	≤5
10	Crushing index(%)	4.0	—	—	≤20

Test results of admixtures indicate respectively that: two kinds of set retarding superplasticizers HA-JC & JM-Ⅱ(C) and new type JM-2000 concrete air entraining admixture can meet first-class requirements in GB 8076—1997.

6.3.3.2 RCC Preparation Strength

Because SL admixture mixed with slag and rock powder has above properties, it is a favorable technical approach to replace cement in same amount and to realize sustainable development. Through multi-disciplinary demonstration, SL admixture mixed with slag and rock powder in accordance with proportion 50:50 is adopted as admixture in RCC in Gelantan Hydropower Station. Meanwhile, to use Jingyu P·O 42.5 cement to carry out mix proportion test.

According to technical articles and design requirement in tender documents, design index for RCC is listed in Table 6.22, strength calculation value for RCC is in Table 6.23.

Table 6.22   Design performance index of RCC

Location	Design requirements	Gradation	Maximum particle size of aggregate (mm)	90 d limiting extended value ($\times 10^{-4}$)	The maximum admixture mixing amount (%)	The maximum allowable water-cement ratio (%)	Assurance factor of strength (%)	Type of aggregate
Inside the dam	C_{90}15W4 F50	Ⅲ	80	0.75	60	0.60	80	Limestone
Upstream impervious layer	C_{90}20W8 F50	Ⅱ	40	0.80	50	0.55	85	Limestone

Table 6.23   The concrete preparation strength

Standard value of strength, $f_{cu,k}$ (MPa)	Strength standard deviation, σ (MPa)	Assurance factor of strength, P (%)	Probability coefficient, t	Preparation strength $f_{cu,0}$ (MPa)
C_{90}15	3.5	80	0.84	17.9
C_{90}20	4.0	85	1.04	24.2

6.3.3.3 The Selection of Mix Design Parameters of RCC with SL Admixture

Ⅰ The selection of gradation proportion for coarse aggregate

RCC with limestone aggregate is Ⅱ or Ⅲ gradation. Choose optimum combination proportion for coarse

Chapter 6 Research and Application of RCC Admixture

aggregate based on maximum vibrated volume-weight. Test results of different combination proportion for coarse aggregate are as follows: the close packing density of Ⅱ gradation with small stone: middle stone of 50:50 and that of Ⅲ gradation with small stone, middle stone and big stone of 40:30:30 is the greatest. Therefore, above proportions are chosen as the proportion for aggregates of RCC of Ⅱ and Ⅲ gradations.

Ⅱ Choosing admixture mixing amount

Water reducing agent: the mixing amount of HA-JC set retarding superplasticizer is 0.8%, and mixing amount of JM-Ⅱ(C) set retarding superplasticizer is 0.6%.

Air entraining agent: the air entraining admixture JM-2000 is mixed with HA-JC and JM-Ⅱ(C) respectively, mixing amount can be controlled on basis of 4%–6% air content in concrete after wet sieving.

Ⅲ Choosing of water consumption per unit and sand rate

Under the condition that RCC can meet requirements of workable consistency, water consumption per unit are mainly influenced by stone gradation and cement type. VC value of RCC mixture is controlled within the range of 5–8 s, water consumption per unit for Ⅱ gradation RCC is 101–104 kg/m^3. The water consumption per unit for Ⅲ gradation RCC is 89–94 kg/m^3.

Sand rate is mainly influenced by coarse aggregate type, gradation, sand fineness modulus and water-cement ratio, etc. The best sand rate for RCC can be determined by the test. Choosing the best sand rate, RCC easy to be compacted can be obtained. The sand rate of Ⅱ gradation RCC with water cement ratio of 0.50 and 0.55 is 37%–38% through test. The sand rate of Ⅲ gradation RCC with water cement ratio of 0.55 and 0.60 is 34%–35% through test.

Ⅳ The choosing of water-cement ratio and admixture mixing amount

According to design index requirement for RCC and regulations in relevant specifications, as well as refer to test data in other projects. The preliminary choice for water-cement ratio of Ⅱ gradation RCC is 0.50 and 0.55, and admixture mixing amount is determined preliminarily as 50%. The preliminary choice for water-cement ratio of Ⅲ gradation RCC is 0.55 and 0.60, admixture mixing amount is determined preliminarily as 50% and 60%.

Ⅴ The selection test of different proportion between slag and rock powder, as well as different mixing amount

According to requirements in task statement, choose the admixture with different proportion of slag:rock powder (40:60, 50:50 and 60:40) to carry out concrete mix design test, and then discuss with designers to determine the proportion of slag and rock powder in next step according to test results.

Select the same water cement ratio, to select admixture with proportion of slag:rock powder (40:60, 50:50 and 60:40) respectively, admixture mixing amount is set as 50%, to carry out preliminary mix design test for RCC with Jinggu P·O 32.5 and Jinggu P·O 42.5 cement respectively, admixture workability (VC value) in RCC is controlled 5–8 s, and air content is controlled 4%–6%. Test results refer to Table 6.24. It indicates that in Table 6.24:

1. When water cement ratio, admixture mixing amount and water reducing agent mixing amount are same, and water consumption is same in same gradation RCCs with different proportion of slag:rock powder (40:60, 50:50 and 60:40). But mixing amount of air entraining admixture will increase with increasing of slag proportion, and will decrease with increasing of rock powder proportion.
2. Under the same condition, water consumption in mixed RCC with Jinggu P·O 42.5 is 1 kg/m^3 less than that with Jinggu P·O 32.5.
3. Under condition with certain water cement ratio, both 90 d compressive strengths of RCCs with these two kinds of cements can meet strength requirements in $C_{90}20$, and Ⅱ gradation RCC with different proportions of slag:rock powder (50:50 and 60:40) can reach 90 d limiting tensile value 0.8×10^{-4}.
4. Under the condition with same cement type, same admixture mixing amount and same water cement ratio,

Table 6.24 Properties of RCC with admixture in different proportions of slag and rock powder

No.	Cement type	Admixture		Additive		Water content (kg/m³)	VC value (s)	Air content (%)	Compressive strength (MPa)			Static compressive modulus of elasticity (GPa)			Axial tensile strength (MPa)			Ultimate tensile value ($\times 10^{-4}$)		
		Mixing amount (%)	Slag: rock powder	HA-JC (%)	JM-2000 (‰)				7 d	28 d	90 d	7 d	28 d	90 d	7 d	28 d	90 d	7 d	28 d	90 d
N1	Jinggu P·O 32.5	50	40:60	0.8	0.50	104	7.0	4.4	14.1	22.8	30.2	20.4	26.0	33.7	1.38	2.12	2.76	0.50	0.65	0.79
N2		50	50:50	0.8	0.60	104	5.0	5.0	15.0	23.4	31.4	20.1	26.9	35.1	1.44	2.18	2.73	0.58	0.78	0.87
N3		50	60:40	0.8	0.70	104	5.8	5.0	16.7	24.7	31.9	21.8	29.0	37.8	1.45	2.54	3.18	0.52	0.81	0.96
N4	Jinggu P·O 42.5	50	40:60	0.8	0.40	103	5.5	5.0	14.5	21.1	28.6	24.4	26.0	31.6	1.55	1.90	2.63	0.60	0.66	0.81
N5		50	50:50	0.8	0.50	103	7.0	4.4	15.5	23.5	29.5	26.0	30.3	35.5	1.46	1.98	2.83	0.61	0.68	0.82
N6		50	60:40	0.8	0.60	103	5.0	4.8	16.0	23.8	30.5	26.2	30.6	37.3	1.58	2.43	3.13	0.61	0.77	0.91

Note: The maximum aggregate particle size is 40 mm, therein, the proportion of small stone: middle stone is 50:50.

compressive strength, static compressive modulus of elasticity, axis tensile strength and limiting extended value of RCC will increase with increasing of slag mixing amount.

In conclusion, considering disadvantage factors, for example, limiting tensile value is required to be more than 0.8×10^{-4} in design, the admixture in fixed amount with higher proportion of slag will cause the increasing of air entraining admixture, hydration heat will rise and the cost will increase. Through analysis and comparison, select SL admixture with the proportion of slag: rock powder 50∶50 to carry out mix proportion design for RCC used in dam and upstream impervious area.

6.3.4 The Performance Test for RCC Mixed with SL Admixture

6.3.4.1 Compressive Strength Test Result

Based on the selection of main parameters of the concrete mix proportion and the test results mentioned above, under the precondition of meeting the consistency and air content of concrete mixture, the mix proportion test of RCC is carried out by using SL admixture with 50∶50 ratio of slag to rock powder. There are three kinds of cements. See Table 6.25 for the test result of RCC mixed with SL admixture. It indicates that in Table 6.25:

1. Under the same condition, the compressive strength of concrete decreases with the increase of water-cement ratio, decreases with the increase of admixture amount and increases with the increase of age.
2. When the water cement ratio is 0.50 – 0.55 and the amount of admixture is 50%, the 90 d compressive strength of Ⅱ gradation RCC with the cement of Jinggu P·O 32.5, Jinggu P·O 42.5 and Jianfeng P·O 42.5 reaches 28.5 – 32.4 MPa, meeting the preparation strength of C_{90}20 24.2 MPa.
3. When the water cement ratio is 0.55 – 0.60 and the amount of admixture is 50% – 60%, the 90 d compressive strength of Ⅲ gradation RCC with cement of Jinggu P·O 32.5, Jinggu P·O 42.5 and Jianfeng P·O 42.5 reaches 25.8 – 30.4 MPa, meeting the preparation strength of C_{90}15 17.9 MPa.
4. It indicates that in Table 6.25: the primary and secondary order on 28 d and 90 d compressive strengths of RCC is A (cement type)→C (Amount of admixture)→B (Water binder ratio). The compressive strength of the concrete with Jinggu P·O 32.5 cement is 1.9 MPa higher than that of the concrete with other two kinds of cement at 28 d, 90 d and 180 d; when the amount of admixture increases from 50% to 60%, compressive strengths at 28 d, 90 d and 180 d decreases by 1.5 MPa on average. The water cement ratio increases from 0.55 to 0.60. The compressive strength at 28 d, 90 d and 180 d decreases by 1.1 MPa on average. Both the blank column and range reflecting test error are small, showing that the test precision is high.

6.3.4.2 Compression Modulus, Ultimate Tensile Strength and Tensile Strength

Test results of compressive elastic modulus and ultimate tensile of RCC are listed in Table 6.26. It indicates that in Table 6.26:

1. Under the same condition, the static compressive elastic modulus of concrete decreases with the increase of water cement ratio, decreases with the increase of admixture amount and increases with the increase of age.
2. The static compressive elastic modulus of the Ⅱ gradation RCC with three kinds of cements and water cement ratio of 0.50 – 0.55, admixture amount of 50% and using Jiangxi Pingxiang HA-JC retarding and high range water reducing agent reaches 30.9 – 33.8 GPa in 28 d. The static compressive elastic modulus at 90 d is 34.5 – 37.3 GPa. The static compressive elastic modulus of 180 d is 38.8 – 40.2 GPa, being 1.1 times of that at 90 d on average; the ultimate tensile value at 28 d is $0.74 \times 10^{-4} - 0.87 \times 10^{-4}$ and the ultimate tensile value at 90 d is $0.91 \times 10^{-4} - 0.97 \times 10^{-4}$. The ultimate tensile value at 180 d is $1.00 \times 10^{-4} - 1.05 \times 10^{-4}$; the 28 d tensile strength is 2.55 – 2.98 MPa; the 90 d tensile strength is 2.86 – 3.43 MPa; the 180 d tensile strength reaches 3.13 – 3.43 MPa. The 90 d ultimate tensile value of RCC in upstream seepage control area meets the demand, not less than 0.8×10^{-4}.

Table 6.25 Mix proportion test results of RCC with SL admixture

No.	Cement type	Water-cement ratio	Admixture amount (%)	Sand ratio (%)	Maximum particle size of aggregate	Additive HA-JC (%)	Additive JM-2000 (‰)	Water	Cement	Slag	Rock powder	Sand	Small-sized stone	Medium-sized stone	Large-sized stone	VC (s)	Air content (%)	Unit weight (kg/m³)	7 d	28 d	90 d	180 d
JM		0.50	50	37	40	JM 0.6	1.00	105	105	53	53	790	670	670	—	6.7	4.8	2 420	16.5	25.0	32.4	36.5
N8		0.50	50	37	40	0.8	0.40	104	104	52	52	793	672	672	—	5.0	5.0	2 420	17.8	25.8	31.5	35.4
N9		0.55	50	38	40	0.8	0.40	104	95	47	47	821	667	667	—	6.0	5.0	2 420	17.4	25.0	30.2	33.6
N10	Jinggu P·O 32.5	0.55	50	34	80	0.8	0.40	92	84	42	42	752	582	436	436	6.2	5.0	2 450	18.0	25.5	30.4	33.7
N11		0.55	60	34	80	0.8	0.55	94	68	51	51	749	579	434	434	6.7	5.2	2 448	15.5	24.8	29.6	32.9
N12		0.60	50	35	80	0.8	0.40	92	77	38	38	779	576	432	432	5.5	5.0	2 458	17.3	25.0	30.0	32.5
N13		0.60	60	35	80	0.8	0.55	94	63	47	47	775	574	430	430	5.0	4.2	2 445	14.9	23.3	28.6	31.6
N14		0.50	50	37	40	0.8	0.45	103	103	52	52	794	673	673	—	7.0	4.4	2 420	18.0	25.3	29.7	33.6
N15		0.55	50	38	40	0.8	0.40	102	93	46	46	824	670	670	—	5.8	5.0	2 422	17.5	24.6	28.5	32.0
N16		0.55	50	34	80	0.8	0.40	90	82	41	41	755	584	438	438	5.4	4.2	2 440	17.7	25.0	29.1	32.3
N17	Jinggu P·O 42.5	0.55	60	34	80	0.8	0.55	92	67	50	50	751	581	436	436	6.0	5.0	2 446	14.5	22.9	26.9	30.2
N18		0.60	50	35	80	0.8	0.40	90	75	38	38	781	578	434	434	6.4	4.8	2 448	16.8	23.0	28.5	31.8
N19		0.60	60	35	80	0.8	0.55	92	61	46	46	778	576	432	432	5.3	5.0	2 442	13.7	22.2	25.8	29.7
N20		0.55	50	38	40	0.8	0.35	101	92	46	46	825	671	671	—	5.6	4.4	2 423	17.3	24.4	29.2	32.4
N21		0.55	50	34	80	0.8	0.35	89	81	40	40	756	585	439	439	6.0	5.0	2 440	17.4	24.6	29.5	33.0
N22	Jianfeng P·O 42.5	0.55	60	34	80	0.8	0.55	90	65	49	49	754	584	438	438	6.0	5.0	2 450	15.7	21.9	28.0	31.1
N23		0.60	50	35	80	0.8	0.37	89	74	37	37	783	579	435	435	5.1	4.8	2 452	16.6	22.6	27.3	30.8
N24		0.60	60	35	80	0.8	0.50	91	61	46	46	779	577	433	433	5.0	5.0	2 444	14.3	21.5	26.6	30.0

Chapter 6 Research and Application of RCC Admixture

Table 6.26 Static compressive elastic modulus and ultimate tensile value of RCC

No.	Cement type	Water-cement ratio	Admixture amount (%)	Kind of water reducing agent (%)	Amount of cementing materials (kg/m³)	Maximum particle size of aggregate (mm)	Static compressive elasticity modulus (GPa)				Tensile strength (MPa)			Ultimate tensile value ($\times 10^{-6}$)		
							7 d	28 d	90 d	180 d	28 d	90 d	180 d	28 d	90 d	180 d
JM	Jinggu P·O 32.5	0.50	50	JM-II (C)	211	40	23.4	29.5	35.7	39.0	2.13	2.82	3.20	0.66	0.75	0.93
N8		0.50	50	HA-JC	208	40	26.6	33.8	37.3	40.2	2.68	2.93	3.22	0.84	0.97	1.05
N9		0.55	50	HA-JC	189	40	25.7	33.4	36.3	39.5	2.55	2.86	3.13	0.81	0.96	1.02
N10		0.55	50	HA-JC	168	80	—	—	—	—	2.77	3.02	—	0.87	0.98	—
N12		0.60	50	HA-JC	153	80	23.1	32.5	36.5	41.7	2.69	3.00	3.19	0.83	0.90	0.94
N13		0.60	60	HA-JC	157	80	22.0	31.2	35.5	39.5	2.60	2.90	3.50	0.72	0.91	1.01
N14		0.50	50	HA-JC	207	40	26.6	33.2	36.5	39.3	2.48	2.95	3.36	0.72	0.87	1.03
N15	Jinggu P·O 42.5	0.55	50	HA-JC	185	40	26.2	32.5	36.1	38.8	2.84	3.21	3.40	0.87	0.93	1.00
N18		0.60	50	HA-JC	151	80	26.0	32.8	36.2	40.0	2.55	2.98	3.05	0.75	0.87	0.98
N20		0.55	50	HA-JC	184	40	24.2	30.9	34.5	39.6	2.59	3.04	4.43	0.74	0.91	1.05
N21	Jianfeng P·O 42.5	0.55	50	HA-JC	161	80	24.5	31.8	34.7	39.3	2.82	3.08	3.28	0.78	0.86	0.92
N23		0.60	50	HA-JC	148	80	22.9	30.2	33.3	37.5	2.55	3.02	3.17	0.78	0.88	0.96

3. When the water cement ratio is 0.55 – 0.60 and the admixture amount is 50% – 60%, the 28 d static compressive elastic modulus of the Ⅲ gradation RCC made by three kinds of cements reaches 30.2 – 32.8 GPa; 90 d static compressive elastic modulus is 33.3 – 36.5 GPa; the 180 d static compressive elastic modulus is about 1.1 times of that of 90 d; the ultimate tensile value at 28 d is $0.72 \times 10^{-4} - 0.87 \times 10^{-4}$ and the ultimate tensile value at 90 d is $0.86 \times 10^{-4} - 0.98 \times 10^{-4}$. The ultimate tensile value at 180 d is $0.92 \times 10^{-4} - 1.01 \times 10^{-4}$; 28 d tensile strength is 2.55 – 2.82 MPa and the 90 d tensile strength is 3.02 – 3.50 MPa. The 90 d ultimate tensile value of RCC inside dam body meets the requirements, not less than 0.75×10^{-4}.

6.3.4.3 Dry Shrinkage Deformation and Autogenous Volume Deformation

The dry shrinkage of concrete refers to the axial length deformation caused by drying and wetting under the condition of no external load and constant temperature. Dry shrinkage curves of the Ⅱ gradation and Ⅲ gradation RCC is shown in Figure 6.44. It can be seen from Figure 6.44 that the dry shrinkage curve of RCC is consistent. In the first 28 d, the dry shrinkage rate is higher and it tends to be stable in the later stage. The average dry shrinkage rate of RCC in 90 d is 200×10^{-6} and the average shrinkage rate of 180 d is 216×10^{-6}. Due to the low water consumption, small amount of cement and large amount of admixture in RCC, the dry shrinkage rate is much smaller than normal concrete.

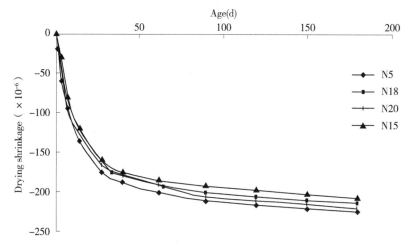

Figure 6.44 *Dry shrinkage curve of RCC*

Autogenous volume deformation of concrete refers to the volume deformation only caused by hydration of cementing materials under the condition of absolute humidity and constant temperature. The autogenous volume deformation curve of RCC is shown in Figure 6.45. It can be seen from Figure 6.45 that: The autogenous volume deformation of the Ⅲ gradation RCC made by Jinggu 32.5 cement with water cement ratio of 0.6 and admixture amount of 50% is the process of expansion first and then shrinkage. The first 3 d is expansion process and the next 3 d is contraction process. The maximum shrinkage deformation is 8×10^{-6} and the authigenic volume remains stable in the later stage. However, the autogenous volume deformation of RCC is obviously less than that of normal concrete, this is because the amount of cementing materials used in RCC is less.

6.3.4.4 Durability of RCC

Frost resistance and impermeability test results of RCC are listed in Table 6.27, showing that:
1. 90 d impermeability grades of the Ⅱ gradation RCC made by three kinds of cements with water cement ratio of 0.55 and admixture amount of 50% are all higher than W8, meeting the design requirements. The relative permeability coefficient is $(4.32 - 5.65) \times 10^{-10}$ cm/s; 90 d impermeability grades of the Ⅲ gradation RCC made by three kinds of cements with water cement ratio of 0.60 and admixture amount of

Chapter 6 Research and Application of RCC Admixture

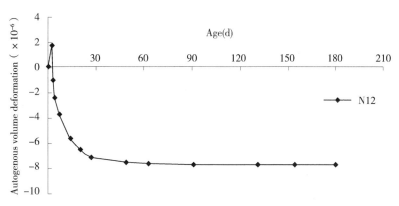

Figure 6.45 *Autogenous volume deformation of RCC*

Table 6.27 Durability test results of RCC

No.	Cement type	Water cement ratio	Admixture amount (%)	Maximum particle size of coarse aggregate (mm)	90 d frost resistance				90 d impermeability	
					Relative dynamic elastic modulus (%)		Weight loss rate (%)		Grade of impermeability	Relative permeability coefficient ($\times 10^{-10}$ cm/s)
					F50	F100	F50	F100		
N9	Jinggu P·O 32.5	0.55	50	40	93.4	89.0	0.2	0.5	>W8	4.32
N12		0.60	50	80	92.2	88.2	0.3	0.7	>W4	5.40
N13		0.60	60	80	85.0	—	0.8	—	>W4	9.65
N15	Jinggu P·O 42.5	0.55	50	40	94.4	90.2	0.1	0.4	>W8	5.06
N18		0.60	50	80	91.5	—	—	—	>W4	8.27
N20	Jianfeng P·O 42.5	0.55	50	40	91.1	86.2	0.1	0.6	>W8	5.65
N23		0.60	50	80	88.6	—	—	—	>W4	6.87

50% - 60% are all higher than W4, meeting the design requirements. The relative permeability coefficient is $5.40 - 9.65 \times 10^{-10}$ cm/s.

2. When water reducing agent HA-JC and air entraining agent JM-2000 are mixed inside, the air content of concrete mixture is controlled at 4% - 6%, and the water cement ratio is 0.55 and the amount of admixtures is 50%, after the II gradation RCC made by three kinds of cements are frozen and thawed by 100 times, the relative dynamic elastic modulus is more than 86.2% and weight loss is less than 1%; When the water cement ratio is 0.60 and the amount of admixtures is 50% - 60%, after the III gradation RCC made by three kinds of cements are frozen and thawed by 50 times, the relative dynamic elastic modulus is more than 85.0% and the weight loss is less than 1%. The result shows that the frost resistance grade of RCC meets the design requirements.

6.3.4.5 Setting Time of RCC

The setting time of RCC is the basis to determine the allowable interval time of roller compacted layers. The setting time of RCC is shown in Table 6.28. The result shows that: The water cement ratio is 0.55 - 0.60; the initial setting time of RCC with 50% admixture amount is about 10 h 10 min; and final setting time is about 15 h 22 min.

Table 6.28 Setting time of RCC

No.	Cement type	Water cement ratio	Admixture amount (%)	Maximum particle size of coarse aggregate (mm)	Time of setting (h: min) Initial setting	Time of setting (h: min) Final setting
N9	Jinggu P·O 42.5	0.55	50	40	10:40	15:30
N12	Jinggu P·O 42.5	0.60	50	80	10:02	15:25
N15	Jinggu P·O 42.5	0.55	50	40	11:10	15:37
N20	Jianfeng P·O 42.5	0.55	50	40	10:45	15:22
N23	Jianfeng P·O 42.5	0.60	50	80	10:25	14:48

6.3.4.6 Shear Strength Test of RCC

The shear strength test of RCC includes the shear strength of internal concrete (inside layer) and the shear strength of concrete layer surface. 150 mm × 150 mm × 150 mm cube specimens are used in the test. The shear specimens used for interlayer bonding are cast in two times, 75 mm thick concrete at the bottom layer is cast first, then the upper layer concrete is cast at different intervals. After 90 d of standardized curing, the test is carried out with direct shear apparatus. The shearing test of RCC layer surface is divided into two working conditions: I. The interval of layered pouring is 6 h and the layer surface is not treated at all; II. 24 h after the bottom layer is poured, the layer surface is not treated, bedding mortar (water cement ratio of mortar is 0.48; the ratio of cement to sand is 1:3; the amount of admixture is 50%, and the dosage amount of water reducing agent HA-JC is 0.4%), and then upper layer of concrete is cast.

No. N12 (III gradation RCC made by Jinggu P·O 32.5 cement with water cement ratio of 0.6 and 50% of admixture amount), N15 (III gradation RCC made by Jinggu P·O 42.5 cement with water cement ratio of 0.55 and admixture amount of 50%). The results of shear strength (calculated with c) test are listed in Table 6.29.

Table 6.29 Test results of shear strength of RCC

S/N	No.	Normal stress (MPa)	Shear strength (MPa)	Friction coefficient f'	Cohesive force c' (MPa)	Remarks
1	N12	2.5	6.2	1.07	3.59	Inside concrete
2	N12	2.5	5.1	1.27	1.91	The interval between layers is 6 h and the layer surface is not treated
3	N12	2.5	5.9	1.22	2.79	The interval between layers is 24 h and the layer surface is not treated; bedding mortar is cast; the upper layer of concrete is cast then
4	N15	2.5	6.5	1.14	3.72	Inside concrete
5	N15	2.5	5.3	1.29	2.20	The interval between layers is 6 h and the layer surface is not treated
6	N15	2.5	6.1	1.24	2.98	The interval between layers is 24 h and the layer surface is not treated; bedding mortar is cast; the upper layer of concrete is cast then,

Chapter 6 Research and Application of RCC Admixture

It indicates that in Table 6.29: when the normal stress is 2.5 MPa, the shear strength of N12 and N15 RCC (in the layer) is 6.2 MPa and 6.5 MPa respectively; After 6 h interval between layers, the layer surface is not treated, the shear strength between layers decreases by about 18%; the interval between layers is 24 h and the layer surface is not treated, bedding mortar is cast and the upper layer of concrete is cast then. The shear strength between layers decreases by 5.5% on average. This shows that the interval time between layers of RCC should be controlled before the initial setting time. If the interval time exceeds the initial setting time or even the final setting time of concrete, before casting the upper layer of concrete, a layer of cement mortar is cast first.

6.3.4.7 Thermal Properties of RCC

The thermal properties of concrete vary with the type of aggregate, concrete mix proportion, moisture content, temperature, etc. At normal temperature, thermal performance test results of RCC are shown in Table 6.30. It indicates that in Table 6.30: The thermal conductivity of RCC is 0.0033 – 0.0037 m^2/h; specific heat is 0.93 – 0.95 $kJ/(kg \cdot ℃)$; coefficient of linear expansion is $6 \times 10^{-6}/℃$; and thermal conductivity is about 8 $kJ/(m \cdot h \cdot ℃)$.

Table 6.30   Test results of thermal performance of RCC

No.	Cement type	Water-cement ratio	Coefficient of thermal conductivity (m^2/h)	Linear expansion coefficient ($\times 10^{-6}/℃$)	Specific heat [$kJ/(kg \cdot ℃)$]	Thermal conductivity factor [$kJ/(m \cdot h \cdot ℃)$]
N12	Jinggu P · O 32.5	0.60	0.0035	6	0.95	8.15
N15	Jinggu P · O 42.5	0.55	0.0037	6	0.93	8.32
N20	Jianfeng P · O 42.5	0.55	0.0033	6	0.95	7.60

The adiabatic temperature rise of concrete is the temperature change and the maximum temperature rise value of cementing materials (including cement, admixtures, etc.) in concrete in the hydration process under adiabatic condition. The regression equation of adiabatic temperature rise-time of N12 and N15 RCC is shown in Table 6.31. The adiabatic temperature rise process curve of RCC is shown in Figure 6.46. When the content of SL admixture amount is 50%, the correlation coefficients of regression equations are all above 0.999.

Table 6.31   Adiabatic temperature rise of RCC

No.	Cement type	Water-cement ratio	Adiabatic temperature rise value (℃)					Expression of regression equation
			1 d	3 d	7 d	14 d	28 d	
N12	Jinggu P · O 32.5	0.60	7.27	11.95	14.83	16.40	17.22	$18.03t/(1.49+t)$
N15	Jinggu P · O 42.5	0.55	9.72	15.34	18.49	20.06	20.99	$21.90t/(1.28+t)$

6.3.5 Construction Mix Proportion of RCC with SL Admixture

6.3.5.1 Mix Proportion Parameters of RCC with SL Admixture

According to the technical terms of the tender documents, the design tender of RCC is carried out. According to above test results, the contractor carries out the mix proportion test of RCC mixed with SL admixture on site. Test conditions and construction mix proportion parameters are as:
1. Test condition: Taiyu P · O 32.5 cement, 0.8% admixture amount of Kunming Shanfeng retarding and high efficiency water reducing agent, SL admixture, manufactured coarse and fine aggregate of limestone are

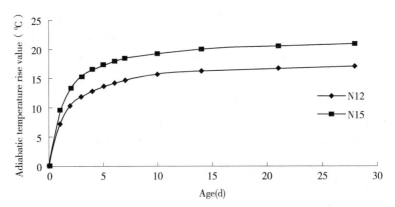

Figure 6.46 *Adiabatic temperature rise process curve of RCC*

used.

2. Mix proportion parameter: the Ⅲ gradation sand ratio of RCC is 35%; the water consumption is 83 kg/m³; Ⅱ gradation sand ratio is 39%; water consumption is 93 kg/m³; and *VC* value is 3 – 8 s.

6.3.5.2 Relationship between RCC Strength and Water Binder Ratio

According to the mix proportion test parameters of RCC, the relationship between RCC strength and water binder ratio is tested. The test results are shown in Table 6.32.

Table 6.32 Test results of the relationship between the strength of RCC and water binder ratio

	Ⅲ gradation (SL admixture 60%)				Ⅱ gradation (SL admixture 55%)			
	Age	Compressive strength (MPa)			Age	Compressive strength (MPa)		
		7 d	28 d	90 d		7 d	28 d	90 d
Kunming Shanfeng is used as additive SFG(0.8%)	Water-cement ratio	0.60 / 9.7	14.0	18.6	Water-cement ratio	0.55 / 10.9	15.0	21.3
		0.55 / 10.6	15.4	19.9		0.50 / 13.1	17.7	24.7
		0.50 / 11.9	17.6	22.8		0.45 / 15.3	21.3	27.9
	$R7 = 6.6x - 1.4 \quad r = 0.9986$				$R7 = 10.9x - 8.8 \quad r = 0.9983$			
	$R28 = 10.8x - 4.2 \quad r = 0.9972$				$R28 = 15.6x - 13.4 \quad r = 0.9997$			
	$R90 = 12.71x - 2.8 \quad r = 0.9866$				$R90 = 16.26x - 8.11 \quad r = 0.9971$			
	Ⅲ gradation (admixture 60%)				Ⅱ gradation (admixture 55%)			
	Age	Compressive strength (MPa)			Age	Compressive strength (MPa)		
		7 d	28 d	90 d		7 d	28 d	90 d
The additive is Shanxi Huanghe UNF-2C(0.8%)	Water-cement ratio	0.60 / 9.3	13.9	18.5	Water-cement ratio	0.55 / 10.4	15.8	20.8
		0.55 / 11.1	15.6	20.3		0.50 / 11.8	18.0	24.7
		0.50 / 12.1	17.8	22.7		0.45 / 15.2	22.3	27.6
	$R7 = 8.3x - 4.3 \quad r = 0.9768$				$R7 = 12.0x - 11.7 \quad r = 0.9841$			
	$R28 = 11.7x - 5.6 \quad r = 0.9998$				$R28 = 16.2x - 13.9 \quad r = 0.9920$			
	$R90 = 12.62x - 2.57 \quad r = 0.9996$				$R90 = 16.69x - 9.24 \quad r = 0.9899$			

The result indicates: The accuracy of correlation coefficient between compressive strength and water binder ratio of RCC with SL admixture is high and has a high correlation.

6.3.5.3 Construction Mix Proportion of Dam RCC

According to the test results of the relationship between the strength and water binder ratio of RCC with SL admixture, see Table 6.33 for mix proportion parameters of RCC and GEVR. The performance test results of RCC and GEVR are shown in Table 6.34. See Table 6.35 for construction mix proportion of dam RCC.

6.3.6 Conclusion

1. The physical properties and chemical composition of "Taiyu" P·O 32.5 cement, Jinggu P·O 32.5 cement, Jinggu P·O 42.5 cement produced by Jinggu cement plant and "Yexiang" 42.5 cement produced by Jianfeng cement plant meet the technical requirements of national standard *Silicate Cement, Ordinary Portland Cement* GB 175—1999.
2. The indexes of slag used in the test meet the technical requirements of S75 grade specified in *Ground Granulated Blast Furnace Slag Used for Cement and Concrete* GB/T 18046—2000. The content of calcium oxide (CaO) in limerock powder is about 55%; The fineness (0.08 mm) is 29.2%; it can be used as admixture.
3. The physical properties of limestone and andesite coarse and fine aggregate meet the requirements of *Specifications for Hydraulic Concrete Construction* DL/T 5144—2001.
4. Under the condition with same cement type, same admixture mixing amount and same water cement ratio, compressive strength, static compressive modulus of elasticity, axis tensile strength and limiting extended value of RCC will increase with increasing of slag mixing amount.
5. From the technical and economic analysis and comparison, in the Project of Gelantan Hydropower Station, SL admixture with the ratio of slag to rock powder of 50:50 is used as the admixture of RCC for dam.
6. The water reducing agent adopts SFG retarding and high efficiency water reducing agent and the admixture amount is 0.8%. JM-2000 is used as air entraining agent and the admixture amount of F50 is 0.04%. The air content is 3%–4%; The admixture amount of F100 is 0.05%; and air content is 4%–5%. The VC value of RCC is controlled as per 3–8 s.
7. Results of mix proportion test of RCC dam construction show that the RCC with SL admixture has good workability, plasticity, stability and compactness; The 90 d compressive strength, impermeability and deformation properties meet the design requirements. At the same time, with the prolongation of age, the strength increases significantly; Meanwhile, the construction shows that the paving quality of SL admixture RCC is good and can meet the purpose of continuous and rapid construction of RCC.

Table 6.33　Mix proportion parameters of RCC and GEVR of dam

No.	Mix proportion strength grade	Concrete classification	Cement type (Taiyu)	Water-cement ratio	Admixture amount (%)	Sand ratio (%)	Unit volume (m³/kg)							Additive(%)			VC (s)	Air content (%)	Volume ratio of mortar to sand PV
							Cement	Extender	Sand	Small-sized stone	Medium-sized stone	Large-sized stone	Water	Retarding high-range water reducing agent	Air entraining agent JM-2000			
R1	$C_{90}15W4F50$	Roller compacted concrete	P·O 32.5	0.55	60.0	35.0	61	90	757	435	580	435	83	SFG (0.8)	0.04	3–8	3–4	0.38
R2	$C_{90}15W4F50$	Grout enriched vibrated mortar	P·O 32.5	0.53	50.0	—	575	575	—	—	—	—	608	SFG (0.8)	—	—	—	—
R3	$C_{90}15W4F50$	Roller compacted concrete	P·O 32.5	0.50	60.0	34.0	66	100	731	439	585	439	83	SFG (0.8)	0.04	3–8	3–4	0.40
R4	$C_{90}15W4F50$	Grout enriched vibrated mortar	P·O 32.5	0.48	50.0	—	608	608	—	—	—	—	583	SFG (0.8)	—	—	—	—
R5	$C_{90}20W8F100$	Roller compacted concrete	P·O 32.5	0.50	55.0	39.0	84	102	812	523	785	—	93	SFG (0.8)	0.05	3–8	4–5	0.40
R6	$C_{90}20W8F100$	Grout enriched vibrated mortar	P·O 32.5	0.48	40.0	—	733	483	—	—	—	—	583	SFG (0.8)	—	—	—	—
R7	$C_{90}20W8F100$	Roller compacted concrete	P·O 32.5	0.45	55.0	38	93	114	784	527	791	—	93	SFG (0.8)	0.05	3–8	4–5	0.42
R8	$C_{90}20W8F100$	Grout enriched vibrated mortar	P·O 32.5	0.43	40.0	—	783	517	—	—	—	—	558	SFG (0.8)	—	—	—	—

Note: The grout enriched vibrated neat mortar is calculated as 6% of the volume of RCC.

Chapter 6 Research and Application of RCC Admixture

Table 6.34 Performance test results of RCC and GEVR

No.	Compressive strength (MPa)			Splitting tensile strength (MPa)			Impermeability strength (90 d)	Frost resistance (90 d)	Compression modulus (GPa)		Ultimate tensile $\varepsilon(\times 10^{-4})$		Time of setting (h:min)	
	7 d	28 d	90 d	7 d	28 d	90 d			28 d	90 d	28 d	90 d	Initial setting	Final setting
R1	10.6	15.4	19.9	0.61	1.18	1.78	>W5	>F50	18.2	21.7	0.61	0.75	15:00	20:20
R2	12.6	17.2	21.8	0.73	1.38	2.02	>W5	>F50	20.8	23.0	0.69	0.82	—	—
R3	11.9	17.6	22.8	0.72	1.30	2.04	>W5	>F50	20.3	24.6	0.67	0.81	—	—
R4	14.7	19.7	23.8	0.82	1.47	2.24	>W5	>F50	21.9	25.4	0.70	0.87	—	—
R5	13.1	19.7	24.7	0.82	1.52	1.88	>W9	>F100	21.3	25.8	0.72	0.85	14:10	19:40
R6	13.5	19.6	24.9	0.93	1.78	2.40	>W9	>F100	23.4	26.7	0.74	0.93	—	—
R7	15.3	21.3	27.9	1.12	1.98	2.44	>W9	>F100	24.6	29.0	0.73	0.90	—	—
R8	16.0	23.1	29.4	1.20	2.17	2.46	>W9	>F100	25.8	29.2	0.78	0.94	—	—

Table 6.35 Mix proportion of RCC and GEVR in dam construction

No.	Mix proportion strength grade	Concrete classification	Cement type (Taiyu)	Water-cement ratio	Admixture amount (%)	Sand ratio (%)	Unit volume (m³/kg)							Additive (%)		VC (s)	Air content (%)	Volume ratio of mortar to sand PV
							Cement	Extender	Sand	Small-sized stone	Medium-sized stone	Large-sized stone	Water	Water reducing agent SFG	Air entraining agent JM-2000			
R3	$C_{90}15$ W4F50	Roller compacted concrete	P·O 32.5	0.50	60.0	34.0	66	100	731	439	585	439	83	0.8	0.04	3–8	3–4	0.40
R4	$C_{90}15$ W4F50	Grout enriched vibrated mortar	P·O 32.5	0.48	50.0	—	608	608	—	—	—	—	583	0.8	—	—	—	—
R7	$C_{90}20$ W8F100	Roller compacted concrete	P·O 32.5	0.45	55.0	38	93	114	784	527	791	—	93	0.8	0.05	3–8	4–5	0.42
R8	$C_{90}20$ W8F100	Grout enriched vibrated mortar	P·O 32.5	0.43	40.0	—	783	517	—	—	—	—	558	0.8	—	—	—	—

Chapter 7

Research and Utilization of Rock Powder in RCC

7.1 Overview

Rock powder refers to the fine particles of rock processed by machine with the particle size less than 0.16 mm (refers to particles less than 0.075 mm in other countries). It includes fine particles with particle size less than 0.16 mm in manufactured sand and specially ground rock powder. The morphology of rock powder is similar to that of cement particles. It is a polyhedron with irregular shape and an inactive admixture. When it is mixed into concrete, the particle gradation of fine powder can be improved. It has filling effect and the mechanical binding force between mortars can be improved. In the mix proportion design of RCC, the content of rock powder generally refers to the percentage of rock powder in the manufactured sand. Because the amount of cementing material and water in RCC is less, when the manufactured sand contains an appropriate amount of rock powder, because it is similar to the fineness of the admixture, rock powder can replace some admixtures in mortar. Together with cementing materials, it can fill voids and wrap sand surface, equivalent to the increase of cementing material mortar. The greatest contribution of rock powder is to increase the volume ratio of mortar to sand. It can significantly improve the workability of RCC with less mortar, improve the homogeneity, density and impermeability of concrete, improve the strength and fracture toughness of concrete, improve the adhesive performance of construction layer, reduce the amount of cementing materials and reduce the adiabatic temperature rise. Meanwhile, the content of rock powder in sand is increased, so that it can increase the production of manufactured sand, reduce the cost, improve technical and economic benefits. Scientifically and reasonably increasing the content of rock powder in RCC is one of the main measures to effectively improve the quality of RCC.

People pay more and more attention to the role of rock powder as admixture in RCC. The rock powder has become an indispensable part of RCC materials now. The content of rock powder in manufactured sand directly affects the performance of RCC. People are more and more aware of the rock powder content in many RCC dams, such as Puding, Fenhe Reservoir 2, Jiangya, Dachaoshan, Mianhuatan, Linhekou, Shapai, Baise, Suofengying, Longtan, Guangzhao and Jin'anqiao. When the content of rock powder is about 18%, the performance of RCC is obviously improved. Especially in recent years, the manufactured sand has been produced by dry process and semi dry and semi wet process. It is beneficial to increase the content and yield of manufactured sand and rock powder. However, dust produced by dry process is easy to cause environmental pollution. At present, the manufactured sand in dry process adopts green environmental protection measures for production. In other words, the semi dry and semi wet process has been successfully applied to dry process production. Recently, large closed workshops are used for dry sand production and the pollution problem has been solved. For example, Suofengying semi dry and semi wet green environmental protection method, dry production in Guanyinyan large-scale closed workshop, etc., have successfully solved the dust pollution problem of dry production.

The mortar content of RCC is much lower than that of normal concrete. In order to ensure its rollability, liquefaction and bleeding, interlayer bonding, compactness and a series of other properties, increasing the content of rock powder in sand is a very effective measure. According to the latest version of *Construction Specification for Hydraulic Roller Compacted Concrete* DL/T 5112, the content of rock powder ($d \leq 0.16$ mm particles) in manufactured sand should be controlled at 12% – 22%. The best rock powder content shall be determined by test. In recent years, in the tender documents of large and medium-sized projects, the content of rock powder in manufactured sand in RCC dam is generally specified to be 15% – 22%. Study results show that the content of rock powder can be further increased to 22%, even breaking 22% ceiling limit. The rock powder, especially the micro rock powder less than 0.08 mm, plays an important role in RCC. Project practice has proved that the micro rock powder less than 0.08 mm has become an important part of RCC admixture. RCC dam construction technology in foreign projects also attaches great importance to the content of rock powder, for example, Koudiata Cerdoune RCC dam in Algeria. The dam is 120 m high. RCC is mixed with limerock powder and the quality requirement is also less than 0.08 mm.

The content and particle size distribution of rock powder in manufactured aggregate are related to the production equipment, production technology and rock types. The main types of aggregate used in most projects are at present as follows: Limestone, granite, diabase, dolomitic limestone, basalt, sandy slate, etc. The rock powder content of manufactured sand produced by different types of rocks is also different. Their chemical composition is also different. Most of them are non-active materials. Although some rock powders have certain activity and they participate in the hydration reaction of cementing materials, their activity is very low. Compared with fly ash, iron ore slag, phosphorous slag and pozzolan, the activity of rock powder has little effect on pozzolanic activity. However, rock powder can play the role of micro aggregate effect, has the function of filling and compacting and increases the volume of cementing materials. In the manufactured sand with broken edges and corners, the shape of the rock powder is polygonal particle. At the early stage of hydration, when a large amount of net hydrated calcium silicate gel (C – S – H) is generated, the fulcrum interlocking action of rock powder is beneficial to the generation of hydrate products. Therefore, the greatest contribution of rock powder is to increase the volume ratio of mortar-sand of RCC, obviously improving the millability and interlayer binding quality of RCC and beneficial to improving various performances of concrete. For example, diabase aggregate is used in the RCC main dam project of Baise Multipurpose Dam Project. Affected by production technology, production equipment and rock types, the rock powder content of manufactured sand is large, accounting for 20% – 24%. Among them, the micro-rock powder less than 0.08 mm accounts for 40% – 60% of the rock powder; the concrete aggregate of Jiangya Dam is produced by Italian crusher. The rocks are limy dolomite, limestone and dolomitic limestone. The content of rock powder in the manufactured sand accounts for about 18.9% of the manufactured sand, in which, the particles smaller than 0.075 mm account for about 13.9%. PL-8500 vertical shaft crusher is used to produce aggregates in Suofengying Hydropower Station. The aggregate is limestone and the content of rock powder is 17% – 21.8%, 18.3% in average. The content of rock powder smaller than 0.08 mm is 11.6% – 14.4%, 12.8% in average.

Particles of stone power in the manufactured sand are very fine. Due to electrostatic action, the content of rock powder tested by drying method is low according to the test regulations. The rock powder content of the same manufactured sand detected by water washing method is different from that detected by drying method, there are big differences. Because the rock powder content of manufactured sand is measured by water washing method, which is in line with the actual situation of the project, so it has been adopted by a large number of projects at present. For example, in the RCC construction of Baise main dam project, through statistical analysis of the inspection result of diabase manufactured sand, diabase manufactured sand is inspected in accordance with *Test Code for Aggregates of Hydraulic Concrete* DL/T 5151—2001. That is to say, dry screening is adopted for detection. The content of rock powder is 16% - 20%. FM = 2.7 - 3.0; diabase manufactured sand is tested by water washing wet screen, that is to say, 500 g diabase manufactured sand which is dried and weighed is sieved in water by a 0.16 mm sieve, then drying and testing. The content of the rock powder obtained is 20% - 24%, FM = 2.7 - 2.9, and the coarse and fine particles are polarized, the gradation is not within the range of particle gradation curve. The content of rock powder exceeds the standard. Manufactured sand in Baise Project is unqualified according to the standard water washing method, but by using and research, the problem of high rock powder content of diabase manufactured sand is solved. In existing standards, the same standard is adopted for the grain gradation of natural sand and manufactured sand, without considering the influence of rock properties. In fact, the most important feature of manufactured sand is its particle shape and can hardly be summed up by any test standard. Practice of RCC construction of Baise main dam project proves that gradation of manufactured sand is discontinuous. The content of rock powder exceeds the standard and breaks through past experience and standards. The construction proves that RCC mixed with diabase manufactured sand with high rock powder content is easy to bleeding and has good millability. According to the core sample of RCC drilling in the main dam, the interlayer bonding of RCC is very dense and has good quality. According to the rock powder test method of manufactured sand, the grain gradation and the function of high rock powder content in RCC, it is necessary to put forward a new viewpoint for re-understanding, establish the manufactured sand testing standard for RCC. It needs further research and verification that the rock powder content of manufactured sand is tested by drying method or water washing method.

In this chapter, the research and application of rock powder of manufactured sand and admixture rock powder in RCC are introduced. From the microscopic analysis of rock powder, the rock powder content of manufactured sand, the applications of rock powder replacing sand, rock powder replacing fly ash, etc., the influence of rock powder on the performance of RCC is discussed, so that the function and mechanism of rock powder in RCC can be brought into full play.

7.2 Limestone Rock Powder

Limerock powder is the most important admixture in RCC, whether the rock powder is added externally or internally or used as an integral part of the admixture, it plays an important role in RCC. As limestone is an important raw material for cement production, therefore, when limerock powder is used as admixture of cement, its performance has good secondary reaction effect with cement. In order to study the properties of rock powder better, it is also necessary to deeply understand and master the properties of limestone, so as to provide scientific and technical support for the application of limerock powder in RCC.

7.2.1 Limestone

Limestone is one of the most widely distributed minerals in the crust. According to its sedimentary area, limestone is divided into marine sediments and continental sediments, the former is the most; According to formation causes, limestone can be divided into three types: biological sediments, chemical sediments and secondary sediments. According to the different components contained in limestone, limestone can be divided into siliceous limestone, clayey limestone and dolomitic limestone.

Distribution of resources: China is rich in limestone mineral resources, there are more than 800 limestone deposits used as cement, solvent and chemical industry. They are distributed all over the country. All provinces, municipalities and autonomous regions can use local limestone near the industrial zone.

Limestone minerals were deposited in every geological era and were distributed in every geological structure development stage, but good quality and large-scale limestone deposits often occur in certain horizons. Taking limestone for cement as an example, the limestone of Majiagou Formation of Middle Ordovician in Northeast China and North China is an extremely important horizon. Carboniferous, Permian and Triassic limestone are widely used in Central South China, East China and Southwest China. Silurian and Devonian limestone are widely used in Northwest China and Tibet. Ordovician limestone in East China, Northwest China and the middle and lower reaches of Changjiang River is also an important layer of cement raw materials. See Table 7.1 for temporal and spatial distribution of limestone resources in China.

Table 7.1 Temporal and spatial distribution of limestone resources in China

Ore-bearing geological age	Limestone distribution area	Main lithology
Early Proterozoic Era	Inner Mongolia, Heilongjiang, central Jilin, Xinyang and Nanyang in southern Henan	Marble
Middle and late Proterozoic Era	Liaodong Peninsula, Tianjin, Beijing, northern Jiangsu, Gansu, Qinghai and Fujian	Siliceous limestone and flint limestone
Cambrian period	Shanxi, Beijing, Hebei, Shandong, Anhui, Jiangsu, Zhejiang, Henan, Hubei, Guizhou, Yunnan, Xinjiang, Qinghai, Ningxia, Inner Mongolia, Liaoning, Jilin and Heilongjiang	Olitic limestone, pure limestone, edgewise limestone, thin dolomitic limestone
Ordovician	Heilongjiang, Inner Mongolia, Jilin, Liaoning, Beijing, Hebei, Shanxi, Shandong, Henan, Shaanxi, Gansu, Qinghai, Xinjiang, Sichuan, Guizhou, Hubei, Anhui, Jiangsu and Jiangxi	Thin-lift, thick pure limestone, dolomitic limestone, tabby limestone, gravelly limestone, etc.
Silurian	Xinjiang Toksun, Qinghai Golmud, Gansu, Inner Mongolia and so on	Muddy limestone, siliceous limestone, crystalline limestone, etc.
Devonian	Guangxi, Hunan, Guizhou, Yunnan, Guangdong, Heilongjiang, Xinjiang, Shaanxi and Sichuan	Thick pure limestone, dolomitic limestone, crystalline limestone, thin limestone, muddy limestone, etc.
Carboniferous	Jiangsu, Zhejiang, Anhui, Jiangxi, Fujian, Guangxi, Guangdong, Sichuan, Hubei, Henan, Hunan, Shaanxi, Xinjiang, Gansu, Qinghai, Yunnan, Guizhou, Inner Mongolia, Jilin and Heilongjiang	Thick pure limestone, thick limestone mixed with sandy shale and dolomitic limestone, marble, crystalline limestone, etc.
Permian	Sichuan, Yunnan, Guangxi, Guizhou, Guangdong, Fujian, Zhejiang, Jiangxi, Anhui, Jiangsu, Hubei, Hunan, Shaanxi, Gansu, Qinghai, Inner Mongolia, Jilin, Heilongjiang and so on	Thick limestone, cherty limestone, siliceous limestone, dolomitized limestone, marble
Triassic	Guangxi, Yunnan, Guizhou, Sichuan, Guangdong, Jiangxi, Fujian, Gansu, Qinghai, Zhejiang, Jiangsu, Anhui, Hunan, Hubei, Shaanxi	Muddy limestone, thick limestone, thin limestone
Jurassic	Zigong area, Sichuan	Continental lacustrine sedimentary limestone
Tertiary	Xinxiang in Henan Province and Suburbs of Zhengzhou	Muddy limestone, loose calcium carbonate

7.2.2 Properties of Limestone Ore

The limestone is mainly composed of calcite, dolomite, magnesite and other carbonate minerals, as well as some other impurities. Magnesium appears as limestone and magnesite. Silicon oxide is free quartz. The chalcedony and opal are distributed in the rock. Aluminum oxide and silicon oxide are synthesized into alumina-silicate (Clay, feldspar, mica); iron compounds exist as carbonates (Magnesite), pyrite (iron pyrite), free oxides (Magnetite, hematite) and hydroxide (hydrogoethite); as well as glauconite. Some types of limestone include organic matter such as coal and asphalt, and sulfate such as gypsum and anhydrite, and compounds of phosphorus and calcium, alkali metal compounds and strontium, barium, manganese, titanium, fluorine and other compounds, but the content is very low.

The main mineral characteristics and chemical properties of limestone are as follows:

Chemical molecular formula	$CaCO_3$
Crystal system	Trigonal system
Crystalline form	Irregular equiaxed grains, or rhombohedral crystal, or scalenohedron and rhombohedral aggregation, cylinder and scalenohedron and rhombohedral aggregation
Color	Colorless, white, grayish yellow, light red or blue-green when containing impurities
Striation	Colorless
Gloss	Vitreous luster
Cleavage	Eminent cleavage
Fracture	Staggered shape
Hardness	3 (f coefficient is generally 8 – 12)
Density	2.715 g/cm^3
Humidity	Generally less than 1%
Theoretical component content	Ca 56.4%; CO_2 43.96%
Tensile and compressive strength	The vertical bedding direction is generally 58.8 – 1 370 MPa and the parallel bedding direction is generally 49 – 117.6 MPa.
Bulk factor	1.5 – 1.6

Limestone has good machinability, polishing property and good cementing capacity, is insoluble in water, easily soluble in saturated sulfuric acid. It can react with strong acids to form corresponding calcium salts and release CO_2. When limestone is calcined at temperature above 900 ℃ (generally 1 000 – 1 300 ℃), it is decomposed into lime (CaO) and gives off CO_2. Quicklime deliquesces when it meets water and immediately forms hydrated lime (Ca(OH)$_2$). The hydrated lime can be adjusted after being dissolved in water and hardens easily in air.

7.2.3 Process Characteristics and Main Uses

7.2.3.1 Process Characteristics

Limestone has excellent properties such as thermal conductivity, firmness, water absorption, air impermeability, sound insulation, polishing, good bonding capacity and machinability. The raw ore can be directly utilized and can also be used after deep processing.

7.2.3.2 Main Purposes

Limestone is an important industrial raw material in metallurgy, structure materials, chemical industry, light industry, construction, agriculture and other special industrial sectors. With the development of steel and cement industry, the importance of limestone will be further enhanced. The main uses of limestone are shown in Table 7.2.

Table 7.2　　　　　　　　　　　　　　　　Main uses of limestone

Field of application		Accounting for the total output of limestone (%)	Main purposes
Metallurgical industry		60.56	Used as calcium oxide carrier in steelmaking, so as to combine coke ash and unnecessary associated elements such as silicon, aluminum, sulfur and phosphorus, then turned into fusible slag and discharged out of the furnace
Chemical industry		8.99	Filling agent for rubber industry; Extender for paper making and coating; raw material for making calcium carbide; it is also widely used in the production of bleaching agents, alkali production, magnesia extraction of seawater, nitrogen fertilizer, plastics, organic chemicals and calcium carbide
Construction industry		9.16	Producing lime mortar for structure, various types of lime, crushed stone, asphalt ingredient during road construction, etc
Agriculture			Calcined lime is used to neutralize acidic soil and as feed
Structure material industry	Cement	17.31	Main raw material of Portland cement
	Glass		The main raw material introducing CaO with the main function as a stabilizer in glass
	Ceramic		Main raw material for introducing CaO
	Refractory material		The lime cream is used as mineralizer to obtain a solid blank and accelerate the process of transforming quartz into tridymite and cristobalite when baking silica bricks
Sugar industry		0.05	Filter aid in sugar production
Plastics industry			Important ingredients for nylon production
Environmental protection			Treatment of industrial wastewater

7.2.4　Quality Requirements of Calcareous Raw Materials for Cement

General industrial indexes of limestone for cement: $CaO \geqslant 48\%$　$MgO \leqslant 3\%$　$K_2O + Na_2O \leqslant 0.6\%$　$SO_3 \leqslant 1\%$　$SiO_2 \leqslant 4\%$.

Calcareous raw materials for cement production. The quality requirements are listed in Table 7.3. For amorphous limestone, the particle size is required to be 30 – 80 mm.

Table 7.3　　　　　　　　Quality requirements of calcareous raw materials for cement

Grade and category		Grade(%)				
		CaO	MgO	R_2O	SO_3	Flint or quartz
Mudstone		35 – 45	<3.0	<1.2	<1.0	<4.0
Limestone	First grade	>48	<2.5	<1.0	<1.0	<4.0
	Second grade	45 – 48	<3.0	<1.0	<1.0	<4.0

7.3 Influence of Rock Powder on Performance of RCC

7.3.1 Rock Powder Content and *PV* Value of Grout-mortar Ratio

The *PV* value of mortar/sand ratio is one of the most important parameters in RCC mix design. It has the same function as water-cement ratio, sand ratio and water consumption. The characteristic values of early RCC mix design mainly refer to foreign experience, generally, α, β and *PV* value are adopted. α is the filling factor of mortar, generally takes 1.1 – 1.3, reflecting the situation that water, cement and admixture fill the sand gap; β is the filling coefficient of mortar, generally takes 1.3 – 1.5, reflecting the situation that water, cement, admixture and sand fill the gap of coarse aggregate; The *PV* value of grout-mortar ratio is the ratio of the volume of mortar (water + cementing material + 0.08 mm micro-rock powder) to the volume of mortar, generally should not be lower than 0.42.

Because the calculation of α and β values is complicated, with a wide amplitude range, it is difficult to be controlled accurately. In recent years, with the research and deepening understanding of the content of rock powder, more and more attention is paid to *PV* value of grout-mortar ratio in RCC mix design. According to the practical experience of full-section RCC damming in recent years, when the content of rock powder in manufactured sand is controlled at about 18%, the grout-mortar ratio is generally not lower than 0.42. Thus it can be seen that the *PV* value of grout-mortar ratio intuitively reflects a proportional relationship between RCC materials. It is an important index to evaluate the construction performance of RCC, such as millability, liquefaction and bleeding, interlayer bonding, anti-aggregate separation and so on.

Rock powder content has great influence on *PV* value of RCC grout-sand ratio. The *PV* value of RCC grout-mortar ratio is directly related to the interlayer bonding, imperviousness performance and overall performance of RCC, generally, *PV* value of pulp-sand ratio should not be lower than 0.42. When RCC adopts natural sand or the stone power in the manufactured sand has a low content, and RCC with low-grade less cement, increasing the rock powder content of manufactured sand or increasing the content of admixture can realize the purpose of improving the *PV* value of RCC grout-mortar ratio and the performance of the mixture. The following is an analysis of the influence of the relationship between rock powder content and grout-mortar ratio *PV* value in several projects as follows:

1. Dachaoshan Project. Dachaoshan Hydropower Project adopts basalt manufactured sand and aggregate. Rock powder content in manufactured sand is particularly important in Dachaoshan RCC. According to the expert's advice and through test and verification, $d \leq 0.15$ mm (old standard), the content of rock powder must reach more than 15% and the rock powder $d \leq 0.08$ mm must reach more than 8%. The rock powder content is low in early stage. Rock powder recovery facilities are added to the production system in later stage. Good results have been achieved. The content of rock powder is obviously improved. When individual rock powder content index does not reach the set value and is lower than 7%, PT (admixture of phosphate rock slag and tuff) admixture is used for over-substitution. The *PV* value of grout-mortar ratio is increased to 0.43 by calculation, ensuring that the PT admixture RCC has good working performance.

2. Mianhuatan Project. Coarse biotite granite aggregate is used in Mianhuatan Hydropower Station Project. The manufactured sand and aggregate are produced by dry method. The main crushing equipment of the system adopts the equipment of Swedish Svedara Company. The system has high output and good aggregate shape, high content of rock powder, saving water, etc. Due to the adoption of dry production, it avoids the loss of a

large amount of rock powder in sand. The rock powder in sand is over 17% in average. Fine particles (≤ 0.08 mm) account for about 30% of the rock powder. The PV value of grout-mortar ratio is calculated to be 0.42 – 0.45. Technological tests show that the rock powder content of Barmac sand is related to the opening and rotating speed. The rock powder content of manufactured sand can be adjusted by adjusting the rotating speed.

3. Linhekou Project. Limestone manufactured sand and aggregate are used in Linhekou Hydropower Station Project. Reasonable rock powder content of manufactured sand is one of the key factors affecting the quality of RCC. At the initial stage of operation of the sand and aggregate production system, the aggregate less than 20 mm washed by the spiral sand washer directly enters Barmac for dry sand making by stoning, causing rock powder package. The content of rock powder in dry sand fluctuates greatly and affects the sand production. After organizing special technical research, firstly, water-washed aggregates smaller than 20 mm are dewatered in the pile and then enter Barmac. After the above process adjustment, the problem of rock powder package in the production process is solved. The content of rock powder is basically controlled between 15% and 22%. After calculation, the PV value of grout-mortar ratio of Ⅱ and Ⅲ gradations of RCC ranges from 0.44 – 0.45. Because the mix proportion design determines the reasonable amount of cementing material, fine aggregate rich in rock powder is used. The field construction shows that cementing materials with high fly ash content have larger volume due to the high content of manufactured sand and rock powder. The RCC of Linhekou Arch Dam has achieved very good results in terms of millability, liquefaction and bleeding, interlayer bonding and anti-aggregate separation.

4. Guangzhao Project. The RCC gravity dam of Guangzhao Hydropower Station is a world-class high dam with a height of 200.5 m. RCC adopts limestone manufactured sand and aggregate. Due to the tight schedule, in the process test of RCC in late August 2005, normal concrete sand is used. The rock powder content of manufactured sand is only 7% – 13%. Because manufactured sand is produced by wet method, as a result, a large amount of precious micro-rock powder below 0.08 mm is lost with water. During the first RCC process test, the mix proportion of RCC is as follows: Ⅱ gradation $C_{90}25F100W10$, water-cement ratio 0.45, water 86 kg/m³, fly ash 45% used at dam imperviousness area; Ⅲ gradation $C_{90}20F100W10$, water-cement ratio of 0.48, water of 77 kg/m³, fly ash of 50% and aggradation of small stone: middle stone: big stone = 30: 35: 35 inside the dam. Due to the adoption of normal concrete sand, the mix proportion design is not very reasonable and the rock powder content in the manufactured sand is low. After calculation, the PV value of RCC grout-mortar ratio is only 0.35. Meanwhile, the water content of manufactured sand greatly exceeds the control index of 6%. Therefore, in the first field process test, the RCC mixture has serious bleeding, aggregate separation, poor millability and poor interlayer bonding. Drilling and coring are carried out at the 90 d RCC for the first time. Core sample has honeycomb pits and has obvious faults at interlayer hot joints. The first RCC process test failed.

The second RCC process test was carried out in mid-December, 2005. The problems existed for the failure of the first RCC process test were analyzed carefully before the test. The mix proportion of RCC was adjusted first. Under the condition of keeping the water binder ratio of the original mix proportion unchanged, the fly ash content of RCC in dam impervious area and inside dam area were increased to 50% and 55% respectively. In view of the low stone power content in the sand for normal concrete, the scheme that fly ash replaces sand by 4% was adopted. The aggregate gradation was adjusted to small stone: middle stone: big stone = 30: 40: 30. The actual water consumption of Ⅱ and Ⅱ gradations after adjustment were 96 kg/m³ and 86 kg/m³ respectively. VC value was controlled in the range of 3 – 5 s and the water content of sand was strictly controlled within 3%. In this way, the PV value of grout-mortar ratio of RCC was increased from 0.35 to 0.44. The second RCC process test was very successful. The liquefaction and bleeding of RCC on the placing surface were sufficient, with good rollability and interlayer bonding. The RCC was drilled for core sample in 10 d age. The surface of RCC core sample was smooth and compact. Interlayer bonding was unrecognizable and won unanimous praise and recognition from consulting experts. The RCC of Guangzhao Dam was constructed and poured on schedule in February 2006.

A large number of RCC dam construction technologies and practices at home and abroad prove that, the strength of RCC almost exceeds the designed strength and the surplus is relatively large, thus the elastic modulus of concrete has been improved. The temperature stress of the dam body is increased and is unfavorable in anti-cracking. The most important quality control of RCC is the performance of mixture and construction performance. These properties are directly related to the PV value of RCCgrout-mortar ratio. Therefore, PV value of grout-mortar ratio shall be monitored closely in mix proportion design. According to the Author, the PV value of RCC grout-mortar ratio shall be listed as one of the important indexes to evaluate the performance and rollability of RCC mixture.

7.3.2 Influence of Rock Powder Content on Performance of RCC

7.3.2.1 Influence of Rock Powder Content on Water Consumption, Air Content and Setting Time

With the increase of rock powder content in RCC, it plays a significant role in improving the workability of RCC, liquefaction and bleeding, interlayer bonding and concrete setting performance. However, the high content of rock powder also has disadvantages. A large number of test studies show that no matter rock powder is mixed internally or externally, the unit water consumption of RCC increases with the increase of rock powder content. For every 1% increase in rock powder content, the unit water consumption increases by about 2 kg/m^3 generally; in addition, the content of rock powder has great influence on the air content of fresh RCC. Generally, with the increase of rock powder content, the air content of RCC decreases obviously. Therefore, the air entrainment of RCC with high rock powder content is much more difficult than that of normal concrete, therefore, the content of air-entraining agent in RCC is often several times higher than that of normal concrete; at the same time, it also has a certain influence on the setting time which is shortened.

7.3.2.2 Influence of Replacing Part of Fly Ash ith Rock Powder on Performance of RCC

Test studies show that while keeping the cement content unchanged, rock powder replaces part of fly ash, that has little influence on the workability of RCC in a certain range. Baise Project practice shows that when diabase powder replaces fly ash in the range of 20–40 kg/m^3, that has little influence on the setting performance of concrete, but when the substitution amount of diabase powder exceeds 40 kg/m^3 fly ash, the strength and other performance of RCC are obviously reduced. It shows that the excessive amount of rock powder replacing fly ash is unfavorable to the performance of concrete.

7.3.2.3 Influence of Rock Powder Content on Hardening Performance of RCC

The increase of rock powder content and the adoption of small VC value can obviously improve the compactness of the RCC. Adding a proper amount of rock powder (whether mixed internally or externally) is beneficial to improving the strength and impermeability of RCC, especially, having obvious effects on improving frost resistance, ultimate tensile value and reducing elastic modulus. However, when the content of rock powder is gradually increased from 18% to 24%, its compressive strength decreases with the increase of rock powder content. 28 d splitting tensile strength also shows a downward trend. Research results also show that: With the increase of rock powder content, the dry shrinkage value of concrete at different ages increases. The dry shrinkage value gradually increases with the extension of the age. It shows that the high content of rock powder is unfavorable to the dry shrinkage of RCC.

7.3.2.4 Coarse Aggregate Shows the Influence of Rock Powder Coated on Aggregate on the Strength of RCC.

In RCC of Mianhuatan and Golantan dams, etc., coarse aggregates are coated by rock powder. The influence of rock powder coated on the surface of coarse aggregate on the strength of RCC is studied. Test Results: rock powder coated on the surface of coarse aggregate will reduce the compressive strength, splitting tensile strength and axial tensile strength of RCC, especially, the splitting tensile strength decreases greatly. With the increase of age, the influence of rock powder coated on the surface of coarse aggregate on the compressive strength of RCC is weakened. This is because the rock powder coated on the surface of coarse aggregate causes the adhesion between the mixing mortar and the surface of coarse aggregate to decrease (or the mechanical snap-in force decreases). With age increases, the cementing material is further hydrated, the newly generated C-S-H binder continuously penetrates into the surface of coarse aggregate, improves the interface structure between coarse aggregate and mortar, enhances the mechanical snap-in force between mortar and coarse aggregate. Therefore, the influence on compressive strength is weakened.

7.4 Project Examples of RCC with Rock Powder Replacing Sand

7.4.1 Technical Scheme of Replacing Sand with Rock Powder in Jin'anqiao RCC

7.4.1.1 Foreword

The quarry of Jin'anqiao Hydropower Station is composed of basalt and weakly weathered basalt. The aggregate must be basalt and weakly weathered basalt which can meet the project needs. Total amount of concrete in main works such as the dam, discharge and powerhouse is about 5.28 million m^3. The required sand and aggregate is about 11.8 million t, among them, gravel is 7.3 million t; sand for RCC is 2.3 million t; sand for normal concrete is 2.2 million t. Early design companies, relevant scientific research units and construction companies all found in the concrete mix proportion test: Basalt aggregate used in concrete of Jin'anqiao Project, different from other aggregates. Basalt aggregate has high density, rough surface and strong absorbability. The characteristic of basalt aggregate lead to a sharp increase in concrete water consumption, moreover, the workability of freshly mixed RCC is poor. It is related to the low content of stone power of basalt manufactured sand. In addition, Jin'anqiao Project is located in Yunnan-Guizhou Plateau which has typical plateau climate characteristics, i.e. great temperature difference between day and night, strong illumination, dry climate and large evaporation. It is unfavorable to RCC construction.

According to the characteristics of basalt aggregate of Jin'anqiao dam and the unfavorable factors of low content of rock powder in the manufactured sand, through the test study on the mix proportion of basalt aggregate of RCC, the mix proportion design adopts the technical route of adding rock powder instead of sand and increasing the admixture content and low VC value, which improves the working performance of RCC mixture, effectively improves the liquefaction and interlayer bonding quality of basalt aggregate of RCC, ensures the rapid construction of RCC dam.

The basalt manufactured sand used in Jin'anqiao project is low in rock powder content and cannot meet the construction requirements. Therefore, RCC adopts the scheme of adding rock powder admixture instead of sand, select limerock powder processed by a cement plant as the rock powder. The fineness of limerock powder is controlled according to grade II fly ash. See Table 7.4 for the test results of rock powder.

Chapter 7 Research and Utilization of Rock Powder in RCC

Table 7.4   Test results of limerock powder in Jin'anqiao

Rock powder factory	Fineness (0.08 mm sieve residual)	Loss on ignition	SiO_2	Al_2O_3	Fe_2O_3	CaO	MgO
Yongbao cement plant	13.2	41.84	0.42	0.08	Not detected	55.82	0.38
Liude cement plant	13.0	42.34	3.25	1.6	Not detected	50.14	1.97

7.4.1.2 Selection of RCC Rock Powder Content

The content of rock powder has great influence on the ratio of mortar to sand of RCC. When the rock powder content of manufactured is low, the volume ratio of RCC mortar to sand can be increased by adding rock powder instead of sand or fly ash instead of sand. Because the rock powder content of basalt manufactured sand in Jin'anqiao is far lower than the design index of 15% - 22%, in order to study the influence of different rock powder content on the performance of RCC, the technical scheme of replacing sand with rock powder is studied. Test condition: Yongbao 42.5 moderate heat cement, Grade II fly ash, limerock powder, basalt aggregate, manufactured sand FM = 2.78, rock powder content 11.8%, II gradation. Mix Proportion for Testing: water-cement ratio 0.50, fly ash 55%, sand ratio 37%, 1.0% of water reducing agent, air content 3% -5%, VC value 3 -5 s, VC value controls water consumption.

See Table 7.5 for test results of influence of rock powder content on RCC performance. The results show that with the increase of rock powder content in manufactured sand, the water consumption increases regularly. When the content of rock powder in sand is controlled in the range of 18% -20%, the volume ratio of RCC mortar to sand is calculated to be 0.44. The liquefaction and bleeding of RCC mixture is obviously accelerated. After the VC value test is finished, the surface of the specimen is smooth and dense, and the mortar is sufficient. The performance of RCC is effectively improved. Because the compactness is improved, the strength of concrete is improved accordingly.

7.4.1.3 Quality of RCC with Rock Powder instead of Sand

Excellent mix proportion is the basic guarantee of rapid construction and quality of RCC. The mix proportion of basalt aggregate RCC of Jin'anqiao dam has been studied repeatedly and tested at production site. According to the characteristics of basalt aggregate RCC and the low content of rock powder in manufactured sand, the technical measure of replacing sand with rock powder is adopted to control the content of rock powder in sand to 18% -19%.

The VC value of RCC is dynamically controlled. The VC value control of the machine outlet is based on the principle of good rollability of the placing surface. The VC value of the machine outlet is controlled for 1 - 3 s during the day. The VC value of warehouse surface is controlled according to 3 -5 s; The VC value of the machine outlet for 2 -5 s at night and controlled according to the lower limit. Accurate mixing was adopted for the content of RCC rock powder of Jin'anqiao Dam. The amount of sand replaced by rock powder and VC value was dynamically controlled. RCC was controlled in whole processes from mixing, transportation, pouring, paving, rolling and spraying and moisturizing, etc. RCC had sufficient mortar, good rollability, non-separation of aggregates, sufficient surface liquefaction and bleeding, and elasticity, so that the upper aggregate could be embedded into the rolled concrete of the lower layer, and ensuring the interlayer bond quality of the RCC. The acquisition rate of dam core sample drilled was high and the first ultra-long core sample of 16.49 m at home and abroad was obtained. The water pressure test was less than the design requirements of 1.0 Lu and 0.5 Lu. In-situ shear resistance parameter showed that the friction coefficient f' was bigger than 1.3, cohesive force

Table 7.5 Test results of influence of rock powder content on RCC performance

Test No.	Rock powder content (%)	Water consumption (kg/m³)	ZB-1RCC15 (%)	ZB-1G (1/10 000)	VC value (s)	Air content (%)	Mixture properties	Compressive strength (MPa)			
								7 d	28 d	60 d	90 d
RSF2-1	12	97	1.0	30	3.5	4.2	Aggregate is coated poorly with coarse specimen surface	12.8	16.0	18.2	22.5
RSF2-2	14	100	1.0	30	3.8	3.8	Aggregate is coated poorly with coarse specimen surface	12.3	16.3	22.0	23.5
RSF2-3	16	102	1.0	30	4.2	3.5	Aggregate is coated poorly with coarse specimen surface	11.6	16.4	20.7	23.1
RSF2-4	18	103	1.0	30	4.2	3.4	Aggregate is coated generally and the surface of the specimen is dense	11.8	17.0	21.9	24.0
RSF2-5	20	105	1.0	30	3.8	3.2	The aggregate is well coated and the surface of the specimen is smooth and dense	12.4	18.7	21.8	24.3
RSF2-6	22	107	1.0	30	3.8	3.0	The aggregate is wrapped very good and the surface of the specimen is smooth and dense	11.2	15.8	18.6	22.0

c' was bigger than 1.6 MPa and the homogeneity was good. The quality control test results showed that the RCC of Jin'anqiao Dam adopted the technical route of replacing sand with rock powder and low VC value, which is scientific and reasonable, getting good technical and economic benefits.

7.4.2 Technical Measures for Replacing Sand with Rock Powder in Kalasuke RCC

7.4.2.1 Foreword

Xinjiang Kalasuke Multipurpose Dam Project is 528.5 km long from National Highway 216 to Urumqi City. The main task of the project is to ensure and improve the social and economic development and ecological environment water consumption in the Eerqisi River Basin, supply water to Urumqi Economic Zone and give attention to power generation and flood control also. The Multipurpose Dam Project consists of RCC gravity dam, power generation and diversion system, power house and auxiliary dam. RCC gravity dam is divided into overflow dam section and non-overflow dam section. The maximal dam height is 121.5 m; The maximum water head is 119.6 m. Total concrete is 2 890 400 m³, in which RCC is 2 522 400 m³ and normal concrete is 368 000 m³.

The annual average temperature at the dam site is 2.7 ℃; The extreme maximum temperature is 40.1 ℃; The extreme minimum temperature is −49.8 ℃. The annual average precipitation is 183.9 mm. The annual average evaporation is 1 915.1 mm. The annual average wind speed is 1.8 m/s. The maximum wind speed is 25 m/s. The maximum snow depth is 75 cm. The maximum frozen soil depth is 175 cm. The average water temperature from May to October is 13.1 ℃ and the maximum water temperature is 25 ℃.

Chapter 7 Research and Utilization of Rock Powder in RCC

Kalasuke Project is located in the north of Xinjiang which has typical complex climate characteristics of "cold, hot, windy and dry". For harsh weather conditions, five different grades of RCC were designed according to different parts of the main dam, especially RCC above the dead water level at the upstream of the dam and outside the water level change area at the downstream. The frost resistance grade is F300 and is a super frost resistance grade. The fine aggregate of RCC of the main dam is natural sand mined in Gobi desert and the natural sand has large mud content. The content of rock powder less than 0.16 mm of the natural sand washed with water is very small, resulting in poor workability of RCC, and not meeting the requirements of RCC liquefaction, rollability, interlayer bonding quality and rapid construction. Therefore, high frost resistance and little content of rock powder in the natural sand are the difficulties and key points in the mix proportion design of RCC for the main dam.

According to the suggestions made in "Expert consultation meeting on RCC mix proportion field test in Kalasuke project" in October 2006, the quality inspection center of the owner, the laboratory of the construction unit and the scientific research units were on the site of Kalasuke conducted the mix proportion test of RCC and checked again the normal concrete of the main dam according to the test result of RCC mix proportion of Kalasuke project submitted by previous scientific research units.

7.4.2.2 Design Requirements of RCC for Main Dam

See Table 7.6 for the design requirements of RCC for Kalasuke main dam. It can be seen from the Table 7.6 that the maximum frost resistance grade F300 is rare in China. RCC strength guarantee rate $P = 80\%$, apparent density $\geq 2\,400$ kg/m^3, and specific requirements raised for setting time also.

Table 7.6   Concrete design index table of each area

Zone No.	I-1	I-2	II-1	II-2	II-3
Location	Concrete above upstream dead water level and outside water level change area at downstream side	External concrete below upstream dead water level	External concrete above the water level change area at downstream side	RCC above 650 m internal elevation	RCC below 650 m internal elevation
Concrete grading	II	II	III	III	III
Design label	R_{180}20W10F300	R_{180}20W10F100	R_{180}20W6F200	R_{180}15W4F50	R_{180}20W4F50
Guarantee rate coefficient t	0.84	0.84	0.84	0.84	0.84
Standard deviation σ(MPa)	4.0	4.0	4.0	3.5	4.0
Preparation strength (MPa)	23.4	23.4	23.4	17.9	23.4
Ultimate tensile($\times 10^{-4}$)	>0.78	>0.78	>0.70	>0.70	>0.70
VC value(s)	5-8	5-8	5-8	—	5-8
Initial setting time (h)	12-17	12-17	12-17	12-17	12-17

7.4.2.3 Raw Material

I Cement

Tianshan P·O 42.5 portland cement was used for the test and the quality of cement was tested according to *Silicate Cement and Ordinary Portland Cement* GB 175—1999. The results showed that all physical indexes of cement met the specifications.

II Fly ash

Manas fly ash was used in the test and the quality was tested according to *Technical Regulation for Mixing Fly Ash with Hydraulic Concrete* DL/T 5055—1996. The results showed that Manas fly ash met grade II ash standard.

III Aggregate

The fine aggregate was natural washed sand from Kalasuke Gobi desert. Coarse aggregate was manufactured crushed stone of gneiss granite mined in C12 quarry. The quality of natural washed sand was tested according to *Test Regulations for Sand and Aggregate of Hydraulic Concrete* DL/T 5151—2001. Test results are shown in Table 7.7 and Table 7.8. The test results of natural sand showed that the content of rock powder less than 0.16 mm was only 1.6%, almost not available.

Table 7.7　　Quality inspection results of natural sand

Requirements in specification	Apparent density (kg/m³)		Water absorption (%)		Bulk density (kg/m³)		Porosity (%)		Clay content (%)	Clay lump content (%)	Mica content (%)	Light substance (%)	Solidity (%)	Organic matter
	Dry sand	Saturated surface dry	Dry sand	Saturated surface dry	Loose pile	Tight pile	Loose pile	Tight pile						
Measured value	2 640	2 610	1.16	1.14	1 422	1 588	46	40	2.5	0	0.1	0.02	4.43	Lighter than the standard color
DL/T 5144—2001 DL/T 5112—2000	≥2 500	—	—	—	—	—	—	—	≤3	Not allowed	≤2	≤1	≤8	Lighter than the standard color

Table 7.8　　Gradation test results of natural sand

Grain size (mm)	>5	5–2.5	2.5–1.25	1.25–0.63	0.63–0.315	0.315–0.16	<0.16	FM
Content (%)	0.78	6.79	13.75	13.90	49.42	13.44	1.6	2.44

Coarse aggregate is manufactured crushed stone of gneiss granite. Test results showed that the apparent density of gneiss granite was 2 700 – 2 720 kg/m³; the dry water absorption rate of the saturated surface was 0.84% to 1.4%, showing that gneiss granite had higher water absorption rate. The crush index of manufactured crushed stone was high also. According to aggregate quality, Kalasuke gneiss granite belongs to weathered rocks; rock strength is generally about 45 MPa; there are some fluctuations in crushing index and firmness. All of them have a certain impact on the performance of concrete.

IV Additive

PMS-3 retarding and high efficiency water reducing agent, NF-1 high efficiency water reducing agent and PMS-NEA3 air-entraining agent were used in the test. Test results showed that admixtures all met the standard requirements of *Concrete Admixtures* GB 8076—1997.

V Rock powder

There were two kinds of rock powder: Rock powder was processed and ground from stone chips in Tianshan

Cement Plant. The content of rock powder with particle size bigger than 0.08 mm was less than 10%; Rock powder with particle size below 0.16 mm was collected from precipitated waste water after washing and screening in aggregate production system. See Table 7.9 for particle analysis results of two kinds of rock powder.

Table 7.9   Grain gradation analysis results of grinding rock powder and collecting sieved rock powder

Type	>0.16 mm Content of grain (%)	>0.08 mm Content of grain (%)	>0.045 mm Content of grain (%)	<0.045 mm Content of grain (%)
Grinding rock powder	0.62	10.24	17.54	71.60
Collected sieved rock powder	0	30.04	32.16	37.80

7.4.2.4   Mix Proportion Test of RCC with Rock Powder instead of Sand

Ⅰ   Test scheme of RCC mix proportion

Test condition: the raw materials were Tianshan P · O. 42.5 cement, Manas fly ash, natural washed sand of Gobi desert, manufactured crushed stone of gneiss granite, PMS-3 retarding and high efficiency water reducing agent and PMS-NEA3 air entraining agent. Considering the dry and high evaporation climate conditions, the VC value of machine output was controlled according to 1 – 3 s; because of the high demand on durability and frost resistance grade, the air content of machine outlet was designed according to different frost resistance levels. The air content was controlled according to F50 3% – 4%, F100 4% – 5%, F200 4.5% – 5.5%, and F300 5% – 6%. Two kinds of rock powder with particle size less than 0.08 mm and less than 0.16 mm were used to replace natural sand equally.

Design scheme of RCC with rock powder instead of sand. When the water-cement ratio was less than or equal to 0.47 and the cementing material was more than 200 kg/m^3, the scheme of not mixing rock powder was adopted. When the water-cement ratio was bigger than 0.50 and the cementing material was less than 170 kg/m^3, sand replacement scheme with rock powder below 0.08 mm and sand replacement scheme with rock powder below 0.16 mm was adopted. See Table 7.10 for test parameters of RCC mix proportion of main dam. See Table 7.11 for performance test results of RCC mixture. See Table 7.12 for the test results of mechanical properties of RCC. See Table 7.13 for test results of ultimate tensile value and elastic modulus of RCC. See Table 7.14 for the test results of frost resistance and impermeability of RCC.

Ⅱ   Test results of RCC mixture

1. VC value. The VC value of machine outlet is small, 15 min and 30 min test results meet the design requirements.
2. Air content. The air content of the machine outlet is large, 15 min or 30 min test results meet the design requirements. The air content loses quickly within 15 min and the loss rate is slow after 15 min and the measured value tends to be stable.
3. Apparent density. The apparent density of RCC is generally low. Analysis shows that it is mainly caused by low density of cement, coal ash, aggregate and high air content of concrete. Test results show that every 1% increase in air content reduces the apparent density by about 25 kg/m^3. Accord to the test result of quality inspection center and construction unit, in the case that internal RCC is within 30 min – 1 h, air content basically tends to a stable range, the apparent density of concrete can meet the design requirements of more than 2 400 kg/m^3.

Table 7.10 Parameter table of recheck test for RCC mix proportion of main dam

No.	Part of works	Design index	Gradation	Mix proportion parameter									Material amount (kg/m³)			
				Water-cement ratio	Sand ratio (%)	Water reducing agent (%)	Air entraining agent (%)	Cement (%)	Fly ash (%)	Rock powder added (%)	Water content	Cement	Fly ash	Rock powder
KF1-2	Over 650 mat upstream face	R₁₈₀20F300W10	II	0.45	35	1.0	0.12	60	40	0	98	131	87	0
KF9-1	Upstream water level change area	R₁₈₀20F200W6	III	0.45	32	0.9	0.07	50	50	0	90	100	100	0
KF4-1	Outside dam body below dead water level	R₁₈₀20F100W10	II	0.47	36	1.0	0.06	50	50	0	95	101	101	0
KF5-1	Dam inside below 650 m	R₁₈₀20F50W4	III	0.52	33	0.8	0.04	35	65	0	95	64	119	0
KF6-1		R₁₈₀20F50W4	III	0.54	33	1.0	0.04	40	60	8	90	67	100	57
KF7-1		R₁₈₀15F50W4	III	0.56	33	1.0	0.04	40	60	8	90	65	96	57
KF8-1		R₁₈₀15F50W4	III	0.54	32	0.8	0.04	35	65	0	95	62	114	0
KF7-3	Dam inside above 651 m	R₁₈₀15F50W4	III	0.56	32	0.9	0.03	40	60	16	90	65	96	111
KF7-4		R₁₈₀15F50W4	III	0.56	32	0.9	0.03	40	60	14	90	65	96	96

Chapter 7 Research and Utilization of Rock Powder in RCC

Table 7.11   Performance test results of RCC mixture

Test piece No.	Concrete design index	Gradation	VC value (s)			Air content (%)			Unit weight (kg/m³)			Time of setting (h: min)	
			Machine outlet	15 min	30 min	Machine outlet	15 min	30 min	Machine outlet	15 min	30 min	Initial setting	Final setting
KF1-2	$R_{180}20$ F300W10	II	0.7	1.7	4.1	>10	6.8	4.2	2 221	2 314	2 414	26:20	34:19
KF9-1	$R_{180}20$ F200W6	III	—	1.2	—	—	4.6	4.8	—	2 363	2 356	28:19	37:34
KF4-1	$R_{180}20$ F100W10	II	1.7	3.2	4.8	7.6	6.2	5.3	2 291	2 284	2 342	34:36	41:19
KF5-1	$R_{180}20$ F50W4	III	0.6	1.7	3.2	8.2	5.2	4.0	2 249	2 336	2 373	33:25	43:25
KF6-1	$R_{180}20$ F50W4	III	—	1.0	2.3	—	8.0	5.3	—	2 269	2 342	33:29	43:16
KF7-1	$R_{180}15$ F50W4	III	—	1.6	3.5	—	5.8	4.4	—	2 334	2 371	33:46	44:35
KF8-1	$R_{180}15$ F50W4	III	—	1.0	3.2	—	6.4	5.4	—	2 308	2 373	42:19	53:07
KF7-3	$R_{180}15$ F50W4	III	0.7	0.9	1.2	5.8	4.4	3.4	2 327	2 356	2 386	32:33	56:45
KF7-4	$R_{180}15$ F50W4	III	0.9	1.4	—	5.6	4.9	—	2 313	2 334	—	—	—

Table 7.12   Test results of mechanical properties of RCC

Test piece No.	Concrete design index	Compressive strength (MPa)				Splitting tensile strength (MPa)		
		7 d	28 d	90 d	180 d	28 d	90 d	180 d
KF1-2	$R_{180}20$F300W10	18.3	30.2	41.5		—	3.06	
KF9-1	$R_{180}20$F200W6	13.8	24.8	33.2		1.83	2.51	
KF4-1	$R_{180}20$F100W10	10.4	29.5	39.4		—	2.62	
KF5-1	$R_{180}20$F50W4	6.8	14.2	27.6		1.34	2.26	
KF6-1	$R_{180}20$F50W4	6.2	15.7	25.0		1.49	1.73	
KF7-1	$R_{180}15$F50W4	5.6	17.3	30.7		0.97	2.11	
KF8-1	$R_{180}15$F50W4	7.1	15.8	27.2		1.13	1.87	
KF7-3	$R_{180}15$F50W4	6.8	13.8			1.15		
KF7-4	$R_{180}15$F50W4	7.4	14.4			1.35		

4. When rock powder with particle size less than 0.08 mm replaces 8% of natural sand, the newly mixed RCC has good workability, fast liquefaction and bleeding, good filling effect of sand pores with rock powder. The compaction performance of RCC is obviously improved.

5. 12%, 14% and 16% natural sand are replaced by rock powder below 0.16 mm in same quantity. The results show that: when sand is replaced with rock powder by 12%, the mill ability and filling effect of RCC are not ideal; when sand is replaced with rock powder by 14%, RCC has good mill ability and filling effect; When sand is replaced with rock powder by 16%, RCC has good mill ability and filling effect. The mix proportion test results of rock powder replacing sand by 14% and 16% are shown in numbers KF7-3 and KF7-4 respectively.

Table 7.13 Test results of ultimate tensile value and elastic modulus of RCC

Test piece No.	Concrete design index	Axial tensile strength (MPa)		Axial tensile elastic modulus (GPa)		Ultimate tensile value (10^{-6})		Axial compressive strength (MPa)		Compression modulus (GPa)	
		90 d	180 d	90 d	180 d	90 d	180 d	90 d	180 d	90 d	180 d
KF1-2	R_{180}20F300W10	2.23		29.6		84.3		29.1		36.0	
KF9-1	R_{180}20F200W6	2.26		33.8		80.5		25.6		31.3	
KF4-1	R_{180}20F100W10	1.96		32.5		78.7		29.2		34.9	
KF5-1	R_{180}20F50W4	2.00		28.8		79.5		21.0		33.6	
KF6-1	R_{180}20F50W4	1.88		30.0		71.4		29.2		29.8	
KF7-1	R_{180}15F50W4	1.94		26.1		82.2		15.9		27.9	
KF8-1	R_{180}15F50W4	1.86		29.5		75.5		23.6		29.8	
KF7-3	R_{180}15F50W4										
KF7-4	R_{180}15F50W4										

Table 7.14 Test results of impermeability and frost resistance of RCC

Test piece No.	Concrete grade	Grade of impermeability (>)	Frost resistance (age 90 d)												Frost resistance grade (>)
			Mass loss rate (%)						Relative dynamic modulus (%)						
			50	100	150	200	250	300	50	100	150	200	250	300	
KF1-2	R_{180}20F300W10	W10	0.1	0.2	0.3	—	—	—	94.7	92.9	89.3	85.2	79.2	74.3	F300
KF9-1	R_{180}20F200W6	W6	—	—	—	—	—	—	92.2	91.3	90.2	83.2	71.2	—	F250
KF4-1	R_{180}20F100W10	W8	0.1	0.2	—	—	—	—	91.1	88.9	—	—	—	—	F100
KF5-1	R_{180}20F50W4	W4	0.8	—	—	—	—	—	86.3	—	—	—	—	—	F50
KF6-1	R_{180}20F50W4	W4	0.4	—	—	—	—	—	82.7	—	—	—	—	—	F50
KF7-1	R_{180}15F50W4	W4	0.4	—	—	—	—	—	80.0	—	—	—	—	—	5F0
KF8-1	R_{180}15F50W4	W4	0.1	—	—	—	—	—	75.5	—	—	—	—	—	F50

Ⅲ Test results of hardened RCC

1. Concrete strength. 90 d RCC strength of the main dam meets the design grade. According to projections, the strength performance index of each mix proportion can meet the requirement of corresponding grade with a large surplus.

2. Axial tensile strength. Because RCC is mixed with high content of coal ash, the axial tensile strength of 90 d RCC of main dam is generally low, being 1.8 − 2.3 MPa, similar to the splitting tensile strength.

3. Ultimate tensile value. The ultimate tensile value of 90 d RCC of main dam ranges from $(70 - 85) \times 10^{-6}$, meeting the design requirements.

4. Axial compressive strength. The axial compressive strength of 90 d RCC of main dam ranges from 15 − 30 MPa, lower than cube compressive strength, but they all meet the design strength.

5. Static compressive modulus of elasticity. The static compressive elastic modulus of 90 d RCC of main dam ranges from 27 − 37 GPa, being moderate.

6. Frost resistance. The air content of RCC is strictly controlled within the design range. The air content of

F300 at machine outlet is controlled to be ≤7.5% and ≥5.0% at site; The air content of F200 at machine outlet is ≤7.0% and ≥4.0% at site. The frost resistance completely meets the design requirements.

7.4.2.5 Conclusion

The Kalasuke RCC adopts the technical scheme of adding rock powder instead of sand in fine aggregate natural sand. The mix proportion test results show that the workability, mill ability and construction performance of RCC after replacing sand with rock powder are obviously improved. The properties of hardened concrete meet the design requirements. See Table 7.15 for the recommended construction mix proportion of RCC for the main dam of Kalasuke Multipurpose Dam Project.

In actual construction of Kalasuke RCC, 29 kg/m^3 of sand was replaced by rock powder; VC value output was controlled by 1-3 s; the air content was strictly controlled in accordance with the design requirements. Frost resistance test results of hardened RCC show that: bubbles on the concrete specimen after formwork removal were tiny and evenly distributed. The hardened concrete had compact internal structure and uniform pore distribution. Bubble diameter and bubble spacing were relatively small. The frost resistance of F300 RCC with high frost resistance grade was effectively improved.

Dam core sample obtaining rate was high and two extra-long core samples larger than 16 m were taken out. The core sample had smooth and compact appearance, good interlayer bonding, all of facts above fully reflected that reasonable and proper technical scheme of replacing sand with rock powder had great importance and practical significance to improve the performance of RCC and ensure rapid construction when the natural sand had little rock powder content.

7.5 Study on Utilization of Baise Diabase Manufactured Sand and Rock Powder in RCC

Baise Multipurpose Dam Project is the backbone project in the flood control engineering system of combining dikes and reservoirs of Yujiang River determined in the flood control planning of Pearl River Basin. It is a flood control project and gives attention to power generation. It is a large-scale multipurpose dam projects with comprehensive utilization of irrigation, shipping and water supply. Baise Multipurpose Dam Project is located in Youjiang river section 22 km upstream of Baise city. The rainwater collection area above the dam site is 19 600 km^2. The annual average runoff is 263 m^3/s; The annual runoff is 8.29 billion m^3. The reservoir has a total storage capacity of 5.6 billion m^3, in which the flood control capacity is 1.64 billion m^3. The Multipurpose Dam Project consists of RCC gravity dam, underground powerhouse system, two auxiliary dams and navigation structure. The maximum dam height of RCC gravity dam is 130 m. Total length of dam crest is 720 m. There are 4 surface holes, 3 middle holes and 1 bottom hole on the dam. The downstream of the main dam is a reinforced concrete stilling basin, 127 m long, 88 m wide. The riverbed downstream of stilling basin is provided with apron. There are about 2.61 million m^3 of concrete quantities in this project. Among them, the RCC of the main dam is 2.1 million m^3, cofferdam RCC is 80 000 m^3, the normal concrete of stilling basin floor and guide wall is about 240 000 m^3. The duration of the RCC main dam of Baise Multipurpose Dam Project was from January 1, 2002 to June 30, 2006. The total construction duration was four and a half years, totally 54 months.

Diabase aggregate is used in RCC main dam of Baise Multipurpose Dam Project. Diabase belongs to hard rock according to rock classification, has high strength, high hardness and brittleness. Due to the lithologic characteristics of diabase itself, the gradation of diabase manufactured sand (hereinafter referred to as diabase sand) is discontinuous (lack of intermediate grade). The content of rock powder is high (particles with size less than 0.16 mm account for 20% -24%). There are many micro particles in the rock powder (the particles

Table 7.15 Recommended construction mix proportion of RCC for main dam of Kalasuke Multipurpose Dam Project

S/N	Pouring location	Concrete grade	Gradation	Mix proportion parameter							Material amount (kg/m³)								Grout-mortar ratio PV
				Water-cement ratio	Sand ratio %	Water reducing agent (%)	Air entraining agent (%)	Cement (%)	Fly ash (%)	Rock powder (%)	Water content	Cement	Fly ash	Rock powder	Sand	Coarse aggregate			Water reducing agent	Air entraining agent	
																Small-sized stone	Medium-sized stone	Large-sized stone			
K1	Water level change area above 650 m of upstream face of the dam	R₁₈₀20 F300W10	II	0.45	35	1.0	0.120	60	40	0	98	131	87	0	708	526	789	0	2.183	0.262	0.41
K2	Above the upstream water level change area	R₁₈₀20 F200W6	III	0.45	32	0.9	0.070	50	50	0	90	100	100	0	663	423	564	423	1.800	0.140	0.41
K3	Outside the dam below the upstream dead water level	R₁₈₀20 F100W10	II	0.47	36	1.0	0.060	50	50	0	95	101	101	0	744	529	794	0	2.020	0.121	0.39
K4	Dam inside below 650 m	R₁₈₀20 F50W4	III	0.52	33	0.8	0.040	35	65	0	95	64	119	0	678	432	576	432	1.464	0.073	0.41
K5		R₁₈₀20 F50W4	III	0.54	33	1.0	0.040	40	60	8.0	90	67	100	57	654	433	577	433	1.667	0.067	0.43
K6		R₁₈₀15 F50W4	III	0.54	32	0.8	0.040	35	65	0	95	62	114	0	680	434	578	434	1.408	0.070	0.40
K7	Dam inside above 650m	R₁₈₀15 F50W4 (0.08 rock powder)	III	0.56	33	1.0	0.040	40	60	8.0	90	65	96	57	656	434	579	434	1.607	0.064	0.42
K8		R₁₈₀15 F50W4 (0.16 rock powder)	III	0.56	32	0.9	0.030	40	60	16	90	65	96	111	580	441	587	441	1.446	0.048	0.45

Note: Tianshan P · O 42.5 cement, manasi fly ash, natural washed sand of Gobi desert, manufactured crushed stone of gneiss granite, PMS-3 retarding and high efficiency water reducing agent, PMS-NEA3 air entraining agent were used. 29 kg/m³ of sand is replaced by rock powder with particle size less than 0.08 mm.

Chapter 7 Research and Utilization of Rock Powder in RCC

with size less than 0.08 mm account for 40% - 60% of the rock powder). In the test and study of RCC mix proportion of main dam project, the Joint Venture of Minjiang-Yellow River Hydropower Project found that: the water consumption, setting time, working performance, strength and drying shrinkage of RCC were different from those of RCC in other projects. Especially, the setting time of RCC was seriously shortened. This was unprecedented in the RCC projects built and under construction in China. After investigation and analysis item by item: Diabase aggregate was the main cause of the above problems of RCC. Due to the application of diabase aggregate in mass RCC, Baise project is the first case no matter at home and abroad and there is no experience to learn from. Therefore, it is necessary to conduct an in-depth test study on the performance of diabase RCC and the utilization of rock powder in RCC.

"The research on the utilization of diabase manufactured sand and rock powder in RCC" was undertaken by Sinohydro Bureau 4 and Yangtze River Research Institute. After a year of test research, the microscopic analysis of diabase powder, various influences of rock powder on RCC performance, setting time of RCC in high temperature climate, feasibility of using diabase sand and rock powder, RCC with diabase sand mixed with river sand and GEVR mixed with fiber were studied respectively. A large number of test data had been obtained to understand the microstructure characteristics and properties of diabase and its rock powder. The setting time problem of RCC construction in high temperature climate was solved and various optimization schemes of diabase aggregate RCC were put forward, which provided a scientific basis for the high-strength and rapid construction of the RCC main dam of Baise Multipurpose Dam Project.

7.5.1 Study on Diabase Sand and Rock Powder

7.5.1.1 Diabase Sand

In this part of the test study, sand samples with four kinds of different rock powder content were used for inspection, among them, No. 1 was washed to remove rock powder; No. 4 was used in construction (rock powder content 21.6%); No. 2 and No. 3 were obtained by consciously sampling from different parts (the content of rock powder was 13.2% and 17.1% respectively). The quality inspection results of diabase manufactured sand are shown in Table 7.16.

Table 7.16 Quality inspection results of diabase manufactured sand

S/N	Rock powder content (%)	Apparent density of dry sand in saturated surface (kg/m^3)	Dry water absorption of saturated surface (%)	Organic matter content (colorimetry)	Solidity (%)
1	0	2 950	0.8	—	—
2	13.2	2 910	1.7	—	—
3	17.1	2 890	1.9	—	—
4	21.6	2 830	2.1	Lighter than the standard color	1.2

As can be seen from the data in the Table 7.16, the content of rock powder has great influence on the saturated dry water absorption and saturated dry density of diabase manufactured sand. This is because water becomes the binder between diabase manufactured sand particles and micropowder particles in rock powder in the process of continuously mixing samples during the inspection. With the evaporation of free water on the surface, under the action of external mechanical force, micro-powder particles keep moving closer to the sand particles, finally are wrapped on the surface of sand grains, become part of the sand. Therefore, the measured dry water absorption rate of saturated surface, in addition to the increase of surface area of manufactured sand caused by micro powder particles, also includes film water and capillary water needed by the micro powder adhering to sand particle. Therefore, the dry water absorption rate of saturated surface increases with the increase of rock powder content. Then, the dry density of saturated surface decreases with the increase of rock powder content.

I Basic parameters of parent rock

The parent rock specimen is gray-green block. Table 7.17 lists the test results of mechanical and physical properties of diabase which were inspected according to *Standard for Tests Method of Engineering Rock Mass* GB/T 50266—1999.

Table 7.17   Test results of mechanical and physical properties of diabase

Mechanical test results					Natural moisture content (%)	Water absorption (%)	Indentation hardness value (MPa)	Platts hardness grade	Density (g/cm³)
Test condition	Force direction	Tensile strength (MPa)	Compressive strength (MPa)	Softening coefficient					
Air drying	Arbitrary	6.9	65.2	0.79	0.12	0.26	6 925	11	3.10
Saturation	Arbitrary	5.1	51.4						

II Lithofacies identification

Microscopic identification was carried out by two times of sampling, 1# sample and 2# sample respectively. 1# sample was taken from quasi-extra-large stone in diabase manufactured aggregate and 2# sample was taken from diabase parent rock sample.

1# sample is gray-green block which is mainly composed of plagioclase, pyroxene and its alteration chlorite, zoisite and other minerals. Plagioclase is subhedral lath-shaped and the particle size is mostly 0.02 mm × 0.1 mm – 0.005 mm × 0.25 mm. There are alterations such as zoisite and sericite in the particles; The refractive index $N_g = 1.541$, $N_p = 1.531$, belonging to acid plagioclase. Pyroxene are heteromorphic particles and the grain size is large. The majority is 0.5 – 4 mm. The refractive index $N_g = 1.724$, $N_p = 1.700$. the extinction angle $C \wedge N_g = 52.2°$, belonging to augite. Some of them have been eroded into hornblende and chlorite. Plagioclase is often embedded in pyroxene, forming an embedded crystal containing long structure. Chlorite becomes green scales and particle size is 0.01 – 0.15 mm. It is a pyroxene alteration. The hornblende is green needle-shaped and particle size is 0.02 mm × 0.05 mm – 0.1 mm × 0.36 mm. It is often embedded in pyroxene grains or their edges and belongs to secondary hornblende. Ilmenite is heteromorphic and semi-autogenous granular; Particle size is 0.01 – 1.2 mm; the majority is 0.5 – 1 mm. Plagioclase is often embedded in larger particles. Under oxidation, some have been decomposed into hematite and white titanite. Zoisite is fine granular and sericite is in the form of microscopic scales; particle size is ≤0.002 – 0.02 mm. Calcite is in the shape of tiny veins and interspersed in rocks. The vein width is 0.03 – 0.25 mm. Auxiliary minerals such as apatite and magnetite are fine grains, about 0.02 – 0.1 mm, in scattered distribution. The rock structure is an embedded crystal ophitic texture.

2# sample is dark green block and is mainly composed of plagioclase, pyroxene and its alteration zoisite. Plagioclase is lath-shaped and particle size is mostly 0.2 mm × 1 mm – 1.4 mm × 9 mm, widely becomes Zoisite and sericitization. The refractive index $N_g = 1.540$, $N_p = 1.530$, it belongs to An5 plagioclase. Pyroxene is brown heteromorphic and semi-autogenous short columnar, particle size is (0.5 – 2.5) mm × 5 mm, mostly 13 mm. The refractive index $N_g = 1.724$, $N_p = 1.700$. The extinction angle $C \wedge N_g = 47°$, belonging to augite. Zoisite is petrosiliceous and particle size is 0.002 – 0.005 mm. Limonite is heteromorphic granular, about 0.01 – 0.3 mm. It is the oxidation product of magnetite and ilmenite. Sericite is in the form of

Chapter 7 Research and Utilization of Rock Powder in RCC

microscopic scales and particle size is ≤0.002 − 0.02 mm. Apatite is needle-columnar and particle size is 0.01 mm ×0.17 mm − 0.24 mm ×0.3 mm. Calcite is heteromorphic granular and particle size is 0.01 − 1 mm, local filling distribution. The grain size of this specimen is coarser than that of common diabase. It is a transitional rock between diabase and gabbro. Its structure is residual hypidiomorphic texture.

See Table 7.18 for the main mineral components observed by microscope.

Table 7.18 Main mineral components observed by microscope

Sample No.	Mineral name	Content (%)	Mineral name	Content (%)	Mineral name	Content (%)
1# sample	Plagioclase	39	Ilmenite	2	Calcite	0.2
	Augite	35	Sericite	1	Limonite and hematite	0.2
	Chlorite	14	Magnetite	0.1	Apatite	<0.1
	Hornblende	2	Leucoxene	0.3		
	Zoisite	6	Biotite	≤0.1		
2# sample	Plagioclase	32	Limonite	4	Calcite	0.3
	Pyroxene	34	Sericite	1	Actinolite	0.2
	Zoisite	28	Apatite	≤0.5	Leucoxene	≤0.1

Ⅲ Grain composition

See Table 7.19 for the grain gradation of diabase manufactured sand.

Table 7.19 Grain gradation of diabase manufactured sand

	Grain composition							Fineness module	Rock powder content (%)	
Screen size(mm)	10.0	5.0	2.5	1.25	0.63	0.315	0.16		Dry process	Water washing
Sub-total screen residue(%)	0	6.5	25.9	15.5	12.1	14.1	8.2	2.83	17.7	21.6
Accumulated screen residue(%)	0	6.5	32.5	48	60.1	74.2	82.3			

Its particle distribution characteristics are as follows: grain gradation is discontinuous and is polarized, lack of intermediate grade. The content of rock powder is large. The content of rock powder is 17.7% when it is screened by dry method and is 21.6% when screened with water washing.

Ⅳ Rock powder content

Three kinds of rock powders (<0.16 mm particles) are used in this part of the test.

The first one is directly obtained by dry screening from diabase manufactured sand. Rock powder content is 17.7%, called as diabase powder. Among them, the micro-powder particles may be less than the actual content. What needed to be noted is that, diabase powder can be obtained by dry method or water washing, but when washing it with water, a large number of suspended particles remain in the sieve washing liquid and are difficult to precipitate after several days. Moreover, diabase manufactured sand and rock powder of Baise Project and rock powder of other projects, such as granite powder and limerock powder, in addition to the above-mentioned high content of micropowder, this kind of rock powder is soaked in water and then left still to dry, then can be formed into hard blocks with considerable strength. It is hard to break by hands and feet kicking. In addition, at the early stage of the project, when hand grasps wet diabase manufactured sand and is washed with water, there is obvious greasy feeling on hands. In order to keep the possible hydration components in diabase powder as much as possible, dry screening was selected.

The second rock powder is based on the fact that it has a great influence on the concrete performance. It should be micropowder particles smaller than 0.08 mm in rock powder. In order to deeply understand the physical and chemical properties of diabase, the micropowder is prepared by ultra-fine grinding, 2.3% sieve residue of 0.045 mm sieve, called as the ground diabase powder.

Furthermore, considering that the loss in the process of dry sieving rock powder is basically the micropowder in the rock powder. According to 21.6% rock powder of manufactured sand by water washing method, a third kind of rock powder is prepared from one or two kinds of rock powder, called as prepared diabase powder. The grain gradation of this kind of rock powder should be close to the actual situation in diabase manufactured sand.

In order to explain the characteristics of diabase powder, in this section, fly ash used in Baise Project is also taken as a reference to continue comparative analysis.

Ⅴ Micro particle content in rock powder

Rock powder is direct dry sieved and the micro particle content in rock powder is 40% – 60%.

Ⅵ Other basic parameters (water soluble products)

As mentioned above, at the beginning of the project, when hand graspes wet diabase manufactured sand and is washed with water, there is obvious greasy feeling on hands. Therefore, it is considered whether there is any substance in diabase powder that can be dissolved in water quickly, which affects the workability of RCC thereby. With the project continuing, this phenomenon no longer exists. According to the analysis of the early strength test and hydration heat test results of cement mortar, in addition to the micro-particle interstitial effect of rock powder, there may be some factors affecting the early physical properties of diabase powder mortar, and this kind of substance can dissolve quickly after meeting water. Therefore, the extraction solution of diabase powder and purified water mixed and stirred according to different proportions (weight ratio) was analyzed (because there are too many suspended substances in the solution, it needs to stand for three days before extraction). Test conditions are listed in Table 7.20. Wherein the test result has been processed with the blank sample. Figure 7.1 is the corresponding dissolution curve.

Table 7.20   Test results of dissolution physical properties of diabase powder

S/N	Diabase powder: Water	Ion content (mg/L)					
		Ca^{2+}	Mg^{2+}	Cl^-	SO_4^{2-}	K^+	Na^+
1	0.5:1	27.86	2.11	3.36	36.13	34.75	29.47
2	1:1	38.30	9.50	23.62	67.65	11.74	40.99
3	1.5:1	38.30	7.29	21.20	105.94	13.06	56.87
4	2:1	41.79	8.45	21.78	129.90	13.86	61.63

Because the dissolution rate changes greatly, it is more active and should have a strong influence on early performance. It is observed that the dissolution of SO_4^{2-} and Na^+ ions among them is changed the most. But SO_4^{2-} can adjust the setting time in cement-water system as the function of delaying hydration process. While Na^+ can accelerate the early hydration of cement and affect the adaptability of cement and admixture, therefore, the influence of Na^+ is analyzed emphatically.

With reference to *Test Method for Water Requirement of Normal Consistency, Setting Time and Soundness of the Portland Cement* GB/T 1346—2001, three groups of tests on setting time of clean mortar are carried

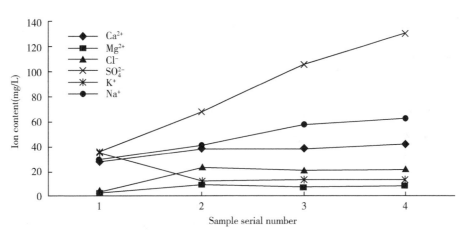

Figure 7.1 *Ion dissolution curve of diabase powder in water*

out. Although the initial setting time is difficult to grasp in the test, it can be thought that the Na^+ concentration in water increases within a certain range, and has little effect on the initial setting time of cement mortar and cement fly ash mortar mixed with 0.8% ZB-1RCC15. The final setting time is shortened.

7.5.1.2 Rock Powder and Fly Ash

Ⅰ Microscopic analysis and X diffraction analysis of particle morphology and mineral composition

1. Microscopic analysis of particle morphology and mineral composition of rock powder. For being convenient for comparison, morphology photos of Youjiang 525[#] moderate heat Portland cement particles, rock powder of limestone manufactured sand and rock powder of granite manufactured sand under scanning electron microscope are provided. Four kinds of micropowder are all polygonal granules, diabase powder, limerock powder and cement have similar particle shapes (see Figure 7.2 to Figure 7.4, Figure 7.7 and Figure 7.8). Granite powder has the clearest edges and corners (see Figure 7.10 to Figure 7.12). According to the micrograph, diabase powder particles are very fine, most of them are micron (see Figure 7.2, Figure 7.3, Figure 7.5 to Figure 7.8). Diabase powder screened in the original state has a tendency to attract and bond with each other (see Figure 7.2 and Figure 7.3) due to the influence of surface molecular force (van der Waals force) or electrostatic force (charge generated in the production process). Diabase powder screened after wetting with water bond into spherical particle (see Figure 7.5 to Figure 7.7) due to the external mechanical action, molecular adhesive force, capillary adhesive force and internal friction force are generated among particles. However, limerock powder and granite powder have no such phenomenon.

Figure 7.2 *SEM photos of diabase powder, very fine micro powder particles*

Figure 7.3 *SEM photos of diabase powder, extremely fine polygonal micro powder*

Figure 7.4 *SEM photos of Youjiang No. 525 moderate heat cement particles*

Figure 7.5 *SEM photos of diabase powder, polygonal particles*

Figure 7.6 *SEM photos of polygonal particles of wet screened diabase powder and solid spherical particles formed by micro powder*

Figure 7.7 *SEM photos of diabase powder by wet screening, forming polygonal particles and solid spherical particles by micro powder*

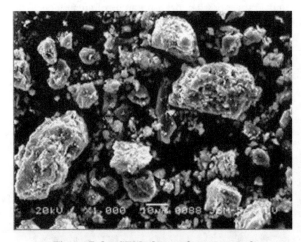

Figure 7.8 *SEM photos of wet screened limerock powder*

Figure 7.9 *SEM photos of wet screened limerock powder*

Figure 7.10 *SEM photos of granite powder, particles with sharp edges and corners*

Figure 7.11 *SEM photos of granite powder, particles are independent of each other*

2. X-ray diffraction identification. Figure 7.12 is the X-ray diffraction pattern of diabase. The main minerals are augite, albite, potash feldspar, hornblende, tremolite, mica and chlorite etc.

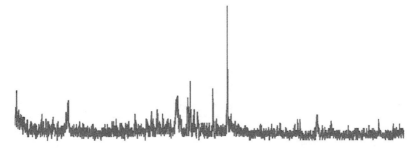

Figure 7.12 *X-ray diffraction pattern of diabase*

Ⅱ Chemical composition

Chemical composition of diabase powder is shown in Table 7.21.

Table 7.21 Chemical composition of diabase

Sample name	Chemical composition(%)									
	Loss	SiO_2	Al_2O_3	Fe_2O_3	CaO	MgO	SO_3	K_2O	Na_2O	R_2O
Diabase	2.00	45.31	14.80	14.62	9.48	6.29	0.14	0.95	2.82	3.45

Note: Alkali content $R_2O = Na_2O + 0.658 K_2O$.

Ⅲ Particle distribution

See Table 7.22 for particle distribution measured by sieve analysis method. See Table 7.23 for particle weight frequency distribution and Table 7.24 for weight cumulative distribution measured by X-ray particle size tester. See Figure 7.13 and Figure 7.14 for the corresponding figures.

Table 7.22 Particle distribution measured by sieve analysis method

Screen size (mm)	0.16	0.125	0.10	0.08	0.061	0.045	0.038	0.031	<0.031
Sub-total screen residue(%)	0	1.2	3.7	9.7	15.6	10.2	9.1	8.8	41.7
Sub-total screen residue(%)	0	1.2	4.9	14.6	30.2	40.4	49.5	58.3	100

Table 7.23 Particle weight frequency distribution measured by X laser particle size tester

Grain size range (μm)	Weight frequency distribution(%)				
	Qujing Grade II fly ash	Panxian Grade II fly ash	Youjiang 525# moderate heat portland cement	Fine grinding diabase powder	Diabase powder
0.50	1.79	2.04	2.07	4.50	3.14
1.60	4.67	5.28	5.22	11.48	8.15
2.38	5.67	6.55	5.72	12.25	8.56
3.53	7.59	9.11	6.45	12.60	7.94
5.24	11.44	13.82	8.83	15.94	8.25
7.78	14.19	16.75	10.99	17.96	7.53
11.55	14.91	17.02	13.30	15.53	7.12
17.15	13.98	15.02	15.74	8.14	7.94
25.46	11.56	10.22	15.46	1.60	8.79
37.79	8.29	3.76	10.88	0	9.73
56.09	4.57	0.44	4.63	0	10.08
83.26	1.28	0.01	0.70	0	7.67
123.59	0.07	0	0	0	3.69
183.44	0	0	0	0	1.29
272.31	0	0	0	0	0.12

Table 7.24 Particle weight cumulative distribution measured by X laser particle size tester

Grain size range (μm)	Weight cumulative distribution (%)				
	Qujing Grade II fly ash	Panxian Grade II fly ash	Youjiang 525# moderate heat portland cement	Fine grinding diabase powder	Diabase powder
0.50	1.79	2.04	2.07	4.50	3.14
1.60	6.46	7.22	7.29	15.98	11.29
2.38	12.13	13.87	13.01	28.23	19.85
3.53	19.72	22.98	19.46	40.83	27.79
5.24	31.16	36.80	28.29	56.77	36.04
7.78	45.35	53.55	39.28	74.73	43.57
11.55	60.26	70.57	52.58	90.26	50.69
17.15	74.24	85.59	68.32	98.40	58.63
25.46	85.80	95.81	83.78	100	67.42
37.79	94.09	99.57	94.66	100	77.15
56.09	98.66	100	99.29	100	87.23
83.26	99.94	100	100	100	94.90
123.59	100	100	100	100	98.59
183.44	100	100	100	100	99.88
272.31	100	100	100	100	100.00

IV Density

The density of diabase powder is 3.09.

V Water demand

The water demand test is carried out with reference to the water demand ratio test method in *Technical Code for*

Chapter 7 Research and Utilization of Rock Powder in RCC

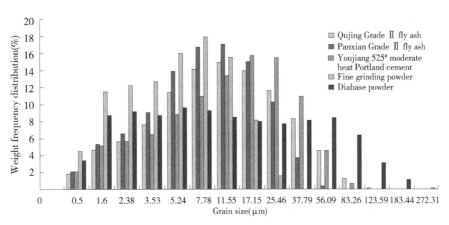

Figure 7.13 *Particle weight frequency distribution measured by X laser particle size tester*

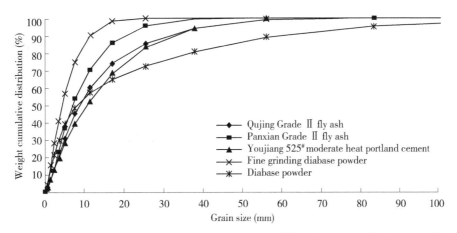

Figure 7.14 *Particle weight accumulation curve measured by X laser particle size tester (local)*

Mixing Fly Ash in Hydraulic Concrete DL/T5055—1996. The amount of diabase powder is the same as that of fly ash and the test results are shown in Table 7.25.

Table 7.25　　　Influence of diabase powder and fly ash on water demand of materials

Sample name	Youjiang 525# moderate heat portland cement	Fine grinding diabase powder	Diabase powder	Qujing Grade II fly ash	Panxian Grade II fly ash	Shimen Grade II fly ash	Xuanwei Grade II fly ash
Water content (mL)	129	138	135	126	132	123	129
Mobility (mm)	125	126	125	127	126	125	127
Water demand ratio	100	107	105	98	102	95	100

Ⅵ　Pozzolanic activity

1. Strength of mortar. The influence of diabase powder and fly ash on the strength of mortar is investigated by means of equal water-cement ratio and equal fluidity of mortar. See Table 7.26 and Table 7.27 for molding mix proportion and strength performance of mortar at various ages.

According to the test results, under the conditions of the same amount and water-cement ratio, the 7 d strength of mortar mixed with fine diabase powder is higher than that of mortar mixed with fly ash; 28 d strength is basically same; The 90 d and 180 d strength are lower than those mixed with fly ash, and there is no increase in strength during this period. The 7 d strength of mortar mixed with diabase powder is basically equivalent to that of mortar mixed with fly ash; 28 d strength is slightly lower; the 90 d strength is not only lower than the mortar mixed with fly ash and but also slightly lower than that mixed with fine diabase powder. 180 d strength is equal to the mortar mixed with fine diabase powder.

Table 7.26   Mix proportion of mortar molding

No.	Extender Variety	Extender Mixing amount(%)	Forming water-cement ratio	Remarks
BS1	Youjiang 525# moderate heat Portland cement	0	0.50	Tests are conducted according to GB/T 17671—1999
BS3	Fine grinding diabase powder	30	0.50	
BS4	Qujing Grade II fly ash	30	0.50	
BS5	Panxian Grade II fly ash	30	0.50	
BS6	Shimen Grade II fly ash	30	0.50	
BS22	Xuanwei Grade II fly ash	30	0.50	
BS23	undisturbed diabase powder	30	0.50	
BS11	Youjiang 525# moderate heat Portland cement	—	0.43	Tests are conducted according to GB 177—77
BS13	Fine grinding diabase powder	30	0.46	
BS14	Qujing Grade II fly ash	30	0.42	
BS18	undisturbed diabase powder	30	0.45	
BS31		30	0.43	

Table 7.27   Test results of mortar strength

No.	Bending strength(MPa)					Compressive strength(MPa)				
	3 d	7 d	28 d	90 d	180 d	3 d	7 d	28 d	90 d	180 d
BS1	5.8	7.0	10.2	10.9	10.6	29.0	36.3	61.1	70.2	70.8
BS3	3.9	5.2	7.9	8.4	8.7	20.7	23.4	40.2	44.5	43.3
BS4	3.4	4.9	8.7	10.4	11.5	20.1	22.5	42.5	57.8	68.4
BS5	3.7	4.9	7.7	11.2	11.3	21.3	23.0	39.5	57.5	67.2
BS6	3.5	4.3	7.0	9.4	10.5	18.8	19.6	34.1	48.4	61.0
BS22		5.1	7.7	10.8	11.8		21.6	44.3	51.8	63.5
BS23		4.7	6.8	9.0	8.4		22.0	37.2	41.6	43.2
BS11		7.7	10.2	11.3	11.3		45.5	68.2	72.8	73.7
BS13		4.9	6.6	8.1	8.1		23.4	38.7	42.4	44.3
BS14		5.1	7.7	9.5	10.8		26.5	44.9	54.6	62.4
BS18		4.6	7.0	7.6	7.9		21.0	35.0	39.8	40.8
BS31		4.7	7.1	8.1	8.6		22.0	33.7	40.8	43.8

Under the condition of the same amount and fluidity of mortar, the 7 d strength of mortar mixed with fine ground diabase powder and diabase powder is lower than that of fly ash, but the gap is not big, almost equal. With the increase of age, the gap between strength of mortar mixed with diabase powder and mortar mixed with fly ash gradually increases. After 90 d, the mortar mixed with diabase powder is as same as cement mortar; The strength increases slowly. However, the fly ash mortar continues to maintain its growth momentum.

Figure 7.15 and Figure 7.16 show that development trend curve of flexural and compressive strength of mortar under the conditions of the same amount and water-cement ratio; Figure 7.17 and Figure 7.18 show development trend curve of flexural and compressive strength of mortar under the conditions of the same amount and the same fluidity of mortar. It is observed that no matter what kind of diabase powder is mixed, 28 d later, its strength curves are nearly parallel to the development of cement mortar. Therefore, it can be affirmed that diabase powder mainly plays a role in filling dense matrix structure in mortar, but its influence on the early strength of mortar may be more than that.

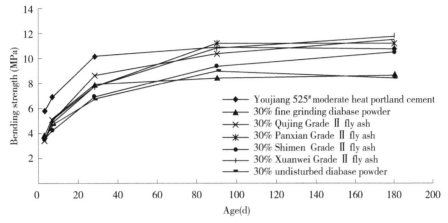

Figure 7.15 *Development curve of flexural strength with equal water-cement ratio*

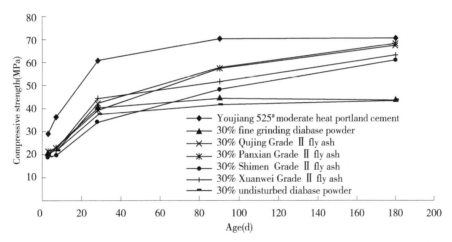

Figure 7.16 *Development curve of compressive strength with equal water-cement ratio*

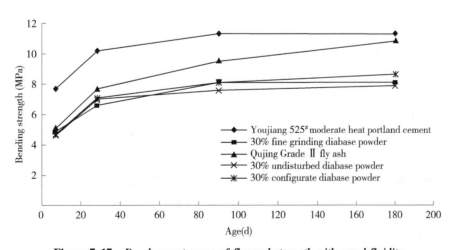

Figure 7.17 *Development curve of flexural strength with equal fluidity*

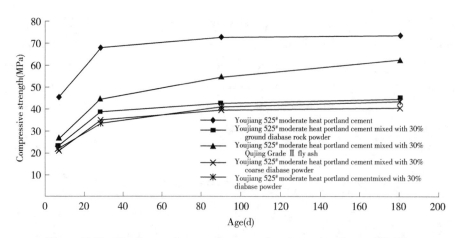

Figure 7.18 *Development curve of compressive strength with equal fluidity*

2. Hydration heat of cementing materials. The hydration heat test is conducted according to *Cement Hydration Heat Test Method (Direct Method)* GB 2022—80. The effects of rock powder and fly ash on hydration heat are investigated by using water consumption according to standard consistency and equal water-cement ratio. See Table 7.28 for test conditions and results. From Figure 7.19 and Figure 7.20, the whole process of hydration heat release within 7 d age can be observed.

Table 7.28　　　　　Test results of hydration heat of cementing materials

Sample name	Hydration heat (kJ/kg)						
	1 d	2 d	3 d	4 d	5 d	6 d	7 d
Youjiang moderate heat portland cement W/C 0.27	188	228	250	264	275	284	291
Youjiang moderate heat cement is mixed with 30% fine-ground diabase powder W/C 0.29	150	174	188	201	211	220	226
Youjiang moderate heat portland cement mixed with 30% Qujing fly ash W/C 0.262 5	138	161	175	185	193	199	204
Youjiang moderate heat portland cement mixed with 30% diabase powder W/C 0.28	131	153	169	180	188	194	199
Youjiang moderate heat cement is mixed with 30% Qujing fly ash W/C 0.3	141	167	180	189	197	205	211
Youjiang moderate heat cement is mixed with 30% fine-ground diabase powder W/C 0.3	150	174	188	201	211	220	226
Youjiang moderate heat cement is mixed with 30% diabase powder W/C 0.3	146	174	188	199	209	217	223
Youjiang moderate heat portland cement mixed with 30% diabase powder W/C 0.3	138	163	175	185	193	200	205

When molding with the same amount and water demand of standard consistency, the hydration heat of fine diabase powder is always higher than that of Qujing fly ash within 7 d age and the hydration heat of diabase powder is basically the same as that of Qujing fly ash.

When molding with the same amount and water-cement ratio, the hydration heat of ground fine diabase powder is still higher than that of Qujing fly ash within 7 d age. The hydration heat of diabase powder is similar to that of Qujing fly ash. The hydration heat measured by adding diabase powder is still higher than that by adding Qujing fly ash.

The above hydration heat test results show that the mixing of diabase powder has certain influence on the early hydration of concrete.

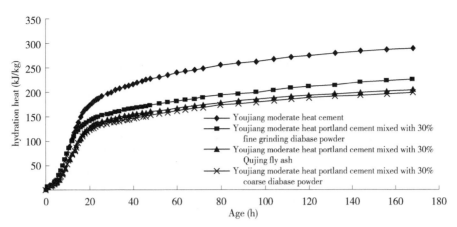

Figure 7.19 *Curve of hydration heat test results measured by standard consistency molding*

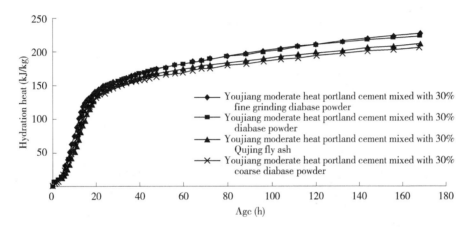

Figure 7.20 *Curve of hydration heat test results measured by equal water-cement ratio molding*

3. Time of setting. Setting time test is conducted according to *Test Method for Water Requirement of Normal Consistency, Setting Time and Soundness of the Portland Cement* GB/T 1346—2001. See Table 7.29 for test conditions and results.

Table 7.29   Setting time test results

No.	Extender		Water consumption for standard consistence (%)	Time of setting (h: min)	
	Variety	Mixing amount (%)		Initial setting	Final setting
1#	—	0	27.0	3:00	4:40
4#	Fine grinding diabase powder	30	29.0	2:13	3:50
5#	Qujing Grade II fly ash	30	26.25	3:00	4:40
6#	Diabase powder	30	23.0	2:00	3:27

The test results show that diabase powder has significant influence on setting time. This also confirms the speculation that there may be some factors affecting the early physical properties of diabase powder mortar from another aspect in Section 1.2.6.2.

Ⅶ Hydrated products (SEM microscopic image analysis)

Figure 7.21 to Figure 7.28 age are SEM photos of various mortar at the age of 28 d age. Figure 7.29 to Figure 7.31 are SEM photos of 90 d age; Figure 7.32 to Figure 7.38 are SEM photos of 180 d age.

As can be seen, after 28 d, 90 d to 180 d ages, the surface of a part of fly ash particles is gradually hydrated; the other part of fly ash particles play the role of micro-aggregate filling. However, no hydrate is found on the surface of diabase powder. It plays the role of fulcrum (see Figure 7.33 and Figure 7.38) and micro-aggregate filling in the process of hydration product extension.

According to the SEM analysis of the microstructure of mortar, it can be seen that the function of diabase powder particles in mortar is mainly micro-aggregate filling. At the same time, it plays a supporting role in the extension process of hydrated calcium silicate binder.

Figure 7.21 *SEM of hydration products of cement mortar in* 28 d

Figure 7.22 *SEM of cement mortar in* 28 d

Figure 7.23 *SEM of cement diabase powder mortar in* 28 d

Figure 7.24 *SEM of cement diabase powder mortar in* 28 d

Figure 7.25 *SEM of cement diabase powder mortar in* 28 d

Figure 7.26 *SEM of Qujing grade* Ⅱ *fly ash cement mortar in* 28 d

Figure 7. 27 *SEM of cement Qujing grade* Ⅱ *fly ash mortar in* 28 d

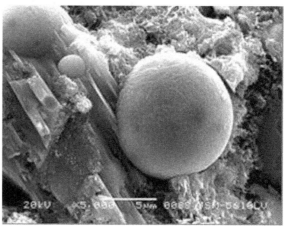

Figure 7. 28 *SEM of cement Panxian grade* Ⅱ *fly ash mortar in* 28 d

Figure 7. 29 *SEM of cement mortar in* 90 d

Figure 7. 30 *SEM of cement diabase powder mortar in* 90 d

Figure 7. 31 *SEM of cement diabase powder mortar in* 90 d

Figure 7. 32 *SEM of cement mortar in* 180 d

Figure 7.33　*SEM of cement mortar in* 180 d

Figure 7.34　*SEM of rock powder diabase cement mortar in* 180 d

Figure 7.35　*SEM of cement diabase powder mortar in* 180 d

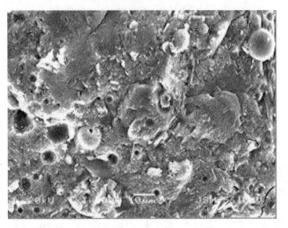

Figure 7.36　*SEM photos of cement Qujing grade* Ⅱ *fly ash mortar in* 180 d

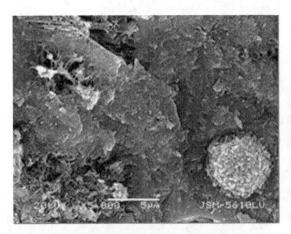

Figure 7.37　*SEM of cement Shimeng grade* Ⅱ *fly ash mortar in* 180 d

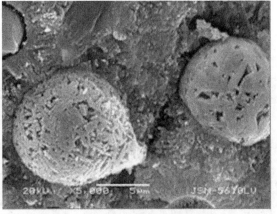

Figure 7.38　*SEM of cement Shimeng grade* Ⅱ *fly ash mortar in* 180 d

Fly ash particles not only play the role of micro-aggregate filling in the mortar, its surface layer can also participate in hydration, generating needle-like or large-particle hydration products which enhance the interfacial strength between particles in mortar. Therefore, the difference between the strength of cement diabase powder mortar and cement fly ash mortar at different ages in Section 1.2.6.1 can be explained as follows: at the early stage of hydration, when the network hydrated calcium silicate binder [C-S-H (II)] is produced in large quantities, fulcrum function of diabase powder is favorable for extending and overlapping hydration products. Therefore, the early strength of diabase powder mortar is higher than that of fly ash mortar. Later, because the surface layer of fly ash particles participates in hydration, hydration products enhance the interfacial strength between particles in mortar. At this time, diabase powder mainly plays the role of micro-aggregate filling. Therefore, the later strength of fly ash mortar is higher than diabase powder mortar.

7.5.1.3 Result Analysis

1. Diabase powder particles are very fine and most of them are micron. Its unique phenomenon is as follows: diabase powder screened in the original state tends to attract and bond with each other due to the influence of surface molecular force (van der Waals force) or electrostatic force (charge generated in the production process). Diabase powder screened after wetting with water bond into pellets due to external mechanical action, the molecular adhesive force, capillary adhesive force and internal friction force generated between particles.

2. Under the conditions of the same amount and water-cement ratio, the 7 d strength of cement mortar mixed with ground fine diabase powder is higher than that of cement mortar mixed with fly ash. The 7 d strength of cement mortar mixed with diabase powder is basically equal to that of cement mortar mixed with fly ash. The long-age strength is obviously lower than that of fly ash. Under the condition of the same amount and fluidity of mortar, the 7 d strength of cement mortar mixed with finely ground diabase powder and diabase powder is lower than that of cement mortar mixed with fly ash, but the difference is not big; with the increase of age, the difference is gradually widening. The strength development trend of cement mortar mixed with diabase powder after 28 d is basically consistent with that of No. 525 moderate heat cement. After 90 d, the strength growth trend is very limited and belongs to cement strength increase. This point is obviously different from the late reaction enhancement of fly ash. Therefore, the interstitial action of diabase powder particles can be determined.

3. Scanning electron microscope analysis shows that: the function of diabase powder particles in mortar is mainly micro-aggregate filling. It plays also a supporting role in the extension process of hydrated calcium silicate binder. Fly ash particles not only play the role of micro-aggregate filling in the mortar, its surface layer can also participate in hydration, generating needle-like or large-particle hydration products which enhance the interfacial strength between particles in mortar. Therefore, the difference of strength between cement diabase powder mortar and cement fly ash mortar at different ages can be explained as follows: At the early stage of hydration, when the network hydrated calcium silicate binder [C-S-H (II)] is produced in large quantities, fulcrum function of diabase powder is favorable for extending and overlapping hydration products. Therefore, the early strength of diabase powder mortar is higher than that of fly ash mortar. Later, because the surface layer of fly ash particles participates in hydration, hydration products enhance the interfacial strength between particles in mortar. At this time, diabase powder mainly plays the role of micro-aggregate filling. Therefore, the later strength of fly ash mortar is higher than diabase powder mortar.

4. When molding with the same amount and water demand of standard consistency, the hydration heat of fine diabase powder is always higher than that of Qujing fly ash within 7 d age. The hydration heat of diabase powder is basically the same as that of Qujing fly ash. When molding with the same amount and water-cement ratio, the hydration heat of grinding fine diabase powder is still higher than that of Qujing fly ash within 7 d age. The hydration heat of diabase powder is basically the same as that of Qujing fly ash. The hydration heat measured by adding diabase powder is also higher than that by adding fly ash. Setting time test also shows that fly ash has little influence on setting time and diabase powder has significant influence on setting time. It can be inferred from this point that in addition to the particle interstitial effect of rock

powder, there may be some factors affecting the early physical properties of diabase powder mortar.

5. Among the ions that can be dissolved in water, the dissolution amount of SO_4^{2-} and Na^+ ions changes the most. They should have a strong influence on the early performance of mortar. But SO_4^{2-} can adjust the setting time in cement-water system and delay hydration process, while Na^+ can accelerate the early hydration of cement and affect the adaptability of cement and admixture. Test shows that concentration of Na^+ increases within a certain range, it has little influence on the initial setting time of cement pure mortar and cement fly ash mortar mixed with 0.8% ZB-1A RCC ±15. The final setting time is shortened.

6. The admixture materials are divided into active and inactive types. The active admixture mainly depends on pozzolanic activity to produce hydraulic property and inactive admixture mainly plays a filling role. Judging the activity of admixtures mainly from the following four indicators:

a) The burning loss of artificial pozzolanic admixture shall not exceed 10%.

b) Sulfur trioxide content shall not exceed 3%.

c) Volcanic ash activity must be qualified.

d) 28 d compressive strength ratio of cement mortar shall not be less than 65%.

Those who meet the first two indexes are inactive admixture materials and those who meet four items can be used as active admixture.

Diabase belongs to igneous rock. The micro-rock powder is a very fine powder and the average particle size is 2 – 80 μm. It has the function of particle interstitial. According to the chemical composition table of diabase powder, diabase powder has higher silicon and oxygen compounds, and meet the requirement of the first two indexes. The pozzolanic activity of diabase powder is qualified, only the compressive strength ratio of cement mortar for 28 d does not meet the requirements. Therefore, diabase powder is an inactive admixture material.

7.5.2 Study on the Influence of Rock Powder on the Performance of RCC

August 2002, according to the mix proportion test of RCC in Baise Project conducted by Min-Huang Joint Venture Test Center, the setting time of RCC was too fast, too early and seriously shortened. The initial setting time was 2 – 4 h. This situation had attracted great attention from Youjiang company, design, supervision and joint venture. In early September 2002, Youjiang company and Min-Huang joint venture invited authoritative experts of RCC to the site for guidance. The Min-Huang joint venture test center, Guangxi Institute of Water Conservancy, and Guixiang Supervision Office of Baise Project jointly conducted the test also. The test results further confirmed that the setting time of RCC with diabase aggregate was seriously shortened. In view of the serious shortening of setting time of diabase aggregate RCC, after experts carefully analyzed and studied the issue, a detailed test plan was worked out to find out the reason by adopting different combinations of cement and fly ash varieties, aggregate types, admixture varieties and amount, etc. The influence of diabase aggregate on the performance of RCC was studied by using diabase manufactured sand with different rock powder content.

7.5.2.1 Analysis and Research on Factors Influencing Setting Time

Ⅰ Influence of rock powder content on setting time

In view of the above situation, the effect of rock powder content in diabase sand on the setting time of RCC was tested. See Table 7.30 and Table 7.31 for test results.

Result in Table 7.30 shows that: when RCC adopts manufactured sand with rock powder content of 20% and 10%, the initial setting time is 2 h 37 min, 4 h 22 min (low VC value). It shows that when the content of rock powder in manufactured diabase sand powder is low, setting time is prolonged. However, the influence of rock powder content on setting time has no qualitative change.

Chapter 7 Research and Utilization of Rock Powder in RCC

Table 7.30   Test results of setting time of diabase coarse and fine aggregates

No.	Concrete category	Water-cement ratio	Fly ash (%)	Sand ratio (%)	ZB-1RCC15 (%)	DH_9 (%)	Water (kg/m³)	VC value (s)	Time of setting (h: min) Initial setting	Time of setting (h: min) Final setting	Rock powder content (%)
NR2-0	Roller compacted concrete	0.50	60	35	—	—	125	2	3:52	7:05	20
NR2-1	Roller compacted concrete	0.50	60	38	0.6	0.015	100	3	2:37	6:30	20
NR2-2	Roller compacted concrete	0.50	60	42	0.6	0.015	100	3	4:22	7:05	10

Note: Youjiang No. 525 moderate heat cement, Qujing grade II fly ash, sand washing (residual on 5 mm sieve) + rock powder = manufactured sand with fixed rock powder content, diabase artificial coarse aggregate, ZB-1RCC15, volume weight 2 500 kg/m³, 3 – 5 s (Joint test by test center, water research institute and supervisor).

Table 7.31   Influence of artificial sand with different rock powder content on performance of RCC

Test No.	water-cement ratio	Fly ash (%)	Sand ratio (%)	Rock powder content (%)	15ZB-1 RCC15 (%)	DH_9 (%)	Water content (%)	VC value (s)	Volume weight of test piece (kg/m³)	Time of setting (h: min) Initial setting	Time of setting (h: min) Final setting	Compressive strength (MPa) 3 d	Compressive strength (MPa) 7 d
Y + XS04 – R1	0.5	60	38	14	0.6	0.015	100	7	2 662	3:00	5:30	5.0	7.0
Y + XS04 – R2	0.5	60	38	16	0.6	0.015	100	7	2 649	2:48	5:38	5.2	6.6
Y + XS04 – R3	0.5	60	38	18	0.6	0.015	100	7	2 602	2:40	5:48	5.2	6.8
Y + XS04 – R4	0.5	60	38	20	0.6	0.015	100	10	2 578	2:25	5:10	5.9	8.2
Y + XS04 – R5	0.5	60	38	22	0.6	0.015	100	13	2 622	2:08	5:25	4.6	6.0
Y + XS04 – R6	0.5	60	38	24	0.6	0.015	100	16	2 590	2:05	4:40	4.8	5.7

Note: Youjiang No. 525 moderate heat cement, Qujing grade II fly ash, diabase artificial aggregate are used; unit weight 2 500 kg/m³.

Test results in Table 7.31 show that: when manufactured sand with rock powder at 14% – 24% is used in RCC, with the increase of rock powder content, the VC value increases correspondingly and the setting time tends to shorten, but the initial setting time is still from 2 h 5 min h to 3 h and the volume weight is reduced. When the rock powder exceeds 22%, the volume weight and strength decreased obviously. When the content of rock powder increases by 1%, the water demand is correspondingly increased by 2 kg/m³ and sand ratio is correspondingly reduced by 0.5%.

II Influence of cement, fly ash, river sand and pebble on setting time

The influence of raw materials on setting time of RCC was investigated by using three kinds of cement: Tiandong No. 42.5 ordinary silicon cement, Tiandong No. 525 moderate heat cement and Shimen No. 525 moderate heat cement, Qujing and Panxian grade II fly ash, Diabase coarse and fine aggregate, Youjiang river sand and pebbles. Test condition: II gradation RCC and 0.50 water-cement ratio are used. The water consumption is 100 kg/m³; the rock powder content of diabase artificial sand is 20%; the amount of ZB-1RCC15 is 0.6%. The mix proportion parameters are basically unchanged. The test results are shown in Table 7.32.

Table 7.32   Test results of setting time of cement, fly ash, river sand and pebble

Test No.	Cement 40% + fly ash 60%	Sand variety	Coarse aggregate variety	Additive(%) ZB-1 RCC15	Additive(%) DH₉	Time of setting(h:min) Initial setting	Time of setting(h:min) Final setting	VC value (s)	Compressive strength (MPa) 3 d	Compressive strength (MPa) 7 d
NR-1	Tiandong Portland cement + Qujing	River sand	Diabase	0.6	0.015	6:05	9:26	4	5.9	7.1
NR-2	Tiandong Portland cement + Qujing	Manual	Diabase	0.6	0.015	1:35	4:08	8	3.8	4.8
NR-3	Tiandong moderate heat cement + Qujing	River sand	Diabase	0.6	0.015	6:12	9:48	4	5.6	7.8
NR-4	Tiandong moderate heat cement + Qujing	Manual	Diabase	0.6	0.015	1:58	3:56	7	4.3	5.6
NR-5	Shimen moderate heat cement + Qujing	Manual	Diabase	0.6	0.015	2:00	4:45	7	3.0	4.4
NR-6	Tiandong moderate heat cement + Qujing	Manual	Diabase	0.6	0.02	2:22	4:50	5	5.0	5.4
NR-7	Tiandong moderate heat cement + Panxian	Manual	Diabase	0.6	0.015	1:54	3:25	3	3.1	5.2
NR-8	Shimen moderate heat cement + Panxian	Manual	Diabase	0.6	0.015	2:38	5:22	5	3.1	4.3
NR-9	Tiandong moderate heat cement + Qujing	River sand	Pebble	0.6	0.015	8:20	15:30	3	—	—

Note: The ZB-1RCC15 was an unimproved product in August 2002.

The results of NR-2, NR-4 and NR-5 show that: when Tiandong No. 42.5 ordinary silicon cement, Tiandong No. 525 moderate heat cement and Shimen No. 525 moderate heat cement are used, the water consumption, bleeding and VC value of RCC do not change obviously. The initial setting time of RCC is 1 h 35 min, 1 h 58 min and 2 h 20 min respectively. The setting time of RCC with Tiandong Portland cement is short and the setting time of RCC mixed with two kinds of moderate heat cements is basically the same.

The results of NR-4 and NR-7, NR-5 and NR-8 show that: when Tiandong and Shimen moderate heat cement are combined with Qujing and Panxian fly ash, the initial setting time of RCC is 1 h 58 min, 1 h 54 min and 2 h, 2 h 38 min respectively. It shows that moderate heat cement and fly ash have little influence on setting time of RCC.

No. NR-1 and NR-3 adopt Tiandong ordinary silica cement and moderate heat cement mixed with Qujing fly ash. When replacing diabase artificial sand with Youjiang river sand, the VC value of RCC decreases obviously and the mixture has good fluidity and quick liquefaction andbleeding. When pouring the vibrated mixture out of the volume cylinder, RCC has smooth surface and the initial setting time was 6 h 5 min and 6 h 12 min respectively. The result indicates: the setting time of RCC with river sand is obviously prolonged.

No. NR-9 uses river sand + pebble natural aggregate instead of diabase coarse and fine aggregate. The setting time of RCC is obviously prolonged; the VC value continues to decrease; the liquefaction and bleeding is faster; the initial setting time reaches 8 h 20 min. Compared with RCC mixed with river sand, the initial setting

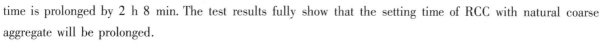

time is prolonged by 2 h 8 min. The test results fully show that the setting time of RCC with natural coarse aggregate will be prolonged.

The above test results show that: Diabase aggregate is the main factor affecting the setting time of RCC.

Ⅲ Influence of admixture on setting time of RCC

In order to analyze the influence of admixture on setting time of RCC, parallel comparative tests were carried out on the influence of different admixtures on the setting time of RCC. Test condition: Ⅱ gradation RCC mix proportion, Tiandong No. 525 moderate heat cement, Qujing grade Ⅱ fly ash, diabase aggregate, 20% rock powder content were used. Selected retarding and high efficiency water reducing agent include: modified ZB-1RCC15 (Zhejiang Longyou), JM-Ⅱ (Jiangsu Academy of Structure Science and Technology), MTG (Yunnan Green), SW (Kunming Shengwei), FDN-04 (Sichuan Panshuang), DH4A (Shijiazhuang), BD-V (liquid, Fujian Sanming) and other additives. See Table 7.33 for the test results of the variety and content of admixtures on the setting time of RCC. The test results show that:

Table 7.33　　　　　　　Test results of setting time of additive variety and content

Test No.	Water-cement ratio	Fly ash (%)	Sand ratio (%)	Rock powder content (%)	Retarding high-range water reducing agent		Air entraining agent DH_9 (%)	VC value (s)	Time of setting (h:min)	
					Variety	Mixing amount (%)			Initial setting	Final setting
NR4-1	0.50	60	38	20	ZB-1RCC15	0.8	0.015	6	5:50	9:00
NR4-2	0.50	60	38	20	ZB-1RCC15	1.0	0.015	4	7:15	11:25
NR4-3	0.50	60	38	20	ZB-1RCC15	1.3	0.015	5	8:30	13:00
NR4-4	0.50	60	38	20	ZB-1RCC15	1.5	0.015	5	10:12	15:50
NR4-5	0.50	60	38	20	JM-Ⅱ	0.8	0.015	4	5:48	8:50
NR4-6	0.50	60	38	20	DH4-A	0.8	0.015	5	2:40	4:30
NR4-7	0.50	60	38	20	FDN-04	0.8	0.015	6	3:20	5:20
NR4-8	0.50	60	38	20	SW	0.8	—	7	2:10	3:50
NR4-9	0.50	60	38	20	MTG	0.8	—	7	2:30	4:05
NR4-10	0.50	60	38	20	BD-V (liquid)	1.6	—	6	3:40	5:45

Note: ZB-1RCC15 and JM-Ⅱ were the improved products in September 2002.

When the amount of retarding and high efficiency water reducing agent is 0.8%, the initial setting time of RCC from short to long is as follows: SW, MTG, DH4-A, FDN-04, BD-V (liquid), setting time was 2 h 10 min, 2 h 30 min, 2 h 40 min, 3 h 20 min and 3 h 40 min respectively. With the content of ZB-1RCC15 and JM-Ⅱ at 0.8%, the initial setting time was 5 h 50 min and 5 h 48 min respectively. The retarding effect was obvious.

The results of NR4-1 – NR4-4 show that: When the content of ZB-1RCC15 is increased from 0.8% to 1.0%, 1.3% and 1.5%, the initial setting time of RCC is 5 h 50 min, 7 h 15 min, 8 h 30 min and 10 h 12 min respectively. The result shows that increasing the content of ZB-1RCC15 can obviously delay the initial setting time of RCC.

The above-mentioned comparative investigation results of raw materials fully show that cent and fly ash have little influence on the setting time of RCC. Diabase aggregate is the main factor causing serious shortening of setting time of RCC. Choosing the admixture suitable for diabase aggregate and properly increasing its content can delay the setting time of RCC.

7.5.2.2 Diabase Manufactured Sand with Different Rock Powder Content

Due to the lithology of diabase, the rock powder content of diabase sand produced is very high. In order to study the influence of rock powder on the performance of RCC, diabase sand with different rock powder contents was prepared. Several different tests were used to prepare diabase sand with different rock powder content, such as negative pressure dry screening method, rock powder mortar setting and drying method, washing undisturbed diabase sand and other test methods. All of them are not ideal. On the basis of several test methods, by analysis and comparison, it is easy to make water wash sand from original diabase sand, that is, the original diabase sand and water are put into a mixer according to a certain quantity, stirred according to the set time, then slowly pour out the muddy water of rock powder from the mixer. When washing with water and stirring twice, the rock powder smaller than 0.16 mm is basically washed away. In this way, the original diabase sand is washed with water to make diabase sand which is basically free of rock powder. Then, the washed sand and the original diabase sand are mixed according to a certain proportion. Diabase sand with different rock powder contents had been prepared.

See Table 7.34 for the gradation test results of diabase sand with different rock powder contents. Results show that: With the decrease of rock powder content, the fineness modulus increases accordingly. It can be found from the grading curve that diabase sand, regardless of the content of rock powder, is lack of intermediate gradation. Particles larger than 2.5 mm are generally more than 35%. It fully shows that due to the characteristics of diabase, the gradation of diabase manufactured sand produced by Barmac dry method is not ideal.

Table 7.34 Test results of gradation of diabase sand with different rock powder content

Content of rock powder of mix proportion (%)	Screen residual at all sieve meshes (%)							Fineness module	Content of rock powder measured (%)
	10.0 (mm)	5.0 (mm)	2.5 (mm)	1.25 (mm)	0.63 (mm)	0.315 (mm)	0.16 (mm)		
Original sand	0	4.48	23.34	12.52	14.48	13.62	6.58	2.6	25.0
Washed sand	0	6.60	33.12	15.14	16.98	15.96	7.40	3.4	4.8
10	0	6.56	30.32	14.78	16.56	15.16	6.18	3.2	10.4
12	0	6.64	30.08	15.04	16.04	14.26	5.24	3.2	12.7
14	0	9.45	32.12	12.61	14.20	10.86	6.80	3.1	14.0
16	0	6.55	32.56	14.28	14.52	10.15	5.63	3.1	16.3
18	0	8.22	29.89	13.06	14.09	12.05	4.80	3.0	17.9
20	0	7.91	28.68	13.13	13.04	12.11	5.03	2.9	20.1
22	0	4.57	25.23	14.19	14.93	13.42	5.80	2.7	21.9
24	0	4.20	22.09	13.22	15.67	13.83	7.20	2.6	23.8

7.5.2.3 Study on Different Rock Powder Content and RCC Performance

In order to study the influence of diabase aggregate on the properties of RCC, determine the optimum rock powder content of diabase RCC, test studies on the properties, mechanical properties, durability, deformation properties and drying shrinkage of RCC mixtures mixed with diabase sand with different rock powder contents (14%, 16%, 18%, 20%, 22%, 24%) were carried out.

Test condition: Tiandong No. 525 moderate heat cement, Qujing Grade II fly ash, Diabase coarse and fine aggregate, retarding high efficiency water reducing agent ZB-1RCC15 are used. The air entraining agent is DH_9; aggregate gradation: small stone: middle stone: quasi large stone = 30:40:30. See Table 7.35 for test mix proportion parameters of quasi-III gradation RCC.

Chapter 7 Research and Utilization of Rock Powder in RCC

Table 7.35   Test mix proportion parameters of quasi-III gradation RCC

Test No.	Level	Water-cement ratio	Sand ratio (%)	Fly ash (%)	ZB-1 (%)	DH$_9$ (%)	Water content (kg/m^3)	Unit weight (kg/m^3)	VC value (s)	Rock powder content (%)
KF1-9	R$_{180}$15S2D50	0.60	34	63	0.8	0.015	106-94	2 650	3-8	14-24
KF2-9	R$_{180}$15S2D50	0.60	34	63	1.0	0.015	103-91	2 650	3-8	14-24

I Influence of different rock powder content on water consumption, VC value and air content of RCC

Adopting diabase manufactured sand with different rock powder content, the effect of different rock powder content on water consumption and VC value of RCC was tested. See Table 7.36 and Table 7.37 for test results. Test data Nos. kf1-1, KF1-2, KF1-3, KF1-4 and KF1-5 show that: With 0.8% of ZB-1RCC15, the content of rock powder is reduced from 24% to 16%; the water consumption is correspondingly reduced from 106 kg/m^3 to 94 kg/m^3. Test data Nos. KF2-1, KF2-2, KF2-3, KF2-4 and KF2-5 show that: with 1.0% of ZB-1RCC15, the content of rock powder is reduced from 24% to 16%; the water consumption is correspondingly reduced from 103 kg/m^3 to 91 kg/m^3. The result shows that: the rock powder content of manufactured sand has great influence on the water consumption of RCC. With the decrease of rock powder content of manufactured sand, the total surface area of RCC decreases correspondingly and the water consumption decreases regularly. That is, every 1% reduction in rock powder content of diabase sand, the water consumption of RCC is correspondingly reduced by about 1.5 kg/m^3.

According to the test data Nos. KF1-7, KF1-8 and KF1-9, when the original diabase manufactured sand is used and the content of rock powder is 23.7%, the amount of ZB-1RCC15 is 0.8%. When the water consumption is reduced from 106 kg/m^3 to 100 kg/m^3, the VC value increases from 4.0 s to 8.0 s correspondingly. The test results of Nos. KF2-7, KF2-8 and KF2-9 also show that when content of ZB-1RCC15 is 1.0% and the water consumption is reduced from 97 kg/m^3 to 91 kg/m^3, the VC value increases from 4.0 s to 7.6 s. The result indicates: the water consumption has great influence on VC value. When the VC value of RCC increases or decreases by 1 s, the water consumption is correspondingly reduced by about 1.5 kg/m^3.

Test data Nos. KF1-1, KF1-2, KF1-3, KF1-4 and KF1-5 show that: when the content of rock powder decreases from 24% to 16% and the water consumption is correspondingly reduced from 103 kg/m^3 to 91 kg/m^3, the content of RCC mortar is decreasing, but the air content increases from 1.6% to 1.8%. It is fully explained that high diabase powder will reduce the air content of RCC, which is unfavorable to RCC.

II Effect of different rock powder content on setting time

See Table 7.36 and Table 7.37 for test results of RCC mixture with different rock powder content and setting time.

Test results in Table 7.29 show that: the amount of ZB-1RCC15 is 0.8%. The content of rock powder is reduced from 24% to 16%. When the indoor temperature of RCC was 16-21 ℃, the initial setting time was prolonged from 4 h 50 min to 9 h. When outdoor natural temperature is 21-31 ℃, the initial setting time was prolonged from 3 h 55 min to 6 h 12 min.

Test results in Table 7.30 show that: when the content of ZB-1RCC15 is 1.0%, the content of rock powder is reduced from 24% to 16%. When the indoor temperature is 16-21 ℃, the initial setting time is prolonged from 8 h 45 min to 10 h 55 min. When outdoor natural temperature is 21-31 ℃, the initial setting time is prolonged from 5 h to 7 h 20 min.

The result shows that: the diabase sand and powder has certain influence on the setting time of RCC. When the content of rock powder is reduced from 24% to 16%, the initial setting time is correspondingly

Table 7.36 Test results of mixture and setting time of RCC with different rock powders
(ZB-1RCC15 0.8%)

Test No.	Rock powder content (%)	Water content (kg/m³)	VC value (s)	Temperature (℃)	Concrete temperature (℃)	Air content (%)	Unit weight (kg/m³)	Liquefaction Bleeding	Condition	Temperature (℃)	Initial setting	Final setting
KF1-1	24	106	6.6	21.5	22	1.6	2 673	Quite good	Indoor	16 – 21	4:50	14:30
								General	Natural	21 – 31	3:55	8:44
KF1-2	22	103	6.3	20	20.5	1.6	2 674	Quite good	Indoor	16 – 21	6:00	14:25
								General	Natural	19 – 31	5:14	10:00
KF1-3	20	100	5.5	19	19.5	1.7	2 678	Quite good	Indoor	16 – 21	6:20	16:45
								General	Natural	16 – 31	5:50	12:35
KF1-4	18	97	6.0	19	20	1.7	2 679	Quite good	Indoor	16 – 21	7:18	15:08
								Quite good	Natural	16 – 31	6:00	13:05
KF1-5	16	94	6.6	22	23	1.8	2 684	Quite good	Indoor	16 – 21	9:00	17:25
								Quite good	Natural	16 – 31	6:12	17:56
KF1-7	23.7	106	4.0	23	24	—	—	—	Indoor	16 – 22	5:10	18:08
									Natural	17 – 30	5:13	15:27
KF1-8	23.7	103	6.0	22	23	—	—	General	Indoor	19 – 20	4:52	16:54
								Slow	Natural	20 – 33	4:35	10:25
KF1-9	23.7	100	8.0	23	24	—	—	—	Indoor	19 – 20	4:29	11:10
									Natural	20 – 33	4:15	8:30

Table 7.37 Test results of mixture and setting time of RCC with different rock powders
(ZB-1RCC15 1.0%)

Test No.	Rock powder content (%)	Water content (kg/m³)	VC value (s)	Temperature (℃)	Concrete temperature (℃)	Condition	Temperature (℃)	Initial setting	Final setting
KF2-1	24	103	5.0	22	23	Indoor	18 – 21	8:45	20:55
						Natural	19 – 31	5:00	15:45
KF2-2	22	100	5.0	23	24	Indoor	18 – 21	9:25	21:00
						Natural	19 – 32	5:42	20:00
KF2-3	20	97	5.0	23	24	Indoor	18 – 21	10:00	24:25
						Natural	19 – 32	6:08	19:32
KF2-4	18	94	6.0	22	24	Indoor	18 – 21	10:35	23:45
						Natural	16 – 32	7:00	20:30
KF2-5	16	91	6.6	20	21	Indoor	18 – 21	10:55	26:40
						Natural	16 – 31	7:20	21:05
KF2-7	23.7	97	4.0	23	24	Natural	19 – 32	9:40	25:30
KF2-8	23.7	94	5.5	22	24	Natural	19 – 32	8:16	23:20
KF2-9	23.7	91	7.6	20	24	Natural	19 – 32	7:21	18:05

prolonged by about 2 h, that is, every 1% decrease in the content of rock powder, the initial setting time is prolonged by about 15 min.

Chapter 7 Research and Utilization of Rock Powder in RCC

Ⅲ Influence of different rock powder content on strength

In order to test the influence of rock powder content on the strength of RCC, the effect of different rock powder content (24% - 14%) on the strength of RCC is tested. Test results are shown in Tables 7.31 and 7.32.

Test results in Table 7.38 show that: the amount of ZB-1RCC15 is 0.8%. When the content of rock powder is reduced from 24% to 16%, compressive strength is increased from 4.9 MPa to 6.9 MPa in 7 d. compressive strength is increased from 8.8 MPa to 11.3 MPa in 28 d. Compressive strength is increased from 19.0 MPa to 21.2 MPa in 90 d. Compressive strength is increased from 25.6 MPa to 28.8 MPa in 180 d; When the content of rock powder is reduced to 14%, the strength begins to decline.

Table 7.38　　Test results of RCC strength with different rock powders (RCC 0.8%)

Test No.	Rock powder content (%)	Water content (kg/m³)	Compressive strength(MPa)				Splitting tensile strength(MPa)	
			7 d	28 d	90 d	180 d	28 d	180 d
KF1-1	24	106	4.9	8.8	19.0	25.6	0.76	2.26
KF1-2	22	103	5.4	8.8	20.5	28.1	0.62	2.36
KF1-3	20	100	6.1	10.6	20.2	27.8	0.76	2.11
KF1-4	18	97	6.3	11.1	21.4	29.6	0.72	2.39
KF1-5	16	94	6.9	11.3	21.2	28.8	0.84	2.20
KF1-6	14	91	6.7	11.0	20.5	27.8	0.76	1.78

Test results in Table 7.39 show that: the amount of ZB-1RCC15 is 1.0%. When the content of rock powder is reduced from 24% to 16%, compressive strength is increased from 3.9 MPa to 6.7 MPa in 7 d. compressive strength is increased from 9.8 MPa to 11.5 MPa in 28 d. Compressive strength is increased from 20.2 MPa to 24.6 MPa in 90 d. Compressive strength is increased from 27.2 MPa to 30.4 MPa in 180 d; when the content of rock powder is reduced to 14%, the strength also begins to show a downward trend.

Table 7.39　　Test results of RCC strength with different rock powders (ZB-1RCC15 1.0%)

Test No.	Rock powder content (%)	Water content (kg/m³)	Compressive strength(MPa)				Splitting tensile strength(MPa)	
			7 d	28 d	90 d	180 d	28 d	180 d
KF2-1	24	103	3.9	9.8	20.2	27.2	0.67	2.31
KF2-2	22	100	4.5	10.0	21.4	29.3	0.73	2.39
KF2-3	20	97	5.3	10.3	22.4	30.2	0.69	2.45
KF2-4	18	94	6.1	11.0	24.0	30.4	0.87	2.40
KF2-5	16	91	6.7	11.5	24.6	30.4	0.88	2.48
KF2-6	14	88	5.9	10.9	23.6	29.8	0.80	2.38

The result also shows that: different rock powder content has little effect on splitting tensile strength of RCC. At the same time, with the decrease of rock powder content, the water consumption of RCC also decreases regularly.

Above results show that when the content of rock powder decreases from 24% to 16%, with the decrease of rock powder content, the strength of RCC is on the rise. When the content of rock powder is less than 14%, the strength begins to decline. The result shows that the content of rock powder has great influence on the strength. When the content of rock powder is in the range of 16% - 18%, compactness and strength of RCC are optimal.

Ⅳ Influence of different rock powder contents on impermeability and frost resistance

See Table 7.40 and Table 7.41 for 180 d impermeability and frost resistance test results of RCC with different rock powder contents. Results show that: the content of rock powder is 24% – 14%. When the admixture content is 0.8% and 1.0%, the frost resistance and impermeability indexes of RCC at 180 d age, meets the design requirements.

Ⅴ Influence of different rock powder contents on ultimate tensile and elastic modulus

See Table 7.40 and Table 7.41 for the ultimate tensile and elastic modulus test results of RCC with different rock powder contents. The test results show that:

Table 7.40   Test results of impermeability, frost resistance, extreme tension and elastic modulus of RCC with different rock powders (ZB-1RCC15 0.8%)

Test No.	Rock powder content (%)	Water content (kg/m³)	Freezing resistance 180 d	Fenetration resistance 180 d	Ultimate tension ε_p ($\times 10^{-6}$)		Compression modulus (GPa)	
					28 d	180 d	28 d	180 d
KF1-1	24	106	>D50	>S2	48	78	13.6	38.4
KF1-2	22	103	>D50	>S2	51	90	10.0	40.2
KF1-3	20	100	>D50	>S2	56	92	11.6	43.1
KF1-4	18	97	>D50	>S2	59	95	16.6	43.7
KF1-5	16	94	>D50	>S2	67	93	15.8	45.6
KF1-6	14	91	>D50	>S2	75	73	18.3	44.0

Table 7.41   Test results of impermeability, frost resistance, extreme tension and elastic modulus of RCC with different rock powders (ZB-1RCC15 1.0%)

Test No.	Rock powder content (%)	Water content (kg/m³)	Freezing resistance	Fenetration resistance	Ultimate tension ε_p ($\times 10^{-6}$)		Compression modulus (GPa)	
					28 d	180 d	28 d	180 d
KF2-1	24	103	>D50	>S2	48	82	12.8	36.7
KF2-2	22	100	>D50	>S2	53	91	11.9	38.4
KF2-3	20	97	>D50	>S2	60	94	21.8	40.2
KF2-4	18	94	>D50	>S2	67	97	24.1	44.7
KF2-5	16	91	>D50	>S2	71	96	23.7	46.2
KF2-6	14	88	>D50	>S2	72	88	24.1	45.1

When the content of rock powder is reduced from 24% to 14%, when the content of ZB-1RCC15 is 0.8%: the ultimate tensile value in 28 d increases from 48×10^{-6} to 75×10^{-6}; the tensile elastic modulus increases from 22.9 GPa to 26.6 GPa in 28 d. The ultimate tensile value in 180 d decreases from 78×10^{-6} to 73×10^{-6} and the tensile elastic modulus increases from 35.8 GPa to 37.2 GPa in 180 d; the 28 d compressive elastic modulus increases from 13.6 GPa to 18.3 GPa and 180 d compressive elastic modulus increases from 38.4 GPa to 44.0 GPa.

When the content of rock powder is reduced from 24% to 14%, when the content of ZB-1RCC15 is 1.0%: the ultimate tensile value in 28 d increases from 48×10^{-6} to 72×10^{-6} and the ultimate tensile value in 180 d increases from 82×10^{-6} to 88×10^{-6}. The 28 d tensile elastic modulus increases from 22.5 GPa to

25.3 GPa and the tensile elastic modulus increases from 36.7 GPa to 45.1 GPa in 180 d.

The above research results show that: With the decrease of diabase sand powder content, the strength and ultimate tensile value of RCC increase and the elastic modulus also increases correspondingly. When the content of rock powder is in the range of 16% – 18%, RCC has high strength, ultimate tensile value and elastic modulus. It further reflects that the rock powder content of diabase sand has great influence on the performance of RCC.

According to statistics, the ultimate tensile relationship of RCC with different rock powder contents is as follows:

$y_{28} = -2.657x + 109.819$ $y_{180} = -0.786x + 30.043x - 191.171$ (ZB-1RCC15 0.8%)

$y_{28} = -2.586x + 110.962$ $y_{180} = -0.451x + 16.448x - 53.150$ (ZB-1RCC15 1.0%)

The relation of elastic modulus is:

$y_{28} = -0.656x + 26.775$ $y_{180} = -0.640x + 54.660$ (ZB-1RCC15 0.8%)

$y_{28} = -1.346x + 45.302$ $y_{180} = -0.999x + 60.856$ (ZB-1RCC15 1.0%)

Ⅵ Study on high rock powder content and dry shrinkage performance of RCC

A large number of test data show that when the manufactured aggregate has high content of rock powder, workability of concrete is improved and bleeding is obviously reduced. However, when the humidity is low, the temperature is high and the air is dry, due to the rapid evaporation of water on the concrete surface, if without timely replenishment, the surface will produce plastic shrinkage and there are many cracks with irregular distribution. After cracks appear, the evaporation of water in concrete is further accelerated, so the cracks spread rapidly, especially when the wind speed exceeds 1.0 m/s, plastic shrinkage of concrete surface increases sharply. Therefore, for the concrete with high rock powder content, plastic shrinkage must be strictly monitored. On the concrete surface with high rock powder content, measures such as curing, moisture retention and covering must be strengthened, so as to prevent it from drying.

The rock powder content in diabase aggregate used in RCC of Baise project is high. It will adversely affect the dry shrinkage of RCC. Therefore, an test study on dry shrinkage performance of RCC with different rock powder content is carried out. See Table 7.42 for test results. Results show that: When the content of rock powder decreases from 24% to 14%.

Table 7.42 Test results of dry shrinkage performance of RCC with different rock powder contents (ZB-1RCC15 0.8%)

Test No.	Gradation	Water-cement ratio	Rock powder content (%)	Water content (kg/m³)	Drying shrinkage(×10⁻⁶)						
					3 d	7 d	14 d	28 d	60 d	90 d	180 d
KF1-1	Quasi Ⅲ	0.60	24	106	80	186	325	451	538	577	610
KF1-2		0.60	22	103	73	172	305	431	504	544	584
KF1-3		0.60	20	100	66	165	290	409	461	514	560
KF1-4		0.60	18	97	59	145	284	389	428	482	521
KF1-5		0.60	16	94	59	139	263	370	409	443	482
KF1-6		0.60	14	91	52	125	250	349	389	408	441

The 3 d dry shrinkage rate is reduced from 80×10^{-6} to 52×10^{-6}. The dry shrinkage rate is reduced by 28×10^{-6}. Every 1% reduction of rock powder will reduce the drying shrinkage by 3.5×10^{-6}.

The 7 d dry shrinkage rate is reduced from 186×10^{-6} to 125×10^{-6}. The dry shrinkage rate is reduced by 61×10^{-6}. Every 1% reduction of rock powder will reduce the drying shrinkage by 8×10^{-6}.

The 14 d dry shrinkage rate is reduced from 325×10^{-6} to 250×10^{-6}. The dry shrinkage rate is reduced by 75×10^{-6}. Every 1% reduction of rock powder will reduce the drying shrinkage by 9×10^{-6}.

The 28 d dry shrinkage rate is reduced from 451×10^{-6} to 349×10^{-6}. The dry shrinkage rate is reduced by 102×10^{-6}. Every 1% reduction of rock powder will reduce the drying shrinkage by 12.5×10^{-6}.

The 60 d dry shrinkage rate is reduced from 538×10^{-6} to 389×10^{-6}. The dry shrinkage rate is reduced by 149×10^{-6}. Every 1% reduction of rock powder will reduce the drying shrinkage by 18.6×10^{-6}.

The 90 d dry shrinkage rate is reduced from 577×10^{-6} to 408×10^{-6}. The dry shrinkage rate is reduced by 169×10^{-6}. Every 1% reduction of rock powder will reduce the drying shrinkage by 21.1×10^{-6}.

The 180 d dry shrinkage rate is reduced from 610×10^{-6} to 441×10^{-6}. The dry shrinkage rate is reduced by 169×10^{-6}. Every 1% reduction of rock powder will reduce the drying shrinkage by 21.1×10^{-6}.

The above research shows that: the content of diabase powder has great influence on the dry shrinkage performance of RCC. With the decrease of rock powder content, the dry shrinkage of RCC decreases regularly. With the extension of age, the dry shrinkage of RCC increases regularly. According to relevant data: the dry shrinkage of normal concrete is generally between 200×10^{-6} and 300×10^{-6}. The dry shrinkage of RCC generally does not exceed 300×10^{-6}; the drying shrinkage of Baise diabase manufactured aggregate RCC is relatively large and this is unfavorable to crack resistance of mass RCC.

R18015S2D50 test results of dry shrinkage performance of RCC with different rock powder content (ZB-1RCC15 0.8%)

7.5.2.4 Summary

The above research results show that:
1. Diabase aggregate is the main factor affecting the setting time of RCC and directly leads to the increase of admixture consumption, water consumption and binder consumption.
2. Due to the characteristic of diabase, lack of intermediate gradation of manufactured sand, that is the main reason to affect the workability of concrete.
3. Every 1% increase or decrease of rock powder content of diabase sand increases or decreases the water consumption of RCC by about 1.5 kg/m³ correspondingly.
4. Every 1 s increase or decrease in VC value of RCC reduces the water consumption correspondingly by about 1.5 kg/m³.
5. Every 1% reduction of diabase powder content prolongs the initial setting time by about 15 min.
6. When the rock powder content of diabase aggregate is in the range of 16% – 18%, RCC has better performance, high compactness and high strength.

When the content of rock powder is too high, the dry shrinkage of RCC increases obviously. Excessive drying shrinkage is unfavorable to the crack resistance of RCC.

7.5.3 Test Study on Diabase Powder Admixture

Diabase aggregate is used in RCC of Baise Project. Due to the characteristic of diabase aggregate, the rock powder content of manufactured sand is as high as 20% – 24%. It is far higher than the control index of 10% – 22% of rock powder content in fine aggregate of RCC as specified in *Construction Specification for Hydraulic Roller Compacted Concrete* DL/T 5112—2000. The gradation is discontinuous, coarse particles above 2.5 mm account for about 35% and 0.16 – 2.5 mm particles only account for about 40%. The coarse and fine particles are polarized. Due to the characteristics of diabase aggregate, the setting time of RCC is seriously shortened. The amount of admixture, water and binder increase sharply. In addition, the gradation of diabase manufactured sand is not ideal. The content of rock powder is high. The liquefaction of RCC mixture is slow. At the same time, the dry shrinkage of RCC is large. Due to the application of diabase aggregate in mass RCC, Baise Project is the first case no matter at home or abroad and there is no experience for reference.

Chapter 7 Research and Utilization of Rock Powder in RCC

Diabase belongs to igneous rock and has strong pozzolanic reaction and microstructure change. Baise diabase manufactured sand not only has high rock powder content and there are many micro powder particles in the rock powder. Micro-powder particles with particle size less than 0.08 mm (hereinafter referred to as micro-rock powder) account for 40% – 60% of the rock powder. The manufactured sand is made by mechanical processing. The surface of the stone chips is rough and porous, therefore, the mechanical snap-in force is strong. When cement is hydrated, alkaline solution enters the hole and a series of physical and chemical changes have taken place, therefore, it has good interfacial adhesion with cement and stone. The micro-rock powder is a very fine powder and the average particle size is 2 – 80 μm. Specific gravity is 3.09 g/cm^3.

Because of the high content of rock powder in diabase manufactured sand, the micro-rock powder has large specific surface area, has strong moisture adsorption capacity, has great influence on the workability of concrete, and can effectively reduce concrete bleeding. Fine rock powder can fill and compact concrete. In the test, rock powder is obtained by water washing. The washed rock powder mortar is settled and dried in the sun. The rock powder is bonded into sheets and blocks which have certain strength. It is not easy to break by hand or foot.

According to the characteristics of high content of micro-rock powder in diabase stone sand and its filling, compacting and reinforcing effect on RCC, in view of improving project quality and reducing project cost, it is necessary to conduct test research and verification on the utilization of diabase stone sand as admixture.

7.5.3.1 Study on Mortar Properties of Rock Powder Admixture

Ⅰ Pure mortar performance of rock powder admixture

There are many micro-rock powders in diabase sand. The micro powder particles with the particle size less than 0.08 mm account for 40% – 60% of the rock powder. In order to verify the feasibility of using micro-rock powder (particles below 0.08 mm) as admixture instead of fly ash, the pure mortar performance test of micro-rock powder admixture was carried out. Tiandong No. 525 moderate heat cement, Qujing Grade Ⅱ fly ash were used. The total amount of binders was 400 g (cement + fly ash + micro rock powder). Wherein 160 g of cement account for 40% of the binder and 240 g of admixture (fly ash + micro rock powder) accounts for 60% of the binder. The test is conducted according to the cement pure mortar test method to verify the influence of micro-rock powder as admixture on water consumption and setting time. See Table 7.43 for test results. Test data shows that: with the increase of the amount of micro-rock powder admixture (corresponding reduction of fly ash), when the content of micro-rock powder increases from 0, 24 g … 144 g, the water consumption is correspondingly increased from 100 g to 104 g. The initial setting time of pure mortar is shortened from 2 h 10 min to 0 h 32 min. The result shows that: with the increase of rock powder content, the water consumption increases and setting time is shortened.

Table 7.43 Test results of pure mortar performance of rock powder admixture

Test No.	Cement (g)	Admixture (g) 240		ZB-1 (%)	Water content (g)	Time of setting (h: min)	
		Fly ash	Micro rock powder			Initial setting	Final setting
CJ1-1	160	240	0	0.5	100	2:10	14:42
CJ1-2	160	216	24	0.5	101	2:00	12:40
CJ1-3	160	192	48	0.5	101	1:45	11:05
CJ1-4	160	268	72	0.5	102	1:40	4:25
CJ1-5	160	144	96	0.5	102	1:18	2:53
CJ1-6	160	120	120	0.5	103	0:40	2:00
CJ1-7	160	96	144	0.5	104	0:32	1:29

II Performance of mortar of rock powder admixture

In order to compare the performance of rock powder and fly ash, according to the test method of cement mortar, the performance comparison test of rock powder and fly ash mortar was carried out. Tiandong No. 525 moderate heat cement, Qujing Grade II ash, standard sand, diabase micro rock powder were used in the test. Cementing material 540 g, standard sand 1 350 g. Comparing 30% fly ash admixture and 30% micro-rock powder admixture with reference mortar, the water consumption is controlled by fluidity. See Table 7.44 for test results. The test results show that:

Table 7.44 Comparative test results of performance of rock powder mortar and fly ash mortar

Test No.	Cement (g)	Admixture(30%)		Water content (g)	Water demand ratio (%)	Time of setting (h: min)		Compressive strength (MPa)				Bending strength (MPa)			
		Coal ash (g)	Rock powder(g)			Initial setting	Final setting	3 d	7 d	28 d	90 d	3 d	7 d	28 d	90 d
CJ4-1	540	—	—	233	100	3:54	5:10	32.0	45.0	63.3	72.1	6.60	8.18	9.43	10.92
CJ4-2	378	162	—	228	97.9	4:29	5:55	20.6	29.2	48.6	70.3	4.90	6.45	9.38	10.78
CJ4-3	378	—	162	260	111.6	3:31	4:23	14.1	19.2	30.2	37.8	3.88	5.15	7.15	7.83

1. Water demand ratio: the water demand ratios of cement mortar, fly ash mortar and micro-stone mortar are 100%, 97.9% and 111.6% respectively. The results show that the water demand ratio of diabase micro rock powder is large.
2. Time of setting: the initial setting time of cement mortar, fly ash mortar and micro stone mortar is 3 h 54 min, 4 h 29 min and 3 h 31 min respectively. The result shows that fly ash can delay the setting time and rock powder can shorten the setting time.
3. Compressive strength: the compressive strength of three kinds of cement mortar, mortar mixed with fly ash and mortar mixed with micro-rock powder decreased in turn in 3 d, 7 d, 28 d and 90 d. It is considered after analysis that the large water-cement ratio is related to the large water demand ratio of rock powder.
4. Bending strength: the flexural strength of cement mortar, mortar mixed with fly ash and mortar mixed with micro-rock powder in 3 d, 7 d, 28 d and 90 d also decreased in turn, but the decrease is small. The analysis shows that the surface of rock powder is rough. Strong snap-in force and dense filling effect are beneficial to flexural strength.

III Performance test of rock powder admixture mortar

To further verify the possibility of using rock powder as admixture, Tiandong No. 525 moderate heat cement, Qujing Grade II ash, diabase stone sand, diabase micro-rock powder were used to test mortar performance with rock powder admixture. Test according to cement 216 g, 324 g of fly ash (accounting for 60% of the cementing material), diabase manufactured sand 1 350 g, 250 g of water as the benchmark; Compared with diabase mortar after replacing 10% and 20% fly ash. In which, the replacement amount is directly counted by the content of micro-rock powder in manufactured sand. Mortar performance is carried out according to hydraulic concrete test method and mortar consistency shall be controlled according to 5 – 7 cm. The test results are shown in Table 7.45. The test results show that:

1. Water demand ratio: with the increase of the amount of micro-rock powder replacing fly ash, the total surface area of mortar decreases. The water consumption of benchmark mortar, replacing 10% fly ash mortar and replacing 20% fly ash mortar is gradually reduced.
2. Time of setting: setting time of benchmark mortar, replacing 10% fly ash mortar and replacing 20% fly ash mortar is slightly shortened. Rock powder can shorten the setting time.
3. Compressive and flexural strength: under the condition of the same consistency, after replacing fly ash, the water consumption decreases, the water-cement ratio is reduced. Compressive strength and flexural strength

are improved.

Table 7.45　　　　　　　　Performance test results of rock powder admixture mortar

Test No.	Cement (g)	Admixture (30%)		ZB-1 RCC15 (%)	Water content (g)	Consistency (cm)	Time of setting (h: min)		Compressive strength (MPa)				Bending strength (MPa)			
		Coal ash (g)	Rock powder (g)				Initial setting	Final setting	3 d	7 d	28 d	90 d	3 d	7 d	28 d	90 d
CJ3-1	216	324	—	0.5	250	5–7	8:25	12:38	1.2	5.6	21.1	49.8	0.50	1.83	4.80	8.80
CJ3-2	216	310.5	13.5	0.5	240	5–7	8:24	12:18	2.6	8.2	24.0	48.6	0.72	1.93	5.52	8.77
CJ3-3	216	297	27	0.5	235	5–7	8:13	12:02	2.7	8.6	21.6	48.1	0.82	2.20	5.55	8.73

As the micro-rock powder has the function of filling and compaction, under the influence of reduced water-cement ratio, compressive strength and flexural strength increase according to benchmark mortar, replacing 10% fly ash mortar and replacing 20% fly ash mortar in turn between 3 d and 7 d to 28 d.

The active function of fly ash mainly concentrates at time after 60 d, therefore, the compressive strength and flexural strength of 90 d decrease with the increase of the amount of substitute fly ash.

Ⅳ　Performance of mortar with different rock powder contents

Diabase sand with different rock powder contents (0, 14%, 16%, 18%, 20%, 22%, 24%) is used for mortar performance test. Tiandong No. 525 moderate heat cement was used in the test, Qujing Ⅱ Grade fly ash. The amount of cementing material is 540 g, in which 40% of cement, 60% of fly ash. The amount of diabase sand with different rock powder content is 1 350 g, and mortar consistency is used to control water consumption; consistency 3–4 cm. See Table 7.46 for test results. Test data shows that:

Table 7.46　　　　　　Performance test results of cement mortar with different rock powder content

Test No.	Cement (%)	Fly ash (%)	Rock powder content(%)	ZB-1 (%)	Water content (g)	Compressive strength (MPa)				Bending strength (MPa)			
						3 d	7 d	28 d	90 d	3 d	7 d	28 d	90 d
CJ2-1	40	60	0	0.5	250	1.2	5.6	21.1	45.8	0.50	1.83	4.80	8.80
CJ2-2	40	60	14	0.5	235	2.6	8.2	24.0	48.6	0.72	1.93	5.52	8.77
CJ2-3	40	60	16	0.5	240	2.7	8.6	22.3	48.1	0.82	2.20	5.55	8.73
CJ2-4	40	60	18	0.5	246	2.4	8.0	21.6	46.1	0.98	2.22	5.45	8.37
CJ2-5	40	60	20	0.5	250	2.4	7.4	19.6	43.7	0.93	2.27	5.52	8.60
CJ2-6	40	60	22	0.5	250	2.2	7.2	19.3	39.0	0.93	2.18	5.63	8.52
CJ2-7	40	60	24	0.5	255	2.1	6.6	19.3	31.6	0.90	2.12	5.03	7.97

With the increase of rock powder content, the water consumption increases correspondingly and compressive strength decreases with the increase of rock powder content. When the content of rock powder is more than 22%, compressive strength decreases obviously, however, it has little effect on the flexural strength.

7.5.3.2　Performance of RCC with Rock Powder Admixture

This paper discusses the possibility, substitution quantity and application conditions of micro-rock powder in diabase sand as admixture in RCC, as a technical scheme to solve the problem of high rock powder content in diabase sand, improve the workability, mill ability, strength, drying shrinkage and other rollability compacted concrete, also achieve the purpose of reducing water consumption and binder consumption.

Under the condition of high rock powder, RCC adopts fly ash and micro-rock powder admixture so as to change the specific surface area and working performance of RCC. Specific surface area is the relationship between surface area and volume of particles. The larger the ratio of surface area to volume, the larger the specific surface area. This is a very important factor in the proportion of concrete. Every surface of every particle of aggregate must be wrapped by cement mortar so as to be bonded together. Different particle shapes affect the content of loose pores in aggregate, thereafter affect the proportion, mechanical properties of concrete admixture and economy of concrete. Because the ratio of surface area to volume of diabase rock powder particles is too high, the amount of fine aggregate has obvious influence on the total surface area of concrete mixture. In order to make the specific surface area reach an acceptable value, the grain shape of fine aggregate should be as close to square as possible. If the particles are not square or of equal size, this will lead to the increase of fine materials (rock powder) in sand gradation. The increase of fine particles significantly increases the specific surface area of the manufactured sand in whole, and the demand of water and cement is correspondingly increased, and the quality of concrete is reduced thereby. It is considered after analysis that when rock powder is used as admixture to replace part of fly ash, the surface area and porosity of RCC are reduced. When the water consumption remains unchanged and the sand particles are fully wrapped by cement mortar, RCC mixture liquefies quickly, rollability is good and the compactness is improved.

Performance test of RCC mixture with rock powder as admixture adopts quasi-III gradation and II gradation RCC mix proportion. Every 1% rock powder is used as admixture to replace 4 kg/m^3 fly ash (about 800 kg/m^3 sand for RCC. The micro-rock powder in the rock powder accounts for about 50%. According to the calculation, the micro-rock powder below 0.08 mm in 1% rock powder is about 4 kg/m^3). The amount of rock powder admixture is 0, 4 kg, 8 kg…40 kg to replace fly ash. The effect of different amount of rock powder admixture on the rollability compacted concrete is studied to demonstrate the feasibility and optimum content of rock powder as admixture. What needed to be explained here is that, as the rock powder is used as admixture to replace part of fly ash, it is necessary to increase diabase sand in corresponding quality, so as to ensure the density of RCC unchanged.

See Table 7.47 for test parameters of RCC with rock powder admixture. See Table 7.48 and Table 7.49 for performance test results of RCC with rock powder admixture. Results show that:

Table 7.47 Test parameter table of RCC with rock powder admixture

Test No.	Level	Water-cement ratio	Sand ratio (%)	Admixture (%) (fly ash + rock powder)	ZB-1 (%)	DH_9 (%)	Water content (kg/m^3)	Unit weight (kg/m^3)	VC value (s)
KT-1 – 10	$R_{180}15S2D50$	0.60	34	63	0.8	0.015	106 – 102	2 650	3 – 8
KT-11 – 21	$R_{180}20S10D50$	0.50	38	58	0.8	0.015	116 – 112	2 600	3 – 8

1. Water content: when 0.8% admixture is added to RCC, the VC values are similar and the water consumption decreases with the increase of the dosage of the rock powder admixture. Every 8 kg/m^3 increase of rock powder admixture reduces the water consumption by 1 kg/m^3.
2. VC value: when the substitution amount is 0, 4 kg, 8 kg…32 kg, the VC value of RCC decreases gradually with the increase of rock powder substitution. When the replacement amount of rock powder reaches 24 kg/m^3, the VC value of RCC starts to increase, workability is reduced and the setting time is shortened obviously.
3. Time of setting: under the natural condition of 22 – 26 ℃, when the fly ash is replaced by the rock powder admixture from 0, 4 kg, 8 kg … 40 kg, the initial setting time of quasi-III gradation RCC is shortened from

Chapter 7 Research and Utilization of Rock Powder in RCC

Table 7.48 Test results of setting time of RCC with rock powder admixture

S/N	Grade	Gradation	Water-cement ratio	Fly ash content (%)	Cement (kg/m³)	Fly ash (kg/m³)	Rock powder admixture (kg/m³)	W (kg/m³)	VC value (s)	Initial setting	Final setting
T-0							0	106	6	5:35	12:45
T-2							8	106	5	5:45	13:12
T-4	R₁₈₀15S2D50	III	0.60	65	60	110	16	106	5	5:18	12:32
T-6							24	106	5	4:20	10:00
T-8							32	106	6	4:25	10:15

Note: Tiandong No. 525 moderate heat cement, Qujing Grade II fly ash, on-site diabase aggregate, manufacturer improved additive ZB-1 and field additive JM-II.

Table 7.49 Performance test results of RCC with rock powder admixture

Test No.	Gradation	Water-cement ratio	Fly ash + micro rock powder (kg/m³)	Water consumption (kg/m³)	Concrete temperature (℃)	VC value (s)	Air content (%)	Liquefaction and bleeding
KT-0			112	106	18.5	6	1.5	General, general
KT-1			108+4	106	18.5	6	1.5	General, general
KT-2			104+8	106	18.5	5.6	1.5	Better, faster
KT-3			98+12	105	18	6.6	1.5	Better, faster
KT-4			94+16	105	18	6	1.6	Better, faster
KT-5	III	0.60	89+20	104	18	6.6	1.6	Better, general
KT-6			85+24	104	18	6.2	1.6	General, general
KT-7			80+28	103	18	6.7	1.6	General, general
KT-8			76+32	103	18	6.1	—	General, general
KT-9			71+36	102	18	6.6	—	General, general
KT-10			67+40	102	18	6.6	—	General, general
KT-11			135	116	18	6	—	General, general
KT-12			131+4	116	18	6	—	General, general
KT-13			127+8	116	18	5.6	—	Better, faster
KT-14			121+12	115	18	6.5	—	Better, faster
KT-15			117+16	115	18	6.0	—	Better, faster
KT-16	II	0.50	112+20	114	18	6.7	—	Better, general
KT-17			108+24	114	18	6.2	—	General, general
KT-18			103+28	113	18	6.6	—	General, general
KT-19			99+32	113	18	6.2	—	General, general
KT-20			94+36	112	18	7.0	—	General, general
KT-21			90+40	112	18	6.5	—	General, general

Note: Adopting Tiandong No. 525 moderate heat cement, Qujing Grade II fly ash and admixture, Diabase aggregate on the construction site. The powder content of diabase sand is 23.7%. The volume weight of Grade II materials is calculated as 2 600 kg/m³ and that of Grade III materials is calculated as 2 650 kg/m³.

5 h 35 min to 4 h 25 min, about 50 min shortened. It shows that with the increase of amount of rock powder, the fly ash consumption is reduced and the setting time of RCC is gradually shortened.

4. Gradation: Quasi-III gradation: small stones : medium stones : quasi-large stones = 30 : 40 : 30; Grade II gradation: small stones : medium stones = 45 : 55.

7.5.3.3 Mechanical Properties of RCC with Rock Powder Admixture

The mechanical test results of RCC with rock powder admixture are shown in Table 7.50. The test results show that: with the content of ZB-1RCC15 0.8%, under the condition of constant unit water consumption, the replacement amount of rock powder admixture ranges from 0, 4 kg, 8 kg...40 kg. The 7 d compressive strength of quasi-III gradation RCC is reduced from 5.2 MPa to 3.8 MPa. The 28 d compressive strength is reduced from 9.3 MPa to 7.9 MPa. The 90 d compressive strength is reduced from 19.6 MPa to 14.8 MPa. The 180 d compressive strength is reduced from 26.6 MPa to 19.3 MPa; the 7 d compressive strength of II gradation RCC is reduced from 9.4 MPa to 8.0 MPa. The 28 d compressive strength is reduced from 15.8 MPa to 14.1 MPa. The 90 d compressive strength is reduced from 29.4 MPa to 19.1 MPa. The 180 d compressive strength is reduced from 37.8 MPa to 23.8 MPa. It shows that with the increase of the dosage of rock powder admixture, the compressive strength of RCC decreases.

Table 7.50   Test results of mechanical properties of RCC with rock powder admixture

Test No.	Gradation	Water-cement ratio	Fly ash + micro rock powder (kg/m^3)	Water consumption (kg/m^3)	Compressive strength (MPa)					
					7 d	28 d	28 d splitting strength	90 d	180 d	180 d splitting strength
KT-0	Quasi III	0.60	112	106	5.2	9.3	0.66	19.6	26.6	2.26
KT-1			108 + 4	106	4.7	9.3	0.65	19.1	26.6	2.14
KT-2			104 + 8	106	5.3	9.1	0.70	18.6	26.4	2.14
KT-3			98 + 12	105	5.1	9.1	0.76	18.6	25.6	1.98
KT-4			94 + 16	105	4.6	9.0	0.64	18.4	24.8	1.72
KT-5			89 + 20	104	5.2	9.0	0.78	18.2	24.0	1.87
KT-6			85 + 24	104	4.5	8.8	0.72	17.6	22.8	1.56
KT-7			80 + 28	103	5.0	8.6	0.68	17.1	21.4	1.46
KT-8			76 + 32	103	3.9	8.2	0.62	16.2	20.8	1.43
KT-9			71 + 36	102	4.7	7.9	0.67	16.0	20.0	1.46
KT-10			67 + 40	102	3.8	7.9	0.58	14.8	19.3	1.38
KT-11	II	0.50	135	116	9.4	15.8	1.22	29.4	37.8	3.06
KT-12			131 + 4	116	9.0	15.8	1.18	29.0	37.5	2.98
KT-13			127 + 8	116	8.8	15.6	1.15	28.6	37.2	3.00
KT-14			121 + 12	115	8.7	15.7	1.06	28.2	36.6	2.78
KT-15			117 + 16	115	8.8	15.7	1.09	27.8	35.4	2.82
KT-16			112 + 20	114	8.8	15.4	1.18	27.0	34.7	2.80
KT-17			108 + 24	114	7.9	15.0	1.10	26.1	33.8	2.71
KT-18			103 + 28	113	7.7	15.0	1.13	25.3	31.8	2.28
KT-19			99 + 32	113	8.3	14.7	1.02	23.0	29.2	2.53
KT-20			94 + 36	112	8.0	14.3	1.20	21.7	26.4	2.40
KT-21			90 + 40	112	8.0	14.1	1.18	19.1	23.8	2.02

Note: Adopting Tiandong No. 525 moderate heat cement, Qujing Grade II fly ash and admixture, Diabase aggregate on the construction site. The powder content of diabase sand is 23.7%. The volume weight of Grade II materials is calculated as 2 600 kg/m^3 and that of Grade III materials is calculated as 2 650 kg/m^3.

7.5.3.4 Impermeability and Frost Resistance of RCC with Rock Powder Admixture

See Table 7.51 for test results of impermeability of RCC with rock powder admixture. The test results show that: the frost resistance and impermeability of RCC with rock powder admixture meet the design requirements.

Chapter 7 Research and Utilization of Rock Powder in RCC

Table 7.51   Durability test results of RCC with rock powder admixture

Test No.	Level	Gradation	Fly ash + micro rock powder (kg/m³)	Rock powder admixture (kg/m³)	Water consumption (kg/m³)	Endurance quality	
						Freezing resistance	Fenetration resistance
KT-0			112	0	106	>D50	>S2
KT-1			108 + 4	4		>D50	>S2
KT-2			104 + 8	8		>D50	>S2
KT-3			98 + 12	12	105	>D50	>S2
KT-4			94 + 16	16		>D50	>S2
KT-5	$R_{180}15S2D50$	Quasi Ⅲ	89 + 20	20	104	>D50	>S2
KT-6			85 + 24	24		>D50	>S2
KT-7			80 + 28	28	103	>D50	>S2
KT-8			76 + 32	32		>D50	>S2
KT-9			71 + 36	36	102	>D50	>S2
KT-10			67 + 40	40		>D50	>S2
KT-11			135	0	116	>D50	>S10
KT-12			131 + 4	4		>D50	>S10
KT-13			127 + 8	8		>D50	>S10
KT-14			121 + 12	12	115	>D50	>S10
KT-15			117 + 16	16		>D50	>S10
KT-16	$R_{180}20S10D50$	Ⅱ	112 + 20	20	114	>D50	>S10
KT-17			108 + 24	24		>D50	>S10
KT-18			103 + 28	28	113	>D50	>S10
KT-19			99 + 32	32		>D50	>S10
KT-20			94 + 36	36	112	>D50	>S10
KT-21			90 + 40	40		>D50	>S10

Note: Adopting No. 525 moderate heat cement, Qujing Grade Ⅱ fly ash and ZB-1RCC15 admixture, diabase aggregate.

7.5.3.5 Ultimate Tensile and Elastic Modulus Properties of RCC with Rock Powder Admixture

See Table 7.52 for ultimate tensile and elastic modulus test results of RCC with rock powder admixture. The test results show that: similar to the strength change pattern, with the increase of the dosage of rock powder admixture, the ultimate tensile value decreases accordingly, but the change is not big.

7.5.3.6 Dry Shrinkage Performance of RCC with Rock Powder Admixture

See Table 7.53 for dry shrinkage test results of RCC with rock powder admixture. It is found from the Table that: when the dosage of rock powder is increased from 0 to 40 kg/m³ (the amount of fly ash is reduced equally), the 3 d dry shrinkage of quasi-Ⅲ gradation RCC is reduced from 93×10^{-6} to 40×10^{-6}. The 7 d drying shrinkage rate decreases from 205×10^{-6} to 109×10^{-6}. The 14 d drying shrinkage decreases from 338×10^{-6} to 208×10^{-6}. The 28 d drying shrinkage decreases from 456×10^{-6} to 296×10^{-6}. The 60 d drying shrinkage decreases from 569×10^{-6} to 464×10^{-6}; The 90 d dry shrinkage rate decreases from 596×10^{-6} to 494×10^{-6}. The 180 d dry shrinkage rate decreases from 662×10^{-6} to 543×10^{-6}.

When the dosage of rock powder admixture increases from 0 to 20 kg/m³ (the amount of fly ash decreases equally), the 3 d dry shrinkage rate of Ⅱ gradation RCC is reduced from 99×10^{-6} to 70×10^{-6}. The 7 d drying shrinkage rate decreases from 218×10^{-6} to 170×10^{-6}. The 14 d drying shrinkage decreases from 356×10^{-6} to 280×10^{-6}. The 28 d drying shrinkage decreases from 475×10^{-6} to 379×10^{-6}. The 60 d drying shrinkage decreases from 594×10^{-6} to 546×10^{-6}. The 90 d dry shrinkage rate decreases from 614×10^{-6} to 566×10^{-6}. The 180 d dry shrinkage rate decreases from 673×10^{-6} to 619×10^{-6}.

Table 7.52 Test results of ultimate tensile and elastic modulus of RCC with rock powder admixture

Test No.	Level	Gradation	Fly ash + micro rock powder (kg/m³)	Rock powder admixture (kg/m³)	Water consumption (kg/m³)	Ultimate tensile value $\varepsilon_p(10^{-6})$		Compression modulus (GPa)	
						28 d	180 d	28 d	180 d
KT-0	$R_{180}15S2D50$	Quasi III	112	0	106	54	96	12.2	36.5
KT-1			108 + 4	4		50	97	17.5	35.5
KT-2			104 + 8	8		52	93	21.6	34.5
KT-3			98 + 12	12	105	47	94	13.8	33.5
KT-4			94 + 16	16		50	93	17.0	31.5
KT-5			89 + 20	20		33	88	16.2	31.0
KT-6			85 + 24	24	104	48	85	20.2	29.0
KT-7			76 + 28	28		46	78	17.5	27.0
KT-8			71 + 32	32	102	32	80	13.7	22.5
KT-9			67 + 36	36		39	76	18.5	20.0
KT-11	$R_{180}20S10D50$	II	135	0	116	43	90	21.6	44.5
KT-12			131 + 4	4		51	96	13.5	43.4
KT-13			127 + 8	8		47	95	14.5	38.8
KT-14			121 + 12	12	115	46	92	20.8	36.2
KT-15			117 + 16	16		67	95	16.8	35.9
KT-16			112 + 20	20	114	69	90	21.5	33.9
KT-17			108 + 24	24		56	86	21.5	31.6
KT-18			103 + 28	28	113	61	86	18.5	31.1
KT-19			99 + 32	32		63	80	13.5	30.0
KT-20			94 + 36	36	112	56	78	14.0	27.2

Note: Adopting Tiandong No. 525 moderate heat cement, Qujing II grade fly ash and admixture, diabase aggregate on the construction site.

Table 7.53 Dry shrinkage test results of RCC with rock powder admixture

Test No.	Level	Gradation	Fly ash + micro rock powder (kg/m³)	Rock powder admixture (kg/m³)	Water consumption (kg/m³)	Drying shrinkage ($\times 10^{-6}$)						
						3 d	7 d	14 d	28 d	60 d	90 d	180 d
KT-0	$R_{180}15S2D50$	Quasi III	112	0	106	93	205	338	456	569	596	662
KT-1			108 + 4	4		90	200	328	448	558	593	657
KT-2			104 + 8	8		86	178	310	435	544	584	643
KT-3			98 + 12	12	105	80	179	293	393	539	559	605
KT-4			94 + 16	16		70	160	281	381	527	547	587
KT-5			89 + 20	20	104	59	138	257	356	514	534	580
KT-6			85 + 24	24		50	148	248	346	505	524	574
KT-7			80 + 28	28	103	50	139	238	327	495	515	656
KT-8			76 + 32	32		40	128	228	316	484	504	554
KT-9			71 + 36	36	102	40	119	218	306	475	495	544
KT-10			67 + 40	40		40	109	208	296	464	494	543
KT-11	$R_{180}20S10D50$	II	135	0	116	99	218	356	475	594	614	673
KT-15			117 + 16	16		80	180	299	398	560	587	633
KT-16			112 + 20	20	114	70	170	280	379	546	566	619

Note: Adopting Tiandong No. 525 moderate heat cement, Qujing Grade II fly ash and admixture, diabase aggregate on the construction site. The powder content of diabase sand is 23.7%. The volume weight of Grade II materials is calculated as 2 600 kg/m³ and that of Grade III materials is calculated as 2 650 kg/m³.

The test results show that rock powder is the main factor affecting the dry shrinkage of concrete. With the increase of the dosage of rock powder admixture (the equivalent reduction of fly ash), total fine particles (particles less than 0.08 mm) in concrete are relatively reduced; The total specific surface area decreases. The corresponding water demand is reduced; moisture hydration and loss are relatively reduced; the drying shrinkage decreases.

7.5.4 Utilization of Rock Powder in RCC

7.5.4.1 Field Application of Rock Powder in RCC

Test study on performance of RCC with rock powder admixture shows that the technical scheme is feasible and meets the design and construction requirements. In April 2003, at the placing entrance of the 5th dam of the 3B-4B dam section of RCC main dam (chainage Ba 0+181.512-0+267, Elevation EL 111-114.72) and dam section 5-6 (chainage Ba 0+267.5-0+298.75, Elevation EL 111-114) close to the downstream side, and in late September 2003, at 7-8 dam section (chainage Ba 0+334-0+448, Elevation EL 132-135, on the downstream side of RCC, with volume of 25 343 m^3), the field construction of RCC with equivalent replacement of fly ash by rock powder was verified.

In April 2003, during the rolling construction of the main dam section 3B-4B, the content of rock powder reached 22%. The air temperature was high (higher than 25 ℃). Part of the particles of Ⅱ gradation RCC on the leveling surface were bound to be spherical, hard and astringent. The liquefaction of RCC after vibration and rolling was slow. This happened frequently when the content of rock powder was high as shown in Figure 7.39. Therefore, according to the research results of rock powder utilization, it was decided to adopt the scheme of replacing fly ash with rock powder in this dam section for field test. Ⅱ gradation RCC replaced fly ash equally with 20 kg rock powder (that is, 20 kg/m^3 fly ash was reduced, adding 20 kg/m^3 manufactured sand). After replacing fly ash with rock powder, the *VC* value of RCC decreased correspondingly; the aggregate was evenly distributed; the liquefaction was accelerated; the bleeding was good after rolling; the interlayer bond of RCC was improved as shown in Figure 7.40. See Table 7.54 for sampling and testing results of RCC with equivalent replacement of fly ash by rock powder.

Figure 7.39 *RCC leveling surface with high rock powder content*

Figure 7.40 *RCC leveling surface with replacement of fly ash by rock powder admixture in same quantity*

In mid-April, 2003 and late September, 2003, on-site rolling tests of quasi-Ⅲ gradation RCC rock powder were carried out at the downstream side of dam section 5-6 and dam section 7-8 respectively. The amount of rock powder replacing fly ash was 20 kg/m^3. After replacement, the *VC* value decreased; the liquefaction and bleeding were fast; the rollability was obviously improved. The results of sampling inspection at that time. Test

Table 7.54 Field inspection results of RCC with rock powder admixture replacing fly ash equally

Location	Chainage and elevation	Grade of RCC	Substitution amount of micro-rock powder (kg)	Temperature(℃)		VC value (s)	Compactness (%)	Compressive strength(MPa)			
				Temperature	Concrete temperature			7	28	90	180
Dam 5 placing entrance at 3B–4B dam section	Dam 0+181.512 – 0+267, 0–10.75 above the dam – 0+95.492 below the dam, EL 111 – 114.72	R$_{180}$20 S10D50	20	28	28	6	>98	10.0	13.8	27.2	33.2
Dam 5 placing entrance at 3B–4B dam section	Dam 0+181.512 – 0+267, 0–10.75 above the dam – 0+95.492 below the dam, EL 111 – 114.72	R$_{180}$20 S10D50	20	25	26	6	>98	10.8	14.0	25.7	30.8
5 and 6 dam section	Dam 0+267.5 – 0+298.75, 0+4.6 – 0+32.5 above dam, EL 111 – 114	R$_{180}$15 S2D50	20	27	25	4	>98	4.1	9.2	15.5	23.4
7 and 8 dam section	Dam 0+334 – 0+448, EL 132 – 135	R$_{180}$15 S2D50	20	34	24	3	>98	—	7.2	—	17.7
7 and 8 dam section	Dam 0+334 – 0+448, EL 132 – 135	R$_{180}$15 S2D50	20	34	24	3	>98	—	7.4	—	16.7

Note: II gradation RCC, water-cement ratio 0.5, fly ash admixture 58%, 108 kg/m^3 water, sand ratio 38%, binder 216 ($C=91$, $F=125$) kg/m^3, volume weight 2 600 kg/m^3.

Quasi-III gradation RCC, water-cement ratio 0.6, fly ash admixture 63%, water 96 kg/m^3, sand ratio 34%, binder 160 ($C=59$, $F=101$) kg/m^3, volume weight 2 650 kg/m^3.

results show that: After replacing fly ash with rock powder, the specific surface area of RCC is reduced and the performance of the mixture is obviously improved. The compressive strength of II gradation RCC reaches 13.8 – 14.0 MPa in 28 d, 27.2 – 25.7 MPa in 90 d and 33.2 – 30.8 MPa in 180 d; the compressive strength of quasi-III gradation RCC reaches 9.2 – 7.2 MPa in 28 d, 15.5 MPa in 90 d and 23.4 MPa in 180 d, all meet the design requirements.

7.5.4.2 Economic Benefit of Using Rock Powder in RCC

In RCC main dam project of Baise multipurpose dam project from December 28, 2002 to the end of October, 2003, about 560 000 m^3 of RCC and normal concrete had been cast, among them, RCC was about 430 000 m^3. There were about 1 120 groups of compression test pieces of RCC. The 28 d compressive strength of II gradation RCC was about 9 – 13 MPa and 180 d compressive strength was 25 – 30 MPa; the 28 d compressive strength of quasi-III gradation RCC was 7 – 9 MPa and 180 d compressive strength was about 22 – 25 MPa, meeting the design requirements ad the strength was far beyond the guarantee value.

In September 2003, a total of 295.6 m RCC core samples were taken. The longest core sample was 12.1 m and it was the longest core sample of RCC dams built in China at that time. The appearance of RCC core sample of main dam was compact and smooth in whole. The aggregate was evenly distributed and the interface combination was good. See Figure 7.41 and Figure 7.42 for details.

The research on the utilization of rock powder in RCC and the on-site rolling construction inspection

Figure 7.41 *RCC core sample of Baise main dam (length 12.1 m, diameter 250 mm)*

Figure 7.42 *Appearance of Baise RCC core sample*

proves that the technical scheme of replacing fly ash with rock powder is not only technically feasible, but also gets certain economic benefits. After analysis and comparison, the unit price of fly ash in Baise project was 155 CNY/t and 12 – 20 kg/m³ fly ash was replaced by rock powder in RCC. 1.59 – 2.64 CNY/m³ (after deducting 0.27 – 0.46 CNY/m³ of supplementary manufactured sand) was saved. According to the substitution amount of 16 kg, 2.12 CNY/m³ could be saved in RCC. There were 1 700 000 m³ of RCC that had not been poured. According to this calculation, CNY 3.604 million could be saved. The technical scheme of replacing fly ash with rock powder had obvious technical and economic effects.

7.5.4.3 Summary

1. According to the research results, Baise diabase had high content of rock powder and had filling enhancement effect. After replacing part of fly ash with micro-rock powder admixture, the specific surface area of the total volume of RCC was decreased and the workability of RCC was improved, meeting the design and construction requirements.
2. After replacing fly ash with diabase micro-rock powder, the *VC* value of RCC can be reduced. The liquefaction can be accelerated. The roll ability of RCC can be improved. The strength, frost resistance, impermeability, ultimate tensile value, axial tensile strength, static elastic modulus and other properties of RCC meet the design requirements also. Especially after replacing fly ash with rock powder admixture in same quantity, the dry shrinkage of RCC is reduced, this is beneficial to improve the crack resistance of concrete.
3. When the content of rock powder is more than 20% and the content of micro-rock powder is higher, the technical scheme of replacing fly ash with rock powder can be adopted. The equivalent replacement of fly ash by rock powder in quasi Ⅲ gradation and Ⅱ gradation RCC is calculated at 12 – 20 kg/m³ and diabase sand of corresponding quality is supplemented at same time. See Table 7.55 for construction mix ratio of

RCC with rock powder admixture. See Table 7.56, Table 7.57 and Table 7.58 for the recheck test results of mix proportion of RCC with rock powder admixture.

Table 7.55 Mix proportion of RCC with rock powder admixture

S/N	Part of works	Design requirements	Gradation	Water-cement ratio	Fly ash (%)	ZB-1 (%)	DH$_9$ (%)	VC value (s)	Material amount (kg/m³)				
									Water	Total amount of rubber material	Cement	Coal ash	Micro rock powder
TR-1	RCC in dam	$R_{180}15$ S2D50	Quasi III	0.60	63	0.8	—	3–8	98	163	60	103	0
TR-2						0.8	—	3–8	98	140	60	83	20
TR-3						1.0	—	3–8	98	140	60	83	20
TR-4	RCC on dam water front face	$R_{180}20$ S10D50	II	0.50	58	0.8	0.015	3–8	108	216	91	125	0
TR-5						0.8	0.015	3–8	108	190	91	99	26
TR-6						1.0	0.015	3–8	108	190	91	99	26

Note: 1. No. 525 moderate heat cement, grade II fly ash, diabase coarse and fine aggregate, sand FM = 2.8 ± 0.2, content of rock powder more than 20%.

2. Sand ratio 34% according to III gradation and 38% according to II gradation. Adopting density method design, 2 650 kg/m³ for III gradation and 2 600 kg/m³ for II gradation.

3. The principle that micro-rock powder replaces fly ash in equal quantity: when the content of rock powder is more than 20% and the content of micro-rock powder is high, quasi-III gradation and II gradation are considered according to the substitution amount of 12–20 kg/m³, the diabase sand of corresponding quality shall be supplemented.

Table 7.56 Test results of mechanical properties of RCC with rock powder admixture

Test No.	Gradation	Water-cement ratio	Fly ash + micro rock powder (kg/m³)	Water consumption (kg/m³)	VC value (s)	Compressive strength (MPa)					
						7 d	28 d	28 d splitting strength	90 d	180 d	180 d splitting strength
TR-1	III	0.60	103 + 0	98	8	9.4	17.4	1.14	—	34.8	2.60
TR-2			83 + 20	98	7	5.5	10.6	0.64	21.4	23.7	1.78
TR-3			83 + 20	98	3	5.9	10.4	0.56	21.2	23.8	1.84
TR-4	II	0.50	125 + 0	108	5	9.2	16.6	1.25	33.6	39.1	2.93
TR-5			99 + 26	108	5	8.0	16.6	0.71	18.0	32.8	2.22
TR-6			99 + 26	108	3	7.8	15.2	1.02	—	30.8	2.44

Table 7.57 Test results of ultimate tensile, elastic modulus and durability of RCC with rock powder as admixture (180 d)

Test No.	Level	Gradation	Fly ash + micro rock powder (kg/m³)	Rock powder admixture (kg/m³)	Water consumption (kg/m³)	Ultimate tensile $\varepsilon_p(10^{-6})$	Modulus of elasticity (10^4 MPa)	Freezing resistance	Fenetration resistance
TR-1	III	0.60	103 + 0	98	98	98.5	4.11	≥D50	>S2
TR-2			83 + 20	98		83	3.83	≥D50	>S2
TR-3			83 + 20	98		91	3.87	≥D50	>S2
TR-4	II	0.50	125 + 0	108	108	110	4.45	≥D50	>S10
TR-5			99 + 26	108		99	3.60	≥D50	>S10
TR-6			99 + 26	108		98.5	4.04	≥D50	>S10

Table 7.58   Dry shrinkage test results of RCC with rock powder admixture

Test No.	Level	Gradation	Fly ash + micro rock powder (kg/m³)	Rock powder admixture (kg/m³)	Water content (kg/m³)	Dry shrinkage value ($\times 10^{-6}$)						
						3 d	7 d	14 d	28 d	60 d	90 d	180 d
T-0	III	0.60	103 + 0	0	98	80	206	346	479	593	612	659
T-1			99 + 4	4		79	212	352	505	598	617	664
T-2			95 + 8	8		79	232	370	516	609	629	669
T-3			91 + 12	12		79	232	377	529	629	642	675
T-4			87 + 16	16		93	237	388	533	645	652	678
T-5			83 + 20	20		93	251	404	542	661	668	688
T-6			79 + 24	24		99	258	411	556	669	675	695
T-7			75 + 28	28		99	257	409	580	666	685	705
T-8			71 + 32	32		99	244	408	593	679	692	725
T-9			67 + 36	36		93	246	419	598	685	698	731
T-10			63 + 40	40		92	251	416	600	680	699	732

4. After on-site inspection, it is feasible to replace fly ash with rock powder, having obvious technical and economic effect.

7.5.5 Conclusion

1. Rock powder particles of diabase manufactured sand are very fine and most of them are micron. Its unique phenomenon is as follows: Diabase powder screened in the original state tends to attract and bond with each other due to the influence of surface molecular force (van der Waals force) or electrostatic force (charge generated in the production process). Diabase powder screened after wetting with water bond into pellets due to external mechanical action, the molecular adhesive force, capillary adhesive force and internal friction force generated between particles.

2. Scanning electron microscope analysis shows that: the function of diabase powder particles in mortar is mainly micro-aggregate filling. It plays also a supporting role in the extension process of hydrated calcium silicate binder.

3. Rock powder content has great influence on *PV* value of RCC cement-sand ratio. The *PV* value of RCC cement-sand ratio is directly related to the interlayer bonding, imperviousness performance and overall performance of RCC, generally, the ratio of cement to sand should not be lower than 0.42. For low-grade RCC with low cement consumption, the *PV* value of cement-sand ratio can be improved by increasing the content of manufactured rock powder or the amount of admixture. The function of diabase powder particles in mortar is mainly micro-aggregate filling. It plays also a supporting role in the extension process of hydrated calcium silicate binder. The content of rock powder has great influence on the quality of RCC. Too high rock powder is unfavorable for drying shrinkage.

4. When rock powder replaces fly ash, with the increase of substitution amount, its compressive strength and splitting tensile strength show a downward trend. When the substitution amount is not more than 25%, the intensity decreases little.

5. Under the condition of rich binder with a large amount of fly ash, a proper amount of rock powder replaces fly ash (not more than 20%) and has little influence on compressive strength, axial tensile strength, elastic modulus and ultimate tensile strength. Under the condition of keeping the water-cement ratio and the dosage of binder unchanged and the content of rock powder is within a certain range, it has little influence on rolling compressive strength, ultimate tensile strength and elastic modulus. When a certain value is

exceeded, with the increase of rock powder content, compressive strength, ultimate tensile strength and elastic modulus of RCC show a downward trend.

6. Through the test study on the action of rock powder in diabase manufactured sand in RCC, it shows that the content of rock powder has great influence on the quality of RCC. In the southwest region rich in hydropower, fly ash resources are in short supply, and transportation is difficult and the cost is high. It is necessary to study the role of rock powder in RCC. The research and utilization of rock powder is particularly important in terms of technology, economy and especially quality.

Chapter 8

Construction Technology of GEVR

8.1 Overview

GEVR is mortar-rich concrete formed by spreading mortar used in RCC paving construction so that it has a certain slump. GEVR is vibrated and compacted by vibrator. GEVR is mainly used in these locations such as the surface of dam seepage control area, area around formwork, water stop, bank slope, corridor, opening, slope toe rolled by inclined layer, embedded monitoring instrument and reinforced parts, etc. GEVR was first applied to the upstream face and places, where the water-stop and embedded parts were set, of Rongdi and Puding RCC dams. After adopting GEVR, the interference to RCC construction was obviously reduced.

Full-section RCC damming technology and GEVR are an important technological innovation in RCC damming technology in China. It is a new material and process with unique characteristics and intellectual property rights in the development of RCC dam construction technology in China. Puding Project in Guizhou, China pioneered the full-section RCC damming technology in 1993. The dam body adopts RCC which can prevent seepage. It is the first to break the traditional impervious structure habit of "RCD" in RCC damming technology in China. II gradation RCC with the maximum aggregate size of 40 mm is used as the impervious layer of the dam, which is built synchronously with III gradation RCC of dam body and rolled in the same layer. The proper amount of grout (about 4% – 6% of the concrete volume) is added to RCC at the edge of formwork, corridor, side wall of vertical shaft, rock slope, etc. RCC is modified into GEVR with slump of 2 – 4 cm and then vibrated with immersion vibrators. At present, almost all upstream faces of RCC dams have generally adopted this GEVR and II gradation RCC together to form impervious bodies. Practices which have been operated by dam body for many years prove that impervious effect is no less than normal concrete. Since the late 1990s to now, RCC dam basically adopts full-section damming technology and GEVR, greatly simplifying the construction, accelerating the progress of the project, reducing the duration. The advantages of RCC rapid damming technology has been brought into full play.

The impervious system of RCC dam has experienced such impervious structures as "RCD", reinforced concrete face slab impervious, asphalt mixture impervious, RCC self-seepage and GEVR imperviousness. Among the various imperviousness structures mentioned above, the type of "RCD" is basically no longer used in China. Concrete face plate and asphalt mixture are rarely used for seepage control. The early imperviousness system of RCC dam adopts "RCD" type. Because there is a big difference between the performance of "RCD" 2 – 3 m normal concrete and RCC, two kinds of concrete have different strength, leading to different performances of concrete such as deformation, temperature control and shrinkage. Under the action of thermal constraint, the normal concrete of "RCD" is prone to cracking, moreover, the phenomenon of "two skins" is easy to form between the normal concrete and RCC on the upstream face of the dam. Therefore, the "RCD" imperviousness system is easier to have leakage. For example, two meters thick CVC (normal concrete) was used in the upstream and downstream surface of Zirdan RCC dam which was started in Iran in 2002. As the impervious structure of the dam, the impervious system of "RCD" was adopted. The actual dam seepage control effect was not as tight as imagined, especially the CVC with 2 m thickness was easier to crack, therefore, only CVC could not ensure water tightness.

In recent years, with the continuous innovation and perfection of RCC rapid damming technology, maturity of mix proportion design, RCC adopts high rock powder content, low VC value, and cementing material-enriched groutgrout. RCC has become semi-plastic concrete without slump, mill ability is obviously improved. The anti-sliding stability and impervious problems of the joint layer (seam) between RCC layers have been thoroughly solved. The progress in construction technology and the application of GEVR effectively promote the rapid development of full-section RCC damming technology. II gradation cementing material-enriched grout RCC (RCC) is widely used in the upstream impervious structure of RCC dam. Generally, the ratio of impervious layer thickness to water head is 1/15 – 1/12. Depending on the dam height, its thickness is about 2 – 10 m. II gradation cementing material-enriched RCC and GEVR impervious structure are convenient to construct. The quality is easier to be guaranteed.

In the process of RCC construction, GEVR has many advantages of flexibility and simplified construction. For example, the VC value of RCC fluctuates too much or too little, the working degree VC value is beyond the control range of the standard, is unable to meet the rolling construction. In view of this situation, the RCC shall not be simply treated as waste material. The RCC can be transported and paved to the periphery of formworks, bank slope or corridor at the downstream of the dam, and can be treated by the construction method for GEVR; when slope rolling is adopted, the toe of slope often becomes the weak plane of slope rolling. The toe of slope is difficult to be compacted. Before paving the upper RCC, grout can be spreaded on RCC at the toe of slope and can be treated by the construction method for GEVR, the effect is very good.

Due to the restriction of dam design and formwork construction, its thickness tends to be thicker and thicker in some parts of the dam. For example, the thickness of GEVR is affected by dam slope ratio (generally 1:0.75 at downstream and 1:0.02 – 1:0.3 at upstream lower part of high dam), formwork bracing, etc. Its thickness develops from 30 cm and 50 cm to 100 cm. The thicker GEVR is not necessary the better. Excessive thickness of GEVR is unfavorable to temperature control, drying shrinkage and surface cracking prevention. The optimum thickness of GEVR needs further study.

The quality of GEVR construction is directly related to the impervious performance of dam, therefore, it has attracted people's attention. The quality control of GEVR needs to be standardized from the aspects of grout mixing ratio, grout uniformity, grouting method (hole making), grouting amount (slump), vibrating time, transitional lap between GEVR and RCC, etc. The quality of GEVR depends more on grout quality, grouting method and good vibrating in time. The uniformity of grout and the way of grouting shall be through continuous technological innovation, improving the quality of GEVR. The key of GEVR quality is an issue of responsibility and management.

The grout used for GEVR is made by mixing cement, flyash, additives and water. The water-cement ratio shall not be greater than that of RCC of the same kind. The main characteristic of grout is its poor stability. Because the apparent densities of cement and flyash in the grout are different, when the grout stays still for a short time, the grout will naturally produce a lot of precipitation, resulting in a significant decrease in the density (concentration) of the upper part of the grout. If grout with different density (concentration) is added to GEVR, it will directly cause the change of water-cement ratio of GEVR. Therefore, the uniformity of grout has always been one of the main factors affecting the quality of GEVR.

Due to the importance of grout in GEVR, grout mixing ratio test shall be carried out for all grouts used in the project. The grout mixing ratio test mainly includes grout density, setting time, water separation rate, uniformity and compressive strength, providing mix proportion for on-site mixing of grout. See Table 8.1 for the mixture ratio of GEVR grout in some domestic projects. As can be seen from Table 8.1, the water consumption of grout is generally 500 – 600 kg/m³ and the amount of flyash is 45% – 60%. The admixture adopts the same retarding high-efficiency water reducing admixture as same of RCC with the amount is generally 0.5% – 0.7% and the density of grout is generally 1 600 – 1 800 kg/m³. Grout density is an important index of grout quality control and is mainly related to the apparent density of cement and admixture. The density of grout is mainly tested by mud hydrometer at site for controlling the uniformity of grout. The essence of controlling the density (concentration) of grout is to control the water-cement ratio, that is, the water-cement ratio of grout shall not be greater than that of RCC of the same kind.

Table 8.1　　Statistical table of GEVR grout mix proportion in some domestic projects

S/N	Project name	Design index of RCC	Mix proportion parameter				Material amount (kg/m³)				Grout density (kg/m³)
			Water-cement ratio	Fly ash (%)	Water reducing admixture (%)	Air entraining admixture (%)	Water	Cement	Fly ash	Additive	
1	Mianhuatan	$R_{180}100$	0.65	65	0.6	—	615	331	615	5.7	1 567
2		$R_{180}150$	0.60	65	0.6	—	597	348	647	6	1 598
3		$R_{180}200$	0.50	55	0.6	—	560	504	616	6.7	1 687
4	Longshou	$C_{90}20F300W8$	0.45	40	0.7	—	531	401	779	8	1 719
5	Linhekou	$R_{90}200D50S8$	0.51	50	0.7	0.007	590	578	579	8	1 755
6	Baise	$R_{90}20S10D50$	0.50	58	0.6	0.03	550	462	638	6	1 656
7	Zhaolaihe Hydropower Station	$C_{90}20F150W8$	0.48	50	0.6	—	523	545	545	6.5	1 613
8	Yixing auxiliary dam	$R_{90}200W8F100$	0.45	50	0.6	—	505	561	561	6.7	1 634
9	Longtan	$C_{90}25W12F150$	0.40	50 Grade Ⅰ	0.4	—	497	621	621	5	1 744
10	Guangzhao	$C_{90}25W12F150$	0.45	50	0.7	—	523	581	581	8	1 693
11		$C_{90}20W10F100$	0.50	55	0.7	—	574	517	631	8	1 730
12		$C_{90}15W6F50$	0.55	60	0.7	—	594	432	648	7	1 681
13	Jin'anqiao	$C_{90}20W8F100$	0.52	50	0.5	—	574	552	552	5	1 683
14	Jufudu	$C_{90}20W8F100$	0.47	55	0.7	—	572	548	669	8.5	1 798
15	Gelantan	$C_{90}20W8F100$	0.43	40 Slag rock powder	0.8	—	583	607	607	8	1 805
16		$C_{90}15W4F50$	0.48	50 Slag rock powder	0.8	—	558	779	519	10	1 866
17	Kalasuke Hydropower Station	$R_{180}20F300W10$	0.44	30	0.7	—	534	850	364	8.5	1 757
18	Gongguoqiao	$C_{180}20W10F100$	0.46	40	0.7	—	558	728	486	8	1 780

Continued Table 8.1

S/N	Project name	Design index of RCC	Mix proportion parameter				Material amount (kg/m³)				Grout density (kg/m³)
			Water-cement ratio	Fly ash (%)	Water reducing admixture (%)	Air entraining admixture (%)	Water	Cement	Fly ash	Additive	
19	Guandi	$C_{90}25W6F100$	0.45	50	0.7	—	500	555	555	7.8	1 618
20		$C_{90}20W6F50$	0.48	50	0.7	—	500	521	521	7.3	1 549
21		$C_{90}15W6F50$	0.50	50	0.7	—	500	500	500	7.0	1 507
22	Lianhuatai	$C_{180}20W6F200$	0.47	60	0.6	—	500	426	638	8	1 572
23	Xiangjiaba	$C_{180}25W10F150$	0.42	50	0.4 Grade I	—	497	592	591	4.7	1 685

Note: Grout is added according to 4% – 6% of the volume of RCC, that is, the amount of grout added is 40% – 60 L/m³; Control slump 2 – 4 cm.

8.2 Mix Proportion Test of GEVR

The GEVR is economical, practical and simple in pouring process, and is widely used in RCC. As GEVR has the same slump and fluidity as normal concrete, the bleeding can be realized by normal vibration, making the appearance surface of the dam smooth, neat and beautiful. The GEVR also has good impermeability and can be combined with RCC very well.

Laboratory test results and field quality control results of GEVR show that: The GEVR formed by adding grout to RCC has obviously prolonged setting time. The formwork shall not be removed prematurely in low temperature season or period. Due to the addition of 4% – 6% grout in GEVR, it has the same slump as normal concrete. However, rich grout is not concealed by un-compacted technology. The GEVR with low water-cement ratio is more in line with the water-cement ratio rule. Its strength, durability, ultimate tensile value and other properties are obviously superior to those of RCC of the same grade. Moreover, the strength development coefficient increases significantly in the later period. For example: The GEVR was systematically tested in the RCC main dam of Baise multipurpose dam project. The test is briefly described as follows:

8.2.1 Mix Proportion Test of GEVR Grout

Except that the water-cement ratio of grout used in GEVR should not be greater than RCC with the same strength grade, it should not be too thick in actual use. See Table 8.2 for mix proportion and test results of GEVR grout. Test results show that: The mix proportion of grout – 2 and grout – 3 is ideal, and meets the standard requirements, convenient for construction and use.

8.2.2 Mix Proportion Test of GEVR

The dosage of grout in GEVR is 4% – 6% of the volume of RCC. The amount of grout added is related to VC value. The larger the VC value, the amount of grout shall be increased accordingly. When the VC value of RCC is 3 – 5 s and the grout content is 6%, the slump can meet the construction requirements of 2 – 4 cm. See Table 8.3 for test parameters of GEVR mixture ratio; see Table 8.4 for performance, strength and splitting tensile test results of GEVR mixture. The ultimate tensile value, elastic modulus and durability test results of GEVR are shown in Table 8.5. The test results show that all properties of GEVR meet the requirements of design and construction. All properties of hardened concrete are obviously higher than those of RCC with the same strength grade, especially, the ultimate tensile value is obviously improved.

Table 8.2　　Test results of grout mix proportion for GEVR

Test No.	Test parameters					Performance index	
	Grout	Water-cement ratio	Fly ash (%)	ZB-1Rcc15 (%)	Water (kg/m³)	Grout density (kg/m³)	Grout condition
Grout-1	Grout	0.44	58	0.6	550	1 540	Thick, fast precipitation, good viscosity
Grout-2	Grout	0.50	60	0.6	550	1 500	Slightly thick, slow precipitation and good uniformity
Grout-3	Grout	0.55	60	0.6	550	1 500	Slightly thin, slow precipitation and good uniformity
Grout-4	Grout	0.60	60	0.6	550	1 480	Dilute, fast precipitation, slightly poor viscosity

Note: Tiandong No. 525 moderate heat cement, Qujing grade II fly ash are used.

Table 8.3　　Parameters of GEVR mix proportion test

Test No.	Concrete design index	Test parameters					Material amount (kg/m³)						VC value/ Slump
		Gradation	Water-cement ratio	Fly ash (%)	Sand ratio (%)	ZB-1 Rcc15 (%)	Water	Cement	Fly ash	Sand	Stone	Apparent density	
BRV-1	$R_{180}20$	II	0.50	60	38	1.2	102	82	122	871	1 421	2 600	3–5 s
	S10D50	Grout	0.50	60	—	0.6	550	440	660	—	—	1 672	2–4 cm
BRV-2	$R_{180}15$	Quasi III	0.60	63	34	1.5	92	63	97	817	1 585	2 650	3–5 s
	S4D50	Grout	0.55	60	—	0.6	550	400	660	—	—	1 657	2–4 cm
BRV-4	$R_{180}15$	Quasi III	0.60	65	34	1.5	92	53	100	817	1 585	2 650	3–5 s
	S4D50	Grout	0.60	65	—	0.6	550	321	596	—	—	1 572	2–4 cm

Note: The grouting amount is 6% of the volume of RCC, that is, adding 60 L grout per cubic meter.

Table 8.4　　Performance, strength and splitting tensile test results of GEVR mixture

Test No.	Water-cement ratio	Slump (cm)	Initial setting time (h:min)	Apparent density (kg/m³)	Compressive strength (MPa)			Splitting tensile strength (MPa)		
					28 d	90 d	180 d	28 d	90 d	180 d
BRV-1	0.50	3.1	19:50	2 575	14.4	23.2	33.0	1.07	1.97	2.58
BRV-2	0.60	2.6	18:30	2 612	10.2	18.5	25.1	0.96	1.73	2.36
BRV-4	0.60	3.5	18:45	2 595	9.3	15.4	23.5	0.83	1.62	2.16

Table 8.5　　Ultimate tensile value, elastic modulus and durability test results of GEVR

Test No.	Gradation	Water-cement ratio	Fly ash (%)	Ultimate tensile value ($\times 10^{-6}$)		Modulus of elasticity (GPa)		Grade of impermeability S		Frost resistance grade F	
				28 d	180 d	28 d	180 d	90 d	180 d	90 d	180 d
BRV-1	II	0.50	60	86	104	27.4	34.1	>S10	>S10	>D_{50}	>D_{50}
BRV-2	Quasi III	0.60	60	79	98	18.7	32.6	>S4	>S4	>D_{50}	>D_{50}
BRV-4	Quasi III	0.60	65	75	100	16.1	31.2	>S4	>S4	>D_{50}	>D_{50}

8.2.3 Development Coefficient of Compressive Strength of GEVR at Different Ages

The development coefficient of concrete strength refers to the proportional relationship of concrete strength at different ages. See Table 8.6 for comprehensive results of compressive strength development coefficient of GEVR at different ages. Statistics show that: The compressive strength of GEVR in 90 d is 161% – 171% of that in 28 d. The compressive strength of 180 d is 229% – 246% of that of 28 d. It shows that the strength of GEVR increases greatly in the later period and is consistent with the high strength of fly ash in later stage. The later strength growth rate of GEVR is higher than that of RCC with the same grade. It is mainly due to the fact that GEVR increases the amount of grout cementing material and obviously prolongs the retarding time.

Table 8.6   Development coefficient of compressive strength of GEVR at different ages

Concrete classification	Concrete design index	Gradation	Compressive strength of each age is (%) of 28 d age.			
			7 d	28 d	90 d	180 d
GEVR	$R_{180}20S10D50$	II	54	100	161	229
	$R_{180}15S4D50$	Quasi III	56	100	171	246

8.3 Construction Technology of GEVR

8.3.1 Grout Adding Method and Technical Innovation

8.3.1.1 Grout Adding Method

Grout adding method is the most critical technology affecting the construction quality of GEVR. The way of grouting has experienced many methods, mainly surface grouting, layered grouting, groove grouting and hole grouting.

1. Grouting on the surface. In the early stage, the grouting method of GEVR is mainly surface grouting which has the advantages of convenience, not interfering with the pouring construction. However, this method is unable to make grout penetrate into concrete evenly. After vibrating, grout floats up and the defect of uneven grouting on the surface layer is exposed. When the amount of GEVR is increased, the application range is wider and wider. This limitation will become more and more prominent.

2. Grouting in layers. RCC is paved in two layers. Pave the first layer by about 17 cm; spread grout on it, then, pave a second layer to be level with the RCC paving layer; spread grout on the surface for the second time. Although this method can spread grout evenly, it takes time and effort, not conductive to on-site construction organization and improving construction speed.

3. Groove method. It can only solve the problem of uniformity at a certain depth, cannot solve the problem of GEVR pouring width. Because the depth and width of groove are limited, grouting in such a limited groove is difficult for grout to penetrate into the whole casting layer, and after the grouting is finished, the groove after grouting shall be covered in time. Heavy manual workload is required and grout tends to float up when vibrating is applied, therefore, it cannot meet the requirement of uniformity of the whole pouring layer of GEVR.

4. Hole grouting. Also known as stamp punching method, mainly making hole first and adding grout into the hole then. The way of hole grouting is composed of hole making and grout adding. The hole depth and

spacing are the key factors affecting the uniformity of grouting. A certain hole depth and spacing is equivalent to reserving uniformly distributed low-constraint grout penetration channels. After the grout is added into the hole permeates for about 10 minutes, the bottom is already relatively uniform. After the vibration is started, the grout will uniformly float up along the reserved low constraint channel, so that the middle and upper grout layers are uniformly distributed. The quality of GEVR can be easily guaranteed. It shall be said the application prospect of hole grouting method in GEVR is very good if following strictly the standard that holes larger than 20 cm, and the spacing between holes should not be greater than 25 cm.

8.3.1.2 Innovation of Hole-making Technology

The quality of GEVR construction is directly related to dam seepage control performance, therefore, people pay close attention to it. The key to affect the construction quality of GEVR is the uniformity of grout in concrete. In order to ensure the uniformity of the grouting quality of GEVR, the grouting method is very critical. A large number of project practices show that the application of hole grouting technology is relatively mature at present and has become the mainstream way of grouting.

The hole grouting method is to make holes on the paved RCC surface by puncher in diameter of $\Phi 40 - \Phi 60$ mm. At present, hole-making mainly adopts manual foot-stepping puncher to punch holes. Manual hole making is laborious and time-consuming and the effect is poor. The hole depth is difficult to meet the requirement of $\geqslant 20$ cm depth and the grout in the hole often cannot penetrate the bottom and the periphery, therefore, the hole-making depth is the key to affect the construction quality of GEVR.

Based on years of construction experience, the mechanized jacks need to be studied by technical innovation. Mechanized puncher can learn from the principle of portable vibrating rammer. The end of the vibrating rammer is transformed into a puncher and a single-rod or multi-rod puncher is arranged at the end of the tamping head, therefore, the hole-making depth and hole-making efficiency can be effectively improved, and labor intensity is reduced. Improvement of hole-making quality and efficiency provide a basic guarantee for adding grout toGEVR. It can obviously improve the construction quality of GEVR.

8.3.1.3 Technical Innovation of Grout Uniformity

Uniformity and amount of grout are a key construction technology in GEVR construction and directly affects the quality of GEVR. In order to ensure the uniformity of grout, grout shall be prepared in strict accordance with the grout mixing ratio. A grout station shall be set up at the construction site for centralized grout making. The placing time of grout from the beginning of mixing to the end of use should be controlled within 1 h. Prepare it when using it.

Due to the properties of grout, cement is prone to precipitation, resulting in uneven grout concentration. To prevent grout from settling, ensure uniformity of grout, it is necessary to carry out technical innovation on the carriage of the grout truck. To study the feasibility of installing agitator in grout compartment, grout truck shall be technically reformed according to the working principle of agitator. The grout in the grout compartment is stirred by a stirrer before adding grout every time. After the grout is homogenized, grouting is carried out on the RCC on which the holes have been made. At present, the grouting method on GEVR is mainly manual on-site grouting. It can't achieve the exact grouting amount and uniformity as the GEVR mixed in the mixing plant. Therefore, special personnel shall be responsible for grouting operation at site. Grouting amount shall be spread strictly according to the unit volume of RCC. Generally, GEVR after grouting shall be stopped for 10 min it is convenient for grout to fully penetrate into RCC, then start vibrating.

8.3.2 Construction of GEVR

8.3.2.1 Vibrating and Lapping of GEVR

GEVR has been widely used in many domestic projects with better effect. According to construction practice, the laying thickness of the grouted RCC is generally the same as that at the leveling and spreading position. The amount of manual work can be reduced and the construction efficiency can be improved. To ensure the construction quality of GEVR, the concrete can be spread with manual assistance.

Grout shall be used in strict accordance with specification. The grout is added in the morphological change area or in the range of 30 – 50 cm from the rock surface or formwork. The deviation of grouting amount per unit volume of concrete shall be controlled within the allowable range. The amount of grout added shall be determined by tests according to specific requirements, mainly related to the VC value of RCC. Generally, the grouting amount is controlled according to the requirements of GEVR slump of 2 – 4 cm. To ensure quality, accurate calibration of amount measuring tools and corresponding grouting area shall be carried out and construction shall be carefully organized. Mechanical grouting is beneficial to controlling grout amount, to ensure the uniformity of grout.

According to the different construction conditions, the vibrating sequence of GEVR mainly adopts spreading, grouting, vibrating and finally rolling. It is also possible to spread grout first, rolling and vibrating then. No matter what vibrating sequence is adopted for GEVR, all joints need to be treated separately by vibrating roller.

When rolling GEVR and RCC, it is necessary to overlap a certain width. Only in this way can a good transitional combination between the morphological change area and the rolling area be ensured. Strong vibration is a necessary measure to ensure the quality of GEVR uniformity, the joint of upper and lower layers and the connection with rolling area, GEVR must be vibrated strongly to ensure uniformity and joint of upper and lower layers. When vibrating, the vibrator should be inserted into the lower concrete for about 5 cm and the overlapping width between adjacent areas and morphological change areas should be greater than 20 cm when rolling concrete. Two kinds of concrete should be poured crosswise at the joint of GEVR and RCC. GEVR should be vibrated and compacted before initial setting. RCC should be rolled within the allowable interval between layers.

8.3.2.2 Machine-mixed GEVR

At the some complicated areas in the construction of RCC dam, the cushion concrete of bank slope, large-volume GEVR around corridor and formwork made by normal concrete or GEVR as required in technical standard, on the premise of not affecting RCC construction, the mixing plant could be used to directly mix GEVR instead of the GEVR added with grout on site, thereby not only reducing the amount of grout added for field GEVR, greatly improving the construction intensity, and also improving the quality of GEVR casting.

Machine-mixed GEVR has been widely used in Mianhuatan, Baise, Longtan, Guangzhao and Jin'anqiao in recent years. Because many parts of RCC dam are constructed with GEVR, such as cushion, bank slope, corridor, upstream and downstream surface parts, etc., are often large in quantity and concentrated, according to the construction method of field grouting, the workload of grouting is too large and the grouting uniformity is difficult to ensure, that directly affects the construction quality of GEVR. In actual construction, as long as the construction conditions permit, at the positions where automobile transportation can be used for direct pouring, the GEVR can be directly mixed by the mixing plant. Machine-mixed GEVR is normal concrete with slump of 2 – 4 cm when the RCC is added with cement and fly ash grout at specified proportion during mixing, that is to say, 40 – 60 L grout is added to the RCC. The machine-mixed GEVR is convenient to construct, not only simplifies the operation procedure of GEVR, but also can ensure the quality of mass GEVR.

8.4 Test Study on GEVR Mixed with Fiber in Impervious Area

8.4.1 Foreword

Diabase manufactured aggregate was used in the RCC main dam of Baise multipurpose dam project. The application of diabase manufactured aggregate in hydraulic mass concrete was the first time at home and abroad. Diabase manufactured aggregate was featured with high density, more than 3.0 g/cm^3, great hardness, high elastic modulus and difficult processing. Especially, the rock powder in diabase manufactured sand had high content, poor gradation and water demand ratio much higher than other kinds of manufactured aggregates adopted in China. It used 30 – 40 kg/m^3 more water than concrete mixed with general manufactured aggregate. In order to improve the impermeability and crack resistance of RCC water front face, new technologies, new materials and new processes should be considered. On the basis of drawing lessons from the measures of adding polypropylene fiber into normal concrete, in this paper, test research and discussion are carried out on RCC and GEVR mixed with polypropylene fiber with II gradation imperviousness area of water front face.

Concrete is an excellent structure material and its development and application have a history of more than 170 years. It has excellent performance in many aspects. But it has one major defect: poor tensile strength, shear strength, impact resistance, explosion resistance and toughness. After years of research: adding fiber into concrete can improve the above performance of concrete. At present, the commonly used fibers are divided into two types: one is fiber with high elastic modulus, comprises glass fiber, asbestos fiber and steel fiber; other one is fiber with low elastic modulus, such as nylon, rayon, vegetable fibers and polymer fibers. Polypropylene fiber is a new type of concrete fiber and is called "secondary reinforcing steel bar" of concrete. After ten million pieces of polypropylene fibers are added into concrete, they can effectively inhibit the plastic shrinkage and cracking of concrete; reduce the occurrence and development of primary micro cracks caused by concrete segregation, bleeding, shrinkage and other factors; reduce the number and scale of primary micro cracks; significantly improve the ultimate tensile strength, bending toughness, fracture toughness, impact resistance and fatigue resistance of concrete; reduce the internal stress concentration of concrete.

Rolling was started on December 28th, 2002 for the RCC main dam of Baise multipurpose dam project. From the study of the influence of rock powder on the performance of RCC, it was found that: high content of rock powder led to increased shrinkage of concrete. In order to improve the impermeability and crack resistance of RCC water front, new technologies, new materials and new processes should be considered. On the basis of drawing lessons from the measures of adding polypropylene fiber into normal concrete, it was proposed to add a proper amount of polypropylene fiber into II gradation RCC in the imperviousness area of the water front, so as to better improve the anti-cracking and imperviousness performance of RCC and GEVR in the imperviousness area. This was an important technological innovation to improve the impermeability and cracking prevention of RCC in impervious area.

8.4.2 Raw Materials in Test

8.4.2.1 Cement

After adopting Tiandong No. 525 moderate heat Portland cement provided by the owner, the test results showed that: physical and chemical indexes of cement met the standard.

8.4.2.2 Fly Ash

The fly ash used in the test was Yunnan Qujing Grade Ⅱ fly ash and Guizhou Panxian Grade Ⅱ fly ash provided by the owner. Testing of fly ash were carried out in accordance with *Technical Standard of Flyash Concrete for Hydraulic Structures* DL/T 5055—1996. Test results showed that: both kinds of fly ash met the requirements of Grade Ⅱ fly ash index.

8.4.2.3 Aggregate

The fine aggregate was diabase manufactured sand which was produced by Barmac dry method. The results of manufactured sand quality inspection and particle grading test showed that the fineness modulus of diabase manufactured sand was 2.67 – 2.77; the content of rock powder was 20.6% – 21.9% and the coarse and fine particles were polarized; The gradation was not ideal; coarse aggregate was diabase manufactured crushed stone and the density was 3.0 g/cm^3. The maximum particle size of aggregate was 60 mm and adopted quasi-Ⅲ gradation. Test results of RCC mix ratio showed that when RCC mixed with diabase aggregate was used, the apparent densities of the quasi-Ⅲ gradation and Ⅱ gradation reached 2 650 kg/m^3 and 2 600 kg/m^3 respectively, such a large density of concrete was rare.

8.4.2.4 Additive

Through tests, RCC adopted the retarding high efficiency water reducer ZB-1Rcc15 made by Zhejiang Longyou Wuqiang Concrete Admixture Co., Ltd. and the air entraining admixture was DH$_9$ from Hebei Concrete Admixture Factory.

8.4.2.5 Polypropylene Fiber

The test research adopted "Haoyite" brand polypropylene fiber produced by Sichuan Huashen Structure Materials Co., Ltd. The specifications were 19 mm, 8 mm, 6 mm and 4 mm respectively. The quality results of polypropylene fiber are shown in Table 8.7. Test results show that: all indexes of polypropylene fiber are qualified.

Table 8.7　　　　　　　　　　Quality results of polypropylene fiber

Test Items	Control index	Inspection result
Water absorption (%)	—	Nil
Specific gravity	0.9 – 1.0	0.91
Melting point(℃)	155 – 165	160 – 170
Flash point(℃)	≥550	590
Heat conduction performance(W/km)	≤0.5	Low
Acid-base impedance	0.99 – 1.01	High
Tensile strength(MPa)	≥500	560 – 770
Young's elastic coefficient(MPa)	≥3 500	3 500

8.4.3　Test Study on GEVR Mixed with Fiber

8.4.3.1 Polypropylene Fiber Grout

The grout mixed with polypropylene fiber is made of cement, fly ash, admixture and polypropylene fiber in proportion. According to relevant information, generally, polypropylene fiber is 0.5 – 1.0 kg per cubic meter of concrete. 0.8 kg/m^3 polypropylene fiber was added to the GEVR in this test and the volume of fiber-doped grout was controlled according to 5% – 8% of the volume of RCC, calculated by 6%. The content of polypropylene fiber in grout was 13.3 kg/m^3.

The grout mixed with polypropylene fiber adopted the RCC mixture ratio of the dam front water surface. See Table 8.8 for details. The water-cement ratio of grout enriched vibrated grout was 0.50; the water consumption was 550 kg/m^3; the content of ZB-1Rcc15 was 0.3%; The polypropylene fiber was 13.3 kg/m^3. The raw materials of grout were composed of Tiandong No. 525 moderate heat cement, Qujing Grade II fly ash, additive ZB-1 RCC15, Huashen polypropylene fiber and water.

Table 8.8   Mix proportion of RCC on dam upstream surface

Design requirements and location in the project	Gradation	Water-cement ratio	Sand ratio (%)	Fly ash (%)	ZB-1Rcc15 (%)	DH$_9$ (%)	Water content (kg/m^3)	Apparent density (kg/m^3)
R$_{180}$20S10D50	II	0.50	38	58	0.8	0.015	108	2 600
Upstream surface RCC	Grout	0.50	—	58	0.3	—	550	1 653

Test method for adding microfiber into polypropylene fiber grout: Firstly, the weighed cement, the fly ash and the microfiber were dried and evenly stirred, then water and additive were added and stirred evenly continuously. In this test, four kinds of microfibers with different length, such as 19 mm, 8 mm, 6 mm and 4 mm, were used. Trial mixing of polypropylene fiber grout was carried out. Test results are as follows:

1. Test of microfiber with length of 19 mm: after adding water and additive, fiber pellets appeared; the stirring time was prolonged; the fiber pellets were still unevenly dispersed; there was a small amount of microfiber clusters with a diameter of about 5 – 6 mm; sedimentation was slow; a water film appeared locally near the fiber after storage. Therefore, the 19 mm microfiber grout still existed as pellets after being added into RCC and is not easy to disperse after being vibrated.
2. Test of microfiber with length of 8 mm: after adding water and additive, a small amount of fibers were pellets. The diameter of the microfiber pellets was about 3 – 4 mm. After being added into RCC, the microfibers were unevenly distributed. After being vibrated, there were still a few microfiber pellets.
3. Microfiber test with length of 6 mm: after adding water and additive, there are a small amount of pellets of fibers. The diameter of microfiber pellets is about 2 – 3 mm. After being added into RCC and manual vibration, fibrous pellets can be dispersed and did not exist.
4. Microfiber test with length of 4 mm: after adding water and additive, the microfibers can be dispersed in the grout and spread out well.

8.4.3.2 Grouting Method of Fiber Grout

According to the grouting method of GEVR in RCC in China, the adding methods of fiber grout include surface drilling grouting method, groove grouting method and layered grouting method. Different grouting methods are compared to know the fiber distribution in GEVR. The fiber GEVR grouting test simulates the paving situation of RCC on site. A specimen box with a volume of 60 cm × 60 cm × 35 cm is used for the test. Test results of fiber grout grouting method are analyzed as follows.

Fiber grouts of different lengths: three kinds of fiber grout grouting have great differences of fiber distribution in GEVR. The microfibers with length of 19 mm and 8 mm are unevenly distributed in GEVR. The microfibers with the length of 6 mm and 4 mm are uniformly distributed in GEVR. It shows that fiber length has great influence on fiber distribution in concrete.

The fiber grout adopts different grouting methods and the distribution of fibers is also different:
1. Boring method: the hole diameter is 4 cm and the spacing between holes is 10 cm. The hole goes deep through the bottom. RCC paving thickness is 35 cm. Holes are punched first and then are injected with fiber

grout. Then the hole opening is blocked by RCC. The vibrating rod vibrates the mortar until the mortar has bleeding, then the fiber GEVR specimen is split longitudinally and transversely for observation. There are no fiber pellets in concrete. There are many dense fibers around the hole from the bottom up, fewer fibers at the bottom and more fibers in the upper part. The fibers in the middle part are slightly less than that in the upper part, so fiber distribution is uneven.

2. Groove method: 25 cm thick RCC is laid at the bottom and grooves are made at areas with this thickness. Groove width is 15 cm and about 15 cm deep. The distance between the grooves is 15 cm. Inject fiber grout into the groove and then cover the groove with RCC. The vibrating rod vibrates the grout until the grout has bleeding. Then the fiber GEVR specimen is split longitudinally and transversely for observation. There are no fiber pellets in RCC and the microfibers are well dispersed. Seen from the longitudinal direction of the groove, there are more fibers in about 15 cm around the groove and few fibers beyond 15 cm. There are fewer fibers at the bottom and more fibers in the middle and upper parts. The fiber distribution is uneven.

3. Interlayer method: RCC with thickness of 8 – 10 cm is paved at the bottom. Fill fiber grout on RCC. Cover the grout with 10 cm thick RCC; spread a layer of fiber grout. Cover the grout with RCC. Finally, total thickness is 35 cm. Then vibrate the concrete until it has bleeding. The specimen is split longitudinally and transversely for observation, There are no fiber pellets in RCC and the fibers are well dispersed. There are fibers distributed above 5 cm from the bottom of the specimen (fibers below 5 cm is less). Two-layer fiber grout paving method is used. The fiber distribution is basically uniform and is better than hole method and groove method.

The above grouting methods of fiber grout in RCC are analyzed comparatively: the fibers at the bottom of three grouting methods are rare. The reason after analysis is that the microfiber itself is very light. Due to the upward movement of microfibers after vibrating, but with the increase of RCC thickness, the resistance of microfibers moving upwards increases and the number of microfibers moving upwards is decreasing. After RCC exceeds a certain thickness, the microfibers can hardly move, only rely on the guiding action of the vibrating rod inserting in and moving out, the microfibers are brought out. Seen from the section after cutting, there are traces displayed when vibrating rods are pulled out obliquely and vertically from bottom to top. There are fibers distributed in vibration from bottom to top and more fibers, but only a small amount of microfibers are distributed at a distance around the vibrating rod. It shows that these fibers are brought out by the guiding action of the vibrating rod during moving out, rather than the result of the movement of the fiber itself during vibrating. According to test results, compared with the other two methods, the interlayer method distributes the fibers more evenly, therefore, the interlayer method is tested again.

The fiber grout is poured twice by interlayer method. RCC thickness is 30 cm and paving thickness of each layer is 10 cm. After the manual vibration is finished, check the specimen after cutting. The fiber distribution is basically uniform, but the quantity of microfibers distributed at the bottom is less, so there are still unevenly dispersed fibers in some places. There are more fibers in some parts and few fibers in other parts. This situation is related to the distance between vibrating rods. If the distance between the vibrating rods is large, there is less fibers distributed in this part of concrete and distributed unevenly.

Therefore, the mechanical mixing test of GEVR with fiber grout was also carried out. The specific method is as follows: II gradation RCC is mixed by forced mixer for 60 s and then stopped. The grout of polypropylene fibers with different length (4 mm, 6 mm, 8 mm and 19 mm) is added respectively. After continuous mixing for 30 s, the RCC becomes fiber GEVR. Take the specimen out of the machine for observation. The workability of the mixture is good and the fibers are evenly distributed. The fiber GEVR is put into a test piece box and vibrated by a vibrating rod, then is cut in longitudinal and transverse direction for observation. Four kinds of

Chapter 8 Construction Technology of GEVR

microfibers with different lengths are uniformly distributed in RCC. Test results show that: if the grout mixed with polypropylene fiber is grouted in three different ways, such as hole method, groove method and interlayer method, polypropylene fibers are not easily distributed evenly in RCC, not suitable for site construction.

8.4.4 Study on Performance of Fiber-reinforced RCC

According to the above test results, considering the feasibility of construction, how to solve the uniformity of GEVR mixed with fibers in RCC is a new subject. Therefore, we must start from another technical route. If we can directly mix polypropylene fibers into II gradation RCC in mixing plant first and then transport the RCC mixed with fibers to the construction site, then add conventional GEVR grout, in this way, the uniformity of fiber GEVR in imperviousness area can be ensured.

Test of II gradation fiber reinforced RCC in impervious area; forced mixer is used. The feeding sequence is as follows: cement, fly ash, polypropylene fibers and aggregate are first dry stirred for 15 s, then add additive solution and water, stir them for 60 s. After leaving the machine, the mixture becomes fiber RCC. Then, the grout of common GEVR is added into fiber RCC and becomes fiber GEVR. According to the observation of the mixture of fiber RCC and fiber GEVR, fibers are distributed uniformly and the workability of the mixture is good. See Table 8.9 for test parameters of RCC mixed with polypropylene fibers in imperviousness area; see Table 8.10 for performance and mechanical performance test results of RCC mixed with polypropylene fiber in impervious area; see Table 8.11 for other performance test results. Results show that:

Table 8.9   Test parameters of RCC with polypropylene fiber

Test No.	Water-cement ratio	Fly ash (%)	Sand ratio (%)	ZB-1 Rcc15 (%)	DH_9 (%)	VC value/Slump	Water (kg/m^3)	Fiber (kg/m^3)	Apparent density (kg/m^3)	Concrete classification
QR-1	0.5	58	38	0.8	0.015	3 – 8 s	106	—	2 600	Roller compacted concrete
QR-2	0.5	58	38	0.8	0.015	3 – 8 s	106	0.8	2 600	Fiber RCC
QR-3	0.5	58	38	0.8	0.015	1 – 3 cm	106	—	2 600	GEVR
QR-4	0.5	58	38	0.8	0.015	1 – 3 cm	106	0.8	2 600	FiberGEVR
Grout	0.5	58	—	0.3	0.003	—	550	—	1 653	Grout

Table 8.10   Test results of mechanical properties of RCC mixed with polypropylene fiber

Test No.	Water-cement ratio	Fiber (kg/m^3)	Apparent density (kg/m^3)	VC value/Slump	Time of setting (h: min)		Compressive strength (MPa)			Splitting tensile strength (MPa)	
					Initial setting	Final setting	7 d	28 d	180 d	28 d	180 d
QR-1	0.5	—	2 600	4 s	6:35	12:55	9.1	14.6	29.7	1.27	2.42
QR-2	0.5	0.8	2 600	6 s	7:28	15:40	8.3	14.5	28.7	1.29	2.72
QR-3	0.5	—	2 600	2.2 cm	16:00	22:48	7.1	13.2	27.0	1.07	2.14
QR-4	0.5	0.8	2 600	2.8 cm	15:30	22:25	7.0	13.0	25.3	1.18	2.61

Table 8.11   Test results of other properties of RCC with polypropylene fiber

Test No.	Grade of impermeability	Ultimate tensile value ($\times 10^{-4}$)		Modulus of elasticity (GPa)		Drying shrinkage ($\times 10^{-6}$)						
		28 d	180 d	28 d	180 d	3 d	7 d	14 d	28 d	60 d	90 d	180 d
QR-1	>W10	0.63	0.76	26.7	36.5	-85	-181	-298	-383	-422	-441	-454
QR-2	>W12	0.67	0.87	20.6	31.7	-65	-162	-259	-343	-382	-394	-402
QR-3	>W10	0.61	0.78	27.4	35.1	-26	-118	-222	-378	-444	-464	-470
QR-4	>W12	0.64	0.95	23.6	32.0	-26	-103	-187	-343	-375	-394	-401

1. Workability. Compared No. QR-2 fiber RCC with QR-1 reference RCC, the VC value increases by about 2 s, initial setting time and final setting time are prolonged for about 0:55 h and 2:45 h. Fibers are distributed uniformly and there is no obvious change in appearance; Compared QR-4 fiber GEVR with QR-3 reference GEVR, there is no obvious change in slump. The setting time is shortened by about 0:30 h. The fibers are distributed uniformly and the appearance is slightly sticky.
2. Strength. Compared fiber RCC and fiber GEVR with the reference GEVR, compressive strength decreases slightly after 7, 28 and 180 d, but the splitting tensile strength is obviously improved.
3. Other properties. According to the design age of 180 d, compared No. QR-2 fiber RCC with QR-1 reference RCC, the anti-permeability grade is improved from W10 to W12; The ultimate tensile value is increased from 0.76×10^{-4} to 0.87×10^{-4}; The elastic modulus is reduced from 36.5 GPa to 31.7 GPa; The drying shrinkage value decreases from 454×10^{-6} to 402×10^{-6}. Compared No. QR-4 fiber GEVR with QR-3 reference GEVR, as same, the anti-permeability grade is improved from W10 to W12; The ultimate tensile value is increased from 0.78×10^{-4} to 0.95×10^{-4}; The elastic modulus is reduced from 35.1 GPa to 32.0 GPa; The drying shrinkage is decreased from 470×10^{-6} to 401×10^{-6}.

In summary: Adding polypropylene fiber obviously improves the performance of RCC and fiber GEVR and reduces the stress of concrete. The impermeability, crack resistance and other properties of RCC and GEVR in impervious area are improved.

8.4.5 Conclusion

1. Test study on polypropylene fibers blended in RCC in impervious area of water front surface is an important technological innovation to improve the impermeability and cracking prevention of RCC.
2. Polypropylene fibers are mixed into grout. Grouting is carried out in RCC in different ways such as hole method, groove method and interlayer method. Polypropylene fibers are not easily distributed evenly in GEVR, the method of adding fiber into grout is not applicable to field construction.
3. RCC mixed with polypropylene fibers in impervious area can be directly mixed in mixing plant. The fiber RCC is transported to the pouring area and conventional grout is added on the pouring area. In this way, fibers in GEVR are evenly distributed and the method is simple and feasible.
4. The test results show that RCC with polypropylene fiber and fiber modified concrete can effectively improve the impermeability and crack resistance of concrete in impervious area.
5. When polypropylene fibers are added into RCC in impervious area, the influence of polypropylene fiber content and specification on RCC performance needs further research.

The recommended mix proportion of polypropylene fiber RCC in impervious area is shown in Table 8.12.

Table 8.12   Mix proportion of polypropylene fiber RCC in impervious area

Design requirements and location in the project	Water-cement ratio	Water (kg/m³)	Fly ash (%)	Sand ratio (%)	Fiber (kg/m³)	ZB-1 Rcc15 (%)	DH$_9$ (%)	VC value/Slump	Apparent density (kg/m³)	Remarks
R$_{180}$20S10D50 Roller compacted concrete	0.50	106	58	38	0.8	0.8	0.015	3 – 8 s	2 600	—
R$_{180}$20S10D50 GEVR	0.50	550	58	—	—	0.3	0.003	1 – 3 cm	1 653	Grout mixing 5% – 7%

8.5 Application of GEVR in Baise RCC Main Dam

8.5.1 Foreword

In construction of RCC main dam of Baise project, the GEVR is fully used in the imperviousness area of upstream front water surface, cushion concrete of dam foundation on both banks, the area surrounding holes, the periphery of formwork, the reinforcement mesh of joints, or the parts where RCC cannot be used. Usually, these areas are constructed by pouring normal concrete. Through the research on the new technology of hole grouting, it is beneficial to control the grouting method and quantity of grouting. For the concentrated parts of GEVR and steel mesh on the bank slope in which the concrete can be directly put into the pouring area by vehicles, the method of mixing GEVR directly in mixing plant is also used, which further promotes the application of GEVR.

Therefore, in the construction of RCC dam, it is necessary to prepare the construction synchronously with normal concrete. This construction method not only affects the pouring speed of RCC, but also is unable to give full play to the characteristics of rapid construction of RCC. Moreover, this construction method increases the preparation works for normal concrete construction, increases the investment in the project, furthermore, the bonding quality between different kinds of concrete is not enough. With the continuous development of RCC dam construction technology, a new concrete is developed and has been widely used in RCC dam construction. In this method, the RCC is transported into the pouring area and vibrated according to the vibration compaction method of normal concrete, so as to replace normal concrete. This new concrete is named as GEVR. So as to further improve the construction speed of RCC and reduce the weak joints between different kinds of concrete, the RCC dam construction technology has stronger vitality.

8.5.2 Construction Technology of GEVR

8.5.2.1 Placing and Spreading

During the paving of GEVR, if the spreader is directly used for the RCC on the large placing surface, the aggregate is concentrated often at the area of distorted concrete, and some areas will be higher than the surface of RCC pouring area. After vibration compaction, the GEVR area is higher than the RCC placing surface. It is not conducive to vibration operation and is easy to cause grout loss of GEVR. Therefore, the GEVR shall be paved and leveled manually. In order to prevent the grout of GEVR from flowing into the RCC placing surface, generally, it is also required that the GEVR area should be paved into a trough about 6 – 10 cm lower than the RCC.

8.5.2.2 Adding Grout

Firstly, in order to meet the physical and mechanical properties of the areas using the GEVR, the mortar used in GEVR shall be tested. The parameters such as the mix proportion of cement, fly ash and various admixtures and the amount of mortar are designed. Secondly, in order to ensure the uniformity of mortar, a mortar station should be set up for centralized mortar mixing. The storage time of mortar should not be too long and should be controlled within 1 hour from the beginning of mixing to the end of use, making sure to mix it as soon as using it; the mortar of GEVR is sprayed manually at present so it is not possible to achieve the same amount and uniformity of grouting as the concrete mixing plant, Therefore, the on-site grouting operation should be strictly controlled and special supervisor shall be responsible for grouting. The special mortar container is used to quantitatively and evenly add mortar according to the area.

There are two main ways to grout, bottom grouting and top grouting. Bottom grouting is to grout on the next layer surface of GEVR and then pave RCC on the grout, and then vibrating. Vibration force can make the mortar penetrate upward until the grout goes up to the top. The advantage of this method has good uniformity, but vibration is very difficult; top grouting is to spread grout on the paved RCC surface then vibrating. The vibration is easy in this way, however, it is difficult for the grout to penetrate downward and is not easy to be uniform. The grout will float on the surface, that is not beneficial.

8.5.2.3 Vibrating Tamping

Generally, the vibration of GEVR is prior to that of adjacent RCC, sometimes it can be carried out after rolling surrounding areas. Generally, the high frequency vibrator with $\Phi 100$ is used. Vibration is generally required to be carried out 15 min after grouting and the vibration time should be controlled between 25 and 30 seconds. When vibrating, the depth of the vibrator inserted into the lower layer shall be more than 10 cm. It is required that the vibration range of high frequency vibrator shall exceed the overlapping range at the overlapping position of RCC, so that two are mixed and compacted with each other. For the local overlap protruding area between the GEVR and RCC, generally, a small vibrating roller is used to supplement and compact the overlapping areas.

8.5.3 Application of GEVR

8.5.3.1 Specific Application Description

As a complete RCC gravity dam, the volume of RCC Baise dam was 2.104 million m^3. During the construction of Baise dam, GEVR had been widely used and the total volume of GEVR was about 163 300 m^3. The GEVR was used around the normal concrete cushion of the whole dam bank slope and some parts of the bank slope dam foundation, upstream and downstream formwork edges, expansion joints, burying places where upstream and downstream water stops were embedded, corridor, elevator shaft periphery, observation cable periphery, joint reinforcement mesh part and parts that cannot be rolled by vibration roller.

8.5.3.2 Grout Mix Proportion of GEVR

The grout used in the GEVR of Baise dam was mixed in a centralized way and the grout station was located at the dam head on the right bank. The production capacity of grout station was 5.0 m^3/h. The pure grout was transported from the mortar station to the placing surface through pipelines. On the placing surface, the grout was loaded to motor dumpers and transported to the place of use. On the paved 30 cm RCC layer, hand bucket was used to spread the grout. The grout amount of GEVR had passed the test. It was determined that 6% cement and fly ash grout were added into RCC, that is, each cubic meter of RCC was added with 60 L grout. At the same time, different kinds of grout were used for different grades of RCC. The mix proportion of grout is shown in Table 8.13.

Table 8.13　　　　　　　　　　　　　　Grout mix proportion of GEVR

Symbol	Gradation	Composition mixed (kg/m^3)			Additive(%)
		Cement	Fly ash	Water	ZB-1Rcc15
Rv15	Quasi Ⅲ gradation	400	600	550	0.6
Rv20	Ⅱ gradation	462	638	550	0.6

The field quality control of grout was mainly by testing its density. If the density failed to meet the requirements, the mortar was regarded as waste mortar and would be completely rejected. See Table 8.14 for density control standards.

Table 8.14　　　　　　　　　　　　　　Field density control index

Item	Grout quality control index of of GEVR		Remarks
Symbol	Rv15	Rv20	Mud hydrometer was used for inspection.
Density (g/cm^3)	≥1.56	≥1.66	Mud hydrometer was used for inspection.

According to the original design requirements, a 1.5 m thick normal concrete cushion was required to be covered on the bedrock surface of Baise dam section 3 – 9. The cushion normal concrete of bank slope and RCC of dam body were raised synchronously. Due to the difference of initial setting time and final setting time of two kinds of concrete, it was difficult to achieve synchronous rising. At the same time, two kinds of concrete were poured simultaneously in the actual construction and there were many disadvantages. In order to simplify the construction procedure of RCC dam and further increase the speed of RCC dam construction, bold innovation was made in the construction of Baise dam. The application scope of GEVR was further extended to the cushion concrete of bank slope bedrock surface and the original 1.5 m thick normal concrete cushion was cancelled. In other words, 1.5 m thick GEVR was directly poured at the bank slope foundation of EL 99.5 – EL 123 of 3B – 4 dam section and EL 124.5 – EL 132 dam section 8B. It was raised synchronously with RCC dam body. The GEVR used in the foundation of bank slope included EL 147 – EL 177 of 2 – 3 dam section on the left bank and EL 163 – EL 177 of 11 dam section on the right bank. The bank slope foundation was covered by 2 m thick GEVR and the original 2 m thick normal concrete cushion was cancelled.

When the GEVR was poured on the cushion layer of bank slope dam foundation, first, a layer of mortar was spread on the bedrock surface, then covered by the RCC. And it was required to roll the RCC dam body first, then the grout was spread and the GEVR of bank slope foundation cushion was poured, two kinds of concrete shall be poured and tamped within 2 h. After GEVR construction of bank slope bedrock surface reached the required thickness, RCC operation could be carried out directly on it.

8.5.3.3 Hole Grouting

As there are some defects in the traditional top and bottom grouting methods, the quality of GEVR cannot be guaranteed. In order to make the mortar evenly penetrate into the RCC, moreover, it can effectively control the amount of mortar added. The grouting process was improved in Baise dam. The hole puncher was designed and the horizontal grouting method was changed to vertical grouting method (See Figure 8.1 for the structure of the puncher). Generally, holes are made on the paved RCC surface by hole puncher with a diameter of 50 mm before grouting and the holes are arranged in the form of plum blossom. The hole spacing is generally 25 cm and the hole depth is more than 20 cm. Then the manual bucket (amount is recorded) is used to spread the mortar. When grouting, a bucket of mortar is poured into the hole specified, so as to control the amount of mortar. The surface of the puncher shall be smooth and the end must be a bullet shaped cone, so as to reduce friction resistance, convenient to go deeper. After use, the surface of the puncher shall be kept clean and bright.

8.5.3.4 Machine-mixed GEVR

The foundation of some bank slopes of Baise dam was made of GEVR. The GEVR was used in the areas where the reinforcement mesh of the joints on the dam surface and the formwork braces were more and more concentrated. Therefore, in places where the amount of GEVR was large and concentrated, such as the bank slope platform, according to the construction method of field grouting, the workload of grouting was too large, moreover, it was difficult to ensure the uniformity of the mortar. In actual construction, the mixing plant is used to directly mix the GEVR at the place where the truck can be used for direct pouring. This new GEVR construction technology not only simplifies the operation procedure of GEVR, but also can ensure the quality of mass GEVR.

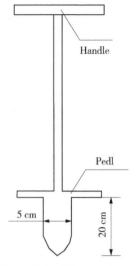

Figure 8.1 *The structure of the puncher*

8.5.4 Quality Inspection of GEVR

The construction period of the first dry season was from December 28, 2002 to June 14, 2003. 450 000 m³ of RCC was poured into the dam, among them, 41 600 m³ of GEVR was poured. The construction period of the second dry season was from September 20, 2003 to June 12, 2004, 870 000 m³ RCC was poured into the dam, among them, 55 900 m³ of GEVR was poured. The third dry season was from September 2, 2004 to September 10, 2005. 580 000 m³ of RCC was poured into the dam, among them, 65 800 m³ of GEVR was poured. The construction quality of GEVR in the first and second dry seasons was inspected. The corresponding physical properties are shown in Table 8.15.

Table 8.15 Physical properties of GEVR core samples

Test Indicators		Machine mixing		On site grouting	
		Rv20 II gradation	Rv15 Quasi III gradation	Rv20 II gradation	Rv15 Quasi III gradation
Density(kg/m³)	Range	2 608 – 2 622	2 630 – 2 668	2 586 – 2 634	2 628 – 2 690
	Mean value	2 610	2 655	2 613	2 664
	Design value	2 600	2 650	2 600	2 650
Compressive strength (MPa)	Range	21.6 – 30.2	18.2 – 25.6	20.6 – 28.3	16.7 – 23.2
	Mean value	26.7	20.4	25.4	20.7
	Design value	≥20.0	≥15.0	≥20.0	≥15.0
Elasticity modulus ($\times 10^4$ MPa)	Range	2.21 – 2.44	2.35 – 2.56	2.06 – 2.37	2.13 – 2.40
	Mean value	2.32	2.44	2.24	2.28
Tensile strength (MPa)	Range	2.2 – 2.9	1.8 – 2.3	1.6 – 2.4	1.4 – 2.2
	Mean value	2.4	2.0	2.1	1.7
Ultimate tensile value ($\times 10^{-6}$)	Range	84 – 108	78 – 92	78 – 102	72 – 86
	Mean value	92	84	88	78
	Design value	>75	>70	>75	>70

8.5.5 Conclusions

In Baise RCC main dam project, the hole grouting method was used for GEVR. The RCC (RCC) in the GEVR area was paved and leveled with manual assistance. The phenomenon of large aggregate concentration was avoided. At the same time, in order to prevent the grout of GEVR from flowing into the RCC placing surface, the paving level of GEVR area was lower than that of RCC. A trough of about 6 – 10 cm was created and the quality problem of lap joint by two kinds of concrete was solved then. Baise RCC dam had made some new attempts in the application of GEVR and got satisfactory results. It is worthy of further promotion in RCC construction.

Chapter 9

Key Technology for Rapid Construction of RCC

9.1 Overview

The RCC construction features rapid, continuous and intensive mechanized construction. In case of any failure, incoordination or incompatibility at any step in the whole production system, the project progress, project quality and the rapid construction characteristics of RCC will be affected. "Interlayer bonding, temperature control and cracking prevention" are core technologies for rapid RCC construction. To ensure "rapid" construction, it is necessary to develop scientific and rational construction organization design, construction technical scheme, optimal resource allocation, high-level organization management and operation, skilled construction personnel and reasonable bid winning price in accordance with the *Design Specification for RCC Dams, Construction Specification for Hydraulic RCC, Specifications for Hydraulic Concrete Construction, Test Code for Hydraulic Concrete* and applicable provisions of technical terms of bidding for the project, which is the precondition to ensure the rapid RCC construction, and can give full play to the advantages of rapid RCC construction technology.

"Being rapid" is the greatest advantage of RCC dam construction technology. Generally, a concrete dam higher than 100 m will be completed in 4 – 5 years of duration by using the traditional normal concrete dam construction technology. However, by using the RCC damming construction technology, the dam will generally be completed in only 2 – 3 years. For example, the construction periods of Dachaoshan, Linhekou, Mianhuatan, Baise, Longtan, Guangzhao, Jin'anqiao RCC dams which are 100 – 200 m high are obviously shortened by more than one third of duration compared with those of Longyangxia, Wujiangdu, Baishan, Lijiaxia, Wanjiazhai and Goupitan and other earlier normal concrete dams which are similar in construction scale. Time is money. Shortening the construction period to achieve power generation ahead of schedule can bring significant return on investment, which is the advantage of the rapid RCC dam construction technology.

The fact that the duration can be greatly shortened is why the RCC dam construction technology is adopted. Engineering practice has proved that by adopting the RCC dam construction technology, a small number of operators are required on the grand construction site, and the RCC construction is conducted in an orderly and well organized manner from mixing, transport, placing, spreading, roller compaction, spraying and moisture preservation, giving a natural and unrestrained sense throughout the construction process. In contrast, with regard to the normal concrete dam construction, due to constraints by the pouring intensity and temperature control, the dam body is divided into many pouring blocks by transverse joints and longitudinal joints. The concrete is poured in columns and to alternating blocks, which results in the multiplication of formwork quantities, and there are crowded and busy people on the site. Besides, the jointing requires a large amount of grouting. Therefore, the rapid damming technology with RCC has incomparable advantages over other damming technologies.

Chapter 9 Key Technology for Rapid Construction of RCC

Over the past 20 years, the RCC damming technology has developed from the early exploration period and transition period to the current maturity period. The RCC has also developed from the dry hard concrete to the no-slump semi-plastic concrete. Constant communication, collision and fusion with regard to RCC have been achieved in the process of damming. It has been proved in practice that the RCC damming is not restricted by climatic and geographical conditions. The RCC damming technology is completely applicable for any dam type. Under the appropriate geological and topographical conditions, rapid RCC damming technology can be adopted.

The main difference between RCC and normal concrete only lies in the material composition in the mix and the construction method. RCC has the characteristics of small amount of cement, large amount of admixture (fly ash) and low hydration heat; without longitudinal joints and with few formwork quantities, the rapid construction process with whole-block placement & thin lift roller compaction and continuous lift is adopted; the RCC is well compacted with vibration by the roller, and thus the defects and deficiencies caused by insufficient vibration of the normal concrete will not occur, which is the real charm of rapid RCC construction.

With the full section RCC damming technology, the interlayer bonding quality has always been a concern. The technical route to improve the interlayer bonding quality in the early days mainly relies on laying mortar on the lift/joint surface. This is because the early RCC is super-dry hard concrete or dry hard concrete, the mixture is with very large VC value, and the concrete lift after rolling is with very poor fluidification and bleeding. The dam mainly relies on the normal concrete on the upstream surface for conventional seepage control, i. e. , so called "RCD". Material science is the foundation of rapid construction of RCC. With the rapid development of full section damming technology in the late 1990s, the performance of RCC mixture has undergone qualitative changes. The RCC mixture has developed into no-slump semi-plastic concrete, which features rich cementing paste and long retarding time. This ensures that the aggregate of upper RCC is embedded into the lower concrete after being rolled, and the problem with interlayer bonding is thus solved. Therefore, the development process of RCC can be vividly compared to the development process from pig iron to wrought iron. The hardened concrete has also developed from brittle concrete with high strength, high elastic modulus, low tension, low impermeability and low frost resistance in early days to the current semi-plastic concrete with high tension, low elastic modulus, high impermeability, high frost resistance and excellent crack resistance. Changes in the performance of RCC mixture reasonably solves the technical problems with "interlayer bonding, temperature control and cracking prevention" of the core technology of rapid RCC damming, and provides scientific technical support for rapid damming.

The balanced construction of RCC is quite different from that of the normal concrete dam, which requires close attention in the construction organization design. Because the RCC dam is wider in the substructure and is of relatively simple construction, it is favorable for the rapid construction of RCC. With the dam body rising, the dam superstructure becomes narrower and narrower, and with the overflow surface outlet in the dam superstructure, the headrace penstock of the powerhouse at dam toe or otherstructures to be provided, the dam superstructure becomes more complex. Although the dam superstructure requires small RCC quantities, it is very unfavorable to the rapid construction of RCC. Therefore, the characteristics of rapid construction of RCC and the influence factors of dam structure shall be incorporated in the construction organization design. The balanced construction quantities cannot be evenly distributed on a monthly basis. Instead, the construction intensity of RCC for the dam substructure is much greater than that of the superstructure, which is the biggest difference in balanced construction of dam between RCC and normal concrete.

According to my practice of RCC test and damming technology for many years, I deeply feel that the key technology of rapid RCC construction needs continuous research and technological innovation. In addition, while summarizing and discussing the advantages of RCC damming technology, the optimal scheme for RCC dam shall be selected according to the value engineering method, taking all factors from design, construction to management into account. Everything has both positive and negative effects. Despite the success of RCC damming technology, the problems still exist. It is necessary to examine, summarize and analyze the successful aspects and deficiencies of rapid RCC construction technology. The rapid construction of RCC dam is a system engineering involved with many technologies, especially the key construction technologies and measures which need repeated research and demonstration. Only with continuous technological innovation can the optimal construction scheme be selected.

This chapter is mainly based on *Construction Specifications for Hydraulic Roller Compacted Concrete* DL/T 5112—2009. With the research and case study on the construction procedure of RCC dam, and on key technologies from construction organization design, dam body construction zoning, cushion RCC, mixing, transport and placement, formwork engineering, block surface operation, inclined layer rolling process, joint formation, interlay bonding, increase of rolling layer thickness and synchronous rise of anti-abrasion concrete on overflow surface and RCC, it provides first-hand reference for the rapid construction of RCC.

9.2 Key Technology for Rapid Construction of RCC

9.2.1 RCC Construction Zoning

For the RCC construction zoning or block surface zoning planning, the detailed block surface construction organization design shall be prepared in accordance with the general schedule for construction organization design, which is the basis for rapid construction of RCC and is directly related to the characteristics of block surface, technical requirements and pouring methods, resource allocation, quality and safety assurance measures, etc.

The RCC construction zoning principles for the dam are mainly based on such factors as the dam material zoning elevation, dam structure layout, mixing production capacity, road layout, transport and placement method and pouring intensity, construction zones shall be reasonably divided, fully considering the characteristics of continuous and rapid pouring for large RCC block surfaces.

Recently, the RCC construction zoning in China is dominated by the whole-block placement and 3 m lift, so it is common practice to combine several blocks into a large one for construction. Because the transverse joints in the dam body are divided by seam cutting machine and the inclined layer rolling construction technology is adopted, the joint division and interval time between layers are no longer main restraints for the construction zoning. For the RCC, the large block surface is not only favorable to the efficiency of block surface equipment, but also helpful for reducing the number of formworks used between dam sections. It can also facilitate the block surface management. However, too large block surface also brings disadvantages. Due to the interval time between layers, temperature control, consolidation grouting for dam foundation and other reasons, continuous lift of RCC is prevented, and the idle pouring equipment during the interval will also affect the efficiency of equipment. Moreover, with too large block surface, under exceptional circumstances, the interlayer bonding quality of RCC will be affected when the interval time between layers is longer than the initial setting time.

No matter for single-block construction or whole-block construction, the block surface construction organization design is of great importance. Block surface design is a highly integrated and comprehensive concentration of design drawings and design contents fully understood and digested by the construction technicians, a plan to be implemented for a RCC pouring layer, and it integrates the construction resource allocation and planning, so that the technical personnel can provide technical disclosure to the operators and the operators can easily understand and accept the above contents. Full compliance with the design can avoid the randomness of construction, ensure the implementation of construction technology and improve the construction efficiency. In bidding for projects in China, according to the specific conditions of the project, the concrete construction organization design shall generally incorporate the detailed planning for RCC construction zoning of the dam.

For example, Longtan Dam requires great RCC pouring quantities, which is required to be completed in tight construction period and with complex technologies. Therefore, high-intensity and rapid construction must be carried out. To this end, according to the different pouring elevation, meteorological conditions, capacity of pouring equipment, dam section image requirements, etc. , reasonable block surface planning and block surface process design are proposed for the pouring blocks. For RCC construction in the hot season, 1 - 2 RCC dam sections are taken as a pouring block, which is with a block surface area of 4 000 - 6 500 m^2, while in the cold season, 3 - 4 RCC dam sections as a pouring block with a block surface area of 10 000 - 15 000 m^2. Block surface design is conducted for each pouring block, which shall briefly integrate the characteristics, technical requirements, construction methods, key quality points and resource allocation for the pouring block, in order to provide guidance for the operation team to carry out well-organized and efficient construction.

Take Jin'anqiao Dam for example, the RCC construction is conducted by zones and in layers. While following the pouring schedule, the construction process inthin-lifts, at short intervals and with uniform lift shall be adopted as far as possible. The Jin'anqiao Dam RCC sub-silo is mainly divided according to the dam structure and the mixing intensity of the concrete production system. The concrete for the sediment discharge bottom outlets at dam sections $6^{\#}$ and $12^{\#}$ on the left and right banks is constructed as one block separately according to the elevation of the bottom outlet; the concrete for non-overflow dam sections $2^{\#} - 6^{\#}$ and $16^{\#} - 19^{\#}$ on the left and right banks is constructed as one block integrally; overflow dam section $13^{\#} - 15^{\#}$ are constructed as one block integrally, and the concrete for powerhouse dam sections $7^{\#} - 11^{\#}$ is constructed as one block integrally. Meanwhile, considering the dam structure, concreting elevation, road layout, placement methods and production intensity of mixing plants, when the RCC of dam is poured above EL 1 352 m, non-overflow dam sections and sediment discharge bottom outlets at dam sections on the left bank and powerhouse dam sections $2^{\#} - 11^{\#}$ will be concreted as one whole-block placement integrally; non-overflow dam sections on the right bank and overflow dam sections $19^{\#} - 13^{\#}$ will be concreted as one whole-block placement integrally; the sediment discharge bottom outlets at the dam section $12^{\#}$ on the right bank will be concreted as one block separately. The RCC placement for Jinanqiao Dam is poured into 10 rolling areas, which greatly reduces the workload of cross joint formwork. The layer thickness of poured RCC for the dam is 3. 0 m, while the thickness of the local position can be adjusted to 1. 5 m according to the structure.

RCC slopes can be adopted to transit the elevation difference between adjacent concrete placement of RCC dam, which has been successfully put into practice in Baise, Guangzhao, Kalasuke, Jinanqiao and other projects. See Figure 9. 1 for the layout of RCC slopes between adjacent RCC placements.

9.2.2 Rapid Construction of Blinding RCC

Construction Specification for Hydraulic Roller Compacted Concrete DL/T 5112—2009 stipulates: before laying foundation blocks, apply the mortar on the bedrock surface first, and then pour the bedding layer concrete or GEVR, or directly lay small aggregate concrete or fat mortar concrete on the bedrock surface. Unless otherwise specially required, the thickness shall be convenient for rolling after leveling.

For RCC dam, its bottom width is larger and foundation restraint range is higher because of no longitudinal joints. In order to prevent cracks in foundation concrete, the allowable temperature of foundation shall be controlled. RCC pouring is not convenient as the foundation of the dam is located on an uneven bedrock

Figure 9.1 *Layout of RCC slopes between adjacent RCC placement*

surface. Therefore, it is necessary to pour the blinding concrete with a certain thickness for leveling and consolidation grouting before RCC pouring, and then RCC construction can be performed. Since the blinding is constructed with normal concrete, on one hand, the concrete for blinding is of high strength (generally ⩾ C25), and the cement consumption is large, which is not conducive to temperature control; on the other hand, due to the small placing surface of the concrete for the blinding, large amount of formwork and low construction strength, the concreting for the blinding is one of key factors restricting the rapid construction of RCC dam.

The construction practice in recent years shows that RCC can achieve the same quality and performance as those of normal concrete. Therefore, the bedrock surface is immediately concreted with RCC after being leveled with low-slump normal concrete, which can obviously accelerate the blinding concrete construction. Generally, blinding concrete pouring is performed in winter or at a low temperature, so RCC (instead of normal blinding concrete) is adopted to quickly construct the foundation cushion at a low temperature, which can effectively speed up the consolidation grouting, control the foundation temperature difference and prevent deep cracks in the dam foundation concrete, and is beneficial to temperature control and construction progress.

For example, RCC rapid construction technology is adopted for the foundation cushion of Baise Dam. The foundation cushion of Baise RCC gravity dam was immediately concreted with RCC after being leveled with normal concrete on December 28, 2002. Due to the rapid construction of RCC, the pouring of blinding concrete in the foundation restraint area was completed in the optimal low temperature season (the foundation consolidation grouting was arranged in the high-temperature season (June-August) when RCC was not poured, which increased the costs of drilling operation, but was beneficial to cover consolidation grouting). Temperature control measures were greatly simplified, and penetration cracks in the foundation are also properly prevented. Core samples were taken by drilling to the bedrock in August, 2003, showing that the layers of bedrock, normal concrete and RCC were closely combined with each other, and the strength fully met the design requirements. Up to now, 3 core samples (10 – 12 m long) are still intact, and the bedrock is closely combined with the joint surface of concrete layer, which is unable to be identified. See Figure 9.2 for the core sample of bedrock, normal concrete and RCC.

Figure 9.2 *Photos of core sample of bedrock, normal concrete and RCC*

9.2.3 RCC Mixing

9.2.3.1 Mixing Equipment

Forced mixing equipment and gravity or continuous forced mixing equipment should be selected for RCC mixing. Forced mixers are suitable for dry concrete. According to a large number of project construction practices, the quality of RCC mixed by forced mixers is good, and the mixing time is short. Most of RCC dam projects in China adopt forced mixers in recent years, such as Longshou, Jiangya, Linhekou, Longtan, Guangzhao, Jinanqiao, Guandi and other projects. Gravity or forced continuous mixers can also mix high-quality RCC. For example, gravity mixers have been successfully used in the Three Gorges Phase Ⅲ Cofferdam Project, Baise and other projects. In addition, continuous mixers have been used in multiple projects, such as Shapai, Suofengying, Zhaolaihe, Kalasuke, Songshen Reservoir and other projects.

Continuous forced RCC mixers have a breakthrough innovation based on traditional RCC mixers. Such mixers are characterized by small quantity of civil works, small volume, light weight and quick and easy installation and removal. For example, the MY-BOX continuous forced mixing system cooperated by Sinohydro Bureau 8 Co., Ltd. and Tsinghua University has been successfully applied in the Shapai RCC Arch Dam Project. The quality control and the results of core sample test show that various physical and mechanical indicators meet the design requirements. The 150 m^3/h continuous forced mixers produced by Guangzhou Duowei Construction Machinery Investment Co., Ltd. are used in the RCC construction of Zhaolaihe, Kalasuke and Xinhui Reservoir dam projects. The performance meets the design requirements and RCC construction specifications, and the quality of concrete is good. However, continuous forced mixers also have some disadvantages, the performance of which is similar to that of truck mixers. Upon the startup/shutdown, the uniformity of concrete mixtures at the front and rear is slightly poor. Due to the centrifugation effect, there are more coarse aggregate and insufficient grout wrapping at the front. On the contrary, there are less coarse aggregate and more mortar at the rear. Moreover, the *VC* value of mixtures fluctuates greatly. The solution is to adjust the discharge time difference of each component to the best time.

Nowadays, all mixing systems have been controlled by computers. Due to the changes of RCC mixture performance, forced mixers, gravity mixers and continuous forced mixers can all produce new RCC with good quality, which can fully meet the quality requirements of precise mixing. See Table 9.1 for RCC mixers applied in domestic projects.

Table 9.1　　RCC mixers applied in domestic projects

S/N	Project title	Batching and mixing plant	Product and model	Specification and capacity (m³)	Quantity	Mixing capacity
1	Fenhe Reservoir Ⅱ	Forcing type	JS1500	2×1.5	2	Mixing capacity 185 m³/h
2	Dachaoshan	Gravity type (869 platform)	—	4×3.0	2	Mixing capacity: for RCC or 480 m³/h + 75 m³/h for normal concrete
		Gravity type	—	3×1.5	1	
3	Longshou	Forcing type	HZS75-1500	3×1.5	3	Mixing capacity (nameplate) 3×75 m³/h = 225 m³/h、1×10 m³/h
		Forcing type	FangYuan	1×1.0	1	
4	Shapai	Continuous forced type	—	—	1	Production capacity: 200 m³/h for continuous type Gravity type 60 m³/h
		Gravity type	Zhengzhou mixing plant	—	1	
5	Linhekou	Forcing type	—	1×3.0	1	Mixing capacity (nameplate) 315 m³/h
		Forcing type	—	2×1.5	2	
6	Baise	Gravity type (1#)	—	4×3.0	1	Capacity rating: 470 m³/h RCC or 200 m³/h normal concrete
		Forcing type (2#)	Japanese mixing plant	2×4.5	1	
		Forcing type (3#)	—	2×3.0	1	
7	Suofengying Hydropower Station	Forced type (double-horizontal shaft)	HZ300-2S4000 L	2×4.0	1	Production capacity: Forcing type 250 m³/h、continuous type 200 m³/h
		Continuous forced type	—	—	1	
8	Phase Ⅲ cofferdam of the Three Gorges Project	Gravity type (low linear velocity)	—	4×3.0	2	Mixing capacity (nameplate): 240 m³/h for low linear velocity + 360 m³/h for high linear velocity = 1 000 m³/h
		Gravity type (high linear velocity)	—	4×4.5	2	
9	Pengshui	Gravity type (right bank)	—	4×4.5	2	Mixing capacity (nameplate): 280 m³/h + 180 m³/h on the right bank and 240 m³/h on the left bank
		Forcing type	—	2×3.0	1	
		Gravity type (left bank)	—	—	1	

Continued Table 9.1

S/N	Project title	Batching and mixing plant	Product and model	Specification and capacity (m^3)	Quantity	Mixing capacity
10	Longtan	Forcing type	Japan	2×6	3	Design production capacity: $3×300\ m^3/h + 180\ m^3/h$ = 1 080 m^3/h, with the highest shift output reaching 4 770 m^3.
		Gravity type	—	4×3	1	
11	Guangzhao	Forced type (left bank)	—	2×4.5	2	Mixing capacity (nameplate): 660 m^3/h on the left bank + 180 m^3/h on the right bank = 840 m^3/h
		Forced type (left bank)	—	2×3.0	1	
		Gravity type (right bank)	Zhengzhou	4×3.024	1	
12	Gelantan	Forcing type	Japan IHI	2×4.0	1	480 m^3/h normal concrete or 360 m^3/h RCC, peak month concrete is 100 000 m^3
		Forcing type	Hangzhou	2×3.0	1	
13	Jin'anqiao	Forced type (left bank)	Zhengzhou	2×4.5	2	Mixing capacity (nameplate): Roller compacted concrete $3×300\ m^3/h$, Normal concrete 240 m^3/h, Concrete under temperature control $3×250\ m^3/h + 180\ m^3/h$
		Forced type (right bank)	Zhengzhou	2×4.5	1	
		Gravity type (right bank)	Zhengzhou	4×3.0	1	
14	Kalasuke Hydropower Station	Forcing type	—	2×4.5	2	Comprehensive capacity: 1 020 m^3/h normal concrete or 860 m^3/h RCC Refrigerated concrete 660 m^3/h
		Forcing type	—	2×3.0	1	
		continuous type	DW	—	1	
15	Gongguoqiao	Forcing type	HL320-2S4500L	2×4.5	1	Comprehensive capacity: 560 m^3/h normal concrete or 460 m^3/h RCC Refrigerated concrete 400 m^3/h
		Forcing type	HL240-2S3000L	2×3.0	1	
16	Guandi	Forced type (low linear velocity)	HL360-2S6000 L	2×6	2	Design production capacity: Low linear velocity $2×300\ m^3/h$, High linear velocity 480 m^3/h
		Forced type (high linear velocity)	—	2×6.0	1	
		Gravity type (high linear velocity)	Zhengzhou	4×3.0	1	

9.2.3.2 Feeding Sequence and Mixing Time

The feeding sequence, mixing time and mixing amount of RCC shall be determined by field RCC mixing process test. The practice shows that, different mixers, such as forced mixers, gravity mixers or continuous mixers, have different feeding sequence and mixing time, and the mixing state and performance of RCC are directly related to the feeding sequence of raw materials, and mixing time and mixers. Therefore, the most reasonable feeding sequence and the optimal mixing time shall be selected through tests.

The RCC mixing process test is of significant importance. RCC mixing process test is carried out before concrete construction, so as to obtain various technical parameters needed for RCC mixing and provide effective parameters for construction quality control of RCC for the dam. The mixing process test is mainly to determine the optimal feeding sequence and the minimum mixing time of RCC, so as to fully ensure the uniformity of RCC mixture and save cost. Therefore, before the construction of RCC for the dam, the productive mixing process test shall be conducted. Proper mixing process can not only ensure the good uniformity of the mixed concrete, but also accelerate the production of concrete.

Different feeding sequences have great influence on the uniformity of the mixture, and the mixing time of different types of mixers is also significantly different. The mixing time of a forced mixer is much shorter than that of a gravity mixer. The mixing time of the former is generally 80 – 90 s, while that of the latter is generally 150 – 180 s.

Ⅰ Mixing process test of the forced mixer

Baise RCC is mixed by a forced mixer. A specific feeding sequence is preliminarily selected in the mixing process test with different mixing time; the test is conducted to determine the minimum mixing time that meets the uniformity of the concrete. Then different feeding sequences are conducted respectively under the same mixing time to finally determine the best feeding sequence and mixing time of RCC. See Table 9.2 for the results of the forced RCC mixing process test.

Table 9.2   Results of the forced RCC mixing process test

No.	Feeding sequence	Batching time (s)	Concrete uniformity
D1	Manufactured sand→quasi-large-sized stones→medium-sized stones→small-sized stones→cement + fly ash + water + admixtures	70	Compared difference
D2		80	General
D3		85	Quite good
D4		90	Good
A	Manufactured sand + cement + fly ash→small-sized stones→medium-sized stones→quasi-large-sized stones→water + admixtures	85	Quite good
B1	Manufactured sand + cement + fly ash + water + admixtures→small-sized stones→medium-sized stones→quasi-large-sized stones	75	Quite good
B2		85	Good
B3		90	Good
C	Manufactured sand→small-sized stones→medium-sized stones→quasi-large-sized stones→cement + fly ash + water + admixtures	85	General
E	Cement + fly ash + water + admixtures-manufactured sand→small-sized stones→medium-sized stones→quasi-large-sized stones	85	Quite good

Results of the RCC mixing process test show that:
1. It can be seen from No. D1 – D4 samples that with the same feeding sequence, the longer the mixing time of RCC, the better the uniformity, and when the mixing time is 85 s, the uniformity is good; when the mixing time is 90 s, the uniformity is better. Therefore, the mixing time of RCC is determined to be 90 s.
2. No. A, B1 – B3, C and E samples are tested with different feeding sequences, which has great influence on the uniformity of the mixture. The uniformity of RCC mixed with feeding sequence B is the best, i.e., manufactured sand + cement + fly ash + water + admixtures are charged at the same time to mix the mortar,

and then small-sized stones, medium-sized stones, and quasi-large-sized stones are put into the mortar in sequence for mixing.

Why does the feeding sequence of a forced mixer have so great impact on the uniformity of RCC mixture? The analysis believes that due to the certain gap between the mixing blade and the steel wall of the forced mixer, the feeding sequences for mixing mortar can lubricate the gap. If the sequence of coarse aggregate→cementing materials + water→sand is followed (adopted for most of gravity mixers), the gap between the blade and the steel wall can easily get stuck by the needle-like aggregate, which may affect the mixing quality.

II Mixing process test of the gravity mixer

The RCC for the cofferdam in Phase III is mixed with gravity mixers, and the RCC is poured in athin-lift by continuous rise method. The production capacity of RCC is very high. The mixing time determines the production capacity and placement strength of RCC, and directly affects the rapid construction of the cofferdam. Therefore, it is very important to select a reasonable mixing process.

1. Feeding sequence. According to the structure of the gravity mixers and the experience of concrete mixing process in the Three Gorges Phase II Project, two feeding sequences are adopted in the test:

a) Medium and small-sized stones→cement + fly ash→admixtures + water + sand→large-sized stones.

b) Large, medium and small-sized stones→cement + fly ash→admixtures + water + sand.

2. Batching time. In order to seek the shorter mixing time, the concrete mixing speed is increased in the test, so as to increase the output of the mixing plant, and the uniformity of the concrete mixture with different mixing times is also tested. The 1 st and 2nd feeding sequences are adopted in the crossover test with 120 s, 150 s and 180 s respectively. See Table 9.3 for the test results.

Table 9.3　　　　　　　　The results of the gravity RCC mixing process test

No.	Feeding sequence	Batching time (s)	VC value (s)	Gas content (%)	Compressive strength (MPa)	
					7 d	28 d
T120	Medium and small-sized stones→cement + fly ash→admixtures + water + sand→large-sized stones	120	9.7	0.9	9.1	15.0
T150		150	6.8	2.8	11.6	19.6
T180		180	8.9	2.4	9.4	16.5
T150	Large, medium and small-sized stones→cement + fly ash→admixtures + water + sand	150	6.9	3.7	10.8	20.0
T180		180	8.2	1.5	9.7	16.3

Thetest results show that: When the mixing time is 120 s, the VC value of the concrete mixture is larger, which indicates that the concrete mixing is insufficient and the uniformity of the concrete mixture is poor. The VC value of mixing time of 150 s is less than that of mixing time of 120 s and 180 s, and the air content of the concrete is higher, which is beneficial to improve the frost resistance of the concrete. The deviation rate of 7 d and 28 d compressive strength of the concrete under the 2nd feeding sequence with mixing time of 180 s is small, but the VC value is large and the air content is lower. After comparison, it is determined that the 2nd feeding sequence (large, medium, and small-sized stones→cement + fly ash-admixtures + water→sand) is adopted for RCC mixing, and the mixing time is determined to be 150 s.

III Mixing process of the continuous mixer

The general feeding sequence of the continuous mixer: Coarse aggregate—cement—admixtures and water—sand. The mixing time is generally about 20 s. The mixtures at the front and rear of the plant are prone to be not uniform upon the startup/shutdown of a continuous mixer.

When adjusting the feeding sequence and mixing time, pay attention to changes in the resistance of the mixer shaft. Now the advanced forced mixers with double horizontal shaft reflect the resistance of the mixer shaft through the current. The influence of the adjustment of the feeding sequence on the mixer is shown in the oil pressure-current curve. When adjusting the feeding sequence, observe the changes of RCC state, such as uniform mixing, VC value, or agglomeration (lumps) of cementing material.

9.2.3.3 Concrete Production System on the Right Bank of Longtan Hydropower Station

I Concrete production system

RCC is produced by two concrete production systems, which are arranged on the right bank at an elevation of 360.00 m and 308.50 m respectively.

The concrete production system at the elevation of 360.00 m is equipped with a 2×6.0 m^3 forced mixing plant with double horizontal shaft and a 4×3.0 m^3 gravity mixing plant. The forced mixing plant is mainly used to produce RCC, and can also produce normal concrete. The design capacity of RCC is 300 m^3/h, and that of normal concrete is 250 m^3/h; the gravity mixing plant is mainly used to produce normal concrete, with a design capacity of 180 m^3/h.

The concrete production system at the elevation of 308.50 m is equipped with two 2×6.0 m^3 forced mixing plant with double horizontal shaft, which are mainly used to produce RCC, and can also produce normal concrete. The design capacity of RCC is 2×300 m^3/h, and that of normal concrete is 2×250 m^3/h. See Table 9.4 for detailed technical indexes of both mixing systems.

Table 9.4   Main technical indexes of concrete production system on the right bank of longtan hydropower station

S/N	Item	Unit	Elevation 360.00 m system	Elevation 308.50 m system	Total	Remarks
1	Design capacity	m^3/h	480	600	1 080	
2	Design capacity for pre-cooling concrete	m^3/h	220	440	660	RCC Tmachine = 12 ℃
			150	360	510	Normal concrete tmachine = 10 ℃
3	Processing capacity of screening & washing and dehydration workshop	m^3/h	1 300	1 400	2 700	Coarse aggregate screening and washing
4	Design conveying capacity of finished aggregate	t/h	2 400	2 700	5 100	The sand is 600 and 700 t/h respectively
5	Dump volume of finished aggregate	t	27 000	45 000	72 000	Meet 1 d consumption during the peak period
6	Design cement conveying capacity	t/h	120	120	240	
7	Cement reserves	t	4 500	4 500	9 000	Meet 7 d consumption during the peak period
8	Design conveying capacity of fly ash	t/h	90	90	180	
9	Fly ash reserves	t	1 700	2 550	4 250	Meet 3 d consumption during the peak period
10	Air volume	m^3/min	180	180	360	The reserve capacity is 20 m^3/min respectively
11	Cold blast rate	10^4 m^3/h	56	64	120	The cold blast rate of the secondary air is not counted

Continued Table 9.4

S/N	Item	Unit	Technical indicators			Remarks
			Elevation 360.00 m system	Elevation 308.50 m system	Total	
12	Production capacity of flake ice	t/h	11.25	10	21.25	
13	Production capacity of 5 ℃ cold water	m³/h	145	35	180	Including 120 m³/h of cooling water for the high dam
14	Water demand	m³/h	950	850	1 800	800 and 680 m³/h for the pre-cooling system respectively
15	Cooling system installed capacity	10^4 kW	1.56	1.62	3.18	
		10^4 kcal/h	1 343	1 399	2 742	
16	Gross installed capacity of the system	kW	15 000	14 724	29 724	
17	Working system	Shift/d	3	3	—	
18	Total structure area of the system	m²	2 690	3 518	6 209	Cooling plant is not included
19	Total floor area of the system	m²	53 000	41 000	94 000	

II Feeding sequence and mixing time of RCC mixing by the forced mixing system

1. The feeding sequence of the Level 2 graded RCC is determined as (sand + cement + fly ash) →(water + admixtures) →(small-sized stones + medium-sized stones), with a mixing time of 90 s.
2. The feeding sequence of the Level 3 graded RCC is determined as (sand + cement + fly ash) →(water + admixtures) →(small-sized stones + medium-sized stones + large-sized stones), with a mixing time of 90 s.

III Feeding sequence and mixing time of RCC mixing by the gravity mixing system

1. The feeding sequence of the Level 2 graded RCC: (small-sized stones + admixtures + water) →(cement + fly ash + sand) →medium-sized stones, with a mixing time of 150 s.
2. The feeding sequence of the Level 3 graded RCC: (Medium-sized stones + small-sized stones + admixtures + water) →(cement + fly ash + sand) →large-sized stones, with a mixing time of 150 s.

IV Mixing time of pre-cooling concrete

When other measures such as ice is required in the production of RCC, the mixing time of a forced mixing system is extended by 15 s, and the mixing time of a gravity mixing system is extended by 30 s on the original basis, provided that the feeding sequence is unchanged.

V Strict monitoring of the first three tanks of concrete produced by the mixing plant

Every time the mixing is started, the on-site test and inspection personnel of the laboratory will closely monitor the first three tanks of concrete produced by the mixing plant, so as to ensure that "The first tank of concrete is tested, the second tank of concrete is adjusted, the third tank of concrete is re-adjusted, and the fourth tank of concrete is normal". The first three tanks of concrete can be placed only after passing the test by the quality controller of the laboratory.

9.2.4 Transport and Placement of RCC

9.2.4.1 Way of Warehousing

According to construction practice, dump trucks, belt conveyors, negative pressure chutes (tubes), special vertical chutes and Manguan chutes (tubes) are mainly adopted for the transport and placement of RCC, and cable cranes, gantry cranes and tower cranes can also be used as auxiliary transportation tools.

Generally, the lower part of the middle and low dams and the high dams are mainly placed directly by trucks, and the vertical transportation for the middle and upper parts mostly performed combined with vacuum chutes in recent years. With the development of high damming, steep slope and vertical transportation equipment have also been developed and applied, and the transportation technology of vacuum chute has made new progress. Hundred meters of negative pressure (vacuum) chutes have been adopted in Dachaoshan, Shapai and Linhekou projects, two vacuum chutes have been arranged on the left and right banks of Dachaoshan Hydropower Station, of which the maximum height difference of the vacuum chute on the left bank is 86.6 m and the length of the chute is 120 m, thus such a vacuum chute has the highest conveying height in China at that time, and the conveying capacity of the vacuum chute is 220 m^3/h. The 100 m level vacuum chute is a kind of economic and effective mean to solve the vertical transportation of RCC in high mountains and valleys area and under the condition of high fall, which has greatly improved the progress of RCC construction, enabling the technical advantage of rapid construction for RCC to be sufficiently applied. See Figure 9.3 for combined placing of Suofengying chute + sliding barrel + placement surface aggregate bin.

Figure 9.3 *Combined placing of Suofengying chute + sliding barrel + placement surface aggregate bin*

Starting from the Xiaolangdi and the Three Gorges Projects in late 1990s, because of the introduction as well as development and application of tower conveyor (top conveyor) and tyred conveyor, the horizontal and vertical transport of the concrete is combined into one. The horizontal and vertical transport of RCC in Longtan Project was integrated, high speed feeding line + tower (top) conveyor was used to place the concrete, its pouring strength is increased twice. As a result, for the allocation of various resources of placing surface, no matter the quality or the quantity will be increased obviously, the requirement for the placement management

Chapter 9 Key Technology for Rapid Construction of RCC

level is also improved significantly. The application of tower conveyor (top conveyor) and tyred conveyor has brought the belt conveyor into wide application, enabled the revolution for the conventional mode of RCC, the researched and developed mobile conveyor and telescopic cantilever spreader has been successfully applied in the construction practice of projects.

9.2.4.2 Full Pipe Chute Vertical Conveying

Transport of RCC into placing surface is always one of the key factors restricting rapid construction. A large number of engineering construction practices have demonstrated that direct truck placing is the most effective mode of fast construction, which can greatly minimize the intermediate links, and reduce the flow backward due to temperature rise of the concrete. The time from the mixing plant to the placing surface of the dam generally ranges from 15 – 30 min, the automatic tarpaulin is provided on top of the dump truck for sun protection, the placing temperature increase generally will not exceed 1 ℃.

As the dam was built in a high mountain valley, the elevation difference of the road to go up the dam is too big for the truck to place directly, many kinds of transportation equipment are used for vertical conveying of RCC at the intermediate links, many kinds of placing modes were used successively such as turnover by negative pressure (vacuum) chute, tower conveyor, cable crane, tower crane, aggregate bin, etc..

In recent year, the Guangzhao 200.5 m RCC gravity dam adopted the full pipe chute, the full pipe chute was used to solve the challenge of vertical transport with large elevation difference, that is the combined placing mode of truck (deep-trough belt conveyor) + full pipe chute + placement surface truck. The practices show that combined placing plan by adopting full pipe chute is the transport placement mode that is least in investment, the most simple and efficient way. Due to the increase of the full pipe chute size, it has completely taken the place of conventional negative pressure (vacuum) chute. The cross-section dimension of the full pipe chute now has reached to 80 cm × 80 cm, with vent holes distributed on the top of the full pipe chute, the declination angle generally is 40° – 50°, the aggregate bin on the top of the placement surface is removed, the truck outside the placement is directly unloaded the concrete to the placement surface truck by passing the discharge hoper through the full pipe chute, allowing very simple and fast transport and pouring. As the existing RCC is the semi plastic concrete with high rock powder content and low VC value, the worrying issue of aggregate separation is also readily solved. See Figure 9.4 for the 1 424 m elevation full pipe chute on left bank of Jin'anqiao. The full pipe chute has been successfully applied in Guagnzhao, Baisha, Gelantan, Jinanqiao and other Projects, and good effect has been obtained, which has effectively solved the construction challenges that restrict vertical conveying and placing intensity of the RCC.

Let's take the Guagnzhao Dam with 200.5 m of dam height as an example. The truck and deep-trough high speed belt conveyor were mainly used in horizontal conveying of the RCC, cable crane and full box pipe were used in vertical conveying. Four kinds of placing modes were used as the placing modes of RCC, that are direct placing from dump truck, belt conveyor + placing from full box pipe + dump truck, dump truck + full box pipe, dump truck + cable crane, which has effectively eliminated the challenge of big elevation difference with the dam. See Figure 9.5 for Longtan high speed feeding line + tower (top) conveyor concrete, see Table 9.5 for details of the corresponding modes of RCC placement and engineering quantities.

9.2.4.3 Factors Affecting Placement and Discharge

The dump truck transports the RCC and carries out direct placing, quantities of the placing opening, structure and closing construction methods have very big influence on the construction quality and speed; the dirt, soil, etc. entrained in the wheels affects the cementing quality of the concrete layer; brought-in moisture will change the workability of the concrete and the water-cement ratio, thus impacting the quality of the concrete; Sudden braking and sharp turning of the truck will damage the concrete surface whose strength is not high enough, and affects the cementing of the layer; to ensure the structure and shape of the dam part entering the placement

Figure 9.4 1 424 *elevation full pipe chute on left bank of Jinan Bridge*

Figure 9.5 *Longtan high speed feeding line + tower (top) conveyor concrete*

opening, it is necessary to take the technical measure that crosses the formwork at the entry of placing.

Table 9.5　　　　　　　　　　RCC placement modes and quantities table

S/N	Placement mode	Cementing height	Quantities ($\times 10^4$ m^3)
1	Placing direct form dump truck	556.5 – 622.5	82.5
2	Placing from belt conveyor + 2$^\#$ full box pipe + dump truck	622.5 – 676.5	67.5
3	Placing from belt conveyor + 1$^\#$ full box pipe + dump truck	667.0 – 740.5	77.5
4	Dump truck + 3$^\#$ full box pipe	704.5 – 740.5	11
5	Dump truck + large chute	740.5 – 748.5	1.5

The present configuration of placing route for RCC. Most of the lower part of the dam body is backfilled directly to form a direct placing access to trucks, no matter the placing from the dump truck is before or behind the dam, the first is due care shall be taken to keep the slope ratio of the road not to be more than 12.5%, the second is the mode for the dump truck to cross the formwork shall use steel trestle to ensure that the dam surface is not damaged.

The sunshade and rainproof measures shall be set on the transport equipment of the RCC such as dump truck, belt conveyor, etc., which can minimize the impact on the concrete mixing and physical properties of the RCC by the external environment. The separation of aggregate is prone to be caused during the operation of the belt conveyor, therefore, the separation prevention device shall be equipped at the discharge end, the scraper and cleaning device shall be mounted when the belt conveyor discharges, which can control the loss of the mortar.

When the equipment such as mixing plant discharge opening, negative pressure (vacuum) chute, full pipe chute (pipe), spreader and so on is in use, the concrete speed at the outlet can be from 5 m/s to 10 m/s, and the concrete can be directly discharged from the dump truck, which can cause serious impact and aggregate separation. The rubber discharge port with good buffering performance below the arc door at the outlet of the mixing plant. The straight down elbows and rubber discharge port shall be provided for the site placement tube (chute), this can significantly slow down the speed at the outlet, and function as mixing and uniformity improving, which can effectively prevent impact and separation. The chute adopting the made-to-order rubber hose and other special structure can effectively prevent the aggregate from separation. The continuous mixer shall be equipped with a storage hopper of sufficient volume, which can guarantee the continuous operation of the continuous mixer.

The measures against sediment and bleeding shall be in place when the mortar is transported, which can ensure that the grout transported to the site is uniform. The transit mixer can be used to transport the mortar, and the effect is good.

9.2.4.4 The shortages and Challenges with Transport by Tower Conveyor

Using the tower conveyor as the mean to place the RCC is the integration of horizontal transport, vertical transport and surface spreading, which can realize complete operations from mixing plant to placing surface, has the advantages of fast speed, high efficiency, large coverage area, continuous feeding, direct spreading and so on, therefore, with significant advantages. However, the shortages and challenges have been observed during the transport of RCC by high-speed tower conveyor, conveyor mounted wheel crane, and deep-trough belt conveyor.

One of the most prominent problems is the separation of aggregates, which can seriously impact the construction quality of the RCC if it is not handled properly. After the RCC mixture is transport to certain distance away by means of the belt, the large aggregate will gradually floating up and the fine aggregate will sink gradually due to the vibration caused by the movement of the belt and carrier rollers under it, which results in preliminary separation of the mortar from the aggregate; when the RCC mixture reaches the top of the discharge belt at certain speed and discharges, due to the law of inertia, the forward speed of the coarse aggregate is fast, which causes the separating of coarse aggregate from fine aggregate. The placing of tower conveyor discharge is quick, during the production, the operation of the tower conveyor frequently lags behind the instruction of stopping discharge at placement surface or moving discharge, which results in over-thickness of concrete pouring billet layer, local mound height or disorder, therefore it is prone to the separation of the aggregates.

When the tower conveyor is used to cast the RCC, its temperature rise is much bigger. As the mixture is transported to the placing block from the discharge port through the feeding line, thethin-lift of concrete is spread on the belt, the area contacting with the air is large, the cold and heat exchange is quick, at the same time, the number of times for receiving and transferring of the feeding line is many. According to the actually measured data from Three Gorges Project, when the air temperature is from 25 – 32 ℃ in high temperature period, the temperature of 12 – 14 ℃ pre-cooled concrete from the mixer will increase 4 – 6 ℃, which is far higher than that is directly placed from the truck.

When the tower conveyor is used to cast the RCC, there is hysteresis for the treatment of out-of-spec materials. During the construction of RCC, it is normal for mixing plant to occasionally produce unqualified material. When the tower conveyor is used to cast the RCC, long distance transport by belt conveyor is required, when the nonconforming material arrives at the placing position, there may be nonconforming material on the belt conveyor with several hundred meters in length. If any equipment failure or power outage occurs during the conveying of the mixture, the feeding line may fail to operate normally, which not only cause stop at the placing surface, but also makes it very difficult to treat the concrete of the belt of the feeding line. Therefore, much higher requirement is proposed to the service, maintenance and operation management of the equipment.

9.2.5 Formwork Work

9.2.5.1 Forms of Formwork

The formwork is one of the key equipment that ensures continuous rise and rapid construction of the RCC, the formwork engineering is an important link in RCC dam construction, and its influence is quite prominent. The construction of earlier RCC works was interrupted due to late installation of formwork, sometimes there was a long interval of 2 – 3 d due to removal of the formwork. The influence of formwork installation on construction speed has been paid more and more attention in the engineering sector, through continuous improvement, optimization and simplification of the types of the formworks, the cantilever turn-over steel formwork has been widely applied in our country.

The special stipulation to the formwork was made in the hydraulic RCC construction specification. The formwork is the important equipment in the construction of the RCC, which has significant impact on various aspects of the RCC such as the appearance, quality, construction progress, cost and so on, the formwork shall meet the requirements of strength, rigidity and stability, reliably bear the various loads in vibration rolling construction, and ensure that the design shape and dimensions of the structure are correct, the deformation is within the allowable range.

The selection of the formwork is as important as the allocation of the mechanical equipment, the formwork design shall meet the requirements of rapid and continuous construction of the RCC, in order to facilitate the surrounding paving work, it is not suitable to set up diagonal brace or minimize the diagonal brace. During the concrete placement, full-time staff shall be arranged to inspect, maintain and adjust the position and form of the formwork so as to prevent the problems such as deformation, leakage of grout, etc.. As the strength of the RCC is much lower at the early stage, the difference between setting time, initial strength, formwork lateral pressure and so on of the RCC and those of the normal concrete is much bigger, therefore, the time for formwork removal and turnover in RCC construction shall be determined through calculation, test and site productivity test, attention shall be paid to reserve certain safety margin.

The form of the formwork shall adapt to the structure, member characteristics, construction condition and casting method; it is advisable to use cantilever formwork, turnover formwork or jack-up formwork, the pre-cast concrete formwork can also be used for back surface. The pre-cast concrete formwork or pre-assembled steelwood formwork can be used in hole structure such as gallery and so on.

The design, fabrication and use of the formwork shall comply with relevant provisions in DL/T 5110—2015. The design of the formwork shall meet the continuous casting characteristics of the RCC. At present, formworks mainly used in RCC dam project in our country are continuous rise cantilever formwork, adjustable cantilever turnover formwork for most of the arch dams, bi-directional adjustable continuous rise cantilever formwork, downstream face continuous rise step formwork, etc.. The gravity dams of Longtan, Guangzhao, Jin'anqiao and so on mainly use large size turnover steel formwork. See Figure 9.6 for the cantilever turn-over formwork in Baise, see Figure 9.7 for gallery precast concrete formwork in Jin'anqiao Dam.

Figure 9.6 *Baise Dam cantilever turn-over formwork*

Figure 9.7 *Jinanqiao Dam gallery precast concrete formwork*

9.2.5.2 Formwork Work

Ⅰ Steel formwork

1. The thickness of the steel formwork panel shall be calculated and determined according to the requirements of structural force, and the thickness shall not be less than 5 mm.
2. All bolts and rivets shall be of countersink.
3. The design of the attachments shall enable the formworks to be assembled reliably, the concrete will not be damaged when removing them.
4. The joints among the metal sheets shall be smooth and tight, any formwork with pit, buckling or other surface blemishes or any formwork with formwork tie hole damaged is not allowed to use.
5. The skeleton frame of the formwork shall be fabricated by using shape steel or steel pipe.

II Slip form

1. The steel lining shall be used for the sliding formwork, the surface of the lining shall be smooth and flat.
2. The sliding formwork shall be oriented by guide rails, which allows precise positioning and continuous movement.
3. The sliding formwork shall have sufficient height, which can prevent the concrete under the sliding formwork from collapse when the formwork slides.
4. The sliding formwork shall have sufficient weight so that the formwork is not affected by lift when concrete is poured.

III Formwork production

The fabrication of the formwork shall meet structure shape of thestructure required by the structure drawing, its allowable fabrication deviation shall not exceed the stipulation in DL/T 5110—2015.

The allowable deviation of special formwork, sliding and traveling formworks and permanent special formwork shall not be more than the stipulation listed in Table 9.6.

Table 9.6   Allowable deviation in formwork fabrication

S/N	Deviation items	Allowable deviation (mm)
I	Wooden formwork	
1	Small formwork: Length and width	±2
2	Large formwork (length and width are more than 3 m) Length, width and diagonal line	±3
3	Height difference between neighboring two plate surface	0.5
4	Local unevenness (to be inspected by a 2 m straightedge)	3
5	Panel gap	1
II	Steel formwork, composite formwork and bakelite formwork	
1	Small formwork: Length and width	±2
2	(length and width exceeding 2 m) Length, width and diagonal line	±3
3	Local unevenness on formwork surface (to be inspected by a 2 m straightedge)	2
4	Hole position of connection fittings	±1

IV Formwork installation

For the installation of the formwork, the survey and setting out shall be carried out according to the detail drawing of the concrete works, more control points shall be set for important structure so as to facilitate the inspection and correction. During the installation of the formwork, sufficient temporary fixtures shall be set to prevent overturn.

The brace of the formwork shall not be bent, the diameter shall be more than 8 mm, the connection of the brace with the anchor ring shall be reliable. The embedded anchoring parts (bolts, reinforcement ring, etc.) shall have sufficient anchoring strength when bearing loads.

Joints between formwork must be smooth and tight. In case of layered construction of the structure, the deviation between the upper and lower layers shall be corrected layer by layer, the lower end of the formwork shall not have any "faulting of slab ends". The allowable deviation of formwork for concrete and reinforced concrete structure shall comply with the stipulations in GB 50204, the allowable deviation for the formwork of concrete in mass shall comply with the stipulations in DL/T 5144. The allowable deviation for formwork installation shall not be more than the values specified in Table 9.7.

Table 9.7   Allowable deviation of integral structure formwork installation

Item No.	Item	Allowable deviation (mm)
1	Axis location	5
2	Foundation	±10
	Wall	+4, −5
3	Height difference between neighboring two plate surface	2
4	Surface evenness (to be inspected by a 2 m straightedge)	5

Special form release agent shall be coated on the formwork panel, the form release agent shall be the same product made by the same manufacturer. The form release agent shall be coated before erection of the reinforcement frame. It is not allowed for the form release agent to contact with the reinforcement and concrete surface.

The contact surface of the precast concrete formwork that embedded in the concrete with the concrete shall be chipped and cleaned according to the provisions concerning "construction joint treatment" in this technical terms and conditions.

V   Formwork removal

In addition to meeting the regulations in the construction drawing, following stipulations about the time limit of the formwork removal shall be complied with:

1. Removal of formwork on non-load-bearing side shall be done when the concrete strength reaches 2.5 MPa and more, the removal can be carried out only after it is assured that no damage occurs on its surface and edges due to removal of formwork.
2. The removal at the locations of piers, walls and columns can be carried out only when its strength is not less than 3.5 MPa.
3. The bottom formwork can be removed only after the concrete strength reaches the values specified in Table 9.8.

Table 9.8   Standard of concrete strength necessary for bottom formwork removal when removing the formwork in cast-in-situ structure

Structure type	Structure span (m)	Percentage of designed concrete strength standard value (%)
board	≤2	50
	>2, ≤8	75
	>8	100
Beam, arch, shell	≤8	75
	>8	100
Cantilever members	≤2	75
	>2	100

VI   Formwork protection

The formwork shall be cleaned before each use, the steel formwork shall be applied with anti-rust protective coating of mineral oil, but the use of any oil agent that may contaminate the concrete is not allowed, the concrete quality shall not be affected. If the concrete surface is contaminated with stains, measures should be taken to remove it in time.

9.2.5.3 Longtan Formwork Work

The dam structure shape within the RCC range on the right bank is relatively regular, most of them belongs to the concrete in mass, therefore, large amount of integral steel formworks were used. The sizes of the large formwork are: 3 m × 3.1 m, 3 m × 1.82 m turnover steel formwork, 3 m × 3.1 m cantilever steel formwork and 3 m × 3.1 m cantilever WISA formwork. According to the structural features at various locations and the relevant requirements in contract documents, see Table 9.9 for the details of the formwork engineering.

1. Dam upstream face, overflow dam section downstream vertical face, navigation dam section face, transverse joint face adopted 3 m × 3.1 m, 3 m × 1.82 m turnover steel formwork.
2. The upstream surface of the navigation wall, gate pier and navigation dam section navigational channel adopted 3 m × 3.1 m cantilever WISA formwork.
3. The overflow dam surface RCC step adopted 1.5 m × 0.6 m set-shaped steel formwork and set-shaped wooden formwork.
4. The underport inlet side wall and breast wall used the special-shaped steel formwork.
5. The of dam body bracket, navigation dam section construction joint face, gate slot, well, chamber, collecting well and secondary concrete used composite steel-wooden formwork.
6. The gallery used the concrete prefab formwork, the composite steel formwork was used at the locations where it is not convenient for installation.
7. The elevator shaft uses the interior cylinder formwork (set-shaped steel formwork with a height of 3.0 m).

Table 9.9   Formwork usage condition statistical table

S/N	Description	Size (length × width) (m)	Quantity (blocks)	The parts used	Remarks
1	Turnover formwork	3.0 × 3.1	702	Dam upstream and downstream faces and transverse joint face	
2		2.0 × 3.1	72	Dam upstream and downstream faces and transverse joint face	
3		3 × 1.82	760	Dam upstream and downstream faces and transverse joint face	
4		2 × 1.82	12	Dam upstream and downstream faces and transverse joint face	
5		1.5 × 0.6	96	Dam downstream face steel trestle placing area	
6	Cantilever form	3.0 × 3.1	280	Dam upstream and downstream faces, transverse joint face	
7	Cantilever formWISA	3.0 × 3.1	140	Guide wall and gate pier	
8	Special-shaped formwork	3.0 × 3.1	60	Underport inlet side wall and breast wall	
9	Circular formwork	R = 1.0	10	Underport inlet bottom slab	
10	Cylinder formwork		6	2 sets for each of elevator shaft, stair shaft, cable shaft	
11	Step formwork	1.5 × 0.6	320	Overflow dam surface RCC step	
12	Steel formwork	P3015	1 000	Locations of holes, corner filling, embedded parts, etc.	

Note: The usage amount of set-shape wooden formwork at locations such as uncounted gate slot, navigation dam section construction joint, overflow dam surface step, etc..

9.2.6 Placing Surface Rolling Construction

9.2.6.1 Placing Surface Support Equipment

The construction process of whole-block placement and continuous thin lift rolling is generally used in the full-section RCC construction, the placement surface leveling machine, joint cutter, vibration roller, placement surface crane and sprayer, material and method of pre-embedded cooling water pipe, construction process of pre-embedded parts, etc. That are used will develop with the development of the RCC construction technology, the placement surface support equipment is the foundation to ensure high strength and continuous rolling construction of the RCC.

For example, RCC placement surface equipment of Longtan Dam is configured according to simultaneous placing of two blocks of about 10 000 m^2, maximum daily casting capacity is up to from 20 000 m^2 to 25 000 m^2. See Table 9.10 for placement surface support equipment of Longtan Dam.

Table 9.10　　Placement surface support equipment of Longtan Dam

S/N	Equipment description	Specification/Model	Unit	Quantity	Single unit production efficiency (m^3/h)
1	Vibrating roller	BW202AD—2	Set	13	70 – 80
2	Small vibration roller	BW75S—2	Set	3	
3	Small vibration roller	SW200	Set	3	
4	Spreading machine	CATD3GLGP	Set	3	130 – 150
5	Spreading machine	SDl6L	Set	5	100 – 120
6	Crawler joint cutter	R130 LC—5	Set	2	
7	Vibrator	EX60	Set	2	
8	High pressure water roughening machine	GCHJ50B	Set	15	50
9	Vehicle-mounted high pressure water roughening machine	WLQ90/50	Nr.	1	50
10	Sprayer	HW35	Set	16	
11	Pneumatic stirring and storage truck	—	Nr.	1	
12	Hydraulic stirring and storage truck	—	Nr.	1	
13	Placement surface crane	8t、16t、25t	Nr.	10	
14	Nuclear densimeter	MC—3	Set	4	

9.2.6.2 Discharge and Leveling

Before paving RCC on the placement surface that passes the acceptance, a layer of mortar shall be paved on the old concrete layer, the pavement of the mortar shall be carried out together with the spreading of the RCC. The allowable intermission time of the rolling layer varies with the air temperature and humidity, if it is necessary to pave the RCC continuously when the time duration of allowable interval time for the rolled concrete is exceeded, the casting can be done only after treatment, motor or grout (cement + fly ash) is generally spread on the layer surface of the RCC with allowable interval time exceeded.

The spreading of the RCC shall be done strip by strip in fixed direction, which will enable the gradation is distinct, the time interval between layers is easy to control, and facilitate orderly construction. Discharge, spreading, leveling strip shall be parallel to the dam axis. If the construction is restricted by the cutting of transverse corridor, hole bottom part and so on, the leveling strip shall be spread in parallel to the direction of water flow, however, leveling and rolling direction shall be parallel to the dam axis within the range of 3 − 5 m of the upstream surface (including the dam surface under the downstream water level), the reason why the leveling direction is parallel to the dam axis is to avoid the imperviousness weak zone in the direction of downstream water flow from being formed at important locations. Based on project practice, if the fresh RCC is placed on a spread but not rolled layer surface, it can act as obvious buffering, and in favor of prevention of aggregate separation, then carry out leveling according to the leveling thickness to allow the spreading zone to extend forward, See Figure 9.8 for the spreading and leveling schematic diagram.

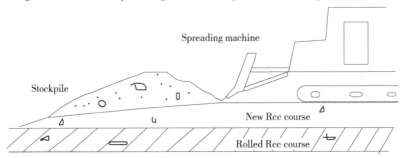

Figure 9.8 *Spreading and leveling schematic diagram*

If the compaction thickness of the RCC is 30 cm, the leveling thickness generally is about 34 cm. To uniform the leveling thickness, it is generally necessary to mark layer thickness scale and its elevation on the placement surface formwork. The discharge of the dump truck on placement surface shall use plum-type overlying discharge or backspread method two-point discharge, 1/3 is discharged first, then 2/3 is discharged after moving forward for 1 m more or less, the plum-type or two-point discharge can reduce the height of heap, help to eliminate the separation of aggregate from the dump truck, and allowing the aggregate concentration at the bottom of the heap to be dispersed. In the actual construction, among various zones, the aggregate separation still occurs due to the leveling, which needs to be treated manually. Leveling and rolling shall be done in a timely manner after discharging, the actual situation often is the rolling lag far behind the leveling.

When the truck drives smoothly and slowly when driving on the rolling compacted concrete surface so as to avoid any sudden braking, sharp turning, etc., otherwise, the constructed RCC quality will be damaged. In discharging, the discharge and leveling of the strip close to the formwork, the distance from the edge of the dump heap to the formwork shall be not less than 30 cm. At the same time, attention shall be paid to the boundary between II and III gradations of RCC when discharging and leveling.

9.2.6.3 Rolling

I Vibrating roller

The vibration roller is the equipment that is very important to the rolling construction. The selection of the type of vibration roller shall take into consideration of the rolling efficiency, exciting force, drum size, vibration frequency, amplitude, traveling speed, maintenance requirement and reliability of operation. At present, the vibration rollers mainly are German BMW, American Hummer, Japan Sakai and domestic vibration rollers. For example: the German BMW BW202AD vibration roller adopts articulated soft connection structure, double steel wheel self-propelled, specific performance parameters are as follows:

Total weight:	10 724 kg		
Front axle load:	5 364 kg	Static linear loading:	25.10 kg
Rear axle loading:	5 360 kg	Static linear loading:	25.10 kg
Vibration angle:	±8 degrees	Steel wheel diameter:	1 220 mm
Turning radius:	4 873 mm	traversing speed:	0 − 6 km/h

Maximum grade ability:	37 degrees		
Vibration frequency	30 – 45 Hz	Vibration breadth:	43 – 91 mm

The engine is Model BF4L913 manufactured by Deutz WC

System rated power:	70 kW	rotation speed:	2 150 r/min
system rated voltage:	12 V		
Fuel tank capacity	129 L	Sprinkler tank volume	830 L

Ⅱ　Rolling operation

It is specified in the specification of hydraulic RCC that the traveling speed of the vibration roller shall be controlled from 1.0 – 1.5 km/h. In the rolling construction, the traveling speed of the vibration roller has direct impact on the rolling efficiency and compaction quality, too fast of the traveling speed can result in bad compaction effect, too slow speed is easy to cause the vibration roller to be trapped and reduce the construction strength. In recent years, because of the change in RCC mixture performance, the traveling speed of the vibration roller has beenincreased significantly, the construction practices demonstrate that, when the slope rolling process is used, the uphill speed has exceeded 1.5 km/h. The author believes that, with regard to the traveling speed of the vibration roller, the principle is that it shall be finally determined by means of site test so as to meet the apparent density and the relative compactness required by the design.

The large sized vibration roller carries out rolling close to the formwork, therefore the construction efficiency is high and the quality can be assured easily, but the displacement of the formwork shall be controlled. For the placement end locations and locations where it is not easy for the formwork to access, the smaller sized vibration roller can be used for compaction by rolling, its number of times of rolling shall be determined in such a way that the apparent density is met.

Within a range of 3 – 5 m on the dam body upstream surface, the rolling direction shall be parallel to the direction of the dam axis. If the rolling direction that is parallel to the dam axis, the water seepage channel formed due to improper contact of rolling strip can be avoided, therefore, the rolling direction shall be parallel to the direction of dam axis within a range 3 – 5 m on the upstream surface. When the inclined-layer placing method is used for rolling, there shall be practical measures in place. To assure the compaction quality of rolling strips overlapped with each other, in case of many vibration rollers working simultaneously, measures shall be taken to avoid any missed rolling. The overlap width between the rolling operation strips is 10 – 20 cm, the overlap width at the end shall be 100 cm more or less. The length overlapped at the end part shall ensure that front and rear wheels of the vibration roller can enter the overlap range, the overlap length can be determined according to the wheelbase of the selected vibration roller. The small vibration roller shall be used to carry out rolling in the overlap area.

After completion of the operation for each rolling strip, the apparent density of the concrete shall be inspected according to the grid points in a timely manner, if the result is lower than the specified indicators, reinspection shall be carry out immediately, and take treatment measures after finding out the causes. The test result of the apparent density is the main indicator that indicates whether the concrete is compacted or not, when the measured value is lower than that of the specified indicator, the number of times of rolling needs to be increased; if it still fails to meet the specified indicator, analyze the causes and take corresponding measures. The project practices demonstrate that, as long as the apparent density can meet the requirements, spring concrete phenomenon occurred in roller compaction is conducive to interlayer bonding, because the RCC conforms to the rule of concrete water-cement ratio.

After completion of the construction for one lift of RCC, after the layer surface or cold joint that is necessary to stop as a horizontal construction joint reaches the number of times of rolling and the apparent density till the RCC is not finally set, the vibration roller is used to carry out vibration-free rolling finish, which can prevent the concrete surface from generating crack and shrinkage, and help to bridge the fine crack on the surface.

The leveling and rolling shall be completed as soon as possible after the RCC is placed. The maximum time duration from mixing and adding water of the RCC to rolling completion is called as the "permissible interval time", the permissible interval time shall be specifically determined according to different seasons, weather condition and variation rule of VC value, through tests or analogy with other engineering examples, which should not be exceed 2 h. The purpose of timely rolling and shortening the interval time is to avoid any quality issues such as poorrollability of concrete, bad interlayer bonding caused by too long placement time of the mixture. As a result, the time from mixing of the RCC mixture to completion of rolling is restricted. The practice of many projects in our country such as Mianhuatan, Longshou, Baise, Longtan, Jin'anqiao Projects, etc. has demonstrated that, if the rolling is completed within 2 – 3 h from the start of RCC mixing to the placement surface, the liquefaction bleeding of the RCC layer surface is fast, the performance is good, which effectively improves the quality of interlayer bonding. Meanwhile, the interval time for the hot joints of upper and lower layers of the RCC shall also be controlled within the permissible interval time that meets the interlayer bonding quality, the specific maximum allowable interlayer interval time shall be determined by site tests according to actual condition of the project.

For example: the interlayer interval time in high temperature season specified in Longtan Project is not more than 4 h, and the interlayer interval time in different seasons and under different air temperature are specified specifically. The strip overlap method is used in rolling operation, the rolling direction is perpendicular to the direction of water flow, the overlap width between the rolling strips is 15 – 20 cm, the GEVR shall be paved at the locations where the rolling is impossible, the immersion vibrator shall be used to vibrate and compact manually. The traveling speed of rolling shall be controlled from 1.4 – 1.6 km/h, the number of times of rolling shall be controlled with the mode of "vibration-free + vibration + vibration-free" as per times of "2 + 8 + 2". The German BOMAGBW-202AD and BW-201AD rolling equipment is selected, the model BW-201. AD vibration roller is used to roll the positions close to the edge of the formwork. At the peak of construction, the placing intensity can be met only 15 vibration rollers are in the placement (the efficiency of the vibration roller is counted as 65 m^3/h per unit). There are so many rolling equipment, together with spreading equipment, transport equipment, formwork hoisting equipment, joint cutting equipment, etc., the placement surface is busy and crowded, to prevent the equipment construction from mutual interference, the rolling strip spreading and the running passages of the truck shall be arranged in a reasonable manner, the operating scope of the equipment shall be defined so as to connect the spreading with the rolling organically, and maximize the benefits of equipment. In addition to the operation according to the specification, attention shall be paid to following points during the rolling process:

1. The rolling layer surface shall be fully bled, the location where bleeding does not occur shall be excavated and rolled after resurfacing with the fine aggregate.
2. The height difference between two rolling strips caused by rolling operation shall be treated by 1 – 2 times of vibration-free pressing at low speed.
3. After the start of each rolling operation, workers shall be sent to timely spread the fine aggregate of the RCC in the area where the coarse aggregate is concentrated locally to eliminate local aggregate concentration and off-contact.
4. After completion of each layer of rolling operation, the compacted bulk density of the concrete shall be inspected according to the grid points in a timely manner, if the measured compacted bulk density is less than the specified indicator, it shall be additionally rolled in time.

9.2.7 Inclined Layer Rolling Construction Process

9.2.7.1 Features of Inclined Layer Rolling Construction

There are two large control indicators for RCC construction: one is to control the time from mixing of the RCC mixture to completion of placement surface rolling within 2 h (which can be properly extended to 4 h depending on temperature condition and setting delay time), the other is to control the interval time of the concrete interlayer within the initial setting time of the concrete (generally take 6 – 8 h).

The full-section RCC dam technique is used, because the improvement of joint cutting technology, many placing positions can be merged into one whole-block placement construction, which can effectively reduce the quantity of the transverse joint formworks and the workload, the placing surface area generally is up to 5 000 – 10 000 m², placing surface area of some projects are often more than 10 000 m², even more than 20 000 m². For example: the placing area of whole-block placement on the left bank of Jin'anqiao 1 350 m elevation 11[#] dam section reached 14 000 m², See Figure 9.9 for details.

Figure 9.9 *The placing area of whole-block placement on the left bank of Jin'anqiao* 1 350 m *elevation* 11[#] *dam section is* 14 000 m²

For such a large placement surface, if the flat-rolling technique is used, the block rolling shall be carried out, that is the entire dam surface is divided into several blocks that are spread and rolled separately, only in this way, two large control indicators of permissible placing time and interlayer interval time can be met. If the rolling construction is carried out according to maximum placement surface, the workload of placement surface treatment will be increased by multiplied, and the configuration of construction equipment such as mixing, transport, leveling, rolling, etc., is required to add, the investment shall also be increased correspondingly.

How to solve the challenge that it is difficult to guarantee the whole-block placement and thin lift spreading, too large placement surface and rolling continuously rise, starting from Jiangya and Fenhe Reservoir in our country, the technical innovation has been made to the placement surface rolling, the construction of RCC adopts inclined-layer placing method in the spreading and rolling process, which is also called as "inclined rolling", and successful experience has been gained.

The inclined rolling process has become the mainstream way of RCC construction. The successful application of the inclined rolling process is an important technical innovation for fast construction of the RCC, it has provided an effective mean for solving large placement surface construction of the RCC under the condition of limited resources allocation. The inclined rolling process not only saves the input of large amount of resources, but also provide more reliable guarantee for improving the interlayer bonding quality of the RCC, ensuring the RCC construction quality in rainy season and sub-high temperature season. See Figure 9.10 for inclined layer rolling construction process of dam RCC.

Figure 9.10 *Inclined layer rolling construction process of Jin'anqiao Dam*

The purpose to adopt inclined rolling mainly is to reduce the operation area of spreading and rolling, and shorten the interlayer interval time. The significant advantages of the inclined rolling is to cast large placement area under certain resources allocation, convert the large placement surface into small ones that the area is basically same, in this case, the interlayer interval time of RCC can be greatly shortened, the interlayer bonding quality of the concrete is improved, the mechanical equipment, staffing and large amount of capital are saved. Especially in seasons with much higher air temperature, the effect of using inclined rolling construction technique is significant. This also helps the construction in rainy season and sub-high-temperature season, allowing controlling the interlayer interval time of the RCC within its permissible range.

Why can the inclined rolling significantly shorten the interlayer rolling interval time? It can be seen from the calculation below that the advantage for using the inclined rolling to control allowable time duration and interlayer interval time. If the single vibration roller's rolling strip width generally is 2 m (with 15 cm of overlap width is deducted), the rolling layer thickness is 0.3 m, the rolling efficiency is 65 m^3/h/unit, then:

Lift height Flat or inclined layer takes 3 m;

Rolling layer thickness 0.3 m;

Slope ratio of inclined rolling Height: Length = 1:10 – 1:15;

Calculated, slope length 30.1 – 45.1 m;

Rolling area calculated according to 6 strips per roller 361.2 – 541.2 m^2(30.1 m ×2 m ×6 – 45.1 m × 2 m ×6);

Rolling strength calculated according to 6 strips per roller 108 – 162 m^3[361.2 m^2 × (0.3 – 541.2) m^2 × 0.3 m];

Maximum allowable time duration and interlayer interval time 1.7 – 2.5 h ≈ (108 – 162) m^3/65 m^3/h.

The calculation results show that: if the rolling efficiency of each vibration roller takes 65 m^3/h/unit and the rolling workload takes 6 strip rolling, the allowable time duration and interlayer interval time by using inclined rolling generally does not exceed 2.5 h, that is the interlayer interval = (placing surface area × rolling layer thickness)/placing intensity. It is obvious that the advantages by using inclined rolling are very significant.

According to the practice in many projects such as Jiangya, Baise, Longtan, Guangzhao, Jin'anqiao Projects and others, when the inclined layer slope is controlled between 1:10 and 1:15, the normal construction can be carried out, the slope rolling effect is good. Under the condition that the coverage before initial setting is met, the smaller the slope ratio, the better, because the smaller the slope ratio, the bigger placing area, which is conducive to rapid construction. When the slope is less than 1:10, as the slope is too steep, the uniformity of the spread material and the rolling quality cannot be assured with ease, and the speed shall be accelerated when the vibration roller goes uphill, otherwise, the sliding may easily occur and cause it difficult to go uphill, making the assurance of rolling quality to be not easy. For example: the RCC of Guangzhao Dam adopts large placement area whole-block casting (maximum placement surface is up to 2.2×10^4 m^2), this is to avoid the time for interlayer coverage of large whole-block placement flat rolling is too long, big loss of *VC* value, bad rollability, too big input of mechanical equipment and personnel, reduce the feed strength of the RCC and improve the bonding quality between layers. The inclined layer rolling construction technique is widely applied in the rolling construction process, the inclined rolling placing direction for the dam with 662.5 m or less elevation is from the downstream to upstream, the inclined rolling placing direction is from the right bank to the left bank for the dam with 662.5 m and more, the volume of the inclined compacted concrete accounts for 92% of the total volume of RCC of the dam, and the inclined layer rolling construction process is continuously deepened in the construction, the in-depth study was carried out on the issues of rolling direction, slope angle processing, addition of horizontal bedding, slope gradient, etc., which enables the inclined rolling construction technology to become increasingly mature. By adopting inclined rolling process, daily placing intensity of the RCC reaches 13 582 m^3, the monthly placing intensity of the RCC reaches 22.25×10^4 m^3, the inclined rolling construction process has been fully affirmed by all the participants in the construction, and very good illustration has been provided for its wide application.

9.2.7.2 Key Points of Inclined Layer Rolling Construction

The requirements of inclined rolling construction process flow is same as that of the flat rolling construction process flow, the key is the control of the three main points (slope, thickness and slope toe processing) of the inclined layer. Inclined layer rolling construction process flow is: prepare the instruction drawing of the placement, the instruction drawing of the placement shall describe the construction scheme in detail, it shall be prepared by the engineering technicians before each placing of the concrete, the construction shall be organized according to the requirements of the instruction drawing of the placement, details are as follows:

1. The placement surface commander organizes the implementation according to the requirements of *Method Statement* and instruction drawing of placement surface casting.
2. Inclined plane slope control range: 1:10 – 1:15.
3. Setting of placing area: the placing area shall be set according to the production capacity of the casting equipment, the width of the placing surface shall not be less than 15 m so as to assure minimum requirement of flow operation for mechanized construction.
4. Formation and propulsion of the inclined plane: The spreading length of the first layer of the RCC shall be determined depending upon the slope ratio, which shall not be less than 30 – 45 m. From the second layer on, the indentation at one end rises up gradually, the end of forward placement surface will form a inclined plane till one end raises to 3 m, the other end forms the inclined plane with pre-defined slope. 1 – 2 cm thickness of mortar shall be paved at the leading-edge when the inclined layer is advanced, then spread the RCC and compact by rolling, the inclined plane RCC construction can be carried out after that.
5. Discharge on placement surface: the discharge is commanded by the placement surface conductor, the backspread method is used for discharge in succession, and piled up in plum shape in succession, 1/3 is discharged first, 2/3 is discharged at the position after moving about 1 m. To minimize the separation of the

Roller Compacted Concrete Rapid Damming Technology

aggregate, the dump truck discharges the concrete on the step at the spreading front edge of the pavement layer, then the concrete is pushed down from the step by the leveling machine and the flat push type placement is carried out. The discharge shall be uniform possibly, if partially separated aggregate occurs at the bottom of the slope, it shall be spread evenly on the unrolled concrete surface by workers.

6. Spreading and levelling: prior to pouring, measure and draw the leveling line of each layer on the surrounding formwork and the old concrete surface at the bottom. The thickness of each leveling layer shall be 34 cm more or less, if any location is found through inspection that the specified value is exceeded, carry out leveling again, the location which is locally uneven shall be spread out manually. The leveling direction shall start from top to bottom of the slope, the slope ratio shall not be less than 1∶10, the sharp corner formed at the slope toe and slope crest shall be rolled and compacted by using small vibration roller. The aggregate concentrated on both sides during the leveling shall be scattered on the strip manually. For the inclined paving construction, the bottom of the slope shall not exceed the outer edge of the flat layer, which generally is backspread for 30 cm.

7. Rolling: the large size double drum high frequency vibration roller is used, the rolling is carried out by using overlapping process, the overlap width is 20 cm, the traveling speed of the vibration roller is 1.0 – 1.5 km/h, the number of times of rolling is according to the mode of "vibration – free + vibration + vibration – free": 2 + 8 + 2, if the concrete still does not bleed after 6 times of rolling with vibration, the water tank equipped with the vibration roller shall be opened for watering and compensation of the moisture; if it still fails to bleed due to too large VC value, the large placement area whole-block & thin lift casting shall be used on the next layer of the concrete, which mainly is inclined layer rolling, the factors such as the rigidity of large turnover formwork, easy installation and difficult uphill of the vibration roller due to too steep slope shall be taken into consideration, the gradient of the inclined rolling shall not be less than 1∶10, the normal construction can be carried out only in this case.

Avoiding to forming thin-lift sharp angle at the toe of slope and strictly removing any secondary pollution are two main problems for guaranteeing the construction quality by inclined-layer placing method. As the aggregate at the location of the thin-lift sharp angle is easy to be crushed, stretching out a section is an effective method to avoid the formation of thin-lift sharp angle. When the construction is carried out by using the inclined-layer placing method, attention shall be paid to that the height of each inclined lift shall correspond to the interval of the embedded cooling water pipe required for temperature control. The height of each inclined lift in Longtan Project is 1.5 – 3.0 m in high-temperature season, 3.0 – 6.0 m in low temperature season.

When the inclined-layer placing method is used, marking shall be made on the formwork according to the proposed slope of the inclined layer and spread out strictly according to the marking. The spreading is carried out along the dam axis from the crest to the toe of the slope to avoid the aggregate at the slope toe is crushed, a "wedge" shape shall be formed while spreading, the "wedge" thickness is same as the layer thickness, the "wedge" length is 3 – 4 times of the layer thickness.

9.2.7.3 Several Problems in Inclined Plane Rolling

1. Placement surface contamination problem. As the inclined plane rolling technology requires that no truck is allowed to drive on the placed RCC, that is end-back placing is used, during the transport and discharge of the truck, as the placing surface area is small, the layer surface interval time is short, the rolled layer surface is easy to be destabilized by the placing truck, the layer surface is subject to more significant damage.

2. The slope toe of inclined plane rolling. The toe cutting process is used to treat slope toe of the inclined plane rolling, the most of the concrete cut out from the toe cutting process needs to be disposed of as waste due to the longer shelving time, the grout is required to be spread for the construction of GEVR.

3. Joint cutting of inclined plane rolling The joint cutting of the inclined plane rolling may produce

unidirectional offset while the vibration roller carries out rolling due to inclined plane rolling operation, the joint from joint cutting may form a kind of vertical fold line joint surface, which has certain deviation from the design line surface.

4. Placement surface of inclined plane construction. When finishing the placement surface of inclined plane construction, there is a transition problem between the inclined plane and the leveling, and the area of the placement surface becomes smaller and smaller, the placement tends to be too fast in the construction, which causes it is difficult to ensure the quality of the layer thickness in rolling operation.

9.2.7.4 Disposal of Out-of-spec Materials during Construction

For the disposal of out-of-spec materials during construction of the RCC, the GEVR mode is used, which has many advantages during the construction of RCC.

The slope toe of the inclined layer rolling often becomes the weak surface of slope rolling, it is difficult to compact the slope toe by rolling. For the 20 – 30 cm width edge of the slope toe reserved when paving the upper layer of the RCC by using the inclined-layer placing method, first cut the part where the thickness at the slope toe is less than 15 cm with the advance of the inclined plane and spread it on the concrete surface, the mortar shall also be paved on the edge of the reserved slope toe and rolled with the next strip when paving mortar on the next construction joint surface. Or, the grout is paved on the concrete surface at the slope toe, after it is treated by using the mode of GEVR, then the RCC shall be paved on it.

The unqualified materials of the RCC is inevitable, it is much easier for the disposal of the out-of-spec materials at the outlet of the mixing plant, which can be disposed of as waste materials. However, because of the impact by various factors, it usually is that after the RCC has been placed, the VC value is found too big or too small or the VC value is out of the range of construction control due to the rolling not in a timely manner, it becomes the hard concrete or normal concrete that is impossible to roll. In this case, it shall not be treated as a waste, the out-of-spec RCC can be discharged around the locations such as dam downstream formwork, bank slope or gallery, etc. ; if the VC value is too big and out of the required range, the mode of construction that paves the GEVR spreading grout is used for treatment; if it is the normal concrete, direct vibration is okay; for the out-of-spec materials that fails to be excavated, the GEVR treatment mode is also used, carry out vibration after spreading the grout. As the RCC complies with the rule of water-cement ratio, the GEVR treatment mode is used, the practices demonstrate that the effect is good. See Figure 9.11 for spreading grout at the slope toe of the slope rolling.

9.2.8 Joint Formation by Joint Cutter

9.2.8.1 Modes of Joint Formation

There are many modes of joint formation for the RCC, its joint formation has experienced many kinds of modes such as preset joint partition by slotting, induction hole by filling dry sand after hole-drilling, joint cutting by jointcutter, etc. . The joint cutter has also experienced the technological innovation process of the joint cutter such as joint cutting by modified plate vibrator, filling colored striped cloth for separation joint, joint cutting by modified high frequency vibrator, filling colored strip, joint cutting by modified tamper, filling colored striped cloth and joint cutting by special hydraulic joint cutter, filling colored strip in the joint, etc. . See Figure 9.12 for using the vibro-rammer type joint cutter equipment.

The application of induction hole in RCC arch dam is much more. When the thin-lift continuous placement construction is used, the induction hole can be drilled by using pneumatic drill after the concrete is rolled, the hole shall be filled with dry sand after drilling so as to avoid the induction hole is filled by concrete while construction on upper layer. In case of intermittent construction, the holes can be drilled at the interval time between layers. As the holes are made by using pneumatic drill, the configuration of the air compressor is

Figure 9.11 *Spreading grout at the slope toe of the slope rolling*

Figure 9.12 *Cutting joint by using simple joint cutter modified by vibro-rammer*

required, which causes that the construction becomes complicated and disturbance is big.

In recent years, the joint cutter is mainly used to form joint for the RCC. The joint formation by joint cutter is good in effect, flexible, small in construction disturbance, and small in damage on the RCC surface. To guarantee the formation of the joint surface, it is specified that the joint formation area for each layer shall meet the design requirement. The materials such as galvanized sheet iron, PVC (colored strip) and so on can be used as joint filler.

At present, most of the joints are cut by using vibro-rammer modified joint cutter, which can assure the formation quality of the cut joint and the working efficiency, the cost also is effectively controlled. The vibro-rammer modified joint cutter is easy to operate, the weight of the cutter is about 35 kg, only 2 – 3 operators are required to operate one joint cutter, the joint cutter is transferred to the working face by using a trolley in the construction, the joint cutting construction is carried out by two workers, the other worker fill the joint separation material, therefore, the operation is very convenient. Prior to the joint cutting, first is to make marks according to the pile number and guy wire at both ends of the construction joint, then the joint cutting can be carried out. The joint cutting depth shall not be less than 2/3 of the paving layer thickness, that is it shall not be less than 25 cm, the spacing of the joint filler shall not be more than 10 cm, the height shall be 3 – 5 cm lower than the compaction depth. The purpose is to guarantee the joint forming area, the height is specified so as not to affect the compactness of concrete and the bonding quality between layers.

9.2.8.2 Modes of Joint Cutting

The formation mode of the transverse joint mainly uses the joint cutter to cut the joint, this mode does not take a linear time limit for a project and does not affect the construction in block, which is adapted to fast construction of large placement surface. Based on the experience from past construction, the joint formation effect in the order of joint cutting after rolling followed by joint riding rolling is good. There are several modes of joint cutting by RCC joint cutter: "cutting after rolling" "cutting before rolling" and "rolling after joint rolling", etc. .

The first mode is suitable to lager joint cutting machines such as joint cutting construction of hydraulic joint cutter, because the large joint cutting machine has enough exciting force to smash the aggregate in the concrete encountered by the blade during the joint cutting and cut it into the concrete, thus the joint is formed according to the designed line, the quality can be assured. "Cutting after rolling" is the mainstream mode at present, which mainly is small in construction disturbance, especially is that the joint cutting construction can be carried out only after the rolling of the slope layer rolling is completed. The position for the transverse joint area of the RCC is located by measuring and setting out, if HP913-C hydraulic joint cutter is used to cut the joint, the continuous joint with a width of 12 mm is cut out, the transverse joint formation area is equal to the design joint formation area, and the joint shall be filled by using the materials indicated in the construction drawing. See Figure. 9.13 for the schematic diagram of HP913-C hydraulic joint cutter.

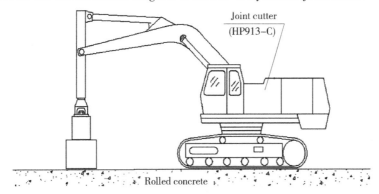

Figure 9.13 *Joint cutter cutting schematic diagram*

The second mode is suitable to the construction by small joint cutting machine. For example: the joint formation efficiency for vibro-rammer modified simple joint cutter is higher if the rolling after cutting mode is used due to its smaller power. However, the joint cutting shall be carried out after completion of the entire leveling, although it is easy to cut joint, the disturbance to the placement surface is big, especially is that the spreading of the strip for inclined layer rolling is not allowed to use the mode of rolling after cutting.

The third mode is especially adapted to the simple joint cutter modified from the vibro-rammer, two times of rolling without vibrations in construction, then the remaining number of times of rolling operation is carried out. From the viewpoint of joint formation in construction, the joint surfaces have effective connection, the cutting effect is better than that of the "cutting after rolling" method. It is mainly because that the placement surface is not compacted by rolling, when the blade encounters with large aggregate in the concrete, the aggregate is moved away from the joint surface position by vibration of the joint cutter, this is easy to guarantee the depth of design line of the joint formation and the joint filler, but the disturbance to the construction is also big.

For example: the hydraulic joint cutter (a joint cutting blade is added in hydraulic backhoe) is used to cut and form the cross expansion joint for the whole-block placement dam section of Longtan Project, the colored strip is used to fill the joints. The method is that: the colored strip is cut into blocks with a length of 28 cm and width of 1 – 2 cm smaller than that of the joint cutting blade in advance, the joint cutting can be started after each layer of RCC has been rolled, the width of joint cutting is 24 mm. Insert the cut colored strip into the joint immediately after the completion of the joint cutting. To ensure that the position for joint cutting is accurate, when the block number of the RCC is prepared, mark the position of the structural joint on the upstream and downstream formeworks by measuring and setting out, draw lines on the concrete surface by using the mark line before joint cutting, cut the joint by approaching the blade along the line to enable that the cut joint agrees with the design drawing. Its construction procedure is: Survey and set out→Draw position line of structural joint→Joint cutting→Insert colored strip→Inspection and acceptance→Vibration roller rolls 1 – 2 times without vibrations.

9.2.9 Key Technique to Improve Interlayer Bonding Quality

"Interlayer bonding, temperature control and cracking prevention" is the core key technology for quick damming of the RCC. The physico-mechanical indexes of the RCC is not inferior to that of normal concrete, according to the experiences from the built RCC dams, if the design of mix proportion is reasonable, the construction speed, construction process, and construction quality control are assured, the interlayer bonding quality can fully meet the design requirements. In recent years, the performance features of the full-section RCC mixture has been highlighted on thedesign of RCC mix proportion, its mixture has the properties of cementing material enriched (high rock powder content, big grout to sand ratio), low *VC* value and significant retarding effect, which enables the initial setting time of the fresh RCC is long and the rollability is good, assuring the rolling construction of the upper layer of RCC can be completed before the initial setting time of the RCC. As the RCC layer surface is more sensitive to the environmental impact, when the air temperature is too high, solar radiation and winds are bigger, the surface is very easy to become white due to water loss, which impacts the interlayer bonding quality. According to the successful experiences from many projects in China, the interlayer bonding quality can be improved if following measures are taken:

1. Strengthen the construction organization and management, achieve rapid placing and spreading, rolling in a timely manner;
2. Carry out dynamics control of the *VC* value, the low *VC* value is used in construction under the circumstance that the roller is not trapped;
3. The measure of moisture retention by spray is taken for the placement surface, control the temperature increase of the concrete, improve the rollability and interlayer bonding quality.

9.2.9.1 Timely Rolling as the Key of Ensuring the Interlayer Bonding Quality

The interlayer bonding quality of the RCC dam is closely related to the design of mix proportion, the carefully designed mix proportion plays an important role in the interlayer bonding, after rolling 2 – 5 times by the vibration roller with good construction mix proportion, the layer surface can fully bleed. After the RCC is mixed, transported and placed, it shall be spread and leveled timely, rolled and covered for moisturizing in a timely manner, that is it is often said that the faster the RCC construction, the better. As the timely rolling and moisturizing by spray can change microclimate, which can ensure the rollability of the RCC, liquefied bleeding and interlayer bonding quality, therefore, quickening the placement speed of the RCC and timely rolling is the key techniques to assure the interlayer bonding quality control.

Timely rolling is the key to ensuring the quality of interlayer bonding after paving RCC. The traveling speed of the vibration roller at site has direct impact on the rolling efficiency and compaction quality, the construction practice shows that the rolling effect is good if the traveling speed of the vibration roller is controlled within a range of 1.5 km/h, too fast of the traveling speed can result in bad compaction effect, too slow speed is easy to cause the vibration roller to be trapped and reduce the construction strength.

Most of the vibration rollers used in our country are German BMW, American Hummer, Japan Sakai and domestic rolling equipment, large amount of construction practices demonstrate that the rolling strength of an individual vibration roller is about 65 – 75 m^3/h of the RCC, by taking into consideration of the impact by factors such as spreading, leveling, curing, transfer and so on, an individual vibration roller can actually roll about 1 000 – 1 400 m^3 RCC each day.

Taking the Jin'anqiao Project as an example, the American Hummer HD130 double drum vibration roller is used, the working width of front and rear drums is 2 140 mm, with the overlap with of 20 cm is subtracted, the actual rolling strip width is 1.94 m (2 140 – 200 mm), by taking into consideration of the backward traveling, the traveling speed of the vibration roller takes 1.25 km/h, 10 times of rolling, that is the mode is 2 + 6 + 2 of the vibration-free + vibration + vibration-free, the rolling thickness per layer is 30 cm, the calculated rolling volume of each vibration roller is about 73 m^3/h. (1 250 ÷ 10) × 1.94 × 0.3 = 73 m^3/h.

The placement surface of the Jin'anqiao Dam adopts large whole-block rolling construction, 3 – 5 dam sections are usually connected together as an entire placement surface, 5 vibration rollers on the placement surface fail to guarantee that the rolling is synchronous with the leveling, the spread RCC can not be rolled in a timely manner, and delayed for a long time. As the number of rolling equipment can not meet the site placing intensity requirements, it not only restricts the rapid construction of the RCC, but also adversely affects the quality of interlayer bonding of RCC.

To improve the rolling efficiency of the placement surface, the first is to allocate sufficient rolling equipment according to the spreading and rolling strength, the second is to allocate vibration rollers with bigger mass and exciting force to reduce the number of times for rolling, the rigidity requirement of the formwork shall also be met.

9.2.9.2 *VC* Value Dynamics Control is the Key to Guarantee the Rollability

The *VC* value has a significant impact on the performance of RCC. In recent years, a large number of construction practices prove that the key points for the control of on-site RCC are *VC* value of mixtures and setting time, and the dynamic control of *VC* value is the key to ensure the rollability and interlayer bonding of RCC. The *VC* value dynamics control is to adjust the *VC* value dynamics control at the outlet in due time according to the conditions of air temperature, wind speed and evaporation, the *VC* value at the outlet is generally controlled from 1 – 5 s, it is much better if the *VC* value at site is controlled within a range of 3 – 6 s, under the condition that roller trapping does not occur and the normal rolling at site is satisfied, the site *VC* value shall use small value as possibly. The rolled rolling surface shall be full bleeding and elastic, there shall be elastic relief before and after the drums of the vibration roller. The test and construction personnel shall inform the mixing plant of the bleeding condition and measured *VC* value after rolling operation at site, it is the test personnel that is responsible for adjusting the *VC* value in a timely manner.

Spray just changes the microclimate on the placement surface so as to achieve the purpose of moisturizing and cooling, the liquefied bleeding of the RCC is to extract the grout from the concrete liquidation under the action of the rolling of the vibration roller, this thin-lift of the liquefied bleeding is the key to guarantee the interlayer bonding quality, the liquefied bleeding has been used as an important criterion to evaluate the rollability of the RCC. Under the premise of satisfying the rollability, the main factors of the RCC mix proportion affecting the liquefied bleeding are whether it is rolled in time after pouring and gradual loss of *VC* value caused by temperature.

The rollability of the RCC, liquefied bleeding and interlayer bonding quality can be effectively improved by the most effective technical measure to improve the liquefied bleeding, keeping the mix proportion parameter unchanged, using different admixture dosage according to the temperature at different time period, decreasing the VC value and delaying the setting time under the double layer superposition of the admixture and retarding.

9.2.9.3 Placement Surface Moisturizing by Spray and Interlayer Bonding Quality

The moisturizing of RCC placement surface by spray is particularly important, which is directly related with the rollability of the RCC site construction, interlayer bonding and prevention against temperature increase and flow backward, significantly changes the microclimate on the placement surface, reduces the temperature by 4 – 6 ℃, is very favorable for temperature control; if the placing temperature increases by 1 ℃, the temperature in the dam generally raises by 0.5 ℃, therefore the moisturizing by spray is an important measure for temperature control and cracking prevention, it also is one of the important technical measures to improve the interlayer bonding quality, which is absolutely not neglectable, there are many lessons in this regard.

Ⅰ Spray by sprayer

Advantages and disadvantages of spraying equipment have great influence on the effect of the moisturizing by spray, there are many successful experiences and lessons of failures in this regard. Therefore, first thing first, the selection of the spraying equipment can ensure the effect of moisturizing by spray. China Gezhouba (Group) Corporation, Minjiang Engineering Bureau, etc.. have conducted a series of test investigation on the spray, by optimizing the nozzle, adding fan and high pressure pump to form the spray set, a kind of novel sprayer has been developed and put into production successfully. The spraying distance can reach 25 – 30 m, when the sprayer carries out 120° rotation, its spraying area can reach 650 m^2.

For example: the Baise Multipurpose Dam Project is located in subtropical region, it is inevitable to carry out construction in high temperature season, to improve the placement surface construction, Minjiang Engineering Bureau has independently designed a mobile placement surface sprayer, see Figure 9.14 and Figure 9.15 for Baise mobile sprayer and site spraying. The power of this sprayer is 37 kW, maximum flow rate is 9.7 m^3/s, modified from tunnel fan of rated pressure 2 400 Pa, angle adjustable, the spraying radius is up to 30 m, available 120° swing, swing frequency is 5 times/min. The microclimate on the placement surface is effectively changed due to good atomization effect of the sprayer.

Figure 9.14 *Baise project mobile sprayer*

Ⅱ Spraying by spray gun

Using the handheld spray gun manually for placement surface spraying is flexible, which is the most common

Figure 9.15 *Placement surface spraying and moisturizing by mobile sprayer*

practice for moisturizing of placement surface by spray, the spraying of the spray gun can form white mist and the effect is very good. The spray gun is mostly replaced by the roughening machine in the past, the nozzle diameter of the most spray gun spraying equipment is 0.5 mm and more, the diameter of the high pressure sprayer nozzle of the roughening machine reaches 1.4 mm, the size of the droplet diameter is between 200 – 500 μm (just like the drizzle). Because of poor atomization effect and little evaporation of water, not only the cooling effect is not obvious, but also the water cement ratio is changed because of the water accumulation on the placement surface due to dropping of mist droplets on the upper concrete surface, which reduces the quality of the concrete on the contrary. Through the modification to the nozzle diameter of the roughening machine, the modified roughening machine is used in most projects, the sprayingeffect is good due to big flow rate and pressure.

If the spray gun with very small droplet size is used, the spraying effect is poor due to very close distance. When the droplet is small to certain degree, under the condition of same initial spray speed, the smaller the droplet size, the smaller its kinetic energy, the shorter the spray distance, the smaller the controlled area. If the droplet is as small as 20 – 30 μm, the sprayed mist just looks like white smoke, which is evaporated after floating for about 0.5 m, which fails to achieve the purpose of cooling and moisturizing by spraying. Furthermore, the water quality on the site generally is worse, very small nozzle is easily blocked, the micro-mist in case of too small nozzle diameter is not adapted to the moisturizing by spraying on large placement surface. The engineering practices show that, when the nozzle diameter of the spray gun is between 50 – 100 μm, the effect of moisturizing by spray is good. The engineering practices show that, the spray effect of the sprayer is not only greatly affected by the nozzle, but also significantly affected by its flow rate and pressure.

Taking the Longtan Project as an example, to achieve the most optimal atomization effect, 3 kinds of nozzles made by 3 manufacturers are selected at the structure site of Longtan Project for test, the droplet is between 30 – 100 μm respectively, and the test is carried out in two working conditions of with/without blower. Test results show that the spraying distance is closely related to the wind force and wind direction at site, if there is change with the former, the change of spraying distance changes significantly. The sunlight intensity and size of wind force also have great influence on evaporation of mist drops when measuring temperature within the range of spray, it is easy for the thermometer to be covered by water mist, there is certain influence on the measurement result, and the measurement of the droplet evaporation capacity is impossible. According to the test results, if the type 1/4 MKB 80014S303-RW hollow conical micromist nozzle manufactured by Japan Ikeuchi is used, the atomization effect is good, the spray distance is far, and optimal to the spray on placement surface, and this type of nozzle has the advantages of convenient assembly and removal, easy cleaning and good blocking resistance.

9.2.9.4 Examples of *VC* Value Dynamics Control and Moisturizing by Spray

Taking the Xinjiang Kalasuke Project as an example, the *VC* value is dynamically controlled with regard to the climatic characteristics of dry, windy and strong evaporation in the dam site area. The principle to determine the working degree control of the RCC for this project is: The *VC* value at the outlet of mixer is generally controlled between 1 - 3 s, the operation temperature of the site placement surface RCC is controlled between 4 - 6 s. In case of rolling operation at site, the *VC* value of the working temperature at the outlet of the RCC is dynamically adjusted according to different temperature, wind speed and sunshine conditions so as to guarantee the timely bleeding of the placement surface rolling and the quality of interlayer bonding.

At the same time, control the intermittent time between layers strictly, the water loss during the RCC construction is serious due to the climatic feature of the dam site, to assure the rollability of the RCC and the interlayer bonding quality, the interval time specified in this project construction is generally controlled within 4 h, and fastplacing, fast spreading, fast rolling and fast covering are carried out, this measure can assure the interlayer bonding quality well.

In case of placing concrete in high temperature season, the outdoor temperature is higher, the monthly average maximum temperature over the years occurs in July, which is 31.8 ℃, the river-water temperature is 17.7 ℃. According to the engineering experience and climatic feature in the dam site, the guide thinking for the compensation of the water loss in this Project is: first is to start from the viewpoint of improving the air temperature around the placement surface, make use of the low temperature mist formed by low temperature river water and high pressure wind to change the micro environment in the block, reduce the air temperature in the block, improve the air humidity in the placement surface, minimize the evaporation, which can achieve the purpose to effectively prevent the flow backward due to temperature change and reduce the loss of *VC* value. The second is to adopt the mode of spray by combination of fixed spray with mobile spray for the placement surface to solve the requirement of water loss compensation of large placement surface. The third is the water loss compensation of the spread RCC is completed by the spraying equipment, the water loss compensation of rolled layer surface is completed by spraying equipment together with the water replenishment by vibration rolling.

This Project needs to allocate 4 remote fixed sprayers and 3 roughening machines with 6 handheld spray guns, which is specially used for moisturizing of placement surface by spraying, two 200 t/h, 5 - 10 ℃ water chilling units are also allocated to provide cooling water for sprinkler system, spraying system and concrete water cooling system. The practice has demonstrated that, the fixed remote sprayer has the advantage of automatic operation, but its relocation is not convenient, and the effect is not good in heavy wind season. The handheld spray gun is featured with flexibility, big control range, small disturbance to the placement surface construction. This Project has effectively combined the fixed spray with the mobile spray, the operation flexibility is strong, the operation effect is good, which has very good cooling and moisturizing effect to the placement surface, basically achieves the purpose of reducing the temperature of micro environment in block by 6 - 8 ℃, the 80% and more of the placement surface humidity, maintaining the placement surface temperature no more than 23 ℃ more or less, the requirement for environment temperature in concrete placing is met.

9.2.10 Technical Innovation for Improving Rolling Layer Thickness while Minimizing Layer Joint Surface

The features of full-section RCC construction is whole-block & thin lift spreading and continuous rolling up, the interlayer thickness after each layer which is rolled is 30 cm, a RCC dam with a height of 100 m has 300 layers of joint surface, such a large number of interlayer joint surface can easily become channels for leakage, which is adversely to the impermeability, interlayer shear and the overall performance. In the early period, some Chinese scholars had doubts and disputes about the interlayer bonding and imperviousness of RCC dams, the challenge of interlayer bonding quality has been the main research topic in the industry for many years, the interlayer joint tensile strength and shear strength have played a leading role in the design of RCC dam, especially in the high earthquake incidence area, therefore, the design has taken the minimum coefficient of friction f' and cohesive strength c' of the RCC dam layer joint surface as the design control indexes.

Construction Specification for Hydraulic Roller Compacted Concrete DL/T 5112—2009 stipulates: The rolling thickness and the number of times of rolling used in the construction should be determined by the test, which is taken into consideration together with of the factors such as comprehensive productive capacity and so on. According to the different conditions of climate, paving methods, etc., the different rolling thickness can be selected. The rolling thickness should not be less than 3 times of maximum aggregate size of the concrete. The construction practices demonstrate that the compacted thickness varies with different vibration rollers, the number of times of rolling for a mixture with same mix proportion required by different vibration rollers. The rolling thickness and the number of times of rolling can be determined through site test together with the comprehensive productive capacity, if different rolling thickness is used in the construction according to the conditions, it is expedient to meet the requirement for intermittent time between layers. If the rolling thickness is less than 3 times of maximum aggregate size, the aggregate with maximum size will impact the compaction effect or the aggregate is crushed.

RCC dams have many layer joints, and rolling is time-consuming and power-wasting, which seriously restricts the rapid construction of RCC. Therefore, it is very necessary to carry out technical innovations in the thickness of the rolling layer and the time of rolling, to study whether to improve the thickness of the rolling layer, reduce the rolling joint surface and increase the lifting layer height of RCC construction. For example, from the traditional thickness of 30 – 50 cm, 70 cm or even more thickness, it is generally possible to increase the 3 m lifting layer to 6 m or higher. As the impermeability of the RCC body is not inferior to that of normal concrete, the increase of the rolling layer thickness can significantly accelerate the sped of rapid RCC damming, and effectively reduce the joint surface between RCC dam layer.

With regard to the technical innovation for improving the RCC layer thickness, the comparison tests of different layer thickness of 50 cm, 75 cm and 100 cm were carried out in October of 2006 at the site of Huanghuazai RCC double curvature arch dam in Guizhou, which explored to break through 30 cm rolling thickness limit in current standard. The test results show that: the SD451 vibration roller made by Japan Sakai is used, the paving layer thickness of the RCC at site is 100 cm, the number of times of rolling for the rolling strip is controlled according to the mode of "vibration-free + vibration + vibration-free" of "2 + 8 + 2, 2 + 10 + 2" and "2 + 12 + 2", the compactness tests of 60 cm and 90 cm from the rolling surface are carried out respectively. The relative compactness at 60 cm from the rolling surface is more than 97%; The relative compactness for the number of times of strip rolling of "2 + 8 + 2" at 90 cm from the rolling surface is less than 97%, which is not in conformance with the requirement of the specification; the relative compactness for the number of times of strip rolling of "2 + 10 + 2" and "2 + 12 + 2" at 90 cm from the rolling surface is more than 97%, which is in conformance with the requirement of the specification. In the last ten-day of November of

2008, the special tests for the rolling layer thickness of 50 cm and 75 cmwere carried out to the RCC dam of Yunnan Mazushan Hydro Power Station.

The enlightenment from the successful site test on improving RCC layer thickness:
1. The RCC spreading layer thickness can completely break 30 cm specified in the standard;
2. The rolling layer thickness can be increased to 40 cm, 50 cm, 60 cm. Even much thicker layer thickness;
3. Different rolling thickness can be used for the top, middle and bottom of the dam;
4. The increase of the thickness of the rolling layer requires in-depth research on the matching of the rigidity and stability of the formwork, the quality of the vibrating roller and the excitation force.

As the thickness of the RCC layer increases, GEVR jacking, grouting, and vibrating will become the construction difficulties. According to the practice experiences from the RCC dams of Baise, Guangzhao, Jin'anqiao and so on, the author solves the construction problem of layer thickness by using the machine-mixing GEVR. The machine-mixing GEVR is a kind of low slump concrete made by adding specified proportion of cement fly ash grout when mixing the RCC at the mixing plant. The pouring of machine-mixed GEVR is the same as that of RCC. After pouring, the concrete is vibrated with a vibrator, which not only simplifies the operation procedure of GEVR, but also guarantees the quality of large-volume GEVR.

At the time being, to accelerate the RCC construction speed, several, even tens of dam sections are connected together as a large placement surface, the spreading and rolling of the RCC basically uses the inclined-layer placing construction method, the actual inclined-layer placed RCC spreading often exceeds the 30 cm limit, the layer thickness of 40 – 50 cm layer has become common, therefore, the limit of 30 cm rolling layer thickness currently specified in the standard shall be modified.

9.2.11 The Technology for Synchronous Rise of Abrasion-resistance Concrete and RCC on Overflow Surface

The abrasion-resistance construction on overflow surface of overflow dam section often becomes one of the main factors affecting the rapid speed construction of RCC. By taking into consideration of the construction quality and construction difficulty in the design, the construction of anti-abrasion concrete of the overflow surface and the dam body RCC is usually divided into two stages. The high performance abrasion-resistance concrete of the overflow surface layer is poured after the RCC rises to a certain height first (which may be a certain stage) in some dam body, the RCC is poured into the top of the dam, and then the high performance abrasion-resistance concrete of the overflow surface is poured back in some other dam, in this case the high performance abrasion-resistance concrete of the overflow surface becomes a bottleneck that restricts the construction progress, the construction difficulty is bigger, the effect is not good because the formwork sliding or formworkdrawing is usually used for construction rush, which causes dam duration to be extended or delayed. In addition, it is also easy to occur that the deformation and stress of high performance abrasion-resistance concrete are not consistent with that of RCC, and two kinds of skin is easily formed. It is well known that the overflow dam section is a flood releasing structure, the overflow surface is closely related to the flood control of the project safety, and also directly related to the target of water storage of the gate under the dam. With regard to the technology for synchronous rise of abrasion-resistance concrete and RCC on overflow surface of overflow dam section, this construction method has been used in the projects such as Baise, Guangzhao, Gelantan, Kelasuoke Projects and others undertaken by Minjiang Engineering Bureau, and successful experience has been obtained.

For example, the overflow dam section of Gelantan Dam adopts the construction process of the one-time-formwork-erection concreting shaping, the overflow surface and the dam RCC are poured and raised simultaneously to enable rapid forming, 3 months more or less of construction period is saved, the passive situation that the schedule of overflow dam section lags behind has been changed, which is expedient to the integrality, overall progress and appearance coordination of the dam body. One formwork-erection, one joint bar and manual chiseling treatment are also eliminated in dam concrete construction, which has simplified the construction, reduced the cost of project, prevention against secondary pollution of the placement surface, and the realization of safe and civilized construction on site.

9.2.11.1 Advantages of Synchronous Rise

As the overflow surface rises synchronously with the RCC, the dam shaping can be assured only a formwork is erected outside of the overflow surface, which prevents the RCC from sliding formwork in abrasion-resistance concrete area due to erection of one formwork. The one-time formwork-erection and manual dabbing of the concrete is eliminated, thus the cost is saved. Meanwhile, as the formwork is only erected on the overflow surface, the concrete is one-time forming, the phenomenon of faulting of slab ends between neighboring sliding formwork concrete during the formwork sliding process is avoided. Where the greatest advantage is to eliminate the unfavorable bonding between the new and old concrete layers of high performance abrasion-resistance concrete and the RCC.

9.2.11.2 Disadvantages of Synchronous Rise

Because of synchronous rise, the temperature control of the high performance abrasion-resistance concrete becomes tighter. And the condition that the joint construction of many kinds of concretes such as RCC, normal concrete, etc. may occur in concrete placing, which is easy to cause the blending problem of the concrete placing, prominent sign is required for placing concrete.

9.2.11.3 Appearance Quality of Overflow Surface

To assure that the appearance shape of the overflow surface concrete is accurate, and complies with the requirements of standard and design, the error for measuring and setting out of the formwork-erection shall be tightly controlled within 5 mm, careful measuring shall be carried out to the deformation, rigidity and so on of the formwork after the formwork is erected till they are acceptable. After the concrete is placed and the formwork is removed, measure the concrete shape, carry out analysis and summary according to the measured data, and propose improvement requirement to the process of the next time of formwork-erection.

The overflow surface concrete and main dam concrete of the Gelantan Dam rises synchronously, one-time forming of the dam overflow surface concrete not only assures that the site is clean and tidy as well as civilized construction, but also the shape of the overflow surface concrete regular, pleasing to the eyes, the most important is that the performance of the abrasion-resistance concrete is improved, especially the interlayer bonding performance. The abrasion-resistance performance of the abrasion-resistance concrete itself can fully meet the design requirements, however, because that the two stages of construction is used for the abrasion-resistance concrete, the age of the bottom layer of the concrete exceeds 28 d, the fresh-old concrete is formed by the bottom layer concrete and the abrasion-resistance concrete, which becomes the area of strong constraint, two skin situation has been formed actually. The overflow surface outlet, during the flood discharge, especially under the action of cavitation erosion by high speed water flow, can easily produce the cavitation erosion at the layering of anti-abrasion concrete and bottom RCC, there are too many lessons to be learned in this regard.

To assure the quality of the anti-abrasion concrete, the construction process has to be changed, using the construction process of synchronous rise with the RCC is not only a matter of rapid construction and saving capital, but also directly related to the quality of the anti-abrasion concrete, as well as the safety of flood discharge and control and the dam quality. The construction technology for synchronous rise of abrasion-resistance concrete and RCC on overflow surface is worth popularizing and applying vigorously in the future construction.

9.3 Jin'anqiao Dam Roller Compacted Concrete Fast Construction Key Technique

9.3.1 Project Overview

The Jin'anqiao Hydropower Station is located on the Jinsha River midstream reach in Lijiang area, Yunnan Province, is the fifth stage power station in the "1-reservoir 8-cascade" cascade hydropower station planned on the midstream reach of the Jinsha River, total installation of power station is 2 400 MW. The barrage of Jin'anqiao Hydropower Station is of roller compacted concrete gravity dam, dam crest elevation is 1 424.00 m, maximum dam height is 160 m, the dam crest length is 640 m. The barrage is composed of left-right $0^\#$ key slot dam section, $1^\# - 5^\#$ left bank non-overflow dam section, $6^\#$ left bank sand flushing underport dam, $7^\# - 11^\#$ powerhouse dam section, $12^\#$ right bank flood discharge and sand flushing underport dam section, $13^\# - 15^\#$ right bank overflow surface outlet dam section and $16^\# - 20^\#$ right bank non-overflow dam section, 21 dam sections in all. See Figure 9.16 for full view of Jin'anqiao Hydropower Station, See Figure 9.17 dam roller compacted concrete construction.

Figure 9.16 *Full view of Jin'anqiao Hydropower Station*

Figure 9.17 *Dam roller compacted concrete construction*

$0^\# - 5^\#$ dam sections, full length is 192 m, dam crest width is 12 m. The sand flushing dam section on the left bank is the $6^\#$ dam section with the length of 30 m, and the sand flushing penstock extends from the underneath of the powerhouse erection bay to the downstream. The powerhouse dam sections at the dam toe in the middle of the river is the $7^\# - 11^\#$ dam sections, with the total length of 156 m and the dam crest height of 26 m. The discharging and sand flushing bottom outlets on the right bank is the $12^\#$ dam section, which is arranged closely next to the right end wall of the powerhouse dam section, the length of dam section is 26 m, the dam crest height is 21 m and the discharging and sand flushing bottom outlet with 2 outlets is arranged. The overflow surface outlets dam sections on the right bank are the $13^\# - 15^\#$ dam sections, the total length is 93 m, the dam crest height is 31 m, and five surface outlets are arranged with energy dissipation by hydraulic jump. The non-overflow dam sections on the right bank are the $16^\# - 20^\#$ dam sections, with the total length of 183 m and the dam crest height of 12 m.

The single-unit and single-tunnel headrace mode is adopted for four units of the powerhouse at the dam toe with the pipe diameter of 10.5 m, it is the half-back penstock at the dam toe, with 1.5 m thick reinforced concrete wrapped. The water inlet of the power station is the vertical water inlet on the dam surface, with the elevation of 1 370.00 m. The dimensions of the main powerhouse is 213 m × 34 m × 79.2 m (length × width × height), with the 4 × 600 MW mixed-flow hydraulic turbine-generator unit inside.

Total weight of dam concrete is 329×10^4 m^3, including 251×10^4 m^3 of roller compacted concrete, 78×10^4 m^3 of normal concrete. The dam concrete is normal concrete at the areas with complex structure such as power station intake, trash rack pier, jointing beam, gate slot, discharge hole, gate pier, etc., other locations are roller compacted concrete. The dam concrete is zoned by taking 1 350.00 m elevation as the boundary, which is divided into dam upper part and dam lower part, the lower part is $C_{90}20$ Ⅲ gradation roller compacted concrete, the upper part is $C_{90}15$ Ⅲ gradation roller compacted concrete, the upstream surface adopts $C_{90}20$ Ⅱ gradation roller compacted concrete for imperviousness. Depending upon different acting head, the upstream face adopts different impervious layer thickness, which is 5 m, 4 m and 3 m Ⅱ gradation imperviousness roller compacted concrete respectively from lower part of the dam to the dam crest.

Depending upon dam concrete construction requirement, this project arranges two 30 t translational cable crane, which undertakes the dam normal concrete pouring and the lifting of the dam metal structure, the roller compacted concrete is mainly solved by means of the modes such as belt conveyor, full pipe, chute and placing from truck. The placement of dam roller compacted concrete started from May 2, 2007, up to the end of December 2009, 259×10^4 m^3 of roller compacted concrete was placed (including 17×10^4 m^3 of right flush and discharge slot foundation roller compacted concrete).

9.3.2 Features, Important Points and Difficulties of Construction

1. The Jin'anqiao Hydropower Station Dam is featured with large project scale, complicated technology, high construction intensity, tight duration, many contributing factors and so on. Subject to the influence by factors such as construction layout, construction methods and construction environment, etc., much bigger difficulty is brought to the construction organization of the dam.
2. There are many holes on the dam body (12 holes), 5 flood discharge surface outlets, 3 sand flushing and flood discharge underports, 4 powerhouse intakes are arranged respectively. The construction of powerhouse section is the key to control the progress of dam concrete construction for this project, especially the installation of diversion pressure steel pipe is cross or parallel operation with the concrete placement, therefore, the disturbance is big, the accuracy and quality of the installation is the important points and difficulties.
3. There are many longitudinal and transverse galleries arranged with the dam body, the galleries are arranged in five layers: foundation layer, EL 1 297 m, EL 1 320 m, EL 1 359 m and EL 1 388 m in criss-cross, the cumulative length of the gallery is more than 5 000 linear meters. The gallery formwork mainly used the form of precast concrete formwork, however, the hoisting of large amount of precast concrete fromworks has brought certain difficulty to placing preparation and rolling, the placement preparation period is long, which weakens the advantages of fast damming construction. How the layout and optimization of the galleries on the roller-compacted concrete gravity dam reflects the features of fast construction is worthy the research of engineering designers.
4. The dam body is not divided into longitudinal joints, the roller compacted concrete is whole-block & thin lift spreading and casting, the casting area is large, the intensity is high (maximum intensity of the roller compacted concrete was up to 22.43×10^4 m^3/month in December, 2007), the peak duration is long. In addition to ensure the mixing capacity of concrete, the requirement for control of temperature at the mixer outlet is strict, the transportation and placing intensity is high, shortening the intermittent time between layers of the roller compacted concrete, strengthening the placement surface mist spray and moisturizing, curing by covering and water cooling are the key to assure the "interlayer bonding, temperature control and

cracking prevention" of the roller compacted concrete, it is also the important points of the construction.
5. According to the total project progress scheduling, the roller compacted concrete dam was originally planned for pouring in November 2006, but the dam roller compacted concrete pouring was postponed to the beginning of May in 2007 due to the influence by the factors such as complicated geological condition at the dam site, etc. , which caused big change with peak period and placing intensity of originally planned roller compacted concrete placing. To meet the requirement for water retaining and safety flood discharging of the dam body in the flood period of 2008, the roller compacted concrete construction is always in the high intensity continuous construction stage, in this course, linkage of the important process such as left and right sand flushing underport normal concrete, steel pipe and gate installation and financial difficulties make the situation of tight schedule more complicated and tense. High intensity continuous construction also is the important feature of the construction in this project.

According to the construction features, important points and difficulties of this project, by utilizing the technical advantages of roller compacted concrete rapid damming, the key technologies such as the basalt aggregate roller compacted concrete mix proportion optimization, dam body material zoning, transportation placing, placement surface zoning planning, interlayer bonging quality, temperature control and cracking prevention, etc. , carry out scientific and reasonable optimization and technological innovation, assuring the time limit target is achieved on schedule at major nodes such as flood period, etc. .

9.3.3 Basalt Aggregate Roller Compacted Concrete Mix Proportion Optimization

9.3.3.1 Behaviors of Basalt Aggregate Roller Compacted Concrete

The mix proportion design for roller compacted concrete plays an decisive role in roller compacted concrete damming, superior quality, scientific and reasonable mix proportion is the basic assurance for guaranteeing roller compacted concrete fast construction and dam quality, is one of the key technologies among the roller compacted concrete fast damming technologies, it has much higher technical content, is directly related to the success of rapid damming, which can get twice the result with half the effort and obtain obvious technical and economic benefits.

For the roller compacted concrete for Jin'anqiao Dam, early-stage design unit, relevant research and development institution and the construction organization found in the roller compacted concrete mix proportion tests that: The basalt aggregate used in Jin'anqiao Project concrete was different from other aggregate, the density of basalt aggregate is big, the surface is rough, which has very strong adsorbability. The water consumption for concrete has increased sharply due to the behavior of the basalt aggregate, the unit water consumption is about 20 – 30 kg/m^3 higher than that of limestone aggregate roller compacted concrete, the apparent density of II gradation and III gradation concrete reaches 2 630 kg/m^3 and 2 600 kg/m^3 respectively. Its behavior is very close to that of igneous rock aggregate roller compacted concrete that the diabase and basalt is used in Baise and Guandi Projects, the high water consumption and high density also exists. It was also found in the test that, the performance of roller compacted concrete mixture mixed by manufactured basalt sand is very poor, the liquidation and bleeding is slow, main cause is that the content of manufactured basalt sand powder is low, it fails to meet the design requirement of 15% – 22% of rock powder content. The basalt aggregate is not good for construction of roller compacted concrete due to its behavior.

9.3.3.2 Mix Proportion Design Optimization

After a large number of tests and studies, by the comparative test analysis of basalt aggregate roller compacted concrete and site productivity process tests, the mix proportion of roller compacted concrete is primarily selected as the dam mix proportion. The primary mix proportion control indicators: rock powder content: 16% –

20%, VC value at mixer outlet: 3 – 5 s, retardation type superplasticizer dosage: 1.0%, air entraining agent dosage: 25 – 30 (1/10 000), in this way, the requirement of air content of 3% – 4% can be met. The application of primarily selected mix proportion started in May, 2007, it was found in the construction placing that the rollability of the fresh roller compacted concrete is worse, the liquidation and bleeding was insufficient, pitted surface often occur on the surface after rolling, which has influence on interlayer bonding, the quality fluctuation is bigger. According to the analysis, the fluctuation in the behavior of basalt aggregate roller compacted concrete and rock powder content of manufactured sand are the main factors that have influence on the workability of the fresh roller compacted concrete, what is more, the placement surface is large, the loss of VC value of the roller compacted concrete with time is inevitable, the control of VC value of 3 – 5 s at the mixer outlet is slightly big.

According to the initial condition of roller compacted concrete construction, it becomes necessary to adjust and optimize the mix proportion of basalt aggregate roller compacted concrete. The optimized mix proportion allows optimal determination of rock powder content by tests, the test results show that: the rock powder contents of III gradation $C_{90}20$ and $C_{90}15$ III gradation roller compacted concrete are 18% and 19% respectively, the working performance of the mixture is good. In this case, the calculation of admixture of rock powder and sand adopts dynamics control, which is the calculation is carried out according to the difference between actually tested manufactured sand powder content and determined rock powder content, the influence by the fluctuation of the rock powder content is minimized. For the condition of the VC value loss with time, adjust the VC value at the mixer outlet to 1 – 3 s, it is okay if the control of VC value at the mixer outlet allows good placement surface rollability. For the high water consumption and poor liquidation and bleeding with the basalt aggregate roller compacted concrete, the challenge of poor liquidation and bleeding is effectively solved by increasing the dosage of retardation type superplasticizer. At the same time, the III gradation $C_{90}15$ graded roller compacted concrete is optimized, the cement amount is reduced from 72 – 63 kg/m³, the temperature rise due to hydration heat is effectively decreased. After the mix proportion is optimized, the performance of roller compacted concrete mixture is significantly improved, the pitted surface occurs on the roller compacted concrete after rolling compaction is eliminated, which significantly improves the interlayer bonding quality of the roller compacted concrete. See Table 9.11 for roller compacted concrete construction mix proportion of Jin'anqiao Dam.

Table 9.11 Roller compacted concrete construction mix proportion of Jin'anqiao Dam

Design index and position	Mix proportion parameter							Material amount (kg/m³)				
	Gradation	Water-cement ratio	Fly ash (%)	Sand ratio (%)	Water content (kg/m³)	ZB-1 Rcc15 (%)	Air entraining agent (/10 000)	Cement	Fly ash	Sand	Stone	Apparent density
$C_{90}20W6$ F100 Lower part of dam	III	0.47	60	33	90	1.2	20	76	115	775	1 574	2 630
$C_{90}15W6F100$ Upper part of dam	III	0.53	63	33	90	1.2	15	63	107	782	1 588	2 630
$C_{90}20W8F100$ External imperviousness area	II	0.47	55	37	100	1.2	20	96	117	846	1 441	2 600

9.3.4 Dam Body Material Zoning and Optimization

9.3.4.1 Non-overflow Dam Section Material Zoning and Optimization

The zoning of dam body material has much bigger influence on roller compacted concrete fast damming, for temperature control and cracking prevention of the dam, the cement amount shall be reduced possibly, and the temperature rise due to the hydration heat is minimized. The original design *Roller Compacted Concrete Construction Zoning and Layering Diagram* of Jin'anqiao Dam shows that, the $0^{\#}$ key slot dam section, $1^{\#} - 5^{\#}$ left bank non-overflow dam sections and $16^{\#} - 20^{\#}$ right bank non-overflow dam sections are roller compacted concrete below EL 1 413 m, the range 11 m from the dam crest EL 1 424 m elevation is designed as normal concrete with same indicators. By comparison of technical proposals, the zoning of dam body materials of the original design is optimized, that is the zoning range of roller compacted concrete at non-overflow dam section is raised from EL 1 413 m elevation to EL 1 422.5 m elevation, in this case the zoning of roller compacted concrete at non – overflow dam section is increased by 9.5 m approximately, the roller compacted concrete can replace 35×10^4 m^3 of the normal concrete.

9.3.4.2 Intake Dam Section Material Zoning and Optimization

$7^{\#} - 10^{\#}$ dam sections are intake dam sections, the width of various dam sections is 34 m, 4 power station intakes are arranged in all. In the original design, the construction of roller compacted concrete at $7^{\#} - 10^{\#}$ intake dam sections is up to EL 1 366 m, then changes to normal concrete above EL 1 366 m, according to the requirement of intake structure construction and convenient condition of roller compacted concrete, the design adjusts the roller compacted concrete construction up to EL 1 367.8 m, the roller compacted concrete adopts the mode of placing from truck, which reduces the pressure of high intensity of dam normal concrete pouring.

From elevation EL1 367.8 m till dam crest elevation of EL 1 424 m at the intake dam section is normal concrete, the zoning of raw materials are II gradation $C_{28}25W8F100$ graded concrete, the dosage of cementing material is 300 kg/m^3. If it is placed only by the II gradation concrete, the concrete amount is up to 225 kg/m^3, which causes the temperature rise due to the concrete hydration heat is very high, cracks on the concrete surface after pouring is much more, and it's not good for the structure. According to the zoning condition of peripheral concrete of the penstock at powerhouse dam sections of large hydropower stations such as the Three Gorges, Lijiaxia, Wanjiazhai, Wudu, etc., the original concrete design indexes are optimized. Except that the II gradation $C_{28}25W8F100$ graded concrete is casted at the locations with intensive reinforcement within a range of 2 m in periphery of the steel pipe, orifice and gate slot, the concrete design index at other locations is adjusted to III gradation $C_{90}30W8F100$, the design age is 90 d, which increases the dosage of fly ash, reduces the dosage of cementing material, and minimizes the temperature rise of cement hydration heat and concrete heat insulation. The result of optimization test of $C_{90}30W8F100$ III gradation concrete mix proportion carried out in Jin'anqiao Project is that the water-cement ratio is 0.47, dosage of fly ash is 30%, the amount of cementing material is reduced from original 300 – 245 kg/m^3, the cement amount is reduced from 225 – 172 kg/m^3 correspondingly.

During practical use in this project, the concrete zoning adjustment is carried out for intake dam section elevation of EL 1 384 m above, the III gradation $C_{90}30W8F100$ concrete is added, 10 m width on the left and right sides of each intake dam section is adjusted to $C_{90}30W8F100$ III gradation concrete, the 14 m wide gate slot area is kept as $C_{28}25W8F100$ II gradation concrete in original design.

9.3.4.3 Zoning and Optimization of Roller Compacted Concrete at $6^{\#}$, $11^{\#} - 15^{\#}$ Dam Sections

Except that the surrounding of left sand flushing underport steel pipe and service emergency gate slot at $6^{\#}$ dam

Chapter 9　Key Technology for Rapid Construction of RCC

section with elevation of EL 1 387 m and less is roller compacted concrete, the dam surface above EL 1 387 m is narrowed and the arranged service emergency gate slot are normal concrete. $11^{\#} - 12^{\#}$ dam section right discharge underport EL 1 333 m to EL 1 346.5 m elevation, service gate slot surrounding, right discharge underport working area is normal concrete, in addition, the mass concrete of EL 1 413 m and less is roller compacted concrete, EL 1 413 m above of the dam surface is narrowed and the arranged service gate slot is normal concrete, $13^{\#} - 15^{\#}$ dam sections are roller compacted concrete from EL 1 393 m and less, EL 1 393 m above is normal concrete construction of surface outlet and gate pier, overflow weir, etc..

The roller compacted concrete of $6^{\#}$ dam section at EL 1 387 m and less basically adopts whole-block construction with neighboring dam section. $11^{\#} - 15^{\#}$ dam sections are rolled to the EL 1 393 m by whole-block after passing through right discharge underport normal concrete area; EL 1 393 m above is normal or roller compacted concrete, its engineering quantity is not too big, the placing mean mainly is cable crane. The dam concrete construction peak mainly is the construction section with EL 1 352 m and less, how to construct the normal concrete in the dam underport area is very important, it is the key location to ensure the roller compacted concrete rapid and uniform lift.

9.3.4.4　Zoning and Optimization of Mass Concrete for Flood Spillway, Right Discharge Slot of Dam Downstream

In the dam downstream flood spillway and right discharge slot foundation concrete, total volume of design is 67×10^4 m^3, the 20.2×10^4 m^3 mass concrete is adjusted to roller compacted concrete, the mode of placing from the truck is used, the layout of gantry crane and tower crane in the original bidding scheme is eliminated, which gives full play to the advantages of rapid construction of roller compacted concrete, significantly reduces the quantity of construction equipment and formworks, the cement amount for the concrete decreases significantly, very favorable to temperature control and cracking prevention, shortening the period of concrete construction, its technical economic benefit also is obvious.

9.3.5　Optimization of Transportation Placing

9.3.5.1　Truck + Full Pipe Chute Transportation Plan

The roller compacted concrete placing transport has always been one of the key factors restricting rapid construction, the dam roller compacted concrete transportation shall be based on the principle of speed, simplicity and economy. There are many kinds of transportation placing plans for roller compacted concrete transportation placing: that is truck transportation, belt conveyor conveying, negative pressure chute, aggregate bin turnover, cable crane or tower crane direct transportation, etc. A large number of engineering construction practices have demonstrated that direct placing of the roller compacted concrete by using truck transportation is the most effective mode of transportation, which can greatly minimize the intermediate links, effectively control the temperature rise of the concrete.

Roller compacted concrete transportation system layout plan in original construction organization design of Jin'anqiao Dam: the horizontal transportation mainly adopts dump truck and deep slot belt conveyor feed line, the vertical transportation mainly adopts negative pressure chute and cable crane. In addition to hoist normal concrete, the cable crane is also used for the ancillary work such as placement preparation, placement equipment transfer, metal structure installation, etc. In the original concrete transportation plan, large amount of placing modes such as belt conveyor feeding line and negative pressure chute, etc.. are used, which will bring great difficulties to the design, arrangement and maintenance of belt conveyor feed line and negative pressure chute during high intensity transportation of roller compacted concrete, and the construction interference is big between them, the layout of transportation plan is complicated, time and labor consumed, vertical transportation of roller compacted concrete by cable crane also does not conform to the actual construction due to poor efficiency of cable crane in our country.

The Jin'anqiao roller-compacted concrete gravity dam is featured with large placement surface, whole-block continuous placing and high intensity construction, the original transportation placing plan is optimized in actual construction, the original transportation layout plan of deep slot belt conveyor and negative pressure chute is removed. The roller compacted concrete transportation optimization plan mainly uses placing direct from dump truck, the vertical transportation adopts box full pipe chute to convey the concrete. The roller compacted concrete from an elevation of 1 348.0 m and less at lower part of the dam adopts the mode of placing direct from truck, the elevation of more than 1 348.0 m at upper part of the dam adopt the combined transportation plan of dump truck + full pipe chute + placement surface truck. In this case, the advantages of dump trucks such as high mobility, high efficiency of full pipe chute and flexibility of placement surface truck are fully utilized, the requirement of rapid construction of dam roller compacted concrete is met.

9.3.5.2 Box Full Pipe Chute

The full pipe chute is latest technical equipment for solving vertical transportation. At the time being, the height of the roller compacted concrete dam becomes higher and higher, for the dam body of narrow river valley, the height difference of the road to the dam is very big, it is impossible for placing direct from truck, the vertical transportation of roller compacted concrete at intermediate links uses full pipe chute for conveying, that is dump truck + full pipe chute + placement surface truck, which has completely replaced the transportation plan of conventional negative pressure chute.

The box full pipe chute structural design is mainly composed of four parts: unloading hopper, full pipe chute body structure, discharge port elbow section and system supporting structure. The problems such as unloading hopper shape, chute section size, system sealing and installation part stability control with the box full pipe chute shall be solved. To assure continuous unloading of the concrete and achieve the effect of full pipe conveying, Jin'anqiao discharge adopts large hopper with a capacity of 20 m^3 more or less, the upper opening size is 3 400 mm × 3 400 mm, the lower opening size is 800 mm × 800 mm, the height is 3 150 mm (to prevent the discharge leakage, a 400 mm high baffle is added on three sides of the upper opening), the thickness of discharge hopper wall steel plate is 8 mm, the hopper is arranged at the gap reserved on the bank slope dam section, which is connected to the foundation through four supporting pillar by foundation bolts. The box type full pipe chute body component includes standard section length of 1.0 m, non-standard section length of 0.55 m, the cross-section dimensions are 800 mm × 800 mm. Each section of box type full pipe chute body adopts flanged bolt connection, the installation and removal are very convenient, the reduction extension section at the outlet is mounted between elbow section and outlet arc gate. The declination angle of the full pipe chute is 40° – 50°, the outlet arc gate and placement surface aggregate bin are removed (the aggregate bin is set on the placement surface for original negative pressure chute). The dump truck unloads the roller compacted concrete directly to the placement surface truck through full pipe chute, it generally takes 15 – 25 s for conveying 9 m^3 roller compacted concrete from the discharge hopper through full pipe chute to completion of receiving by placement surface truck, it is very simple and fast. As the existing roller compacted concrete in recent years is the semi plastic concrete with high rock powder content and low VC value, the worrying issue of aggregate separation is also readily solved. See Figure 9.18 and Figrue 9.19 for the box-type full pipe structure layout drawings.

The Jin'anqiao Dam is basalt aggregate roller compacted concrete, the wear of the full pipe chute is very fast and serious due to the behaviors of basalt aggregate such as big hardness, etc. . . , the bottom of the chute is generally worn through after conveying 3×10^4 m^3 roller compacted concrete, therefore, 20 mm thick steel sheet is welded at bottom and 200 mm height on both sides for reinforcing in the full pipe chute. The box-type full pipe chute systems are arranged at 1 360.0 m and 1 424.0 m elevations on the left bank of the dam and

Figure 9.18 *Right bank $20^{\#}$ dam section full pipe chute*

Figure 9.19 *Left bank $1^{\#}$ dam section full pipe chute*

1 400. 0 m and 1 424. 0 m elevations on the right bank of the dam respectively. The design delivery capacity of single strip box-type full pipe is 500 m^3/h. The box-type full pipe enables rapid transportation of roller compacted concrete, the loss of *VC* value is small, which is very good for improving the bonding between layers, thoroughly solves the difficulty of vertical transportation of concrete, in the actual casting, the full pipe operating state is stable. The Jin'anqiao Dam adopts box-type full pipe to deliver 70×10^4 m^3 more or less of the concrete, the dump truck is matched with box-type full pipe, allowing satisfactory conveying intensity and quality.

9.3.6 Placement Surface Zoning and Rapid Construction

9.3.6.1 Principle of Placement Surface Zoning

The Jin'anqiao Hydropower Station Dam roller compacted concrete construction is big in intensity and tight in connection of work operations, because the original placing modes such as deep slot belt conveyor, etc. are canceled, the mode of placing direct from truck is used, it is necessary to study and solve the problem of high intensity and continuity of concrete transportation in the construction. The zoning principle of roller compacted concrete placement surface is to reasonably divide the placement surface mainly according to the dam structure, layout type, mixing system intensity, different casting elevation, casting equipment capacity, node requirements of different dam sections and the transportation and placing mode.

According to the roller compacted concrete construction features of the Jin'anqiao Dam, the large area continuous and fast placement is used to give full play to the advantages of rapid mechanized construction of roller compacted concrete, under the precondition that the intermittent time between layers is met, the larger the placement surface, the higher the construction efficiency. At the time being, the roller compacted concrete spreading and rolling almost adopts inclined-layer placing method for rolling construction, the construction experience of Jin'anqiao Dam is to possibly combine many units into one placement surface as soon as the conditions permit, this allows flow shop, cyclic operation and balanced construction, significantly improves the efficiency of roller compacted concrete construction.

9.3.6.2 Placement Surface Zoning and Transportation Placing

The zoning of dam roller compacted concrete placement surface is closely related with the mode of transportation placing, the placement surface zoning shall be analyzed in combination with the mode of transportation placing, try to combine multiple placing areas into one large area for construction, this can effectively reduce the quantity of the formworks and fully utilize the advantages of rapid construction of the roller compacted concrete. The plan for specific placement surface zoning and transportation placing mode of the dam roller compacted concrete is as follows:

Roller Compacted Concrete Rapid Damming Technology

1. Combine the $7^{\#} - 11^{\#}$ dam sections with EL 1 330 m and less elevation into one big area for construction. Roller compacted concrete placing, mixing plant→right bank 1 320 m elevation road to dam→passing $17^{\#}$ dam section→dam upstream slag filled road→placing direct from truck→casting $7^{\#} - 11$ dam sections. In September, 2007, the right bank spillway berm slope geological slump caused the right bank 1 320.0 m elevation upper dam road disruption, the technical measures was taken later, by using the spillway slope section to arrange "Z" shape road to the dam, this timely connects and guarantees the smooth flow of the road to the dam.

2. Combine the $12^{\#} - 15^{\#}$ dam sections with EL 1 330 m and less elevation into one big area for construction. Roller compacted concrete transportation placing, mixing plant→road to dam→right discharge slot foundation→spillway slope section "Z" road→placing direct from truck at $17^{\#}$ dam section→casting $12^{\#} - 15^{\#}$ dam sections.

3. Combine the $5^{\#} - 6^{\#}$ dam sections with EL 1 313.8 - 1 324 m elevation into one big area for construction. Roller compacted concrete transportation placing, mixing plant→left bank access tunnel→left bank 1 320 m elevation road→dam downstream slag filled road→placing direct from truck at $6^{\#}$ dam section →casting $5^{\#} - 6^{\#}$ dam sections.

4. Combine the $5^{\#} - 11^{\#}$ dam sections with EL 1 330 - 1 348 m elevation into one big area for construction. Roller compacted concrete transportation placing, mixing plant→right bank 1 424 m elevation access tunnel→upstream road→upstream cofferdam 1 345 m elevation→cofferdam downstream sloe road→dam upstream slag filled road→placing direct from truck from $11^{\#}$ dam section→casting $5^{\#} - 11^{\#}$ dam sections.

5. Combine the $12^{\#} - 17^{\#}$ dam sections with EL 1 330 - 1 348 m elevation into one big area for construction. Roller compacted concrete transportation placing, mixing plant→right bank 1 424 m elevation access tunnel→upstream road→upstream cofferdam 1 345 m elevation→cofferdam downstream sloe road→dam upstream slag filled road→placing direct from truck from $15^{\#}$ dam section→casting $12^{\#} - 17^{\#}$ dam sections.

6. Combine the $5^{\#} - 18^{\#}$ dam sections with EL 1 348 - 1 354 m elevation into one big area for construction. Roller compacted concrete transportation placing, mixing plant→left bank 1 360 m elevation road to dam→$4^{\#}$ dam section 1 360 m elevation→dump truck→$4^{\#}$ dam section full pipe chute (two) → placement surface dump truck→casting $5^{\#} - 18^{\#}$ dam sections.

7. Combine the dam left bank $6^{\#} - 11^{\#}$ dam sections with EL 1 354 - 1 363 m elevation into one big area for construction. Roller compacted concrete placing, mixing plant→left bank 1 360 m elevation road to dam→dam downstream slag filled road→placing direct from truck at $5^{\#}$ dam section→casting $6^{\#} - 11^{\#}$ dam sections.

8. Combine dam right bank the $12^{\#} - 18^{\#}$ dam sections with EL 1 354 - 1 375 m elevation into one big area for construction. Roller compacted concrete transportation placing, mixing plant→right bank 1 400 m elevation road to dam→dump truck→$19^{\#}$ dam section full pipe chute→placement surface dump truck→casting $12^{\#} - 18^{\#}$ dam sections.

9. Combine the dam left bank $2^{\#} - 11^{\#}$ dam sections with EL 1 422.5 m and less elevation into one big area for construction. Roller compacted concrete transportation placing, mixing plant→left bank 1 424 m road to dam →dump truck→$1^{\#}$ dam section full pipe chute→placement surface dump truck, casting respectively: $7^{\#} - 11^{\#}$ dam sections EL 1 363 - 1 367 m elevation, $2^{\#} - 6^{\#}$ dam section EL 1 363 - 1 381 m elevation, $2^{\#} - 5^{\#}$ dam sections EL 1 381 - 1 422.5 elevation.

10. Combine the dam right bank $11^{\#} - 19^{\#}$ dam sections with 1 422.5 m and less elevation into one big area for construction. Roller compacted concrete transportation placing, mixing plant→right bank 1 424 m road

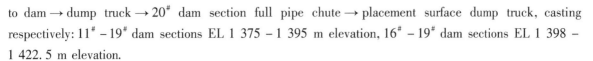

to dam → dump truck → $20^{#}$ dam section full pipe chute → placement surface dump truck, casting respectively: $11^{#} - 19^{#}$ dam sections EL 1 375 – 1 395 m elevation, $16^{#} - 19^{#}$ dam sections EL 1 398 – 1 422. 5 m elevation.

To meet the requirement in flood period in 2008, the dam $5^{#} - 18^{#}$ dam sections EL 1 348 – 1 352 m elevation are combined into one large whole – block for construction, the placement area is up to 2×10^4 m^2 and more (425 m × 55 m), $4.2 \times 10^4 m^3$ roller compacted concrete was cast.

9.3.6.3 Construction In Layers

For the layering of Jin'anqiao Dam roller compacted concrete construction, according to dam concrete temperature control zoning and construction scheme, the lift height of roller compacted concrete is controlled at 3.0 m, the maximum lift of the left bank $6^{#}$ and right bank $12^{#}$ underport dam sections is up to 9 m for construction rush. Based on the dam placement surface zoning plan, specific construction layering is as below:

Powerhouse dam section, $7^{#} - 11^{#}$ dam sections EL 1 268 – 1 366 m elevation, the layering height is 3.0 m, 33 lifts in all;

Overflow dam section, $13^{#} - 11^{#}$ dam sections EL 1 294 – 1 393 m elevation, the layering height is 3.0 m, 33 lifts in all;

Left bank non-overflow dam section, $2^{#} - 5^{#}$ dam sections EL 1 318 – 1 422.5 m elevation, the layering height is 3.0 m, 35 lifts in all;

Right bank non-overflow dam section, $14^{#} - 19^{#}$ dam sections EL 1 324 – 1 422.5 m elevation, the layering height is 3.0 m, 33 lifts in all;

Left flushing bank underport dam section, to $6^{#}$ dam sections EL 1 315 – 1 396 m elevation, the layering height is 3.0 – 9.0 m, 23 lifts in all;

Right flushing dam section, $12^{#}$ dam sections EL 1 294 – 1 327 m elevation, the layering height is 3.0 m, 11 lifts in all; EL 1 345 – 1 411 m elevation, the layering height is 3.0 – 9 m, 19 lifts in all.

9.3.7 Key Technique for Interlayer Bonding

9.3.7.1 Factors Affecting Interlayer Bonding

The interlayer bonding quality is particularly important in roller compacted concrete construction, which is directly related to the imperviousness, anti-sliding stability and overall performance of the dam. The key point of quality control at roller compacted concrete site is the interlayer bonding. There are many factors affecting the interlayer bonding, the factors such as the mix proportion, rock powder content, VC value, setting time, joint surface mortar spreading, rolling intermittent time between layers, interlayertreatment, etc. After the working performance of the fresh roller compacted concrete meets the requirements of rollability and liquidation and bleeding, to assure good bonding between concrete layers, the intermittent time between layers shall be strictly controlled. After flushing treatment of the concrete joint surface and its acceptance by the supervisor is acceptable, a layer of 1.5 cm mortar layer is scraped and spread first (strength classes of the mortar is one class higher than that of the roller compacted concrete) before spreading and rolling the concrete, then immediately spread the roller compacted concrete on it, and complete the rolling before the initial setting of the mortar. The large area thin-lift continuous placement is used for roller compacted concrete of rolling layer, the rolling layer thickness is 30 cm, the spreading thickness is from 34 – 36 cm. In roller compacted concrete construction, the shorter the intermittent time between upper and lower layers, the better the interlayer bonding quality and interlayer cohesion of the continuous paving layers (hot joints). The intermittent time between layers mainly controls the permissible time for direct spreading of upper and lower layers, that is the spreading time between layers is less than the initial setting time to enable the full bleeding and elastic on the rolled concrete surface, which guarantees the upper layer aggregate is embedded into the lower layer roller compacted concrete after completion of rolling, effectively improves the interlayer bonding quality and anti-slip stability. If the interval time exceeds the initial setting time of the lower layer roller compacted concrete, the plasticity is lost, and the cold joints will appear, the cold joint is that the construction joints that needs to take the measures such as paving mortar, etc.. for the layer treatment, and has influence on the rapid construction of roller compacted concrete. The concrete rolling at various rolling layers shall be controlled according to specified rolling parameters and inspected timely, whether the number of times of rolling is increased or not can be determined according to the condition of bleeding on the concrete surface and the result of test by nuclear density gauge, measures shall be taken timely to deal with the situations such as the non-bleeding, rough surface, aggregate concentration, cold joints and so on.

9.3.7.2 Control of *VC* Value is the Key to the Rollability

The size of the *VC* value has significant influence on the performance of the roller compacted concrete, large amount of project practices in recent years demonstrates that the key point of roller compacted concrete site is the mixture *VC* value and initial setting time, *VC* value dynamics control is the key to guarantee the rollability and interlayer bonding of roller compacted concrete. The *VC* value of Jin'anqiao Roller Compacted Concrete Dam at the mixer outlet was adjusted at any time with the change of season, time period, air temperature, the *VC* value at the mixer outlet is controlled between 1 – 3 s in day time, the placement surface *VC* value is controlled between 3 – 5 s; the *VC* value at the mixer outlet is controlled between 2 – 5 s in night time, the *VC* value at the mixer outlet is controlled at the lower limit, under the condition that the normal rolling at site is satisfied, the control of the *VC* value at the mixer outlet is based on the principle that the roller is not trapped in the placement surface.

The liquefied bleeding of the roller compacted concrete is to extract the grout from the concrete liquidation under the action of the vibrating and rolling of the vibration roller, this thin-lift of the liquefied bleeding is the key to guarantee the interlayer bonding quality, the liquefied bleeding has been used as an important criterion to evaluate therollability of the roller compacted concrete. Under the precondition that the mix proportion meets the rollability, the main factor affecting the liquidation and bleeding is the loss of *VC* value with time and whether the roller compacted concrete is rolled after placement or not.

With respect to the worse liquidation and bleeding condition of basalt aggregate roller compacted concrete for Jin'anqiao, under the condition that the mix proportion parameter is kept unchanged, increase the dosage of the additive, under the double layer superposition of the admixture and retarding, the *VC* value is reduced, the setting time is delayed, which effectively improves the rollability, liquidation and bleeding and interlayer bonding quality of the roller compacted concrete.

9.3.7.3 Inclined Layer Rolling, Shortening Intermittent Time

With the Jin'anqiao Dam, several dam sections were usually combined into one placement surface, the area is very big, the rolling process mainly uses inclined layer rolling, the inclined layer rolling has effectively reduced the spreading operation area of the roller compacted concrete, improved the rolling efficiency, shortened the intermittent time between layers, the effect to improve interlayer bonding quality is significant.

The inclined layer rolling is to carry out rolling to the placed roller compacted concrete after it is spread by inclined-layer placing method. The specific practice for inclined layer rolling is to carry out flat-spread rolling to the placing section first, then shorten the spreading layer from bottom to top for rolling successively, this enables to form a slope on the freshly poured concrete surface, and allows the most part of the concrete at the placement finishing end is spread and rolled according to this slope, the spreading method is basically same as that of roller compacted concrete flat-spread method, the placement finishing end is finished by means of several extended flat layer in succession. The direction inclined-layer placing is parallel to the dam axis in principle, if the pouring is along the direction of upstream and downstream, the inclined layer placing and rolling direction shall be from the downstream direction toward the upstream direction.

Inclined layer slope, lift height and rolling layer thickness are three main parametersof inclined layer rolling, by selecting proper parameters, the intermittent time between layers can be controlled within the initial setting time of the roller compacted concrete. In the roller compacted concrete construction of Jin'anqiao Dam, if the slope of inclined layer rolling is controlled in a range of 1:10 − 1:12, the lift height is 3 m, the rolling layer thickness is controlled at 30 cm, the slope length is about 31 − 36 m, if it is controlled according to maximum zoning width of concrete placement, the slope area is generally controlled in 2 000 m^2, see Photo 2 for inclined layer rolling construction. At present, the inclined rolling has become the mainstream way for rapid construction of roller compacted concrete. Main advantages of inclined layer rolling:

1. it can greatly shorten the intermittent time between layers of the roller compacted concrete, which is shortened from 4 − 6 h of original flat layer rolling to 2 − 4 h, it is very beneficial to improve the interlayer bonding of rolling layer and change the performance of interlayer structure.
2. The balanced casting intensity can be used to carry out rolling on the placement surface with very big coverage area, which reduces the resources allocation of large placement surface casting, comprehensively reduces the equipment input and cost of temporary structure works, enabling the saving of project investment.
3. It enables to carry out large scale cyclic flow shop, minimize the impact due to too small block area casting and the quantity of the formwork works, under the precondition that the casting intensity is not increased, the construction efficiency is improved and the project progress is quickened. Using inclined layer rolling to complete a large placement surface of 10 000 m^2 with a lift of 3 m can completed 3 d advance than that of the flat layer rolling.
4. As the interval time between layer surface is significantly shortened, the upper layer of concrete is quickly covered, the rise of casting temperature of the roller compacted concrete is reduced, at the same time, because the casting area is small, the measures such as insulation by spraying, etc. is easy to implement, therefore it has good adaptability to the construction in high temperature season.
5. In the rainy areas, as the slope allows easy drainage, small casting area enables easy handling, which reduces the influence range and degree of rainfall, so it is also suitable for construction in rainy weather.

9.3.8 Key Technology for Dam Temperature Control and Cracking Prevention

The Jin'anqiao Project is located in the Yunnan-Guizhou Plateau, the altitude at the dam site is high, and has typical characteristics of plateau climate, the temperature difference between day and night is big, the sunshine is intensive, the evaporation capacity is big, the extreme maximum temperature is 40 ℃, the extreme minimum temperature is −6.2 ℃, the temperature drop is frequent due to cold snap in winter and spring, which has much bigger influence on the temperature control and cracking prevention. At the same time, because the roller compacted concrete adopts the casting mode of large whole-block and thin-lift continuous lift without longitudinal joint, the placement surface zoning is too big, the binding effect of the bed rock is strong, it is easy to produce through cracks. And the temperature drop process with the roller compacted concrete is slow, the internal restraint time due to the temperature difference between inside and outside is long, it is easy to produce surface cracks, even causes the vertical cracks, which has influence on the safety of the dam. As a result, the concrete temperature control is one of the key technical issues, the strict temperature control measures shall be taken to the concrete during the construction so as to prevent cracks from occurring and guarantee the integrity of dam body structure.

9.3.8.1 Temperature Control Design Requirement

According to the design requirements, the temperature difference of the internal and outside of dam concrete, allowable maximum placement temperature and permissible maximum temperature of the Jin'anqiao Dam are as follows:

1. Temperature difference of the internal and outside of concrete according to the concrete surface temperature control criteria required by the design, the temperature difference of the internal and outside of concrete is controlled to be no more than 16 ℃, temperature difference of the internal and outside of roller compacted concrete is controlled to be no more than 15 ℃.
2. Allowable maximum placement temperature according to design concrete temperature control zoning, the roller compacted concrete casting conditions are: the distance from the bed rock surface thickness is 0 – 0.2 L, the casting layer thickness is 3 m, the interval of the casting layer: 7 – 10 d, primary water cooling is not be less than 15 d, the high temperature period determined by the design is from April to September, allowable maximum placement temperature: dam cushion restrain zones is 17 ℃, dam lower part 8 zones (EL 1 350 m and less) is 20 ℃, dam upper part 15 zones (EL 1 350 m above) is 22 ℃. The temperature of roller compacted concrete at mixer outlet from April to September correspondingly is controlled as follows: restrain zone is 12 ℃, dam lower part is 14 ℃, dam upper part is 16 ℃.
3. Permissible maximum temperature. High temperature period is from April to September, the permissible maximum temperature of roller compacted concrete is that: restrain zone is 27 ℃, dam lower part Section 8 31.5 ℃, dam upper part Section 15 33 ℃.

9.3.8.2 Lower Temperature Rise of Concrete Hydration Heat

Under the precondition that the mix proportion of the roller compacted concrete meets the design and construction requirements, it possibly reduces the cement amount and unit water consumption, increases the dosage of fly ash, achieves the purpose to effectively reduce the temperature rise due to the concrete hydrous heat.

The Jin'anqiao adopts 42.5 moderate heat Portland cement, the internal control indicators is established for the moderate heat Portland cement: 28 d compressive strength is(47.5 ±2.5) MPa, flexural strength is more than 8.0 MPa, the heat of hydration is less than 283 kJ/kg, the fineness is less than 320 m^2/kg, magnesium oxide content is 3.5% – 5.0%, temperature of cement into tank is not more than 60 ℃. The required water content ratio for the supplied Grade II fly ash is not more than 100%, the temperature of fly ash into tank is not more than 45 ℃. With regard to the issue that the water content of manufactured basalt sand is high, the technical measure of mixing with 20% dry sand is used to effectively reduce the water content of manufactured sand.

The optimization of mix proportion design adopts "two-high-two-low-one-replacement" technicalroute, that is the technical route of high fly ash dosage, high additive dosage, low water consumption, low VC value and replacement of the sand with rock powder. The dosages of fly ash for $C_{90}20$ and $C_{90}15$ III gradation roller compacted concrete are increased to 60% and 63% respectively, the dosage of additive is increased to 1.2%, the water consumption for original III gradation roller compacted concrete is decreased from 100 kg/m^3 to 90 kg/m^3, the VC value at mixer outlet is reduced from 3 – 5 s to 1 – 3 s, at the same time, the technical proposal that is mixed with rock powder instead of sand is used to effectively reduce the temperature rise due to concrete hydrous heat and improve the performance of the roller compacted concrete.

By strict control of raw materials and the optimization of the mix proportion, the temperature rise due to concrete hydrous heat is effectively controlled from the source.

9.3.8.3 Control of Temperature at Mixer Outlet

Ⅰ Concrete refrigeration system

One mixing system is arranged each on left and right of dam downstream, there are 4 mixing plants in all. The left bank mixing system is about 2.7 km from the dam axis with an elevation of EL 1 220 m, the $3^{\#}$ and $4^{\#}$ mixing plants are arranged, two mixing plants are 2×4.5 m^3 forced mixing plant, which mainly mixes the roller compacted concrete, the nameplate production intensity is 500 m^3/h; the right bank mixing system is about 0.75 km from the dam axis with an elevation of EL 1 370 m, the $1^{\#}$ and $2^{\#}$ mixing plants are arranged, of which the $1^{\#}$ mixing plant is 4×3 m^3 gravity type mixing plant, the $2^{\#}$ mixing plant is 2×4.5 m^3 forced mixing plant, this mixing system mainly mixes the normal concrete. The pre-cooling installation of the left and right bank mixing systems is configured according to that the temperature of roller compacted concrete at mixer outlet is not more than 12 ℃, the temperature of normal concrete at mixer outlet is not more than 10 ℃. According to the ambient temperature and concrete outlet temperature requirements, different combinations of primary air cooling, secondary air cooling, addition of cold water and addition of ice for aggregates can be chosen to meet the requirements of temperature at the outlet of pre-cooling concrete mixer.

Ⅱ Pre-cooling concrete

1. Reduce the temperature of aggregates. To produce the qualified pre-cooling concrete, it is necessary to take cooling measures to concrete materials, so as to reduce the temperature at the outlet of concrete mixer. Based on the temperature at the outlet of concrete mixer, it is found that the temperature of stones among all kinds of materials has the largest influence on the temperature at the outlet of concrete mixer. If the temperature of stone is reduced by 1 ℃, the temperature at the outlet of concrete mixer can be reduced by about 0.6 ℃. For roller compacted concrete, the moisture content of finished sand should be strictly controlled and the primary and secondary air cooling should be carried out for coarse aggregate, so that the minimum temperature at the outlet of concrete mixer in the hot period from April to September live up to the requirement of 12 ℃.

2. Mixing with cold water and ice. Mixing with cold water: water for mixing shall be supplied from the screw chilling units. If the water temperature is reduced by 1 ℃, the temperature at the outlet of concrete mixer can be reduced by about 0.2 ℃. Mixing with ice: some water can be replaced with flake ice for mixing. With the ice melting, 80 kcal/kg of latent heat will be absorbed, which will further reduce the temperature at the outlet of concrete mixer. According to the calculation, if the amount of ice added per cubic meter accounts for 25% and 50% of water consumption, the concrete temperature can be reduced by 2.8 ℃ and 5.7 ℃ respectively.

9.3.8.4 Temperature Control During Concrete Pouring

Ⅰ Temperature control during transportation

In order to reduce the temperature rise of concrete in the process of transportation, it is necessary to strengthen the management in the construction, speed up the placement of concrete, reduce the exposure time and the number of handling, so as to reduce the temperature rise in the transportation process. The roller compacted concrete of Jin'anqiao dam is completely transported by dump truck. The sliding sunshading tarpaulin is set on the top of the carriage of the dump truck to effectively prevent the temperature rising during the transportation. It generally will take 15 – 20 minutes to convey concrete from the mixing plant to the dam surface by dump truck. The statistics show that the temperature rise of roller compacted concrete is generally 0.6 – 1.1 ℃.

Ⅱ Temperature control during concrete pouring

1. Speed up the rolling on the surface. During the construction, it is necessary to strictly control the

transportation time and the exposure time before closing the pouring layers for the roller compacted concrete, speed up the placement and covering of concrete, and reduce the pouring temperature of concrete, so as to reduce the maximum temperature of dam. After the roller compacted concrete is placed, it shall be leveled and rolled in time. In Jin'anqiao dam, the roller compacted concrete is mainly poured continuously and in slant layers. The pouring of roller compacted concrete is generally completed within 4 h from feed out of the mixer to the completion of rolling. The interval time between layers is controlled within 6 – 8 h.

2. Reduce temperature and keep moisture by spraying water. In the dry days or high-temperature period, in order to improve the crushability, prevent initial setting and temperature inversion of the concrete, it is necessary to spray water over the surface with the spraying gun in the process of concrete pouring and rolling. In this way, the moisture can be maintained and the concrete surface can be prevented from drying and whitening; in addition, the microclimate on the surface can be effectively improved, and the temperature on the surface can be reduced by 4 – 6 ℃.

3. Cover concrete and spray water for curing. For every belt of closed surface that has been compacted and tested to have meet the compactness requirements, the closing surface shall be covered with PVC coil immediately; after the roller compacted concrete becomes final setting, it is necessary to spray water on the finished concrete continuously to keep its moisture, until the roller compacted concrete of upper layer is poured.

With various temperature control measures in the transportation and pouring process, the temperature rise of concrete from the outlet of mixer to the completion of rolling can be generally controlled for not more than 5 ℃. In this way, the pouring temperature of roller compacted concrete can be basically controlled to meet the design requirements.

9.3.8.5 Water Cooling of Dam

I Layout of cooling water pipe

The cooling water pipe for RCC of the dam body shall be laid with the horizontal spacing of 1.0 m in the upstream impervious area and 1.5 m outside the impervious area, and the vertical spacing of 1.5 m. The cooling water pipe shall be HDPE plastic pipe, and laid along the rolling strip. Each layer of cooling water pipe shall be laid on the first rolling layer of opening surface and fixed with U-clamp, which shall be strictly prevented from mechanical rolling before covering with the roller compacted concrete. The cooling water pipe shall be laid according to the specified row spacing, 1.0 – 1.5 m from the upstream and downstream dam surface, and 1.0 – 1.5 m from the joint surface and the periphery of holes. Each pipeline shall have the total length of not greater than 300 m. After the cooling water pipe is led into the gallery or behind the dam according to the design of cooling pipeline and pipe leading planning, they shall be arranged orderly and marked well. The pipe orifice shall be properly protected to prevent blockage.

II Phase-I water conveying cooling

Phase-I water conveying cooling is one of the effective measures to reduce the temperature rise by concrete hydration heat. In the parts with strict temperature control requirements, the Phase-I water cooling must be carried out to reduce the peak hydration heat of concrete. According to the observation data, the peak temperature of roller compacted concrete appears in the 4 – 9 d after the concrete pouring.

The Phase-I water conveying cooling includes the supply with river water and 10 ℃ cooling water. In the low-temperature season, the river water can be directly used for cooling; while in the high-temperature season, the cooling water must be used. Such cooling water must be recycled. The temperature difference between the cooling water and the concrete of dam shall not exceed 25 ℃, and the temperature drop of dam body shall not exceed 1 ℃/d. The volume of water in the phase-I water cooling is controlled at 1.2 m^3/h. From the beginning of concrete pouring to the water supply, the direction of cooling water shall be changed every 24 h, so that the concrete of dam body cool evenly. The Phase-I water conveying cooling shall be dynamically controlled. When the temperature in the internal of concrete is rising, the monitoring for internal temperature shall be strengthened. If necessary, the water supply intensity can be increased or the temperature of cooling water can be reduced, to ensure that the internal temperature of concrete is controlled in the allowable range. When the internal temperature of concrete reaches its peak, the water supply requirements can be relaxed to avoid unnecessary super-cooling. The Phase-I water conveying lasts for about 20 d, which can reduce the peak temperature by 2 – 4 ℃.

9.3.9 Conclusion

1. Jin'anqiao dam makes full use of the advantages of rapid construction technology of roller compacted concrete, optimizes and innovates the rapid construction technology scientifically and reasonably, ensuring the realization of construction period target of major nodes such as flood season on schedule.
2. In view of the characteristics of basalt aggregate, the RCC mix design optimization is correctly carried out, which effectively reduces the temperature rise of concrete hydration heat, and obviously improves the working performance of RCC.
3. With reasonable material zoning of dam body, the advantages of rapid construction of roller compacted concrete can be given full play to, which is beneficial to temperature control and cracking prevention, and has remarkable technical and economic benefits.
4. The transport scheme of dump truck + full-pipe chute + truck transport for block surface is the fastest, most efficient and flexible, which can effectively reduce the rise of concrete temperature and accelerate the construction progress.
5. According to the construction experience of Jin'anqiao, if the conditions permit, several dam sections can be combined into a large surface as far as possible, which can realize flow operation, circulation operation, balanced construction, and improve the construction efficiency of roller compacted concrete.
6. The key to the quality control of roller compacted concrete is the combination of layers. The *VC* value should be strictly controlled and the concrete shall be compacted in time. With the inclined rolling method, the interval time between layers can be effectively shortened, the allocation of resources for surface can be reduced, which is very beneficial to improve the quality of interlayer bonding and the performance of interlayer structure.

9.4 RCC Construction of Longshou Arch Dam in Cold and Dry Area

9.4.1 Foreword

9.4.1.1 Project Overview

Longshou Hydropower Station is located at the mouth of Yingluo gorge in Heihe River Basin, northwest of Zhangye City, Gansu Province, 41 km away from Zhangye highway. The hydropower station is a medium-sized third-class project with a total installed capacity of 52 MW and a storage capacity of 13.2 million m^3. The main structures of Longshou Hydropower Station are advanced in design and complex in structure. The main dam is an ultra-thin dome dam, with the maximum dam height of 80 m, the crest width of 5 m, the bottom width of 13.5 m, the thickness-height ratio of 0.17, the total length of 217.3 m and the crest elevation of 1 751.5 m. The water retaining structure is constructed with a full section of RCC to have a volume of 207 300 m^3. The

RCC design indexes are very strict, especially that the frost resistance grade shall reach F300.

The dam site of Longshou Hydropower Station is located in a area with frequent earthquake, where it is the typical continental climate, extremely hot in summer, with little rainfall and large evaporation; the freezing period is as long as four months, and the maximum frozen soil depth is 1.5 m. The mean annual precipitation is 171.6 mm, and the mean annual evaporation is 1 378.7 mm, which is more than 8 times of precipitation. The mean annual temperature is 8.5 ℃, the absolute maximum temperature is 37.2 ℃, the absolute minimum temperature is −33.0 ℃, and the average daily temperature difference is more than 20 ℃.

The construction period of Longshou Hydropower Station is very tight. The project is required to be completed within 2 years from the commencement to the operation of first generating unit. Therefore, RCC construction must be carried out both in hot days withhigh temperature period and cold days with low temperature, to achieve the total duration objective. Longshou project was the first RCC arch dam built in Northwest China at that time, so there was no experience to learn from. Under that circumstance, many difficulties and challenges will be encountered while adopting the full-section RCC dam construction technology in dry and severe cold areas.

The Longshou project was started on April 18, 1999, and the RCC construction was commenced on March 11, 2000. The workload for 16 months was completed in 10 months. In April 2001, the goal of impoundment was achieved half a year in advance.

9.4.1.2 Characteristics of RCC Design Index

The main design indexes of RCC construction for arch dam of Longshou Hydropower Project are shown in Table 9.12. It is shown that the design age of RCC is 90 d, the design strength is 20 MPa, the guarantee rate is 95%, and the strength of prepared RCC strength is 28.25 MPa. Although the design strength of RCC is not very high, the frost resistance grade F300 is the highest in China. It reflects that the durability requirement of arch dams is very strict in the design.

Table 9.12   Main design indexes of RCC for arch Dam of Longshou Hydropower Project

The parts used	Design index	Assurance rate P (%)	Apparent density (kN/m^3)
II gradation above ▽ 1 736 m of upstream face	$C_{90}20F300W8$	95	≥24
II gradation below ▽ 1 736 m of upstream face	$C_{90}20F200W8$	95	≥24
III gradation for the inside of arch dam	$C_{90}20F100W6$	95	≥24
III gradation for the inside of gravity dam	$C_{90}20F50W6$	95	≥24
II gradation for GEVR	$C_{20}F300W8$	95	≥24
III gradation for GEVR	$C_{20}F150W6$	95	≥24

9.4.2 Raw Material

9.4.2.1 Cement

It is tested for many times that "Qilianshan" P. II 42.5 Portland cement in Yongdeng of Gansu has the compressive strength of 53.5 − 64.8 MPa at 28 d, with a large margin.

9.4.2.2 Fly Ash

By testing the grade-II fly ash in Yongchang Power Plant for many times, the average is as follows: loss on ignition of 5.3%, water demand ratio of 98%, fineness of 15.8%, and sulfur trioxide of 0.6%.

9.4.2.3 Expansive Admixture

MgO (magnesium oxide) from Haicheng in Liaoning Province was used as the expanding agent to compensate

Chapter 9 Key Technology for Rapid Construction of RCC

the shrinkage deformation of concrete. The content of MgO was 4.3% of cementing material.

9.4.2.4 Additive

NF-A composite set retarding superplasticizer with the dosage of 0.9% was used; the content of air entraining agent NF-C is determined according to different frost resistance grades.

9.4.2.5 Aggregate

The natural sand and gravel are the aggregates of RCC for Longshou Project. The quarry is located 2 km downstream of the dam site, where the reserve of sand and gravel is large. In the early stage, the aggregate was produced in traditional wet process, generating too much loss of fine powder less than 0.16 mm in sand (the fine powder only accounts for 5%). Therefore, since March 2000, the aggregates were produced with dry method. Due to the dry climate in Longshou, the sand and gravel are dry and loose. By taking winnowing measures, the large silt content in sand and gravel is effectively reduced, and the content of fine sand is increased to 14%. The result indicates: the natural sand has small fineness modulus, small porosity of pebble, hard texture and good quality.

9.4.3 Construction Mixing Proportion

9.4.3.1 Optimization and Adjustment of Construction Mix Proportion

For Longshou Project, the RCC frost resistance is high, and the dry and severe coldenvironment had adverse impacts on the construction. Technology roadmap of construction mix proportion design: firstly, select a smaller water-cement ratio and a larger amount of fly ash as far as possible to reduce the temperature rise of concrete and make up for the low content of natural sand and rock powder; secondly, use low VC value and add high quality water-reducing retarder and air entraining admixture in RCC to prolong RCC setting time and increase air content, so as to adapt to large evaporation in Longshou, effectively improve the interlayer bonding and high frost resistance of RCC; thirdly, add proper amount of expansion agent MgO in RCC to compensate the temperature drop and shrinkage deformation of concrete and reduce the temperature crack of concrete.

From November 5, 1999 to November 9, 1999 and from January 11 to 18, 2000, two productive RCC process tests were carried out for Ⅱ gradation and Ⅲ gradation aggregates of RCC, with the volume of 603 m^3 and 1 800 m^3. The RCC mix proportion of productive test was carried out completely according to the designed mix proportion, showing that: the workability of RCC is not good, the separated aggregates are more, and it is obvious that the aggregate on the concrete surface is not wrapped by mortar after rolling; the surface of RCC is not full of grout, and the volume density of rolling and mixing are small. It is analyzed that the low water consumption and small sand ratio in the mix proportion as well as the high VC value submitted by the designer make it difficult for the mortar to fully wrap the coarse aggregate. Based on the rolling situation on the site, the compactability is good when the VC value is less than 8 s, and the rollability is poor when the VC value is more than 10 s. In addition, it is found that the air content of concrete does not reach the original design value.

After careful study and analysis by all parties involved in the construction, it is considered that the water consumption and sand rate of the mix proportion submitted by the designer are low, and the content of fine particles below 0.16 in natural sand is too small, only accounting for 5%. These factors lead to poor workability and rollability of RCC. Therefore, the construction unit optimizes and adjusts the mix proportion of the original design on the basis of the RCC mix proportion parameters submitted by the designer. The main parameters of RCC mix proportion after adjustment are shown in Table 9.13.

It can be seen from the data comparison in Table 9.13 that the adjustment of Ⅱ gradation aggregates in RCC focuses on the improvement of α value and mortar ratio P_V, that is, the improvement of mortar content of RCC; The adjustment of Ⅲ gradation aggregates focuses on the improvement of β value, that is, the increase of mortar proportion. According to the calculation, for Ⅱ gradation and Ⅲ gradation aggregates in RCC, the α

Table 9.13   Comparison of main parameters of RCC mix proportion after adjustment

Mixing proportion No.	Water-cement ratio	Water (kg/m³)	Cement (kg/m³)	Fly ash (kg/m³)/%	Sand ratio (%)	α	β	Grout-mortar ratio P_V	Unit weight (kg/m³)
Original design 1#, 2#	0.43	83	91	102/53	32	1.40	1.34	0.41	2 327
Adjustment 1	0.43	88	96	109/53	32	1.48	1.35	0.43	2 400
Adjustment 2	0.43	91	99	112/53	32	1.54	1.35	0.45	2 420
Original design 3#, 4#	0.48	78	55	108/66	28	1.43	1.23	0.40	2 368
Adjustment 1	0.48	82	58	113/66	30	1.41	1.28	0.39	2 450
Adjustment 2	0.48	85	60	117/66	30	1.47	1.28	0.41	2 450

value falls between 1.41 – 1.54, the β value falls between 1.28 and 1.35, and the grout and mortar ratio is between 0.39 – 0.45. By properly increasing the proportion of grout and mortar in RCC mix proportion, the workability of concrete is improved, which is beneficial to air entrainment, improve the frost resistance of RCC, and improve the rollability and bonding quality of RCC.

9.4.3.2 Analysis of Parameters of Construction Mix Proportion

1. water-cement ratio: the water-cement ratio directly affects the physical properties and durability indexes of RCC. Considering the influence of fly ash content on RCC performance, on the premise of not changing the water-cement ratio of RCC mix proportion of the original design unit, it is demonstrated that the water-cement ratio of RCC is as follows: II gradation aggregate: 0.43, III gradation aggregate: 0.48.

2. Sand ratio: according to the porosity of natural sand and pebble aggregate, the optimal sand ratio of RCC is adjusted to 32% for II gradation aggregate and 28% for III gradation aggregate. In addition, the washable natural sand is replaced by the natural sand produced in dry process, so that the content of fine powder less than 0.16 mm in the sand is increased to 14%. In this way, the ratio of grout and mortar is effectively increased.

3. Water content: in the original design, the water consumption of II gradation and III gradation aggregate is 83 kg/m³ and 78 kg/m³ respectively, which can't adapt to the RCC construction in dry and evaporating days in Longshou. After optimization, the water consumption of II gradation and III gradation aggregate in the RCC construction mix proportion are 88 kg/m³ and 82 kgm³ respectively. In order to ensure RCC construction in high temperature period in summer, the water consumption increased by 3 kg/m³, that is, the water consumption of II gradation and III gradation aggregate reached 91 kg/m³ and 85 kg/m³ respectively.

4. Fly ash: on the premise of meeting the requirements of durability and frost resistance, the content of fly ash content shall be increased as much as possible. The II gradation RCC on the upstream and downstream surfaces of the arch dam are mixed with fly ash of 53%, and the RCC inside the arch dam is mixed with fly ash of 66%, which is beneficial to crack resistance and reduce temperature rise.

5. Grout-mortar ratio: in the design of RCC mix proportion, people pay more and more attention to the grout and mortar ratio, that is, the ratio of the absolute volume of grout (including the volume of particles with grain size less than 0.08 mm) to the absolute volume of mortar. According to the practical experience of full section RCC dam construction in recent years, the grout and mortar ratio should not be lower than 0.42. The grout and mortar ratio reflects a kind of proportion relationship between RCC materials. It is an important

construction performance index to evaluate the rollability, liquefaction and bleeding, interlayer bonding, and aggregate separation resistance of RCC mixture. According to the calculation, the grout and mortar ratio of RCC should be 0.40 – 0.43.

6. Gradation: good aggregate gradation could contribute to the maximum apparent density of concrete. In addition, two factors should be considered: One is the balance between the optimal gradation and natural aggregate; the other is the separation resistance of RCC aggregate. According to the test results and comprehensive consideration, the aggregate gradation is selected as follows: II gradation gradation, medium stone: small stone = 60: 40; III gradation, big stone: medium stone: small stone = 35: 35: 30.

7. VC value VC value has a significant impact on the performance of RCC. It is proved by the construction practice that, with the small VC value, the workability of RCC mixture can be greatly improved, improve the rollability, liquefaction, bleeding and interlayer bonding can be improved, and the adverse construction effects of RCC in dry, high temperature, large evaporation and severe cold weather conditions can be solved. According to the special climate conditions in Longshou area, the relationship between VC value of RCC out of mixer and VC value of concrete placed is studied, providing a basis for dynamic control of VC value. The VC value shall be controlled dynamically. The VC value out of mixer shall be 0 – 3 s, and that of concrete placed to surface shall be 3 – 7 s.

8. Gas content According to different frost resistance grades and different air contents, the air content of 4.5% – 5.5%, 4.0% – 5.0%, 3.5% – 4.5% and 2.5% – 3.5% can be controlled by F300, F200, F100 and F50 respectively. The content of corresponding air entraining agent NF-C is 0.45%, 0.40%, 0.07% and 0.05%. In order to meet the requirements of F300 frost resistance grade, the air content of RCC shall be controlled in the range of 4.5% – 5.5%, resulting in a large amount of air entraining agent NF-C (accounting for 0.45%), which is about 30 – 40 times of normal concrete. According to the analysis, it is concluded that: firstly, RCC is the concrete without slump, and its air content is very low, so it is difficult for air introduction; secondly, the gas content is determined under the simulated RCC compaction state, and the loss of gas content is fast and large; thirdly, the content of fly ash is high; the fly ash has a strong adsorption of bubbles. Therefore, when the air content of RCC is controlled at 4.5% – 5.5%, the amount of air entraining agent needs to be increased.

9. Apparent density: the apparent density of RCC is 2 400 kg/m^3 for the II gradation aggregate and 2 420 kg/m^3 for the III gradation aggregate.

9.4.3.3 RCC Construction Mix Proportion of Longshou Arch Dam

The RCC construction mix proportion of Longshou arch dam is shown in Table 9.14. It is shown that the construction performance of RCC can be effectively improved by using smaller water-cement ratio and VC value, high fly ash content, and proper increase of water consumption and admixture content. Meanwhile, the performance of RCC meets the design requirements.

Table 9.14

Design index	Gradation	Water-cement ratio	Water consumption (kg/m^3)	Fly ash (%)	Sand ratio (%)	Water reducing agent NH-A (%)	Gas-attracting agent NF-C (%)	MgO (%)	VC value (s)	Gas content (%)	Apparent Density (g/cm^3)
$C_{90}20F300W8$	II	0.43	88	53	32	0.9	0.45	4.3	3 – 7	4.5 – 5.5	2 400
$C_{90}20F200W8$	II	0.43	88	53	32	0.9	0.40	4.3	3 – 7	4.0 – 5.0	2 400
$C_{90}20F100W6$	III	0.48	82	66	30	0.9	0.07	4.3	3 – 7	3.4 – 4.5	2 420
$C_{90}20F50W6$	III	0.48	82	66	30	0.9	0.05	4.3	3 – 7	2.5 – 3.5	2 420

9.4.4 RCC Construction in Dry and High Temperature Period

9.4.4.1 Measures to Control RCC Temperature

On March 11, 2000, the RCC pouring for main dam of Longshou Hydropower Project began. It is proven that on the construction site: due to the high content of fine powder of natural sand produced by dry process, with a small VC value, the newly mixed RCC has good rollability and sufficient liquefaction and bleeding performance, which will produce sufficient bleeding when rolling for 3 – 4 times.

For Longshou Hydropower Project, after entering June, the temperature in the dam site area is very high in the daytime, the maximum temperature reaches 38 ℃, the natural temperature of aggregate is about 35 ℃, and the temperature of river water is from 17 – 19 ℃. The pouring temperature of arch dam and gravity dam shall be controlled below 20 ℃ and 24 ℃, so the temperature of RCC out of the mixer shall be controlled below 16 ℃ and the temperature of block surface shall be reduced to 30 ℃ to meet the concreting temperature requirements.

In order to effectively control the concrete temperature, the cooling water system and heat supply system shall be set up in the mixing plant. With these two systems, the concrete outlet temperature can be reduced in the high temperature period, and the concrete outlet temperature can be increased in the low temperature period.

9.4.4.2 Control the Temperature of Aggregate

The temperature of raw materials has a great influence on the RCC temperature at the outlet of mixer. The aggregate occupies a large proportion in concrete. It is an effective way to reduce the concrete temperature by controlling the aggregate temperature. Therefore, there are three ways to reduce the temperature of aggregates, including sun shed, cold water spray and cooling water pipe.

In Hexi Corridor, the temperature difference between day and night is large. The block surface temperature under sunlight can reach above 45 ℃ and the block surface aggregate temperature can reach about 40 ℃. For this reason, a sunshade shall be set up on the aggregate bin for the aggregates stacked in the open air. In addition, a cold water spray device shall be provided in the shed. This method can effectively reduce the temperature in the shed and the temperature of the block surface aggregate. It has been found that the temperature of the aggregate can be reduced by 8 – 11 ℃.

In addition, the air (water) supply pipes shall be arranged in the finished aggregate bin and batching bin, which are composed of Φ 108 mm steel pipes. These pipes are laid with the purpose to supply the 1 – 3 ℃ circulating cold water in the high temperature period to cool the aggregates; and supply heat in the low temperature period to increase the temperature of aggregates.

Due to the influence of air temperature during the transportation of aggregate, a sunshade (high temperature period) and heat preservation shed (low temperature period) shall be set up on the belt conveyor for transporting finished aggregate. In addition, pipelines shall be laid in the shed to supply cold water in summer and supply heat in winter.

9.4.4.3 Control the Temperature of Cementing Materials

The block surface temperature of cement and fly ash storage tank can rise to about 45 ℃ under the sunlight. In order to prevent the temperature rise of cement and fly ash, a sunshade shall be set up around the storage tank, and a Φ 38 rigid plastic pipe shall be tightly wrapped on the cone part of the lower part of the storage tank, with an external thermal insulation. Then, the artificial cooling circulating water is flowing through the pipe to reduce the temperature of the cementing material.

9.4.4.4 Control the Temperature of Mixing Water

The water for mixing the concrete of Longshou Project is from Heihe River, which mainly sourced from the snow

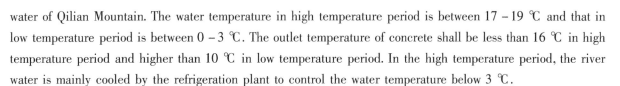

water of Qilian Mountain. The water temperature in high temperature period is between 17 – 19 ℃ and that in low temperature period is between 0 – 3 ℃. The outlet temperature of concrete shall be less than 16 ℃ in high temperature period and higher than 10 ℃ in low temperature period. In the high temperature period, the river water is mainly cooled by the refrigeration plant to control the water temperature below 3 ℃.

9.4.4.5 Control the Temperature Rising during Transportation, Leveling and Rolling

Due to the limitation of construction road and other conditions in Longshou Project, RCC is generally placed in three ways, namely, directly pumped by truck, conveyed by pipe in negative pressure and hanging tank by truck. In the high temperature period, a key to control the temperature is how to shorten the exposure time of RCC from the outlet of the mixer to be placed into form for rolling. In order to prevent the temperature rising from the sun, a sunshade is set on the dump truck and negative pressure chute. In addition, the RCC on the transportation vehicle sis covered with thermal insulation to reduce water evaporation and block surface water loss. It is necessary to strictly shorten the transportation time, and control the exposure time of RCC fed from the mixing plant to the working block surface within 2 h as far as possible.

9.4.4.6 Preserve Moisture and Lower Temperature by Spraying and Covering

There are several functions of spraying water on the RCC block surface: firstly, it is to humidify and retain water on the RCC of the block and prevent it from drying and whitening, thus to avoid affecting the internal quality of RCC; secondly, it is to change the microclimate on the block surface and to form atomization area, so as to reduce the block surface temperature and increase the humidity; thirdly, by spraying and moistening the previous concrete block surface, the interlayer bonding quality can be enhanced.

Because of the climatic characteristics of Longshou region, the block surface of RCC is easy to dry and white after the material is placed into the form. In this case, the concreteshall be sprayed with water for moisture and cooling, and then rolled. Otherwise, the aggregate that is dry and white will be crushed easily, and the layer will not be bleeding. Spraying shall be combined with block surface coverage. Through field tests, sack bags were used to keep moisture for block surface. As the sack has good water absorption, the continuous praying on block surface can be ensured, and there will not be stagnant water on the RCC block surface due to continuous spray. In this way, the microclimate on the block surface can be effectively improved, the air temperature can be reduced, and the moisture on the RCC block surface can be preserved.

There are usually three opportunities for spraying. The first is to spray before laying. It is purposed to moisten the bottom RCC and good for interlayer bonding. The second is to spray before rolling in the RCC paving process. It is to improve the loss of water and whitening on the RCC block surface due to large evaporation, reduce the disadvantage of large loss of VC value, and increase moisture. At this time, the spray gun shall be upward, so that the whole RCC block surface can be covered by the atomizing area and be evenly moisturized. The third is to spray water on the rolled RCC block surface. It is mainly purposed to reduce the block surface temperature and preserve the moisture on the entire RCC block surface. In this case, the block surface shall be covered and the spraying shall be continuous. The spray gun shall be upward to maximize the atomization area. In the whole process of spraying, the weather changes shall be also considered. In the nighttime and rainy days, the intermittent spraying can be carried out; in the daytime with high temperature and strong sunlight, the continuous spraying can be carried out.

In high temperature period, the RCC shall be covered with wet sack bags to prevent water evaporation and temperature rising due to sunlight and high temperature. After rolling or closing, the RCC block surface shall be covered, and the water shall be sprayed in time according to the evaporation of the water in the sack. During the construction of the next layer of RCC, the sack bag shall be lifted up while discharging along with the paving strip. In this way, the large water evaporation of RCC in dry and high temperature period can be dealt with.

The RCC block surface after rolling shall be fully bleeding with 0.5 – 1.0 cm mortar layer on the block surface. If people feel elastic when walking on the block surface, it indicates that the rolling effect is good.

From mid June to early September of 2000, the temperature falls in the range of 26 – 36 ℃ from 11:00 a.m. to 18:00 p.m., so the spraying and wet sack bags can be used for covering. When the temperature reaches above 28 ℃, the spraying shall be implemented every 5 – 8 minutes. The temperature in the form basically stays at 19 – 23 ℃, and the relative humidity is about 70%, which reduces the loss of VC value and ensures the quality of RCC construction.

The temperature of RCC can be kept at 18 – 26 ℃, so the cooling and moisturizing effect is very obvious. The moisture on the RCC block surface can be effectively maintained. In addition, the temperature and humidity on the block surface covered by wet sack bags can meet the requirements of RCC setting time.

9.4.4.7 Dynamic Control of VC Value

The Longshou Project is located in Hexi Corridor, where the annual rainfall is rare, and the evaporation is 8.03 times of the rainfall. Therefore, during the RCC construction in the high temperature period, the water evaporation in RCC is extremely fast, which brings great difficulties to the construction. In the process of RCC transportation, placment, paving and rolling, the water evaporation and VC value loss are very serious. The loss of VC value will increase with time, temperature, especially sunlight. According to the field test, when RCC is not covered, the VC value has a difference of 5 – 7 s from that not been covered within 90 min. In the placement of concrete process, the most intuitive performance is that the block surface of RCC turns white. This phenomena only appears about 10 minutes after RCC closing.

When the VC value of RCC on the paving block surface is large, if the layer is still not bleeding after three times of rolling by the vibration roller, the water can be supplemented by the vibration roller before re-rolling, but the effect is not ideal. The field operation shows that when the VC value is greater than 10 s, the rolling effect is poor, and there are overhead aggregate and insufficient bleeding on the block surface.

The main factor lies in the VC loss and water evaporation in high temperature period. Therefore, a fundamental method to reduce VC loss is to spray water, make covering and replenish water properly. Due to wind and sunshine, the concrete loses water quickly before compaction. Appropriate water replenishment measures can be adopted to reduce the VC value of RCC. Generally, water consumption increases or decreases by $1.0 - 1.5$ kg/m^3 for every 1 s increase or decrease of VC value. During the high temperature period, the VC value out of the mixer shall be controlled within 0 – 3 s, and the VC value of the bin block surface shall be controlled within 3 – 7 s; at the same time, the RCC pouring shall be accelerated and the transfer times shall be reduced as much as possible; the poured RCC shall be sprayed and covered in time. All the above measures significantly improve the rollability and liquefaction of RCC, and effectively improve the interlayer bonding quality.

9.4.4.8 Non-vibration Rolling after Rolling

After rolling, the RCC can be rolled without vibration within 2 – 4 h, which can reduce the micro cracks on the block surface and delay the setting time of RCC. This method is not only suitable for high temperature period, but also suitable for other periods when the pouring block surface is large and the pouring speed is slow.

9.4.5 RCC Thermal Storage Method Construction in Severe Cold Period

9.4.5.1 RCC Construction with Heat Accumulation Method

It is the first time for RCC construction of Longshou Hydropower Project in severe cold area. The construction time of Longshou Project in normal temperature period is relatively short, only 7 months lasting from April to October every year. At that time, the construction unit asked to make impoundment half a year in advance. Therefore, the construction unit must carry out RCC construction in the low temperature and severe cold period, so as to ensure the realization of the goal of impoundment in advance. From November to December of 2000, the average temperature of Longshou dam site is $-2 - -15$ ℃. How to deal with RCC construction in the low temperature period is a new problem encountered in the Longshou Project, and there is no experience to learn from. Therefore, according to the construction experience in the cold region of Qinghai, the construction unit decided to carry out the RCC construction with the heat accumulation method.

The main measures for construction with heat accumulation method are as follows: firstly, the temperature of raw materials is increased by heat accumulation method to ensure that the temperature of fresh concrete out of the mixer reaches the specified temperature; secondly, a certain amount of antifreeze is mixed in RCC to prevent the concrete from freezing damage in the early stage; thirdly, the warm shed is used to ensure the normal temperature of the bedrock block surface and concrete layer; fourthly, the poured concrete shall be covered and provided with thermal insulation timely, so as to meet the needs of cement hydration reaction under negative temperature conditions, and ensure the concrete strength and durability.

9.4.5.2 Heat Accumulation for Raw Material

Heat accumulation method: according to the characteristics of construction site and aggregate pile, the mixing water and aggregate of Longshou Project can be heated with heat accumulation method. How to heat the water: the mixing water is heated by steam. The steam pipe is inserted into the water tank. The heating temperature of water is controlled in the range of 40 – 60 ℃ to prevent the concrete from false setting. How to heat the sand and stone aggregate: steam pipes with a diameter of 108 mm are embedded in the sand and stone piles, and the diameter of the pipes is 108 mm and laid in epsilon type. When heated by steam, the heat on the pipe wall makes the surrounding air form a thermal cycle. Through heat transfer, the sand and stone aggregate can be preheated.

Thermal insulation measures: in the periphery of aggregate silo, 3 cm thermal insulation materials are attached and the external insulation quilt is hung. The conveyor belt of aggregate is fully enclosed to minimize the heat loss of aggregate.

9.4.5.3 Measures for Mixing Antifreeze into RCC

When the temperature falls in the range of $-3 - -10$ ℃, in order to prevent early freezing damage of RCC, 4% DH8 antifreeze can be added into RCC to improve the antifreeze performance of concrete, reduce the freezing point of mixing water in concrete, and make the liquid phase of concrete not freeze in a certain negative temperature range, so that RCC will not suffer from freezing damage, and the hydration reaction of cement can continue, and the concrete will continue to harden.

With the addition of anti-freezing agent in RCC, the concrete without thermal insulation measures can be protected against freezing damage, but the concrete strength is slightly lower than the standard curing strength. The results show that the initial hydration reaction is not complete. That is to say, the fresh concrete mixed with antifreeze develops slowly to the solid phase hardening under negative temperature. Due to the small volume of the specimen, it is easy to be affected by temperature, but the mass concrete is slightly affected by the temperature.

9.4.5.4 Thermal Insulation Measures for RCC Construction

By preheating all raw materials, the outlet temperature of mixture is about 12 ℃. Due to the low external temperature, part of heat is carried away by the air and transport machinery in the transportation process, so that the mixture has a large heat loss. The distance between the mixing plant and the construction site is about 3 km. On average, it will take about 30 minutes to place the concrete of each truck. The longer the concrete placement time is, the more heat loss will be. Therefore, the key for heat loss is to shorten the placement time, so the concrete placement time shall be controlled within 30 minutes.

Thermal insulation measures for placement block surface: according to the requirements of construction specifications, the temperature of bedrock block surface before construction must reach normal temperature. Therefore, a warm shed shall be set up on the placement block surface, and the furnace shall be used to raise the temperature, so that the temperature inside reaches the positive temperature, the newly poured concrete can still meet the temperature required for strength growth under the external negative temperature.

In the low temperature and severe cold period, in order to prevent the block surface RCC from freezing, water evaporation and temperature reduction, the RCC block surface that has been rolled or leveled but not rolled shall be covered with colored fiber strip cloth in time, and then covered with 5 cm thermal insulation quilt. When the lower layer of RCC construction is carried out, the material shall be unloaded along with the paved strip.

The temperature of concrete after rolling is not only related to heat loss during transportation and rolling, but also related to part of heat absorbed by formwork and thermal insulation materials after concrete is placed. As the concrete contacts the air in the rolling and leveling process, the heat loss is large. The minimum temperature is reduced to 3 ℃. Due to the high content of fly ash in RCC, low heat of hydration and low early strength, the early thermal insulation of concrete must meet the requirements of strength growth.

The poured block surface shall be covered with thermal insulation materials to preserve the preheat in raw materials and hydration heat of cement in the concrete, so as to obtain the early strength of concrete and improve the frost resistance.

9.4.6 Conclusion

The research and application of RCC for Longshou arch dam in dry and cold areas has experienced the pain of failure and the joy of success. In the early stage, the RCC process test and production test of Longshou were carried out according to the mix proportion submitted by the original design unit. Since the original mix proportion test was conducted in the southern region with mild and humid climate, the unit water consumption, sand rate, admixture content, *VC* value and other parameters of the mix ratio cannot adapt to the windy and dry climate with large temperature difference, high evaporation and long cold period in Longshou dam site area, leading to the failure of the process test and production test. According to the climate characteristics in Longshou, new problems and challenges will be encountered in the RCC dam construction. Through continuous exploration, summary, research and innovation, the constructionunit has successfully solved the RCC construction of Longshou Project under the harsh weather conditions, providing valuable first-hand information for RCC construction in cold and dry areas.

Roller Compacted Concrete Rapid Damming Technology

(Volume Ⅲ)

中文作者 田育功
译　者　张宝瑞　赫庆彬　等

Chinese Writer　Tian Yugong
Translators　Zhang Baorui　He Qingbin et al

黄河水利出版社
·郑州·

Abstract

This book describes the theory, method, experience and engineering practice of rapid damming technology with roller compacted concrete(RCC). Citing a large number of rich first-hand RCC test and research results, tender/bidding documentation and engineering examples of rapid damming, with full and accurate contents, this is a practical technology book with very high theoretical level and rich engineering practice.

The book includes:13 topics of development level of RCC damming, design and rapid construction of RCC dam, examples of raw materials and projects, mixing proportion design with examples, research and examples of RCC performance, admixture research and application, the role of rock powder in RCC, construction technology of grout enriched vibrated concrete(GEVR), key technology for rapid construction of RCC, temperature control and cracking prevention, quality control of RCC, core sample pressurizing water and in-situ shear resistance and RCC cofferdam work.

This book is a valuable reference for vast engineering and technical personnel engaged in the structure, design, research, construction and supervision in water conservancy and hydropower engineering sector, as well as teachers and students of related majors in colleges and universities.

图书在版编目(CIP)数据

碾压混凝土快速筑坝技术 = Roller Compacted Concrete Rapid Damming Technology：全三册/吴正桥等译：田育功著. —郑州：黄河水利出版社，2021.8

ISBN 978-7-5509-3069-8

Ⅰ.①碾… Ⅱ.①吴…②田… Ⅲ.①碾压土坝-混凝土坝 Ⅳ.①TV642.2

中国版本图书馆 CIP 数据核字(2021)第 165756 号

出　版　社：黄河水利出版社　　　　　　　　　网址：www.yrcp.com
　　　　　地址：河南省郑州市顺河路黄委会综合楼 14 层　邮政编码：450003
发行单位：黄河水利出版社
　　　　　发行部电话：0371-66026940、66020550、66028024、66022620(传真)
　　　　　E-mail:hhslcbs@126.com
承印单位：广东虎彩云印刷有限公司
开本：787 mm×1 092 mm　1/16
印张：42
字数：1 500 千字
版次：2021 年 8 月第 1 版　　　　　　　　　印次：2021 年 8 月第 1 次印刷
定价(全三册)：198.00 元

Translator's Preface

Roller compacted concrete (RCC) gravity dam has shown rapid construction speed, low cost and good engineering quality and other advantages. In the recent year, the rapid development of RCC dams in China, especially the completion of high DAMS over 100m, damming RCC materials and construction technology of China is indeed a kind of technical concept with unique characteristics of China. Carefully analyzing various technologies, their contents are very rich, and the connotation is also deep, and the application is very convenient and handy, to be easy to master. What more valuable is that, whenever a large RCC dam is completed, there are almost always innovations and improvements, so that the development of the technology is constantly rooted in the source of project construction. For technical progress in various aspects, many engineering technicians have written a large number of special treatises or monographs, well welcomed by industry peers. In order to better promote the design and construction concept of Chinese RCC dam in the world, Wu Zhengqiao and others translated the book with the permission of the author.

The translation results are divided into three volumes, division of labor is as follows:

Chief editor: Wu Zhengqiao.

Other editor:

(1) Volume I: Wu Zhengqiao, Yu Yang.

(2) Volume II: Yang Haiyan, Li Jianqiang, Yao Desheng, Zhao Chuntao, Zhou Hongmin, Bao Dixiao, Li min, Sun Shunan, Zhao Jianli.

(3) Volume III: Zhang Baorui, He Qingbin, Peng Xiaochuan, Li Miao, Xu Lizhou, Pei Xianghui, Xu Ting, Liu Zhuo, Zhang Kai, Zhao Lin.

In this process of translating this book, Tian Yugong gave great support and help, thank you!

In addition, in the process of publishing this book, I would like to thank Wang Xiaohong and Yu Ronghai from the editorial department of China Water Resources Bei Fang investigation, Design & Research Co. LTD and the Yellow River Water Publishing House for their strong support.

Wu Zhengqiao

August 2021

Preface

In recent years, China has successively completed 200-metre roller compacted concrete(RCC) gravity dams and several 100-metre level thin double-curvature arch RCC dams. In construction process, all these dams have shown rapid construction speed, low cost and good engineering quality and other advantages. These high dam construction achievements signify that, China's RCC damming technology has made a breakthrough in overall development in a great number of engineering practices, through earnestly summing up experience, actively carrying out exploration and test study, continuous innovation and development. The technology level is getting higher and higher, getting more mature. These new technical achievements and engineering achievements have also brought many beneficial effects to international dam industry.

Looking back to early days, due to hazy knowledge, in RCC dam projects in China, "drying hard" concrete commonly applied in foreign countries was generally applied. As a result, the lack of bleeding at pouring face often caused poor inter-layer bonding, forming a "multi-layer steamed bread" phenomenon. This caused doubts and reservations about the quality of RCC mass of dam. In view of this condition, with a spirit of tenacity, many engineering technical personnel and researchers are continuing to explore, adjust and improve materials of concrete and construction technical actions, finally, a technology mode of "sub-plastic state" RCC with good bleeding performance, being easy to compact and being able to prevent vibration-press subsidence has been formed, widely used nowadays. It has better solved the problem of concrete, hot joint bonding, thoroughly improved impervious performance and shear resistance of course face, solved the safety and impervious problem for the construction of 200- meter high RCC gravity dam. Compared with some foreign RCC damming technologies, damming RCC materials and construction technology of China is indeed a kind of technical concept with unique characteristics of China. Carefully analyzing various technologies, their contents are very rich, and the connotation is also deep, and the application is very convenient and handy, to be easy to master. What more valuable is that, whenever a large RCC dam is completed, there are almost always innovations and improvements, so that the development of the technology is constantly rooted in the source of project construction. For technical progress in various aspects, many engineering technicians have written a large number of special treatises or monographs, well welcomed by industry peers.

The greatest advantage of RCC damming technology is its rapid speed, which has strong vitality. Generally, for concrete dam with a height of over 100 m, it can be completed for 2 – 3 years adopting RCC damming technology. Compared with normal concrete damming technology, the construction duration can be shortened by one-third and more. Engineering practice has proved that, RCC damming technology is credible and reliable in quality, of which the advantage is beyond any doubt.

The design is the key to technical innovation of rapid RCC damming. Technical innovation not only requires a solid and scientific technical foundation, but also allows failure and discussion, and also requires the courage to bear failure and responsibility, to achieve the objective of technical innovation through practice.

"Interlayer combination, temperature control and cracking prevention" is the core technology of rapid RCC damming. In recent years, with the increase in the height and volume of RCC dams, it has become common practice to continuously pour RCC in high-temperature seasons in order to crash the schedule or shorten the duration. As a result, temperature control measures for RCC have become more and more stringent,

which it has no different from normal concrete, to make temperature control measures of RCC show a trend more and more complicated. In this respect, RCC dams in Thailand, Cambodia, Laos, Myanmar and other countries located in subtropical region shall be learnt from. The RCC dams in these countries do not adopt cooling water pipe temperature control measures, and the design indicators adopt a single strength index and with a single aggregate gradation, the design concepts and temperature control measures of these dams are worth learning and thinking about.

The construction of RCC dam has the characteristics of one-time, and it is particularly important that construction quality is being under control all the time. Especially, for on-site RCC construction, dynamic control of VC value, timely rolling, mist spray moisturizing and covering for curing and other construction links concern directly the quality of interlayer bonding and the performance of temperature control and cracking prevention, which must be raised to the height of quality problem to understand their importance.

The Author of this book has collected a large number of RCC dam engineering practical data and multifaceted scientific test and research results in China, with full and accurate contents. While sorting out and analyzing these achievements, combined with personal exploration and study of RCC dam participated in the construction of RCC dams personally, the Author has carried out in-depth display and interpretation, and has more realization and understanding of "plastic RCC", to expand the discussion in the book. RCC damming is still a developing technology. For all concrete materials, construction process, dam engineering design, construction management, temperature control and cracking prevention, there is still room for improvement and development. It is beneficial for technical development to constantly summarize, analyze, study and exchange. After reading this book, I have benefited a lot and expanded my eyesight.

Former Director of the RCC Damming Professional Board of China Society for Hydropower Engineering

Wang Shengpei
Beijing, March 2010

Foreword

Roller compacted concrete(RCC) damming technology is a major technical innovation in the history of world damming. RCC damming technology is favored by global dam industry because of its rapid construction speed, short duration, low investment, safe and reliable quality, high degree of mechanization, simple construction, strong adaptability and green environmental protection, etc. In particular, the damming period of can be significantly shortened by one-third and more compared to similar normal concrete dams, to show strong advantage of vigorous development, and inject a fresh air and vitality into dam construction, which is in line with the development direction of good, rapid and economical.

The "rapid" is the greatest advantage of RCC damming technology, which is its strong vitality. Although the adoption of RCC damming technology has only been for 20 years more, the speed of damming and the large number of dams are unable to be matched by other damming technologies. As of the end of 2008, there were 180 RCC dams(including cofferdams) completed or under construction in China, ranking first globally. The greatest charm of RCC damming technology is its compatibility. RCC has the characteristics of concrete, conforming to the rules of water-cement ratio, and its cross-section design is the same as that of normal concrete dam, regardless of RCC gravity dam or arch dam; meanwhile, its construction has the characteristics of rapid construction of earth-rock dam. A large number of engineering practice has proved that, RCC dam has become one of the most competitive dam types.

Using RCC damming technology, the dam is bold and generous and real. The internal quality of the dam is good, and the appearance quality is beautiful, so its quality is not inferior to normal concrete dams. Although the RCC damming is bold, the construction is not rough at all. RCC damming technology is very delicate, to be a well-organized and standardized construction. The construction site is sparsely staffed, and the concrete roller compacting is carried out in an orderly manner from placing, spreading to rolling. On the contrary, for the pouring of normal concrete dam, due to that the normal concrete is restricted by pouring strength and temperature control and other factors, dam body is divided into blocks by transverse and longitudinal joints. For concrete construction, columnar pouring is adopted, which results in a large amount of formwork workload, and the amount of merging joint grouting is great and of long period. The dam construction often presents a large number of people on placement surface and appears busy, lacking of well-organized, unified and bold of RCC construction.

RCC damming is a systematic project integrating the research, design, construction, quality control and other aspects. What mostly reflects the characteristics of rapid RCC damming is whole-block placement and thin-lift concreting. Because of the change in damming technology, the design concept of complex layout and dam engineering structure has been changed. The complex layout design shall not only consider starting from meeting the requirements of rapid construction of RCC and simplifying the layout of the dam, but also require carrying out in-depth study on dam structure, temperature stress and overall performance. The adoption of RCC damming is a promotion to the design. The design concept must advance, the dam layout shall be different from normal concrete dam, to start from thin-lift concreting and whole-block placement, simple and rapid damming technology, combined with the characteristics of RCC itself, the simpler the dam's structural layout, the more obvious its advantages.

"Interlayer bonding, temperature control and cracking prevention" is the core technology of RCC rapid damming. Through a large number of tests and studies and engineering examples, for problems of great or low fluctuations in artificial sand and gravel powder content of RCC, the successful application admixing fly ash or admixing rock powder instead of sand, precise control of rock powder mixing, dynamic control of VC value and whole-block placement inclined layer rolling and other technologies have effectively improved the rollability, liquefaction and bleeding and interlayer bonding quality of RCC, promoting the development of RCC rapid damming technology. The "grout-mortar ratio" has become one of important parameters in the design of RCC mixing proportion, with the same important effect as three major parameters of water-cement ratio, sand ratio and water consumption. Rock powder has got more and more attention in RCC, become an indispensable part of RCC materials. Hydraulic RCC has been developed into non-slump semi-plastic concrete. The worrying problem of interlayer bonding quality has been well solved, significantly improved the ultimate tensile value and frost resistance.

Looking back at test history of drilling and coring of RCC and on-site water pressurizing tests, core samples at early stage with the longest length of 60 cm more have been developed into ones of ultra-long 16 m more, and core samples longer than 10 m are not uncommon. Drilling and coring have fully proved that the maturity of rapid RCC damming technology gets gradually maturing, reflecting the change in the properties of RCC from one side, which is indeed very beneficial to improve the quality of interlayer bonding and impervious performance of dam body.

RCC also has its dual nature. Because RCC adopts low cement content and high admixtures, it is required to adopt advanced design concepts to constantly innovate and deepen the study of its frost resistance, ultimate tensile value and carbonization and other properties(compared to normal concrete).

Over 20 years since 1988 when starting studying dam dam technology of left auxiliary dam RCC test at Longyangxia Dam, the Author has personally witnessed the development and growth of China's RCC damming technology, who is one of major participants, implementers, researchers and promoters of RCC dam technology of China, mastered a large number of first-hand test and research results, with rich engineering practical experience, having novel viewpoints in RCC damming technology. This book is achievement data of the test and research, construction technology, technical consultation, water storage acceptance and construction management in Longshou, Linhekou, Baise, Guangzhao, Jin'anqiao, Kalasuke and other RCC dams participated in and presided over by the Author in recent years, compiled on the basis of over 40 papers (thesis) published at home and abroad, in which many of valuable first-hand data are disclosed for the first time, being one of windows to understand and master RCC damming technology of China.

RCC damming technology of China ranks among global leading level, which is inseparable from continuous efforts and innovations of vast research, design, technical consultation, supervision, and construction personnel of a large number of water conservancy and hydropower projects in China, especially massive, scientific and hard works and strong supports of the RCC Damming Profession Board. Here, I would like to express my highest respect and heartfelt gratitude to elder Wang Shengpei, who has contributed his whole life and made outstanding contributions to China's RCC damming technology, and colleagues of the RCC Damming Profession Board! I would like to express my heartfelt gratitude to leaders and colleagues who have supported my technical work for long term!

In view of the characteristics of hydraulic RCC, this book makes a quite detailed analysis and research from the mechanism and application of RCC rapid damming. Through the discussion of rapid RCC damming technology and the analysis on typical engineering examples, the Author summarize in time, and think more calmly, to discuss and research from both the pros and cons, always insisting on putting the quality and safety of the dam in the important position of the first priority, so that the RCC damming technology is more viable,

safer and more perfect.

This book is compiled in the intention of constructing best and first-class RCC dam in the world. Because there are many research topics on RCC damming technology, restricted by the Author's capability and limited time, it is inevitable to make mistakes and mistakes for sorting out, analyzing and studying and giving examples for so many topics, therefore, readers are expected to criticize and correct!

The Author
July 2010

Contents

Translator's Preface

Preface

Foreword

Chapter 1 General .... (1)
   1.1 Development Level of RCC Damming Technology in China .... (1)
   1.2 Development History of RCC Dam .... (10)
   1.3 Popularization and Application of RCC Damming Technology .... (12)
   1.4 Discussion on Key Rapid Damming Technologies of RCC .... (16)
   1.5 Innovation and Reflection on RCC Rapid Damming Technology .... (23)

Chapter 2 Design and Rapid Construction of RCC Dams .... (27)
   2.1 Overview .... (27)
   2.2 Design and Rapid Construction of RCC Dam .... (28)
   2.3 Design Index of RCC and Material Zoning of Dam .... (40)
   2.4 Introduction to RCC Dam Design .... (42)
   2.5 Conclusion .... (64)

Chapter 3 Raw Materials and Project Examples .... (66)
   3.1 General .... (66)
   3.2 Cement Properties and Project Cases .... (68)
   3.3 Performance and Quality Test of Fly Ash .... (78)
   3.4 Performance of Hydraulic Concrete Admixtures .... (88)
   3.5 Aggregate Properties and Project Cases .... (95)
   3.6 Conclusion .... (106)

Chapter 4 RCC Mix Proportion Design and Example .... (108)
   4.1 Overview .... (108)
   4.2 RCC Mix Proportion Parameter Selection .... (109)
   4.3 Design Basis and Content of Mix Proportion .... (133)
   4.4 RCC Mix Proportion Design Method .... (134)

Chapter 5 Research and Application of RCC Performance .... (141)
   5.1 Overview .... (141)
   5.2 Properties and Influencing Factors of RCC .... (142)
   5.3 Study on Relationship of Admixture, VC Value, Temperature with Setting Time .... (169)
   5.4 Autogenous Volume Deformation, Adiabatic Temperature Rise and Creep Tests .... (174)
   5.5 Test Study on Properties of RCC with Manufactured Sand and River Sand .... (186)

Chapter 6 Research and Application of RCC Admixture .... (193)
   6.1 Overview .... (193)
   6.2 Micro Analysis and Study of RCC Admixture .... (195)
   6.3 The Research and Application of SL Admixture in Gelantan Project. .... (232)

Chapter 7 Research and Utilization of Rock Powder in RCC (253)
    7.1 Overview (253)
    7.2 Limestone Rock Powder (255)
    7.3 Influence of Rock Powder on Performance of RCC (259)
    7.4 Project Examples of RCC with Rock Powder Replacing Sand (262)
    7.5 Study on Utilization of Baise Diabase Manufactured Sand and Rock Powder in RCC (271)

Chapter 8 Construction Technology of GEVR (315)
    8.1 Overview (315)
    8.2 Mix Proportion Test of GEVR (318)
    8.3 Construction Technology of GEVR (320)
    8.4 Test Study on GEVR Mixed with Fiber in Impervious Area (323)
    8.5 Application of GEVR in Baise RCC Main Dam (329)

Chapter 9 Key Technology for Rapid Construction of RCC (334)
    9.1 Overview (334)
    9.2 Key Technology for Rapid Construction of RCC (336)
    9.3 Jin'anqiao Dam Roller Compacted Concrete Fast Construction Key Technique (374)
    9.4 RCC Construction of Longshou Arch Dam in Cold and Dry Area (389)

Chapter 10 Temperature Control and Cracking Prevention of Roller Compacted Concrete (399)
    10.1 Overview (399)
    10.2 Basic Information and Standards of Temperature Control (401)
    10.3 RCC Temperature Control Measures (413)
    10.4 Technical Innovation and Discussion on Temperature Control and Anti-cracking (425)
    10.5 The Temperature Control in RCC Gravity Dam in One Hydro-junction Project (430)

Chapter 11 RCC Quality Control and Project Cases (456)
    11.1 Overview (456)
    11.2 Quality Control and Evaluation Regulations (458)
    11.3 Other Quality Control Measures and Discussions of RCC (469)
    11.4 Application of Nuclear Densimeter in RCC (475)

Chapter 12 Core Drilling, Pump-in and In-situ Shear Tests (482)
    12.1 Overview (482)
    12.2 Core Drilling of Dam RCC (484)
    12.3 Field Pump-in Test of RCC (493)
    12.4 Performance Test of RCC Core Samples (500)
    12.5 In-situ Shear Test of RCC Dams (504)
    12.6 Core Drilling and Pump-in Test of Dam RCC (508)
    12.7 Performance Test of RCC Core Sample of Jin'anqiao Dam (515)
    12.8 On-site In-situ Shear Test of RCC Dams (519)

Chapter 13 RCC Cofferdam Construction and CSG Damming Technology (531)
    13.1 RCC Cofferdam (531)
    13.2 Cemented Sand and Gravel (CSG) Damming Technology (532)
    13.3 Design and Rapid Construction of Longtan RCC Cofferdam (535)
    13.4 CSG Mix Proportion Design and Application of Upstream Cofferdam of Gongguoqiao Dam (544)

Appendix A   National Method Hydraulic Concrete Mix Proportion Test Method ········· (561)
 A.1 Foreword ········· (561)
 A.2 Characteristics of Construction Method ········· (562)
 A.3 Applicable Scope ········· (562)
 A.4 Technological Principle ········· (562)
 A.5 Construction Process Flow and Key Points of Work ········· (562)
 A.6 Material and Equipment ········· (569)
 A.7 Quality Control ········· (572)
 A.8 Safety Measures ········· (572)
 A.9 Environmental Protection Measures ········· (573)
 A.10 Benefits Analysis ········· (573)

Appendix B   National Construction Method GEVR Construction Method in RCC Dam Construction ········· (575)
 B.1 Foreword ········· (575)
 B.2 Characteristics of Construction Method ········· (575)
 B.3 Applicable Scope ········· (576)
 B.4 Technological Principle ········· (576)
 B.5 Construction Technology and Operation Points ········· (576)
 B.6 Mechanical Equipment Configuration ········· (578)
 B.7 Quality Control ········· (578)
 B.8 Safety Measures ········· (579)
 B.9 Environmental Protection Measures ········· (579)
 B.10 Technical and Economic Analysis ········· (579)
 B.11 Project Examples ········· (579)

Appendix C   Example of Construction Methods: Construction Method of RCC for Dam of a Certain Project ········· (581)
 C.1 General Principles ········· (581)
 C.2 Normative References ········· (581)
 C.3 Terminology ········· (582)
 C.4 Concrete Compaction Process Flow Chart ········· (585)
 C.5 Control and Management of Raw Materials ········· (586)
 C.6 Selection of Roller Compacted Concrete Mix Ratio and Issuing of Charger Sheet ········· (588)
 C.7 Inspection and Acceptance before Placement Construction of Roller Compacted Concrete ········· (588)
 C.8 Concrete Mixing and Management ········· (591)
 C.9 Concrete Transport ········· (593)
 C.10 Construction Management inside Block ········· (594)
 C.11 Construction under Special Climatic Conditions ········· (604)
 C.12 Quality Control Management ········· (605)
 C.13 Management of Placement Surface after Rolling and Placement Finishing ········· (609)

Main References ········· (610)

Chapter 10

Temperature Control and Cracking Prevention of Roller Compacted Concrete

10.1 Overview

Although the amount of cement used in RCC is small and the amount of fly ash and other admixtures is large, the calorific value of concrete is greatly reduced, but the exothermic process of RCC hydration heat is slow and lasts for a long time. The roller compacted concrete construction is carried out on full section in whole-block and thin-lifts, and continuous in rolling. There is no longitudinal joint in the dam body, and the transverse joints are not exposed. It relies on the block surface for heat dissipation, so the heat dissipation process is greatly prolonged. It was ever believed that it is unnecessary to make temperature control for RCC dam. Later, a large number of construction practice and research results showed that the problem of temperature stress and temperature control also existed in the RCC dam. As the cooling process of roller compacted concrete lasts for a long time, the roller compacted concrete inside the dam body will be in a high temperature state for a long time, It will take decades to reduce the temperature inside the dam to the stable temperature.

Concrete is a kind of poor thermal conductivity material. The increase rate of cement hydration heat is far greater than the thermal diffusivity, so the internal temperature of concrete increases. Due to the thermal expansion and cold shrinkage of materials, the internal concrete shrinks with time. However, the shrinkage is constrained by the surrounding concrete and cannot occur freely, resulting in tensile stress. When the tensile stress exceeds the tensile strength of concrete, cracks will occur. Therefore, in order to reduce the temperature cracks in concrete, it is necessary to control the maximum temperature rise of concrete inside the dam. A large number of test results show that the higher the concrete pouring temperature is, the faster the chemical reaction of hydration heat of cement is. The influence of temperature on the reaction rate of concrete hydration heat further aggravates the temperature cracks. Therefore, the temperature is the main cause of cracks in dam concrete.

In addition, the annual temperature change and cold wave are also important reasons for dam cracks. They have the same impact on RCC and normal concrete. In fact, cracks also appear in RCC dams built in China. Therefore, it is important to prevent block surface cracks caused by environmental temperature change and excessive temperature difference between inside and outside of RCC dam.

To control the temperature of RCC is to prevent cracks caused by temperature stress due to excessive temperature difference inside and outside the dam. Because the dam is an important water retaining structure, leakage is not allowed. Cracks will have a very adverse impact on the safety and long-term durability of the dam. Since the dam concrete is poured, it has to suffer from its own hydration heat and the external environment temperature. Under the action of thermal expansion and cold contraction, the displacement and deformation of any point in the concrete are constantly changing. If it is constrained by the external and internal conditions, the temperature stress will generate. The temperature stress is caused under the constraint conditions. If the temperature stress exceeds the ultimate strength of concrete or the strain exceeds the ultimate tensile value of concrete, the concrete dam structure will have cracks. If the cracks develop seriously, the bearing capacity of the dam will be weakened or even destroyed. For decades, the temperature control and cracking prevention of dams have always been a major concern in engineering and technical circles.

The temperature stress and temperature control of roller compacted concrete dam has its own characteristics. The temperature control should be studied according to the material performance, structure size, climate conditions, paving thickness, pouring temperature, rolling layer lifting and intermittent mode, and combined with the cooling and heat dissipation measures of block surface. In addition, the construction period shall be reasonably arranged and the temperature control mode shall be simplified. Only by fully considering and mastering these characteristics, the temperature control, cracking prevention of roller compacted concrete and rapid dam construction technology can be perfectly combined.

Concrete dams are divided into low dam, medium dam and high dam. Dams with the height of less than 30 m are low dam; dams with the height of 30 – 70 m are medium dam; dams with the height of more than 70 m are high dam. One of the advantages of roller compacted concrete (RCC) is to simplify or cancel the temperature control. Most of the early RCC dams are low and medium dams, which are constructed in low-temperature season and low-temperature period. No temperature control measures are taken in most of these dams. However, in recent years, due to the increase of the height and volume of RCC dam, continuous pouring of RCC in high temperature season and high temperature period has become a common practice in order to speed up the construction or shorten the construction period. In this way, the temperature control measures of roller compacted concrete are more and more strict. The technical route of temperature control is mainly to copy that of normal concrete, making the roller compacted concrete temperature control more and more complex. Regardless of low dam, medium dam, high dam or high temperature period or low temperature period, some roller compacted concrete dams are designed with cooling water pipes. The uniform temperature control has a certain negative impact on the simple and rapid construction of roller compacted concrete in the most favorable low temperature period.

For example, in the subtropical regions including Vietnam, Thailand, Cambodia, Laos and Myanmar, the cooling water pipes are not laid in the dam body for temperature control. Generally, a single strength index and a single aggregate gradation are adopted. The maximum size of aggregate is generally 40 mm, 50 mm or 63 mm. For example, for the Bolegolon RCC gravity dam in Vietnam, the 180 d compressive strength is 20 MPa and the maximum particle size of aggregates is 40 mm; for the Tatan RCC gravity dam in Thailand, the 91 d compressive strength is 15 MPa and the maximum particle size of aggregates is 63 mm; for the Yeywa RCC gravity dam in Myanmar, the 365 d compressive strength is 20 MPa and the maximum particle size of aggregates is 40 mm. The temperature control measures and design concept of these dams are worth learning.

The temperature load of roller compacted concrete dam is different from that of normal concrete dam, so the design of roller compacted concrete dam is complicated. Due to the adoption of full-section thin-lift andwhole-block placement and continuous rolling of RCC dam, the temperature stress caused by hydration heat temperature rise of RCC during construction period will affect the stress of RCC dam for a long time due to the long process of temperature drop. Temperature load is a special load caused by the temperature change of dam. The five main loads of dam body include temperature load, water (sand) pressure, self weight, seepage pressure and seismic force. Temperature load has some particularity on the one hand, when the concrete cracks due to excessive temperature pressure, the constraint conditions will change and the temperature stress will be eliminated or relaxed; on the other hand, temperature stress depends on many factors, especially many factors in RCC construction, such as pouring period, construction progress, temperature control measures, construction technology, etc.. Therefore, there is a certain difference between the calculated temperature stress and the

Chapter 10 Temperature Control and Cracking Prevention of Roller Compacted Concrete

actual temperature control effect.

Cracks are a common problem in concrete dams. The dam with cracks has long troubled people. For a long time, people have taken a series of measures to prevent and resist cracks of concrete dams, in terms of design, construction and management, including dam body joint and block, improvement of concrete crack resistance performance, temperature control, surface maintenance, etc. , but almost all dams still have cracks in actual situation. The temperature and temperature stress of RCC are different from those of normal concrete dams. Although RCC dams have low hydration heat and fewer cracks on the dam surface in the early stages, the high temperature in the RCC dam lasts for a long time, so that the surface of the dam is weathered by low temperature, cold wave, exposure, drying and wetting, causing the temperature difference between the inside and outside of the dam, and easy appearance of surface cracks. The period of more surface cracks in RCC dams is usually after the dam is built. Surface cracks can develop into deep cracks or even penetrating cracks. Therefore, it is very important to strengthen the concrete curing on the surface.

The temperature control cost for roller compacted concrete is large, which has become one of the key factors restricting the rapid construction of roller compacted concrete. It is necessary to put forward new views on the temperature control standard and technical route of temperature control, carry out technical innovation research, break the deadlock and passive situation of temperature control, and combine the RCC rapid construction and temperature control and cracking prevention measures in a best manner, so as to provide scientific and reasonable technical support for rapid construction of RCC dam.

Academician Tan Jingyi made a speech at the "Academic Exchange Meeting on Concrete Materials and Temperature Control of Hydraulic Dams" in July 2009: Due to the small margin of the factor of safety against cracking and some uncertain factors in concrete crack resistance, some effective measures should be taken in construction management, cooling system and cooling technology. With the guiding ideology of " small temperature difference, early cooling and slow cooling", the temperature gradient and temperature difference in the process of cooling and temperature drop should be reduced as much as possible to reduce the thermal creep stress. In addition, the surface protection shall also be provided to make the dam have a greater safety against cracking.

Through the research and discussion of RCC temperature control standard, basic data, temperature control and cracking prevention measures, technical innovation and engineering examples, this chapter comprehensively expounds the temperature control and cracking prevention of roller compacted concrete dam in terms of theory, technology and practice, providing more detailed first-hand reference materials for temperature control and cracking prevention of roller compacted concrete dam.

10.2 Basic Information and Standards of Temperature Control

10.2.1 Temperature Difference, Temperature Stress and Crack

10.2.1.1 Temperature Difference of RCC Dam

The early hydration heat of RCC is low, the elastic modulus is small, and the creep is large. However, due to the rapid rising of the dam body and the long duration of high temperature inside the dam body, it is very slow to have the concrete dropped from the maximum temperature to stable temperature. Such process always takes several decades. Therefore, it is more likely to have crack due to the internal and external temperature difference. It can be seen that the temperature difference is the main reason for the temperature stress of the dam and the cracks in the dam.

Most of the cracks in concrete dams are caused by temperature stress, which is generated by temperature difference under the constraint conditions. The key to prevent cracks is to control the temperature stress of concrete, that is, to control the temperature difference of concrete. The temperature difference of RCC dam mainly includes the temperature difference of foundation, the temperature difference inside and outside the dam body, and the temperature difference between fresh and old concrete.

1. Foundation temperature difference. The temperature difference of foundation refers to the difference between the maximum temperature and the stable temperature of concrete within the constraint range of the dam foundation. The temperature difference of foundation is an important index to control the deep cracks of dam foundation concrete, which changes with the performance of roller compacted concrete, the height-length ratio of pouring block, the length of longer side of pouring block, the elastic modulus ratio between concrete and bedrock, and the climatic conditions in dam site area. For RCC gravity dam, there is no longitudinal joint, its bottom width is larger, and the foundation restraint is also high. In order to prevent cracks in the foundation concrete, the allowable temperature difference of foundation should be controlled.
2. Temperature difference inside and outside the dam body. The temperature difference inside and outside the dam body generates due to the slow heat dissipation of hydration heat of roller compacted concrete, long time to reach the maximum temperature after RCC pouring, and the long cooling process. The roller compacted concrete inside the dam body will be in a high temperature state for a long time. When winter comes after construction or when the temperature drops suddenly in case of cold tide, the external concrete will be cooled, which is easy to form a large internal and external temperature difference. Such temperature difference has become a main control factor. Therefore, it is an important issue of cracking prevention for RCC to prevent the surface cracks caused by the temperature change of climate environment and the large temperature difference between inside and outside.
3. Temperature difference between upper and lower layer. It refers to the temperature difference between the fresh and old concrete. Due to the adoption of full-section whole-block rolling, large cross-section of the dam body and construction block surface, the large temperature stress caused by temperature difference between fresh and old concrete will always generate, which has become a kind of temperature difference to be controlled.

10.2.1.2 Temperature Stress

In the concrete structure of dams, the development of concrete temperature stress can be divided into three stages: early stress stage, medium stress stage and late stress stage.

1. Early stress stage. This stage starts from the beginning of concrete pouring to the end of cement exothermic action, generally lasting for about one month. For the roller compacted concrete, the period is 40 – 90 d. This stage has two characteristics. First, a large amount of hydration heat is released due to the hydration of cement, which leads to the sharp change of temperature field.
2. Middle stress stage. This stage starts from the end of cement exothermic action to the final stable temperature of concrete. In this period, the temperature stress is caused by the cooling of concrete and the change of external temperature. These stresses are superimposed with the temperature stress produced in the early stage. During this period, the elastic modulus of concrete has some slight changes.
3. Late stress stage. This is the operation period after the concrete is completely cooled. In this period, the temperature stress is mainly caused by the changes of the external air temperature and water temperature. These stresses and the residual stresses in the early and middle stages are superimposed each other to form the late stress of the concrete.

10.2.1.3 Cracks in Concrete Dams

Most of the cracks in RCC dams are surface cracks. Under certain conditions, the surface cracks can develop into deep cracks or even penetrating cracks. Therefore, it is very important to strengthen the concrete curing on the surface. Due to the low strength of roller compacted concrete in early stage, the sudden drop of temperature is one of the most unfavorable factors causing surface cracks of roller compacted concrete, and the large temperature difference between inside and outside in winter is also one of the reasons for surface cracks of roller compacted concrete. Therefore, attention should be paid to the temperature protection measures of roller compacted concrete during the period of sudden temperature drop and in winter. According to the *Design Specification for Concrete Gravity Dams* SL 319, the concrete cracks in dam can be divided into three categories.

1. Surface crack. The cracks with the width of less than 0.3 mm, the depth of no more than 1 m, and the length of less than 5 m in regular shape mostly formed due to temperature impact and poor thermal insulation during the period of temperature drop, which has a slight impact on the structural stress, durability and operation safety.

   At present, a certain thickness of grade-II RCC and normal concrete impervious layer are set on the upstream face of RCC dam. The cement content of impervious layer is high, the adiabatic temperature is increased, and there is a certain gap between the mechanical and thermal properties of the impervious layer and that of the internal roller compacted concrete. These characteristics contribute to the greater tensile stress of the impervious layer, so it is prone to have cracks. Surface cracks can also cause local stress redistribution, affecting the dam strength. It should be noted that surface cracks may develop into deep or penetrating cracks in wedge shape.

2. Deep crack. The cracks with the width of no more than 0.5 mm, the depth of no more than 5 m, and the length of more than 5 m in regular shape mostly formed due to the large internal and external temperature difference or the impact of large temperature drop and poor thermal insulation, which has a certain impact on the structural stress and durability. Once it is expanded, it will be more harmful.

3. Penetrating cracks. The cracks with the width of more than 0.5 mm, the depth of more than 5 m, the length of more than 5 m on the side view, running through the entire surface or a block on the plane are formed because the temperature difference of foundation exceeds the design standard or it is suffered from sudden drop of ambient temperature in the restraint area of foundation and further develop in late cooling period, which reduces the structural stress, durability and safety factors to the critical value or below, and damages the physical properties and stability of structures. Both the penetrating cracks and deep cracks are harmful, which must be avoided.

10.2.2 Basic Design Information for Temperature Control

The temperature control standards and measures of RCC dam are closely related to the climate and other natural conditions of the dam site. It is necessary to carefully collect and analyze the temperature, water temperature and foundation temperature of the dam site, which can be served as the basic basis for the temperature control design of the dam. In addition, there are many factors affecting the water temperature of the reservoir. The water temperature of the upstream reservoir can be determined by referring to the water temperature of similar reservoirs.

The design of dam temperature control is closely related to the mechanical, thermal and deformation properties of RCC. The engineering practice shows that the design of temperature control has not only simply analyzed the temperature field, temperature stress and cooling measures, but also paid attention to the properties of roller compacted concrete materials. For example, increasing the ultimate tensile value of roller compacted concrete, selecting aggregates with low thermal expansion coefficient, compensating temperature shrinkage by self volume deformation and creep of roller compacted concrete, and considering the cracking prevention of roller compacted concrete from RCC materials. With the maturity of RCC mix design and the change of RCC definition, the ultimate tensile value, frost resistance and inter layer sliding stability of RCC have been significantly improved in recent years.

10.2.2.1 Air Temperature and Water Temperature

See Table 10.1 for the average air temperature and water temperature in RCC dam site area of some projects in China.

10.2.2.2 Ultimate Tensile and Elastic Modulus

See Table 10.2 for ultimate tensile value and elastic modulus of RCC in some projects in China.

10.2.2.3 Thermal Properties of Concrete

The thermal properties of concrete generally include thermal conductivity coefficient α, thermal conductivity coefficient λ, specific heat c, linear expansion coefficient α and adiabatic temperature rise. In large and medium-sized project, the thermal properties of concrete are determined by tests. Because the thermal properties of concrete depend on the thermal properties of water, cement and coarse aggregate, the thermal properties of concrete can be estimated with weighted average method according to the amount and characteristics of materials in concrete mix proportion. See Table 10.3 for thermal properties of RCC in some projects of China, and Table 10.4 for thermal properties of various concrete materials.

10.2.3 Criterion of Temperature Control

10.2.3.1 Requirements of Temperature Control Standard

The temperature standard of RCC dam is simulated based on the specification and temperature control design. The stable temperature at different parts of the dam body is determined according to the code and the simulation results of temperature control design, which is taken as the temperature control standard at different parts of the dam body. The temperature control standards of dam body mainly include the control of temperature difference of foundation, the control of temperature difference between fresh and old concrete, the control of temperature difference between internal and external concrete of dam, the control of allowable maximum temperature and the control of height difference among adjacent blocks.

1. Foundation temperature difference. It refers to the difference between the maximum temperature and the stable temperature of concrete within the constraint range of dam foundation.
2. Temperature control standard for fresh and old concrete. When pouring on the old concrete surface with an interval of more than 28 d, the newly poured concrete within $1/4L$ above the old concrete surface shall be controlled according to the temperature difference between the fresh and the old concrete.
3. Temperature control standard for surface concrete. The internal and external temperature difference of concrete shall not exceed the designed standard.
4. Permissible maximum temperature. The allowable maximum temperature of concrete block of dam body shall not exceed the design allowable maximum temperature.
5. Control of height difference among adjacent blocks. In the RCC construction, each dam block should rise evenly. The height difference of adjacent dam blocks, the interval time of adjacent dam blocks, and the height difference between the highest and lowest dam block should be controlled according to the specifications and design requirements.

Chapter 10 Temperature Control and Cracking Prevention of Roller Compacted Concrete

Table 10.1 Average air temperature and water temperature in RCC dam site area of some projects in China

Unit: ℃

S/N	Project title	Description	January	February	March	April	May	June	July	August	September	October	November	December	Annual average
1	Jiangya	Temperature	4.9	6.1	11.1	16.5	21.2	—	—	—	23.1	17.6	11.9	6.8	—
2	Longshou	Temperature	—	—	3.3	10.1	15.7	21.0	23.0	21.6	16.7	8.6	-0.4	-6.3	—
3	Baise	Temperature	13.3	15.1	19.2	23.7	26.6	28.0	28.5	27.9	26.6	22.2	18.6	14.8	22.1
4		Water temperature	14.9	16.1	18.9	23.0	26.2	27.1	27.4	26.7	25.9	23.7	20.5	16.9	22.3
5		Radiation temperature rise of dam surface	3.06	3.53	3.59	3.84	4.04	4.13	4.49	4.30	4.00	3.44	3.21	2.81	3.71
6	Zhaolaihe Hydropower Station	Temperature	4.6	6.3	10.6	16.4	21.1	24.8	27.5	27.2	22.6	17.4	11.8	6.6	16.2
7		Water temperature	7.5	8.6	12.1	16.1	19	21.6	23.2	24.7	21.7	18.1	14.2	9.7	16.4
8	Dahuashui	Temperature	3.6	4.7	9.6	14.5	18.4	21.2	23.2	22.4	19.4	15.0	10.1	5.5	14.0
9		Water temperature	9.1	9.9	12.7	16.8	19.7	21.2	23.0	23.4	21.7	18.3	14.7	11.0	16.8
10	Jinghong	Temperature	16.0	18.0	21.1	24.3	25.7	25.8	25.4	25.1	24.5	22.6	19.4	16.2	22.0
11		Water temperature	13.1	14.5	16.9	18.6	20.4	21.7	22.2	22.2	21.0	19.7	16.9	14.0	18.4

Continued Table 10.1

S/N	Project title	Description	January	February	March	April	May	June	July	August	September	October	November	December	Annual average
12	Longtan	Temperature	11.0	12.6	16.9	21.2	24.3	26.1	27.1	26.7	24.8	21	16.6	12.7	20.1
13		Water temperature	14.5	15.2	18.0	21.7	24.2	24.7	25.1	25.6	24.9	22.0	19.4	16.1	21.0
14	Gelantan	Temperature	12.4	13.8	16.8	19.7	21.7	22.4	22.2	22.0	21.0	19.1	15.9	12.8	—
15	Wudu Yinshui	Temperature	5.1	6.8	11.3	16.4	21.0	23.8	25.5	25.2	21.0	16.7	11.4	6.6	15.9
16		Water temperature	5.1	7.4	10.8	14.0	15.7	17.9	18.9	19.1	17.1	14.1	11.1	7.8	13.3
17	Jin'anqiao	Temperature	13.3	16.9	20.6	23.4	25.1	25.0	24.1	23.7	21.7	19.7	15.4	12.5	20.1
18		Water temperature	8.2	9.9	12.6	14.9	16.8	18.5	19.1	19.3	17.6	15.4	11.0	8.5	14.3
19	Kalasuke Hydropower Station	Temperature	-20.6	-17.6	-6.7	7.2	14.9	20.3	22.0	20.0	13.7	5.1	-6.8	-17.5	2.7
20	Gongguoqiao	Temperature	7.6	9.6	12.9	16.2	19.6	21.6	21.4	20.7	19.5	16.5	11.7	8.1	15.4
21		Water temperature	5.6	7.6	9.9	13.2	16.6	18.6	18.4	17.7	16.5	13.5	9.7	6.1	12.8
22	Guandi	Temperature	11.0	14.6	19.1	21.8	22.8	22.7	23.0	22.9	20.6	18.7	14.6	11.1	18.6
23		Water temperature	7.2	9.6	12.3	15.7	17.6	18.0	18.4	18.5	16.8	15.2	11.3	8.2	14.1

Chapter 10 Temperature Control and Cracking Prevention of Roller Compacted Concrete

Table 10.2　　Ultimate tensile value and elastic modulus of RCC in some projects of China

S/N	Project title	Design index	Mix proportion parameter				Ultimate tensile value ($\times 10^{-4}$) 90 d	Modulus of elasticity (GPa) 90 d
			Gradation	Water-cement ratio	Cement (kg/m^3)	Amount of rubber material (kg/m^3)		
1	Yantan	$R_{90}150$	Level III	0.566	55	159	0.70	—
2	Puding	$R_{90}150$	Level III	0.55	54	153	0.72	41.2
3		$R_{90}200$	Level II	0.50	85	188	0.81	39.8
4	Jiangya	$R_{180}100$	Level III	0.61	46	153	0.77	—
5		$R_{180}200$	Level II	0.53	87	194	0.86	—
6	Mianhuatan	$R_{180}150$	Level III	0.60	51	147	0.73	—
7		$R_{180}100$	Level III	0.65	48	136	0.72	—
8		$R_{180}200$	Level II	0.50	90	200	0.75	—
9	Dachaoshan	$R_{180}150$	Level III	0.50	67	168	0.74	—
10		$R_{180}200$	Level II	0.50	94	188	0.86	—
11	Longshou	$R_{90}20$	Level III	0.48	62	177	0.78	34.2
12		$R_{90}20F300$	Level II	0.43	96	205	0.87	29.6
13	Shapai	$R_{90}200$	Level II	0.53	115	192	0.136	16.65
14		$R_{90}200$	Level III	0.50	93	186	0.135	16.65
15	Linhekou	$R_{90}200$	Level II	0.47	74	185	0.91	42.7
16		$R_{90}200$	Level III	0.47	66	177	0.82	41.3
17	Baise	$R_{180}15$	Quasi III gradation	0.60	59	160	0.78	31.5
18		$R_{180}20$	Level II	0.50	89	212	0.90	33.9
19	Suofengying Hydropower Station	$C_{90}15$	Level III	0.55	64	160	0.71	34.5
20		$C_{90}20$	Level II	0.50	94	188	0.86	37.8
21	Zhaolaihe Hydropower Station	$C_{90}20$	Level II	0.48	88.5	177	0.93	31.0
22		$C_{90}20$	Level III	0.48	70.3	156	0.88	31.3
23	Longtan (Xuanwei fly ash)	Lower part $C_{90}25$	Level III	0.41	85	193	0.86	43.9
24		Central part C9020W6F100 $C_{90}20$	Level III	0.45	67	173	0.75	36.9
25		$C_{90}15$ Upper part	Level III	0.48	56	165	0.72	35.9
26		Impervious area $C_{90}25$	Level II	0.40	98	217	0.96	37.9

Continued Table 10.2

S/N	Project title	Design index	Gradation	Water-cement ratio	Cement (kg/m³)	Amount of rubber material (kg/m³)	Ultimate tensile value (×10⁻⁴) 90 d	Modulus of elasticity (GPa) 90 d
27	Guangzhao	Lower part $C_{90}25$	Level III	0.45	83	180	0.90	45.3
28		Upper part $C_{90}20$	Level III	0.50	70	177	0.87	44.2
29		Upper part $C_{90}15$	Level III	0.55	3 357	164	0.81	42.3
30		External $C_{90}25$	Level II	0.45	92	207	0.92	45.4
31		External $C_{90}20$	Level II	0.50	77	195	0.88	43.1
32	Gelantan	$C_{90}15$	Level III	0.50	66	166	0.91	34.5
33		$C_{90}20$	Level II	0.45	93	207	0.90	37.3
34	Jin'anqiao	Lower part $C_{90}20$	Level III	0.47	76	191	0.79	38.8
35		Upper part $C_{90}15$	Level III	0.53	63	170	0.78	35.2
36		Impervious area $C_{90}20$	Level II	0.47	96	213	0.83	34.6
37	Kalasuke Hydropower Station	$R_{180}200F300W10$	Level II	0.45	131	218	0.84	36.0
38		$R_{180}200F100W10$	Level II	0.47	91	202	0.81	31.3
39		$R_{180}200F200W6$	Level III	0.45	100	200	0.79	34.9
40		$R_{180}150F50W4$	Level III	0.56	61	161	0.79	33.6
41	Gongguoqiao	$C_{90}15$	Level III	0.50	81	180	0.82	24.2
42		$C_{90}20$	Level II	0.46	109	218	0.89	30.5
43	Lianhuatai	$C_{180}15$	Level III	0.55	54	155	0.74	27.8
44		$C_{180}20$	Level II	0.47	81	202	0.78	29.7
45	Guandi	Lower part $C_{90}25$	Level III	0.45	82	205	0.81	35.4
46		Central part C9020W6F100 $C_{90}20$	Level III	0.48	77	192	0.77	32.7
47		Upper part $C_{90}15$	Level III	0.51	63	180	0.75	31.5
48		Impervious area $C_{90}25$	Level II	0.45	102	227	0.85	39.4
49		Impervious area $C_{90}20$	Level II	0.48	96	213	0.82	35.1

Table 10.3　　Thermal properties of RCC in some projects of China

S/N	Project title	Type and position of concrete	Coefficient of thermal conductivity $\alpha(m^2/h)$	Thermal conductivity factor $\lambda[kJ/(m \cdot h \cdot ℃)]$	Specific heat $c[kJ/(kg \cdot ℃)]$	Linear expansion coefficient $\alpha(10^{-6}/℃)$	Density (kg/m³)	Poisson's ratio μ	Adiabatic temperature rise (at 28 d) (℃)
1	Puding	RCC Level II	0.003 268	7.942 3	0.966 9	5.097 6	2 475	—	22.95
2		RCC Level III	0.003 836	8.090 1	0.884 8	5.824 2	2 481	—	16.05

Chapter 10　Temperature Control and Cracking Prevention of Roller Compacted Concrete

Continued Table 10.3

S/N	Project title	Type and position of concrete	Coefficient of thermal conductivity $\alpha(m^2/h)$	Thermal conductivity factor $\lambda[kJ/(m \cdot h \cdot ℃)]$	Specific heat $c[kJ/(kg \cdot ℃)]$	Linear expansion coefficient $\alpha(10^{-6}/℃)$	Density (kg/m^3)	Poisson's ratio μ	Adiabatic temperature rise (at 28 d) (℃)
3	Longshou	RCC Level Ⅱ	0.003 6	8.088	0.924	10.2	2 400	—	20.3
4		RCC Level Ⅲ	0.004 7	8.292	0.849	10.5	2 400	—	17.8
5	Baise	RCC Level Ⅱ	0.003 039	7.668	0.94	5.823	2 600	0.167	20.33
6		RCC Quasi Ⅲ gradation	0.003 039	7.668	0.94	6.744	2 650	0.167	14.50
7		Normal cushion	0.002 87	6.46	0.90	7.0	2 530	—	—
8		Bedrock	0.003 19	6.87	0.77	7.0	2 980	0.25	—
9	Suofengying Hydropower Station	RCC Level Ⅱ	0.003 3	7.76	0.963	5.67	2 468	—	17.75
10		RCC Level Ⅲ	0.003 5	8.04	0.960	5.61	2 450	—	16.79
11	Zhaolaihe Hydropower Station	Dam body RCC	0.005 8	12.69	0.903	7	2 403	—	17.7
12		Bedrock (limestone)	0.006 9	14.25	0.76	10	2 437	—	—
13	Dahuashui	RCC Arch dam	0.003 6	8.22	0.942	6.5	2 400	0.170	19.3
14		RCC gravity dam	0.003 5	7.88	0.931	6.5	2 400	0.166	17.9
15	Yujian River	Normal concrete	—	13.56	0.945	8.0	—	0.167	22.4
16		RCC Level Ⅱ	—	7.67	0.9549	6.48	—	0.163	14.8
17	A project	RCC Level Ⅱ	0.002 87	6.46	—	7.00	2 400	0.167	20.3
18		RCC Level Ⅲ	0.003 04	7.67	—	6.74	2 420	0.167	13.5
19	Longtan	Normal concrete	0.003 704	8.776	0.9672	7	2 450	0.167	24.42
20		RCC Level Ⅲ	0.003 941	9.27	0.9672	7	2 400	0.163	15.08、18.5
21	Wudu Yinshui	RCC Level Ⅱ	0.002 99	9.289	1.25	—	2 450	—	13.6
22		RCC Level Ⅲ	0.002 98	9.151	1.24	—	2 450	—	11.8
23	Jin'anqiao	RCC Level Ⅱ	0.002 401	7.38	0.966 2	8.0	2 600	0.198	18.6
24		RCC Level Ⅲ	0.002 511	7.39	0.926 9	8.21	2 630	0.211	16.8

Continued Table 10.3

S/N	Project title	Type and position of concrete	Coefficient of thermal conductivity $a(m^2/h)$	Thermal conductivity factor $\lambda[kJ/(m \cdot h \cdot ℃)]$	Specific heat $c[kJ/(kg \cdot ℃)]$	Linear expansion coefficient $\alpha(10^{-6}/℃)$	Density (kg/m^3)	Poisson's ratio μ	Adiabatic temperature rise (at 28 d) (℃)
25	Kalasuke Hydropower Station	$R_{180}20W10$ F300 Level II	0.003 8	8.49	0.951	9.25	2 400	—	24.29
26		R18020W4 F50 Level III	0.003 8	8.34	0.884	8.96	2 450	—	17.61
27	Gongguoqiao	RCC Level II	0.004 52	10.96	1.012 7	11.4	2 420	0.17	20.9
28		RCC Level III	0.004 3	10.66	0.977 3	11.0	2 440	0.17	19.1
29	Guandi	$C_{90}25$ Level II	0.003 0	7.76	0.96	7.54	2 630	—	20.9
30		$C_{90}20$ Level II	0.002 9	7.58	0.94	7.61	2 630	—	19.6
31		$C_{90}25$ Level III	0.003 0	7.34	0.90	7.70	2 660	—	18.1
32		$C_{90}20$ Level III	0.003 0	7.54	0.95	7.72	2 660	—	17.4
33		$C_{90}15$ Level III	0.002 9	7.71	0.98	7.73	2 660	—	16.9
34	Xinsong Reservoir	RCC Level II	0.003 172	7.31	0.962	8.0	—	—	21.0
35		RCC Level III	0.003 061	6.85	0.919	8.6	—	—	17.8
36		Foundation bedding	0.003 172	7.31	0.962	8.0	—	—	22.6

Note: RCC grade-II-impermeable area on the upstream surface of the dam, and RCC grade-III-the internal part of the dam.

Table 10.4 Thermal properties of concrete materials

S/N	Material	Thermal conductivity factor $\lambda[kJ/(m \cdot h \cdot ℃)]$				Specific heat $c[kJ/(kg \cdot ℃)]$			
		21 ℃	32 ℃	43 ℃	54 ℃	21 ℃	32 ℃	43 ℃	54 ℃
1	Water	2.160	2.160	2.160	2.160	4.187	4.187	4.187	4.187
2	Ordinary cement	4.446	4.593	4.735	4.865	0.456	0.536	0.662	0.825
3	Quartz sand	11.129	11.099	11.053	11.036	0.699	0.745	0.795	0.867
4	Basalt	6.891	6.871	6.858	6.873	0.766	0.758	0.783	0.837
5	Dolomite	15.533	15.261	15.014	14.336	0.804	0.821	0.854	0.888
6	Granite	10.505	10.467	10.442	10.379	0.716	0.708	0.733	0.755
7	Limestone	14.528	14.193	13.917	13.657	0.749	0.758	0.783	0.821
8	Quartzite	16.910	16.777	16.638	16.475	0.691	0.724	0.758	0.791
9	Rhyolite	6.770	6.812	6.862	6.887	0.766	0.775	0.800	0.808

10.2.3.2 Allowable Temperature Difference of Foundation

The temperature difference of foundation is an important index to control the deep cracks in the dam foundation, which mainly changes with the performance of roller compacted concrete, the height-length ratio of poured block, the length of long edge of poured block, the elastic modulus ratio between concrete and bedrock, and the climatic conditions in dam site area. As there is no longitudinal joint, the bottom width is large and the restraint range of foundation is large in the RCC dam, in order to prevent the concrete cracks in foundation, it is necessary to control the allowable temperature difference of foundation.

According to the investigation of cracks in some concrete dams in China, the foundation has cracks under the following circumstances:

1. The thin-lift of poured block on the bedrock has been maintained for a long time, leading that the constraint stress of the thin concrete layer overlaps with the stress caused by the internal and external temperature difference. Then, the tensile stress in the middle of the block is far greater than the tensile strength of the concrete, forming a penetrating crack.
2. The rock surface fluctuates greatly, and there are deep pits or protruding sharp corners in some parts, which leads to uneven thickness of concrete blocks and local stress concentration, forming the concrete cracks in the foundation.
3. As the openings are reserved on the dam for diversion or flood flowing in the construction period, when the concrete temperature is high, the concrete cracks occur in the foundation due to the cold impact of water.

Therefore, the allowable temperature difference of foundation and temperature control standards for RCC dam should conform to the tender documents, and be in accordance with the structure joint, block size and allowable maximum temperature of concrete shown in the construction drawings as well as the *Design Specification for RCC Dams* SL 314. The allowable temperature difference of foundation for RCC gravity dam shall refer to the *Design Specification for Concrete Gravity Dam* DL 5108—1999 or SL 319—2005 and be converted based on the ultimate tensile value of roller compacted concrete (RCC) of 0.70×10^{-4}. The allowable temperature difference of foundation recommended in the specification is shown in Table 10.5.

Table 10.5　　Allowable temperature difference of foundation of RCC gravity dam　　Unit: ℃

Height from foundation surface h	Length of long side of poured block L		
	Below 30 m	30 – 70 m	Above 70 m
0 – 0.2L	18 – 15.5	14.5 – 12	12 – 10
0.2L – 0.4L	19 – 17	16.5 – 14.5	14.5 – 12

Because many factors are involved in the allowable temperature difference of foundation, and the characteristics of roller compacted concrete are different from that of normal concrete, the specific conditions of each project are also very different, considering the allowable temperature difference of foundation is an important index leading to deep cracks in dam, the allowable temperature difference value of foundation for high and medium RCC dams should be determined according to the specific conditions of the project and must be determined after the design of temperature control.

10.2.3.3 Pouring Temperature, Allowable Maximum Temperature

Both the pouring temperature and maximum temperature rise of concrete shall meet the requirements of construction drawings specified in the design. In the construction, the relationship between the temperature at the outlet of concrete and the temperature at the pouring site shall be established through tests, and the effective measures shall be taken to reduce the temperature rise in the process of concrete transportation. Generally, the outlet temperature is determined by the allowable pouring temperature on site, that is, about 5 ℃ lower than the allowable pouring temperature. See Table 10.6 for temperature difference standard and allowable pouring temperature of some RCC dams in China.

Table 10.6　　　Temperature difference standard and allowable pouring temperature of some RCC dams in China　　　Unit: ℃

S/N	Project title	Length of long side of poured block L (m)	Allowable temperature difference of foundation Standard Height from foundation surface h		Allowable pouring temperature		Temperature difference inside and outside the dam body	Permissible maximum temperature
			$0 - 0.2L$	$0.2L - 0.4L$	Constrained zone	Unconstrained zone		
1	Puding	—	14	17	—	—	17 – 19	—
2	Jiangya	>70	13	15	15	18	18 – 23	—
3	Longshou	30 – 70	14	16	20	24	17 – 19	38
4	Baise	>70	10	21	16	22	Controlled according to different maximum temperature	36
5	Suofengying Hydropower Station	>70	14	17	18	22	20	36.5
6	Zhaolaihe Hydropower Station	30 – 70	14	16	16	20	22	35
7	Longtan	>70	16	19	17	22	20	35
8	Guangzhao	>70	16	18	20	20	15	38
9	Jin'anqiao	>70	12	13.5	17	22	15	33
10	Pengshui	>70	12	15	15	17	18 – 20	35
11	Gelantan	20 – 70	13	15	19.5	20	16	38
12	Wudu Yinshui	>40	14	19	22 without precooling	30 without precooling	—	36
13	Kalasuke Hydropower Station	>70	12	14.5	15	18	16	34
14	Gongguoqiao	20 – 70	13	15	17	19	15	34
15	Guandi	>70	12	13	17	17	15	33

Chapter 10 Temperature Control and Cracking Prevention of Roller Compacted Concrete

10.3 RCC Temperature Control Measures

10.3.1 RCC Temperature Control Characteristics

As stipulated in the *Design Specification for RCC Dam* SL 314, the high and medium RCC gravity dams shall be designed for the temperature control, and the temperature control standards and cracking prevention measures shall be proposed. The temperature control design for high and medium dams can refer to the *Design Specification for Concrete Gravity Dams* SL 319 or DL 5108. For the high and medium dams, the temperature field and temperature stress of dam body should be analyzed with the finite element method. The RCC concrete shall be designed for cracking prevention and temperature control according to the construction characteristics of whole-block and thin-lift, full section and continuous rolling rise.

Temperature control and cracking prevention is an important task in the design and construction of RCC dams. To prevent concrete crack is mainly to control the temperature stress by temperature control measures and structural measures. The greatest difference between RCC and normal concrete lies in the mix proportion design and construction method. There are some differences in the performance of mixture, mechanical property, deformation performance, durability and thermal performance between RCC and normal concrete, in particular that the adiabatic temperature rise of RCC is obviously different from that of normal concrete. Because the longitudinal joint has been cancelled, the construction of RCC is carried out in whole-block and thin-lifts, in full section and continuous rolling rise. Therefore, RCC temperature stress and temperature control measures have their own characteristics. In the temperature control, the influence factors of RCC material characteristics and construction method shall be fully considered.

After selecting the appropriate joint spacing, the specific maximum temperature creep stress is directly related to the maximum temperature difference of the dam. That is to say, if the maximum temperature rise of the dam is controlled, the maximum temperature creep stress can be controlled. Therefore, it is crucial to adopt effective temperature control measures for concrete. In the design of temperature control measures, the reasonable joint and block division, layered rolling height and the interval time shall be determined firstly. Apart from the restriction of initial setting time and pouring capacity of concrete, the design of dam joint is mainly related to the requirements of dam temperature control. Reasonable joint and block division can not only speed up the construction progress, but also make the necessary release of temperature stress to prevent and reduce cracks. By simplifying construction technology, reducing construction interference, speeding up the dam pouring progress and improving the integrity of the dam body, through the calculation and analysis of the temperature stress of dam body, with certain temperature control measures, the longitudinal joints are basically not needed. So, the RCC gravity dam has a large bottom width. At present, according to the temperature control design and calculation of RCC dam, the spacing of transverse joint of gravity dam should be controlled at about 20 m. If more than 25 m, a special demonstration shall be carried out. Generally, when the width of the dam section is more than 25 m, a short joint with a depth of 3 - 4 m is basically set in the middle of the upstream face of the dam, which can prevent the occurrence of surface cracks and even split cracks on the upstream face of the dam. Taking the RCC gravity dams in Baise, Longtan, Guangzhao and Jin'anqiao as the examples, the short joints are set on the upstream face of the dam when the transverse joint exceeds a certain width, proving a good effect in the practice.

In addition to the normal construction method and quality control, the most important thing of mass concrete is to control the temperature rise of mass concrete to reduce the temperature stress and cracks. With the rapid construction of RCC, although the rate of hydration heat release in the first stage is slow, the temperature inside the dam body will rise higher than the pouring temperature with the hydration reaction. The maximum temperature rise of RCC dam body depends on the outlet temperature of mixture, cement dosage, exothermic character of cement, dam section size and pouring speed, surrounding temperature conditions, thermal characteristics of concrete and artificial cooling. In order to prevent cracks in RCC, the comprehensive measures must be taken for structural design, raw material selection, mix proportion design, construction scheme, construction quality, temperature control, curing and surface protection, so as to control the maximum temperature rise of concrete within the allowable range of design.

The temperature control measures in the construction process of RCC shall conform to the provisions of *Specifications for Hydraulic Concrete Construction* DL/T 5144, including: ① Lower temperature rise of concrete hydration heat; ② Lower concrete pouring temperature; ③ Supply water for cooling and reduce the temperature difference inside and outside the dam body; ④ Surface curing and heat preservation; ⑤ Temperature monitoring, etc..

10.3.2 Lower Temperature Rise of Concrete Hydration Heat

10.3.2.1 Control the Temperature of Raw Material

In order to prevent cracks in concrete dams, it is necessary to control the temperature rise of hydration heat from the source of concrete constituent materials. In recent years, more and more attention has been paid to the temperature of concrete raw materials in the projects of China. The detailed control indexes have been proposed for cement fineness, magnesium oxide content, hydration heat, mineral composition, water demand ratio for fly ash, water reduction rate of admixture, etc.. In addition, the temperature of cement and fly ash to the construction site is specially specified. Generally, the temperature of cement shall not exceed 60 ℃ and that of fly ash shall not exceed 45 ℃. These provisions are very beneficial to reduce the temperature rise of hydration heat of concrete. Therefore, under the condition that the concrete indexes are satisfied, the following factors shall be considered in the selection of raw materials:

1. Cement with low hydration heat temperature rise shall be selected as far as possible, such as medium low heat cement, low heat cement or low heat micro expansive cement.
2. The aggregate with small thermal expansion coefficient α shall be selected. For example, limestone aggregate has the minimum thermal expansion coefficient α.
3. The high-quality admixtures shall be selected, and the admixtures with small water demand ratio, such as grade-I fly ash, shall be chosen as far as possible.
4. The admixtures shall be the retarded superplasticizer with high water-reducing rate and meeting the setting time.
5. The ultimate tensile value of RCC shall be increased and the elastic modulus shall be reduced as possible.
6. The self-grown volume deformation and shrinkage of concrete shall be controlled to the minimum extent.
7. The creep of RCC shall be large as far as possible so that the relaxation coefficient K is small.

10.3.2.2 Optimize Mix Proportion Design

Cement is the main cementing material of concrete. When cement reacts with water, a lot of hydration heat will generate. Therefore, the temperature in concrete mainly depends on the hydration heat, dosage and pouring temperature of cement. The concrete prepared by cement has poor thermal conductivity. The heat increase rate of cement is far greater than the thermal diffusivity. A large number of test researches show that the higher the concrete pouring temperature, the faster the chemical reaction of cement hydration heat. The faster hydration heat reaction rate is the main factor leading to higher temperature inside the concrete.

Therefore, in the design of concrete mix proportion, in addition to meet the main indexes such as strength, frost resistance, impermeability, ultimate tensile value and construction performance of concrete, it is necessary

Chapter 10 Temperature Control and Cracking Prevention of Roller Compacted Concrete

to optimize the mix proportion design and reduce the unit cement consumption scientifically and reasonably. This contributes to obvious practical significance for reducing the temperature rise of concrete hydration heat.

On the condition that the concrete design requirements and raw materials requirements, the main technical routes for RCC mix proportion optimization are as follows: select high quality additives to improve water reduction rate and reduce unit water consumption; increase the amount of admixtures and reduce the amount of cement; reasonably select the size and gradation of aggregate, and reduce the porosity; increasing the content of rock powder in sand can effectively improve the workability of RCC and reduce the cementing materials. In addition, whether the RCC design indexes match each other and the adoption of long age will have certain influence on reducing cement consumption, reducing temperature rise of hydration heat and preventing cracks.

10.3.3 Lower Concrete Pouring Temperature

10.3.3.1 Reduce the Temperature of Aggregate in Silo

Ⅰ Cooling of aggregate in silo

The aggregate temperature has a great influence on the concrete outlet temperature, so there are special provisions on the measures to reduce the temperature of finished materials in the *Specifications for Hydraulic Concrete Construction*. The stacking height of aggregate in finished material silo should not be less than 6 m, and there shall be sufficient reserve. Besides, the sunshade and rainproof shed shall be set up for the finished material silo to avoid the increase of temperature under sunlight or the moisture content of aggregate beyond specification. When the stacking height is more than 6 m, the internal temperature of the aggregate pile that has been stored for more than 5-7 d can be close to the monthly average temperature. In order to obtain a lower aggregate temperature, a gallery shall be set up under the aggregate pile. The gallery for taking materials from the ground and discharging materials from the ground shall be partially or fully buried in the ground. When using the aggregate, special attention shall be paid to the opening of the radial gate of each hopper, so that the aggregate which has been stored for several days can be used. In addition, the aggregate that has just been processed or transported to the mixing plant with higher temperature in the clean material yard cannot be fed to the mixing plant.

In the high temperature season, when the aggregate in the stock yard is cooled by spraying water, the characteristics of low unit water consumption of RCC shall be considered. The moisture content of aggregate shall be strictly controlled, not exceed the standard, so that the *VC* value of RCC mixture is in the controllable range.

Ⅱ Cooling of aggregate by air

The most effective measure to reduce the concrete outlet temperature is to reduce the aggregate temperature, because the aggregate accounts for more than 80% of the concrete mass, and the coarse aggregate accounts for more than 60%. The cooling of aggregate is usually carried out in the storage bin or aggregate conveying gallery of the mixing plant. The water content of the air-cooled aggregate decreases slightly in the cooling process, and the material temperature can be maintained or reduced when the mixing plant is shut down. The most common way to cool the aggregate below zero is to cool the aggregate by air and ice. The aggregate of 5 – 20 mm should not be cooled by cold air at negative temperature. If it has been cooled by water, only the cold air is used to keep the original feed temperature in the storage bin of mixing plant. In order to overcome the wind resistance of small stone, the fan with smaller air volume and higher air pressure can be used. The primary air cooling and secondary air cooling are mainly adopted to reduce the temperature of RCC aggregate, having significant effect. The parameters of air cooling of aggregate are generally as follows:

Primary air cooling: 5 – 20 mm aggregate, inlet air temperature of 0 ℃, material temperature of 8 ℃.

20 – 40 mm aggregate, inlet air temperature of – 5 ℃, material temperature of 5 ℃. 40 – 80 mm aggregate, inlet air temperature of – 5 ℃, material temperature of 5 ℃. In this way, the average temperature of the three aggregates can reach below 6 ℃.

Secondary air cooling: 5 – 20 mm aggregate, inlet air temperature of 0 ℃, material temperature of 4 ℃. 20 – 40 mm aggregate, inlet air temperature of – 10 ℃, material temperature of 0 ℃. 40 – 80 mm aggregate, inlet air temperature of – 10 ℃, material temperature of 0 ℃. In this way, the average temperature of the three aggregates can reach below 1 ℃.

The main features of the two air cooling process are as follows: the secondary air cooling is the continuation of primary air cooling, unlike air cooling to maintain temperature after water cooling; in addition, the moisture content of aggregate decreases after air cooling, so more ice can be added to concrete mixing. Air cooling time of aggregate: when the cooling range is 70% – 80%, 45 – 50 min for large stones; about 40 – 45 min for the medium stone; about 35 – 40 min for small stones.

For example, the mixing system on the left and right side of Jin'anqiao has enough refrigeration capacity and has a large margin. The primary air cooling and secondary air cooling were carried out on the aggregate. The inlet air temperature of primary air cooling is – 5 – 0 ℃, and the outlet temperature of aggregate is below 6 ℃. Under two working conditions: 0 – 3 ℃ wind temperature suitable for 5 – 20 mm aggregate, and – 5 – – 2 ℃ wind temperature suitable for 20 – 40 mm and 40 – 80 mm aggregates. In this way, the average final temperature of primary air-cooled aggregate can be below 6 ℃. As long as the special pressure regulating valve is equipped in the primary cooling workshop, different liquid supply pressure can be realized; Like primary air cooling, the secondary air cooling also operates under two working conditions 0 – 3 ℃ wind temperature suitable for 5 – 20 mm aggregate, and – 12 – – 8 ℃ wind temperature suitable for 20 – 40 mm and 40 – 80 mm aggregates. The temperature of 5 – 20 mm aggregate is about 4 ℃, and the temperature of 20 – 40 mm aggregate is about 0 ℃. It is feasible that the temperature of 40 – 80 mm aggregate is at negative temperature. It shall be emphasized that the intermittent cooling method shall not be adopted, otherwise the aggregate temperature will fluctuate.

10.3.3.2 Mixing with Ice

I Flake ice

In recent years, one of the most common control measures to reduce the concrete temperature is to add flake ice. The thickness of flake ice is generally 1.5 – 2.5 mm, and the specific surface area per ton of flake ice is about 1 700 m^2/t, so the ice added in the concrete is easy to melt. As long as the flake ice is kept dry and supercooled, it can be stored and transported.

Under normal pressure, the density of ice made of pure water is 917 kg/m^3, the melting point is 0 ℃, the melting heat is usually 335 kJ/kg; when the ice temperature is – 50 – 0 ℃, the thermal conductivity is 2.326 W/(m·K); when the ice temperature is – 20 – 0 ℃, the average specific heat is 2.093 kJ/(kg·K). The flake ice is generally 1.5 – 2.5 mm thick and irregular.

II Cooling effect by adding ice

The main factors affecting the cooling are the rate of ice addition and the degree of supercooling and drying of ice surface. Other factors such as ice temperature and icicle shape also have certain influence. According to the cooling effect of flake ice in the projects of China, adding 10 kg/m^3 of ice in concrete can reduce the concrete temperature by 1.2 – 1.4 ℃. Due to the small water consumption of RCC, the amount of ice added is about 20 – 30 kg/m^3.

After adding flake ice, the mixing time shall be appropriately extended. The extension time depends on the performance of mixing equipment. The mixing time of RCC with ice shall be determined by test. In order to improve the cooling effect of concrete, ensure the uniformity of mixing and add more ice as much as possible, the water content of sand and gravel and the water for admixture shall be strictly controlled. The amount of ice added is directly related to the moisture content of sand and aggregate.

A lot of engineering practice has proved that, the moisture content of sand shall be strictly controlled below 6%, 5 – 20 mm aggregate shall be controlled below 1%, 20 – 40 mm aggregate shall be controlled below 0.5%, 40 – 80 mm aggregate shall be controlled below 0.3%. If living up to these requirements, the ice content of RCC can be increased to more than 30 kg/m^3, generally reducing the concrete temperature by 2.5 ℃.

10.3.3.3 Reduce the Temperature of Automobile Transportation

I Sunshade of automobiles reduces the temperature rises

RCC is mainly transported by dump truck. In general, it will take 15 – 30 min to transport the RCC from mixing plant to the placement surface. A sunshade shall be set on the top of dump truck. According to the measurement results of Jin'anqiao, the temperature rise of RCC is generally 0.4 – 0.9 ℃, which is very beneficial to reduce the temperature rise. Similarly, if transporting the RCC by the dump truck without a sunshade, the RCC temperature will rise up to 2 – 5 ℃ in the sun.

II Spraying water for cooling by dump trucks

If the dump truck transports concrete and return to the mixing plant in empty, the dump truck shall be sprayed with water in front of the mixing plant for cooling well. The spray device can be erected on both sides of the road of 10 – 25 m long before entering the mixing plant, slightly higher than the dump truck compartment, to form a atomizing environment. When the vehicle is waiting in front of the structure, spraying not only reduces the temperature of the compartment, but also avoid direct sunlight, playing a great role in reducing the temperature of the concrete.

In addition, it is good for cooling, keeping the car clean and discharging smoothly by setting the washing table to wash the dump truck carriage frequently.

10.3.3.4 Spraying Water to Keep Moist and Change Microclimate

During the construction in high temperature season, windy and dry climate conditions, the surface water of RCC evaporates rapidly, and the surface is easy to whiten and rise in temperature. Spraying and moisturizing can form a insulation layer over the placement surface, so that the direct sunlight intensity of the concrete during the pouring process is reduced. It is a very important measure to reduce the environmental temperature and the concreting temperature. Spraying and moisturizing is not only directly related to the grind and intergration of RCC, but the most important thing is to change the microclimate of the surface (just like a person stands in the rain), which can effectively reduce the temperature of the surface by 4 – 6 ℃. This is very beneficial to temperature control. Generally, if the pouring temperature rises by 1 ℃, the temperature in the dam rises by 0.5 ℃.

Spray can be carried out either by manual or by machine. Whether it is artificial or mechanical spray, the nozzle of spray gun is very important. The droplets sprayed shall fall in the range of 40 – 100 μm, ensuring that the surface contains white mist. If using the hair sprayer instead of spray gun, it is easy to form rain on the placement surface, and easy to have stagnant water on the surface of concrete. If a hair sprayer is used, it is necessary to refit the nozzle, and install the spray nozzle for spraying.

Spraying mist is an important procedure and guarantee measure to control the interlayer bonding and temperature of RCC. It must be highly valued. In the process of construction, due to the weak quality awareness of individual construction units and the supervision engineer did not pay high attention, the costly lessons can be learnt from. Spraying and moisturizing on RCC surface is not a simple quality problem. It must be aware of at a certain height.

10.3.3.5 Timely Rolling and Covering to Prevent Temperature Rise

I   Timely rolling

Rolling of RCC in time after placement can not only effectively control the temperature rise of the surface, but also is the key to ensure the quality of roller compaction. The travel speed of the vibration roller directly affects the rolling efficiency and compaction quality. The construction practice shows that, if the travel speed of the vibration roller is generally controlled within the range of 1.0 - 1.5 km/h, the compaction effect is good. If the travel speed is too fast, the compaction effect is poor. If the travel speed is too slow, the vibration roller is easy to sink and the construction intensity will decrease. In the actual construction, the strip of vibration roller usually rolls for 10 times. The rolling efficiency of a single vibratory roller is about 70 m^3/h. Due to the whole-block and thin-lift, full-section and continuous rolling rising in RCC construction, the RCC on the surface can not be rolled in time after being paved, which delays for a long time, failing to guarantee the synchronization of rolling and leveling. If the spraying and moisturizing is not in place, the temperature rise of RCC pouring and the quality of interlayer bonding will be seriously affected. Therefore, after mixing, transportation and warehousing, the RCC must be paved, rolled, sprayed, moisturized and covered in time. That is to say, the sooner the RCC construction, the better. Therefore, the construction of RCC shall be accelerated and rolled in time. It is the key to ensure the quality of interlayer bonding and prevent the rise of pouring temperature, which must be attached great importance to.

II   Timely covering

For the roller-compacted concrete, the concrete temperature rebounds rapidly under the sun during the daytime, so timely covering the newly poured concrete placing surface is the key to prevent the temperature rebounding. The site construction level shows that roller-compacting has become the main factor restricting rapid construction. As for the shortest inclined layer roller-compacting, it usually takes more than 2 h to complete one layer lifting, which is very unfavorable to the temperature control of roller compacted concrete (RCC) in high temperature season. When the sunshine is strong, the temperature rebounding rate of the poured concrete can reach 30%.

During the high temperature season or high temperature period, there are two kinds of temperature changes during the concrete pouring process: one is the temperature rebounding during transportation. According to the statistics of the measured data of many projects, the concrete temperature rebounding is generally not more than 1 ℃ when the concrete is directly poured to the concrete placing area by dump truck; secondly, the temperature of the concrete placing surface rebounds, which is the main factor. The temperature rebounding value varies with the duration from the concrete poured to the concreting placing area to the upper layer covered with new concrete. Generally, the rebounding rate is 20% at intervals of 1 h, 35% at intervals of 2 h, and 45% at intervals of 3 h. Judging from the previous construction level, the time from roller-compacted concrete poured to the concrete placing area to upper layer of concrete covered is generally as long as 2 - 4 h, which is very unfavorable to the concrete temperature control during high temperature period, and the rebounding rate usually reaches 30% - 70% during strong sunshine.

For example, in the Jin'anqiao Project, the temperature of the roller-compacted concrete was measured at 15:00 p.m., and the pouring temperature was 17 ℃, and the concrete placing surface was not sprayed and covered. At that time, the sun was shining strongly and the temperature was 30 ℃. The temperature was measured again at 16:00, that is, 1 hour later. At this time, the pouring temperature quickly rose to 22 ℃, and the temperature rose as high as 5 ℃. The temperature measurement results show that if the surface of the roller-compacted concrete is not covered in time, it will be very unfavorable to control the pouring temperature rebounding and temperature flows backwards.

The concrete temperature rebounding of the concrete placing surface is a very important issue in

Chapter 10 Temperature Control and Cracking Prevention of Roller Compacted Concrete

temperature control. Laying heat preservation quilt is a convenient and effective measure to controlthe concrete temperature rebounding of the concrete placing surface. In recent years, the heat preservation material for the concrete placing surface generally is thermal insulation quilt, which is made of two layers of 1 cm thick polyethylene thermal insulation coiled material and plastic woven striped cloth. This kind of heat preservation quilt is mainly made of high pressure polyethylene foam plastic of the third generation foam plastic product, the thermal diffusivity of the material is small, that is, the thermal insulation performance is good, soft and foldable; it has high tensile strength, good waterproof performance (with independent bubble structure), low density, good elasticity, external luster, good impact resistance, good chemical aging resistance, non-toxic, odorless, low temperature resistance, oil resistance and fire resistance, so it can be applied to any shape of uneven concrete surface in the placing area as a covering, and close to the concrete surface to achieve temperature insulation effect. This heat preservation quilt has the characteristics of light weight (one labor can fully undertake the heat preservation of one concrete placing area), convenient use and reuse (not only the heat preservation quilt can be repeatedly used in one concrete placing area, but also after it is washed it can be repeatedly used in another placing area after this concrete placing area is poured and can be used for several years).

In order to prevent the concrete temperature of concrete placing surface from rebounding too fast, the newly poured concrete must be covered and insulated during the pouring process when pouring temperature-controlled concrete in high temperature season or high temperature period. During the construction of roller compacted concrete, each concrete placing area is required to be equipped with a heat preservation quilt with an area not less than 50% of the concrete placing surface area before the concrete is poured in the concrete placing area, and the concrete that has been roller-compacted shall be immediately covered. That the paved roller-compacted concrete that has not been roller-compacted in time and has been placed for too long time shall be recovered too is one of the most favorable temperature measures to control the temperature rebounding of concrete poured under strong sunshine.

The actual measurement to Three Gorges Project in summer confirms that when the new concrete with quilt is compared with the concrete without quilt, the temperature of the concrete in the depth of 10 cm is 5 ℃ lower at intervals of 1 h, 5.5 ℃ lower at intervals of 2 - 3 h and 6.75 ℃ lower at intervals of 4.5 h. Therefore, it can be seen that when the sun is direct and the temperature is 28 - 35 ℃, the pouring temperature can be reduced by 5 - 6 ℃ by covering the heat preservation quilt. In addition, when the measured temperature reaches 33 - 36 ℃, the temperature of the poured concrete covered with the heat preservation quilt at the depth of 15 cm after 1 - 3 h rises less by 2 - 4 ℃, 4 - 5 ℃ and 7 - 8 ℃ respectively than the concrete without heat preservation quilt. In other words, the concrete temperature rises only 0.5 - 2.0 ℃ after the concrete has been conversed with heat preservation quilt for 3 - 4 h, which fully shows that the covering of the heat preservation quilt plays a significant role in reducing the concrete temperature rise rate.

For the timely roller-compacting and timely covering pre-cooled RCC after being poured to the concrete placing area, it is a very effective and simple technical measure to prevent the pouring temperature rebounding. In the temperature control of controlling the concrete pouring temperature, we must not follow the old path of taking tonic and laxative, which does more harm than good!

10.3.4 Supply Water for Cooling and Reduce the Temperature Difference inside and outside the Dam Body

10.3.4.1 Effect of Embedded Cooling Water Pipe in Dam

A large number of temperature control simulation calculations and engineering water cooling results show that for the cooling water pipes buried inside the dam body, generally, the horizontal spacing between the internal water pipes is 1.5 m × 1.5 m, and the vertical spacing between the upper and lower layers is 3 m. Water shall be supplied one day for cooling after concrete is poured, and the water supplying time will generally be about 20 d, the highest temperature of the dam body can be reduced by 2−5 ℃, which shows that it is effective to take water cooling measures to reduce the internal temperature of the dam body.

Main functions of cooling water pipe embedded in dam concrete: Reducing the first-stage hydration heat temperature rise of concrete placing area and reducing the internal temperature of concrete during overwintering, so as to control the highest temperature of concrete and the temperature difference of foundation, reduce the internal and external temperature difference, and change the temperature distribution of dam body during construction.

The water cooling process of water pipes is generally divided into two stages, namely, the first stage cooling (initial water cooling) and the second stage cooling (later water cooling). The first-stage cooling is to supply water for cooling after the roller compacted concrete is poured. Water cooling should be strictly controlled during the roller-compacted process to prevent the cooling water pipe from being rolled and cracked, and leaking. There are many lessons in this respect. The function of first-stage cooling is to reduce the hydration heat temperature rise of concrete and control the maximum temperature of concrete. The purpose of the second-stage cooling is to reduce the maximum temperature of dam body concrete, so that the temperature of the upstream surface of the dam body can be reduced to the dam surface temperature required when the sluice is lowered for water storage, and cracks caused by excessive temperature difference on the upstream surface of the dam due to reservoir water cooling strike can be prevented.

10.3.4.2 Technical Requirements for Cooling Water Pipes

For the concrete of dam body to be cooled by water, cooling water pipes shall be buried according to the design requirements, and pressurized cooling water or natural river water shall be supplied to the cooling water pipes embedded in the concrete in order to cool the concrete of dam body. The temperature difference between the temperature of the dam body concrete and the cooling water should not be too large during water supply. Generally, the temperature difference should not exceed 25 ℃ and should not be greater than 0.5 ℃ every day, so as to avoid cracks caused by excessive internal temperature difference and too fast cooling speed of the dam body due to water supply. The engineering lesson in this respect is profound.

At present, container cooling units are mostly used for dam cooling. Black iron water pipes are basically eliminated from cooling water pipes, and a large number of high-density polyethylene plastic water pipes are used. Requirements are put forward for the design of cooling water pipes:

1. The plastic cooling water pipe is HDPE high density polyethylene plastic water pipe with an inner diameter of 28 mm and an outer diameter of 32 mm, and its indexes are shown in Table 10.7.
2. Before the cooling water pipe is buried in concrete, the inner and outer walls of the water pipe shall be clean and free of scale. The joint of the water pipe shall be expansive waterproof joint (binding type is mostly used). The length of a single circulating cooling water pipe is generally not more than 250−300 m. Special attention should be paid to the fact that the pre-arranged cooling water pipe cannot cross the transverse joint.
3. Cooling water pipes shall be arranged perpendicularly to the water flow direction. The horizontal spacing of cooling water pipes should be 1.5 m, and the vertical spacing should be the same as the pouring layer thickness. Water pipes are arranged at the bottom of each placing area. Before concrete pouring, water tests shall be conducted to check whether the water pipes are blocked or leaking. Water pipes shall be carefully protected to prevent the cooling water pipes from being displaced or damaged during concrete pouring or other work after concrete pouring, as well as during pipe tests. Pipe heads extending out of concrete shall be covered with caps or protected by other methods or by methods satisfactory to the supervisor.

Chapter 10 Temperature Control and Cracking Prevention of Roller Compacted Concrete

Table 10.7　　　　　　　　　　　Index of cooling HDPE plastic water pipe

Item		Unit	ndicator
Thermal conductivity factor		kJ/(m·h·K)	≥1.0
Tensile yield stress		MPa	≥20
Longitudinal dimensional shrinkage		%	<3
Destruction of internal hydrostatic pressure		MPa	≥2.0
Hydraulic test	Temperature 20 ℃ Time: 1 h Reversing stress: 11.8 MPa	Uncracked No leakage	
	Temperature: 80 ℃ Time: 170 h Reversing stress: 3.9 MPa	Uncracked No leakage	

4. During concrete pouring, circulating water at a pressure of not less than 0.18 MPa shall be supplied to the cooling water pipe to see if there is water seepage. Pressure gauge and flowmeter shall be used to indicate the resistance during concrete pouring at the same time. Water seepage and blockage shall be repaired before concrete pouring. If the cooling water pipe is damaged during the concrete pouring process, the concrete pouring shall be stopped immediately until the cooling water pipe is repaired and passes the test before pouring.

5. After the cooling water pipe was used, grouting and backfilling shall be carried out as required. The water pipe butt exposed out of the dam surface shall be cut off, and the left orifices shall be completely filled with mortar immediately.

10.3.4.3　General Requirements for Water Cooling

The first-stage water cooling is one of the measures to reduce the hydration heat temperature rise of the pouring layer. Refrigeration water is generally used for the first-stage water cooling, which can generally reduce the hydration heat temperature rise of concrete by 2 – 4 ℃. For the dam body poured with pre-cooled roller-compacted concrete, when the maximum temperature of the concrete may still exceed the designed allowable maximum temperature, the first-stage water cooling should be carried out to reduce the maximum temperature of the concrete and ensure that the maximum temperature of the dam body is controlled within the allowable range.

1. The temperature of the first-stage cooling water shall be changed according to the temperature of the concrete, so that the maximum temperature of the concrete does not exceed the allowable maximum value. Refrigeration water at 6 – 8 ℃ should be used for cooling, and river water can also be used if the cooling requirement is not high.

2. The water flow rate of the single water pipe in the first-stage water flowing shall not be less than 18 L/min, and the water flowing shall be started after compacting of the concrete and leveling of concrete. The water cooling time shall be controlled at about 20 d, the cooling water direction shall be changed once 24 h, and the cooling speed of the water shall not be greater than 1 ℃/d. At present, on the issue of cooling speed, according to the cooling situation of domestic projects, it is required that the water cooling speed should not be greater than 0.5 ℃/d, otherwise internal cracks will easily occur due to too fast cooling speed.

3. The second-stage water cooling temperature can be determined according to the dam joint grouting time and dam joint grouting temperature. Refrigeration water can be used when the water cooling time is short and the

dam joint grouting temperature is low, while river water can be used when the water cooling time is long and the dam joint grouting temperature is high. The difference between cooling water temperature and concrete temperature shall be controlled at 20 – 25 ℃.

4. Cooling water shall be clear water with little sediment content, and its flow rate and velocity shall ensure turbulence in the pipe. For water pipes with a diameter of 28 mm, the flow rate should be about 18 – 25 L/min.
5. In order to fully grasp the cooling and temperature-reducing situation of concrete, a proper amount of thermometers should be buried in the pouring block, and cooling water pipes can also be used in a planned way to measure the temperature of stuffy water. The stuffy time is generally 5 – 7 d.
6. Cooling water shall be kept clean and free of mud and debris. All necessary protective measures shall be implemented to prevent any part of the cooling system from being blocked or unusable due to other reasons.

10.3.5 Curing and Surface Protection to Prevent Surface Cracks

10.3.5.1 Curing Requirements

Roller-compacted concrete shall be cured according to design requirements or a combination of methods suitable for local conditions. The continuous curing time of roller-compacted concrete should not be less than 90 d or 180 d of the design age, so that the roller-compacted concrete can be kept at appropriate temperature and humidity within a certain period of time, which can cause good hardening conditions for the concrete and is the necessary measures to ensure the strength growth of the concrete and avoid surface cracking.

After the concrete is poured, the concrete surface and all sides shall be sprinkled with water for curing in time to keep the concrete surface moist frequently. To cure the concrete surface by flowing water is one of the effective measures to reduce the maximum temperature of concrete and it can reduce the early maximum temperature of concrete by about 1.5 ℃.

After the concrete is poured, it shall avoid sunlight exposure in the early stage, and the concrete surface should be covered. Generally, the concrete should be cured within 12 – 18 h after it is poured, but it should be cured in advance in hot and dry climate.

For the top surface concrete, after the concrete can resist the damage of water, immediately cover the concrete with water-holding material or use other effective methods to keep the concrete surface moist. The formwork and concrete surface shall be kept moist before and during the removal of the formwork by allowing the curing water to seep from the top of the concrete to the joints between the formwork and concrete to keep the surface moist. All these surfaces shall be kept moist until the formwork is removed. The concrete shall be cured with water continuously after the formwork is removed. The permanently exposed surfaces of the concrete shall be cured with flowing water for long term. The concrete shall be cured continuously. Sometimes dry and sometimes wet curing methods shall not be adopted during the curing period.

Specialized teams or special personnel shall be set up for curing concrete, and maintenance records shall be carefully made to truly and comprehensively record the concrete curing process and situation.

10.3.5.2 Surface Protection to Prevent Surface Cracks

Ⅰ Surface protection requirements

Practical experience shows that most of the cracks generated by mass concrete are surface cracks, but some of them will later develop into deep or penetrating cracks, which will affect the integrity and durability of the dam and cause great harm. The causes of surface cracks are dry shrinkage and temperature stress. Surface cracks caused by dry shrinkage can reach a depth of several centimeters, which can be mainly solved by curing. The temperature factors causing surface tensile stress are air temperature change, hydration heat and initial temperature difference. Temperature changes mainly include: sudden drop in temperature, annual and daily changes in temperature. No matter in the north or the south of our country, the sudden drop in temperature is an important cause of surface cracks. In the cold north, due to the low temperature in winter, the annual change of temperature is also an important cause of surface cracks and even deep cracks. The daily change of temperature has less influence due to the short change period. In the process of concrete construction, sometimes some gaps shall be reserved for water supplying and flowing. After the concrete contacts with low-temperature water, cracks often appear at the bottom and both sides of the gaps.

Both theoretical and practical experience show that surface protection is the most effective measure to prevent surface cracks, especially when the internal temperature of the first-stage concrete pouring is relatively high. The equivalent heat release coefficient value that the insulation layer shall reach after the concrete surface is insulated can be determined through calculation according to the sudden drop of temperature and annual change of temperature at the dam site.

According to the experience of the Three Gorges Project, the concrete surface protection requirements are as follows:

1. Thermal insulation material: materials with good thermal insulation effect and convenient construction shall be selected, and the equivalent heat release coefficient value β of the selected thermal insulation materials shall be checked. Equivalent heat release coefficient of thermal insulated concrete: mass concrete $\beta \leqslant 2.0 - 3.0 \ W/(m \cdot K)$; concrete of hole structure such as bottom hole and deep hole $\beta \leqslant 1.5 - 2.0 \ W/(m \cdot K)$.
2. For the concrete poured on the permanently exposed surface and in low temperature season (such as november to march), the concrete surface must be insulated after the formwork is removed.
3. Before the low temperature season comes every year, the entrances and exits of bottom holes, deep holes, draft pipes, vertical shafts, corridors and all other holes shall be shielded and blocked.
4. When the daily average temperature continuously drops by more than 6 ℃ within 2 - 3 d, the concrete surface (top side) must be insulated.
5. In low temperature season, if the temperature drop on the concrete surface may exceed the sudden drop period of 6 - 9 ℃ after formwork removal, the formwork removal time shall be postponed, otherwise other surface protection measures shall be taken immediately after formwork removal.
6. When the temperature drops below freezing point, the concrete with an age of less than 7 d shall be covered with high foamed polyethylene foam or other qualified thermal insulation materials as temporary thermal insulation layer.

Ⅱ Surface insulation material

Surface thermal insulation materials should be selected according to the calculation results of temperature stress. During construction, thermal insulation materials should be tightly attached to the concrete surface to prevent cold air convection. The dam surface of hydropower project has a short thermal insulation period and belongs to temporary project, but its requirements for wind resistance and water resistance are higher, therefore, different construction technology measures should be taken to protect the dam surface according to the different properties of different materials during thermal insulation construction. The thermal insulation materials used in hydropower projects in China include perlite, fiberboard, polyethylene, polystyrene, etc. At present, polyethylene coiled material and polystyrene thermal insulation board are most commonly used for concrete surface protection.

The construction methods of concrete surface protection materials can be divided into three types: spraying, internal pasting and external pasting. Spraying is to directly spray the thermal insulation material on the concrete surface with a spray board, and to form a thermal insulation layer with a certain thickness by foaming the material to form the surface protection of the concrete. The so-called internal pasting means that the thermal insulation material is pasted or fixed on the innerside of the erected formwork, and after the concrete is poured and the formwork is removed, the surface protection of the concrete is formed. The so-called external pasting means that after the concrete is poured and removed, the thermal insulation material is riveted or pasted on the concrete surface to form the protection in a time of duration.

Generally speaking, internal pasting is simpler than external pasting, and avoids aerial work, which is conducive to improving the quality of concrete surface protection. However, for concrete surface with higher flatness requirements, overflow dam surface or the concrete surface with other special requirements, external pasting is appropriate. In addition, when concrete poured in high temperature season needs surface protection only before low temperature season or cold wave comes, external pasting method should be adopted.

Ⅲ Surface insulation effect

Due to the adoption of thin-lift roller-compacting, the temperature of roller compacted concrete poured into the placing area rises rapidly. For example, if the pouring temperature in the south is 15 ℃ and the concrete is poured into the placing area, after it has been covered for 4 h, the pouring temperature during the daytime in summer can reach 26 – 27 ℃ and it can rise by more than 10 ℃. Therefore, it is very necessary to use spray, cover thermal insulation quilt and quickly cover to reduce the pouring temperature in high temperature season.

Insulation of concrete placing area during pouring in high temperature season is an important measure. Engineering practice shows that when 1.5 cm thick polyethylene thermal insulation quilt is used, the thermal insulation effect can be equivalent to 0.5 – 0.6 m thick concrete when the thermal insulation quilt does not hold water. After the concrete is rolled, the heat preservation quilt shall be used for heat preservation. After the concrete temperature is higher than the ambient temperature that the radiant heat is considered, the heat preservation quilt shall be removed to utilize surface heat dissipation. The maximum temperature of concrete poured in summer can be reduced by about 2 ℃ by using the placing surface insulation quilt.

10.3.6 Temperature Monitoring

If there are only temperature control measures and no necessary temperature measurement and monitoring methods, the effect of temperature control cannot be evaluated and the cause of cracks are not conveniently analyzed. Therefore, in order to ensure the effects of temperature control measures, it is required to carry out temperature observation in the whole construction period, to monitor the temperature control measures and to observe the internal structure of placed concrete. There must be special personnel to engage in these works in large and middle projects. The temperature observation includes observation in construction period, concrete highest temperature observation and the observation for the temperature changing process inside dam body.

The temperature observation in construction period. The temperature observation in construction period includes the observation for precooled aggregates, the observation for concrete just mixed out of the mixer, the observation for concrete just placed into formwork and the observation for concreting temperature. The observation for precooled aggregates is used to assess the precooling performance of aggregates, generally, to use infrared thermometer or thermometer to measure the temperature. The temperature observation for concrete and concreting temperature are also used to assess whether design requirement are met or not. Only when the above mentioned temperature controls in four links meet requirements, is the highest temperature of concrete can be controlled within the range of design requirements.

The observation for highest concrete temperature. Generally, instruments embedded inside the dam are used to observe the highest temperature of concrete, and temperature instruments include the differential resistance thermometer, optical fibre temperature detection. If the embedded instruments are insufficient, steel pipe can be embedded into concrete during placement, and then to put temperature thermometer inside the steel pipe to measure and observe the temperature until the concrete placement in the next step is finished.

10.4 Technical Innovation and Discussion on Temperature Control and Anti-cracking

10.4.1 The Discussion on Canceling the Water Cooling Pipe inside Dam

One of the advantages of RCC is to simplify or cancel temperature control. Early RCC dams have a low height, and make full use of low-temperature seasons and low-temperature periods for construction, and most of them do not take the temperature control measures. However, with the increasing of RCC dam height and volume in recent years, it has already been an routine construction activity to carry out continuous the placing works of RCC in high temperature period to accelerate the construction or to shorten the construction duration, in this case, the temperature control measures are becoming more increasingly strict, the temperature control measures in some RCC dam have no different from that in common concrete dam and are becoming more and more complicated. There are a large number of cooling water pipes embedded inside dam, which have negative impacts on the simple and fast construction for RCC.

Temperature measuring in large quantity of projects indicate that, generally, the concreting temperature is 3 – 5 ℃ higher than that out of mixing machine, the internal temperature rising inside concrete caused by hydration heat can be about 10 ℃ higher. For there are often incomplete formwork temperature measures and negligent management during construction, the precooled RCC loses its advantage and the temperature of RCC placed in formwork rises quickly, just like that cotton blanket in the sun can produce accumulation of heat. Therefore, concreting temperature during construction must be controlled strictly, and the RCC placed must be rolled, covered and cured timely to avoid temperature rising and flowing backward and to prevent placed concrete absorbing heat from ambient environment . It has been proved by the practices in projects that it is an important measure to preserve wet with spraying on block face to improve the combing quality between layers and to avoid concreting temperature rising and flowing back, and this measure can change the block face microclimate and can reduce temperature of block face by 4 – 6 ℃, which is beneficial to temperature control.

Among the measures to prevent concrete from cracking, it is an important method to promote concrete anti-cracking properties by increasing the limiting extended value. However, due to the properties of RCC, it is more harm than good to promote concrete anti-cracking properties by increasing the limiting extended value. Because the RCC has lower strength degree, and increasing the limiting extended value of RCC means reducing the water-cement ratio, and means increasing cement consumption and strength. But the increasing of cement consumption also causes the increasing of hydration heat rise accordingly, and the increasing in strength brings about corresponding increasing in concrete temperature stress and elasticity modulus, which are not good for temperature control. For the cement consumption has the direct relation with the adiabatic temperature rise of the concrete, generally, the hydration heat of cement is 293 kJ/kg, the adiabatic temperature rise is about 1 – 1.5 ℃ with every adding 10 kg/m^3 cement into concrete. For the concrete has the property of fragile material, the rising in limiting extended value does not has the direct relation with the rising in concrete strength, therefore, there is great difference between achieving higher anti-cracking property by increasing limiting extended value and the actual results. The Author holds that, it is appropriate to select the value of later strength on the long period for the calculation of anti-cracking.

The temperature control in RCC has comprehensive relations with anti-cracking, either factor cannot be emphasized partially, they are complementary with each other. For instance, there are differences in temperature control measures between that in low temperature and that in second-rate high temperature in Baise project. In low temperature condition, to take measures on the basis that the highest temperature in dam is 38 ℃, to place concrete into formwork with natural temperature control measures, and to cancel the cooling water pipe in low temperature condition. It is worth learning to take temperature control measures without imposing one single solution in RCC temperature control.

Cooling water pipes embedded in formwork have great interference in the fast and simple construction of RCC, and it is challenging to take temperature control measures without cooling water pipes inside dam. The cancellation of the cooling water pipe does not mean absence of control to the temperature of the RCC, but the technical route of the temperature control measures is different. The temperature control measures are mainly taken in the temperature control of the RCC before placing and during rolling, to strictly control the placing temperature to comply with the standard. The main technical route to cancel the cooling water pipe temperature control measures:

Firstly, do not pour RCC in high-temperature seasons and periods, try to use low-temperature seasonsor periods to pour RCC, to effectively reduce the maximum temperature rise of dam concrete.

Secondly, strictly control the temperature at the outlet of the RCC. The control of temperature at the outlet is conducted with the conventional temperature control method, that is, control the temperature of cement and fly ash into the tank, pre-cool aggregates, and mix with cold water or ice. . Although there is a certain cost to control the temperature of the RCC at the outlet, it can significantly reduce the interference to the rapid construction of the on-site RCC.

Thirdly, it is required to control the temperature rise in RCC strictly, the key point to prevent temperature rising is to roll timely and to change microclimate by preserving block face wet with spraying in time. To cure the concrete and to cover insulation materials timely after rolling concrete, so as to prevent temperature rising and concrete cracking, these are very effective measures.

The author has profound understanding that, it is very important to carry out temperature control for RCC before rolling, which can greatly reduce the construction interference caused by cooling water pipes. It has been proved by many practices in a large number of projects that, preserving wet with spraying on block face can change microclimate clearly and can obviously reduce temperature by 4 – 6 ℃. The observation documents in Jin'anqiao indicate if wet preservation with spraying on block face is not covered in time or is not covered at all, the RCC will accumulate heat in large quantity in high temperature period or in the sun exposure, which can increase the concreting temperature quickly, and the RCC will exceed standard requirement seriously. The observation data indicates that the temperature inside RCC dam in which concreting temperature is beyond standard is 3 – 5 ℃ higher than that of RCC placed in low temperature period or after preserving wet with spraying.

The temperature control has turned into one of the key factors which constrains the fast construction of RCC dam. It needs study and analysis for the temperature control standards and technical methods for RCC dam, it is required to break the temperature control deadlock and to change the passive situation, and it is a new task to be solved to find out one best combined point between temperature control standard and fast construction.

10.4.2 The Impacts of Construction Quality on Temperature Control and Anti-cracking

The construction quality is particularly important for temperature control and anti-cracking in RCC dam, thus, it is key to carry out construction in accordance with temperature control design requirements in temperature

Chapter 10 Temperature Control and Cracking Prevention of Roller Compacted Concrete

controlling. Main temperature control measures taken in RCC include optimizing the mix proportion, reducing cement content, increasing limiting extended value and reducing elasticity modulus. Strictly control the concreting temperature, precooled aggregates, mixing with ice, concrete temperature out of mixing machine, increasing placing intensity, rolling inclined layers, to control the temperature rising again in concrete transportation, placing, rising of concreting temperature. Preserve wet with spraying on block face to change microclimate in block face, and to cover and cure timely to prevent cracking on surface. Reduce the maximum temperature rising insidedam with water pipe cooling and to control the temperature difference between dam inside and outside.

In the process of RCC construction, due to unreasonable construction period, or in order to accelerate the construction, the cooling systems are insufficient, or the precooled concrete cannot meet design requirements. The cracking causes huge loss, though the temperature control measures can save some investments. However, it is too late and the consequence is serious, the dam stress due to temperature increase rapidly. For instance, some arched RCC dam in China has cracking due to temperature control measures out of control, and the lesson is profound in this respect.

The RCC is rolled with thin-lift and continuous ascending in concrete construction in China, which has broken through the restriction that it is not appropriate to construct RCC in high temperature period. Therefore, in recent years, the embedded cooling water pipes are used in most RCC dam, the cooling water pipe made of HDPE (High Density Polyethylene) can reduce the highest concrete temperature by 2 – 4 ℃. To take advantage of the favorable time in low temperature season to place the RCC, in which concreting temperature can be reduced, thus the highest temperature of the dam concrete can be reduced, and the investment into the concrete temperature control measures can be saved.

The pouring method of the RCC has certain influence on the dam temperature. When the RCC is rolled with thin-lifts in continuous way or there is bigger lifting height in concrete construction, it has lower heat dissipation effect, the hydration heat rising in RCC is close to adiabatic temperature rise, and the highest temperature inside dam is higher accordingly. However, placing concrete in this way can reduce works in layer surface handling and can promote the construction speed. When the RCC is rolled with thin-lifts at short intervals and there is lower lifting height in concrete construction, it has higher heat dissipation effect (it is not suitable to take this method in high temperature period for preventing the temperature from flowing back), the hydration heat rising in RCC and the highest temperature inside dam can be reduced. However, placing concrete in this way can increase works in layer surface handling and can reduce the construction speed. The RCC has heat dissipation in the interval time between two layers construction, which has close relation with concreting temperature, climate conditions and temperature. The heat dissipation effect in the high temperature period in the daytime is not good, and temperature flowing back is often produced, which is bad for temperature control. Therefore, the placing method of the RCC must be determined under the overall consideration of block face size, concreting temperature, construction season, construction progress requirements, construction equipment, available temperature control measures and temperature control analysis.

The temperature features in impermeable area on upstream face. The cement consumption in the II gradation RCC and the GEVR in impermeable area on upstream face of dam is 30 – 50 kg/m^3 higher than that in the concrete inside dam. Test results indicate that the concrete adiabatic temperature rise in impermeable area is obviously higher than that inside dam, the heat dissipation rate is high, the highest temperature in temperature control calculation often occurs in the range of 2 – 8 m close to the upstream water-side face. If water cooling measures were not taken, the temperature in the impermeable area on upstream water-side face is 3 – 5 ℃ higher than that in center location.

Considering the thoughts about spreading mortar on the impermeable area on upstrean face, the technical measure of spreading mortar or grout on every layer is taken to improve the binding between layers and impermeable performance in some projects. The temperature rise of hydration heat in concrete is increased invisibly, which is harmful to temperature control and anti-cracking and also is not economical. The bonding and impervious technology between layers is mainly of the fully bleeding on the concrete surface after rolling, which can enable the aggregates in upper layer can embed into the lower rolled concrete, this is the best technical measure to guarantee the interlayer bonding and imperviousness.

10.4.3 Technical Innovation in Surface Anti-cracking by "To Dress Dam with Clothes"

Cracks are a common problem in concrete dams. The dam with cracks has long troubled people. For a long time, we take a series of measures for anti-cracking and crack resistance in concrete dam from design, construction to management, including blocking of concrete dams, improving concrete anti-cracking properties (by using moderate heat cement, mixing with flyash, additives, multi-gradation, etc.), temperature control and surface curing, etc.. However, the reality is still that "All dams have cracking".

The temperature and temperature stress of RCC are different from those of normal concrete dams. Although RCC dams have low hydration heat and fewer cracks on the dam surface in the early stages, the high temperature in the RCC dam lasts for a long time, so that the surface of the dam is weathered by low temperature, cold wave, exposure, drying and wetting, causing the temperature difference between the inside and outside of the dam, and easy appearance of surface cracks. The period of more surface cracks in RCC dams is usually after the dam is built. The RCC dam is a layer joint structure with multiple layers and low tensile strength between layers. In case of water storage or low water temperature of the RCC dams, and cold shock of water or temperature drop, the temperature difference between the inside and outside of the dam is large and easily causes layer or horizontal cracks, even vertical cracks.

In the thesis "To end the history of *All Dams Have Cracking'* with *Overall Temperature Control and Heat Preservation in Long Time* written by Academician Zhu Bofang, scientific analysis is conducted for the fundamental reason of "All dams have cracking", the author holds that insufficient protection on the dam is one important reason, and there is no sound protection for the exposed upstream and downstream surface of dam for long time.

Considering that the high temperature inside RCC can last for long time and the dam has a large quantity of layers, in order to prevent big temperature difference from causing surface cracking, in accordance with the top coating imperviousness material used on the RCC dam upstream surface in China in recent years and the inspiration about overall temperature control from Academician Zhu Bofang, the author holds that it is required to conduct technical innovation for dam surface protection, namely "To dress dam with clothes", the polymer cement flexible waterproofing coatings is adopted to bring the dam into overall protection, which can get two folds results with half the effort.

The polymer cement waterproofing coatings is one new kind of waterproofing material started in developed countries, such as European countries, the USA, Japan and the South Korea, this kind of material has excellent properties in water resisting, durability and weather resistance, and it is attractive in market. The polymer cement waterproofing material is made of special cement modified by organic polymer emulsion and is supplemented with a variety of additives, and it is produced with special technology. The compound design mechanism of this coating is based on that the organic polymer emulsion turns into elastic film with cohesiveness and continuous property after water loss, meanwhile, the cement will absorb the water in the emulsion and turns into hardened cement product after hydration reaction, the flexible polymer film interlaces with hardened cement product and tightly coheres into one solid and flexible waterproofing layer. The hardened cement product can effectively change the deficient property of polymer film swelling with water, enables the waterproofing layer possess organic material property with high toughness and high resilience, and possess the inorganic material property with good waterproofing and high strength.

Prevention of cracks on the surface of RCC dams shall be focused on, because most of cracks of RCC dams are surface cracks. Under certain conditions, they can develop into deep cracks that are often uneasily dealt with. Therefore, strengthening the surface protection of concrete dams is very important. Coating the surface of the dam with a polymer cement flexible waterproof material with a thickness of 3 – 5 mm can play a good role in surface protection, that is, it can prevent seepage and cracks, keep moisture and insulation, also can enhance the durability of concrete, with simple and fast construction and low cost. Compared with conventional XPS insulation plate for concrete surface protection, the polymer cement flexible waterproof material has great advantages.

10.4.4 Brief Summary on Temperature Control and Anti-cracking Innovation

Compared with the temperature control measures used in common concrete, the measures used in RCC are simplified, but also they are complicated in some aspects. To summarize the researches and applications in recent years as follows:

1. The RCC has lower adiabatic temperature rise, which is conducive to control temperature. However, it has low ultimate tensile, particularly the resistance to tension between layers is low, which goes against anti-cracking.
2. Regarding that the RCC has slow hydration heat rise and low strength in early period, it is supposed to make full use of the strength in long term. It is appropriate to take the 180 d or 360 d for the design age of concrete, meanwhile, to match the design index as much as possible (such as the limiting extended value).
3. The temperature control standard is focused on the temperature difference between inside and outside, the restriction area in dam upstream face is influenced by the temperature difference between inside and outside and the foundation temperature difference, the foundation temperature in middle and lower stream region can be relaxed appropriately.
4. Currently, the temperature control measures are becoming more and more complicated, all the successful temperature control measures used in common concrete dam are almost adopted in RCC dam.
5. Rolling RCC in oblique layer is mainly for the pouring capacity, for the rolling in oblique layer can reduce the covering time, which reduce the heat flowing back and can be effective in temperature control. The simulation analysis results indicate that if the covering time of rolling in oblique layers was reduced from 4 h to 2 h, the highest temperature can be reduced by 1 – 2 ℃ in the condition that temperature control measures are not taken in block face.
6. The RCC has the advantage of simplifying or canceling temperature control, and it is a new issue to be solved how to find the best combined point between temperature control and fast construction.
7. The method of "To dress dam with clothes" with polymer cement flexible waterproofing coatings can provide all-round protection on the dam surface in simple and fast way, and it is an effective technical measure to prevent the cracking on dam surface.

10.5 The Temperature Control in RCC Gravity Dam in One Hydro-junction Project

10.5.1 Project Overview

The main dam in one hydro-junction project is RCC gravity dam, the dam crest is 1 570 m in length, including: terrace dam section at left bank, bank slope dam section at left bank, dam section on riverbed, bank slope dam section at right bank, terrace dam section at right bank, totally 87 dam sections, the maximum dam height is 121.50 m and the concrete volume is 2 892 000 m^3. $1^{\#}$ and $2^{\#}$ auxiliary dams are of earth dam with clay core, which are located on the Level IV terrace platform on the right bank. The maximum height of $1^{\#}$ auxiliary dam is 14.5 m with dam crest 150 m in length, and the maximum height of $2^{\#}$ auxiliary dam is 12.0 m with dam crest 485 m in length. The release structure is located in the dam section of riverbed, in which surface outlet, bottom outlet and middle outlet are arranged from left side to right side. The surface outlet has 4 holes with weir length 12 m on each hole, the elevation of weir crest is 730.00 mm, totally 5 dam sections are occupied. The middle outlet has one hole, the elevation of inlet floor is 690.00 m, the hole size is 5 m × 5 m, totally one dam section is occupied. The bottom outlet has one hole, the elevation of inlet floor is 660.00 m, the hole size is 4 m × 4 m, totally one dam section is occupied. The continuous flip bucket is adopted in release structure to carry out energy dissipation.

The region where this project is located has the perennial mean temperature 2.7 ℃, the extreme highest temperature is 40.1 ℃, and the extreme lowest temperature is −49.8 ℃, and the annual average precipitation is 183.9 mm, the perennial mean evaporation capacity is 1 915.1 mm, the average wind velocity in many years is 1.8 m/s with maximum wind velocity 25 m/s, the maximum snow thickness is 75 cm, the maximum frozen earth depth is 175 cm. There are only 7 months effectively available for construction in one year, the weather and climate conditions are very bad.

The key technology in engineering design of RCC main dam is about the dam temperature control and anti-cracking measures, and this is a challenging technology in the dams of the same kind at home and abroad because of its technical difficulty. Through survey, researching, theoretical analysis and many scientific researches, thermal insulation material is paved on the upstream and downstream dam faces of the dam body, and the above-mentioned problems are solved, which provide successful experience for the temperature control and anti-cracking for RCC in severe cold region.

10.5.2 The Mix Design of RCC in Construction

The Contractor makes further optimization testation for the actual raw material on the basis of the mix design recommended by the Research Institutes in the early stage, it proposes the technical route of "To reduce sand rate, to add rock powder, to reduce *VC* value", and carries out site raw material test and concrete construction mix design test respectively, meanwhile, the quality testing department from the Project Owner also conducts review test. Following the submitted RCC mix design, the Construction Organization has carried out rolling technology test two times on site. Through the technology test, the properties of the mixed RCC from the batching plant is tested, including the value loss and separation during RCC transportation, offloading and spreading, as well as the rolling availability and compactness on site. The Contractor makes improvement for the problems found on site and to determine the mix design of RCC in construction. According to the sections of main dam concrete in the design, the design indexes of RCC see Table 10.8, and the mix design of RCC in dam construction see Table 10.9.

Table 10.8   The concrete design indexes of RCC dam

Zone No.	I-1	I-2	II-1	II-2	II-3	III
Location	The area above the dead water level in upstream and the outside of water level changing area in downstream	External concrete below upstream dead water level	External concrete above the water level change area at downstream side	RCC above 650 m internal elevation	RCC below 629 m internal elevation	The concrete in the foundation pad and GEVR in bank slop pad
Gradation	II.	II.	III.	III.	III.	III.
Design index	$R_{180}200W10F300$	$R_{180}200W10F100$	$R_{180}200W6F200$	$R_{180}150W4F50$	$R_{180}200W4F50$	$R_{28}200W8F100$
Density (kg/m^3)	$\geq 2\,380$	$\geq 2\,380$	$\geq 2\,380$	$\geq 2\,380$	$\geq 2\,380$	$\geq 2\,380$
Ultimate tensile	$>0.78 \times 10^{-4}$	$>0.78 \times 10^{-4}$	$>0.70 \times 10^{-4}$	$>0.70 \times 10^{-4}$	$>0.70 \times 10^{-4}$	$>0.80 \times 10^{-4}$
VC value	5-8	5-8	5-8	5-8	5-8	5-8
Initial setting (h)	12-17	12-17	12-17	12-17	12-17	12-17

Note: The design indexes of GEVR in each part can reach the design indexes of adjacent concrete.

10.5.3 The Simulation Calculation of Temperature Control for RCC

This project is located in one area of middle and high latitude, and in the back-land of Eurasia, where the weather conditions are harsh. It is very cold in winter and the winter lasts long, and it is hot and dry in summer, it winds frequently at ordinary times. The extreme lowest temperature is -49.8 ℃, and the extreme highest temperature is 40.1 ℃, the perennial mean temperature is 2.7 ℃. The annual evaporation capacity is 1 915.1 mm. The main environmental factors restricting the temperature control include "Cold, Heat, Wind and Dry" in this project. The difficulties in temperature control in RCC dam mainly are as follows:

1. For there is long time of minus temperature, the time available to carry out RCC construction is short, and it is inevitable to construct in high temperature period, which causes difficulties not only in construction intensity, but also in the RCC temperature control in high temperature working period.
2. According to the need of flood control, the breach flow must be reserved on the dam for flood control, which has disadvantage influence on the temperature control in overflowing section in the dam.
3. Based on the arrangement of construction duration, the dam construction period will cover two winters, it is also the difficult part to carry out temperature control and protection in winter.
4. For the temperature inside RCC is unable tot be reduced for long time, and the dam is in the status of higher temperature. Thus, it is required to control the low environment temperature influence on dam temperature stress to prevent the harmful cracking from occurring in construction period and in a considerably long operation period.

With regard to the above-mentioned difficult problems, the Project Owner and Design Institute have authorized Guiyang Hydropower Design Institute, Xi'an University of Technology, Hehai University and China Institute of Water Resources and Hydropower Research etc. to carry out researches in temperature control, and

Table 10.9 Mix proportion of dam RCC construction

Pouring location	Design requirements	Gradation	Water-cement ratio	Fly ash (%)	Sand ratio (%)	Rock powder (%)	Water reducing agent (%)	Gas-attracting agent (%)	Material amount (kg/m³)								Density (kg/m³)	VC value (s)	Gas content (%)
									Water	Cement	Fly ash	Rock powder	Sand	Coarse aggregate					
														5–20	20–40	40–80			
I-1	$R_{180}200F300W10$	II	0.45	40	35	4	0.85	0.12	98	131	87	29	690	533	800	0	2 370	1–3	5–6
I-2	$R_{180}200F100W10$	II	0.47	55	35.5	7	0.85	0.07	95	91	111	52	684	534	802	0	2 370	1–3	4–5
II-1	$R_{180}200F200W6$	III	0.45	50	32	6	0.85	0.10	90	100	100	40	634	430	574	430	2 400	1–3	4.5–5.5
II-2	$R_{180}150F50W4$	III	0.56	62	30	10	0.85	0.06	90	61	100	64	580	451	601	451	2 400	1–3	3–4
II-3	$R_{180}200F50W4$	III	0.53	62	30	10	0.85	0.06	90	65	105	64	577	449	599	449	2 400	1–3	3–4
III	$R_{90}200F100W8$	III	0.45	30	29	8	0.85	0.08	90	140	60	49	562	449	599	449	2 400	1–3	4–5
III*	$R_{28}200F100W8$	III	0.45	30	30	0	0.75	0.004	112	174	75	0	606	423	565	423	2 380	3–5	4–5

Note: 1. The cement used in the Dam Zone I and Zone II in 2007 is P·O 42.5 ordinary Portland cement with PMS-3 retarding high-range water reducing agent. The P HSR 42.5 sulphate resistant cement and PMS-3 A retarding high-range water reducing agent are used in the blinding concrete in Zone III. PMS-NEA 3 air entraining agent and fly ash from Manasi Power Plant are adopted in all mix proportions.

2. The cement used in the concrete of Zone I and Zone II in 2008 is the special P·O 42.5 ordinary Portland cement (the specific surface area $(310\pm10)\,\text{m}^2/\text{kg}$). P HSR 42.5 sulphate resistant cement is used in the blinding concrete in Zone III. FDN retarding high-range water reducing agent and PMS-NEA 3 air entraining agent are used in the dam concrete.

3. The concrete in Zone III is the normal concrete, which is conducted to level the concrete cushion, with P HSR 42.5 sulphate resistant cement and NF-1 high-range water reducing agent adopted, and the number in the VC value column is the slump value.

Chapter 10 Temperature Control and Cracking Prevention of Roller Compacted Concrete

has authorized China Institute of Water Resources and Hydropower Research to conduct simulation, tracking and feedback calculation analysis in construction period to guide project construction. The main research includes: the RCC material research; the research of measures for temperature control structures of RCC main dam; the research of the temperature stress simulation calculation of the RCC main dam; the research of the construction process, temperature control measure and the temperature control standard in different construction periods; the research of the protection measures of the overwintering layer; the research of the long-term thermal insulation measure of the main dam; the research of the temperature monitoring of the dam body; and the analysis and research of the temperature control tracking and feedback of the RCC main dam.

Through the above researches, the temperature control standard of the RCC main dam and the temperature control measure adopted are determined initially, and the effectiveness of the temperature control measures can be verified by the prototype monitoring, tracking and feedback research.

10.5.3.1 Structure Measures to Prevent Temperature Cracking

1. In order to prevent the vertical crack, the dam section length of the main riverbed dam section and the bank slope dam section is 15 m, while the length of the terrace dam section are respectively 15 m or 20 m according to the different dam height.
2. The length-width ratio of the weak constrained foundation layer is adopted to broaden the allowable temperature difference of the concrete in the foundation restraint area, so that a longitudinal joint is set in the foundation layer of the main river bed dam section, which can reduce the fundament block length from 98 m to 49 m, and the joints are combined by galleries.
3. Based on the experience of the horizontal cracks generated in the overwintering layer of Guanyinge RCC Gravity Dam, a piece of metal horizontal water stop is set in the upstream side of the overwintering layer to strengthen the horizontal seepage control.

Above structure measures have positive effects to prevent the harmful cracks, lower the foundation constraint stress and strengthen the imperviousness performance.

10.5.3.2 Simulation Analysis of Temperature Field and Stress Field during Construction and Operation Period of Main Dam

Ⅰ Calculation model

The $31^{\#}$ dam section is taken as the typical dam section for analysis and calculation, the whole dam section placed in 2007, 2008 and 2009 is taken as the research object, and the calculation model is taken along the dam axis for 15 m. The foundation scope is as follows: taking 100 m in the upstream of the dam heel and the downstream of dam toe respectively, and taking 100 m in depth.

The origin of coordinates of the overall computation coordinate system is in the dam heel of the dam section, x-axis is the water flow direction, and the forward direction points from upstream to downstream; y-axis is the vertical flow direction, and the forward direction points from the right bank to left bank; the forward direction of z-axis is plumbs upward.

When calculating the stress field, the foundation bottom surface of the calculation model is the fixed bearing, the foundation in the upstream and the downstream direction is the simple support along x-axis, and both boundaries of the foundation along the dam axis is the simple support along y-axis.

3D finite element model used in calculation can be seen in Figure 10.1.

Ⅱ Construction and calculation boundary of typical dam section

①Concrete construction of typical dam section

The $31^{\#}$ dam section is taken as the typical dam section, and the placing part placed in 2007 has an elevation of 624.00 – 645.00 m; The placing part in 2008 is: Elevation 645.00 – 696.00 m; The placing part in 2008 is: 696.00 – 745.50 m elevation. The placing thickness of each layer is 3 m with the maximum placing

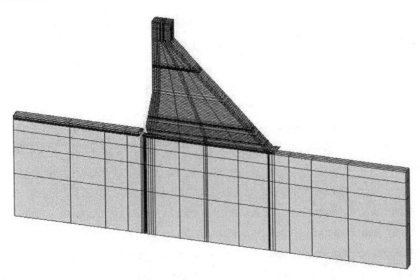

Figure 10.1 *Calculation model diagram of $31^{\#}$ dam section placing block*

temperature as 22 ℃ for total placing time of 886 d. The formal placing is started on April 20, 2007.

②Concrete material properties

The thermal parameters of the concrete material are shown in Table 10.10, due to the lack of corresponding test data, the thermal parameter and adiabatic temperature rise of the sulphate resistant RCC refer to these of the concrete in Zone Ⅰ -1. The RCC adiabatic temperature rise is shown in Table 10.11, and the allowable stress of dam body concrete in different ages are shown in Table 10.12.

Table 10.10   **Test results of concrete thermal performance**

Mixing proportion No.	Concrete positions and strength grades		Pecific heat (kJ/(kg·℃))	Hermal conductivity factor (kJ/(m·h·℃))	Oefficient of thermal conductivity (m²/h)	Inear expansion coefficient (10⁻⁶/℃)
1	External RCC	Region Ⅰ-1 R_{180}20W10F300	0.951	8.49	0.003 8	9.25
2		Region Ⅰ-2 R_{180}20W10F100	0.918	8.23	0.003 7	9.19
3		Region Ⅱ-1 R_{180}20W6F200	0.902	8.38	0.003 8	9.12
4	Internal RCC with an elevation above 650	Region Ⅱ-2 R_{180}15W4F50	0.897	8.57	0.003 5	9.01
5	Internal RCC with an elevation under 650	Region Ⅱ-3 R_{180}20W4F50	0.884	8.34	0.003 8	8.96
6	Region Ⅲ R_{28}20W8F100 Foundation cushion concrete		0.982	8.59	0.003 6	9.38

Chapter 10 Temperature Control and Cracking Prevention of Roller Compacted Concrete

Table 10.11 Test results of adiabatic temperature rise of dam body RCC

Mixing proportion No.	Concrete positions and design indexes		Adiabatic temperature rise value (℃)							Final adiabatic temperature rise	Fitting formula
			1 d	3 d	5 d	7 d	14 d	21 d	28 d		
1	External RCC	Region Ⅰ-1 $R_{180}20W10F300$	0.0	5.3	7.5	18.8	20.5	21.6	23.1	24.29	$T = 24.29\,d/(2.06+d)$
2		Region Ⅰ-2 $R_{180}20W10F100$	0.9	4.5	6.0	17.1	19.0	20.4	21.5	22.68	$T = 22.68\,d/(2.08+d)$
3		Region Ⅱ-1 $R_{180}20W6F200$	0.5	3.1	5.2	16.3	17.9	19.5	20.8	21.9	$T = 21.9\,d/(2.20+d)$
4	internal RCC	Region Ⅱ-2 $R_{180}15W4F50$	5.4	9.2	10.7	12.3	13.9	15.2	16.1	17.42	$T = 17.42\,d/(2.84+d)$
5		Region Ⅱ-3 $R_{180}20W4F50$	5.5	9.3	10.8	12.5	14.1	15.3	16.3	17.61	$T = 17.61d/(2.82+d)$
6	Ⅲ Foundation cushion concrete $R_{28}20W10F100$		8.6	17.1	22.0	24.6	27.1	28.2	29.1	31.6	$T = 31.6\,d/(2.37+d)$

Table 10.12 Allowable stress of concrete in different ages

Concrete positions and design indexes	Allowable horizontal stress or the primary stress (MPa)			Allowable vertical tensile stress on upstream surface (MPa)		
	28 d	90 d	180 d	28 d	90 d	180 d
External concrete of the water level fluctuation area in Zone 1-1 $R_{180}200W10F300$	1.33	1.76	2.24	1.01	1.38	1.61
External concrete under the upstream dead water level in Zone 1-2 $R_{180}200W10F100$	1.19	1.67	2.01	0.95	1.18	1.40
External concrete above the downstream water level fluctuation area in Zone 11-1 $R_{180}200W6F200$	1.27	1.71	2.15	—	—	—
Internal Ⅲ gradation concrete of the dam body in Zone 11-2 $R_{180}150W4F50$	0.90	1.33	1.71	—	—	—
Internal Ⅲ gradation concrete of the dam body in Zone 11.3 $R_{180}200W4F50$	0.95	1.41	1.81	—	—	—
Normal concrete for foundation leveling layers in ZoneⅢ $R_{90}200W8F100$	1.88	—	—	—	—	—
Normal concrete for outlet structures in ZoneⅣ $R_{90}250W8F200$	2.16	—	—	—	—	—

Ⅲ Calculation of boundary conditions

The following boundary conditions are adopted in the calculation of 31[#] dam section placing block:

1. The initial temperature of the foundation is the measured ground temperature of 31[#] dam section in April.
2. The transverse joint surface is treated as the adiabatic boundary.
3. The placing temperature of the concrete in 2007 is the measured on-site data, and the placing temperature of the concrete in 2008 and 2009 refer to the placing temperature of the concrete at the same time in 2007.

4. After placing, the upstream and downstream surface of the placing block in 2007 have been covered with 2 cm thick polyurethane foam quilts for temporary thermal insulation before backfilling, which is considered according to the Class III boundary conditions; the upstream and downstream surfaces of the placing block of 2008 and 2009 are also covered with 2 cm thick polyurethane foam for temporary thermal insulation before permanent thermal insulation.
5. The placing surface is covered with 2 cm polyurethane foam quilts for temporary thermal insulation, and cured by "spraying" method from May to September each year during construction. According to the measured data in 2007, the water temperature of "spraying" was about 15 ℃ in April, 18 ℃ in May, and 24 ℃ from June to August, and 18 ℃ in September.
6. The concrete placed from May to October in 2007, 2008 and 2009 shall be cooled by river water (stage I) one day after the placing for 15 d. The concrete placed from June to August each year shall be cooled by water (stage II) from October 1 – 15 in the same year.
7. The first backfilling time of the upstream surface was in the middle of October 2007, for the concrete placed before the middle of October, 5 cm thick XPS boards shall be pasted on the upstream surface before backfilling the slope deposits for thermal insulation; After the concrete placed in late October in 2007, the slope deposits backfilled on the upstream surface shall reach to the elevation of 645.00 m. In addition, the upstream surface within the range of 2 cm under the elevation 645.00 m shall be pasted with 10 cm thick XPS boards for thermal insulation.
8. The upstream surface and the downstream surface of the placing blocks of 2008 and 2009 shall be pasted with 10 cm XPS boards for permanent thermal insulation. The pasting method of the permanent insulation boards on the upstream and downstream in 2008 and 2009: the insulation boards shall be pasted on the upstream and downstream surfaces of the concrete placed before the end of August, and for the concrete placed in September and thereafter, the permanent insulation boards shall be pasted immediately after placing.
9. The 26 cm thick cotton quilts are used on the overwintering surface every year for thermal insulation, and the equivalent heat emission coefficient is 15.37 kJ/($m^2 \cdot d \cdot K$). Before the placing of the concrete in the next year, the thermal insulation quilts of the overwintering surface were gradually removed, and the detailed procedures are shown in Table 10.13.

Table 10.13　Gradual removal process of the thermal insulation quilt on overwintering surface

Date	Number of the layer of removed thermal insulation quilts	Thickness of the removed quilt (cm)
2008(2009)-03-08	1	2
2008(2009)-03-17	1	2
2008(2009)-03-20	1	2
2008(2009)-03-23	1	2
2008(2009)-03-27	2	4
2008(2009)-03-29	2	4
2008(2009)-04-01	1	2
2008(2009)-04-03	1	2

The remaining thermal insulation quilts have been completely removed in April 5.

10. Water temperature. the underground seepage temperature from October 2007 to September 2008 is shown in Table 10. 14; the water temperature after impounding in September 2008 is calculated as the reservoir water temperature specified by the specification.

Table 10. 14   Monthly underground seepage temperature

Month	10	11	12	13	1	2	3	4	5	6	7	8	9
Water temperature(℃)	7.0	5.0	5.0	5.0	5.0	5.0	5.0	7.0	9.4	13.5	17.3	17.2	12.0

11. Temperature. the calculation period is from April 20, 2007 to January 11, 2062. In which the temperature from April 20, 2007 to December 4, 2010 is the daily average temperature measured on site in 2007, and the air temperature of other periods was the ten-day average air temperature.

10.5.3.3 Analysis of Calculation Results

Ⅰ Typical profile temperature field

The distribution of the maximum temperature during dam construction can be seen in Figure 10. 2.

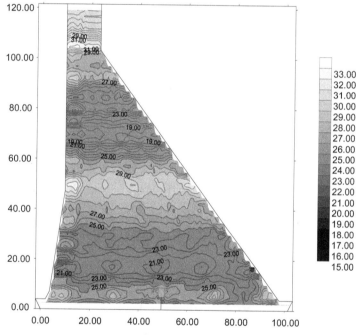

Figure 10. 2 *Envelope diagram of the maximum temperature of dam profile during construction* (℃)

The total construction duration of the dam is 3 years, in concrete placing positions from June to August in each year, there are 3 corresponding high-temperature areas in the dam body with the maximum temperature about 30 – 32 ℃ in the local area of each high-temperature area, this is because the outside temperature during this period is higher, the placing temperature can reach to 20 – 22 ℃, although the cooling measures like pipe cooling, surface covering and spraying are taken, it is still difficult to avoid the high-temperature area.

Ⅱ Typical profile stress field

The envelope diagrams of the daily maximum stress during the dam construction (April 20, 2007 to March 30, 2010) are shown from Figure 10. 3 to Figure 10. 5. The envelope diagrams of the maximum stress in the cross section during dam operation (March 31, 2010 to January 30, 2062) are shown from Figure 10. 6 to Figure 10. 8.

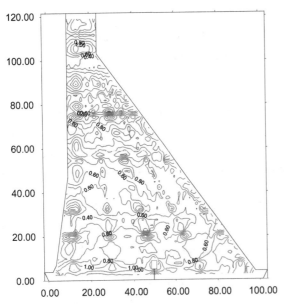

Figure 10.3 *Envelope diagram of maximum horizontal stress σ_x along water flow direction during construction* (MPa)

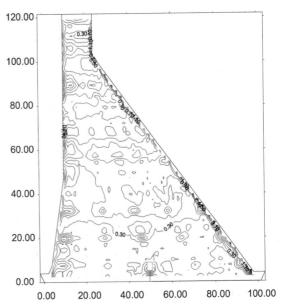

Figure 10.4 *Envelope diagram of maximum horizontal stress σ_y along vertical water flow direction during construction* (MPa)

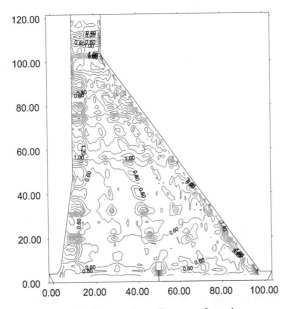

Figure 10.5 *Envelope diagram of maximum vertical stress σ_z during construction* (MPa)

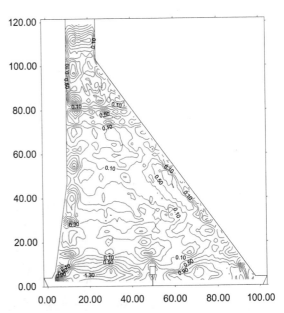

Figure 10.6 *Envelope diagram of maximum horizontal stress σ_x along water flow direction during operation* (MPa)

As we can see from the above envelope diagrams of the maximum temperature and the maximum stress during construction and operation:

1. Stress distribution on the upstream and downstream surfaces of the dam: The stress on the upstream surface is well controlled, and all comprehensive stresses on the upstream surface of the dam is basically controlled within 1.8 MPa, except those in the concrete nearby the upstream dam heel area. Even if in the vicinity of the wintering surface with high stress, the comprehensive stress is basically controlled at about 1.8 MPa, which is close to the allowable stress value. Stresses on the downstream surface of concrete blocks poured in 2008 and 2009 exceeded the standard during the construction period, but it is generally distributed at the depth 50 cm below the surface, which would not impact the safety of the dam.

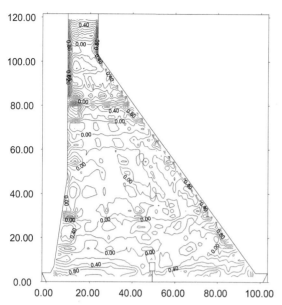

Figure 10.7 *Envelope diagram of maximum horizontal stress σ_y along vertical water flow direction during operation* (MPa)

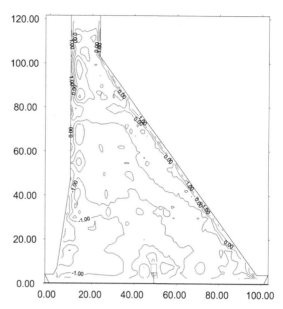

Figure 10.8 *Envelope diagram of maximum vertical stress σ_z during operation* (MPa)

2. At the middle part of the dam, the horizontal stress σ_x of Ⅲ gradation concrete in the direction of water flow is large; Near the upstream and downstream surfaces, the stress perpendicular to the water flow direction σ_y and vertical stress σ_z of the concrete in the Ⅱ gradation area are large. Generally, σ_y is greater than σ_z in the strong constraint area of dam foundation, but σ_z is generally larger than σ_y at the place above the strong constraint area of dam foundation.

3. Stresses near the dam heel and dam toe are very great during construction and operation, mainly because these parts are located in the strong constraint area of the foundation and are influenced by the factor of stress concentration.

Ⅲ Analysis on temperature and stress change processes of typical points during construction and operation

In order to study the temperature and stress change law of the upstream and downstream surface concrete of dam body, and the concrete at the Ⅱ and Ⅲ gradation areas during construction and operation, typical points are selected for analysis.

1. Analysis on temperature and stress change process of typical points of upstream and downstream surface concrete. Figure 10.9 and Figure 10.10 respectively show the change process lines of temperature and stress of typical points on the upstream surface at elevations of 656.00 m and 633.00 m.

    It can be seen from the temperature and stress change process curve of typical points at upstream and downstream surfaces that: temperature and stress change sharply at the initial stage of construction, but during the operation period, the temperature and stress of the upstream surface concrete show periodic changes along with the periodic changes of air temperature or water temperature, and the change of concrete stress has obviously negative correlation with the change of concrete temperature, and the numerical value of stress is also greatly reduced compared with that during the construction period.

    According to the temperature process lines, the concrete near the upstream and downstream surfaces is in quasi-stable temperature field 10 years after pouring, it is no longer affected by the initial factors such as placing temperature and adiabatic temperature rise during construction of dams, but affected only by the periodic change of external air temperature or water temperature. After entering the quasi-stable temperature field, the periodic change curves of the temperature of the parts (underwater and above water level) of the

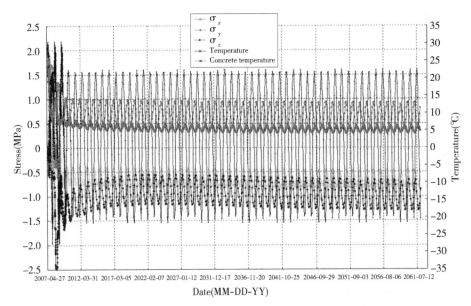

Figure 10.9 *Change process lines of temperature and stress of upstream surface at elevation of 629.00 m (pouring on June 13, 2007)*

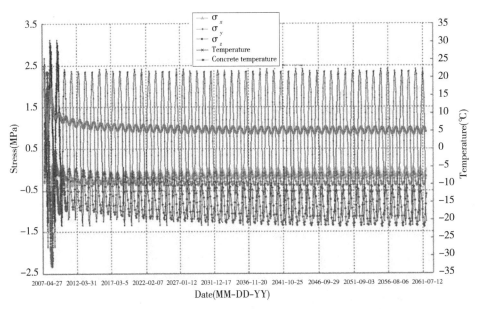

Figure 10.10 *Change process lines of temperature and stress of downstream surface at elevation of 633.00 m (pouring on September 4, 2007)*

upstream and downstream surfaces are respectively with smaller amplitude (that is about 1 ℃ in case that 5 ℃ is taken as a core temperature) and larger amplitude (that is about 5 ℃ in case that about 3 ℃ is taken as a core temperature).

2. Analysis on change processes of temperature and stress of typical points in upstream and downstream Ⅱ gradation concrete areas. Figure 10.11 shows the change process lines of temperature and stress of typical points in upstream Ⅱ gradation concrete area at elevation of 640.00 m.

According to the change process lines of temperature and stress of typical points in upstream Ⅱ gradation concrete area of the dam, Ⅱ gradation concrete on the upstream and downstream surfaces is in quasi-stable temperature field after about 15 years of pouring, after that, the temperature field of Ⅱ gradation concrete area is affected only by the periodic change of external air temperature or water temperature.

Chapter 10 Temperature Control and Cracking Prevention of Roller Compacted Concrete

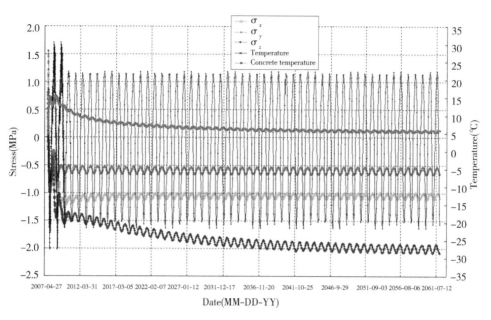

Figure 10.11 *Change process lines of temperature and stress of typical points in upstream* Ⅱ *gradation concrete area at elevation of* 640.00 **m** (*poured on October* 14, 2007)

After the concrete in upstream and downstream Ⅱ gradation area is in quasi-stable temperature field, its stress changes periodically with temperature, and its maximum stress basically does not change with time. The maximum stress near upstream and downstream surfaces fails to exceed 1.0 MPa and 1.5 MPa respectively, which meets the design requirements.

3. Analysis on change processes of temperature and stress of typical points in Ⅲ gradation concrete area inside the dam. Figure 10.12 shows the change process lines of temperature and stress of typical points in Ⅲ gradation concrete area inside the dam.

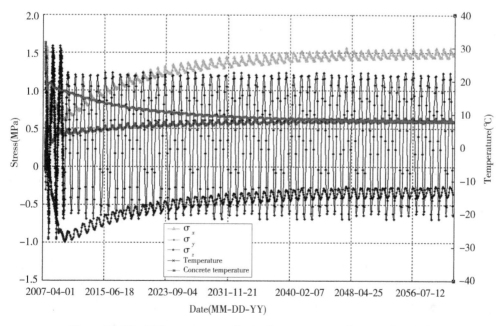

Figure 10.12 ∇*Change process lines of temperature and stress of typical points in* Ⅲ *gradation concrete area in the middle of the dam of* 631.0 **m** (*pouring on July* 29, 2007)

According to the change process lines of temperature and stress of typical points in Ⅲ gradation concrete area in the middle of the dam, the concrete inside the dam is basically in stable temperature field about 50 years after pouring, and the temperature basically does not change. In the process of entering stable temperature field in operation period from construction period, the stress increases gradually with gradual decrease of temperature within the first 25 – 30 years, but its increase rate is very slow due to the effect of concrete creep. However, 30 years later, the stress only fluctuates periodically in a very small amplitude, and basically does not increase.

According to the maximum stress of Ⅲ gradation concrete inside the dam in construction and operation periods, the maximum stress and allowable stress of Ⅲ gradation concrete in other parts of the dam are not large in construction and operation periods except that those of the concrete under strong restraint in the range of 4 m at the bottom are in the same order of magnitude; in operation period, the maximum stress is less than 1.0 MPa, which meets the design requirements.

The main reason for exceeding stress of the concrete in the range of 4 m at the bottom after 5 years of pouring as being in the strong restraint area of foundation is that the concrete of this part is too strongly restrained by the foundation. Therefore, it is necessary to set up longitudinal induced joints in the range of 5 m at the bottom, which has certain effect on reducing the temperature stress at the bottom of placing blocks.

10.5.3.4 Simulation Calculation Conclusion

1. Based on the changes of temperature and stress in construction and operation periods of the dam, the temperature and stress of the upstream and downstream concrete surfaces and the upstream and downstream Ⅱ gradation concretes of the dam, and Ⅲ gradation concrete inside the dam change dramatically in the first two years in construction and operation periods after taking actual temperature control measures on construction site; after that, the temperature and stress changes tend to be stable with passage of time, and the longer the time, the slower the change.
2. The concrete near the upstream and downstream surfaces is in quasi-stable temperature field 10 years after pouring, after that, the periodic change curves of the temperature of the parts (underwater and above water level) of the upstream and downstream surfaces are respectively with smaller amplitude (that is about 1 ℃ in case that 5 ℃ is taken as a core temperature) and larger amplitude (that is about 5 ℃ in case that about 3 ℃ is taken as a core temperature).
3. Ⅱ gradation concrete is in quasi-stable temperature field about 15 years after pouring. After that, the stress fluctuates with temperature, and the maximum stress basically does not exceed previous level in operation period, or 1.0 MPa and 1.5 MPa in upstream and downstream Ⅱ gradation concrete areas respectively, which meets the design requirements.
4. The concrete inside the dam is basically in stable temperature field about 50 years after pouring. From construction period to operation period, the temperature in the dam gradually decreases, and the stress increases slowly within the first 25 – 30 years; the concrete in the range of 4 m at the bottom will show exceeding stress 5 years after pouring as being in the strong restraint area of foundation; the maximum temperature stress inside the dam is basically within 1.0 MPa, but the stress inside the dam tends to be stable 30 years after pouring.

In combination with the above analysis, it can be concluded that, after taking the current temperature control measures, the temperature stress of local parts is slightly larger in construction period of the dam, but, in operation period, the dam stress increases slowly except for the area in the range of 4.0 m at the bottom, and the maximum stress of the upstream surface and the upstream Ⅱ gradation concrete area of the dam, as well as inside the dam is basically controlled within 1.0 MPa. Only the downstream surface of the dam is nonconforming locally in construction period, but the exceeding depth is not large, and the cracks can be controlled within 50 cm; in operation period, the stress of the downstream surface of the dam is basically controlled within 1.5 MPa. Therefore, in construction and operation periods of the dam, the temperature control standards and measures adopted for the dam are basically feasible.

10.5.4 Temperature Control in Construction Period

10.5.4.1 Design of Temperature Control Standards

I Foundation temperature difference

According to the actual method statements and temperature control measures on site, the foundation temperature difference controlled in pouring of RCC dam is shown in Table 10.15.

Table 10.15   Foundation temperature difference of dam (construction stage)

Locations of dam body	Foundation temperature difference of II gradation concrete (5 ℃)	Foundation temperature difference of III gradation concrete (9 ℃)
Strong restraint area of foundation	20.0	19.0
Weak restraint area of foundation	25.0	24.0

II Temperature difference between upper and lower layer

Temperature difference between upper and lower layer: $\Delta T \leqslant 15.0$ ℃.

III Internal and external temperature difference

Temperature difference between inside and outside of concrete: $\Delta T_{inner\ surface} \leqslant 19.0$ ℃.

IV Maximum allowable temperature of the dam is shown in Table 10.16.

Table 10.16   Maximum temperature control index designed for dam concreting   Unit: ℃

Locations of dam body	Maximum temperature of II gradation concrete	Maximum temperature of III gradation concrete
Strong restraint area of foundation	≤26.0	≤25.0
Weak restraint area of foundation	≤31.0	≤30.0
Unconstrained zone	≤32.0	≤31.0

V Placement temperature

Placing temperature designed is shown in Table 10.17.

Table 10.17   Concreting temperature control index   Unit: ℃

Month	10 days	Placement temperature	
		Strong restraint area of foundation	Weak restraint area and non-restraint area of foundation
4	Up	9.0	9.0
	Middle	13.5	13.5
	Down	13.5	13.5
5	Up	15.0	15.0
	Middle	17.0	18.0
	Down	17.0	18.0
6	Up	17.0	19.0
	Middle	17.0	20.0
	Down	17.0	20.0

Continued to Table 10.17   Unit: ℃

Month	10 days	Placement temperature	
		Strong restraint area of foundation	Weak restraint area and non-restraint area of foundation
7	Up	17.0	21.0
	Middle	17.0	22.0
	Down	17.0	22.0
8	Up	17.0	21.0
	Middle	17.0	20.0
	Down	17.0	17.0
9	Up	16.5	16.5
	Middle	16.0	16.0
	Down	15.0	15.0
10	Up	10.0	10.0
	Middle	10.0	10.0
	Down	9.0	9.0

10.5.4.2 Main Measures and Effects of Temperature Control

I Control of placing temperature

In order to control the placing temperature, the following temperature control measures have been taken: take out aggregates from the bottom of the gallery when stacking height is greater than 8 m; carry out secondary air cooling for aggregates; cover charging conveyor with a sunshade; prepare cooling water for concrete mixing in high temperature seasons; strengthen the organization and management of concrete transportation to accelerate concreting speed. In high temperature time, sun-shading measures are taken for carrier vehicles to reduce heat intrusion.

Table 10.18 shows the statistics of measured placing temperature at construction site in high temperature periods in 2007. The average daily placing temperature is lower than the allowable placing temperature, which meets the design requirements. However, when instantaneous temperature is above 30 ℃, the instantaneous placing temperature of corresponding individual points is nonconforming. The above situation shows that the measures taken to control the placing temperature are effective.

Table 10.18   Measured placing temperature in high tyemperature period in 2007   Unit: ℃

Date	Placement surface temperature		Placement temperature	
	Daily average	Maximum value	Daily average	Maximum value
2007-07-10	28.1	31	20.2	21
2007-07-11	27.2	34	20.2	23
2007-07-12	21.3	25	19.9	21
2007-07-17	25.4	28	22.0	24
2007-07-21	26.8	37	19.8	25
2007-07-22	26.6	34	20.9	24
2007-07-27	26.8	32	19.4	22.5

Continued to Table 10.18 Unit: ℃

Date	Placement surface temperature		Placement temperature	
	Daily average	Maximum value	Daily average	Maximum value
2007-07-28	18.7	22	19.4	21
2007-07-29	18.8	19	18.0	18
2007-08-02	26.9	30	18.9	21.5
2007-08-03	26.7	31.5	18.6	22
2007-08-04	23.2	25	17.7	19
2007-08-06	22.5	29	19.0	21
2007-08-07	20.0	20	20.0	20
2007-08-08	24.8	28	18.0	19
2007-08-09	24.0	33	19.2	24
2007-08-10	23.9	33	20.1	24
2007-08-11	26.0	26	20.0	20
2007-08-12	18.6	21	19.1	21
2007-08-13	18.8	26	19.2	22
2007-08-17	19.7	29	19.5	21
2007-08-18	18.0	20	17.7	18
2007-08-19	20.3	27	17.9	21
2007-08-20	22.1	25	17.4	19
2007-08-21	22.0	24	17.6	19
2007-08-22	17.5	20	17.5	19.5
2007-08-23	15.5	16	16.8	17
2007-08-25	27.5	29	17.0	17

Ⅱ Mist spraying to placement surface

The humidity has a great impact on the interlayer bonding of RCC, so the environmental humidity of an area with the dam fluctuates greatly. According to the measured humidity from June 22 to June 25 in 2007, the minimum and maximum humidity reaches 10% – 20% and 70% – 80% respectively. In terms of daily humidity distribution, the humidity reaches approximately more than 40% from 1:00 to before 13:00. The humidity is the maximum from 5:00 to 8:00, reaching more than 50%; the humidity reaches approximately less than 30% from 14:00 to 19:00. In one day, the relative humidity from 14:00 to 19:00 is relatively low.

During concreting of the dam, the environmental humidity is improved by water spraying to placing surface. According to the actual measurement data on site, the placing surface temperature at spraying decreases by about 4 ℃ on average compared with that at non spraying in high temperature seasons. The temperature drop is closely related to the wind speed. The relative humidity of the placing surface can reach more than 40% by using hand-held spray gun. It can be found that the spraying has certain effect on improving microclimate of concreting surface. As the spraying cooling is temporary, the temperature will rise and the humidity will decrease in case of stopping spraying. Spraying is beneficial to replenishing the lost water of the surface paved with concrete, which significantly improves the bleeding effect of the compacted layer surfaces and facilitates the bonding of layer surfaces. The effective range of each hand-held spray gun is about 500 m². Therefore, the high pressure spray gun should be provided for RCC placing surface based on foregoing effective range in high temperature time.

Ⅲ Spraying and polyurethane insulation cover effect

After the surfaces of the concrete rolled compacted layers are finished, a polyurethane insulation cover with thickness of 2 cm is used for temporary thermal insulation. After the initial setting of the concrete, the sprayer (with water jet radius of 30 – 40 m) is used for wet curing immediately. In construction period, "spraying" method is adopted for curing from May to September every year. According to the measured data in 2007, the water temperature of "spraying" was about 15 ℃ in April, 18 ℃ in May, and 24 ℃ from June to August, and 18 ℃ in September; spraying can increase humidity and reduce temperature of the environment.

1. Impact of spraying on humidity. By field measurement, the humidity of the parts near the ground in spraying raindrop control area and in the area adjacent to raindrop control area respectively reaches more than 60% and about 40%, so the concrete curing condition of the dam surface is greatly improved. After 30 min of spraying, the humidity decreases from 41.2% to 36.6%, decreasing by about 5%, which shows that the spraying has a relatively significant effect on improving humidity.

2. Cooling effect of spraying. In high temperature time in high temperature periods, the stop time of spraying equipment should not exceed 30 min, especially in high temperature time from 11:00 to 17:00 every day, the spraying must be strengthened.

    According to the measured results of $35^{\#} - 37^{\#}$ dam sections on July 15, 2007, the concrete surface temperature can be controlled in 25 – 27 ℃ (that is about 6 ℃ lower than temperature) when the temperature is about 33 ℃, and the concrete surface under the insulation cover is with a certain ponding. The concrete surface temperature under the insulation cover can be controlled in 25 – 31 ℃ (that is about 3 ℃ lower than temperature) when the temperature is about 33 ℃, and the concrete surface under the insulation cover is moist.

    The insulation cover can prevent concrete from dry shrinkage and external heat from intrusion in high temperature time and the cold impact on concrete surface by sudden drop of air temperature. According to the field observation data and simulation calculation results during construction in 2007 and 2008, the "spraying + insulation cover" can effectively prevent heat intrusion and surface cracks, and minimize the adverse effects of "heat" "wind" "drying" and cold wave on concrete layer surface.

Ⅳ Water cooling of water pipe

Through simulation calculation and analysis in construction period, it is determined that the concrete poured from May to October in 2007, 2008 and 2009 shall be cooled by setting cooling water pipe (Phase Ⅰ). The spacing of concrete water pipes in Ⅱ gradation concrete area and inside the dam should be 1.0 m (horizontal) × 1.5 m (vertical) and 1.5 m (horizontal) × 1.5 m (vertical) respectively. The cooling water pipes should be made of high-strength polyethylene pipes with inner and outer diameter of 30 mm and 32 mm respectively, and the length of each volume of cooling water pipe is 200 m. The starting time and the duration of water filling are respectively 1 day after completion of pouring and generally 15 d. The water filling rate is about 15 – 20 L/min, and the direction of water filling is reversed once a day.

In order to prevent the cracks around pipe wall caused by cold impact on concrete due to relatively low water temperature in cooling water pipe, it is determined that the water temperature shall be controlled from 15 – 18 ℃ at initial water filling by the simulation calculation.

According to the simulation calculation results, in addition to Phase Ⅰ water filling, the concrete poured

Chapter 10 Temperature Control and Cracking Prevention of Roller Compacted Concrete

in June, July and August of the same year is cooled by river water (Phase II) at the beginning of October every year, so as to lower the temperature gradient of the concrete near upstream and downstream surfaces during overwintering, decrease the temperature stress of the upstream and downstream surfaces, and reduce the risk of cracks on the upstream and downstream surfaces. The water temperature at Phase II water filling is controlled from 15 - 20 ℃.

The monitoring data shows that, after starting water cooling, the average temperature of the concrete around (0.75 m) and near (0.1 m) the cooling water pipe in II gradation concrete area is respectively 34.64 ℃ and 25.05 ℃, with a temperature difference of 9.59 ℃, and the one in III gradation concrete area is respectively 31.64 ℃ and 22.10 ℃, with a temperature difference of 9.54 ℃. The temperature difference shows that the cooling water pipe has a significant effect on absorbing the concrete heat of the dam and reducing the temperature peak. The concrete temperature drop rate is 0.465 ℃/d and 0.19 ℃/d respectively in the case of 15 d of water filling and non-water filling of $35^{\#}$ and $36^{\#}$ dam sections respectively. Based on the measured data, Phase I cooling enables the reduction of internal temperature of concrete by about 4.4 ℃, so it has a significant effect on reducing the internal temperature of concrete.

By simulation analysis of $37^{\#}$ dam section, the concrete temperature drop rate in II gradation and III gradation concrete areas is respectively about 0.5 ℃/d and 0.4 ℃/d during 15 - 21 d of water filling.

Measured results are consistent with the simulation calculation results of temperature drop effect of the cooling water pipe.

10.5.4.3 Comparison of Temperature Monitoring Results with Simulation Calculation Results

In order to evaluate the correctness of the simulation calculation methods and results, 3 thermometers arranged in the key monitoring dam section ($35^{\#}$ dam section) are selected as examples to compare the calculation results with the actual monitoring results. Embedding locations of such 3 thermometers are shown in Table 10.19.

Table 10.19 Statistics on basic information of thermometers selected

Instrument name	T3-6	TB4-1	T3-14
Embedding elevation of thermometers (m)	631.50	641.0	658.0
Distance between thermometer and dam axis (m)	61	-2	18
Remarks	In III gradation concrete, and being embedded on August 29, 2007.	In II gradation concrete of overwintering surface, and being embedded on October 11, 2007.	In III gradation concrete, and being embedded on May 26, 2008

I Comparison of calculated values with measured values

T3-6 thermometer. The process line on comparison of calculated values with measured values is shown in Figure 10.13. The measured values are counted according to actual monitoring frequency, and the calculated values are calculated once a day. At the same time, the change process line of the calculated stress is attached in Figure 10.13.

Table 10.20 shows statistics on measured temperature, calculated temperature and stress on the date with actual monitoring data of T3-6 thermometer. According to the table, the difference between calculated temperature and monitoring temperature is generally less than 2 ℃; the maximum temperature difference is 2.38 ℃. In later period, the calculated temperature is higher than the actual monitoring temperature.

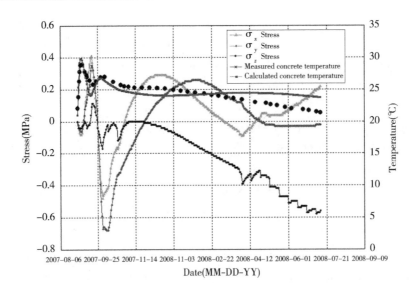

Figure 10.13 *Change process lines of measured temperature, calculated temperature and stress of concrete at the embedding position of T 3-6 Thermometer*

Table 10.20 Statistics of measured temperature, calculated temperature and stress of concrete at the embedding position of T3-6 thermometer

Date	Measured concrete temperature (℃)	Calculated concrete temperature (℃)	Difference value (℃)	Σ x (MPa)	Σ y (MPa)	Σ z (MPa)	Primary major stress (MPa)
2007-08-29	22.05	21.00	1.05	0	0	0	0
2007-08-30	23.80	25.37	-1.57	-0.02	-0.02	-0.04	-0.02
2007-08-31	26.30	27.86	-1.56	-0.05	-0.04	-0.04	-0.04
2007-09-01	27.80	29.13	-1.33	-0.07	-0.05	-0.04	-0.05
2007-09-02	28.80	29.70	-0.90	-0.08	-0.06	-0.04	-0.04
2007-09-04	28.90	29.55	-0.65	-0.07	-0.04	-0.03	-0.03
2007-09-08	27.80	27.40	0.40	0.08	0.08	0	0.09
2007-09-10	27.30	26.35	0.95	0.16	0.12	-0.02	0.16
2007-09-11	27.00	25.82	1.18	0.20	0.14	-0.04	0.20
2007-09-14	26.40	24.13	2.27	0.36	0.23	-0.02	0.36
2007-09-19	25.80	24.71	1.09	0.33	0.32	0.11	0.33
2007-09-23	26.30	25.89	0.41	0.09	0.13	0.04	0.13
2007-09-29	26.90	26.79	0.11	-0.38	-0.50	-0.13	-0.13
2007-10-04	27.00	26.41	0.59	-0.44	-0.67	-0.08	-0.08
2007-10-15	26.20	25.47	0.73	-0.22	-0.45	-0.01	-0.01
2007-10-25	25.70	24.85	0.85	-0.10	-0.30	-0.08	-0.06
2007-10-30	25.50	24.70	0.80	0.00	-0.25	-0.03	0.02
2007-11-5	25.40	24.57	0.83	0.08	-0.19	0	0.09
2007-11-13	25.30	24.42	0.88	0.15	-0.12	0	0.16
2007-11-23	25.30	24.25	1.05	0.23	-0.03	0	0.23

Continued Table 10.20

Date	Measured concrete temperature (℃)	Calculated concrete temperature (℃)	Difference value (℃)	Σx (MPa)	Σy (MPa)	Σz (MPa)	Primary major stress (MPa)
2007-12-03	25.30	24.13	1.17	0.27	0.05	-0.01	0.27
2007-12-13	25.20	24.05	1.15	0.29	0.12	-0.03	0.29
2007-12-23	25.10	24.01	1.09	0.29	0.17	-0.05	0.29
2008-01-03	24.90	24.01	0.89	0.27	0.21	-0.07	0.27
2008-01-13	24.80	24.04	0.76	0.23	0.23	-0.10	0.23
2008-01-22	24.60	24.07	0.53	0.20	0.25	-0.12	0.25
2008-02-01	24.50	24.13	0.37	0.15	0.26	-0.15	0.26
2008-02-13	24.30	24.20	0.10	0.10	0.25	-0.18	0.25
2008-02-22	24.10	24.26	-0.16	0.06	0.23	-0.21	0.23
2008-03-02	24.00	24.32	-0.32	0.03	0.20	-0.23	0.20
2008-03-12	23.70	24.37	-0.67	-0.01	0.16	-0.26	0.16
2008-03-22	23.60	24.41	-0.81	-0.04	0.13	-0.28	0.13
2008-04-02	23.40	24.42	-1.02	-0.09	0.11	-0.39	0.11
2008-04-19	23.00	24.39	-1.39	0.00	0.02	-0.33	0.02
2008-05-03	22.80	24.34	-1.54	0.05	-0.02	-0.34	0.05
2008-05-09	22.60	24.31	-1.71	0.04	-0.02	-0.41	0.04
2008-05-17	22.50	24.27	-1.77	0.04	-0.03	-0.42	0.04
2008-05-28	22.20	24.20	-2.00	0.06	-0.03	-0.47	0.06
2008-06-05	22.00	24.11	-2.11	0.08	-0.03	-0.50	0.08
2008-06-20	21.80	23.95	-2.15	0.13	-0.03	-0.53	0.13
2008-06-28	21.70	23.86	-2.16	0.15	-0.0	-0.56	0.15
2008-07-08	21.50	23.74	-2.24	0.20	-0.02	-0.58	0.20
2008-07-13	21.30	23.68	-2.38	0.22	-0.02	-0.56	0.22

TB4-1 thermometer. The change process lines of measured temperature, calculated temperature and stress at the embedding position of TB4-1 thermometer are shown in Figure 10.14. According to the figure, the measured temperature at the initial stage of concreting fluctuates greatly. After about 10 d, the difference between calculated temperature and monitoring temperature is generally less than 3.6 ℃. In monitoring time, the calculated temperature is always lower than the actual monitoring temperature, and the development trend of the calculated and measured temperature is consistent.

TB4-2 thermometer. The change process lines of measured temperature, calculated temperature and stress at the embedding position of TB4-2 thermometer are shown in Figure 10.15. According to the table, the measured temperature at the initial stage of concreting fluctuates greatly. After about 30 d, the difference between calculated temperature and monitoring temperature is generally less than 2 ℃. In monitoring time, the initial calculated temperature is always higher than the actual monitoring temperature, but the later calculated temperature is lower than the measured temperature.

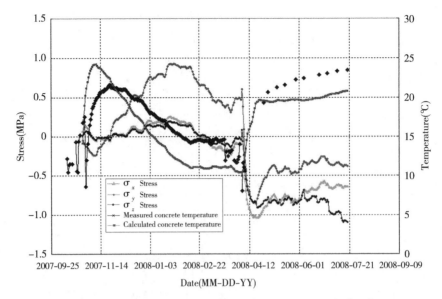

Figure 10.14 *Change process lines of measured temperature, calculated temperature and stress of concrete at the embedding position of TB 4-1 thermometer*

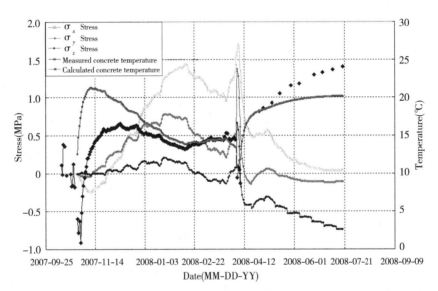

Figure10.15 *Change process lines of measured temperature, calculated temperature and stress of concrete at the embedding position of TB 4-2 thermometer*

Ⅱ　Result analysis

According to the measured and calculated results of T3-6 and TB4-2 thermometers, the calculated results are properly consistent with the measured data, indicating that the simulation analysis model and the calculation method adopted are basically reliable.

10.5.4.4　Evaluation of Temperature Control Measures in Construction Period

1. In construction period, the reasonable temperature control measures adopted for concrete mixtures ensure that the concrete placing temperature meets the specifications.
2. Various temperature control measures in concreting and curing periods ensure the control of environmental temperature and humidity, especially, the artificial spraying measures have a significant effect on thermal insulation and moisture preservation in the event of special "cold, heat, the insulation cover has a positive effect on preventing backward temperature flow and cold wave attack, and is conducive to thermal insulation and moisture preservation.

3. Phase Ⅰ cooling of cooling water pipe enables the reduction of the internal temperature of concrete by about 4 ℃, which has a significant effect on cutting temperature peak and homogenizing temperature field, thereby effectively improving the temperature field distribution of the dam and reducing the temperature stress level.
4. The monitoring data shows that the calculated results are in properly consistent with the measured data, indicating that the simulation analysis model and the calculation method are basically reliable.

10.5.5 Initial Test Results of Permanent Thermal Insulation and Insulation of Overwintering Layer Surface

10.5.5.1 Permanent Thermal Insulation of Upstream and Downstream Surfaces of the Main Dam

The upstream and downstream surfaces of the dam, being exposed for a long time, are insulated by permanent insulation materials to form a permanent insulation layer, achieving thermal insulation and moisture preservation. The specific scheme is described as follows:

Upstream surface: The surfaces above and below the ground are respectively of different thermal insulation and impervious structures, namely, polyurethane impervious coating (with thickness of 2 mm) + sticking XPS board (with thickness of 10 cm) and "polyurethane impervious coating (with thickness of 2 mm) + sticking XPS board (with thickness of 5 cm) + backfilling slope deposits".

Downstream surface: it is of the thermal insulation structure of sticking XPS board (with thickness of 10 cm) + external coating of anti-crack polymer mortar (with thickness of 1 – 1.5 cm).

According to the inspection results of concrete of RCC main dam constructed in 2008, the thermal insulation effect is better, and no cracks are found on the upstream and downstream surfaces of the main riverbed dam section.

10.5.5.2 Thermal Insulation of Overwintering Layer Surface

Before the beginning of winter, the surface where the surface where concreting is stopped and the exposed surface of the dam side shall be subject to thermal insulation and overwintering treatment. According to the simulation calculation results, the specific scheme adopted is as follows:

Layer surface of the dam where concreting is stopped: lay a layer of plastic film (with thickness of 0.6 mm) on the storage layer surface, and then lay two layers of 2 cm thick polyethylene insulation covers on the plastic film, and then lay a cotton quilt on the polyethylene insulation covers. The total thickness of the insulation cover is not less than 26 cm. Finally, 1 layer of three-proofing (waterproof, oilproof and antifouling) canvas is laid on the top.

Protection of side surface of the dam: stick 10 cm thick XPS board and spray 5 cm thick polyurethane on the side surface of the dam.

Thermal insulation of the top of upstream and downstream surfaces (by insulation board): the "corner" where the overwintering surface connects the upstream and downstream surfaces should be protected emphatically because its "two-direction heat dissipation" causes fast temperature drop in winter. The range of 2.6 m under the overwintering surface should be sprayed by 15 cm thick polyurethane rigid foam again based on the thermal insulation of upstream and downstream surfaces.

Ⅰ Determination of the thickness of the insulation cover on overwintering surface

According to research results of simulation calculation, just 20 cm cotton quilts is enough according to the winter temperature changes of the construction site in recent years in the event of only making the stress of overwintering surface conforming during overwintering. However, in order to prevent horizontal cracks on the concrete surface under extremely low temperature, it is necessary to cover the surface with 26 cm thick cotton quilts.

During the thermal insulation by 26 cm thick cotton quilts, the stress changes on the overwintering surface under the extreme condition of −40 °C for 10 d are calculated as follows:

For the upstream corner of the overwintering surface, the temperature drops rapidly in the first 3 d of −40 °C, with a temperature drop by 0.49 °C and 0.33 °C respectively every day. During the last 7 d, the concrete temperature continues to drop, and the daily temperature drop is 0.27 − 0.18 °C. According to the stress change process of the upstream corner during this period, the stress of concrete increases rapidly with the decrease of temperature in the continuous process of low temperature lasting for 10 d, and reaches the maximum value at the end of extreme low temperature, but does not exceed 1.0 MPa.

For II gradation concrete, the stress of concrete increases rapidly with the decrease of temperature in the continuous process of low temperature lasting for 10 d, and reaches the maximum value at the end of extreme low temperature, but does not exceed 1.0 MPa.

For III gradation concrete, if it encounters the low temperature of −40 °C lasting for 10 d in January, the allowable tensile stress of the concrete is about 1.1 − 1.3 MPa because the concrete age is only 2 − 3 months. However, during this period, the horizontal stress of large area of III gradation concrete can reach 1.2 − 1.3 MPa, there is a risk of longitudinal cracks in III gradation concrete of the overwintering layer.

II Removal time of thermal insulation quilts on overwintering surface

The removal time of the overwintering thermal insulation quilts is an important issue of thermal insulation in severe cold areas. The simulation results show that when the overwintering thermal insulation quilts are removed, the temperature difference between inside and outside (temperature difference between inside and outside = concrete temperature-outside temperature) should not be too large; otherwise, the concrete stress will increase too much upon the removal of the quilts, which may lead to cracks on the overwintering surface. The optimum removal time of the thermal insulation quilts is when the outside temperature is higher than or equal to the temperature of the concrete on the overwintering surface, thus eliminating the adverse effects of air "cold shock" during removal. If the quilts must be removed when the outside temperature is lower than the concrete temperature of the overwintering surface, the temperature difference between inside and outside should not exceed 3 °C as per the calculation. The specific quilt removal time in 2008 should be determined according to the actual temperature change in April of that year and the measured temperature of concrete on the overwintering surface.

The results show that it is beneficial for the construction in 2008 to gradually remove the thermal insulation quilts and completely remove the thermal insulation quilts as soon as possible. In order to avoid the cold shock on the overwintering surface, the thermal insulation quilts were gradually removed. The removal time of thermal insulation quilts on the overwintering surface in 2007 is shown in Table 10.21.

III Temperature field monitoring results during overwintering period in 2007 − 2008

Figure 10.16 shows the actual observation data of temperature field after overwintering protection:

1. The temperature at the corner of the dam body on the overwintering surface is low, the minimum temperature is 7.48 °C, and the temperature difference with the outer side of the upstream protective layer is 32.12 °C, which indicates the remarkable thermal insulation effect.
2. The temperature difference between the dam center and the downstream concrete surface is 8.9 °C, which is less than the design allowable temperature difference of 19.0 °C.
3. The temperature field in the dam body is evenly distributed, and the temperature difference has little change.
4. The temperature distribution of dam foundation conforms to the general law, the temperature gradient is small, and the temperature difference between dam body and dam foundation is small, which effectively reduces the constraint stress of the foundation.

Table 10.21   Gradual removal process of the thermal insulation quilt on overwintering surface

Date	Number of layers of thermal insulation quilts removed	Thickness of the removed quilt (cm)	Remarks
2008-03-08	2	2	Cotton quilt
2008-03-17	1	2	Cotton quilt
2008-03-20	1	2	Cotton quilt
2008-03-23	1	2	Cotton quilt
2008-03-27	2	4	Cotton quilt
2008-03-29	2	4	Cotton quilt
2008-04-01	1	2	Cotton quilt
2008-04-03	1	2	Cotton quilt
2008-04-05	3	6	Polyethylene thermal insulation quilt

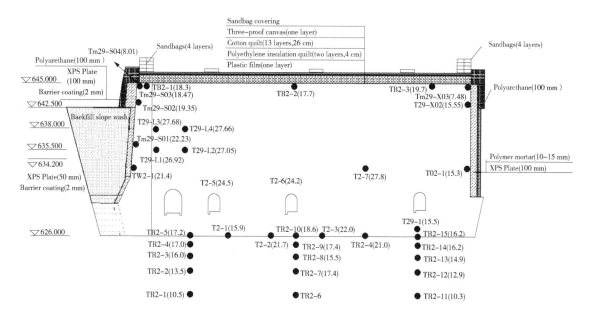

Figure 10.16 *Temperature distribution on dam body and surface of* $29^{\#}$ *dam section on January 22, 2008 (average daily temperature* -24.64 ℃*)*

5. The higher temperature of the overwintering layer is conducive to reducing the temperature difference between the upper and lower layers of the new concrete and the old concrete in the coming year, reducing the constraint stress of the lower layer concrete on the new concrete, and reducing the risk of horizontal cracks.

Above test results show that the permanent thermal insulation and overwintering layer thermal insulation of the main dam are suitable.

10.5.6 Simulation Analysis on Temperature Stress in Gap Section of Dam Body during Temporary Flood Season

According to the construction arrangement, in the main flood season (May – June) in 2009, temporary gaps will be reserved for the main dam to vent the design flood and check flood in the construction period together with the bottom outlet and the middle outlet. The temporary gap has a width of 45 m and bottom elevation of 695.00 m.

Considering that the inside temperature of the temporary gap dam section is still relatively high, in order to prevent the adverse cracks caused by cold shock due to water temperature, the temperature stress of this gap dam section is analyzed, and the maximum temperature stress is shown in Table 10.22.

Table 10.22   Maximum temperature stress of overflow dam gap during flood season

Item	Σx_{max} (MPa)	Location (m)			Occurring time (d)	Σy_{max} (MPa)	Location (m)			Occurring time (d)	Σz_{max} (MPa)	Location (m)			Occurring time (d)
		x	y	z			x	y	z			x	y	z	
Maximum temperature stress	1.06	7.50	31.29	75.00	752	2.21	3.75	21.41	75.00	752	1.88	0.00	7.20	74.80	752
Maximum comprehensive stress	1.05	7.50	31.29	75.00	752	2.21	3.75	21.41	75.00	752	1.87	0.00	38.45	74.80	752

According to the simulation calculation results of unstable temperature field in flood season of temporary gap dam section, it can be seen that the depth of dam body temperature affected by surface water temperature is about 6 m, and the maximum surface cold shock tensile stress occurs in the early stage of flood season; the tensile stress in most areas of the gap during flood season is small, and the general trend is that the surface tensile stress increases from inside to outside.

1. Temperature stress and comprehensive stress of gap dam section during temporary flood season. Temperature stress $\sigma_{x max}$ is 1.06 MPa; $\sigma_{y max}$ is 2.21 MPa; $\sigma_{z max}$ is 1.88 MPa. Comprehensive stress $\sigma_{x max}$ is 1.05 MPa; $\sigma_{y max}$ is 2.21 MPa; $\sigma_{z max}$ is 1.87 MPa.

    It can be seen from the above results that the temperature stress in y and z directions is great, and the concrete on the surface of the gap during flood season shall be protected.

2. The distribution of temperature stress along the height direction of the gap dam section during temporary flood season is basically consistent with the distribution of temperature field, and the temperature stress in the high temperature area is large, while that in the low temperature area is relatively small.

3. It can be seen from the stress envelope diagram of horizontal sections at different elevations that the closer the distribution of temperature stress and comprehensive stress in 3 directions is to the concrete surface, the greater the stress is, and cracks are more likely to appear on the surface, so it is necessary to protect the surface.

4. By comparing the calculation results of temperature stress comprehensively, it can be found that the comprehensive temperature control measures adopted in the gap for temporary flood season can effectively reduce the temperature stress. The stress in most areas of the gap for temporary flood season can be controlled within the allowable stress range, but the tensile stress at some points is larger and exceeds the allowable stress.

    Special attention shall be paid: if the concrete has a short age, the allowable tensile strength of concrete may be lower than the tensile stress, and then the concrete surface may crack. Therefore, the gap dam section shall be constructed as soon as possible to ensure that the increased strength meets the requirements of temperature stress.

10.5.7 Conclusion

1. The results of simulation analysis and preliminary on-site monitoring show that the temperature control scheme and the temperature control measures taken during the construction are basically reasonable.
2. The total construction duration of the dam is 3 years. There are 3 high temperature areas in the concrete poured from June to August every year, and the maximum temperature in local parts of each high temperature area is about 30 – 32 ℃. The monitoring data show that, except for the maximum temperature and foundation temperature difference of the concrete in some dam sections in the strong restraint area of foundation poured at high temperature in 2007, the maximum temperature, foundation temperature difference, upper and lower temperature difference and temperature difference between inside and outside of concrete poured at other periods basically meet the design requirements.
3. The calculation results show that the concrete near the upstream and downstream faces of the dam features a quasi-stable temperature field 10 years after pouring; the II gradation concrete area features a quasi-stable temperature field about 15 years after pouring; the stress field of III gradation concrete features stable state 30 years after pouring and formed a stable temperature field about 50 years after pouring.
4. The upstream surface stress is well controlled, and the comprehensive stress on the upstream surface of the dam is basically controlled within 1.8 MPa except for the concrete near the upstream dam heel area. Even near the overwintering layer with high stress, the comprehensive stress can be controlled at about 1.8 MPa.
5. II gradation concrete is in quasi-stable temperature field about 15 years after pouring. The stress then fluctuates with the change of temperature, and the maximum stress tends to be stable. The stress value near the upstream surface does not exceed 1.0 MPa, and near the downstream surface does not exceed 1.5 MPa, which meets the design requirements. Except for the local tensile stress exceeding the allowable value, the comprehensive stress and tensile stress of III gradation concrete of the dam during construction and operation do not exceed the allowable value of materials, which could meet the specification requirements.

Chapter 11

RCC Quality Control and Project Cases

11.1 Overview

The quality control of RCC is the work directly related to the quality of RCC during and after the construction. The quality control features comprehensiveness (various indexes), whole process (various working procedures), timeliness (various ages) and representativeness (various test frequencies).

RCC is a kind of synthetic dam structure material consisting cement, coarse and fine aggregates (rock powder), admixtures, additives and water formed by mixing, spreading and rolling according to the approved construction mix proportion. The service performance, safety performance, durability and adaptability to the environment of RCC are closely related to its quality, which directly affects the safe operation of dam structures, and further relates to the national safety and people's lives and property. Once a quality accident occurs, the consequences are very serious. Therefore, the RCC construction must follow the policy of "quality first in constructing a project of vital and lasting importance", and the undertaker, designer, supervisor, constructor and the whole society shall attach great importance to RCC quality.

Due to the influence of raw materials, mixing equipment, construction environment and other factors, there is a certain difference between the actually produced RCC mixture and the approved construction mix proportion. The mix proportion is an important basis for the quality control of RCC production. Ensuring the fluctuation of the performance of its mixture is controlled within the allowable range is the prerequisite for ensuring the construction quality and hardening RCC to meet the design requirements.

The test and quality control of RCC are closely related to many procedures, such as raw materials, mixing, transportation, placement, spreading, rolling, spraying moisturizing, and covering curing. Each construction procedure and quality control are closely linked, and problems in any process will not only affect the subsequent procedures, but also affect the quality of the whole RCC, especially the interlayer bonding quality. The interlayer bonding quality of RCC is directly related to the dam's imperviousness performance, interlayer anti-sliding stability and use function, and poor quality may cause serious adverse consequences. Therefore, it is necessary to effectively test and control the whole process of RCC construction quality.

The quality control of RCC is mainly carried out in accordance with the relevant provisions of *The Construction Specification for Hydraulic Roller Compacted Concrete* DL/T 5112—2009 and the technical clauses of the bidding documents. *The Construction Specification for Hydraulic Roller Compacted Concrete* DL/T 5112—2009 is implemented on December 1, 2009, and this is the fourth revision of *The Construction Specification for Hydraulic Roller Compacted Concrete* DL/T 5112—2009. China's RCC construction technology has developed rapidly, and the construction of 200 m dams has been realized. Many new technologies, new materials, new processes and new equipment have been widely used. In order to meet the needs of the RCC rapid damming technology, the revision focuses on the quality and safety of hydraulic RCC construction in the revision content, so as to reflect the new level of the hydraulic RCC construction. Especially, the modification of VC value of RCC mixture reflects the achievements of construction practices for many years, and also reflects the deepening understanding of hydraulic RCC performance. For example, DL/T 5112—2009 stipulates that "The VC value of RCC mixture shall be 2 – 12 s on site." The VC value in outlet should be changed according to the climatic conditions in dynamic selection and control within 2 – 8 s. Meanwhile, canceled the requirement of "spring concrete" treatment in rolling process, and the article states that "as long as the apparent density can meet the requirements, the spring concrete in rolling process is conducive to interlayer bonding and does not need to be treated"; in order to ensure personal and environmental safety, the requirements for the use of nuclear density-moisture gauges have been added in the revision, and relevant standards have been introduced.

In RCC construction, there are many methods of quality test and control, and mathematical statistical analysis is the most basic method for the comprehensive quality inspection and control in the whole process of RCC production. The quality data obtained in each procedure of RCC construction (including the quality inspection results of materials, production process parameters and product quality parameters) shall be analyzed by quality management charts. Currently, with the development of information, dynamic management of information-based construction and quality control has been adopted in RCC construction and quality control.

As the RCC dam is generally formed in one time, it is particularly important to check and control the production process and keep the production under control. The quality management assurance system is an important guarantee for quality test and control. The key to quality control is to have a scientific control procedure and an effective quality assurance operation system. Simply speaking, an effective quality assurance operation system means that every link related to RCC shall be seriously and strictly inspected and approved by a special person or institution according to the standards required by design or specifications, and quality assurance measures shall be implemented. The key of quality control is the proper management, human factors come first, and high quality management must be the scientific, standardized and institutionalized management. The key to the project quality is the quality consciousness of the project manager. In order to treat the project quality above all else, the quality must be controlled and tested in accordance with the terms of bidding documents and bids, regulations and design requirements, so as to implement quality management.

Since the implementation of the bidding system for water conservancy and hydropower projects in China in the 1990s, the project supervision system has been implemented in the project construction, and the whole process of quality supervision and management has been implemented. Through effective control, management, organization and coordination, it has ensured the smooth progress of the projects according to objectives of the construction contract and ensured that the projects could playimportant roles. Therefore, the quality and responsibility of supervision engineers are directly related to the supervision and control of project quality. Also, more and more attention has been paid to the field laboratory in quality control, and the laboratory has been an extremely important technical department of quality control, which can effectively carry out quality control and project quality assurance through tests. Therefore, the field laboratory is specially specified in the terms of the bidding technology contract. In order to ensure the quality of the project, all parties involved in the project shall earnestly perform their duties, the undertaker shall give full play to the leading role of the owner, the supervision company shall play the role of supervision and coordination according to the contract, and the construction contractor shall assume the role of quality assurance according to the contract.

With the rapid development of RCC damming technology, there are also unsatisfactory quality problems. In some projects, due to improper RCC materials, cutting corners, extensive construction technology, weak quality awareness and little knowledge or superficial knowledge of RCC damming technology, project quality problems and accidents have occurred in RCC damming, and there are also many painful lessons in the development process. For example, a RCC gravity dam in Brazil collapsed after impoundment, and RCC dams built in some countries have to be repaired as defective dams. Some RCC dams built in China also have quality problems, especially those with a height below 100 m built on some tributaries. After the completion of some dams, the permeability is high, the recovery rate of core samples is low, and the shear resistance of interlayer bonding is poor. After impounding, the dam leaks seriously, so the dam must be treated by grouting and imperviousness, thus affecting the project quality and causing the loss of benefit. On the one hand, these defective RCC dams suffer management problems, but they are mainly caused by the construction contractor's little understanding of RCC damming technology. Therefore, the safety and durability of the dam are the key directions for the research and development of RCC dams in the future. We shall always keep a clear understanding of the technical characteristics of rapid RCC damming and strictly control the quality.

According to the characteristics of RCC construction and the requirements of specifications and bidding documents, this chapter comprehensively and systematically expounds the key points, difficulties, influencing factors and project examples of RCC quality control, mainly from the aspects of test and control of raw materials, test and control of mixing materials, on-site quality test and quality evaluation.

11.2 Quality Control and Evaluation Regulations

11.2.1 Test and Control of Raw Materials

11.2.1.1 Test and Quality Evaluation of Raw Materials

The test of RCC raw materials is put forward according to the project construction experience and referring to relevant data. The purpose of test of raw materials is to check whether the quality of cement, admixtures, aggregates and additives meets the quality standards, adjust the mix proportion of RCC and improve the construction technology according to the test results, and evaluate the production control level of raw materials. According to the relevant provisions of *The Construction Specification for Hydraulic Roller Compacted Concrete* DL/T 5112—2009, the test and quality evaluation of RCC raw materials are detailed in Table 11.1.

Ⅰ Cement testing

Each batch of cement shall be accompanied with the quality test report of the manufacturer. Sampling and testing shall be carried out for each batch of cement in accordance with relevant national and industrial regulations, as well as chemical composition analysis if necessary. 200 t cement with the same type and grade shall be taken as a sampling unit for testing and sampling, and it shall also be taken as a sampling unit when it is less than 200 t. Mechanical continuous sampling can be used, or the sampling can be carried out cement from 20 different parts by same amount, and mixed sampling is taken as the sample, of which total quantity shall be at least 10 kg. Test items shall include: cement strength grade, setting time, volume stability, consistency, fineness, density and hydration heat and other tests. The quality test of cement shall be carried out in accordance with current relevant national standards and industrial standards.

Ⅱ Test of fly ash and other admixtures

For testing and sampling of fly ash and other approved admixtures, every 200 t is taken as a sampling unit, and that less than 200 t is also taken as a sampling unit. Test items include the fineness, water requirement ratio, loss on ignition and sulfur trioxide and other indexes. The fineness and water requirement ratio shall be inspected at least once a day. In 10 consecutive samples, the difference between the fineness of individual samples and the average value shall not be greater than 10% - 15%. The test of fly ash shall be carried out in accordance with *The Technical Specification of Fly Ash for Use in Hydraulic Concrete* DL/T 5055.

Chapter 11 RCC Quality Control and Project Cases

Table 11.1   Testing items and testing frequency of raw materials

Description		Test Items	Sampling location	Inspection frequency	Test purpose
Cement		Rapid test grade	Cement warehouse of mixing plant	As necessary	Verify the cement activity
Cement		Fineness, stability, water demand of standard consistency, setting time and grade	Cement storehouse	Once every 200 – 400 t❶	Verify the quality of cement delivered
Admixture		Density, fineness, water demand ratio or fluidity ratio, and loss on ignition	Warehouse	Once every 200 – 400 t❶	Evaluate the quality stability
Admixture		Strength ratio or activity index	—	As necessary	Test the activity
Fine aggregate		Fineness modulus, and content of rock powder and particle	Mixing plant, sieving plant	Once a day	The sieving plant controls the production and adjusts the mix proportion
Fine aggregate		Grain composition	Sieving plant	As necessary	The sieving plant controls the production and adjusts the mix proportion
Fine aggregate		Moisture content	Mixing plant	Once every 2 h or when necessary	adjust the water consumption of RCC
Fine aggregate		Clay content and apparent density	Mixing plant, sieving plant	As necessary	the fine aggregate quality
Coarse aggregate	Large-sized stone Medium-sized stone Small-sized stone	Super-inferior size ratio	Mixing plant, sieving plant	1 times per shift	The sieving plant controls the production and adjusts the mix proportion
Coarse aggregate	Small-sized stone	Moisture content	Mixing plant	Once every shift or when necessary	Adjust the water consumption of RCC
Coarse aggregate	Small-sized stone	Content of clay, silt, fines	Mixing plant, sieving plant	As necessary	Inspect the quality of pebble
Additive		Solution concentration	Mixing plant	1 times per shift	Adjusting Additive agent content

Note: ❶When each batch is less than 200 t, it should be tested once.

Ⅲ Testing of admixtures

All kinds of admixtures used in concrete shall be accompanied with the quality certificate of the manufacturer, and shall be tested and identified in accordance with national and industrial standards. If anyone with too long storage time, it shall be re-sampled to test. It is strictly forbidden to use the deteriorated or unqualified admixtures. For the water reducing agent solution concentrate mixed on site, 5 t is taken as the sampling unit; for the air entraining agent, 200 kg is taken as the sampling unit. Measures of preventing precipitation and ensuring uniform concentration shall be taken for the additive solution prepared that shall be tested once or twice a day, and tested at least once per shift. The additive shall be inspected in accordance with the relevant terms of *The Technical Code for Hydraulic Concrete Admixtures* DL/T 5100.

Ⅳ Aggregate quality test

The quality of aggregates shall be tested in the mixing plant according to the following provisions: the percentage of water content in sand and small-sized stone shall be checked at least once every 4 h, and its variable shall be controlled in the range of ±0.5%; when the temperature changes greatly or the water content of aggregate changes suddenly after rain, the check shall be conducted once every 2 h; the FM and the rock powder content of sand shall be checked at least once per shift; when the FM of sand exceeds the control median for ±0.2, it is necessary to adjust the sand ratio in the batching sheet; the aggregates over size and under size and the silt content shall be checked once every 8 h, and the apparent density shall be checked once a day. Other items in Table 11.2 shall be tested once a month.

Table 11.2   Inspection standards for batching and weighing

Material name	Water	Cement and admixture	Coarse and fine aggregate	Additive
Allowable deviation	±1%	±1%	±2%	±1%

Ⅴ Water quality test

For water used for concrete mixing and curing, besides the water quality analysis as required, it is needed to test and compare the water quality through mortar strength when the water source is changed or water quality is doubted. If the compressive strength of mortar made with this water is more than 90% lower than that of 28 d age mortar made with previous qualified water, then this water cannot be used.

Ⅵ Temperature test of raw materials

The temperature of concrete raw materials shall be subject to spot check once or twice per shift depending on the temperature.

11.2.1.2 Rock Powder Content and Percentage of Water Content in Sand-key Control Points of Raw Materials

Ⅰ Control of rock powder content

A large number of test studies and project practices prove that the rock powder has become an indispensable component of RCC. High-quality rock powder can greatly improve the workability and rollability of RCC, which is beneficial to the improvement of liquefied bleeding of layer surface and interlayer bonding quality, enhancing the cementation performance of RCC layer surface. At the same time, the compaction, impermeability and other combination properties of RCC are effectively improved.

Due to characteristics of aggregates, the rock powder content in the manufactured sand processed is always fluctuating. When rock powder content is too low or natural sand is used, the sand is replaced with rock powder, which can effectively improve the mixtures performance and construction performance of RCC. When rock powder content increases or decreases by 1%, the water consumption correspondingly increases or decreases by about 1.5 kg/m^3, which affects the VC value for about 1 s. Generally, the rock powder content of RCC is controlled in 16% - 22%. When the rock powder content of the manufactured sand fluctuates in the range of 16% - 22%, the water consumption correspondingly increases or decreases by about 9 - 12 kg/m^3. Therefore, the rock powder content directly affects the water consumption and the water-cement ratio, further affecting the performance stability of mixtures and the quality of hardened concrete. Therefore, the rock powder content should be controlled as per the specified range.

Ⅱ Strict control of the percentage of water content in sand

The fluctuation of percentage of water content in and FM of the sand may cause changes in water consumption and mixtures performance, so it is necessary to strictly control the percentage of water content in sand and the grain gradation of fine aggregates. Sand accounts for about one third of the components of RCC, and its mass is in the range of 700 – 800 kg/m^3. When the percentage of water content in sand fluctuates by 1.0%, the water content correspondingly fluctuates by 7 – 8 kg/m^3, which may cause the change of VC value for 3 – 5 s. Therefore, the percentage of water content in sand is the main factor that causes the VC value change of fresh RCC.

11.2.2 Test and Control of Mixtures

11.2.2.1 Calibration of Weighing Equipment and Test of Mixing Uniformity

The weighing and batching of RCC, especially the control of water consumption, are stricter than that of normal concrete. If the weighing apparatus is not accurate enough or its batching is not accurate, the quality of RCC cannot be controlled. The mixing and feeding sequence and the mixing time of RCC must be arranged according to the method determined in the test. The mixing time shall be subject to spot check at least twice per shift. If necessary, the mixing uniformity shall be tested.

In the test of RCC mixtures, it is stipulated that the weighing apparatus for batching and weighing of RCC shall be calibrated once a month, and the allowable deviation in batching and weighing is specified in Table 11.2. Before weighing per shift, zero check shall be conducted for weighing equipment. Since the weighing accuracy of various components of each plate of concrete is an important factor affecting the concrete production quality, the mixing equipment shall be tested regularly after being put into operation. When the weighing error of the mixing plant is in "accidental fluctuation range", the workers shall conduct handling as per the opinions of laboratory personnel. When the situation is serious and there is a great impact on the quality of RCC, waste treatment shall be conducted. If the range fluctuates frequently and the quality of RCC is out of control, the causes shall be found out in time for treatment.

The quality control of RCC mixtures largely depends on the experience of test personnel. In order to find out the case of being out of control during mixing in time, experienced personnel can be assigned to regularly observe whether the color of RCC mixtures in outlet is uniform and whether the surface of sand and gravel particles is sticked with grout evenly, and visually estimate whether the VC value of the mixtures is appropriate, whether the test data of the percentage of water content corresponds to the actual storage of sand tank (silo), and whether the rock powder content is within the allowable control range.

11.2.2.2 Test Items and Frequency of RCC

The quality test of RCC can be carried out by random sampling at the mixer outlet. The test items and frequency shall conform to the specifications, and details are shown in specifications in Table 11.3. The test of RCC shall be carried out by sampling in the same plate. The key point of the test is the performance of fresh RCC in outlet. The test purpose is to find out the factors of being out of control in construction and make adjustments.

Table 11.3 Test items and frequency of RCC

Test items	Inspection frequency	Test purpose
VC value	Once every 2 h❶	Controlling working degree change
Gas content	During use of air entraining agent 1-2 times per shift	Adjusting air entraining agent content
Temperature	Once per 2 – 4 h	Temperature control requirements
Compressive strength	Moulding of one group of RCC blocks every 500 m^3 in 28 d and every 1 000 m^3 in design age respectively; at least sampling once per shift in case of being less than 500 m^3	Inspecting mixing quality and construction quality of RCC

Note: ❶The test times shall increase appropriately in case of great climate changes (strong wind, rain and high temperature).

11.2.2.3 Strict Control of *VC* Value of Mixtures

The *VC* value of RCC mixtures shall be dynamically controlled according to the climate and the placing surface construction condition. After the *VC* value of RCC mixtures is determined, the *VC* value in outlet shall be checked once or twice every 4 h, and the allowable deviation of *VC* value in outlet shall be ±3 s. When *VC* value exceeds the control limit, the causes shall be found out in time for adjustment. The mixtures confirmed to be unsuitable for rolling compaction shall be treated properly.

As the mixtures of RCC have higher requirements than that of normal concrete, and the *VC* value increases with the extension of the RCC placement time and is greatly affected by the temperature and meteorological conditions, therefore, required *VC* value shall be different during construction in different seasons and weather conditions, and even during the day or at night. During construction, different *VC* reference values of the mixtures shall be selected according to different conditions for dynamic control. In order to achieve this purpose, a curve of the relationship between *VC* value and time and temperature conditions shall be established by test to be a basis for dynamic control of *VC* value in outlet before commencement of construction of RCC, and the parameter relationship between RCC water consumption and the mix proportion shall be adjusted according to the control range of *VC* value.

11.2.2.4 Strict Control of Air Content of Mixtures

In order to improve the concrete durability of the dam, it is necessary to maintain a certain amount of air content in the concrete. In cold regions in the north or subtropical mild regions in the south, the frost resistance grade has become an indispensable index for RCC durability design. Due to the larger proportion of admixtures mixed into RCC and the "sub-plastic" concrete without slump for RCC mixtures, it is relatively difficult to entrain air for RCC. To reach the same air content as that of normal concrete, it is necessary to increase air entraining agent content several times more than that of normal concrete. In order to ensure the construction quality and improve the durability of RCC, the air content of RCC mixed with air entraining agent shall be strictly controlled, and the fluctuation range of the air content shall be controlled within ±1%.

11.2.2.5 Temperature Detection

During construction of fresh concrete and concrete, it is very important to check and control the temperature.

1. External temperature, and the temperature in shed and of gallery shall be measured at least once every 4 h.
2. Water temperature and aggregate temperature shall be measured at least once every 2 h.
3. The outlet temperature and the placing temperature of the concrete shall be measured at least once every 2–4 h.
4. The internal temperature of the placed blocks shall be especially subject to strengthening observation in 3 d after pouring, and, can be observed regularly later according to the temperature and position conditions. In case of measuring temperature, attention shall be paid to edges and corners that are most easily cooled.

11.2.2.6 Requirements for Sampling and Forming of RCC Specimens

In the test of compressive strength of RCC, the samples are mainly taken from the outlet, and the quantity of the concrete specimens at the same grade shall be determined according to the provisions in DL/T 5112 specification and the bidding documents. Specimens in the same group shall be taken from the same plate of concrete. Specimens shall be randomly sampled at the outlet, but shall not be selected at will. For comparison, the sampling quantity of specimens at the placing location shall be no less than 10% of that at the outlet. The sampling quantity and the compressive strength of concrete specimens at the same strength grade shall comply with the requirements specified in Table 11.3. Besides, the test requirements of tensile strength, ultimate tensile strength, impermeability, frost resistance and others shall be in accordance with *The Provisions of Specifications for Hydraulic Concrete Construction* DL/T 5144.

Compressive strength: moulding of one group of large-volume RCC blocks every 500 m^3 in 28 d and every

1 000 m³ in design age respectively.

Tensile strength: moulding of one group of large-volume RCC blocks every 2 000 m³ in 28 d and every 3 000 m³ in design age respectively.

Impermeability, frost resistance or other major special requirements shall be properly tested after sampling during construction, the quantity can be 1 − 2 groups which are moulded after sampling in the main construction locations every quarter.

11.2.3 Field Quality Inspection and Control

11.2.3.1 Test Items at Paving and Rolling Compaction on Site

In the field test of RCC quality, it is stipulated that the RCC shall be tested according to the provisions in Table 11.4 during paving, and test records shall also be completed.

Table 11.4　　　　　　　　Test items and standards at RCC paving on wite

Test items	Inspection frequency	Control standards
VC value	Once every 2 h	Allowable deviation ± (meeting the requirements in 6.0.4 in DL/T 5112 Specification)
Compressive strength	Being equivalent to 5% − 10% of the sampling quantity at the outlet	Design index
Apparent density	As specified in 8.3.2	The measured apparent density of each paving layer shall all meet the relative compactness index specified in 8.3.4
Aggregate separation	Whole process control	No aggregate concentration is allowed
Interval time between two rolling layers	Whole process control	It is determined on the basis of the allowable interval time between layers under different temperature conditions confirmed by the test
Time for mixing concrete with water until completion of rolling compaction	Whole process control	It is less than 2 h or determined by the test
Placing temperature	Once per 2 − 4 h	Design index

Note: The test times of VC value and pouring temperature shall increase appropriately in case of great climate changes (strong wind, rain and high temperature).

11.2.3.2 Specific Requirements for Quality Control of Placing Surface

1. In order to ensure sound interlayer bonding of upper and lower layers, the allowable interval time between layers must be controlled within the range required in the design, so that the quality of the layer surfaces meets the requirements of shear strength and impermeability. The allowable interval time between layers shall be determined by the test according to different temperature and construction environment conditions. With the development and application of high-range retarding water reducing agent, the setting time of RCC has been greatly extended. Therefore, the allowable time for mixing RCC with water until completion of rolling compaction has exceeded the 2 h control standard.

There are a variety of reasons for poor interlayer bonding caused by control of interlayer bonding quality, where, there are mainly two aspects related to site construction: The intermittent time between layers is too long, or RCC has been set and hardened without treatment, and treatment is improper; improper mix proportion and construction method result in aggregate separation, thereby causing excessive concentration of coarse aggregate separation on the next layer surface.

As for whether the RCC is allowed to be covered on the upper layers of the rolling layer surface and the rolling layer surface treated, the layer surface coverage standard shall be determined according to the specific conditions of the site. In current construction specifications, the initial setting time of RCC is taken as a criterion for layer surface coverage, namely, the so-called "2 h" and "4 h or longer time" criteria for control.

2. When the plastic bonding between layers cannot be guaranteed, construction joints shall be made. However, the treatment method varies with the degree of setting and hardening. The common method is to clean up the set and hardened RCC layer surface before continuing rolling compaction, and cover the RCC mixtures as soon as possible after the joint cushion materials are spread out to ensure sound cohesiveness. At present, mortar or grout is mostly paved to form a cushion, but attention shall be paid to uniform paving and timely coverage after paving, so as to prevent mortar or grout from whitening and drying, resulting in the sandy interlayer between two layers of RCC and the deterioration of joint properties, which is particularly important.

3. The rolling layers shall be prevented from disturbance damage and pollution. Mostly, direct truck placement of RCC is adopted, or dispersed or handled on the placement surface by truck. Attention shall be paid to the control of the running speed and turning radius of the trucks. A special site shall be set up to clean the trucks; before entry of pouring site, the wheels shall be cleaned up, so as to prevent dirt and sludge from being brought into the pouring site. All kinds of machinery in the pouring site shall be strictly prevented from oil leakage. Because of hydrophobicity of oil stain, the layers inevitably may not be bonded. Any oil stain found shall be dug off.

4. When the construction joints are treated with water pressure spray roughening machine, the cleaning time shall be controlled, and the cleaning shall be carried out shortly after the final setting of RCC, so as to avoid the impact on interlayer bonding due to too early roughening.

5. During leveling of spreading machines, the spreading machine shall not run back and forth on the hardened RCC surface, or even rotate in place. The crawler has great destructiveness to the hardened RCC surface. When the exposed stones are loose or broken, the mortar shall be paved first after cleaning up, followed by the RCC. The rolling surface shall be kept clean and pollution-free, besides, it shall also be kept moist until the upper layer of RCC is covered. To prevent layer surface drying, spraying or water sprinkling can be used, and no water drops are formed.

6. The aggregate separation shall be avoided and improved. Since RCC mixtures are made of materials with different particle size and density, it is difficult to avoid aggregate separation in the process of transportation, discharging and leveling. In case of obvious separation, the mixtures shall be manually dispersed in the RCC unrolled. In case of separation due to the change of mix proportion, the causes shall be found out in time and then the mix proportion is adjusted.

7. Methods of improving aggregate separation are as follows: a. Optimize the mix proportion of RCC with good separation resistance; b. reduce the drop and stacking height during discharging and loading; c. set buffer facilities at the outlets of mixer and each intermediate transfer hopper to improve aggregate separation.

11.2.3.3 Apparent Density Test and Relative Compactness Requirements on Site

Ⅰ Apparent density test

The apparent density of RCC is mainly tested with nuclear density-moisture gauges on site. The concrete paved and rolled shall be provided with at least one test point every $100 - 200 \text{ m}^2$, and also with no less than 3 test points in the compacted placing surface in each paving layer. Because of the release process of compaction energy after rolling, the test results of nuclear density-moisture gauges after 10 min upon completion of rolling are taken as the basis for determining apparent density. The nuclear density-moisture gauges shall be calibrated

Chapter 11 RCC Quality Control and Project Cases

before use, and the RCC calibration block shall be calibrated by using the RCC prepared in the same raw material mix proportion as the projects.

II Relative compactness requirements

According to DL/T 5112 specification, the relative compactness of RCC in impervious area of the dam shall not be less than 98%. The relative compactness of internal RCC shall not be less than 97%. The apparent density of RCC shall be more than 97% of the apparent density designed in mix proportion, so as to meet the requirements of RCC compaction and gravity stability of the dam.

III Equirements for use of nuclear density-moisture gauges

Nuclear density-moisture gauge is a kind of testing instrument with radioactive source, and shall be highly concerned during use. The nuclear density-moisture gauges shall be used and maintained by specially trained personnel. It is strictly prohibited to disassemble and assemble the radioactive source in the instrument, and the work shall be in strict accordance with the work procedures. At the same time, the nuclear density-moisture gauges shall be registered and recorded, and stored in a place that meets the safety provisions. In case of any loss or damage to the radioactive source of the instrument, measures shall be taken immediately for proper handling, and a timely report is made for the relevant management department.

The nuclear density-moisture gauges shall be used according to the provisions in *Field Test Methods for Nuclear Density-moisture Gauges* SL 275.

11.2.3.4 Quality Control of Bonding Parts between RCC and Heterogeneous RCC

In the construction of RCC, it is unavoidable to pour heterogeneous RCC on the surrounding bank slope (gallery periphery, etc.). Due to the great differences between performance of heterogeneous RCC and that of RCC, the construction methods are quite different. In order to ensure the construction quality of the bonding parts, a special method shall be adopted for tamping.

Due to the differences in the process of heterogeneous RCC and RCC, the initial setting time of the two kinds of RCC is often different. In order to ensure the synchronous rise of the two kinds of RCC layer surfaces, in addition to paying attention to the control of the paving and rolling quality of the overlapping parts of the two kinds of RCC, attention should also be paid to the adjustment and control of the setting time and the construction sequence of the heterogeneous RCC, so as not to form cold joints and affect the quality.

11.2.3.5 Time Control of Key Working Procedures

The test shows that the quality of RCC is closely related to the time of the following working procedures:

1. Mixing time of RCC (its minimum is determined according to different mixing equipment, and it is generally no less than 90 s for forced mixers)
2. Time from outgoing to completion of rolling compaction of RCC mixtures (generally no more than 2 h).
3. Time from entry of pouring site to commencement of rolling compaction of RCC mixtures (generally no more than 60 min).
4. Time from paving to coverage of cushion materials (for example, mortar) on the layer joint surface (generally no more than 15 min).

11.2.3.6 Supervision and Inspection Measures on Site

Before each mixing and pouring of RCC, the following items shall be tested by special personnel, and the records are kept. During pouring of RCC, the quality test personnel shall conduct patrol test and supervision at any time:

1. Record the pouring process, and adhere to the principle of "person who constructs is required to be responsible" in the construction quality responsibility.
2. Test the RCC mix proportion and the batching sheet, test the raw materials (such as the cement, additives, coarse and fine aggregates, water content and water) comply with the specified requirements, and

immediately adjust the mix proportion or prohibit mixing in case of any change.

3. Test the content of each raw material and the additive content, the spot check shall be no less than 4 times in each shift and make records.
4. Record each parameter about the RCC production process. Such as the mixing time, rock powder content and others.
5. Test the *VC* value of RCC complies with the requirements with random sampling, while it shall be no less than 3 times in each shift.
6. Test and record the temperature during RCC production and the time and temperature change of the RCC transported to the construction site.
7. Test and supervise the whole process of the specimen preparation.
8. Test the specimen curing condition and the test equipment conform to the requirement.

11.2.4 Quality Evaluation

11.2.4.1 Regulations for Sampling and Forming of RCC Specimens

In RCC quality evaluation, it stipulates that the RCC specimen shall be sampled and formed at the mixer outlet. The RCC production quality control is based on the compressive strength of 150 mm standard cubic specimen of standard curing (28 d) or the design age. The compressive strength of specimen of 28 d or design age sampled and molded at the outlet is mainly used to measure and evaluate the quality management level of the RCC mixtures production.

11.2.4.2 Frost Resistance and Impermeability Test Indexes of RCC

The percentage of pass of the RCC frost resistance and impermeability tests shall be no less than 80%. Impermeability, frost resistance or other major special requirements shall be properly tested after sampling during construction, the quantity can be 1 – 2 groups which are formed after sampling in the main construction locations every quarter according to the bidding documents, design requirements or specifications.

11.2.4.3 Standard for Evaluation of RCC Production Quality Level

See Table 11.5 for standard for evaluation of RCC production quality level. It shall be indicated by the standard deviation σ of compressive strength of a batch (at least 30 groups) of concrete of 28 d or design age continuously sampled at outlet.

Table 11.5   RCC Production quality level

Evaluation index		Quality level			
		Excellent	Good	General	Bad
Standard deviation σ (MPa) of concrete strength under different strength grades	$\leqslant C_{90}20(C_{180}20)$	<3.0	$3.0 \leqslant \sigma < 3.5$	$3.5 \leqslant \sigma < 4.5$	>4.5
	$> C_{90}20(C_{180}20)$	<3.5	$3.5 \leqslant \sigma < 4.0$	$4.0 \leqslant \sigma < 5.0$	>5.0
The percentage $P_s(\%)$ of the strength not lower than the standard strength		$\geqslant 90$	$\geqslant 85$	$\geqslant 80$	<80

Note: The calculation method of standard deviation σ of concrete strength and the percentage of the standard strength P_s is shown in the Appendix Table B.2 of Section 12.2.5 (DL/T 5112—2009).

The evaluation of the RCC production quality level is the same as the evaluation standard (DL/T 5144), according to the RCC characteristic, the percentage (P_s) of the strength not lower than the strength standard value, the excellent and good are respectively $\geqslant 90\%$ and $\geqslant 85\%$.

The quality level of RCC production is often described with the average value of strength and the standard deviation, when evaluating the strength quality of a batch of concrete, selecting several groups of specimens randomly in the total test batches to test, and infer the overall quality. The reliability of RCC is related to the variation degree of RCC strength, the variation degree of RCC strength can comprehensively reflect the quality management level of RCC production. According to the RCC strength survey result, the higher the quality management level, the smaller the standard deviation of strength reflecting the strength variation is.

When the number of the statistical data n group is large enough (such as $n > 30$), the result of the assurance rate P and the percentage of pass P_s (i. e., the percentage of strength no less than the specified strength grade) is similar. In order to calculate the quality control of the concrete on site conveniently, P_s is listed as one of the indexes to measure the management production level, so as to avoid that the standard deviation is good while the percentage of pass or the assurance rate is low, and the quality would be measured as a higher level mistakenly.

11.2.4.4 RCC Quality Evaluation Requirements

The calculation formulas and symbols used of RCC quality evaluation are subject to DL/T 5144 and consistent with DL/T 5330 and GBJ 107. According to the RCC characteristic, the RCC average strength is calculated according to the Formula (11-1), and the minimum value are 0.75 and 0.80 according to the coefficient of the formulas (11-2) and (11-3), which are basically consistent with the assurance factor of strength in the specification before revision.

RCC quality evaluation shall be based on the compressive strength of the concrete of design age (90 d or 180 d), and the average value and the minimum value of RCC strength shall meet the following requirements simultaneously:

$$m_{f_{cu}} \geq f_{cu,k} + Kt\sigma_0 \tag{11-1}$$

$$f_{cu,min} \geq 0.75 f_{cu,k} (\leq 20 \text{ MPa}) \tag{11-2}$$

$$f_{cu,min} \geq 0.80 f_{cu,k} (> 20 \text{ MPa}) \tag{11-3}$$

Where $m_{f_{cu}}$ = average concrete strength, MPa

$f_{cu,k}$ = standard strength of the concrete of design age, MPa

K = qualification determination coefficient, which is selected from the Table 11.6 according to the statistics group number of the acceptance batch n

t = probability coefficient, which is shown in Appendix Table B of Section 12.2.4.5

σ_0 = standard deviation of concrete strength in the acceptance batch, MPa

$f_{cu,min}$ = minimum value in n groups, MPa

Table 11.6 K value of qualification determination coefficient

n	2	3	4	5	6 – 10	11 – 15	16 – 25	> 25
K	0.71	0.58	0.50	0.45	0.36	0.28	0.23	0.20

Note: 1. The concrete in the same acceptance batch consists of the concrete with same strength standard, and the basically same mix proportion and production process.

2. When the calculated strength standard deviation of the concrete in the acceptance batch is less than 0.06, it shall be taken as 0.06 $f_{cu,k}$.

11.2.4.5 The Calculation Method of the Concrete Average Strength $m_{f_{cu}}$, the Standard Deviation σ and the Strength Factor of Assurance P

I The concrete average strength ($m_{f_{cu}}$) is determined according to the following formula:

$$m_{f_{cu}} = \frac{\sum_{i=1}^{n} f_{cu,i}}{n} \tag{11-4}$$

Where $m_{f_{cu}}$ = average strength of specimens of n groups, MPa

$f_{cu,i}$ = strength of the specimen of the ith group, MPa

n = group number of specimens

Ⅱ The percentage P_s of the concrete strength standard deviation σ and the strength no less than the designed strength standard value, which is calculated according to the following formula:

①Standard deviation

$$\sigma = \sqrt{\frac{\sum_{i=1}^{n} f^2_{cu,i} - nm^2_{f_{cu}}}{n - 1}} \qquad (11\text{-}5)$$

②Percentage

$$P_s = \frac{n_0}{n} \times 100\% \qquad (11\text{-}6)$$

Where $f_{cu,i}$ = strength of concrete specimen in the ith group during statistical period, MPa

n = number of groups with concrete specimens have the same strength standard value during statistical period

$m_{f_{cu}}$ = average strength of concrete specimens in n groups during the statistical period, MPa

n_0 = number of groups that the strength of the specimens is no less than the standard strength as required during the statistical period

The calculation formula of the concrete strength standard deviation σ_0 of the acceptance batch is the same as that of the σ.

Ⅲ Assurance factor of strength P

①Calculation of the probability coefficient t

$$t = \frac{m_{f_{cu}} - f_{cu,k}}{\sigma} \qquad (11\text{-}7)$$

Where t = probability coefficient

$m_{f_{cu}}$ = average strength of concrete specimens, MPa

$f_{cu,k}$ = design standard strength of the concrete, MPa

σ = Standard deviation of concrete strength, MPa

②The relationship between the assurance rate P and the probability coefficient t can be found in Table 11.7.

Table 11.7 Relationship between the assurance rate P and the probability coefficient t

Assurance rate P (%)	65.5	69.2	72.5	75.8	78.8	80.0	82.9	85.0	90.0	93.3	95.0	97.7	99.9
Probability coefficient t	0.40	0.50	0.60	0.70	0.80	0.84	0.95	1.04	1.28	1.50	1.65	2.00	3.00

11.2.4.6 Quality Evaluation of RCC Core Drilling

The drilling for sampling is a comprehensive method to evaluate the RCC quality. According to relevant strength tests and regulations of *The Specifications for Hydraulic Concrete Construction* DL/T 5144, the cores for mass concrete test can be sampled for 2 – 10 m per 10 000 m³ concrete, and the specified position of drilling for sampling is determined according to the specific circumstance of the engineering construction. The molded standard cube specimen which is sampled at the mixer outlet cannot reflect a series of construction operations after concrete outgoing, including the quality difference caused by the transport, leveling, rolling and curing, therefore, the method of drilling for sampling is often adopted in the field comprehensive evaluation of RCC quality. The drilling for sampling can be conducted after the RCC reaching to the design age. The number and position of the drilling for sampling can be determined according to the bidding documents, design requirements and needs. The content of the drilling for sampling includes:

1. The recovery rate of core samples: it evaluates the RCC homogeneity.

2. The pump-in test: it evaluates the RCC impermeability.
3. The physical and mechanical property test of core samples: it evaluates the RCC homogeneity and mechanical property.
4. The fracture position and morphological description of core samples: it describes the fracture morphology, and makes statistics of numbers of the core sample fractures in different kinds of interlayer bonding positions in the compacted layer, and calculates the ratio of the above numbers to the total fracture numbers, so as to evaluate the interlayer bonding complies with the design requirements.
5. The appearance description of core samples: it evaluates the RCC homogeneity and compaction, and the evaluation standard can be seen in Table 11.8.

Table 11.8   Appearance evaluation standard of RCC core samples

Grade	Surface smoothness	Surface compactness	Aggregate distribution uniformity
Good	Smooth	Dense	Even
General	Basically smooth	Slightly holed	Basically even
Bad	Unsmooth	Several holes	Uneven

Note: The table is applicable to the core samples drilled by a diamond bit.

11.3 Other Quality Control Measures and Discussions of RCC

11.3.1 RCC Defect Treatment

The RCC quality defects include surface failure, surface evenness (faulting of slab ends and distortion), pitted surface, honeycomb, cavity, poor interlayer bonding, poor bonding of heterogeneous RCC, and poor bonding, water seepage and cracks of RCC and the bedrock. According to the defect positions, the defects can be divided into RCC surface defects and the internal defects. During construction, the reason and treatment measures of the RCC quality defect found and the quality after treatment shall be tested and evaluated.

11.3.1.1 RCC Defect Repair

1. The surface quality defects of RCC are: the surface damage, surface evenness (faulting of slab ends and distortion), pitted surface, honeycomb, cavity, and surface crack which can be treated simply according to the general RCC defect treatment method. The common repair methods: backfilling and screeding with cementing materials (cement mortar or epoxy mortar) after the artificial chiseling, or repairing by polishing machines to meet the flatness requirements, and fine cracks and cracks on the surface shall be chiseled and then backfilled. The treatment effect of the surface quality defects shall meet the requirements of relevant regulations and specifications.
2. The RCC internal quality defects include poor interlayer bonding, poor bonding of heterogeneous RCC, and poor bonding, water seepage and cracks of RCC and the bedrock. The internal quality defects have a great impact on RCC quality, which may even endanger the safety of the dam.
a. The RCC internal quality defects, such as poor interlayer bonding, poor bonding of varied RCC, and poor bonding, water seepage and cracks of RCC and bedrock occurred during core drilling, ultrasonic geophysical exploration and pump-in process, are mainly caused by factors like the failure of the cushion cementing material, aggregate separation, the overhead of large aggregates concentration, the interlayer without effective

treatment due to the long interlayer interval time, the polluted layer surfaces, and missing vibration, and feature locality, dispersity and difficulty in treatment. The suspicious placing parts are arranged with boreholes to pump-in, with wet hole pumping water and injecting quantitatively, so as to measure the RCC recovery rate of core samples and volumetric weight, record the drilling process and the appearance of RCC core samples to test the severity and range of RCC internal defects, so that the treatment methods and measures can be determined. Cement grouting is often used for such defects. Sometimes, chemical grouting is also used according to the hazard degree.

b. Treatment of RCC cracks. The cracks can be divided into the surface crack, the deep crack and the penetration crack in form, and desiccation crack, temperature crack and stress crack in cause. The cracks with greater harmfulness destroy the integrity of structures, change the stress state of structures, and cause water seepage, water leakage and steel-bar corrosion, which lower the durability of structures and endanger the safety operation of structures. Therefore, it is necessary to take each crack found seriously, analyze the causes of cracks, and conduct reinforcing treatment in strict accordance with relevant requirements.

The crack detection is the basis to find out the crack shape, analyze the causes and hazards, and formulate the treatment plan. The detection modes include general investigation in low-temperature seasons and the periodical observation for the major cracks. The detection methods include surveying and mapping of the surface, air pressure test, pump-in test, drilling for sampling detection, borehole camera, borehole TV, telerecording and acoustic sounding. The following results shall be proposed by detection: a. the crack position, length, width, depth, tendency, dip separation, water seepage and calcium precipitation from seam; b. the opening change of major cracks and its relationship with the temperature and load; c. joint surface condition, the cracks connect with the RCC overhead accidents, as well as the reserved joints and pipelines of adjacent structures.

The surface cracks of the mass RCC are generally treated simply, while the deep and penetration cracks must be treated seriously. Due to the different distribution of the cracks, their influences on structures are different, and the treatment requirements and standards are not exactly the same. The cracks in important positions must be submitted to the design organization for the design of crack treatment and reinforcement scheme, so as to ensure the safety operation of the structures.

11.3.1.2 Crack Treatment and Reinforcement Measures of RCC

1. The surface treatment measures of the crack surface are laying splicing reinforcement along the crack, caulking of groove chiseled in the joint mouth, pasting or painting the impermeable and leaking stoppage materials. The sealing material of chiseled groove generally is epoxy mortar or pre-shrinking cement motor or slightly expanded pre-shrinking cement mortar and others.
2. The common measures of the crack grouting are cement grouting and chemical grouting.

11.3.2 Quality Control Under Special Environment

11.3.2.1 Quality Control Characteristics of RCC Construction during High Temperature Period

The quality control characteristics of RCC construction during high temperature period are the temperature control and the control of the VC value at the outlet. The temperature control mainly includes the temperature of the RCC at the outlet and the temperature of the block surface. The temperature control at the outlet is the same as that of the normal concrete, by controlling the temperature of the mixtures, i. e., the temperature control can be achieved by lowering the aggregates and mixing water temperature. The control measures of the block surface temperature are taken according the following aspects:

1. Protection of transportation facilities. All equipment transporting the RCC like the dumpers and belt conveyors are provided with sunshades, so as to prevent the temperature flowing backward of RCC during

transportation.
2. Reduction of the block surface temperature. Through spraying on the placing surface, the block surface microclimate can be created, which can make the temperature inside the placed concrete lower than the outside temperature by 4 – 6 ℃, meanwhile, it can keep the humidity of the block surface at 60% – 80%.
3. Timely coverage of block surface. Plastic cloth or other water retaining materials (wet sack and others) are adopted on the block surface for coverage to prevent the RCC from being directly exposed to the sunlight, so as to isolate the temperature and prevent the temperature flowing backward and water evaporation.

The control of the *VC* value is mainly for the characteristic of rapid water evaporation of RCC during high temperature period, it is controlled under 3 s generally, based on the principle of field rolling without sinking, the rollability and the quality of interlayer bonding can be effectively improved.

11.3.2.2 Quality Control Characteristics of RCC Construction During Low Temperature Period

The quality control characteristics of RCC under low temperature are how to increase the temperature of RCC in the outlet and the control of placing temperature, and how to prevent the RCC from freezing. The following measures are adopted to increase the temperature at the outlet:
1. Aggregate preheating. In the aggregate bin and the batch bin, the tubes used for the cooling water in summer are filled with steam to warm the aggregates, and all belt conveyors transported aggregates are filled with warm steam under sealing condition to prevent the temperature loss during transportation.
2. Hot water mixing. The method of heating with steam (supply steam into the water tank) is adopted for the RCC mixing water to increase the mixing water temperature.

The placing temperature is controlled mainly by the method of heat accumulation, mainly including:
1. Thermal insulation by coverage. The thermal insulation quilts are covered during the RCC transportation, and after the RCC leveling or rolling, the thermal insulation quilts shall be covered immediately. The thermal insulation quilts shall be covered for curing after the removal of formwork.
2. The insulated formworks are adopted, and the foam boards with a thickness of 3 – 5 cm are pasted on the inner side (not the permanent surface) or the outside of the formworks, so as to prevent fresh RCC from freezing.
3. Warming of the block surface. The warming by a stove is adopted inside the placed concrete from the preparation of the block unit to the screeding of block unit, so as to keep the temperature of the foundation plane (or the old RCC surface) above 0 ℃.

In order to prevent the RCC from freezing, in addition to the above temperature control, the antifreezing agent with appropriate amount is mainly adopted. The antifreezing agent has a great impact on the later strength of RCC (which can be compensated by adjusting the mix proportion), its content must be strictly controlled and demonstrated by tests before use.

For example, the site of the Longshou Project Dam is in Hexi Corridor, with a typical inland climate, hot summer, cold winter and large evaporation. The temperature in summer can reach above 35 ℃, the lowest temperature in winter can reach – 30 ℃, and the annual evaporation is 8.03 times the precipitation. According to the conventional high temperature period from the middle June to the middle August, and the low temperature period of November and December, the RCC construction cannot be conducted, the effective construction period in a year is only four months. The dam is in the "V" shape valley with difficult placement method and limited placing strength, no matter from the construction period or the overall economic benefit, the limitation that the RCC placement cannot be conducted under the extreme temperature environment is required to be solved. Therefore, under the high temperature period and the ice period (0 – 15 ℃), the RCC construction technology in dry and cold area has been studied, through the strict quality control and the construction measures under cold climatic conditions, the RCC construction can be guaranteed. Now, the

Project has been operated for almost 10 years with good quality.

11.3.3 Quality Management Chart

In order to achieve the whole process and comprehensive quality inspection and control of RCC construction, the mathematical statistic analysis is the most basic method. The new standard requires that the quality data obtained in each procedure of RCC construction (including the quality inspection results of materials, production process parameters and product quality parameters) shall be analyzed for mathematical statistics by quality management charts. The common quality management charts are the followings.

11.3.3.1 Pareto Diagram and Stratification

Pareto diagram is the graphic method adopted to find out the major quality problems or the main reasons influencing quality, it applies the principle of "critical minority and secondary majority" to sort out various problems. Stratification is also called the classification method or the grouping method, and it is the method to classify the data collected according to different aims, and then process the data. This method is often used in combination with other methods, in case of the combination with a paretodiagram, i.e. the "impermeable and leaking stoppage material".

11.3.3.2 Questionnaire Dorms and Cause-and-effect Diagram

A questionnaire form is a chart to collect and collate data and investigate causes, and conduct roughly analysis on this basis. Common questionnaire forms are waste items questionnaire form, defect location questionnaire form and matrix questionnaire form. A cause-and-effect diagram is a diagram that shows the relationship between quality characteristics and causes, it is drew by the stepwise trace ad analysis from many aspects through the quality characteristic phenomenon, based on this diagram, the cause-and-effect of a problem can be clear at a glance.

11.3.3.3 Scatter Diagram

Scatter diagram is also called correlation diagram, showing the relationship between two variables. It judges whether the various factors have an impact on product quality and the degree of impact by establishing a functional relationship between the two.

11.3.3.4 Histogram Method

Histogram is a method to analyze and master the distribution of quality data and estimate the non-conforming product rate in the process by processing and sorting the data.

11.3.3.5 Control Chart Method

Control chart is a method to display quality fluctuations in the process that production changes with time by the chart. It is conducive to analyzing and judging whether the fluctuation is caused by systemic or accidental reasons, so as to remind the operator of making correct judgments in time, taking countermeasures to eliminate the systemic influence, and keeping the process in a stable state.

11.3.4 Quality Management Assurance System

11.3.4.1 Project Manager as the First Person Responsible for Quality

Quality management and assurance system is an important guarantee for quality inspection and control. The project manager of each project is the first person responsible for quality, the deputy manager in charge of production is the person responsible for quality, and the chief engineer is the person responsible for quality technology.

11.3.4.2 Adherence to the "Three-inspection" System

"Three-inspection" system. The three-level quality inspection departments for initial inspection, mutual inspection and final inspection shall have a complete institution and sufficient staff. The "three-inspection" system focuses on initial inspection. Strengthen the quality inspection force of the construction team and shifts, allocate full-time quality inspectors, give full play to the role of the grassroots quality inspectors, and control the first quality inspection.

11.3.4.3 Responsibility of Quality Inspectors

Regularly conduct quality inspector evaluation. Remove the quality inspectors with poor sense of responsibility, low quality, lax review and no competence for quality inspection from the quality inspection team through evaluation. At the same time, establish the work archives of quality inspectors, regularly assess and record the work conditions of quality inspectors, and strengthen the sense of responsibility of quality inspectors, so as to promote the improvement of the overall quality improvement of the quality assurance system.

11.3.4.4 Guaranteed Operation of the Quality Assurance System

Ensure the operation of the quality assurance system by using the ISO 9000 series of standards. On the one hand, strengthen the self-inspection of the quality system, and improve the non-compliance with the requirements of the system documents according to the ISO 9000 series of standard systems. On the other hand, organize and conduct the daily supervision and review of the quality system, and carry out the additional review to the severity of the frequency of quality accidents (defects), so as to promote the continuous and normal operation of the quality system.

11.3.4.5 Standardization and Institutionalization of Quality Work

Under the premise of establishing and improving the quality management and assurance system, the new standard requires that technicians and test and measurement equipment shall be provided, and a set of technical management and quality control system shall be formulated according to the project scale and the needs of quality control and management, to ensure that quality work of each project is standardized and institutionalized, and implemented truly and effectively.

11.3.5 Discussion on the Measurement and Calculation Method of the Percentage of Water Content in Sand

11.3.5.1 Analysis on the Influencing Factors of the Measurement of the Percentage of Water Content in Sand

The percentage of water content in sand is extremely sensitive to the influence of the water consumption of RCC mixtures and is the key on the adjustment of the water consumption of RCC mixtures. Before the concrete was mixed, the test personnel paid great attention to the percentage of water content in sand to measure the percentage of water content. But the VC value or slump of the fresh concrete often has a great fluctuation, and the workability of the mixture is beyond the control range when the measured percentage of water content in sand is used to calculate and adjust the water consumption.

The concrete mixing system is specially set up with various raw material storage tanks (silos) for coarse and fine aggregates, cement, admixtures, additives, water, etc.. The main factors affecting the changes in the measured value of the percentage of water content in sand are related to the amount of storage of sand storage tanks (silos) in the mixing plant. As sand has a very strong filtration effect, if sand tanks (silos) in the mixing plant are fully stored, the percentage of water content at the bottom of the tank will be much higher than the that measured; if sand tanks (silos) in the mixing plant are partially stored (generally less than one-third), the percentage of water content in sand in the tank is much lower than that measured. Under the different storage conditions of the sand tank (silo), if the water consumption is still calculated and adjusted according to the measured percentage of water content in sand, the VC value or slump of the fresh concrete will fluctuate greatly, easily mixing nonconforming materials.

Therefore, the experienced test personnel not only calculate and adjust the water consumption according to the detected percentage of water content in sand, but also observe the storage situation of the sand tank (silo) at any time, and correct the percentage of water content in sand based on experience, to ensure that the VC value or slump of concrete mixtures is controlled within the designed range.

11.3.5.2 Analysis on Calculation Method of the Percentage of Water Content in Sand

The percentage of water content of coarse and fine aggregates of hydraulic concrete involves the percentage of water content of saturated surface dry state of aggregates, which is completely different from the percentage of water content of aggregates in industrial and civil constructions (absolute dry conditions). The percentage of water content in coarse aggregates is small due to characteristics, therefore, the specification or textbook adopts the positive multiplication for the calculation the water content in coarse and fine aggregates. That is: water content of coarse aggregate = mass of coarse aggregates × percentage of water content.

However, the percentage of water content in fine aggregate sand calculated by positive multiplication will cause errors in the water content, which will have certain impact on the VC value or slump of the mixture. Why is there certain error in calculation of amount of water content in sand by the positive multiplication? Different calculation methods of water content in sand are used for contrast, as shown in Table 11.9. If the percentage of water content in sand is calculated as per 5% and two different calculation methods are used, there are differences in the results when the consumption of water and sand is adjusted. The positive multiplication uses more water of 1.97 kg/m³ and less sand of 1.97 kg/m³. However, the reverse division calculation considers that the adjusted sand still has the percentage of water content of 5%, while the positive multiplication method does not consider that, so the two calculation methods are different. In actual construction, the test personnel used the reverse division to calculate and adjust the consumption of water and sand for the percentage of water content in sand, to improve the consistency of the mix proportion and the workability of the mixture.

Table 11.9   Contrast of different calculation methods for water content in sand

Item	Unit dosage	Positive multiplication calculation	Division method
Water content	90 kg/m³	Positive multiplication calculation formula: water content = sand × percentage of water content	Reverse division calculation formula: water content = sand ÷ (1 − percentage of water content)
Concrete sand	750 kg/m³		
Moisture content	5.0%		
Moisture content	kg/m³	37.5 = 750 × 5.0%	39.47 = 750 ÷ (1 − 5.0%) − 750
Sand consumption at the percentage of water content of 5.0%	kg/m³	787.5 = 750 + 37.5	789.47 = 750 ÷ (1 − 5.0%)
Calculate the adjusted unit water consumption when the percentage of water content is 5.0%	kg/m³	52.5 = 90 − 37.5	50.53 = 90 − 39.47
Difference in water consumption	kg/m³	+1.97 = 52.5 − 50.53	
Difference in sand amount	kg/m³	−1.97 = 787.5 − 789.47	
Result analysis		The positive multiplication uses more water of 1.97 kg/m³ and less sand of 1.97 kg/m³	

11.4 Application of Nuclear Densimeter in RCC

11.4.1 Foreword

In the construction, the compactness of RCC is mainly detected by the nuclear density-moisture gauges, so as to control the quality of rolling compaction. In addition to detecting concrete, the nuclear densimeter can also detect the compactness of soil, compound and asphalt. It is fast and convenient, data-based, and highly accurate. In order to adapt to the rapid construction of the RCC main dam of the Baise Multipurpose Dam Project, the compactness quality of RCC is quickly tested by the nuclear densimeter. It has been widely used due to test accuracy and simple and fast operation. The article mainly introduces the application of the nuclear densimeter in RCC construction of the Baise Multipurpose Dam Project for reference.

11.4.2 Working Principle of Nuclear Densimeter

At present, most of the nuclear densimeters used to detect the water content, density or compactness of stone, concrete, asphalt concrete and other materials are surface-type nuclear density-moisture gauges, which are characterized by the small size, light weight, simple operation, and fast detection, so they are especially suitable for the quality inspection of large area compaction density. There are transmission method and backscatter method to measure density. The thickness of RCC is generally above 10 cm, in order to improve the measurement accuracy, the transmission method is used to measure the compactness of RCC. During construction, the compaction density controls the rolling quality of RCC. There are two nuclear density-moisture gauges used by Baise Multipurpose Dam Project to test the compactness of RCC, they are TROXLER3440 made in US and K2030 made in China. The instruments have different characteristics although the working nature and test methods are basically the same. Now the use of two different types of nuclear densimeters in the main concrete construction of the Baise Multipurpose Dam Project is given a brief introduction for reference.

The working principle of the nuclear density-moisture gauges produced at home or abroad is basically the same. Density measurement is that photon rays emitted by the radioactive source penetrate the measured material, some are absorbed and others penetrate the measured material, and finally reach the radioactive source detector; the ray count measured by detector reflects the density of the material within the set measurement time. The radioactive source is usually cesium-137, and the radioactive source detector is usually a Geiger-Müller gas counter or a scintillation detector composed of sodium iodide scintillator and photomultiplier. The moisture is measured with the fast neutron moderation method, that is, the fast neutron emitted by the neutron source enters the measured material. After multiple collisions with the nucleus of the hydrogen substance (for example, water) in the measured material, it is decelerated to form thermal neutrons. The more hydrogen atoms in the measured material, the more thermal neutrons. Within the set measurement time, the thermal neutron count measured by the thermal neutron detector reflects the water content in the material. The neutron source is usually americium-241-beryllium or plutonium-239-beryllium, and the thermal neutron detector is usually helium-3 proportional gas counter, also boron trifluoride proportional gas counter or lithium glass scintillation detector. Because of the moisture count rate is much lower than the density count rate, the longer measurement time shall be used when measuring water content, in order to reduce statistical fluctuation errors and improve the accuracy of measurement.

11.4.3 Operation of Nuclear Densimeter

11.4.3.1 Major Technical Parameters

Main technical parameters of nuclear density-moisture gauges are shown in Table 11.10.

Table 11.10　　Main technical parameters of nuclear density-moisture gauges

Technical parameters	Model TROXLER3440	Model K2030
Maximum depth of density measurement (cm)	30	30
Depth measurement level (cm)	10, 15, 20, 25, 30	10, 15, 20, 25, 30
Density measurement range (kg/m^3)	1 100 – 2 700	1 200 – 2 700
Moisture measurement range (kg/m^3)	0 – 645	0 – 640
Density measurement accuracy (transmission method)	$\sigma \leqslant \pm 0.2\%$	$\sigma \leqslant \pm 0.2\%$
Density measurement accuracy (backscatter method)	—	$\sigma \leqslant \pm 0.6\%$
Moisture measurement accuracy	$\sigma \leqslant \pm 2.0\%$	$\sigma \leqslant \pm 1.0\%$
Saving calibration curve (group)	4	6 + 1
Measuring time (s)	15, 60, 240	15, 30, 60, 120, 240
Operation keys (nos.)	20	8 + 1 (night light)
Display mode	Digital display with LCD 4 × 16	Digital display with LCD 4 × 20
Unit weight (kg)	13.2	16.5

11.4.3.2 Standard Count

Before on-site test or when there is doubt about the work status of the instrument, the standard count of the instrument shall be measured and inspected, and used after being acceptable upon inspection. During standard count measurement, the standard block shall be placed on the dry and flat surface of compacted stone, concrete or other structure materials with density no less than 1.6 g/cm^3 and water content not greater than 0.24 g/cm^3. The instrument is placed on the standard block, and the source rod of the instrument is in a safe position. The surface of the standard block and the bottom surface of the instrument shall be in good contact, without grease and dirt.

Ⅰ　Operation of TROXLER3440 instrument made in US for standard count measurement

1. Press ⟨ON⟩ to turn on, and the instrument will conduct self-inspection for 300 s. Operate the instrument continuously when it is normal after self-inspection. Otherwise, a prompt is displayed and it is automatically turned off after 10 s.
2. Lay the standard block flat, put the instrument after self-inspection between grooves of the standard block, and the right side is close to the metal plate at the bottom of the standard block.
3. After pressing ⟨STANDARD⟩, press ⟨YES⟩ twice continuously, and then the instrument will start measuring and counting for 240 s. It is acceptable if two "P" are displayed after the count is completed. Press ⟨YES⟩ to the next step. It is unacceptable if "F" is shown. Press ⟨NO⟩ to redo this step.

Ⅱ　Operation steps of K2030 instrument made in China for standard count measurement

1. Start operation after turning on the power switch for 10 minutes to stabilize the high voltage.
2. First put the standard block on the packing box of the instrument, and then place the instrument flat on the standard block for alignment.
3. After pressing ⟨Depth/Time⟩, press ⟨Enter⟩ to make adjustment, so that DP = SDP, TM = 240, and then press ⟨Measure⟩ for measurement for 240 s. The SD and SM values displayed at the end of the measurement are compared with the previous standard count value, if the error is less than 2%, proceed to the next step, otherwise, test again.

11.4.3.3 Setting the Optimal Density Value

The optimal density value is generally the maximum wet density, determined by the test in laboratory according

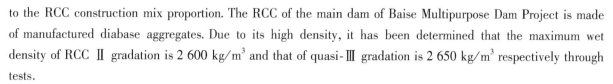

to the RCC construction mix proportion. The RCC of the main dam of Baise Multipurpose Dam Project is made of manufactured diabase aggregates. Due to its high density, it has been determined that the maximum wet density of RCC Ⅱ gradation is 2 600 kg/m^3 and that of quasi-Ⅲ gradation is 2 650 kg/m^3 respectively through tests.

Ⅰ Setting of the optimal density value of TROXLER3440 instrument made in US
1. Firstly select the test object, press ⟨SHIFT⟩, ⟨MODE⟩ and ⟨1⟩ continuously to enter the soil (concrete) test method.
2. Then press ⟨PROCTOR/MARSHALL⟩, ⟨YES⟩ and ⟨1⟩ to select wet density, press ⟨2⟩ (dry density) to enter the maximum wet density value, and press ⟨START/ENTER⟩, ⟨YES⟩ and ⟨NO⟩ to complete.

Ⅱ Setting of the optimal density value of K2030 instrument made in China
1. Press ⟨Function⟩, and then press ⟨Increase⟩ for 4 times. After PG = 04 is displayed, press ⟨Enter⟩.
2. Press ⟨Enter⟩ again to adjust the position of the arrow and select the wet density OM (or dry density OD). After the selection is completed, press ⟨ + ⟩ or ⟨ − ⟩ to set the maximum density value.

11.4.3.4 Setting of Compensation Value

Because of the difference between the measured material and the reference material used by the manufacturer to calibrate the instrument, the density and moisture measured by the instrument are different from the actual density and moisture of the material. Therefore, it is necessary to calibrate the nuclear density-moisture gauges to determine the compensation value, so that the error between test results and density or moisture of material is less than the values required in the specification.

Ⅰ Setting of compensation value of TROXLER3440 instrument made in US
1. Press ⟨OFFSET⟩ and ⟨1⟩ in turn (⟨2⟩ for moisture, ⟨3⟩ for groove) to select wet density compensation.
2. Select and press ⟨YES⟩, ⟨YES⟩, ⟨1⟩ or ⟨2⟩ in sequence. After entering the compensation value, press ⟨ENTER⟩ to finish. Open the compensation value when working.
3. Compensation value = actual value − measured value.

Ⅱ Setting of the compensation value of K2030 instrument made in China
1. Press ⟨Function⟩, and then press ⟨Increase⟩ for twice. After PG = 02 is displayed, press ⟨Enter⟩.
2. Press ⟨Enter⟩ again to adjust the position of the arrow and select the density compensation BD or moisture compensation BM. After the selection is completed, press⟨ + ⟩or⟨ − ⟩to set the compensation value.
3. Compensation value = measured value-actual value.

11.4.3.5 Setting of the Measurement Time

Select the appropriate measurement time according to the test and actual condition, in order to improvethe measurement accuracy and the service life of the instrument after extending the charging.

Ⅰ Setting of test time of TROXLER3440 instrument made in US
Press ⟨TIME⟩, and then select ⟨1⟩, ⟨2⟩ or ⟨3⟩. (⟨1⟩ for 15 s, ⟨2⟩ for 60 s and ⟨3⟩ for 240 s)

Ⅱ Setting of the test time of K2030 instrument made in China
1. Press ⟨DEPTH/TIME⟩ and then press ⟨ENTER⟩ to make the arrow stop at DP.
2. Press and hold ⟨ + ⟩ or ⟨ − ⟩ to select the measurement time required.

11.4.3.6 Setting of Test Depth

The choice of test depth also affects the measurement accuracy. Generally, the concrete rolling thickness of 30 cm is selected as the test depth of the instrument.

Ⅰ Setting of test depth of TROXLER3440 instrument made in US
Press ⟨SHIFT⟩ and ⟨DEPTH⟩ in turn, and then press ⟨2⟩ (⟨1⟩ for manual), to select automatic depth test mode, and then press ⟨ENTER⟩ to complete the setting.

Ⅱ Setting of the test depth of K2030 instrument made in China

1. Press ⟨DEPTH/TIME⟩ and then press ⟨ENTER⟩ to make the arrow stop at TM.
2. Press and hold ⟨ + ⟩ or ⟨ − ⟩ to set the measurement depth.

11.4.3.7 Field Calibration

Under the different tested materials and different measurement conditions, the nuclear densimeter shall be calibrated on site before testing to determine the compensation value, so that the test results of the instrument are consistent with the actual value of the tested material. The field calibration of the instrument can be conducted with the density and moisture reference material method, or the in-situ sampling method. The measurement time of the instrument shall be 240 s during field calibration.

Ⅰ　Density and moisture reference material method

The density and moisture reference material method is to use the measured materials on site, to simulate the onsite construction method according to the construction mix proportion of RCC, to make a concrete reference material of known volume and weight, and calculate its density. Then use the nuclear density-moisture gauges to measure the reference material by transmission method. Compare the measurement result of the instrument with the actual density of the reference material. In case of a difference, the compensation value can be adjusted to make the error between measurement results of the instrument and the density of the concrete reference material within the required range. The dimensions of the reference material shall be no less than 60 cm × 45 cm × 35 cm. The concrete shall be filled and compacted in layers during the production process. The layer thickness can be 5 cm or 10 cm, and the top surface of the reference material shall be flat. The process of the moisture reference material is similar to the density reference material.

Ⅱ　In-situ sampling

The in-situ sampling method is to calibrate the nuclear densimeter at the construction site. For the concrete rolled, select several representative measuring points. First, measuring points are measured by the nuclear densimeter, and then the concrete is sampled at the measuring point in situ, and the density of the concrete at the sampling site is calculated by sand filling method. After comparing the mean of values measured with the instrument and the mean of all in-situ sampling values at all measuring points, the difference between the two is determined as the compensation value. During in-situ sampling, the sampling site shall be between the radiative source rod and the detector, and the sampling depth is the measurement depth of the instrument.

In the Baise Multipurpose Dam Project, the nuclear densimeter is subject to field calibration with the density reference material method and the in-situ sampling method. Before the on-site construction of the RCC, the laboratory produced two reference materials of Ⅱ gradation and Quasi Ⅲ gradation according to the mix proportion, and calibrated the instrument before construction. After the construction of RCC, the instrument is calibrated mainly by in-situ sampling method and once every 3 months, to keep the instrument in normal working condition.

11.4.3.8 On-site Testing

Ⅰ　Point setting

After the rolling for 10 minutes, start to select measuring points. The surface of the measuring points shall be as flat as possible, avoiding start and stop of vibration of the vibrating roller, and arrange a measuring point every 100 – 200 m². Baise Project requires a measuring point in every 100 m², and at least 3 measuring points are required. The measuring points are arranged in a grid manner.

Ⅱ　Drilling hole

When drilling holes, the four corners of the guide plate shall be used to draw the position line of the instrument for measurement, so that the source rod can be inserted smoothly to the measuring hole, the measuring hole shall be perpendicular to the measuring surface, and the depth of the hole is slightly larger than the test depth of the source rod. The size of the measuring hole should not be too large, otherwise it may influence the

Chapter 11 RCC Quality Control and Project Cases

measurement results, when the hole is too large, the source rod of the instrument can be moved gently along the direction of the detector, so as to make the source rod attach to the wall of the measuring hole closely.

Ⅲ Survey

During the field measurement, the compensation value, the maximum wet density, test depth, test time and other relevant parameters shall be set, after setting, the source rod of the nuclear densimeter will be inserted to the measuring hole as the test depth required, and the press of the measuring button can conduct the test of RCC compactness. The control standard of the RCC compactness is: the compactness of internal concrete is greater than 97% and that of external concrete is great than 98%. When an instrument is measuring or performing the standard count, there shall be no radioactive source within 10 m around the instrument, and no large structures within 3 m. The distance between people and the instrument shall be kept more than 4 m.

11.4.4 Application of Nuclear Densimeter in Construction

11.4.4.1 Determination of Measuring Depth and Time

The measuring depth and time of the nuclear densimeter can influence the RCC test results, therefore the tests are conducted for the different measuring depth and time of the instrument respectively, and the test results of the several measuring points can be seen in Table 11.11.

Table 11.11 Compactness test results of nuclear densimeter in different measuring depth and time

Survey point	Gradation	Measuring depth (cm)	Test time (s)	Wet density (kg/m³)	Degree of compaction (%)
A	Quasi Ⅲ	10	15	2 554	96.4
		10	30	2 553	96.3
		10	60	2 561	96.6
		20	15	2 610	98.5
		20	30	2 608	98.4
		20	60	2 603	98.2
		30	15	2 625	99.1
		30	30	2 618	98.8
		30	60	2 630	99.2
B	Quasi Ⅲ	10	15	2 587	97.6
		10	30	2 573	97.1
		10	60	2 581	97.4
		20	15	2 624	99.0
		20	30	2 624	99.0
		20	60	2 622	98.9
		30	15	2 616	98.7
		30	30	2 601	98.2
		30	60	2 601	98.2
C	Quasi Ⅲ	10	15	2 560	96.6
		10	30	2 561	96.6
		10	60	2 557	96.5
		20	15	2 610	98.5
		20	30	2 605	98.3
		20	60	2 599	98.1
		30	15	2 620	98.9
		30	30	2 601	98.2
		30	60	2 607	98.4

The test results show that the deviation between the measuring results with the same point location and depth in different test time is not large; the deviation between the measuring results with the same point location and test time in measuring depth respectively 20 cm and 30 cm is not large. Due to the inclined-layer rolling, in order to be convenient for hole forming and the source rod of the instrument to insert into the samling hole, 2/3 of the RCC thickness is the measuring depth, i. e. , 20 cm, and the measuring time is 15 s.

11.4.4.2 Determination of the Optimal Rolling Times

Under the selected condition, the rolling times of the vibrating roller is related to the concrete compaction density. In case of insufficient rolling times, the compactness cannot meet the design requirements, while the excessive rolling time will damage the internal structure of the concrete, anddecrease the density. In order to improve the use efficiency of the vibrating roller and speed up the construction process, the optimal rolling times must be tested. When the VC value is controlled within 4 – 6 s, other test parameters shall be adjusted, and the statistics of the instrument test results is shown in Table 11.12.

Table 11.12   Relationship between rolling times and compaction density

Rolling form	Compaction times	traversing speed (km/h)	Wet density of compaction (kg/m^3)	Degree of compaction (%)	Gradation
No vibration + vibration	2 + 2	1.0	2 581	97.3	Quasi III
No vibration + vibration	2 + 4		2 642	99.7	
No vibration + vibration	2 + 6		2 678	101.1	
No vibration + vibration	2 + 8		2 691	101.5	
No vibration + vibration	2 + 10		2 682	101.2	
No vibration + vibration	2 + 2	1.2	2 541	95.9	Quasi III
No vibration + vibration	2 + 4		2 594	97.9	
No vibration + vibration	2 + 6		2 658	100.3	
No vibration + vibration	2 + 8		2 677	101.0	
No vibration + vibration	2 + 10		2 665	100.6	
No vibration + vibration	2 + 2	1.5	2 504	94.5	Quasi III
No vibration + vibration	2 + 4		2 566	96.8	
No vibration + vibration	2 + 6		2 639	99.6	
No vibration + vibration	2 + 8		2 672	100.8	
No vibration + vibration	2 + 10		2 686	101.4	

Test results show that when the traveling speed of the vibrating roller is 1.0 km/h, the design requirements can be met in case of "2 + 6" times of rolling, and the compaction density is the maximum in case of "2 + 8" times of rolling; when the traveling speed of the vibrating roller is 1.2 km/h, the compaction density is the maximum in case of "2 + 8" times of rolling; and when the traveling speed of the vibrating roller is 1.5 km/h, the compaction density is the maximum in case of "2 + 10" times of rolling. The optimal rolling times determined are "2 + 6" times, i. e. , after twice of rolling without vibration, then the rolling with vibration for 6 times is conducted.

11.4.4.3 Use of Instrument

The wet density is used to control the rolling quality of the RCC of the main dam of the Baise Multipurpose Dam Project, the test results of the nuclear densimeter only record the wet density and compactness, and no record for the water content. Due to the longer daily high temperature of the Baise, in order to satisfy the field construction of RCC, the smaller VC value is adopted to control. There is lots of fog at night, in addition of the unstable water content test of the instrument, the density test is relatively stable.

The US TROXLER3440 instrument has light weight and long working hours, and it can be used for 150 h after being charged once, however buttons are all in English, the operation is more complicated. The China K2030 instrument is heavy, it can be used continuously for 50 h after being charged once, however it is simple and easy to operate with nigh light on the display screen, which is convenient for the instrument to be used at night.

11.4.5 Nuclear Densimeter Maintenance and Personnel Safety Protection

11.4.5.1 Maintenance of Instruments

1. Turn on the nuclear densimeter 10 minutes in advance before use to stabilize the high pressure.
2. After the use of the instrument in each shift, extend the source rod, clean the cement mortar on the source rod quickly by towels, and also wipe up the entire instrument, then put it into the packaging box of the instrument.
3. After the continuous use of the instrument for one week, wipe up the inner wall of the source rod, and apply the anti-wear hydraulic oil to ensure the lubrication.
4. After the continuous use of the instrument for one months or when the source rod is difficult to be extended, open the shutter access panel at the bottom of the instrument, take out the shielding shutter, brush off the dust and the cement mortar with a brush, then apply oil and reassemble the instrument, so as to ensure the flexibility in use.
5. Handle the instrument with care in use and in the process of movement, and avoid shaking, dampness and water of the instrument.
6. In case of insufficient power, it shall be charged in time, and it is noted that the charging time is 18 h, and do not overcharge.
7. When the instrument is continuously tested and used on site, the field calibration is necessary every 3 – 6 months.
8. The instrument must be kept and used by a person specially assigned.

11.4.5.2 Safety Protection of Personnel

1. The operators must know the performance and structure of the instrument, master the radioactive protection knowledge, and be familiar with the operation instructions before work.
2. When the instrument is not in use, place the handle of the instrument in a safe position.
3. When cleaning the shutter and the source rod, the operators must wear protective clothing, or find a shielding cover, and clean the instrument quickly, so as to reduce the radiation from the radioactive source.
4. When the instrument is not in use for a long time, place it in an unoccupied room with signs, so as to prevent the unauthorized movement by non-operators.
5. The operators shall be replaced regularly, so as to avoid the long-term radiation hazard.

11.4.6 Conclusion

The nuclear densimeter is a special testing instrument, which needs to be returned to the manufacturer for maintenance after damage with high maintenance cost. In addition to enhancing the normal maintenance of the instrument, in order to avoid the unnecessary damage and lower the repair rate of the instrument, and ensure the operators to standardized the operation and take good care of the instrument, the necessary safety protection equipment shall be provided for the operators, so as to make the operators focus on their work, and the instrument be in a stable and normal state for a long time.

Chapter 12

Core Drilling, Pump-in and In-situ Shear Tests

12.1 Overview

RCC construction technology determines the multiple layer joint surfaces in the concrete, and the multiple layer joint surfaces are the weak links, which is unfavorable for the imperviousness performance and anti-sliding stability of the dam. The impermeability and shear resistance capability of RCC body are not inferior to these of the normal concrete, while the structural characteristic of the multiple layer joint surfaces is a major factor to influence the imperviousness performance, and the permeating characteristic is an important index to evaluate RCC quality. If the water flow into the dam body along the layer surface or the weak part, the pore water pressure and uplift pressure of the dam body will be increased, the anti-sliding stability of the dam body will be decreased, and the seepage water will take away the $Ca(OH)_2$ and other ingredients of the concrete structure, which can influence the strength and durability of the concrete. Therefore, the interlayer bonding quality of RCC is always the focus.

The interlayer bonding quality of RCC has been the focus of the research in the field and the content of the national scientific and technological breakthrough topic. Before the national key scientific and technological research for the "Eighth Five-year Plan", the RCC gravity dam was constructed with the dam height no more than 150 m, and for the gravity dam with height over 150 m, the normal concrete was used in the lower part, the RCC is used in the upper part, and they formed a combined gravity dam with the height over 150 m. The result of the national key scientific and technological research for the "Eighth Five-year Plan" is that the RCC can be used to construct a dam with the height of 216.5 m like the Longtan Dam. Practices have proved that the construction of RCC gravity dam with the height of 200 m has been the reality. A series of key technological problems of the high RCC dams have been solved in the national key scientific and technological research for the "Eighth Five-year Plan", and fruitful achievements have been achieved, in which the shear and stability analysis of the layer surface is one of the important achievements.

In early stage, so-called "RCD" method is adopted in the RCC imperviousness system in China, so RCC is defined as super hard or hard concrete, with the VC value of mixtures within 15 – 25 s. Meanwhile, lacking of understanding and research of the rock powder can make the RCC mixtures dry, aggregate separation, and poor cohesiveness and rollability, and the layer joint surface of the concrete after rolling and compaction is easy to be the so-called "layer-cake", which has seriously restricted the rapid development of RCC damming technology for a time. In conclusion, the interlayer bonding quality is the key factor to influence the RCC rapid damming technology.

In 1993, the Puding Arch Dam in China had been groundbreaking for the full-section RCC damming technology, hereafter, the full-section RCC damming technology is widely used in China. Through a large number of test researches and project practices, the technical and engineering personnel makethe hard RCC mixtures gradually transmit to the hypoplastic concrete without slump. Meanwhile, the design of the mix

Chapter 12 Core Drilling, Pump-in and In-situ Shear Tests

proportion is tested in close connection with the performance of mixtures, with the dynamic control of rock powder content, setting time and *VC* value and in-depth study of interlayer bonding quality, the rollability of construction is effectively improved, so that the surface of the concrete after rolling and compaction is in full liquefied bleeding and elastic, which ensures that the aggregates of the upper RCC are embedded into the lower concrete which has been rolled, and the interlayer bonding concerned is also easily solved.

From the review of RCC core drilling, field pump-in and in-situ shear tests, the core samples have been developed form the length of 60 cm by core drilling in early 1986 to the current super long length more than 16 m, and a core sample with a length more than 10 m is common. The overall evaluation of most RCC core drilling and pump-in test shows that the permeability could not meet the design requirements, and the friction coefficient f' and the cohesive strength c' are greater than the designed control indexes. The appearance of core samples is smooth and compact with uniform aggregate distribution, no matter the hot joints caused by continuous spreading and rolling or cold joints, the interlayer bonding is good without obvious layer joint surface, and the core drilling fully proves that the RCC rapid damming technology has been mature. Although the length of the core samples cannot completely represent the imperviousness performance quality of RCC dams, it reflects the change of RCC performance from certain extent, which is very beneficial to improve the imperviousness performance of the dam body.

With the great achievement of RCC damming technology in China, we can also soberly aware the problems of interlayer bonding quality and imperfect points during RCC damming. For example, a RCC high dam has the high recovery rate of core samples, longer length, close interlayer bonding, and low permeability in the pump-in test, however, it may be not always true for all core samples and the pump-in tests. For a specific RCC dam, due to the rough construction technology and extensive construction, which are mainly in the drilled core samples, there are more or less quality problems, such as lots of fractures and pores, overhead aggregates, poor adhesion of the layer joint surface and high permeability in the pump-in test. The conditions of some small and medium dams are especially serious, which may have low recovery rate of core samples and high permeability. For example, in a RCC dam, the permeability is far greater than the design value 1 Lu. After impoundment, the downstream surface of the dam has serious water seepage, so the reservoir has to be emptied and grouted again. There are many lessons in this respect, which need to be paid great attention to and worthy of our deep thought.

Core drilling, pump-in test and in-situ shear test are the most intuitive test methods to evaluate the interlayer bonding quality, imperviousness performance and anti-sliding stability of RCC dams. *Specifications for Hydraulic Concrete Construction* DL/T 5114 clearly points out that proper core drilling and compaction tests shall be carried out for completed concrete structures. Coring and pump-in tests of mass concrete shall be conducted at 2 – 10 m per 10 000 m^3 concrete. Specific core drilling positions, testing items and pump-in test positions, and moisture absorption evaluation standards shall be determined according to the specific conditions of project construction. Drilling, processing and testing of concrete core samples can be carried out according to CECS03. Core drilling is also specified in *The Construction Specification for Hydraulic Roller Compacted Concrete* DL/T 5112—2009: core drilling is a comprehensive method to evaluate the quality of RCC. The drilling for sampling can be conducted after the RCC reaching to the design age. The location and quantity of boreholes shall be determined as required.

In this chapter, we have a deeper understanding of the core technology of RCC rapid damming, i. e. , " interlayer bonding, temperature control and cracking prevention", through the elaboration of RCC core drilling, field pump-in, core sample performance, and in-situ shear tests and project cases.

12.2 Core Drilling of Dam RCC

12.2.1 Specifications of Core Drilling

The molded standard cube specimen which is sampled at the mixer outlet cannot reflect a series of construction operations after concrete outgoing, including the quality difference caused by the transportation, spreading, rolling and curing. Now core drilling is mostly used to comprehensively evaluate the quality of RCC. As the interlayer bonding quality of RCC is very important, in order to better reflect the interlayer bonding situation, the specifications, bidding documents and design all put forward the requirements of core drilling, field pump-in and in-situ shear test for RCC reaching the age.

Core drilling is also specified in *The Construction Specification for Hydraulic Roller Compacted Concrete* DL/T 5112—2009: core drilling is a comprehensive method to evaluate the quality of RCC. The drilling for sampling can be conducted after the RCC reaching to the design age. The location and quantity of boreholes shall be determined as required.

The content of the drilling for sampling includes.

1. The recovery rate of core samples: it evaluates the RCC homogeneity;
2. The pump-in test: it evaluates the RCC impermeability;
3. The physical and mechanical property test of core samples: it evaluates the RCC homogeneity and mechanical property;
4. The fracture position and morphological description of core samples: it describes the fracture morphology, and makes statistics of numbers of the core sample fractures in different kinds of interlayer bonding positions in the compacted layer, and calculates the ratio of the above numbers to the total fracture numbers, so as to evaluate the interlayer bonding complies with the design requirements.
5. The appearance description of core samples: it evaluates the RCC homogeneity and compaction, and the evaluation standard can be seen in Table 12.1.

Table 12.1　　Appearance evaluation standard of RCC core samples

Grade	Surface smoothness	Surface compactness	Aggregate distribution uniformity
Good	Smooth	Dense	Even
General	Basically smooth	Slightly holed	Basically even
Bad	Unsmooth	Several holes	Uneven

Note: The table is applicable to the core samples drilled by a diamond bit.

12.2.2 Relationship between Core Drilling and Interlayer Bonding

From the blasting of the RCC cofferdam at the downstream of Yantan Hydropower Station in 1991 and the partial guide walls of RCC open channels at the side gate in 1992, many layers were obviously exposed in the blasting, resulting in the so-called "smooth plate" (the so-called "smooth plate" means that the layers are not bonded and the plane fails).

Professor Zhang Zhongqing of Guangxi University made field observation on demolition blasting of downstream cofferdam of Yantan Hydropower Station. The cofferdam at the downstream of Yantan is 39.2 m high, with an axis length of 314.58 m, vertical upstream face, slope 1:0.55 – 1:0.66 of downstream face and 119 700 m³ concrete. The concrete grade is C10, the mix proportion is cement: coal ash: water: sand: stone = 47:101:85:669:1 623, the additive content is 2%, the volumetric weight is 2 450 kg/m³, the 28 d tensile

Chapter 12 Core Drilling, Pump-in and In-situ Shear Tests

strength is 1.62 MPa, and the 90 d compressive strength R_c = 21.6 – 22.5 MPa. The downstream cofferdam concreting with RCC was started from January 20, 1988 and was completed in mid-May. The downstream cofferdam was officially demolished in the first half of 1992, drilled with pneumatic drills and blasted with TNT explosives manually. During the explosion, detailed tests were carried out, and a large number of "smooth plates" were found from time to time, with the maximum smooth plate reaching 20 m × 5 m, which was a most obvious weak link. Such smooth plates were mainly caused by cold joints (i.e., construction joints). RCC pouring shall be stopped to erect formworks after every 6 layers of pouring, or stopped as required by the temperature control, and then the concreting can be continued after 3 – 5 d. Although mortar had been laid on the layers, the upper layer of aggregate cannot be embedded into the concrete on the lower layer, and the bonding is poor, resulting in "smooth plates". Currently, although the construction quality has been greatly improved and the core drilling has achieved good results, the representativeness is still dwarfedby the visual "smooth plate". The layers of continuous pouring are generally covered and rolled before initial setting. Although there are embedded aggregates between the upper and lower layers, there are still weak links due to too many layers, resulting in a terrace failure mode. It shall be pointed out that failure along the layer is the weakest link where sliding is most likely to occur. Although the layer joint surface has been treated, failure along the layer is inevitable, which can only occur if it exceeds the safety factor allowed by the code. If the layer is not properly treated, the safety factor will be reduced.

In 1993, Puding concrete arch dam in Guizhou Province wad drilled for sampling. The total length of core samples was 210 m, the core sample recovery rate reached 98% – 99%, and the 4 longest core samples were 4.2 m. From the outside, the surface of core samples was smooth and dense, and no interlayer can be found. The tests found that the volumetric weight, compressive strength and tensile strength all met the requirements. c' reached 2.753 – 3.664 MPa and f' reached 1.822 – 1.412 in the shear test. The shear fracture surface was mostly not on the plane, which reflected the good interlayer bonding.

From April 1997 to June 1999, ϕ150 mm and ϕ250 mm boreholes were drilled for sampling in Jiangya three times. The total length of core samples was 454.2 m, and the longest core sample reached 6.67 m. Similarly, the surface of core samples was smooth and dense, and the position of the interlayer could not be seen clearly, and the rate of fracture on the surface was less than one tenth. In the shear test of core samples, the c' reached 1.27 – 1.4 MPa, and f' reached 1.03 – 1.15.

From November to December, 1999, ϕ171 mm boreholes were drilled for sampling in dam sections $1^{\#} - 3^{\#}$ of Fenhe River II Reservoir, the samples have a diameter of ϕ150 mm, a total length of 76.38 m, and a recovery rate of 99.93%. The sampling results show that the core samples have smooth surface, dense structure and good cementation. There are 5 core samples with a length more than 8 m, and the longest core sample reaches 8.5 m long. Some sample sections have small pores with a diameter of 2 – 5 mm, and two sections have small honeycombs. There were two or three breaks along the horizontal layer on the concrete of which the construction had been postponed in the winter from 1998 – 1999, which reflected that the protection in winter was not good, and the weak layers were formed in new and old concrete construction joints.

Dachaoshan Dam had been drilled for sampling four times from October 1999 to August 2001, with a core sample recovery rate over 97.9%. The total hole depth was 1 266.94 m, and there were 20 core samples over 7 m long, including 3 core samples over 10 m long, and the longest core sample reached 10.47 m. The surface of the core samples was smooth and dense, and the interlayer cannot be identified. The core sample test results met the design requirements, and so did the shear test results.

Baise Dam was drilled for sampling twice from September 2003 to September 2004, with a total length of 791.14 m, and the core sample recovery rate reached 99% and 96.94% respectively, including 3 core samples over 10 m long, and the longest core sample reached 12.1 m long. The surface of core samples of Baise Dam is smooth and dense without surface pores, and the interlayer cannot be identified, which indicates that the manufactured sand with high rock powder content plays a significant role in RCC.

Afterwards, with in-deep research and continuous improvement of RCC damming technology, especially technical route of low cement content, high admixtures, high rock powder content and low VC value has been adopted for the design of mixing proportion, the grout-mortar ratio has been increased, so that the performance of mixture has got a qualitative change, which effectively improves the rollability of RCC construction. Especially in the layer face treatment process, the key is to control allowable interval time well between upper and lower layers, that is, to ensure that the upper layer of RCC is rolled before initially setting of next layer of concrete, so that the upper layer of aggregate can be embedded into lower layer of concrete, which has extremely improved the impermeability and shear resistance capacity between layers.

In recent years, the record of super-long core samples of dam RCC drilling coring has been constantly refreshed. In 2000, in Dachaoshan Project, the 10 m super-long core sample (10.47 m) record was first broken. Since then, super-long core samples longer than 10 m have been continuing to emerge. In 2007, super-long core samples longer than 15 m, i.e., 15.03 m, 15.33 m, 15.85 m and 15.30 m, were taken out of RCC dams in Longtan, Guangzhao, Gelantan and Jinghong, etc., respectively; In 2009, extra-long core samples of 15.73 m, 16.49 m, 16.44 m and 16.55 m were taken from Jin'anqiao and Kalasuke dams respectively, created the world record of the longest core sample of RCC. See Table 12.2 for the statistics of drilling coring in some projects in China, and see Figure 12.2 to Figure 12.5 for super long core samples.

Table 12.2　　　　　Statistics of drilling coring in some projects in China

S/N	Engineering	Coring date	Total length of coring (m)	Core sample acquisition rate (%)	Longest core sample (m)	Core sample appearance and length description
1	Yantan 13 – 22	1996.04	68.59	98.3	4.2	Coring is carried out in $13^{\#}-17^{\#}$, $22^{\#}-23^{\#}$ dam sections, core sample evaluation: excellent 9.5%, common 69% and poor 24.1%
2		2001.12 – 2002.01	60.86	97.1	—	
3	Puding	1993.08 – 1993.06	298.39	98	4.2	The appearance is smooth and dense, no layer plane can be discovered, and there are total four with the length of whole piece longer than 4.0 m
4	Jiangya	1997.08 – 1999.06	412.57	97	7.56	The appearance is smooth and dense, no layer plane can be seen, and there are total ten with the length of whole piece longer than 6.0 m
5	Fenhe Reservoir II	1998.11 – 1999.12	236.6	99.93	8.53	The appearance is smooth and dense, and is well cemented, and there are total 24 with the length of whole piece longer than 7.0 m

Chapter 12 Core Drilling, Pump-in and In-situ Shear Tests

Continued Table 12.2

S/N	Engineering	Coring date	Total length of coring (m)	Core sample acquisition rate (%)	Longest core sample (m)	Core sample appearance and length description
6	Mianhuatan	1999.07 – 1999.12	257.26	97.96	8.38	Average section length is 0.96 m, and there are total six with the length of whole piece longer than 7.0 m, and the excellentrate is 85.88%
7	Gaobazhou	1999.09 – 2003.03	798.03	—	3.35	
8	Dachaoshan (four times)	1999.10 – 2001.08	599.76	97.9	10.47	Smooth and dense appearance; a total of 3 core samples and 20 core samples with length of more than 10.0 m and 7.0 m respectively
9		1999.10 – 2001.08	667.18	96.0		
10	Hongpo Reservoir	2000.07	81.6	97.3	6.3	Dense inner surface and fine interlayer bonding
11	Shapai	1999.10 – 2002.08	126.63	98.5	5.13	Smooth and dense appearance, good interlayer bonding and compact cementation; a total of 4 core samples with length of more than 10 m
13		2003.01 – 2003.02	70.44	99.7	13.15	
14	Linhekou	2001.12 – 2003.06	447.76	98.29	10.57	A total of 3 core samples with length of more than 8.0 m
15	Phase III cofferdam of the Three Gorges Project	2003.08 – 2003.09	62.10	—	12.3	A total of 3 core samples with length of more than 8.0 m
16	Baise	2003.07 – 2003.09 (1st dry season)	305.04	99.0	11.0	The appearance is smooth and dense, no layer plane can be seen, and there are total two with the length of whole piece longer than 8.0 m
17		2004.07 – 2004.09 (2nd dry season)	486.1	96.94	11.98	Smooth and dense appearance of core samples; a total of 3 core samples with the whole piece length of more than 10.0 m
18		2004.07 – 2004.09 (subcontracting 1# – 3# dams on the left bank)	61.75	95.03	2.4	Using the same mix proportion and raw materials; rough construction technology and poorer core samples

Continued Table 12.2

S/N	Engineering	Coring date	Total length of coring (m)	Core sample acquisition rate (%)	Longest core sample (m)	Core sample appearance and length description
19	Jinghong	2004.09 – 2004.11	274.25	99.87	8.16	Using natural aggregate; a total of 8 core samples with length of more than 4.0 m
20		2006.09 – 2006.10	—	98.84	14.13	Using double admixtures containing natural aggregate; smooth and dense appearance; being the longest core sample in 2006
21		2007.09 – 2007.11	800	98.6	15.30	Using natural aggregates; smooth and dense appearance and uniform distribution of aggregates; a total of 4 core samples with length of more than 10.0 m
22	Gelantan	2007.07 – 2007.8	182.03	99.86	15.85	Smooth and dense appearance is smooth and dense; taking 2 core samples with length of more than 13 m from one hole
23	Longtan	1# – 21# dams on the right bank	1 168	99.1	15.03	Smooth and dense appearance; a total of 268 m core sample with the whole piece length of more than 5.0 m, accounting for 25%
24		Dam on the left bank (inlet)	245.3	99.7	12.67	Smooth and dense surface and uniform distribution of aggregates; a total of 2 core samples with length of more than 10.0 m
25	Guangzhao	2007.01 – 2007.3	549.02	99.5	14.73	Smooth and dense appearance; a total of 6 core samples with the whole piece length of more than 10.0 m
26		2007.10 – 2007.12	492.4	99.8	15.33	Smooth and dense appearance, uniform distribution of aggregates, good interlayer bonding and compact cementation

Continued Table 12.2

S/N	Engineering	Coring date	Total length of coring (m)	Core sample acquisition rate (%)	Longest core sample (m)	Core sample appearance and length description
27	Jin'anqiao	2008.05 – 2008.06 (1st time)	171.54	99.56	11.03	Smooth and dense appearance; a total of 3 core samples with the whole piece length of more than 10.0 m
28		2008.12 – 2009.01 (2nd time)	347.97	99.7	15.73	Smooth and dense appearance; a total of 6 core samples with the whole piece length of more than 10.0 m
29		2009.03 – 2009.05 (3rd time)	272.66	97.47	16.49	Smooth and dense appearance; a total of 2 core samples with the whole piece length of more than 10.0 m
30	Kalasuke Hydropower Station	2008.03 – 2009.5 (1st dry season)	118.2	100		Smooth and dense surface of most core samples, and uniform distribution of aggregates
21		2009.03 – 2009.05 (2nd dry season)	416.19	100	16.55	Obtaining 4 long core samples with length of more than 10 m; the intact rate of layer joint surface – 97.82%

Figure 12.1 *Guangzhao RCC core samples of* 15.33 m

Why is it possible to obtain the super-long core samples continuously by RCC core drilling in recent years? On the contrary, there are a few long cores drilled in normal concrete, which is mainly due to the entirely different pouring and vibrating methods of normal concrete compared with the pouring method of

Figure 12.2 *Gelantan RCC core samples of* 15.85 m

Figure 12.3 *Jin'anqiao RCC core samples of* 15.73 m

Figure 12.4 *Jin'anqiao RCC core samples of* 16.49 m

Chapter 12　Core Drilling, Pump-in and In-situ Shear Tests

Figure 12.5　*Kalasuke RCC core samples of* 16.55 **m**

RCC. The compaction of RCC pouring is obtained under the actual rolling compaction of vibrating roller. Some people think that it is the result of the improvement of core drilling technology. In fact, in different parts of the dam, not all the core samples taken are long core samples. Some core samples still have the defects such as poor bonding of layer joint surface and obvious non-compaction, which shows that the improvement of core coring level is not the main factor of obtaining long core samples, and the increasing core sample length essentially reflects the continuous improvement of the RCC damming technology and the quality.

With the continuous improvement of RCC damming quality, people pay great attention to the RCC dam quality inspection technology while constantly exploring construction technology, improving material performance and learning the world's most advanced level. How to correctly test the permeability and quality of RCC, especially, the field precision test of permeability of RCC is related to the correct evaluation of imperviousness effect of RCC. In the quality inspection of RCC dam, Hunan Xiangshui Foundation Engineering Technology Development Co., Ltd., as the main testing unit for core drilling and pump-in test, has explored and researched a set of method, equipment and technology suitable for RCC core sample drilling and field pump-in test from Fenhe River II Reservoir, Jiangya, Dachaoshan, Gaobazhou, Phase III cofferdam of the Three Gorges Project, Linhekou, Jinghong, Guangzhao, Gelantan, Silin, Jin'anqiao and other project practices, which has provided scientific and technical support for quality inspection of RCC of the dam.

12.2.3　Method of Core Drilling

12.2.3.1　Coring Time Determination and Drilling Arrangement

Core sample drilling shall be carried out after RCC is in the design age. Core drilling of the RCC in design age within the specified time may not affect the engineering quality of the dam, but may facilitate obtaining good core samples, so as to test the engineering quality more accurately.

The location and quantity of core drilling shall be determined according to the elevation. In the plane layout, the joint length, distance and position (for example, outlet position) shall also be considered. Generally, 1 – 2 holes are arranged every 30.0 m along the axis of the dam in II gradation concrete area (impervious area) and III gradation concrete area (inside the dam) respectively, but the number of holes in each dam section shall not be less than 2.

When the hole location is arranged, attention shall be paid to avoid failure to the embedded equipment and pipelines in the dam, and there shall be a certain distance from the drainage hole to prevent mutual impact.

12.2.3.2 Technical Requirements for Core Drilling

1. Drilling equipment. Core drill and diamond bit are used for drilling, appropriate diamond bit and reamer are selected according to different positions and borehole diameters, and different rig speed, water supply, pressure and others are determined according to different RCC aggregate gradation, and then the field test is conducted.

    The drill pipe of $\phi 89$ mm or $\phi 114$ mm shall be selected. The bending of drilling tool, drill pipe and other pipe materials shall be less than 0.3%. After thread connection, the coaxiality shall be less than $\phi 0.05$ mm, and the perpendicularity between end face and axis shall be less than 0.10 mm.

2. Setting out of hole location and positioning adjustment of drill. Total station and level are used for setting out. After the drill is in place, it shall be aligned with hole location, its rack is fixed by ground anchor, and it is leveled with theodolite or plumb bob. The clearance between the vertical shaft power head of the drill and the slide rail shall be adjusted properly to ensure that the vertical shaft power head is stable during drilling, and all the factors causing drilling instability of the drill shall be eliminated. The holes can be drilled only after being all set. The drill drills the holes at its minimum speed (generally 40–60 r/min) in the low pressure to ensure the perpendicularity of the hole drilling.

3. Drilling flush fluid (generally clean water). In order to improve the ability of the flush fluid to carry rock powder, avoid the disturbance on core samples caused by deposition of rock powder and disintegrating slag during drilling and timely cool the drill bit to make core samples complete and smooth, the lubricating flush fluid can be used for drilling if necessary. General lubricant can be common detergent, and efficient lubricant can be L-HP and other products.

4. Construction technology requirements. In the dam sections arranged with both pump-in holes and core holes, the inspection holes for pump-in test shall be constructed first, followed by the core holes.

    The location and quantity of core drilling shall be determined according to the elevation. In the plane layout, the joint length, distance and position (for example, outlet position) shall also be considered. Generally, 1–2 holes are arranged every 30.0 m along the axis of the dam in II gradation concrete area (impervious area) and III gradation concrete area (non-impervious area) respectively, but the number of holes in each dam block (or dam section) shall not be less than 2.

    During core drilling, it is required to make uniform numbering according to the sequence of obtaining core samples, draw the borehole histogram, describe the core, fill in number and pack the core samples. During drilling, the core samples shall be obtained to the maximum extent. No matter how long the core samples are, they shall be taken out immediately once they are found to be stuck or worn. For the drilling cycle of 1 m or greater, if the recovery rate of core samples is less than 80%, the cycle depth shall be reduced by 50% for the next cycle until the cycle reaches 50 cm. If the recovery rate of core samples is very low, the drill shall be replaced or the drilling method shall be improved. During drilling, it is required to inspect, observe and record the drilling flush water, drilling pressure, core sample length and other factors that can fully reflect concrete characteristics. No necking core samples are allowed.

12.2.3.3 Core Drilling

1. After the drilling length of the core samples conforms to the requirements, the core samples shall be taken by long core barrel as far as possible. At the bottom of the long core barrel, the clamp spring base and the clamp spring in corresponding specifications are installed to clamp the bottom of the core samples, and then the core barrel is clamped with a clamp plate at the orifice. At the same time, the clamp plate is evenly pressed and fixed by two 5 t jacks, and then the core samples are forcibly pulled off from the bottom of the clamp spring. When the core barrel with clamp spring is lowered into the hole bottom, it is strictly forbidden to lift the core barrel upward to avoid fracture of the core samples. After the core samples are pulled off, the

clearance between the core barrel and the long core samples shall be filled with uniform silty sand with particle size of less than 0.5 mm, so as to protect the core samples during core sample lifting.

If the core samples are with too small diameter and cannot be clamped by the clamp spring, a small amount of clamp material can be added into the notch of 2 mm on the inner surface of the clamp spring to clamp the core samples immediately. Quartz sand or short steel wire with uniform particleswith particle size of about 2 mm can be used as the clamp material for coring.

2. Lifting and storage of the core samples are as follows. The core samples can be lifted by the drill hoist or lifting equipment after being broken. Before lifting, the channel steel with corresponding length is inclined to a support. After the core samples are lifted out, they will be slowly put into the channel steel forward and bound firmly, and then leveled together with the channel steel and the core barrel with core samples. Then, the channel steel and core barrel are lifted into the transport truck by the special lifting beam or crane made of I-beam. The truck is paved with sand in advance and is leveled by triangle sleepers to prevent the core samples from fracture due to uneven road during transportation. After the core samples are transported to the storage place, the clamp spring base is removed. While the silty sand in the core barrel is washed with water, the core samples are being lifted into the channel steel by jack and hoist. In order to keep the core samples for a long time and prevent them from cracking, a layer of transparent protective film can be applied on the surface of the core samples.

12.2.3.4 Backfilling of Core Sample Holes

After drilling for sampling and relevant tests, generally, the core sample holes shall be backfilled by the mortar that is no lower than the concrete grade at this part.

12.3 Field Pump-in Test of RCC

12.3.1 Significance of Pump-in Test

How to correctly test the permeability and quality of RCC, especially, the field precision test of permeability of RCC is related to the correct evaluation of imperviousness effect of RCC. Field pump-in test is an important method to evaluate the permeability of RCC, and its reliability and correctness are very important. For a long time, the evaluation method of RCC is to inspect the RCC by "sampling and moulding in outlet", but the mechanism of stone formation of RCC is different from that of normal concrete. Although the method of measuring the compactness of placing surface after rolling compaction increases, the bonding quality of many layer surfaces formed by continuous rise of RCC still cannot be tested. For high dams, the interlayer bonding quality is particularly important. How to accurately understand the interlayer bonding quality has been the main study topic in the industry for many years.

In the impervious element of RCC dam, in-situ drilling pump-in test is carried out to determine the permeability and provide basic data for quality evaluation. Field drilling pump-in test is a method to evaluate the impermeability of RCC dam. Therefore, the boreholes with different depth are arranged in different positions of the dam for pump-in test, so as to determine the permeability index, thereby understanding and evaluating the overall imperviousness performance of the dam and the constructionquality of the local area of the dam.

12.3.2 Main Drilling Pump-in Testing Equipment

12.3.2.1 Drilling Pump-in Equipment

As the equipment performance and accuracy required by the specifications for drilling pump-in tests in water conservancy and hydroelectric engineering, the tested water supply equipment and test accuracy are far from meeting the requirements for checking concrete imperviousness performance for RCC dams with high impermeability requirements, and the permeability coefficient of RCC is very low, in order to measure the imperviousness performance of RCC more accurately, Hunan Xiangshui Foundation Engineering Technology Development Co., Ltd. has participated in the national "Ninth Five-year Plan" key scientific and technological project 96-220-01-02 special research, developed a set of core drills, precision metering pumps, flow testers, mud pumps and other special equipment that can meet the above requirements, and explored a set of test methods under the guidance of China Institute of Water Resources and Hydropower Research.

According to the requirements for core diameter and drilling pump-in test, the main core drilling and pump-in test equipment is shown in Table 12.3.

Table 12.3   Main core drilling and pump-in test equipment

Equipment description	Model and specifications	Unit	Remarks
Core drilling machine	HGY300、XY-2、XY-1A、XY-4	Set/Unit	Pressure 0 – 1.6 MPa
Precision metering pump		Set/Unit	
Flow tester	H97-2	Set/Unit	
Mud pump	BW-150	Set/Unit	

12.3.2.2 Water Stop Cock

At present, types of the plunger in use include pneumatic, single-pipe jacking, double-tube circulation, etc. However, single-pipe jacking plunger is widely used for its simple operation. At the same time, the length of the plunger shall be greater than 7 times the borehole diameter.

12.3.2.3 Water Supply Equipment

Water supply equipment mainly includes test water pump (precision flow pump and ordinary flow pump), stable air compressor room, water filter, etc. In the pump-in tests, a special rubber tube without joint in the middle is used as the working pipe instead of the drill pipe, which solves the problem of water leakage of the drill pipe joint caused by the common water pump and the drill pipe as the connecting pipe in the conventional pump-in tests.

Test pump shall meet the following requirements:

1. Under the pressure of 1 MPa, the flow can be maintained at 100 L/min. The BW-150 pump selected by the Company can reach up to 150 L/min, while the maximum pressure of precision flow pump used is 1.6 MPa, and the pressure can be adjusted between 0 – 1.6 MPa; the maximum flow is 1.33 L/min, and the flow can be adjusted arbitrarily between 0 – 1.33 L/min.
2. The capacity of the air stabilizing chamber is greater than 5 L.
3. The water filter has two layers of filter screens with pore size less than 2 mm.
4. Stable pressure, even water output and reliable work.
5. Measuring equipment.

Pressure: there are test section pressure gauge and pressure meter for pressure measurement. Generally, pressure gauges are used, but required to be sensitive and the pointer returns to zero after pressure relief. The work pressure of the pressure gauge is kept within 1/3 – 3/4 of the limit pressure value of the gauge.

Flow: the water meter and flow tester are used for measurement, the measurement range is 10^{-5} – 100 L/min, the water seepage in the pump-in test section is greater than 1 L/min, and the water meter is used for measurement. The water seepage in the pump-in test section is less than 1 L/min, and measured by H97-2 flow tester. The principle of the tester is to measure the corresponding volume with a pressure measurement tube, and measure the flow with an electronic watch.

Because the RCC has the large dispersion of permeability, the tester is set with four gears and is assembled by four different measurement pipes, which can meet the accuracy requirements of different flow, and the minimum water seepage is 0. 000 072 L/mim. The flow tester reads using the volume method combined with the electronic watch, and its accuracy reaches the indoor test standard.

12.3.3 Pump-in Test of RCC

12.3.3.1 Testing Method and Status of Permeability of RCC

At present, there are three main methods for studying the permeability of RCC at home and abroad:
1. According to the mix proportion of RCC used in the project, specimens are moulded in the laboratory to determine the permeability of RCC body and layer surface;
2. Core samples are drilled on the poured RCC to determine the permeability of its body and the layer surface;
3. The on-site pump-in test is performed on the poured RCC to determine the actual permeability (including the RCC body, layer surface and special parts of the dam).

The indoor test conditions are quite different from the on-site construction conditions, the indoor test cannot reflect the unevenness of the on-site RCC pouring due to wet sieve, etc.; on-site construction is affected by external factors, and the construction conditions are complex. Although specimens of penetration test for the core samples are from the site, only core samples that can be made into specimens can be measured. If the layer (joint surface) is poorly bonded or with cold joints and less dense when cores are taken, core samples are fractured or broken, and these parts cannot be measured by the core sample penetration test; only through the on-site pump-in test, the insufficiency of the indoor test and the core sample permeability test can be overcome, and the permeability of RCC can be measured more realistically.

At present, the on-site pump-in test of RCC is carried out in accordance with *The Specifications for Drilling Pump-in Tests in Water Conservancy and Hydroelectric Engineering* SL 31—2003, which is suitable for pump-in test of rock mass. Due to the big difference between the two media (rock mass and RCC), although there are structural surfaces similar to rock mass in the RCC layer, the difference is no other structural surfaces except the layer. Especially for RCC with good construction quality, there are very good layer (joint) bounding, low permeability, and high accuracy requirements. So, there are major limitations for the application of current *Specifications for Drilling Pump-in Tests in Water Conservancy and Hydroelectric Engineering* to the on-site pump-in test of RCC.

The on-site pump-in test results of some built RCC dams are 0. 000 or too large, and the permeability coefficients between $10^{-9} - 10^{-3}$ cm/s cannot be tested (permeability coefficient is less than 10^{-9} cm/s, the pump-in time is about 1 h, and the inflow is very small), mainly because the accuracy of the pump-in equipment and measurement devices fails to meet the requirements.

12.3.3.2 Field Pump-in Test Method

I Field pump-in test method

The holes in the pump-in test are drilled with a diamond drill bit, the pore size is generally ϕ75 mm – ϕ91 mm, and the pump-in is blocked by a single plunger from top to bottom. The segment length is determined by the design and contract, which is generally equal to one or a half lift height (3 m or 1.5 m), with or without cold joint layer. In case of great penetration, change the position of the plunger, and then shorten the length of the section, and find the part with great penetration.

Calculate the permeability (Lu) of the test section according to the relevant articles for *Specifications for Drilling Pump-in Tests in Water Conservancy and Hydroelectric Engineering* SL 31—2003. After the pump-in test is completed, the holes are sealed with grouting.

Ⅱ Drill hole layout for pump-in test

Drill hole layout for pump-in test is determined according to RCC part functions, construction technology and material structure.

For the overall impermeability inspection of the dam, the general pump-in test drill holes are arranged in the impermeable area of the RCC dam (Ⅱ gradation), and the drill holes are 1.0 – 3.0 m away from the upstream face of the dam. Each dam block (or dam section) depends on the length of the joint. Generally, 1 – 2 holes are arranged every 30.0 m, but each dam block (or dam section) is no less than 1 hole.

During the layout, avoid damaging the embedded equipment and pipelines in the dam, and have a certain distance from the drainage hole to prevent crossing.

Ⅲ Determination of test section length

The length of the test section in the pump-in test is determined according to the test requirements for RCC. Generally, it is determined according to each lift, with two considerations: one is that each test section length contains a cold joint layer, which is a lift height.

The other is that one lift height is divided into two test section lengths. One section (1/2 lift height) does not contain the cold joint layer surface, which is used to calling the RCC body. The other section (1/2 lift height) contains the cold joint layer surface, and the bottom layer is 0.5 – 0.75 m below the cold joint.

Ⅳ Test pressure of test section

The pressure of the test section is generally 0.3 – 0.6 MPa, and the average pressure of the test with a hole depth of 6.0 – 8.5 m is 0.3 MPa, which shall be subject to no lifting of the RCC surface. The average pressure of each test section below is 0.6 MPa.

12.3.3.3 Requirements for Pump-in Test Process

Ⅰ Drilling and cleaning

The core drill is used for drilling, and the $\phi 75$ mm – $\phi 91$ mm diamond drill bit is used for rotary drilling. The depth of each drilling is the length of the pump-in test section. Before drilling, fix the base of the drill with embedded fasteners to ensure that the vertical axis of the drill and drill hole are vertical.

Lower the drill to the bottom of the hole, and adjust the flow of the water injection pump to the maximum gear. In case of no rock powder in the hole, the reflux is cleaned for the next process.

Ⅱ Test section isolation and equipment installation

Before installing the equipment, check the plunger to see whether the rubber plug is free from deformation, slides flexibly, and the connecting pipe joint is loose. After setting the aforementioned devices as required, place the water stop devices into the borehole by the hoist on the drill. Install the water filter, water supply pump, flow tester, pressure gauge and air stabilizing chamber, etc.

Ⅲ Equipment commissioning

Commission the test devices before each section-time of pump-in test. Carefully observe the stop cocks, connecting pipes and pipeline connections, valves and water supply pumps to ensure that there is no water leakage before the next step. The commissioning order is as follows:

1. Open the precise flow pump for water supply and close the pressure regulating valve, then lift the stop cock to let the water in the boreholes return, and drain the air in the test section and pipelines. After carefully observing and ensuring that there is no rock powder in the reflux, press the rubber plug with the drill to stop the water.

2. When the water is stopped by the compressed rubber plug, turn down the opening of the pressure regulating valve to observe the increasing pressure value on the precise pressure gauge. Stop operating when the pressure reaches or slightly exceeds the design value.

3. Close the water supply valve and open the flow tester. Keep regulating the pressure regulating valve to make

Chapter 12 Core Drilling, Pump-in and In-situ Shear Tests

the test section pressure reach the design value and stable.

Pressure gauge reading shall be calculated according to the following formula:

Pressure gauge reading = test section design pressure − water column pressure of 1/2 of the test section measured by center distance of the pressure gauge.

Ⅳ Observation and readings of pressure and flow

When the pressure reading on the pressure gauge is stable and no leakage is found on the observed parts, observe the flow and time with an electronic stopwatch. According to the technicalrequirements of borehole pump-in test, the flow must be recorded every 5 min. Finish the test when there is no trend of continuous increase in flow and the difference between the maximum and minimum among the 5 continuous flows is less than 10% of the end value.

When there is serious leakage, use a BW − 150 water pump and read on the water meter. Get the readings using the same methods of test process.

Ⅴ Sealing hole

Seal the borehole with grouting and cement mortar after finishing the pump-in test.

12.3.4 Data Compilation for Field Pump-in Test

According to the features of the RCC layering construction, the flow state observed macroscopically could be considered as laminar flow, and no obvious change of crack status may occur during the pump-in test. Therefore, the test is implemented applying the method of one pressure for one stage. Currently, the permeability is calculated according to *Specifications for Drilling Pump-in Tests in Water Conservancy and Hydroelectric Engineering* SL 31—2003. The calculation formula is the following:

$$q = \frac{Q}{L} \cdot \frac{1}{p}$$

Where q = permeability of the test section, Lu

L = length of the pump-in test section, m

Q = flow, L/min

P = pressure value of the test section, MPa

The test results of each section are compiled into a table of pump-in test results of the RCC of each borehole, as Table 12.4 Pump-in test results of RCC Dam of Fenhe River Ⅱ Reservoir.

Comprehensive mathematical statistical analysis is carried out from the test results of each section. The results are generally divided into five levels, namely statistics of $0.000 < q < 0.001, 0.001 \leqslant q < 0.01, 0.01 \leqslant q < 0.1, 0.1 \leqslant q < 1.0$, and $q > 1.0$. Calculate the corresponding percentage and comprehensively analyze the imperviousness conditions of the dam, as Table 12.5 Statistics of permeability (%) of pump-in test in borehole of the Linhekou RCC Arch Dam.

12.3.5 Field Pump-in Test Results in Some Domestics Projects

See Table 12.6 for the field pump-in test results in some domestic projects. The results show that most of the dam body's RCC has a small permeability in the pump-in test, which meets the design requirements. For the high dam, the design permeability is required to be lower than 0.5 Lu for the impervious area, and lower than 1.0 Lu for the inside of the dam. The results also show that grouting has solved the problem of over-permeability for some RCC parts of the dam, which has met the design requirement.

The main reasons for the over-permeability are rough construction technology, weak quality awareness, extensive management and especially the failure in quality control such as the excessive *VC* value, the concrete failing to be poured and rolled in time, and exposed for much time after spreading, the untreated separated

Table 12.4   Pump-in test results of RCC dam of fenhe river II reservoir

Test section No.	Elevation of the borehole test section (m)		Section length (m)	Pressurizing water		
	From	To		Flow (L/min)	Pressure (MPa)	Permeable rate (Lu)
1	906.25	905.20	1.05	0.000 00	0.3	0.000 00
2	905.20	904.15	1.05	0.000 00	0.3	0.000 00
3	904.15	903.10	1.05	0.006 67	0.3	0.021 16
4	903.10	902.05	1.05	0.007 68	0.6	0.012 18
5	902.05	901.00	1.05	0.006 46	0.6	0.010 26
6	901.00	899.95	1.05	0.023 33	0.6	0.037 04
7	899.95	898.70	1.25	0.00333	0.6	0.00444
8	898.70	896.95	1.75	0.002 02	0.6	0.001 92
9	896.95	895.90	1.05	0.000 81	0.6	0.001 28
10	895.90	894.30	1.60	0.001 62	0.6	0.001 68
11	894.30	893.25	1.05	0.006 67	0.6	0.010 58
12	893.25	891.50	1.75	0.020 00	0.6	0.019 05
13	891.50	889.75	1.75	0.003 64	0.6	0.003 46
14	889.75	888.30	1.45	0.004 85	0.6	0.005 57
15	888.30	887.25	1.05	0.020 00	0.6	0.031 75
16	887.25	885.50	1.75	0.013 33	0.6	0.012 70
17	885.50	884.45	1.05	0.006 67	0.6	0.010 58
18	884.45	882.80	1.65	0.000 40	0.6	0.000 41
19	882.80	881.75	1.05	0.002 02	0.6	0.003 21
20	881.75	879.90	1.85	0.000 80	0.6	0.000 72
21	879.90	878.85	1.05	0.000 81	0.6	0.001 28
22	878.85	876.85	2.00	0.016 666	0.6	0.013 89
23	876.85	874.75	2.10	0.003 333	0.6	0.002 65

Note: No.2 dam block, No.2-1 borehole, elevation of the borehole mouth: 912.0 m.

Table 12.5   Statistics of permeability (%) of pump-in test in borehole of the Linhekou RCC Arch Dam.

S/N	Borehole No.	Section-time	Permeable rate (Lu)			
			$10^{-3} \leqslant q < 10^{-2}$	$10^{-2} \leqslant q < 10^{-1}$	$10^{-1} \leqslant q < 1.0$	$q \geqslant 1.0$
1	ysy-1	12	33.3	66.7	—	—
2	ysy-2	30	3.3	93.4	3.3	—
3	ysy-3	30	33.3	60.0	—	6.7
4	ysy-4	12	58.3	41.7	—	—
5	ysz-5	27	25.9	18.5	51.9	3.7
6	ysz-6	27	—	100.0	—	—
7	Total	138	21.0	65.9	10.9	2.2

aggregates, rolling for the time longer than the allowable interval, unpunctuality or failure in spraying to resist solar radiation, and improper moisturizing covering, resulting in poor rollability, liquefied bleeding and compaction. The above-mentioned factors have directly affected the interlayer bonding quality. If the construction is strictly implemented according to construction methods, good interlayer bonding quality could totally be achieved, and the problem of over-permeability for part of the dam section could totally be avoided.

Table 12.6　　　　　Statistics of field pump-in test results in some domestics projects

S/N	Engineering	Date of pump-in test	Section of pump-in test		Permeable rate(Lu)		Remarks
			Length of the borehole (m)	Total section-time	Min.	Maximum	
1	Jiangya	1997.08 – 1999.06	656.85	595	0.007 8	>1.0	RCC of permeability higher than 1.0 Lu accounts for 7.6% – 17.0%
2	Fenhe Reservoir II	1998.11 – 1999.12	320.91	207	0.000 7	0.654 1	Good impermeability
3	Hongpo Reservoir	1998.12 – 2000.07	66.0	21	0	1.000	0.004 2 m/d Permeability coefficient 0.004 2 m/d
4	Mianhuatan	1999.07 – 1999.12	223.3	44	0		The permeability of 5# dam is lower than 1 Lu after grouting
5	Dachaoshan	1999.10 – 2001.08	2 734.9	758	0	<1.0	97.2% is less than 1.0 Lu
6	Hongpo Reservoir	2000.07 – 2000.08	—	—	0.019	1.0	The permeability of the upstream face is lower than 0.20 Lu
7	Shapai	1999.10 – 2002.08	—	33	0	0.43	17-borehole water pump-in test
8	Linhekou	2001.12 – 2003.06	275.8	138	0.002	1.02	More than 1.02 Lu is 1 section-time
9	Baise	2003.07 – 2003.09 (1st dry season)	108.9	26	0	0.38	The pump-in test results meet the design requirements
		2004.07 – 2004.09 (2nd dry season)	181.5	36	0	0.77	The pump-in test results meet the design requirements
10	Jinghong	2004.09 – 2005.10	—	198	0.01	—	RCC of permeability lower than 1.0 Lu accounts for 91.9%. The permeability does not meet the requirement, grouting is needed
11	Gelantan	2007.07 – 2007.08	172.73	33	0.07	—	RCC of permeability lower than 1.0 Lu accounts for 87.9%. The permeability does not meet the requirement, grouting is needed

Continued Table 12.6

S/N	Engineering	Date of pump-in test	Section of pump-in test		Permeable rate (Lu)		Remarks
			Length of the borehole (m)	Total section-time	Min.	Maximum	
12	Longtan	$1^{\#}-21^{\#}$ dams on the right bank	2 170.4	693	0	0.993	The permeability does not meet the requirement, grouting is needed
		Dam sections $22^{\#}-35^{\#}$ at the left bank	1 011.2	342	0	0.738	
13	Guangzhao	2007.01 – 2007.03	320.5	106	0.005	0.218	Good overall impermeability
		2007.10 – 2007.12	212.0	68	0.011	0.536	Good overall impermeability
14	Jin'anqiao	2008.05 – 2008.06 (1st time)	108.0	36	0.02	0.5	Good overall impermeability
		2008.12 – 2009.01 (2nd time)	66	22	0.05	0.8	Good overall impermeability
15	Kalasuke Hydropower Station	2009.04 – 2009.05 (2nd dry season)	97.87	18	0.05	0.95	The permeability does not meet the requirement, grouting is needed

12.4 Performance Test of RCC Core Samples

12.4.1 Regulations on Core Sample Test

Test Code for Hydraulic Concrete SL 352—2006, *Construction Specification for Hydraulic Roller Compacted Concrete* DL/T 5112—2009 have made specific regulations on test of RCC core samples:

Generally, the diameter of the core sample shall be 3 times the maximum particle size of the aggregate, at least not less than twice. The diameter of the core sample for measurement of compressive strength shall be 150 – 200 mm. For the part of the concrete with maximum aggregate particle size less than 80 mm, core samples with diameter more than 200 mm or above shall be adopted. 3 specimens are grouped for the compressive strength and spliting tensile tests, and other performance tests are carried out in accordance with the test code.

Take the core samples of 2.0 height-to-diameter ratio as the standard specimens. See Appendix A of standard DL/T 5112—2009 for the ratio of the compressive strength of core samples with different height-to-diameter ratios and the standard specimens with a height-to-diameter ratio of 2.0. The core samples with a height-to-diameter ratio less than 1.5 shall not be used for compressive strength test. See Appendix A for the conversion relation of the compressive strength between a standard specimen of $\phi 150 \text{ mm} \times 300 \text{ mm}$ and a 150 mm cube specimen. See Table 12.7 for the conversion coefficient of the compressive strength between the various height-to-diameter ratios, and between a cylinder specimen and a cube specimen.

Table 12.7   Conversion coefficient of compressive strength

Strength grade (MPa)	Conversion coefficient of compressive strength of the specimens with various height-to-diameter ratios		ϕ50 mm × 300 mm Compressive strength / ϕ150 mm cube compressive strength
	1.5	2.0	
10 – 20	1.166	1.0	0.775
20 – 30	1.066	1.0	0.821
30 – 40	1.039	1.0	0.867
40 – 50	1.013	1.0	0.910

Note: 1. The conversion coefficient of the height-to-diameter ratio of 1.5 – 2.0 can be obtained by interpolation method.

2. Conversion coefficient of compressive strength of the specimens with various height-to-diameter ratios = $\frac{\text{Compressive strength of the specimens with various height-to-diameter ratios}}{\text{Compressive strength of the specimen with a height-diameter ratio of 2.0}}$.

3. The height-to-diameter ratio of the specimens for the elastic modulus test, axial tensile strength test and tensile deformation test is 2.0 – 3.0.

12.4.2 Content and Processing Method of Core Sample Tests

12.4.2.1 Content of Core Sample Tests

The core samples for the RCC test are classified and subdivided according to the requirements of the design and related procedures and specifications for cutting and milling processing, and tested according to the number and elevation of the boreholes. The core sample tests include apparent density (volumetric weight), compressive strength, tensile strength, shear strength, ultimate tensile value, tensile elastic modulus, static compressive modulus of elasticity, impermeability and frost resistance, etc. Before the test of the core samples, sketch and take pictures as required, and preserve the core samples as per the required designated place and method.

12.4.2.2 Processing Method of Core Sample Tests

The processing methods for testing various properties of RCC core samples are implemented according to the height-to-diameter ratio requirements of the *Test Code for Hydraulic Concrete* SL 352—2006, among which the height-to-diameter ratio of the specimens for tests of compressive strength, compressive modulus of elasticity and tensile strength is 2:1.

The GEVR is processed according to the height-to-diameter ratio requirements of the Technical Specification for *Testing Concrete Strength with Drilled Core Method* CECS 03:88, among which the height-to-diameter ratio of the specimens for compressive strength test is 1:1, and that for static compressive modulus of elasticity and tensile strength is 2:1.

The processing method of core samples for testing impermeability grade is as follows: cut the core sample into an approximate square specimen with a height of 150 mm (the length and width shall fit in the test mould for impermeability). The test mould for impermeability is a truncated cone with an upper opening diameter of 175 mm, a lower opening diameter of 185 mm, and a height of 150 mm. Fill the gap between the specimen and the impermeability test mould with high-grade expansive mortar, and then cure the processed specimen for 14 d after demoulding to ensure good combination between the impermeability specimen and expansive mortar. The processing of the specimen for interlayer impermeability is basically the same as that of the specimen for body impermeability. The only difference is that the layer must be placed longitudinally during molding to ensure that the layer is parallel to the water seepage direction for the impermeability test.

The processing method of core samples for shear test: cut the ϕ150 mm core sample into a cylinder with a height of 200 mm, and ensure that the layer of the specimen for interlayer shear test must be in the middle of the cylinder height. ϕ150 mm cylinder shear boxes are adopted for the shear test.

12.4.3 Result Analysis of Core Sample Performance Test

The author has made statistics on the results of the performance test of RCC core samples in some domestic projects. See Table 12.8 for details.

Table 12.8 Statistics of results of RCC core sample performance test in some domestic projects

S/N	Engineering	Test result statistics	Apparent density (kg/m³)	Compressive strength (MPa)	Ultimate tensile value ($\times 10^{-4}$)	Modulus of elasticity (GPa)	Grade of impermeability (W)	Frost resistance grade	Remarks
1	Yantan	Average	2 475	24.2	—	30.0	—	—	14#、15#、17# dam section
2	Puding	Average	2 497	36.1	0.81	3.98	>S7	—	R_{180}200 Ⅱ gradation
3		Average	2 518	38.0	0.72	4.12	>S5	—	R_{180}200 Three-graded
4	Mianhuatan	Average	2 464	31.3	0.58	28.0	>S8	—	R_{180}200 Ⅱ gradation
5		Average	2 469	30.2	0.53	28.4	>S4	—	R_{180}150 Three-graded
6	Dachaoshan	Average	2 612	24.6	0.76	26.8	>S8	—	R_{90}150 Three-graded
7		Average	2 560	26.6	0.86	32.7	>S10	—	R_{90}200 Ⅱ gradation
8	Hongpo Reservoir	Maximum	2 521	26.1	—	41.0	—	—	Imperviousness of Ⅲ gradation RCC
9		Minimum	2 479	24.3	—	34.4	—	—	
10	Shapai	Average	2 466	27.2	—	16.2	—	—	R_{90}200 Ⅱ gradation
11		Average	2 471	28.6	1.45	15.2	>W8	F150	R_{90}200 Three-graded
12	Linhekou	Average	2 504	29.3	0.6	30.4	—	—	RCC Ⅱ gradation
13		Average	2 507	30.0	0.53	30.6	—	—	RCC Three-graded
14	Baise	Average	2 626	25.7	0.78	30.07	>S10	—	Ⅱ gradation
15		Average	2 646	19.2	0.74	29.8	>S2	—	Quasi Ⅲ gradation
16	Gelantan	Average	2 427	24.5	0.93	—	>W8	F100	C_{90}20 Ⅱ gradation
17		Average	2 438	19.9	0.89	—	>W4	F50	C_{90}15 Three-graded
18	Longtan (right bank)	Average	2 494	32.1	—	74.3	>W6	F50	R Ⅰ C_{90}25 Three-graded
19		Average	2 499	30.8	—	40.2	>W6	F50	R Ⅱ C_{90}20 Three-graded
20		Average	2 488	34.0	0.42	45.0	>W12	F25	R Ⅳ C_{90}25 Ⅱ gradation

Continued Table 12.8

S/N	Engineering	Testresult statistics	Core sample performance test result						Remarks
			Apparent density (kg/m^3)	Compressive strength (MPa)	Ultimate tensile value ($\times 10^{-4}$)	Modulus ofelasticity (GPa)	Grade of impermeability (W)	Frost resistance grade	
21	Longtan (left bank)	Average	2 489	26.10	—	—	>W6	—	Three-graded
22		Average	2 478	35.6	0.56	—	>W12	—	II gradation
23	Sunlight (lower part dam)	Average	2 441	33.6	0.83	34.5	>W12	F100	C_{90}25W12F150 Level II
24		Average	2 491	29.7	0.69	36.5	>W12	F100	C_{90}25W12F150 Level III
25		Average	2 489	21.9	0.57	36.2	>W8	F75	C_{90}20W6F100 Level III
26	Sunlight (higher part dam)	Average	2 405	32.2	0.61	33.1	>W12	F100	C_{90}25W12F150 Level II
27		Average	2 479	31.6	0.59	35.5	>W12	F75	C_{90}20W10F100 Level II
28		Average	2 510	28.0	0.54	28.5	>W10	F75	C_{90}20W6F100 Level III
29		Average	2 539	22.0	0.52	27.5	>W6	F25	C_{90}15W6F50 Level III
30	Jin'anqiao (three times)	Average	2 555	27.9	0.70	30.9	>W8	F75	C_{90}20 W8F100 Level II
31		Average	2 641	24.1	0.58	29.5	>W8	F75	C_{90}20W6F100 Level III
32		Average	2 630	25.6	0.56	30.8	>W6	F75	C_{90}15W6F50 Level III
33	Kalasuke (one dry season)	Average	2 416	36.5	0.81	38.5	>W8	F200	R_{180}200F300W10 Level II
34		Average	2 404	30.4	078	2.63	>W8	F75	R_{180}200F200W10 Level II
35		Average	2 420	28.7	0.72	28.1	>W4	F25	R_{180}200F50W4 Level III

12.4.3.1 Apparent Density

The apparent density (volumetric weight) is the quality of the core sample divided by its volume. The diameter and height of the core samples are measured for 4 times, and calculate the mean value. The apparent density of RCC core samples in some domestic projects is generally 2 404 – 2 646 kg/m^3. The apparent density of core samples is generally higher than the design requirements, and their apparent density is significantly higher than that of the normal concrete, which is affected by higher content of aggregates and good compaction of RCC. Results also show that the variation of the apparent density of the core samples is significant, reflecting the great influence of the properties of different aggregate qualities on the apparent density.

12.4.3.2 Compressive Strength

The compressive strength of the core samples meets the design requirements. Generally, high strength indicates that the strength of RCC has not been the main index of controllability. On the one hand, the high strength of RCC is related to high compaction, and on the other hand, to the high fly ash content and the high strength growth rate of RCC in the later period.

12.4.3.3 Ultimate Tensile Value

Results of the core samples show that due to the small content of cement in RCC, and the fully-graded test and disturbance of the core samples, the ultimate tensile value of the core samples is relatively low and generally fails to meet the design requirements. Although the ultimate tensile values of some core samples are very low, the modulus of elasticity is very high. The analysis shows that this is mainly related to the mix proportion design, rock powder content, cementing mortar and VC value. The performance of aggregate type has especially great influence on the ultimate tensile values.

12.4.3.4 Static Compressive Modulus of Elasticity of Core Samples

The static compressive modulus of elasticity of a core sample truly reflects the stress of concrete. Generally, the main factors affecting the modulus of elasticity of RCC are related to the type of aggregate and the maximum particle size. The mix proportion is also the main factor that affects the modulus of elasticity. For example, the modulus of elasticity of RCC with rich cementing mortar, low VC value and high air content is generally low.

12.4.3.5 Impermeability Grade Test of Core Samples

The results of the core sample impermeability test show that the impermeability of the RCC is good, and all meet the design requirements.

12.4.3.6 Frost Resistance Test of Core Samples

The results of the frost resistance test of core samples do not meet the design requirements, which are as same as the results of the frost resistance test of normal concrete core samples. Because the bubbles in concrete are unstable, they are easy to migrate and merge into the large bubbles during concrete construction, and the air content of the RCC migrates and loses rapidly under the operation of vibrating roller. A great deal of test results show that the air content of a core sample is generally only about 50% of that of the fresh concrete, which is far lower than the air content design requirement. The concrete core samples drilled in the built hydraulic concrete structures for the frost resistance test have much lower test results compared to the specimens for indoor frost resistance test, which is a common phenomenon in most projects.

12.5 In-situ Shear Test of RCC Dams

12.5.1 Explanation for In-situ Shear Test

The RCC is spread in a thin-lift without longitudinal joints and continuous rolling compaction. The general thickness of each rolling layer is 30 cm. A large number of layers tend to cause invisible layered structure, which affects the bonding strength of the layer surfaces. The reasons for this are as follows: the fresh RCC has poor cohesiveness, the aggregate is not well wrapped and easy to be separated; the gap concreting time between the layer surfaces is too long, the layer surface has initially set, which affects the cohesion between the upper and lower layers; the over-high VC value causing poor rollability results in poor liquefied bleeding layer after rolling, and the construction technology and quality control cannot be strictly implemented. All these factors will lead the layer joint surface tobe the weak link.

The shear strength indexes of RCC are mainly set based on the results of in-situ shear test. When the vertical compressive stress on the top surface of the specimen is a fixed value, such as $\sigma = 3$ MPa, the specimen will be failed as the horizontal thrust P is increasing. When the P value is smaller, the specimen is in the elastic stage and there is no sign of failure in the test. As the P value increases, stress concentration occurs at the edge of the specimen. When the stress exceeds the allowable value, a cracking zone can be found. The cracking area gradually moves to the middle as the P value increases, and another cracking area can be found at the other end. Gradually, the crack area expands to the middle as the P value increases. Finally, the left and right cracks are connected and the specimen is failed. The failure process of the specimen under the horizontal thrust is actually very similar to the sliding failure of a gravity dam. For this reason, *Design Specification for Concrete Gravity Dams* DL 5108—1999 has made special provisions on the shear parameters of the concrete layer. See Appendix D and Table 12.9 Shear parameters of concrete layer for details.

Table 12.9　　　　　　　　　　　Shear parameters of concrete layer

S/N	Item	Characteristics	Average and standard values of shear parameters			
			$\mu f_c'$	f_{ck}'	$\mu c_c'$ (MPa)	c_{ck}' (MPa)
1	RCC (layer bonding)	RCC with lean cementing materials and 180 d age	1.0 – 1.1	0.82 – 1.0	1.27 – 1.50	0.89 – 1.05
		RCC with rich cementing materials and 180 d age	1.1 – 1.3	0.91 – 1.07	1.73 – 1.96	1.21 – 1.37
2	Normal concrete (layer bonding)	Age C10 – C20	1.3 – 1.5	1.08 – 1.25	1.6 – 2.0	1.16 – 1.45

Note: RCC with cementing materials less than 130 kg/m^3 is RCC with lean cementing materials; with that equal to or larger than 160 kg/m^3 is RCC with rich cementing materials; and with that within 130 – 160 kg/m^3 is RCC with medium cementing materials.

12.5.2　Procedure of In-situ Shear Test for Layers

12.5.2.1　Procedure of In-situ Shear Test

Test Code for Hydraulic Concrete SL 352—2006 has made specific regulations on "In-situ shear strength test of RCC body and interlayer bonding". Instrument and equipment installation and test methods: the treatment of the specimen-bearing surface, installation, arrangement, commissioning, and inspection of instruments and equipment for loading, force transmission and measurement, and specific test methods shall be carried out according to the provisions of SL 264.

In the in-situ shear test, the representative parts and layers on the structures shall be selected in the construction test section or dam body. The area of the selected test area shall be no less than 2 m × 8 m, and 4 – 5 specimens shall be arranged on the same layer. The shear area of each specimen shall not be less than 500 mm × 500 mm, and the net distance between the specimens shall not be less than 1.5 times the smallest side length of the specimen. The height of the part of specimen above the shear surface is preferable to be slightly longer than 2/3 of the side length of the specimen. Upon arrangement of the test, the direction of the horizontal thrust applied to the surface of the specimen shall be consistent with the stress direction of the structure.

1. The age of specimens excavated shall not be less than 21d. The manual excavation shall be applied to excavate the concrete surrounding the specimen in the test area to prevent the disturbance of the specimen.
2. The excavation depth of the specimens shall reach the test layer surface, but the excavation depth under the horizontal thrust surface shall be deeper than the layer surface to ensure enough space for installing the jack. The jack shall be positioned strictly, its center line shall be parallel to the predetermined shear surface, and the distance from the shear surface shall not be greater than 5/6 of the side length of the specimen in the shear direction. The size error of the specimen after excavation should not be greater than

±2 cm, and the specimen shall be smoothed by the cement mortar of strength similar with the specimen. Upon the shear test for layer joints, leave shear joints of about 10 mm width around the shear surface.

3. After the excavation of specimens and the test area, the test pit shall be filled with water or backfilled with wet sand, and the specimens shall be maintained and preserved. The removal shall not be implemented until the required test age and the start of equipment installation. Meanwhile, measures for keeping the specimen and its shear surface in a water-saturated state shall be continued.

12.5.2.2 Processing of Test Result

1. The test results are mainly the layer surface shear strength parameters, i.e., f' and c' are processed according to Section 4.7.4 in *Test Code for Hydraulic Concrete* SL 352—2006.

2. To organize the test results, it is necessary to collect and provide the test results of the mix proportion of the shear surface concrete, the quality of the mixtures, and the compaction quality; construction layer thickness, spreading and vibrating method, equipment model, compaction technology, layer surface treatment, intermittent time, placement temperature, construction date, climate and layout drawing of test area, description drawing of shear surface and the test device, and pictures of typical shear surface failure conditions, etc.. These results would be adopted for analysis and application.

12.5.3 In-suit Shear Test Results in Some Domestics Projects

The in-situ shear parameters of RCC are the extremely important design parameters for the anti-sliding stability of RCC dams. The field shear test of the RCC layer of the dam body mainly tests the shear load resistance performance of the RCC layer to evaluate its rolling quality, and to provide shear stability parameters to check the dam body. For example, the RCC gravity dam of Longtan Hydropower Station is 216.5 m high. The full-section construction technology is adopted for the RCC dam. In order to verify the workability of RCC mix proportion, construction technology, layer joint bonding measures, high-strength construction requirements for continuous pouring on large placing surface under high temperature, and to ensure the quality of the interlayer bonding, the design attaches great importance to the anti-sliding stability of the interlayer bonding quality of the RCC dam. Two process tests have been carried out on the cofferdam and the test blocks, and the in-situ shear test for a total of 455 blocks of 91 groups have been carried out. It is the first time in China to implement such massive in-situ shear tests. A large number of shear parameters of the anti-sliding stability of the interlayer bonding have been obtained through the in-situ shear test, which provides a scientific basis for the design and rapid construction.

For the general concern about the interlayer bonding quality and shear strength of RCC, many projects conduct indoor, in-situ and core samples shear tests on site according to the actual project materials. See Table 12.10 for results of shear strength and in-suit shear tests of the core samples in some domestics projects. The results show that the values of f' and c' of the shear strength of the concrete interlayer bonding are far greater than the design indexes during the normal construction conforming construction requirements. Even in the middle and low dam projects applying lean concrete with little cementing material, the shear strength can also be higher than the design value. For Longtan and Guangzhao 200-metre high dams, the design index is $f' \geqslant 1.0 - 1.1, c' \geqslant 1.3 - 1.5$ MPa. The test results show that the f' and c' values of RCC poured at normal temperature or high temperature are higher than 1.1 MPa and 1.5 MPa respectively. Meanwhile, the shear result of the core sample shows that the shear strength of the core sample body is better than the interlayer shear strength and is also higher than the in-situ shear strength result.

Chapter 12 Core Drilling, Pump-in and In-situ Shear Tests

Table 12.10　　Test results of shear strength and in-suit shear of core samples in some domestics projects

S/N	Engineering	Test Result Statistics	Design index		Core sample shear result		In-situ shear result		Remarks
			f'	c'	f'	c'	f'	c'	
1	Yantan	Average	—	—	1.12	0.99	1.12	0.839	$14^\#, 15^\#, 17^\#$ dam section
2	Puding	Average	—	—	1.412	3.664	1.517	3.148	$R_{180}200$ II gradation
3		Average	—	—	1.656	3.177	1.882	2.753	$R_{180}200$ Three-graded
4	Jiangya	Average			1.4	1.03		—	Flat layer rolling and compaction
5		Average	—	—	1.27	1.15	—	—	Inclined layer rolling and compaction
6	Mianhuatan	Average	—	—	1.37	2.55	1.38	2.8	$R_{180}200$ II gradation
7		Average	—	—	1.36	2.16	1.2	2.56	$R_{180}150$ Three-graded
8	Dachaoshan	Average	—	—	2.14	4.00	1.483	1.201	Heat rising layer and III gradation
9		Average	—	—	1.88	3.50	1.428	1.001	Cool rising layer and III gradation
10	Hongpo Reservoir	Maximum	—	—	—	—	1.82	3.57	The shear index of the III gradation is the core sample test result
11		Minimum	—	—	—	—	1.3	2.75	
12	Linhekou	Maximum	—	—	—	—	2.12	4.12	The test is the III gradation result
13		Minimum	—	—	—	—	1.99	3.81	
14	Baise	Average	1.1	1.0	1.36	2.71	1.47	1.6	Design dam foundation index and II gradation
15		Average	1.1	0.9	1.29	2.62	1.13	1.09	Design interlayer index and quasi III gradation
16	Pengshui	Average	1.0	1.0	—	—	1.43	2.62	$C_{90}20$ II gradation in upstream face
17		Average	1.0	1.0	—	—	1.14	1.86	$C_{90}15$ III gradation inside the dam
18	Longtan (right bank)	Average	1.1	1.36	1.35	4.36	1.35	2.02	R I . $C_{90}25$ Three-graded
19		Average	1.1	1.35	1.73	3.93	1.29	1.79	R II . $C_{90}20$ Three-graded
20		Average	1.0	1.30	—	—	1.15	1.77	R III . $C_{90}15$ Three-graded
21		Average	—	—	1.53	4.35	1.35	2.02	R IV . $C_{90}25$ gradation

Continued Table 12.10

S/N	Engineering	Test Result Statistics	Design index		Core sample shear result		In-situ shear result		Remarks
			f'	c'	f'	c'	f'	c'	
22	Guangzhao	Average	1.1	1.5	1.539	1.89	1.56	2.02	$C_{90}25$ gradation
23		Average	1.1	1.5	1.328	1.598	1.57	1.68	$C_{90}20$ II gradation
24		Average	1.0	1.28	1.253	1.560	1.31	1.63	$C_{90}20$ Three-graded
25		Average	0.9	0.6	1.153	1.342	1.13	1.29	$C_{90}15$ Three-graded
26	Jin'anqiao	Average	1.1	1.3	1.26	1.77	—	—	$C_{90}20W8F100$ Level II
27		Average	1.1	1.3	1.23	1.62	1.28	1.63	$C_{90}20W6F100$ Level III
28		Average	1.0	1.2	1.13	1.47	1.20	1.79	$C_{90}15W6F50$ Level III
29	Kalasuke (one dry season)	Average	—	—	1.08	1.61	—	—	$R_{180}200F300$ Level II
30		Average	—	—	1.09	1.29	—	—	$R_{180}200F200$ Level II
31		Average	—	—	1.06	142	—	—	$R_{180}200F50$ Three-graded

Note: The shear results of indoor core samples are mainly based on the test results of layer joint surface.

12.6 Core Drilling and Pump-in Test of Dam RCC

12.6.1 Foreword

The total volume of Jin'anqiao Dam concrete is about 3.29 million m³, of which RCC volume is 2.417 million m³, accounting for about 73.3% of the total volume of the dam body concrete. According to the bidding documents and design requirements of *Construction Specification for Hydraulic Roller Compacted Concrete* DL/T 5112—2015 and the provisions of Article 3.0.2 of *Technical Specification for Testing Concrete Strength with Drilled Core Method* CECS 03:88, the core position arrangement principles of core drilling of dam RCC of core drilling and pump-in test of dam RCC are as follows: ①Parts of structures or members with less stress; ②Representative parts of concrete strength and quality; ③Parts convenient for placement and operation of core drilling machine; ④Avoid the position of main reinforcement, embedded parts and pipelines, and try to avoid other steel bars; ⑤When comprehensively measuring strength by drilled core method and nondestructive method, the same measuring area shall be taken as that by nondestructive method.

The $\phi 171$ mm drilling tool is used to drill the $\phi 150$ mm core sample of core drilling in the II gradation area of RCC dam, and the $\phi 219$ mm drilling tool is used to drill the $\phi 200$ mm core sample of core drilling in the III gradation area. The core drilling of dam RCC of Jin'anqiao Hydropower Station has been performed for three times and pump-in test twice. A total of 791.25 m core samples were drilled through core drilling for three times, and 58 section-time pump-in tests were performed. See Figure 12.6 to Figure 12.11 for the core samples and core drilling of dam RCC.

Chapter 12 Core Drilling, Pump-in and In-situ Shear Tests

Figure 12.6 11.30 m *rock sample*

Figure 12.7 11.45 m *rock sample*

Figure 12.8 10.36 m *rock sample*

Figure 12.9 10.92 m *rock sample*

Figure 12.10 *Taking the rock core*

Figure 12.11 *The rock core is taken out*

12.6.2 The First Core Drilling

12.6.2.1 Borehole Position

The first core drilling was performed in May 2008 in dam sections 8#, 10#, 13# and 14#, of which 4 inspection holes (all of which are III gradation RCC), 2 concrete core holes and 2 pump-in inspection holes are arranged at EL 1 317.5 m of powerhouse dam sections 8# and 10#. See Table 12.11 for the specific locations. 3 inspection holes are arranged at the elevation of the EL 1 345.0 m dam surface of the overflow dam section 13#, of which 2 concrete core holes (one for II gradation RCC and one for III gradation RCC, and the III gradation area is taken to the bedrock), and 1 pump-in inspection hole (II gradation RCC area). 4 inspection holes are arranged at the elevation of the EL 1 342.0 m dam surface of the overflow dam section 14#, of which 2 concrete core holes (one for II gradation RCC and one for III gradation RCC), and 2 pump-in inspection holes (one for II gradation RCC and one for III gradation RCC). A total of 6 concrete core inspection holes and 5 pump-in inspection holes are arranged. The age of the RCC at the above-mentioned core position layout has reached or exceeded 90 d.

Table 12.11　　Position of the first core drilling　　Unit: m

No.	Horizontal axis of dam	Longitudinal axis of dam	Crest elevation	Bottom elevation	Depth	Remarks
8 - III - 01	0 + 86.00	0 + 243.00	1 317.5	1 296.5	21.0	
10 - III - 01	0 + 86.00	0 + 311.00	1 317.5	1 296.5	21.0	
13 - II - 01	0 + 03.07	0 + 379.00	1 345.0	1 322.0	23.0	
13 - III - 01	0 + 46.00	0 + 375.50	1 345.0	1 291.5	53.5	Take to the bedrock
14 - II - 01	0 + 3.00	0 + 413.00	1 342.0	1 322.0	20.0	
14 - III - 01	0 + 56.42	0 + 413.74	1 342.0	1 311.0	31.0	
8 - III - 02'	0 + 82.00	0 + 243.00	1 317.5	1 284.0	33.5	
10 - III - 02'	0 + 82.00	0 + 311.00	1 317.5	1 305.5	12.0	
13 - II - 02'	0 + 03.07	0 + 369.00	1 345.0	1 333.0	12.0	
14 - II - 02'	0 + 03.00	0 + 403.40	1 342.0	1 330.0	12.0	
14 - III - 02'	0 + 52.86	0 + 400.19	1 342.0	1 303.5	38.5	

Note: The number in this table is "7 (dam section) - III (Grade) - 01 (01 is the core hole, and 02' is the pump - in hole)".

12.6.2.2 Core Completion

A total of 6 core holes are arranged this time, and 171.29 m RCC core drilling are completed (including 43.25 m II gradation RCC, 127.77 m III gradation RCC and 0.27 m bedrock). For the first time, 43 m core samples taken out from ϕ 150 mm drilling hole, accounting for 6.6% of the total contract amount, 126.5 m core samples taken out from ϕ 200 mm drilling hole, and ϕ 250 mm drilling core was not arranged; 36 section-time pump-in inspection was performed with the drilling hole of 108 m.

There are 159 pieces of core samples, and the main fracture causes of core samples are as follows: ①Mechanical and artificial fracture (tooth-shaped fracture) for 116 pieces, accounting for 73%; ②Layer surface fracture (uneven fracture) for 20 pieces, accounting for 12.4%; ③Joint surface fracture (fracture with sticky mortar) for 3 pieces, accounting for 2%; ④Manual handling fracture (irregular fracture) for 8 pieces, accounting for 5%; ⑤Aggregate is overhead, grout from segregation, and aggregates separation fracture for 6 pieces, accounting for 3.8%; ⑥Large aggregates accounted for 1/3 of the core sample fracture for 6 pieces, accounting for 3.8%. Among the 159 pieces of core samples, there are a total of 53 through the joint surface and 517 through the layer surface, with the joint surface intact rate of 94.34% and the layer surface intact rate of 96.13%.

Core sample in the III gradation area of dam section $13^{\#}$ (taken to the bedrock): the surface of core sample is smooth, the aggregate is evenly distributed, with dense structure and good cementation. The footage of the main hole is 53.7 m, the length of the columnar core sample is 53.47 m. Two single core samples with a whole length of $\phi 200$ mm greater than 10.0 m are obtained, of which the single longest core sample is 11.04 m. 0.27 m into the bedrock, the core sample is relatively complete, the concrete and the bedrock surface are cemented tightly, with good combination.

12.6.2.3 Field Pump-in Test

The arrangement of pump-in inspection holes is based on the layout of the core holes. During construction, the pump-in hole shall be drilled at the same dam section to check the water absorption of the RCC, and then the core holes are drilled to avoid pressurized water penetration. According to the above principle of hole arrangement, 5 holes were arranged for pump-in inspection with borehole footage of 108.0 m, and 36 section-time pump-in inspection was performed.

The diameter of borehole is $\phi 75$ mm, and the pump-in test is performed in sections from top to bottom. The length of pump-in section of RCC is 3.0 m, and the pressure of the first section is 0.3 MPa; the pressure of other sections is 0.6 MPa. Single-point method was used in the test.

The on-site pump-in inspection and test results indicate:

1. The minimum reading of the concrete permeability in the inspection range is 0.02 Lu, and the maximum permeability is 0.90 Lu. The minimum reading of permeability in the II gradation area is 0.04 Lu, and the maximum permeability is 0.90 Lu; the minimum reading of permeability in the III gradation area is 0.02 Lu, and the maximum permeability is 0.47 Lu;
2. A total of 36 section-time pump-in test statistics: 18 section-time tests with the permeability less than 0.1 Lu, accounting for 50.0%, and 17 section-time tests with the permeability less than 0.5 Lu, accounting for 47.2%. 1 section-time test with the permeability greater than 0.5 Lu and less than 1.0 Lu, accounting for 2.8%. Good overall impermeability.

   A total of 8 section-time pump-in test statistics in II gradation area: 2 section-time tests with thepermeability less than 0.1 Lu, accounting for 25.0%, and 5 section-time tests with the permeability less than 0.5 Lu, accounting for 62.5%. 1 section-time test with the permeability greater than 0.5 Lu and less than 1.0 Lu, accounting for 12.5%. Good overall impermeability

   A total of 28 section-time pump-in test statistics in III gradation area: 16 section-time tests with the permeability less than 0.1 Lu, accounting for 57.1%, and 12 section-time tests with the permeability less than 0.5 Lu, accounting for 42.9%. Good overall impermeability.
3. According to the above analysis, the permeability of the RCC III gradation area of dam section $14^{\#}$ is relatively high, with the maximum permeability of 0.90 Lu. Because it is the first section-time pump-in test, the penetration is shallow, and the concrete in some parts of the orifice may not be dense, so this hole is treated with grouting. After grouting, the hole was tested for 4 section-time of pump-in test again, the minimum permeability was 0.05 Lu, and the maximum permeability was 0.12 Lu.

12.6.3 The Second Core Drilling

12.6.3.1 Borehole Position

The second dam core time was from November to December, 2008. At that time, the project image was as follows: dam sections $2^{\#} - 6^{\#}$ were poured to EL 1 369 m, powerhouse dam sections $7^{\#} - 11^{\#}$ were poured to EL 1 367.8 m, and dam sections $13^{\#} - 19^{\#}$ were poured to EL 1 372 m. The second inspection core holes are arranged at the RCC parts of dam section $5^{\#}$ and powerhouse dam sections $7^{\#}-11^{\#}$ on the left bank. See Table 12.12 for specific positions. In this inspection, a total of 44 m core samples taken out from $\phi 150$ mm drilling hole, 302.5 m core samples taken out from $\phi 200$ mm drilling hole, and $\phi 250$ mm drilling core was not arranged; 22 section-time pump-in inspection was performed with the drilling hole of 66 m.

Table 12.12　　　　　　　　　　Position of the second core drilling　　　　　　　　　　Unit: m

No.	Horizontal axis of dam	Longitudinal axis of dam	Elevation	Bottom elevation	Depth	Remarks
5 - III - 01	0 + 47.35	0 + 135.00	1 364.0	1 320.0	44.0	Take to the bedrock
7 - III - 01	0 + 33.00	0 + 185.00	1 367.8	1 323.8	44.0	
8 - III - 01	0 + 33.0	0 + 246.00	1 367.8	1 323.8	44.0	
9 - III - 01	0 + 33.0	0 + 277.0	1 367.8	1 323.8	44.0	
10 - III - 01	0 + 33.00	0 + 311.00	1 367.8	1 323.8	44.0	
11 - II - 01	0 + 02.80	0 + 329.00	1 367.8	1 323.8	44.0	
11 - III - 01	0 + 96.00	0 + 318.00	1 317.5	1 279.0	38.5	Take to the bedrock
11 - III - 02	0 + 33.00	0 + 322.00	1 367.8	1 323.8	44.0	
8 - III - 02'	0 + 28.00	0 + 246.00	1 367.8	1 334.8	33.0	Pump-in hole
11 - II - 02'	0 + 02.80	0 + 322.00	1 367.8	1 334.8	33.0	Pump-in hole

A total of 8 core holes are arranged this time, and 347.97 m RCC core drilling are completed (including 44.07 m II gradation RCC, 303.90 m III gradation RCC and 0.43 m bedrock).

12.6.3.2 Core Sample Condition

The core sample length is 347.97 m, and the recovery rate of core samples is 99.7%. There are 358 pieces of core samples, with a total average length of 0.97 m. 5 single core samples with a whole length greater than 10.0 m were obtained, of which a single whole core sample with the longest $\phi 150$ mm of II gradation concrete was 15.73 m, which was the second length of RCC core samples in the same period in China. Inspection of core samples taken out from borehole: RCC is of good overall quality, with cylindrical core samples and smooth surface. Aggregates are basalt manufactured sand and manufactured gravel, with uniform distribution, dense structure and good cementation.

Core sample in the III gradation area of dam section $5^{\#}$, $11^{\#}$ (taken to the bedrock): the surface of core sample is smooth, the aggregate is evenly distributed, with dense structure and good cementation. The footage of main hole is 40.8 m and 42.45 m respectively, and the length of columnar core sample is 40.8 m and 42.1 m respectively. 0.25 m and 0.18 m respectively into the bedrock, the core sample is relatively complete, the concrete and the bedrock surface are cemented tightly, with good cementation.

There are total 358 pieces of core samples, and the main fracture causes of core samples are as follows: ①Mechanical and artificial fracture (tooth-shaped fracture) for 226 pieces, accounting for 63.1%; ②Layer surface fracture (uneven fracture) for 37 pieces, accounting for 10.3%; ③Joint surface fracture (fracture with sticky mortar) for 12 pieces, accounting for 3.4%; ④Manual handling fracture (irregular fracture) for 61 pieces, accounting for 17.1%; ⑤Aggregate is overhead, grout from segregation, and aggregates separation fracture for 13 pieces, accounting for 3.6%; ⑥Large aggregates accounted for 1/3 of the core sample fracture for 9 pieces, accounting for 2.5%.

12.6.3.3 Field Pump-in Test

The arrangement of pump-in inspection holes is based on the layout of the core holes. During construction, the

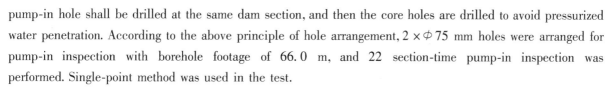

pump-in hole shall be drilled at the same dam section, and then the core holes are drilled to avoid pressurized water penetration. According to the above principle of hole arrangement, $2 \times \phi 75$ mm holes were arranged for pump-in inspection with borehole footage of 66.0 m, and 22 section-time pump-in inspection was performed. Single-point method was used in the test.

12.6.3.4 Analysis of Inspection Results

The on-site pump-in inspection and test results indicate:

1. The minimum reading of the concrete permeability in the inspection range is 0.05 Lu, and the maximum permeability is 0.80 Lu. The minimum reading of permeability in the II gradation area is 0.21 Lu, and the maximum permeability is 0.41 Lu; the minimum reading of permeability in the III gradation area is 0.05 Lu, and the maximum permeability is 0.80 Lu.

2. A total of 22 section-time pump-in test statistics: 1 section-time tests with the permeability less than 0.1 Lu, accounting for 4.5%, and 19 section-time tests with the permeability less than 0.5 Lu, accounting for 86.4%. 2 section-time test with the permeability greater than 0.5 Lu and less than 1.0 Lu, accounting for 9.1%. Overall impermeability is good.

A total of 11 section-time pump-in test statistics in III gradation area: 1 section-time tests with the permeability less than 0.1 Lu, accounting for 9.1%, and 8 section-time tests with the permeability less than 0.5 Lu, accounting for 72.7%. 2 section-time test with the permeability greater than 0.5 Lu and less than 1.0 Lu, accounting for 18.2%. Good overall impermeability.

A total of 11 section-time pump-in test statistics of II gradation area: 11 section-time tests with the permeability less than 0.5 Lu, accounting for 100%. Good overall impermeability.

12.6.4 The Third Core Drilling

12.6.4.1 Borehole Position

The third dam core time was from March to May, 2009. During this core drilling, the construction image of dam project was as follows: dam sections $0^{\#}$ and $1^{\#}$ were poured to EL 1 422.5 m, dam sections $2^{\#} - 6^{\#}$ were poured to EL 1 381 m, powerhouse dam sections $7^{\#} - 10^{\#}$ are constructed at the inlet orifice section, and dam sections $11^{\#} - 19^{\#}$ were poured to EL 1 381 m. See Table 12.13 for the specific locations.

Table 12.13　　　　　　　　　　　Position of the third core drilling　　　　　　　　　　　Unit: m

No.	Horizontal axis of dam	Longitudinal axis of dam	Elevation	Bottom elevation	Depth	Remarks
1 - II - 01	0 + 02.50	0 + 018.00	1 422.5	1 392.50	30.0	Take to the bedrock
11 - II - 01	0 + 03.00	0 + 330.50	1 384.0	1 354.0	30.0	
13 - III - 01	0 + 22.00	0 + 372.00	1 384.0	1 354.0	30.0	
14 - III - 01	0 + 22.00	0 + 431.00	1 374.0	1 354.0	30.0	
15 - III - 01	0 + 22.00	0 + 447.50	1 381.0	1 351.0	30.0	
16 - III - 01	0 + 22.00	0 + 472.00	1 381.0	1 351.0	30.0	
17 - II - 01	0 + 03.00	0 + 499.00	1 381.0	1 351.0	30.0	
17 - III - 01	0 + 22.00	0 + 499.00	1 381.0	1 351.0	30.0	
18 - III - 02	0 + 22.00	0 + 527.00	1 381.0	1 351.0	30.0	

This core inspection is planned at dam section $1^{\#}$ and dam sections $11^{\#} - 18^{\#}$, of which a core inspection hole is arranged at the EL 1 422.5 m of dam section $1^{\#}$ and taken to the bedrock. 8 core inspection holes are arranged in dam sections $11^{\#} - 18^{\#}$.

12.6.4.2 Core Sample Condition

A total of 9 core holes are arranged this time, and 272.66 m core drilling are completed (including 91.10 m for II gradation (ϕ150 mm core sample), 181.16 m for III gradation (ϕ200 mm core sample) and 0.40 m for bedrock).

The core sample length is 272.66 m, and the recovery rate of core samples is 97.47%. There are 261 pieces of core samples, with a total average length of 1.04 m. 2 single core samples with a whole length greater than 10.0 m were obtained, of which a single whole core sample with the longest ϕ150 mm of II gradation concrete was 16.49 m, which is currently the longest core sample in China. Inspection of core samples taken out from borehole: RCC is of good overall quality, with cylindrical core samples and smooth surface. Aggregates are basalt manufactured sand and manufactured gravel, with uniform distribution, dense structure and good cementation.

Core sample of dam section $1^{\#}$ (taken to the bedrock): the surface of core sample is smooth, the aggregate is evenly distributed, with dense structure and good cementation. The footage of the main hole is 30.70 m, and the length of the columnar core sample is 30.39 m. 0.40 m respectively into the bedrock, the core sample is relatively complete, the concrete and the bedrock surface are cemented tightly, with good cementation.

There are 261 pieces of core samples, and the main fracture causes of core samples are as follows: ①Mechanical and artificial fracture (tooth-shaped fracture) for 105 pieces, accounting for 40.0%; ②Layer surface fracture (uneven fracture) for 40 pieces, accounting for 15.0%; ③Joint surface fracture (fracture with sticky mortar) for 25 pieces, accounting for 10.0%; ④Manual handling fracture (irregular fracture) for 36 pieces, accounting for 14.0%; ⑤Aggregate is overhead, grout from segregation, and aggregates separation fracture for 34 pieces, accounting for 13.0%; ⑥Large aggregates accounted for 1/3 of the core sample fracture for 21 pieces, accounting for 8.0%.

12.6.5 Summary of Core Sample Appearance and Pump-in Inspection

The RCC of Jin'anqiao Hydropower Station has been core drilled for three times and pump-in tested twice. Its purpose is to check the construction quality of RCC, which is summarized as follows:

The first inspection was arranged at the RCC parts of overflow dam sections $13^{\#}$ and $14^{\#}$, powerhouse dam sections $8^{\#}$ and $10^{\#}$ on the right bank, and the second inspection was arranged at dam section $5^{\#}$ and powerhouse dam sections $7^{\#}-11^{\#}$ on the left bank. For the second hole arrangement of RCC, it is also considered the parts that it used to be the transportation channel of RCC, and the quality of RCC in this part is greatly disturbed due to truck placement, which may be the weak link of RCC. The third core hole arrangement is a supplement to the insufficient arrangement of the last two.

For the three times before and after the core drilling of the project, a total of 178.42 m core samples taken out from ϕ150 mm drilling hole, and the three inspections accounted for 27.28% of the total contract amount; 612.83 m core samples taken out from ϕ200 mm drilling hole and 1.1 m into the bedrock. The core number of borehole inspection for three times accounted for 64.41% of the total contract amount.

After the appearance inspection of each core sample, it is found that the core sample is cylindrical, and most of the core samples have smooth appearance, dense concrete structure and good cementation; the appearance of some core samples is basically smooth, with a few holes. Aggregates of core samples are basalt manufactured sand and manufactured gravel, with uniform distribution.

Meanwhile, according to the inspection analysis twice of the borehole pump-in of the dam RCC, the permeability of most pump-in sections of RCC in the II gradation area is less than 0.5 Lu, except individual sections. The permeability in the III gradation area is less than 1 Lu. The permeability of most pump-in sections is less than 0.5 Lu, with good overall impermeability.

12.7 Performance Test of RCC Core Sample of Jin'anqiao Dam

12.7.1 Foreword

The core drilling of dam RCC of Jin'anqiao Hydropower Station has been performed for three times and the core sample test was entrusted by the scientific research unit. The core samples are tested according to different parts of RCC of the dam, namely, Ⅱ gradation $C_{90}20F100W8$ in the dam impervious area, Ⅲ gradation $C_{90}20F100W6$ below 1 350 m and $C_{90}15 F100W6$ above 1 350 m in dam interior. The test results of RCC core samples in dam are summarized as follows.

12.7.2 Procedures and Specifications Adopted in the Test

Procedures and specifications adopted in the test are as follows:
Specifications for Hydraulic Concrete Construction SL 352—2006.
Specifications for Hydraulic Concrete Construction DL/T 5150—2001.
Specifications for Hydraulic Concrete Construction DL/T 5144—2001.
Construction Specification for Hydraulic Roller Compacted Concrete DL/T 5112—2000.
Specifications for Rock Tests in Water Conservancy and Hydroelectric Engineering SL 264—2001.

All instruments and equipment used in this test have been certified by Guizhou Metrology and Testing Institute.

12.7.3 Test Content of RCC Core Samples in Dam

This time, $C_{90}20F100W8F100$ Ⅱ gradation RCC (ϕ 150 mm) core sample, $C_{90}20F100W6F100$ Ⅲ gradation RCC core sample (ϕ 200 mm), and $C_{90}15W6F100$ Ⅲ gradation RCC core sample (ϕ 200 mm) were tested. The test items are: appearance description, volumetric weight, compressive strength, static compressive modulus of elasticity, tensile strength, ultimate tensile value, impermeability grade, frost resistance grade, shear strength (body and interlayer) of concrete core samples.

On-site core of RCC core samples of dam is divided into three times, the first sampling time is November 29, 2008, the second sampling time is February 19, 2009, and the third sampling time is June 17, 2009. The RCC core samples drilled on site are sampled and transported back to Guiyang Research Institute.

Among the retrieved core samples, the Ⅱ gradation ϕ 150 mm core samples are divided into dam sections 1 - Ⅱ -01, 11 - Ⅱ -01, 13 - Ⅱ -01, 14 - Ⅱ -01 - 17 - Ⅱ -01, and the total length of the Ⅱ gradation ϕ 150 mm core samples is about 60 m.

The Ⅱ gradation ϕ 200 mm core samples are divided into dam sections 5 - Ⅲ -01, 7 - Ⅲ -01, 8 - Ⅲ -01, 9 - Ⅲ -01, 10 - Ⅲ -01, 11 - Ⅲ -01, 13 - Ⅲ -01, 14 - Ⅲ -01, 15 - Ⅲ -01, 16 - Ⅲ -01, 17 - Ⅲ -01 - 18 - Ⅲ -01, and the total length of the Ⅲ gradation ϕ 200 mm core samples is about 260 m.

12.7.4 Processing and Test Methods of Core Samples

The retrieved ϕ 150 mm and ϕ 200 mm RCC cylindrical core samples were classified and subdivided according to the requirements of the design and related procedures and specifications for cutting processing and end surface grinding. Classification tests were performed according to the grade and different performances of the concrete core samples.

12.7.4.1 Processing Methods of Core Samples

The core samples of RCC of the dam were processed in accordance with the requirements of Article 8.4.7 and Appendix A in *Construction Specification for Hydraulic Roller Compacted Concrete* DL/T 5112—2000 "Article 8.4 Quality Control and Evaluation", among which the volumetric weight, compressive strength, static the compressive modulus of elasticity, ultimate tensile value, and frost resistance of the core samples $\phi 150$ mm and $\phi 200$ mm have the core sample ratio of height to diameter of 2:1.

The processing method of impermeability grade core sample is as follows: because the specifications of concrete impermeability test moulds are truncated cone with upper opening diameter of 175 mm, lower opening diameter of 185 mm and height of 150 mm, process the $\phi 150$ mm core samples into 150 mm high core samples, and fill the gap between core samples and impermeability test moulds with high-strength expansive mortar, and then cure it after demoulding to ensure good combination between the impermeability core samples body and expansive mortar. First cut the $\phi 200$ m core samples into cubes with a side length of 150 mm, so that the core samples can be placed in the impermeability test moulds, and then fill the gap between core samples and impermeability test moulds with high-strength expansive mortar, and then cure it after demoulding to ensure good combination between the impermeability core sample body and expansive mortar.

Processing methods of shear core samples.

1. First cut the $\phi 150$ m core samples into cylinders with a height of 200 mm, and ensure that the shear plane is in the middle of the cylinder height. A 150 mm square shear box (the $\phi 150$ mm cylinder inside, and the $\phi 150$ mm core sample can be directly put in) is used in the shear test.

2. First cut the $\phi 200$ mm core samples into a cylinders with a height of 200 mm, and then place them in a 200 mm × 200 mm × 200 mm square test moulds, fill the gap between core samples and test moulds with high-strength expansive mortar to 1/2 height of the test moulds, after curing the test moulds for 1−2 d, put a layer of 2 mm thick impervious rubber at 1/2 height of the test mould, continue to fill the upper half of the gap with high-strength expansive mortar, and cure it for 14 d after demoulding for shear test, and ensure that the layer surface of the interlayer shear core samples is in the middle of the cylinder height, and the square shear box of 200 mm × 200 mm × 200 m is used in the shear test.

12.7.4.2 Test Methods of Core Samples

Ⅰ The method for determining the age of core samples

The method for determining the age of core samples: the number of days from the concrete pouring date of the placing surface where the elevation of the bottom surface of core samples is to the inspection date of core samples.

Ⅱ Core sample compressive strength

The compressive strength of the core sample is tested in accordance with 4.2 Concrete Cube Compressive Strength Test in the Test *Code for Hydraulic Concrete* SL 352—2006.

Ⅲ Static compressive modulus of elasticity of core samples

The static compressive modulus of elasticity of the core sample is tested in accordance with 4.8 Tests for the compressive strength and modulus of elasticity under static compressive stress of concrete cylinders (axial) in the *Test Code for Hydraulic Concrete* SL 352—2006, with the resistor disc.

Ⅳ Tensile strength and ultimate tensile value of the core sample

Tensile strength and ultimate tensile value of the core sample are tested in accordance with Figure d in Figure 4.5.2-1 Axial Tensile Specimen and Embedded Part of Concrete in 4.5 Axial Tensile *Test of Concrete in the Test Code for Hydraulic Concrete* SL 352—2006. First bond the core sample and the steel pulling plate together with epoxy resin adhesive, and cure it for 3−5 d to make the core sample and the pulling plate adhere firmly, so as to ensure that the core sample is not broken from the pulling plate, and then perform the ultimate tensile value with a tensile tester, with the resistor disc and resistance strain gauge.

Ⅴ　Impermeability grade test of core samples

The core sample impermeability test method is performed in accordance with the 4.21 Concrete Impermeability Test (Gradual Compression Method) in the *Test Code for Hydraulic Concrete* SL 352—2006. For RCC core samples with permeability grades of W8 or W6, the water pressure starts from 0.1 MPa, and each 0.1 MPa is one grade. After the water pressure is increased to 0.9 MPa or 0.7 MPa, if there are less than 3 core samples with surface water seepage of the 6 core samples within 8 h, the core samples of this group shall be assessed to meet the design impermeability grade requirements, and then split the core samples to measure the seepage height.

Ⅵ　Frost resistance test of core samples

The core sample is tested in accordance with 4.23 Concrete Frost Resistance Test in the *Test Code for Hydraulic Concrete* SL 352—2006, with the height of the core sample of 400 mm. When the freeze-thaw quality loss of core sample exceeds 5% of freeze-thaw, the cycle number is the frost resistance grade.

Ⅶ　Test method for volumetric weight of core samples

The volumetric weight is the weight of the core sample divided by its volume. The diameter and height of the core sample are measured at different positions for 3 times, and the average value is the diameter and height of the core sample.

Ⅷ　Test method for shear of core samples

The core sample is tested in accordance with 4.7 RCC Shear Strength Test in the *Test Code for Hydraulic Concrete* SL 352—2006. The maximum normal load is 5.0 MPa, which is applied in 5 grades, from 1.0 MPa to 5.0 MPa.

12.7.5　Conclusion

Through tests and statistical analysis of appearance description, volumetric weight, compressive strength, static compressive modulus of elasticity, tensile strength, ultimate tensile value, impermeability grade, frost resistance grade, shear strength (body and interlayer) of borehole core samples of $C_{90}20F100W8F100$ Ⅱ gradation RCC, $C_{90}20F100W6F100$ Ⅲ gradation RCC, and $C_{90}15W6F100$ Ⅲ gradation RCC of dam of Jin'anqiao Hydropower Station, the results show:

1. From the appearance of Ⅱ gradation and Ⅲ gradation RCC core samples in dams, most of the surfaces are smooth and dense, and aggregates are equally distributed, with good interlayer cementation. The grade of core samples is "good" according to "8.4 Quality control and evaluation" in *Construction Specification for Hydraulic Roller Compacted Concrete* DL/T 5112—2000.
2. The compressive strength of $C_{90}20F100W8F100$ Ⅱ gradation, $C_{90}20F100W6F100$ Ⅲ gradation and $C_{90}15W6F100$ Ⅲ gradation RCC core samples meets the design requirements.
3. As for $C_{90}20F100W8F100$ Ⅱ gradation, $C_{90}20F100W6F100$ Ⅲ gradation and $C_{90}15W6F100$ Ⅲ gradation RCC core samples, the measured values of modulus of elasticity under static compressive stress are normal.
4. As for $C_{90}20F100W8F100$ Ⅱ gradation, $C_{90}20F100W6F100$ Ⅲ gradation and $C_{90}15W6F100$ Ⅲ gradation RCC core samples, the measured values of ultimate tensile value are normal. While the ultimate tensile value of Ⅲ gradation RCC is lower than that of Ⅱ gradation, mainly because the mortar content of Ⅲ gradation RCC is less than that of Grade Ⅱ.
5. The impermeability of $C_{90}20F100W8F100$ Ⅱ gradation, $C_{90}20F100W6F100$ Ⅲ gradation and $C_{90}15W6F100$ Ⅲ gradation RCC core samples meets the corresponding requirements.
6. The frost resistance grade of $C_{90}20F100W8F100$ Ⅱ gradation, $C_{90}20F100W6F100$ Ⅲ gradation and $C_{90}15W6F100$ Ⅲ gradation RCC core samples reaches F75. The large loss of air content of concrete has

certain effect on the frost resistance of RCC after the on-site RCC is subject to transportation, paving and rolling compaction in layers.

7. The shear strength of body and layer surface of $C_{90}20F100W8F100$ II gradation, $C_{90}20F100W6F100$ III gradation and $C_{90}15W6F100$ III gradation RCC core samples meets the design requirements. The shear strength parameter of the body (hot joint) is slightly higher than that of the interlayer (cold joint).

8. The unit weight value of C_{90} 20F100W8F100 II gradation, C_{90} 20F100W6F100 III gradation and $C_{90}15W6F100$ III gradation RCC core samples is slightly big. The main reason is that the concrete aggregate is dense basalt.

9. The statistics results of C_{90} 20F100W8F100 II gradation, C_{90} 20F100W6F100 III gradation and $C_{90}15W6F100$ III gradation RCC core samples are shown in Table 12.14 to Table 12.16.

10. The test and measurement results of core samples did not consider the impact of the case that III gradation RCC core samples (200 mm in diameter) failed to meet the test requirements of three times the particle size (more than 240 mm).

Table 12.14   Statistics of test results of C9020F100F100 II gradation RCC core samples

Category	Compressive strength (MPa) (converted into 150 mm cube)	Axial tensile strength (MPa)	Ultimate tensile value ($\times 10^{-4}$)	Static compressive modulus of elasticity (GPa)	Grade of impermeability	Frost resistance grade	Unit weight (kg/m³)	Shearing strength Body		Interlayer	
								f'	c'	f'	c'
Mean value	29.4	1.12	0.70	30.9	W8	F75	2 555	1.32	1.83	1.26	1.77
Maximum value	47.1	1.52	0.85	36.0	—	—	2 613	—	—	1.28	1.84
Minimum value	18.6	0.92	0.53	26.4	—	—	2 442	—	—	1.24	1.70
Number of test groups	20	8	8	6	4	2	20	1	1	3	3

Table 12.15   Statistics of test results of C9020F100 III gradation RCC core samples

Category	Compressive strength (MPa) (converted into 150 mm cube)	Axial tensile strength (MPa)	Ultimate tensile value ($\times 10^{-4}$)	Static compressive modulus of elasticity (GPa)	Grade of impermeability	Frost resistance grade	Unit weight (kg/m³)	Shearing strength Body		Interlayer	
								f'	c'	f'	c'
Mean value	24.6	1.00	0.58	29.5	W6	F75	2 641	1.30	1.80	1.23	1.62
Maximum value	38.6	1.42	0.77	42.4	—	—	2 729	—	—	1.27	1.68
Minimum value	17.2	0.70	0.47	24.6	—	—	2 573	—	—	1.16	1.54
Number of test groups	40	14	14	8	7	2	40	1	1	3	3

Chapter 12 Core Drilling, Pump-in and In-situ Shear Tests

Table 12.16　　Statistics of test results of C9020F100 Ⅲ gradation RCC core samples

Category	Compressive strength (MPa) (converted into 150 mm cube)	Axial tensile strength (MPa)	Ultimate tensile value (×10⁻⁴)	Static compressive modulus of elasticity (GPa)	Grade of impermeability	Frost resistance grade	Unit weight (kg/m³)	Shearing strength			
								Body		Interlayer	
								f'	c'	f'	c'
Mean value	25.1	0.96	0.56	30.8	W6	F75	2 630	1.32	1.83	1.13	1.47
Maximum value	36.4	1.34	0.70	37.5	—	—	2 718	—	—	1.15	1.52
Minimum value	17.5	0.68	0.47	24.7	—	—	2 522	—	—	1.10	1.38
Number of test groups	20	6	6	4	3	2	20	1	1	3	3

12.8 On-site In-situ Shear Test of RCC Dams

12.8.1 Overview

Jin'anqiao Hydropower Station is the fifth-level power station planned for the "one reservoir and eight cascades" in the middle reaches of the Jinsha River. The project focuses more on power generation and less on flood control. The barrage is a RCC gravity dam with height of 160 m. The amount of RCC in the dam is about 2 648 000 m³, accounting for about 67.5% of the total concrete of the dam. The project is characterized by a large scale, tight construction period, high construction requirements, and continuous construction at high temperature and in rainy seasons, etc.

The total construction duration of the Jin'anqiao Hydropower Station is 87 months. In December 2005, closure was realized as planned, and the first unit would be scheduled to generate electricity in December 2009. The RCC dam began being concreted from May 2007. According to the construction progress of the project, Guiyang Engineering Corporation Limited of Hydrochina Corporation Limited conducted on-site in-situ shear tests of RCC dams, combined with the dam construction schedule upon entrustment by Jin'anqiao Hydropower Station Co., Ltd.

12.8.2 Purpose and Contents of Test

12.8.2.1 Purpose of Test
The field shear test of the RCC layer of the dam body mainly tests the shear load resistance performance of the RCC layer to evaluate its rolling quality, and to provide shear stability parameters to check the dam body.

12.8.2.2 Test Site Layout
The on-site in-situ shear test of RCC dams shall be arranged in 6 groups and divided into two periods according to the contract requirements. The test time of the first period is December 1, 2008 – December 17, 2008, the number of test groups is 3, the test location is the 1 320 m elevation platform at dam toe of 11# dam section, and the concrete design grade is $C_{90}20W6F100$ Ⅲ gradation RCC; the test time of the second period is February 11, 2009 – February 26, 2009, the number of test groups is 3, the test location is the 1 422.5 m elevation platform at dam toe of 1# dam section, and the concrete design grade is $C_{90}15W6F100$ Ⅲ gradation RCC. See Figure 12.12 for the specimen layout.

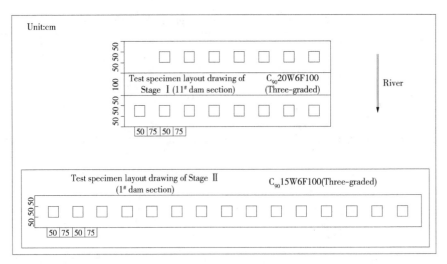

Figure 12.12 *Specimen layout for on-site in-situ shear test of RCC dams*

12.8.2.3 Test Workload

According to RCC dam sections, the on-site in-situ shear test of RCC and workload are shown in Table 12.17.

Table 12.17   Test content and workload for on-site In-situ shear test of RCC dams

Test stage	Test location	Design concrete grade	Layer (joint) surface form	Layer surface treatment measures	Number of test groups
Stage 1	1 320 m-elevation platform behind 11# dam section	$C_{90}20W6F100$ (Ⅲ gradation)	Hot joint	No processing	3
Stage 2	1 422.5 m elevation platform at 1# dam section	$C_{90}15W6F100$ (Ⅲ gradation)	Hot joint	No processing	3

12.8.3   Test Equipment and Methods

12.8.3.1   Main Instruments and Equipment Used in the Test

The test equipment system is mainly composed of three parts: loading, load transfer and measurement.

Loading system: hydraulic jack, high-pressure oil pump, pressure gauge and high-pressure hose.

Load transfer system: load transfer column, counterforce support, bearing plate and roller row.

Measurement system: large-stroke dial indicator, magnetic indicator stand and indicator stand support.

Others: digital camera, recording device, hydraulic oil, geology description and installation and other instruments.

12.8.3.2   Test Methods

The test methods are in line with the relevant regulations on *Specifications for Rock Tests in Water Conservancy and Hydroelectric Engineering* SL 264—2001 and *Test Code for Hydraulic Concrete* SL 352—2006. The test adopts the placing method, and the shear direction is consistent with the actual force direction of the dam.

12.8.4   Test Steps

1. Method of applying vertical load. The maximum normal stress of shear test of on-site RCC is 3.0 MPa. The

on-site shear test adopts multi-point peak method, each group contains 5 specimens, and the maximum normal stress is 3.0 MPa. The normal stress of the 5 specimens is initially set to 0.6 MPa, 1.2 MPa, 1.8 MPa, 2.4 MPa and 3.0 MPa. The vertical load of each specimen is applied in 3 – 5 levels. For application of each level of vertical load, the next level of load can be applied 5 min later upon measurement of the vertical deformation. After application to the predetermined load, it is considered that stability requirements have been met when the difference between two consecutive vertical deformations does not exceed 1‰ mm, and the horizontal shear load can be applied.

2. Method of applying shear load. Begin to apply uniformly and equally in grades according to 8% – 10% of the estimated maximum shear load. When the horizontal deformation caused by the applied load is 1.5 times the deformation of the previous load (or as determined by the specific situation), the load is halved by 4% – 5% applied until sheared. The load application method is controlled by time, about once every 5 min, and the deformation is read before and after the load of each level is applied; closely watch and record the pressure change and the corresponding horizontal deformation (simultaneous observation of pressure and deformation) when it is close to the shearing. The vertical load is always constant during shearing.

3. During the test, record the meter adjustment, meter change, meter contact, jack oil leakage, pressure compensation, concrete loosening, block falling, etc.

4. After being completed, the test block is turned over to describe and measure the physical characteristics of the shear section, such as failure form, fluctuation and shear section area, and take photos.

12.8.5 Test Result Sorting

1. Calculate the stress and corresponding deformation on the shear surface under vertical loads at all levels respectively.
2. Calculate the stress acted on the shear surface as follows:

$$\sigma = P/F$$
$$\tau = Q/F$$

Where σ = normal stress acted on the shear surface, MPa

τ = shear stress acted on the shear surface, MPa

P = all vertical loads acted on the shear surface (including jack output, equipment weight and test block weight, N

Q = horizontal load acted on the shear surface. Its value is equal to the total horizontal load applied minus the friction force of the roller row, N

F = shear area, mm^2

3. Draw the curve of the relationship between shear stress [τ] and shear deformation [μs] and the curve of the relationship between shear stress [τ] and normal stress [σ] under different normal stresses, and determine the peak shear strength parameters f' and c' with graphic method or least square method according to the above various curves.

The shear strength is expressed by the Coulomb formula: $\tau = \sigma f' + c$.

12.8.6 Statistics Analysis of Test Results

According to the contract requirements, 6 groups of on-site shear tests are arranged. The results are shown in Table 12.18 to Table 12.20, the curve and shear plane sketch of results are shown in Figure 12.13 to Figure 12.18, and the shear surface and on-site test pictures are shown in Figure 12.19 to Figure 12.24.

Table 12.18   Results of on-site in-situ shear test of $C_{90}20W6F100$ (III gradation) RCC dams

Test location	Peak parameter of shear strength calculated with c and φ							
1 320 m elevation platform behind 11# dam section	Comprehensive value		Group 1 (τ_I)		Group 2 (τ_{II})		Group 3 (τ_{III})	
	f'	c' (MPa)	f'	c' (MPa)	f'	c' (MPa)	f'	c' (MPa)
	1.28	1.63	1.28	1.65	1.32	1.51	1.24	1.73
Age(d)	Placement surface elevation (m)		Placement opening time		Placement closing time		Test date	
302	1 315.00 – 1 318.70		2008.02.08 PM 9:30		2008.02.19 PM 9:30		2008.12.16 – 2008.12.17	
Strength grade and joint condition of concrete design	$C_{90}20W6F100$ (III gradation) hot joint							
Pouring condition	Weather: drizzle, temperature: 13 ℃, relative humidity: 62%, wind speed: 2.0 m/s							

Table 12.19   Results of on-site In-situ shear test of $C_{90}15W6F100$ (III gradation) RCC dams

Test location	Peak parameter of shear strength calculated with c and φ							
1 422.5 m elevation platform at #1 dam section	Comprehensive value		Group 4 (τ_{IV})		Group 5 (τ_V)		Group 6 (τ_{VI})	
	f'	c' (MPa)	f'	c' (MPa)	f'	c' (MPa)	f'	c' (MPa)
	1.20	1.79	1.20	1.87	1.20	1.71	1.21	1.78
Age (d)	Placement surface elevation (m)		Placement opening time		Placement closing time		Test date	
106	1 420.0 – 1 422.5		2008.11.08		2008.11.10		2009.02.22 – 2009.02.24	
Strength grade and joint condition of concrete design	$C_{90}15W6F100$ (III gradation) hot joint							
Pouring condition	Weather: Temperature:　℃, Relative humidity:　%, Wind speed:　m/s							

Table 12.20 Normal stress and shear stress applied during the on-site in-situ shear test of each group of specimens

S/N	Test No.	Stress (MPa)	Test piece No.				
			1st block	2st block	3st block	4st block	5st block
Group 1	τ_I	Normal stress σ	2.28	1.74	2.99	1.20	0.46
		Shear stress τ	4.54	4.15	5.39	3.00	2.23
Group 2	τ_{II}	Normal stress σ	2.94	0.50	1.09	2.33	1.81
		Shear stress τ	5.13	2.27	2.66	4.99	3.92
Group 3	τ_{III}	Normal stress σ	2.11	0.60	1.17	1.64	2.99
		Shear stress τ	4.19	2.57	3.29	3.55	5.61
Group 4	τ_{IV}	Normal stress σ	0.62	1.23	1.64	2.39	2.81
		Shear stress τ	2.86	3.21	3.49	4.89	5.31
Group 5	τ_V	Normal stress σ	0.68	1.64	1.23	2.39	3.08
		Shear stress τ	2.23	3.63	3.42	5.03	5.03
Group 6	τ_{VI}	Normal stress σ	0.68	1.78	1.03	2.46	2.81
		Shear stress τ	2.44	3.49	3.49	4.61	5.45

Chapter 12 Core Drilling, Pump-in and In-situ Shear Tests

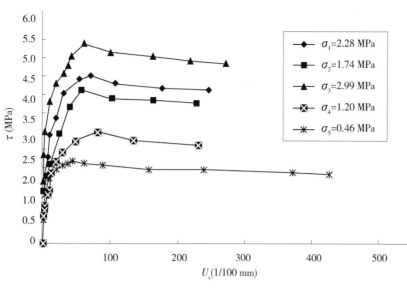

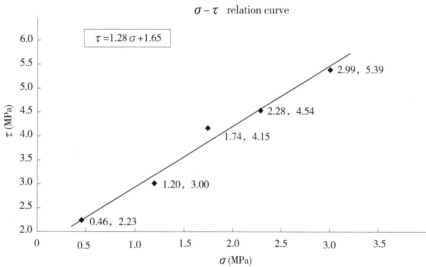

Test No.	Test location		f'	c' (MPa)
τ_{I}	1 320 m-elevation platformbehind 11# dam section	Peak parameter ofshear strength	1.28	1.65
Placement surface elevation (m)	Placement opening time	Placement closing time	Test date	
1 315.00 – 1 318.70	2008. 02. 08PM9: 30	2008. 02. 19PM9: 30	2008. 12. 16 – 2008. 12. 17	
Design strength grade and joint surface working condition	$C_{90}20W6F100$ (Ⅲ gradation)		Hot joint	
Pouring condition	Weather: light rain	Temperature: 13 ℃	Relative humidity: 62%	Wind velocity: 2.0 m/s
Failure plane condition	Shear short or concrete shear through along joint plane. The area of concrete cutting through accounts for about 40%. The sheared surface has a great fluctuation difference, with maximum fluctuation difference of about 7 cm			

Figure 12.13 *The resulting curve graph of* τ_{I}

Roller Compacted Concrete Rapid Damming Technology

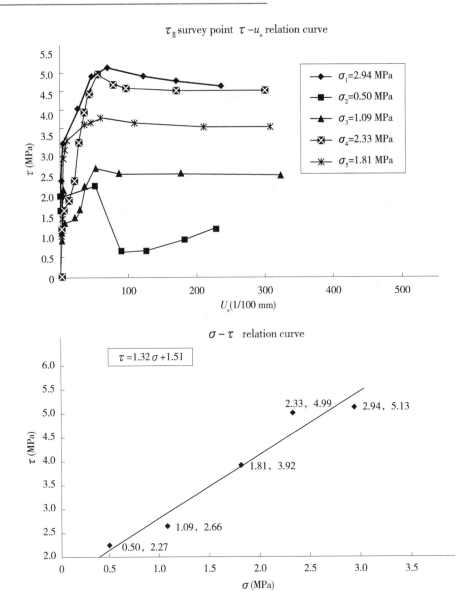

Test No.	Test location		f'	c' (MPa)
τ_{II}	1 320 m-elevation platform behind 11# dam section	Peak parameter of shear strength	1.32	1.51
Placement surface elevation (m)	Placement opening time	Placement closing time	Test date	
1 315.00 – 1 318.70	2008.02.08PM9:30	2008.02.19PM9:30	2008.12.16 – 2008.12.17	
Design strength grade and joint surface working condition	$C_{90}20W6F100$ (III gradation)		Hot joint	
Pouring condition	Weather: light rain	Temperature: 13 ℃	Relative humidity: 62%	Wind velocity: 2.0 m/s
Failure plane condition	Shear short or concrete shear through along joint plane. The area of concrete cutting through accounts for about 25%. The sheared surface has a great fluctuation difference, with maximum fluctuation difference of about 6 cm			

Figure 12.14 *The resulting cunve graph of* τ_{II}

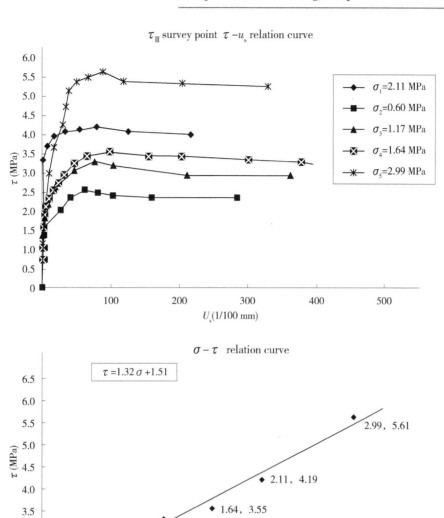

Test No.	Test location			f'	c' (MPa)
τ_{III}	1 320 m-elevation platform behind 11# dam section	Peak parameter of shear strength		1.24	1.73
Placement surface elevation (m)	Placement opening time	Placement closing time		Test date	
1 315.00 – 1 318.70	2008.02.08PM9:30	2008.02.19PM9:30		2008.12.16 – 2008.12.17	
Design strength grade and joint surface working condition	$C_{90}20W6F100$ (III gradation)			Hot joint	
Pouring condition	Weather: light rain	Temperature: 13 ℃	Relative humidity: 62%	Wind velocity: 2.0 m/s	
Failure plane condition	Shear short or concrete shear through along joint plane. The area of concrete cutting through accounts for about 45%. The sheared surface has a great fluctuation difference, with maximum fluctuation difference of about 5 cm				

Figure 12.15 *The resulting curve graph of* τ_{III}

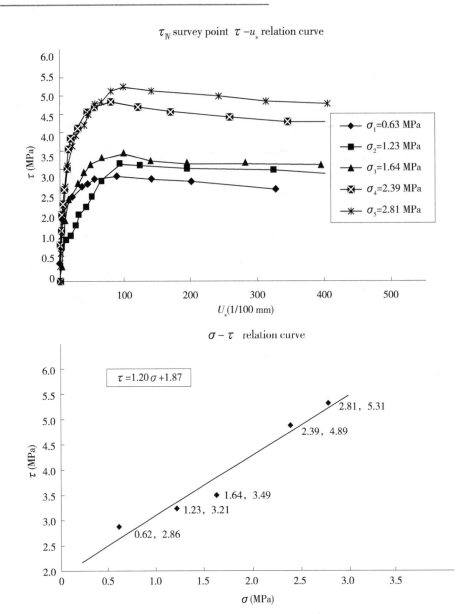

Figure 12.16 *The resulting curve graph of* τ_{IV}

Chapter 12 Core Drilling, Pump-in and In-situ Shear Tests

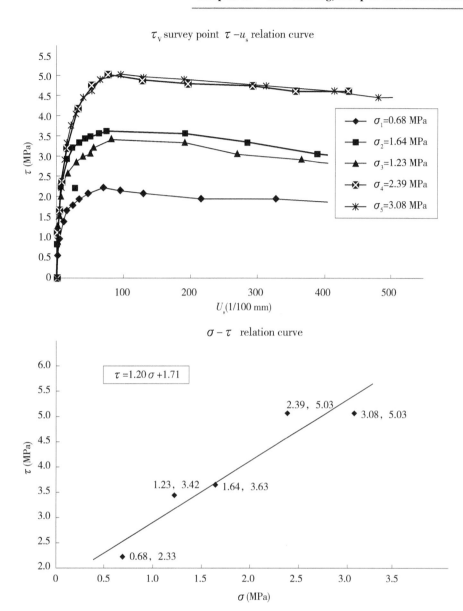

Test No.	Test location	Peak parameter of shear strength		f'	c'(MPa)
τ_V	1 422.5 m-elevation platform behind 11# dam section	Peak parameter of shear strength		1.20	1.71
Placement surface elevation (m)	Placement opening time	Placement closing time		Test date	
1420.00 – 1 422.50	2008.11.08	2008.11.10		2009.02.22 – 2009.02.24	
Design strength grade and joint surface working condition	C_{90}20W6F100 (Ⅲ gradation)			Hot joint	
Pouring condition	Weather: Temperature: ℃ Relative humidity: % Wind velocity: m/s				
Failure plane condition	Cutting off along joint plane or concrete, the joint plane is well cemented, gravels are evenly distributed, and there are a few stones are cut off				

Figure 12.17 *The resulting curve graph of* τ_V

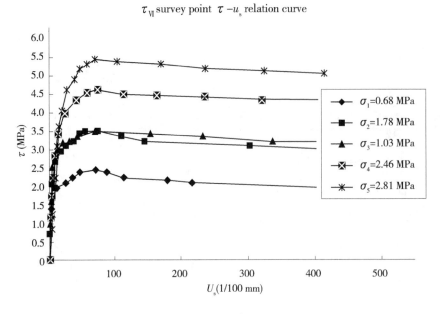

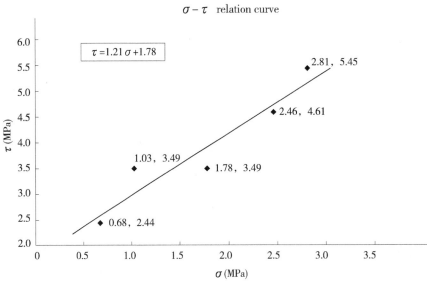

Test No.	Test location		f'	c' (MPa)
τ_{VI}	1 422.5 m-elevation platform behind 11# dam section	Peak parameter of shear strength	1.21	1.78
Placement surface elevation (m)	Placement opening time	Placement closing time	Test date	
1 420.00 – 1 422.50	2008.11.08	2008.11.10	2009.02.22 – 2009.02.24	
Design strength grade and joint surface working condition	C9020W6F100 (III gradation)		Hot joint	
Pouring condition	Weather: Temperature: ℃ Relative humidity: % Wind velocity: m/s			
Failure plane condition	Cutting off along joint plane or concrete, the joint plane is well cemented, gravels are evenly distributed, and there are a few stones are cut off			

Figure 12.18 *The resulting curve graph of* τ_{VI}

Figure 12.19 *After shear of σ_1 specimen under normal stress*

Figure 12.20 *After shear of σ_2 specimen under normal stress*

Figure 12.21 *After shear of $C_{90}15W6F100$ (Ⅲ gradation) specimen*

Figure 12.22 *Excavation of on-site specimens*

Figure 12.23 *Installation of on-site test equipment*

Figure 12.24 *On-site testing*

Group Ⅰ: The No. is $\tau_{\text{Ⅰ}}$, there are a total of 5 specimens, and the design grade of concrete is $C_{90}20W6F100$ (Ⅲ gradation) hot joint; the shear failure surface of the specimen is cut along the joint surface or the concrete, the concrete shear area is about 40%, and the shear surface has a large fluctuation difference (the maximum is about 7 cm), with good cementation and even distribution of stones; the peak shear strength of the group of specimens is $f' = 1.28$ and $c' = 1.65$ MPa.

Group II: The number is τ_{II}, there are a total of 5 specimens, and the design grade of concrete is $C_{90}20W6F100$ (III gradation) hot joint; the shear failure surface of the specimen is cut along the joint surface or the concrete, the concrete shear area is about 25%, and the shear surface has a large fluctuation difference (the maximum is about 6 cm), with good cementation and even distribution of stones; the peak shear strength of the group of specimens is $f' = 1.32$ and $c' = 1.51$ MPa.

Group III: The number is τ_{III}, there are a total of 5 specimens, and the design grade of concrete is $C_{90}20W6F100$ (III gradation) hot joint; the shear failure surface of the specimen is cut along the joint surface or the concrete, the concrete shear area is about 45%, and the shear surface has a large fluctuation difference (the maximum is about 5 cm), with good cementation and even distribution of stones; the peak shear strength of the group of specimens is $f' = 1.24$ and $c' = 1.73$ MPa.

Group IV: The number is τ_{IV}, there are a total of 5 specimens, and the design grade of concrete is $C_{90}15W6F100$ (III gradation) hot joint. Most of the shear failure surface of specimens is cut along the joint surface, and shear surface is relatively flat. The maximum fluctuation difference is 3 cm, with good cementation and evenly distribution of stones, and a few stones are cut. The shear peak strength of the group of specimens is $f' = 1.20$ and $c' = 1.87$ MPa.

Group V: The number is τ_V, there are a total of 5 specimens, and the design grade of concrete is $C_{90}15W6F100$ (III gradation) hot joint. Most of the shear failure surface of specimens is cut along the joint surface, and shear surface is relatively flat. The maximum fluctuation difference is 3 cm, with good cementation and evenly distribution of stones, and a few stones are cut. The 2/3 area of shear surface of the second specimen is made of concrete, which is rough, with relatively large fluctuation (about 8 cm). The shear peak strength of the group of specimens is $f' = 1.20$ and $c' = 1.71$ MPa.

Group VI: The number is τ_{VI}, there are a total of 5 specimens, and the design grade of concrete is $C_{90}15W6F100$ (III gradation) hot joint. Most of the shear failure surface of specimens is cut along the joint surface, and shear surface is relatively flat. The maximum fluctuation difference is 5 cm, with good cementation and evenly distribution of stones, and a few stones are cut. The 1/2 area of shear surface of the first specimen is made of concrete, which is rough, with relatively large fluctuation (about 9 cm). The shear peak strength of the group of specimens is $f' = 1.21$ and $c' = 1.78$ MPa.

12.8.7 Conclusions

1. On-site in-situ shear strength of RCC dams.

    $C_{90}20W6F100$ (III gradation) is $f' = 1.24 - 1.32$ and $c' = 1.51 - 1.73$ MPa. Comprehensive value of shear strength: $f' = 1.28$ and $c' = 1.63$ MPa.

    $C_{90}15W6F100$ (III gradation) is $f' = 1.20 - 1.21$ and $c' = 1.71 - 1.87$ MPa. Comprehensive value of shear strength: $f' = 1.20$ and $c' = 1.79$ MPa.

2. The layer (joint) surface of the RCC is with good cementation and even distribution of stones.

Chapter 13

RCC Cofferdam Construction and CSG Damming Technology

13.1 RCC Cofferdam

Rapid damming technology of RCC has not only been developed rapidly in gravity dams and arch dams, but also widely used in cofferdam projects. Since the RCC cofferdam project was firstly adopted for Yantan Hydropower Station in 1988, nearly more than 30 temporary RCC cofferdam projects have been built in China, giving full play to the advantages of high degree of mechanization, fast speed, cost-effectiveness and safety in the RCC construction. RCC cofferdams are safer than earth-rock cofferdams, and the dam crest can be overflown, especially suitable for overflow cofferdams with high flood control standards; the RCC cofferdam has a small cross section, which can effectively reduce the foundation pit area and be convenient for rapid restoration of foundation pit construction; moreover, the diversion tunnel is significantly shortened, and the diversion standard can be significantly reduced; at the same time, it plays the role of temporary water-retaining power generation and efficiency in advance. Among more than 20 RCC cofferdams built, the Phase III cofferdam of the Three Gorges Project, the Longtan and the Goupitan upstream cofferdam with a height of more than 70 m are high concrete dams. All these cofferdams have been successfully completed within only 4 – 5 months, and put into operation in time in a dry season. Some cofferdams have also withstood the test of large-flow overflow for many times during flood blocking, overflow and flooding, and all cofferdams have no obvious leakage and have been in safe operation. At the same time, the construction technology of the RCC dam is known and the team is cultivated, to lay the foundation for dam construction through the construction of the RCC cofferdam. An example is described as follows.

1. The Phase III cofferdam of the Three Gorges Project adopts concrete gravity cofferdam, with maximum height of 115 m, top width of 8 m and maximum bottom width of 107 m. From December 16, 2002 to April 16, 2003, its placing was completed. It took only 4 months to complete the total volume of 1 100 000 m^3 of RCC and the highest daily and monthly strength of pouring reaches 21 000 m^3 and 475 000 m^3, so that it was completed in advance (55 d earlier than the contract period). The cofferdam was completed in a dry period to solve the construction problem. The storage capacity reached 12.4 billion m^3, to ensure that the Phase III cofferdam of the Three Gorges Project is put into water-retaining power generation on schedule. It is vividly called the Three Hundred Project. In more than 100 d, a dam with a height of more than 100 m and RCC of more than 1.1 million m^3 was built, creating the miracle in damming history, strongly proving that the advantages of "fast" RCC damming technology are incomparable by others.

2. Dachaoshan RCC is mixed with a new type of PT (phosphorus slag + tuff) admixture. In order to master the performance of PT admixture RCC, test construction on the right bank gravity pier was conducted in August 1997. By the end of 1997, the initial setting time, VC value control, fine adjustment of the mix proportion of the content of sand and rock powder and the quality fluctuation of PT admixture and other relevant factors of

PT admixture RCC are mastered basically; during the construction of the RCC arch cofferdam from March to May 1998, the construction characteristics of the large placing surface and the RCC in high-temperature seasons were further found out, to provide the technical support for PT admixture RCC to be under large placing surface construction of dam.

3. The tunnel diversion method of using a cofferdam to block the riverbed at one time is adopted for Longtan Hydropower Station. The RCC cofferdam is used to block water in the initial diversion stage, and one diversion tunnel is arranged on the left and right banks for drainage. The project was closed at the beginning of November 2003, and the RCC cofferdam was constructed under the protection of the upstream and downstream earth-rock cofferdams. The maximum height of the upstream RCC cofferdam is 73.20 m, which is second only to the Phase III RCC cofferdam of the Phase III cofferdam of the Three Gorges Project in the current domestic construction. Longtan RCC cofferdam has a large amount of quantities, tight construction period and high construction intensity. In addition to considering normal water retaining conditions, the overflow protection measures also need to considered for excessive flood. According to model tests, water pre-filling measures for foundation pits such as no water exceeding the weir shoulders on both banks, the gap reserved in the middle of the riverbed, the self-breaking weir set on the gap of the downstream cofferdam, and the slow closing and non-return water valve set on the weir body are adopted. Only an energy dissipation platform with width of 6.00 m is set on the slope behind the upstream cofferdam, and the bank slopes and foundations behind the weir are not protected separately, which greatly simplifies the anti-scouring protection for excessive flood and facilitates construction, to create conditions for rapid construction. The maximum daily and monthly construction height of the cofferdam construction is 1 m and 21 m respectively, and the daily and monthly maximum pouring intensity is 10 500 m^3 and 185 000 m^3 respectively. The RCC cofferdam was completed on schedule in May 2004, creating the conditions for smooth construction of subsequent projects.

4. RCC cofferdams are used in the upstream of the RCC main dam of Baise Multipurpose Dam Project. Through the construction of the upstream RCC cofferdam, the production process test of diabase aggregate RCC is carried out, and the primary parameters are obtained; the stepped formwork of precast concrete block is used as the downstream formwork, which speeds up the construction; in August 2003, it is safe upon cofferdam overflow, ensuring the flood control safety of the dam.

13.2 Cemented Sand and Gravel (CSG) Damming Technology

13.2.1 Characteristics of CSG Dam

CSG (Cemented Sand and Gravel, CSG for short) dam is basically the same as RCC in dammingtechnology with CSG damming technology. The main difference is that the design index, imperviousness performance and mix proportion material composition are different. The first application of CSG in damming technology was proposed in France in 1970. It is a new type of damming technology. Its design concept and construction method combine those of FRD and RCC dam. CSG dam is an extension of RCC damming technology. Its biggest advantage is to broaden the scope of aggregates used, use local materials, NSG mixture, spoil excavated or weathered rock generally not used as far as possible; the amount of cementing materials is small, the general cement is less than 50 kg/m^3, the total amount of cementing materials is no more than 100 kg/m^3; through the CSG mix proportion design, upon mixing, transportation, pouring, paving, rolling compaction and cementation,

Chapter 13 RCC Cofferdam Construction and CSG Damming Technology

the dry and hard dam body with certain strength is formed. The goal of CSG damming construction technology is to achieve zero spoil (i.e., no spoil is generated in the dam site area). This type of dam is usually cost-effective for weak foundation and is beneficial from the perspective of environmental impact.

The design section of CSG dam is mostly symmetrical, with the same slope in upstream and downstream sections, which is also known as the cipollettic dam. Although the symmetrical design section increases the roller compacted volume of the dam body, the total project cost does not increase compared with the RCC dam because the material performance requirements of the symmetrical CSG dam are lower than that of the RCC dam. Another important difference between symmetrical CSG dam and traditional gravity dam is the average shearing stress in the foundation. CSG dam is suitable for rock foundation with low strength or even with shear plane. The vertical load acted on the foundation is quite uniform and will not change greatly with the change of reservoir water level. This is the basic difference between symmetrical CSG dam and traditional gravity dam, which is especially important for rock foundation with low modulus of elasticity. Therefore, when the foundation rock is weaker, it is allowed to build the symmetrical CSG dam in places where it is not suitable to construct the traditional gravity dam.

The deformation of concrete or RCC is lower than that of rock foundation below. When the dam site conditions are not good, as if the rigid dam body is supported by a deformation body, the dam body may generate large structural stresses and cause cracking. From the point of view of structure, it is more advantageous to have less stiffness of damming materials. High permeability and low stiffness are corresponding to the low strength of CSG. In fact, the modulus of elasticity of CSG largely depends on the strength, modulus of elasticity, silt content, mix proportion design, etc. of aggregates. The modulus of elasticity of CSG is expected to be less than 10 GPa.

13.2.2 Advantages of CSG Damming Technology

CSG damming technology is similar to RCC damming technology, and the construction method is basically the same. Since the maximum particle size of coarse aggregates can be 250 mm or 300 mm, the compaction layer can generally 40 – 60 cm thick, which can significantlyaccelerate the speed of rolling compaction. The main advantage of using CSG dam is that many construction requirements are relaxed. An example is described as follows:

1. The layer surface treatment can be reduced to the minimum, because only the shear friction resistance is required, the next layer can be poured without any special treatment for the interlayer or construction joint.
2. The damage caused by aggregate separation is not considered in site rolling compaction because requirements for overall strength and imperviousness performance are relatively low.
3. The seepage of horizontal interlayer joints will not harm the overall stability of the dam, because the CSG dam is similar to the CFRD, and the normal concrete cover is usually used in the upstream and downstream of the CSG dam, so it is unnecessary to lay cushion materials on the construction joint surface and the construction surface in the upstream area of the dam body.
4. The formwork is simple. Because there is no need to set construction shrinkage joints, when the upstream and downstream dam surface slope is relatively large, the upstream and downstream surfaces can adopt concrete precast formwork or movable reinforced concrete precast formwork (namely, the movable extruded reinforced concrete precast formwork for side wall of FRD).
5. No temperature control is required. Due to the low content of CSG cement, its temperature rise is also very low. The temperature stress mainly depends on adiabatic temperature rise and modulus of elasticity of damming materials. Therefore, the temperature stress of CSG dam is lower than that of RCC dam, so it is unnecessary to set transverse joints and control temperature.

Roller Compacted Concrete Rapid Damming Technology

6. Advantages of CSG cofferdam. Most of the concrete and sand and gravel systems have not been put into use in the diversion and closure of general projects. Due to the low impermeability requirements of cofferdams, CSG damming eliminates the sand and gravel screening system, simplifies the mixing process, and has fast construction speed and low investment. CSG dam has strong scour resistance and relatively large water permeability. Therefore, CSG dam is especially suitable for the application in cofferdam projects.

13.2.3 Process of CSG Damming Technology

CSG damming technology was first applied in some temporary projects in Japan. The first project was the upstream cofferdam of Nagashima dam built in 1991 (the dam height is 14.9 m, and CSG of 22 900 m³ is used). The first permanent structure with CSG damming technology is the sand retaining dam in the upstream of Nagashima Reservoir, with a height of 34 m, which was completed in 2000.

The CSG damming technology was firstly applied in the downstream flow measurement cofferdam of Jiemian Hydropower Station and upstream cofferdam of Hongkou Hydropower Station in Fujian Province in 2004. Through the research on CSG damming technology of these two projects, from the CSG mix proportion, material strength characteristics, mixture mixing and rolling compaction process, etc., the research results are applied to project practice in China, so as to take the first step of CSG damming technology in China, accumulate valuable experience and lay a foundation for the further application of the new dam type.

1. The dam of Jiemian Hydropower Station in Youxi County, Fujian Province is a CFRD with a maximum dam height of 126 m. The downstream flow measurement cofferdam of the dam is constructed by CSG technology. The flow measurement cofferdam is also combined with the dam body as the downstream cofferdam. The parameters of the flow measurement cofferdam are as follows: the designed maximum height of 13.54 m, length of 51.93 m, the upstream face slope ratio of 1:0.4, the downstream face slope ratio of 1:0.96, and the total volume of 4 177 m³. The construction of the flow measurement cofferdam began on November 21, 2004 and ended on December 3, 2004. In the construction mix proportion, the amount of cementing materials is 90 kg/m³, the thickness of rolling layer is 60 cm, the maximum aggregate size is 300 mm, and the VC value is controlled within 3-8 s. 3 groups of samples were collected on site. Through the test of compactness and compressive strength, the apparent density of CSG is more than 2 300 kg/m³, and the compressive strength is more than 7.5 MPa. After grouting of impermeable curtain, the water permeability is less than 5 Lu. The project practice of downstream flow measurement cofferdam of the dam of Jiemian Hydropower Station shows that CSG damming technology meets the expected design requirements. Compared with the normal concrete scheme, the construction period is shortened by half and the project cost is reduced by about 25%.

2. The dam of Hongkou Hydropower Station in Ningde City, Fujian Province is a RCC dam with a maximum height of 130 m. The upstream cofferdam of the dam is based on the CSG damming technology. The parameters of the upstream cofferdam are as follows: the designed maximum height of 35.5 m, length of 99.7 m, the crest width of 4.5 m, the upstream face slope gradient of 1:0.3, the downstream face slope gradient of 1:0.75, and the total volume of 32 000 m³. The seepage control of the cofferdam is carried out with the centralized seepage control method of CSG with enriched grout gradation on the upstream face. The enriched grout area is 2.5 m from the bottom of upstream face to 0.5 m at the crest. In the design of CSG mix proportion, the amount of cementing materials is 70 kg/m³, and aggregates are local NSG and part of foundation pit excavation spoil. The thickness of rolling layer is 40 – 60 cm, the density is 2 250 – 2 400 kg/m³, the 28 d compressive strength can reach 4 MPa, and the tensile strength can reach 0.35 MPa. The construction of the upstream cofferdam of Hongkou Hydropower Station is for 3 months from January 10, 2006. Compared with the original RCC cofferdam scheme, the CSG technology applied in the upstream

Chapter 13 RCC Cofferdam Construction and CSG Damming Technology

cofferdam of Hongkou Hydropower Station can reduce the project cost by 28% and shorten the construction period by 25%.

3. In March 2009, the CSG damming technology was adopted in the upstream cofferdam of Lancang River Gongguoqiao Hydropower Station in Yunnan Province. Gongguoqiao Hydropower Station is a RCC gravity dam with the maximum dam height of 105 m. The upstream cofferdam of the dam is based on CSG damming technology. The cross section of CSG cofferdam is subject to trapezoidal design, with the maximum height of 50 m and the cofferdam crest length of 130 m. The upstream face of CSG cofferdam is impervious with normal concrete, with a total volume of 97 000 m^3. The CSG cofferdam in the upstream of Gongguoqiao Hydropower Station was completed in May 2009. On August 5 of that year, the cofferdam suffered from the flood once in 10 years, and the cofferdam was safe. Details are shown in Figure 13.1.

Figure 13.1 *Upstream CSG cofferdam of gongguoqiao hydropower station*

13.3 Design and Rapid Construction of Longtan RCC Cofferdam

13.3.1 Overview

The dam of the Longtan Hydropower Station is a 200 m full-section RCC dam. The tunnel diversion method of using a cofferdam to block the riverbed at one time is adopted for Longtan Hydropower Station. The RCC cofferdam is used to block water in the initial diversion stage, and one diversion tunnel is arranged on the left and right banks for drainage. The project was closed at the beginning of November 2003, and the RCC cofferdam was constructed under the protection of the upstream and downstream earth-rock cofferdams. The maximum height of the upstream RCC cofferdam is 73.20 m, which is second only to the Phase Ⅲ RCC cofferdam of the Phase Ⅲ cofferdam of the Three Gorges Project in the current domestic construction. The cofferdam has a large amount of quantities, tight construction period and high construction intensity. The cofferdam has to experience 2 flood seasons during use. In addition to considering normal water retaining conditions, the protection measures alsoneed to be considered for excessive flood.

The upstream RCC cofferdam of Longtan Hydropower Station Dam Project is designed as a full-section Ⅲ gradation RCC (C10) gravity cofferdam. The upstream face is subject to seepage control with GEVR every 50 cm (width). The cofferdam adopts the layout of heightening the shoulders on both banks and reserving gaps in the middle. The elevation of foundation surface is 199.5 m, the crest elevation of riverbed reserved gap section is 273.2 m, the crest elevation of bank slope section on both banks is 278.6 m, the maximum height is 73.7

m, the crest width is 7 m, the axis length is 386.9 m, and the total amount of concrete is 520 000 m³. The water retaining standard of cofferdam is designed according to the flood once in 10 years of measured hydrological series, and the peak discharge is 14 700 m³/s. At the same time, considering that the RCC cofferdam may overflow in case of excessive flood during its operation, appropriate overflow and energy dissipation protection shall be carried out for the cofferdam, and the water retaining standard of shallers on both banks shall be appropriately improved. The water retaining standard is long-series (i.e. considering historical flood) flood once in 10 years and the flood peak discharge is 16 300 m³/s; the safety standard of excessive flood is checked as per 18 500 m³/s of long-series flood once in 20 years.

From January 23, 2004 when the blinding concrete below the elevation of 201.0 m was poured to May 30, 2004 when the full section of the upstream cofferdam rose to the elevation of 273.2 m, the designed flood crossing elevation of cofferdam in flood season was reached. It took 4 months and 9 d to pour 520 000 m³ concrete, with an average monthly pouring strength of 120 000 m³. The actual maximum monthly pouring strength was 203 000 m³, which occurred in April 2004.

Through the construction of Longtan RCC cofferdam, the following two objectives must be achieved: firstly, conduct rapid construction to meet the requirements of construction period to ensure the safety of foundation pit construction in flood season; secondly, explore the construction mix proportion, construction process parameters and construction technical measures of RCC suitable for the raw materials of dam concrete and special high temperature and low humidity climate characteristics of Longtan Hydropower Station, test the adaptability of admixtures, the supply capacity of raw materials, the operation of equipment, and verify the rollability and construction performance of RCC, to ensure that the quality of dam concrete meet the design requirements, thus accumulating experience for dam RCC construction.

13.3.2 Cofferdam Design

13.3.2.1 Layout And Characteristics of Cofferdam

RCC cofferdams are used for upstream and downstream cofferdams of Longtan Hydropower Station.

The upstream RCC cofferdam is located about 100.00 m in the upstream of the dam axis, the shortest distance from the toe to the dam heel is only about 8.50 m,

The foundation is located on the Luolou Formation and Banna formation, and the cofferdam bedrock is Grade Ⅲ-V; the crest is 368.30 m long, the maximum height is 73.20 m, and the total concrete volume is about 505 000 m³.

The downstream RCC cofferdam is arranged at 370.00 m in the downstream of the dam axis, and the foundation is located on the Banna formation, with good geological conditions; the crest is 273.00 m long, the maximum height is 48.50 m, and the total concrete volume is about 11 000 m³.

13.3.2.2 Design Standard

According to *Specification for Construction Organization Design in Water Conservancy and Hydroelectric Engineering* SDJ 338—1989 and consultation opinions of experts, the water retaining standard of upstream and downstream RCC cofferdams is according to flood once in 10 years and peak discharge of 14 700 m³/s of 43 years measured hydrologic series. Considering that the RCC cofferdam may overflow in case of excessive flood during its operation, appropriate overflow and energy dissipation protection shall be carried out for the cofferdam, and the water retaining standard of shoulders on both banks shall be appropriately improved. The water retaining standard is long-series (i.e. considering historical flood) flood once in 10 years and the flood peak discharge is 16 300 m³/s; the safety standard of excessive flood is checked as per 18 500 m³/s of long-series flood once in 20 years.

Chapter 13 RCC Cofferdam Construction and CSG Damming Technology

13.3.2.3 Shape Design of Cofferdam

Gravity cofferdams are used for upstream and downstream RCC cofferdams. The crest width of cofferdam is 7.00 m. The upstream cofferdam is a vertical plane above upstream face elevation of 220.00 m, the slope ratio below the elevation of 220.00 m is 1:0.2, and the slope ratio of the downstream face below the elevation of 271.30 m is 1:0.75. Considering that the upstream RCC cofferdam is higher and closer to the dam, in order to prevent the damage to the foundation and the dam foundation in case of excessive overflow of the cofferdam, the foundation pit shall be filled with water before the overflow, and appropriate overflow and energy dissipation protection are carried out combined with the cofferdam structure design. The upstream cofferdam adopts the layout of heightening the shoulders on both banks and reserving a gap in the middle. At the same time, an energy dissipation platform with the width of 6.00 m is set at the downstream slope elevation of 240.00 m. The elevation of upstream RCC cofferdam crest is 273.20 m for gap section and 278.60 m for shoulders on both banks. The lowest foundation elevation of cofferdam is 200.00 m and the maximum height is 73.20 m.

The downstream cofferdam is a vertical slope above upstream face elevation of 235.00 m, the slope ratio below the elevation of 235.00 m is 1:0.55, and the slope ratio of the downstream face below the elevation of 235.00 m is 1:0.33. Considering that the foundation pit shall be filled with water in case of excessive flood, a water filling gap of collapsible woven bag clay cofferdam with height of 2.60 m and length of 162.00 m is set in the downstream cofferdam, the elevation of concrete top of cofferdam gap is 242.40 m, and the shoulder on both banks is 247.50 m. The lowest foundation elevation of the cofferdam is 196.50 m, and the maximum height is 48.50 m. At the same time, 6 water return valves are set in the body for recession.

13.3.2.4 Tability and Stress Analysis of Cofferdam

I Upstream RCC cofferdam

The minimum stress and maximum stress of foundation surface in water retaining period are 0.27 MPa and 2.02 MPa respectively; the minimum stress and maximum stress of foundation surface in overflow period are −0.04 MPa and 1.68 MPa respectively; the minimum stress and maximum stress of foundation surface in recession period are 0.27 MPa and 1.46 MPa respectively. The anti-sliding stability of foundation in water retaining period, overflow period and recession period can meet the requirements of the specification.

The anti-sliding stability of cofferdam RCC in water retaining period and overflow period can meet the requirements of the specification. When the cofferdam is not provided with drainage facilities, the minimum stress and maximum stress are −0.02 MPa and 1.78 MPa in the water retaining period, which meets the requirements; the minimum stress and maximum stress are −0.30 MPa and 1.62 MPa in the overflow period, and the maximum tensile stress occurs at the elevation of 220.0 m (i.e. near the break of slope), which not only exceeds the stress control standard, but also has a large distribution range, so the cofferdam shall be provided with drainage. According to the calculation results, the cofferdam is provided with drainage facilities from foundation to the elevation of 235.00 m. The minimum stress and maximum stress are −0.15 MPa and 1.65 MPa respectively in the overflow period, which meet the requirements.

II Downstream RCC cofferdam

The minimum stress and maximum stress of foundation surface in water retaining period are −0.09 MPa and1.22 MPa respectively; the minimum stress and maximum stress of foundation surface in overflow period are −0.06 MPa and 0.60 MPa respectively; the minimum stress and maximum stress of foundation surface in recession period are 0.01 MPa and 0.67 MPa respectively. The anti-sliding stability of foundation in water retaining period, overflow period and recession period can meet the requirements of the specification.

The anti-sliding stability and stress of cofferdam RCC in water retaining period, overflow period and recession period can meet the requirements of the specification.

13.3.2.5 Structural Design of Cofferdam

I Thickness of impervious layer and concrete partition

The impervious layer of the upstream face upstream cofferdam face is made of 0.50 m thick III gradation GEVR (C10, W8), while the upstream face of the downstream cofferdam is not provided with the impermeable layer of GEVR; the cofferdam is made of III gradation RCC (C10, W6). The cofferdam foundation cushion is made of 1.00 m thick III gradation normal concrete (C15, W8), and partial deep groove is backfilled with III gradation normal concrete (C15, W8); cofferdam foundation curtain grouting gallery, drainage gallery and water stop copper strip are no less than 1.00 m around the use of III gradation GEVR (C10, W8), fault grooving concrete plug and water stop copper plate base are backfilled with fine aggregate concrete (II gradation C15, W8).

II Jointing of cofferdam and water stop

In principle, a transverse joint shall be set up at the 30–40 m long cofferdam section and the part with abrupt change of terrain, and a copper water stop shall be set in the joint. Each transverse joint is made of 10 mm thick asphalt fir wood caulking in normal concrete and GEVR. In the RCC, the joint cutter is used for cutting joints, and 10 mm thick polyethylene closed foam joint board is filled.

The upstream RCC cofferdam is 368.34 m long in total, with 10 transverse joints and the length of the maximum cofferdam section of 43.60 m. The downstream RCC cofferdam is 273.04 m long in total, with 7 transverse joints originally set and the length of the maximum cofferdam section of 47.00 m. In the actual construction process, combined with test requirements that the downstream cofferdam is mixed with MgO, the all transverse joints and their water stop copper strips (with thickness of 1.2 mm) and polyethylene closed cell foam board (with thickness of 10 mm) of the downstream cofferdam are canceled, andwhole-block pouring of RCC is conducted.

III Curtain grouting and drainage

The upstream cofferdam is provided with a longitudinal foundation curtain grouting and drainage gallery with the size of 2.50 m × 3.00 m ($b \times h$). The center line is located at 0.75 m in the downstream of the cofferdam axis; 3 transverse galleries are set up to connect with the longitudinal gallery in the deep groove in the river and the bank slopes on both banks. The cofferdam foundation is provided with a single row of curtains, which is 3.00 m below the 5 Lu line of foundation, and the maximum depth of grouting curtain is about 49.00 m. The foundation in the downstream of the curtain is provided with a single row of drainage holes, with a depth of 15.00 m. A row of drainage holes with a spacing of 3.00 m and a pore diameter of D150 are set below the cofferdam elevation of 235.00 m. The water seepage of cofferdam and foundation is drained into the foundation pit by drainage gallery and then pumped.

The downstream RCC cofferdam is not provided with curtain and drainage facilities.

13.3.2.6 Foundation Treatment of Cofferdam

The foundation of upstream and downstream cofferdams is mainly composed of Banna formation sandstone and siltstone with mudslate. The saturated compressive strength of the rock is 50–100 MPa. The angle between the rock stratum strike and the axis of the upstream cofferdam is 32.5°–17.5° and basically parallel to the axis of the downstream cofferdam. The rock stratum inclines to the left bank in the downstream and the dip angle is steep. As for the foundation surface, except that the shoulders on both banks are located in the lower part of the strongly weathered rock mass, most of the foundation is located in the middle and lower part of the strongly and weakly weathered rock mass. The consolidation grouting is only applied to the dense joint zone of the upstream cofferdam, and the integrity of the downstream cofferdam foundation is good, and no other consolidation grouting is applied.

Chapter 13 RCC Cofferdam Construction and CSG Damming Technology

13.3.2.7 Overflow Protection Measures of Cofferdam

Ⅰ Overflow protection of cofferdam

In order to simplify the overflow protection of cofferdam and save the quantity of overflow protection, the engineering measures of heightening the shoulders on both banks of the cofferdam and filling the foundation pit first are adopted. The standard for heightening the shoulders on both banks of the cofferdam is as follows: In case of long-series annual flood once in 10 years ($q = 16\,300$ m^3/s), the shoulders on both banks of the cofferdam will not be filled with water. The upstream cofferdam has a gap with elevation of 273.20 m and width of 231.10 m, and shoulders with elevation of 278.60 m on both banks; the downstream RCC cofferdam has a gap with elevation of 242.40 m and width of 162.00 m, and shoulders with elevation of 247.50 m on both banks. In case of overflow of the upstream RCC cofferdam, the foundation pit is filled with water from the gap reserved in the downstream RCC cofferdam to form a water cushion at the toe during overflow of the upstream RCC cofferdam, to protect the toe and dam heel from being damaged by the overflow flood.

After analysis and calculation, in case of flood once in 10 years ($q = 16\,300$ m^3/s), the water flowing through the upstream cofferdam falls on the cofferdam, and there is a stepped energy dissipator on the downstream face of the cofferdam, and there is a water depth of 20.00–60.00 m under the cofferdam, so no additional protection measures are taken under the cofferdam.

Ⅱ Water filling and recession measures for downstream cofferdam

The downstream cofferdam is provided with a gap with width of 162.00 m and crest elevation of 242.40 m. The upper part of the gap section is a woven bag clay cofferdam with height of 2.60 m. According to the flood forecast, when there is flood exceeding the designed level, the woven bags filled with clay with a height of 2.60 m are removed to fill the foundation pit with water. According to the calculation, it takes about 1.2 h when the foundation pit is filled with water to 235.00 m and 1.4 h when the foundation pit is filled with water to 240.00 m, so as long as the flood forecast time is advanced by 3 h, the water filling time can be guaranteed enough.

After the flood peak, the accumulated water in the foundation pit will be drained from the gap of the downstream cofferdam with a width of 162.00 m (crest level of 242.40 m), so as to ensure that the downstream cofferdam will not be affected by the reverse water pressure as much as possible. Meanwhile, in order to avoid the tensile stress at the toe of the weir due to the difference between the internal and external water heads in the process of draining, 2 rows of 6 slow-closing check valves for water drainage are set at the weir elevation of 230.00 m and 238.00 m respectively.

13.3.3 Design Indexes and Construction Mix Proportion of Cofferdam Concrete

See Table 13.1 for design indexes of cofferdam concrete and Table 13.2 for construction mix proportion.

Table 13.1　　　　　　　　　　Design indexes of cofferdam concrete

S/N	Construction site	Design strength	90 d Assurance rate P (%)	90 d Strength (MPa)	Ultimate tensile (10^{-4})	Grade of impermeability	Remarks
1	Cushion concrete	C15	80	22	0.8	W8	—
2	Concrete of energy dissipation platform	C25	80	34	0.9	—	—
3	Roller compacted concrete	C10	80	15	0.7	W6	—
4	GEVR	C10	80	15	0.7	W8	—

Table 13.2 Mix proportion of cofferdam concrete

Design index and position	Gradation	water-cement ratio	Fly ash (%)	Sand ratio (%)	ZB-1RCC15/JM-II	Material amount (kg/m³)				
						Water	Cement	Fly ash	Sand	Stone
C15 cushion layer	III	0.50	32	30	0.5	101	141	61	688	1 485
C25 energy dissipation platform	III	0.45	31	30	0.5	101	157	67	661	1 493
Roller compacted concrete	III	0.54	65	34	0.6	81	52	98	767	1 500
GEVR grout	—	0.54	50	—	0.3	525	584	584	—	—

13.3.4 Construction of RCC Cofferdam

13.3.4.1 Optimization of Construction of Upstream RCC Cofferdam

The construction period of RCC cofferdam in upstream of Longtan Project is only 4.5 months, which is very tight. In order to ensure the completion of the upstream RCC cofferdam pouring construction on May 31, 2004 and ensure the safety of flood season in 2004, facing the severe and complicated construction conditions, with the support of all participating parties, the following measures have been decided.

I Appropriate adjustment of grouting gallery

Precast members are applied for the side walls and top arch of the grouting gallery, and the steel bars in the gallery are properly adjusted. The grouting gallery is installed with the rise of RCC surface; in order to avoid too much interference to the continuous rise of RCC, the side walls of the slope section of grouting gallery descend into the bedrock, and the trenches are excavated in the weir foundation in advance during the excavation, and the floor concrete is adjusted to GEVR, which is poured together with the RCC with placing surface.

II Precast concrete block

In the downstream face of concrete cofferdam, considering the fast rising speed of RCC with large placing surface, it is conducive to mechanical and rapid construction. The downstream face of the upstream RCC cofferdam is constructed with precast concrete blocks below the energy dissipation platform with an elevation of 240 m, and the precast blocks are part of the weir body; cast-in-place step formworks are adopted in the gap of weir body above the energy dissipation platform with an elevation of 240 m, and precast concrete blocks are still used outside the gap of weir body.

III The downstream slope of the cofferdam adjustment

In order to adapt to the rapid construction of RCC with large placing surface, on the downstream energy dissipation platform, the cofferdam slope angular point and weir crest point are kept unchanged for the upper weir section of 0 + 072.60 − 0 + 324.74, and the downstream slope of the cofferdam is adjusted from 1:0.75 − 1:0.7, correspondingly, the width of the energy dissipation platform with an elevation of 240 m is adjusted from 6 m to 5 m, and the vertical surface height under the platform is adjusted from 8.0 m to 1.9 m.

IV The weir crest gap adjustment

The straight ridge and steep trenches in weir crest gap are adjusted to be connected by slopes with the left bank transitioning from upper horizontal chainage 0 + 303.700 to upper horizontal chainage 0 + 258.70, and the right bank transitioning from upper horizontal chainage 0 + 72.60 to upper horizontal chainage 0 + 117.60, with a slope of 12%.

V Blinding concrete adjustment

The original design of 1.00 m thick normal blinding concrete for foundation is adjusted to leveling concrete, and the thickness is determined according to the foundation surface excavated on site and the requirements of RCC placing surface, i.e., leveling with normal concrete. The normal concrete for the foundation cushion at the bank slope is adjusted to GEVR, which is poured together with RCC.

13.3.4.2 Construction Method

According to the characteristics of cofferdam construction layout, full-section RCC dam damming technology is adopted. 3 placement methods are adopted for upstream cofferdam: direct placement by dump trucks, direct placement by dump trucks + placement by high speed belt conveyor feeding lines and placement by high speed belt conveyor feeding lines. Totally 420 600 m³ concrete is poured by direct placement by dump trucks, accounting for 81.9% of the total volume; totally 94 000 m³ concrete is poured by placement by high speed belt conveyor feeding lines, accounting for 18.1% of the total volume. The downstream cofferdam is totally constructed by direct placement by dump trucks.

RCC is poured by continuous concreting inwhole-block and flat lift, the thickness of the rolling layer is 30 cm, and the permanent transverse joints are made by joint cutters; for normal concrete pouring, due to the placing surface area is small, flat pouring method with a pouring layer thickness of 50 cm, because the foundation blinding concrete and normal concrete of energy dissipation platform are only 1.00 m thick, and the concrete is poured by one layer.

13.3.4.3 Arrangement of Placement Road

The RCC at the lower part of cofferdam is transported by 32 t and 20 t dump trucks, while the RCC at the upper part of cofferdam is mainly supplied by high speed belt conveyors, and when the supply is insufficient, it is supplemented by the direct placement by 32 t dump trucks. A total of 4 placement roads are arranged to meet the needs of direct placement of concrete below elevation of 260 m. The direct placement road filled with clean gravel with dehydrated pavement is 30 – 50 m from the placement entrance, and an automatic truck flushing platform is set at a distance of 30 – 50 m from the placement entrance.

13.3.4.4 Arrangement of High Speed Belt Conveyors

The high speed belt conveyors are arranged with comprehensive consideration into the placement of the dam RCC, and the No. 1 and No. 2 high speed belt conveyor feeding lines for the dam RCC pouring are fully utilized for the RCC pouring of upstream cofferdam in the early stage. The layout of the feeding lines takes into account the needs of cofferdam and dam concrete construction. After cofferdam concreting is completed, the feeding line between the concrete production system and dam is reserved for dam concrete construction, and the feed line between the dam and cofferdam is removed for dam concrete construction.

13.3.4.5 Concrete Production System

The concrete production system with elevations of 360 m and 308.5 m on the right bank arranged by dam concrete pouring is used. The concrete production system with an elevation of 308.5 m consists of two 2 × 6 m³ forced mixing structures, and the concrete production system with an elevation of 360 m consists of one 2 × 6 m³ forced mixing structure and one 4 × 3 m³ gravity mixing structure, which are constructed before the pouring of upstream cofferdam RCC, and can ensure the production of RCC with monthly output of 400 000 m³.

13.3.4.6 Formwork

The upstream RCC cofferdam formwork is divided into: large formwork for upstream face, downstream turnover formwork, downstream precast concrete block formwork, gallery formwork, etc.

Ⅰ Large formwork for upstream face

3 m × 3.1 m (width × height) continuous turnover large formworks are adopted as the formwork for upstream face, which is designed and manufactured by Hubei Gezhouba Dolion Formwork Co., Ltd., and the three floors are continuously turned up. The large formworks were used from the elevation of 201 m to the elevation of 273.2 m, and turned over 23 times in total. Except for a few formworks, the overall deformation of the formworks is small, so they can still be used in the dam project. A 16 t crane is used for hoisting during construction.

Ⅱ Downstream turnover formwork

The downstream turnover formwork is designed based on the following points: ①After the cofferdam elevation exceeds the energy dissipation platform with an elevation of 240 m, there is an overcurrent requirement in the gap section and the cast-in-situ downstream steps are needed; ②After the cofferdam is concreted to the upper part, the placing surface is narrow, which is not conducive to hoisting precast blocks by the crane; ③Manual quick installation. In order to meet the design requirements, the overflow part of the cofferdam is cast-in-place concrete, and the downstream turnover formwork is applied from the elevation of 240 m.

Ⅲ Downstream precast concrete formwork

The downstream precast concrete formworks have simple structure and can be manufactured conveniently and formed in one time, thus rapid and large-scale prefabrication can be achieved; such formworks can be easily installed. The size of a single precast formwork is 150 cm × 100 cm × 60 cm (length × width × height), and the weight is 2.4 t. The crane on the placing surface can meet the lifting requirements. Such formworks have good self-stability and are stable by own weight. The downstream precast concrete formworks are adopted at the full surface below the cofferdam downstream face with an elevation of 240 m, gap section below the elevation of 240 m, and at the placement entrance of dump trucks. The arrangement meets the requirements of mechanical and rapid construction of large placing surface, and is a key construction design adjustment for completion of cofferdam construction according to the general schedule.

Ⅳ Gallery formwork

Precast concrete formworks are adopted for the side walls and top arch of the gallery. The cast-in-place concrete is replaced by precast concrete, the installation of gallery formworks rises with the rise of RCC placing surface, and GEVR with thickness of 50 cm is poured around the gallery formwork, which meets the requirements of mechanical and rapid construction of large placing surface, and saves a lot of consumable supporting materials.

13.3.4.7 Construction of Cofferdam RCC

Ⅰ Discharging, spreading and rolling

The RCC is spread by strip method, with strip direction parallel to cofferdam axis, strip width controlled within 10 – 13 m, and two strips above 267 m elevation.

3 spreading machines are used for spreading. The compacted thickness of each layer of RCC is 30 cm, and the spreading thickness is 35 cm. The spreading machines spread the concrete at one time according to the spreading thickness. During spreading, the layered spreading height lines are drawn on the upstream and downstream formworks every 5 m for monitoring.

9 sets of BW202AD vibrating rollers and 3 sets of BW75S vibrating rollers are adopted for rolling by rolling with no vibration for 2 times and then rolling with vibration for 6 – 8 times. The traveling speed of the rollers is 1.0 – 1.5 km/h, with high frequency and low amplitude adopted for the first time, and then high frequency and high amplitude for the remaining times. Rolling strips must overlap by 15 – 20 cm. When the same strip is rolled in sections, the joint parts overlap by 2.4 – 3.0 m. The height difference between the two strips due to rolling operation shall be flattened by rolling with vibration slowly for 1 – 2 times.

Chapter 13 RCC Cofferdam Construction and CSG Damming Technology

Ⅱ Measures for interlayer bonding of RCC

The duration for RCC mixture from mixing to rolling shall not exceed 2 h; in the design of mix proportion, the retarder with obvious retarding effect is adopted and timely adjusted according to different temperature conditions, so as to ensure that the initial setting time of RCC in high temperature or low temperature season is greater than or equal to 8 h; the GEVR with a width of 50 cm on the upstream face of the weir is set; spraying humidification measures are taken on the placing surface to reduce the environmental temperature and prevent the concrete surface from losing water, which may affect the interlayer bonding; the concrete on the compacted surface shall be comprehensively bled.

Ⅲ Construction of transverse joints

Transverse joints are constructed by grinding first and then cutting. After concrete rolling, a HC-70 hand-held joint cutter is used to cut the joints with a depth of 30 cm, and then the joint material is manually inserted for molding. Joints formed are rolled by a small vibrating roller with vibration for 2 – 4 times to eliminate the local looseness of concrete caused by joint cutting.

Ⅳ Construction under high temperature

Notes during construction under high temperature (daily average temperature ≥25 ℃), strong sunshine and windy seasons: ①The VC value is dynamically controlled, namely the VC value in outlet is reduced by 1 – 3 s to ensure that the VC value of the placing surface is within 5 –7 s; ②Spraying on the placing surface shall be performed to compensate the moisture evaporated on the placed concrete surface, keep the placing surface moist, and control and reduce the environmental temperature of the whole placing surface; ③ Effective temperature control measures shall be taken to reduce the concrete outlet temperature, and ensure that the maximum temperature of RCC does not exceed the maximum temperature allowed in the design.

13.3.5 Conclusion

1. The blinding concrete for upstream cofferdam was poured from January 23, 2004, RCC was poured from February 2, 2004 and poured to the crest elevation on June 10, 2004, which lasted for 139 d, with the highest daily output of 10 500 m^3. The blinding concrete for downstream cofferdam was poured from February 16, 2004, RCC was poured from Thursday, March 11, 2004 and poured to the crest elevation on May 28, 2004, which lasted for 109 d.

2. Longtan cofferdam has been operated well after two years operation since it has been built in May 2004. During the operation of the cofferdam, it resisted the flood seasons in 2004 and 2005. The flood started to rise on July 9, 2004, and reached the maximum peak discharge of 8 890 m^3/s at 17:00 on July 11, 2004, and the flood was discharged to the normal water level on July 18, 2004, which lasted for 10 d. The flood started to rise on June 17, 2005, and reached the peak discharge of 8 090 m^3/s at 17:00 on July 19, 2004, and the flood was discharged to the normal water level on July 26, 2004, which lasted for 10 d.

3. Longtan RCC cofferdam has a large amount of quantities, tight construction period and high construction intensity. In addition to considering normal water retaining conditions, the overflow protection measures also need to considered for excessive flood. According to model tests, water pre-filling measures for foundation pits such as no water exceeding the weir shoulders on both banks, the gap reserved in the middle of the riverbed, the self-breaking weir set on the gap of the downstream cofferdam, and the slow closing and non-return water valve set on the weir body are adopted. Only an energy dissipation platform with width of 6.00 m is set on the slope behind the upstream cofferdam, and the bank slopes and foundations behind the weir are not protected separately, which greatly simplifies the anti-scouring protection for excessive flood and facilitates construction, to create conditions for rapid construction.

13.4 CSG Mix Proportion Design and Application of Upstream Cofferdam of Gongguoqiao Dam

13.4.1 Foreword

13.4.1.1 Project Overview

Gongguoqiao Hydropower Station, located on the west side of Dalishu, Yunlong County, Yunnan Province, is the most upstream power station in the cascade development of the middle and lower reaches of Lancang River. The main purpose of the project is power generation. The reservoir has a normal water level of 1 307 m, a capacity of 316 million m^3 and a regulated capacity of 49 million m^3. It is a daily regulation reservoir. The power station has the installed capacity of 900 MW and the annual power generation of 4 041 million kW·h.

The multipurpose dam is mainly composed of water retaining structures, flood discharge structures, water diversion and power generation structures, among which the barrage is RCC gravity dam; the flood discharge structure is the dam body, which is arranged on the right side of the main riverbed; the headrace system is arranged underground on the right bank. The crest elevation is 1 310 m, the crest length is 356 m, and the maximum dam height is 105 m. 4 × 225 MW units are installed in the underground powerhouse.

The diversion of hydropower station construction is performed by tunnel diversion in dry season andthe combination of overflow cofferdam and diversion tunnel in flood season. The upstream overflow cofferdam of Gongguoqiao Hydropower Station is a CSG cofferdam, with CSG quantity of 97 000 m^3.

Gongguoqiao Hydropower Station was intercepted in November 2008, and the dam met the water storage conditions on September 30, 2011, and met the requirements for power generation by the first generator on November 15, 2011. The project was completed on March 31, 2012.

13.4.1.2 Design Indexes of CSG Materials

The upstream overflow cofferdam of Gongguoqiao Hydropower Station draws lessons from CSG construction technology of upstream cofferdam of Hongkou Hydropower Station. According to Technical Requirements for CSG Materials of Upstream Cofferdam, the design indexes of CSG materials of upstream cofferdam of Gongguoqiao Hydropower Station are shown in Table 13.3.

Table 13.3   Design indexes of CSG materials for upstream cofferdam

Design strength grade	Maximum particle size (mm)	Design age (d)	Grade of impermeability	Shear strength	Unit weight (kg/m^3)	Relative compaction (%)
≥C7.5 P≥80%	≤250	28	>W4	$f' > 0.8 c' > 0.5$ MPa	≥2 300	≥96

13.4.2 Raw Material Test

13.4.2.1 Cement

According to the test requirements of *Upstream Cofferdam CSG Construction Technology*, it is determined that Shilin P.O 42.5 ordinary Portland cement produced by Sanjiang Cement Co., Ltd. is used in the mix proportion test. The test results of cement physical and mechanical properties are shown in Table 13.4. According to the test results, all indexes of the cement meet the requirements of specifications.

Chapter 13 RCC Cofferdam Construction and CSG Damming Technology

Table 13.4   Test for physical and mechanical properties of cement

Test items	Specific surface area (kg/m²)	Time of setting		Stability	Standard consistency (%)	Bending strength (MPa)		Compressive strength (MPa)	
		Initial setting	Final setting			3 d	28 d	3 d	28 d
Standard requirements	≥300	≥45 min	≤600 min	Qualified	—	≥3.5	≥6.5	≥17.0	≥42.5
Test Results	327	154	208	Qualified	27.2	4.9	8.5	19.7	51.9

Note: The indexes of all tested items meet the requirements of *Common Portland Cement* (GB 175—2007) and are qualified.

13.4.2.2 Fly Ash

The fly ash used for CSG mix proportion test is Grade II fly ash produced by Kunming No. 2 Power Plant. The test of fly ash quality and mortar performance is carried out in accordance with *Technical Standard of Flyash Concrete for Hydraulic Structures* DL/T 5055—2007. The quality test of fly ash is shown in Table 13.5. According to the fly ash quality test results, all indexes meet the requirements of specifications.

Table 13.5   Analysis test results of physical properties of fly ash

Variety	Fineness (%)	Water demand ratio (%)	Sulfur trioxide (%)
Grade II fly ash	15.6	98	0.68
DL/T 5055—2007 On the quality indexes of Grade II fly ash	≤20	≤105	≤3.0

13.4.2.3 Water Reducing Agent

SBTJM-II high-range retarding water reducing agent produced by Jiangsu Bote New Materials Co., Ltd. is applied, which is mainly used to improve the performance of concrete to meet the requirements of retarding, water reducing and workability. The quality test is performed as per the *Technical Code for Hydraulic Concrete Admixtures* DL/T 5100—1999. See Table 13.6 for the contents of additive performance test. According to the additive quality test results, all indexes meet the requirements of specifications.

Table 13.6   Performance test parameters of concrete mixed with admixtures

Test Items	Water reducing ratio (%)	Ratio of bleeding rate (%)	Gas content (%)	Time of setting (min)		Compressive strength (%)		
				Initial setting	Final setting	3 d	7 d	28 d
Standard requirements	≥15	≤100	<3	+120 – +240		≥125	≥125	≥120
JM-II Mixing amount 0.6%	20.9	46.2	2.2	+210	+207	153	148	139

Note: The admixture is tested as per the *Technical Code for Hydraulic Concrete Admixtures* (DL/T 5100—1999).

13.4.2.4 Water for Mixing

The water quality analysis is performed according to the *Standard of Water for Concrete* JGJ 63—2006, and the analysis results are shown in Table 13.7.

13.4.2.5 Aggregate

According to the actual situation of riverbed aggregate and the design requirements, samples are taken in riverbed with a sampling quantity not less than 6 m³. Sieving test was carried out first, of which the results are shown in Table 13.8.

Table 13.7   Analysis results of water for mixing

S/N	Test items	Unit	Standard requirements	Test Result	Judgment on single items
1	pH value	—	≥4.5	8.1	Qualified
2	Insoluble substance content	mg/L	≤2 000	Not detected	Qualified
3	Soluble matter content	mg/L	≤5 000	296	Qualified
4	Chloride content (calculated as Cl^-)	mg/L	≤1 000	23	Qualified
5	Sulfate content (calculated as SO_4^{2-})	mg/L	≤2 000	83	Qualified
6	Total alkali content	mg/L	≤1 500	23	Qualified

Note: All performance indexes meet the standard JGJ 63—2006.

Table 13.8   Test results of grain gradation of natural sand and gravel aggregate excavated from riverbed (undisturbed)

Aggregate size (mm)	Sample mass (kg)	Mass of sieve residue (kg)	Screening screen residue performance (%)	Gradation distribution of aggregate >250 mm (%)	Distribution ratio of aggregate >250 mm		Accumulated percentage of residue on sieve (%)
					Grain size range (mm)	Distribution ratio (%)	
>250		12 630	39				39
150–250		3 600	11	18			50
80–150		2 624	8	13	40–250	50	58
40–80	32 622	3 896	12	19			70
20–40		3 316	10	17	5–40	30	80
5–20		2 508	8	13			88
<5		4 048	12	20	<5	20	100
Sampling location	50 m away from the upstream cofferdam						

The quality inspection and physical performance test of coarse and fine aggregates at all levels are carried out. The main contents of the tests mainly include grain gradation test, apparent density test, needle-plate like content test, silt content test, crushing index test, etc.

Ⅰ   Fine aggregate

The quality inspection results of the fine aggregate are shown in Table 13.9 and Table 13.10. It can be seen that all indexes of fine aggregate meet the quality requirements of sand materials specified in *Specifications for Hydraulic Concrete Construction* DL/T 5144—2001.

Table 13.9   Test results of quality indexes and physical properties of fine aggregate

Variety	Fineness module	Apparent density (kg/m³)	Saturated dry apparent density (kg/m³)	Moisture absorption of dry saturated surface (%)	Bulk density (kg/m³)	Close packing density (kg/m³)	Content of silt fines (%)	Clay lump content (%)	Solidity (%)
Natural sand	2.73	2 600	2 520	2.0	1 590	1 810	3.8	Nil	4.1
DL/T 5144—2001	2.2–3.0	≥2 550	—	≤2.5	—	—	≤3	Not allowed	≤10

Chapter 13 RCC Cofferdam Construction and CSG Damming Technology

Table 13.10   Sand gradation test results

Grading situation	Sieve mesh (mm)	Measurement of sieved test specimen (g) 500×2							Grading curve
		Screen residue mass (g)		Sub-total screen residue (%)		Accumulated screen residue (%)			
		1	2	1	2	1	2	Average	
	10.0	0.0	0.0	0.0	0.0	0.0	0.0	0.0	
	5.00	92.0	84.0	18.4	16.8	18.4	16.8	17.6	
	2.50	90.0	90.0	18.0	18.0	36.4	34.8	35.6	
	1.25	39.0	38.0	7.8	7.6	44.2	42.4	43.3	
	0.63	70.0	65.0	14.0	13.0	58.2	55.4	56.8	
	0.315	127.0	138.0	25.4	27.6	83.6	83.0	83.3	
	0.16	48.0	54.0	9.6	10.8	93.2	93.8	93.5	
	Bottom	34.0	31.0	6.8	6.2	100.0	100.0	100.0	
	Fineness module $FM1 = 2.74$				$FM2 = 2.72$				Sieve mesh size (mm)
	Average $FM = 2.73$ Belonging to: medium sand								Grading curve

II Coarse aggregate

The test results of the coarse aggregate are shown in Table 13.11. It can be seen that all indexes of coarse aggregate meet the quality requirements of sand materials specified in *Specifications for Hydraulic Concrete Construction* DL/T 5144—2001.

Table 13.11   Test results of quality Indexes and physical properties of coarse aggregate

Aggregate size (mm)	Saturated dry apparent density (kg/m³)	Apparent density (kg/m³)	Close packing density (kg/m³)	Moisture absorption of dry saturated surface (%)	Needle-plate like (%)	Crushing index (%)	Clay content (%)	Voidage (%)
5–20	2 650	2 670	1 960	1.01	7	6.4	0.5	26
20–40	2 670	2 690	1 810	0.88	14	—	0.3	33
40–80	2 680	2 690	1 710	0.32	12	—	0.2	36
40–250	2 690	2 700			—	—	—	
DL/T 5144—2001	≥2 550	—		≤2.5	≤15	≤13	D20, D40≤1 D80, D150≤0.5	—

13.4.3 CSG Design of Mix Proportion

13.4.3.1 Mix Proportion Design Principles

According to the CSG mix proportion design indexes and the actual natural distribution of riverbed NSG resources, CSG mix proportion design is considered from two aspects.

1. Mix proportion design is carried out from the design of maximum dry volumetric weight and control of CSG compactness;
2. Mix proportion design is carried out from the optimum workability of CSG rolling and control of the compacted volumetric weight.

   In the selection of cementing materials, tests are carried out according to two conditions: adding 30% fly ash and not adding fly ash; in order to obtain the optimum workability of CSG rolling, appropriate water consumption and water reducing agent can be adopted to control CSG materials in the appropriate VC value range.

13.4.3.2 CSG Preparation Strength

According to *Technical Requirements for CSG Materials of Upstream Cofferdam*, the preparation strength of CSG concrete of upstream cofferdam of *Gongguoqiao Hydropower Station is in accordance with Specifications for Hydraulic Concrete Construction* DL/T 5144—2001. The formula for calculating preparation strength is:

$$f_{cu,o} = f_{cu,k} + t\sigma$$

Wherein $f_{cu,o}$ = concrete preparation strength, MPa

$f_{cu,k}$ = standard value of design concrete strength, MPa

t = Probability coefficient

σ = standard deviation of concrete strength for construction, MPa, since there is no statistical data on the standard construction strength value σ of CSG materials currently, refer to the standard value σ of ordinary concrete strength temporarily, and take $\sigma = 3.5$ MPa to calculate the prepared strength

$$f_{cu,o} = f_{cu,k} + t\sigma = 7.5 \text{ MPa} + 0.84 \times 3.5 \text{ MPa} = 10.44 \text{ MPa}$$

13.4.3.3 CSG Compaction Test

Test condition: riverbed NSG is used. Sand and gravel without oversize aggregates with particle size greater than 250 mm is taken as the aggregates in the mix proportion test. The cement is Shilin P.O 42.5 ordinary Portland cement produced by Sanjiang Cement Co., Ltd. the fly ash is Grade II fly ash produced by Kunming No. 2 Power Plant, and the amount of cementing materials is 100 kg/m³. The above materials are mixed and compacted under different percentage of water content to determine the maximum dry volumetric weight and optimum percentage of water content of CSG materials. Combined with the specific indoor conditions, the aggregate with particle size larger than 40 mm is eliminated and then compaction test is carried out. The mix proportion test is carried out according to the maximum dry volumetric weight and optimal water content determined by the compaction test of CSG materials smaller than 40 mm or the converted maximum dry volumetric weight and optimal water content of CSG materials smaller than 250 mm. See Table 13.10 for compaction test results of CSG materials with cement alone, and Table 13.11 for compaction test of CSG materials with 30 fly ash.

It can be seen from the test results in Table 13.12 that the optimum percentage of water content and the maximum dry volumetric weight of CSG materials mixed with cement only with grain size ≤40 mm are 7.25% and 2.16 g/cm³ respectively. When the CSG materials are converted into gravel materials with grain size ≤250 mm, the optimum water content and the maximum dry volumetric weight are 3.50% and 2.41 g/cm³ respectively. The total water content of gravel materials with grain size ≤250 mm per square is about 85 kg/m³ on the basis of the maximum dry volumetric weight. In addition, the average moisture absorption and the total moisture absorption of gravel materials with grain size ≤250 mm are 0.82% and 19 kg/m³ respectively on the basis of the moisture absorption of gravel materials with different grain sizes. In this case, the effective water content that can be combined with cementing materials is about 66 kg/m³, and the calculated water-cement ratio is 0.66.

It can be seen from the test results in Table 13.13 that the optimum percentage of water content and the maximum dry volumetric weight of CSG materials mixed with 30% fly ash with grain size ≤40 mm are 7.30% and 2.14 g/cm³ respectively. When the CSG materials are converted into CSG with grain size ≤250 mm, the optimum water content and the maximum dry volumetric weight are 3.60% and 2.40 g/cm³ respectively. The total water content of CSG with grain size ≤250 mm per square is about 87 kg/m³ on the basis of the maximum

Table 13.12   Compaction test of CSG materials (mixed with cement only)

	≤40 mm CSG Moisture content (%)	≤40 mm CSG Dry density (g/cm³)	Test diagram
Compaction test results of CSG materials with grain size ≤40 mm	5.60	2.14	Correlation curve between the moisture content and dry density of CSG materials with grain size ≤40 mm (mixed with cement only)
	6.10	2.14	
	6.80	2.15	
	7.50	2.17	
	8.10	2.16	
	8.80	2.13	
	Optimum moisture content	Maximum dry density	
	7.25	2.16	
	≤250 mm CSG Moisture content (%)	≤250 mm CSG Dry density (g/cm³)	
Compaction test results of converted CSG materials with grain size ≤250 mm	2.7	2.39	Correlation curve between the moisture content and dry density of CSG materials with grain size ≤250 mm (mixed with cement only)
	2.9	2.40	
	3.3	2.40	
	3.6	2.41	
	3.9	2.40	
	4.3	2.39	
	Optimum moisture content	Maximum dry density	
	3.50	2.41	

dry volumetric weight. In addition, the average moisture absorption and the total moisture absorption of gravel materials with grain size ≤250 mm are 0.82% and 19 kg/m³ respectively. In this case, the effective water content that can be combined with cementing materials is about 68 kg/m³, and the calculated water-cement ratio is 0.68.

13.4.3.4 VC Value Test of CSG Mixtures

Test condition: riverbed NSG is used. Sand and gravel without oversize aggregates with grain size greater than 250 mm is taken as the aggregates in the mix proportion test. The cement is Shilin P.O 42.5 ordinary Portland cement produced by Sanjiang Cement Co., Ltd. the fly ash is Grade II fly ash produced by Kunming No. 2

Table 13.13   Compaction test of CSG materials (mixed with 30 fly ash)

	≤40 mm CSG Moisture content (%)	≤40 mm CSG Dry density (g/cm³)	Test diagram
Compaction test results of CSG materials with grain size ≤40 mm	5.80	2.12	
	6.50	2.13	
	7.10	2.14	
	7.60	2.15	
	8.30	2.13	
	9.00	2.11	
	Optimum moisture content	Maximum dry density	Correlation curve between the water content and dry density of CSG materials with grain size ≤40 mm (mixed with 30 fly ash)
	7.30	2.14	
	≤250 mm CSG Moisture content (%)	≤250 mm CSG Dry density (g/cm³)	
Compaction test results of converted CSG materials with grain size ≤250 mm	2.8	2.38	
	3.1	2.39	
	3.4	2.40	
	3.7	2.40	
	4.1	2.39	
	4.4	2.38	
	Optimum moisture content	Maximum dry density	Correlation curve between the water content and dry density of CSG materials with grain size ≤250 mm (mixed with 30 fly ash)
	3.60	2.40	

Power Plant, and the amount of cementing materials is 100 kg/m³. The above materials are mixed for VC value test under different percentage of water content to determine the water content of CSG materials with a VC value of 5–10 s. VC values are tested according to the RCC VC value determination method in the *Test Code for Hydraulic Concrete* SL 352—2006. CSG concrete VC value test result is as shown in Table 13.14.

It can be seen from the VC value test results of CSG mixtures that if the suitable water content of CSG materials with grain size ≤250 mm without water reducing agent is about 115 kg/m³, and the calculated effective water content is about 96 kg/m³, the water-cement ratio is 0.96; if the suitable water content of CSG materials with 0.8% water reducing agent is about 92 kg/m³, and the calculated effective water content is

Chapter 13 RCC Cofferdam Construction and CSG Damming Technology

Table 13.14   CSG mixture VC value test results

Test No.	Sand-gravel material (kg/m³)	Water (kg/m³)	Cement (kg/m³)	Fly ash (kg/m³)	Additive Variety	Additive Mixing amount (%)	VC value (s)	Measured volumetric weight (kg/m³)	Effective water consumption (kg/m³)	Calculated water-cement ratio
VC-1	2 300	100	100	—	—	—	14.2	2 420	81	0.81
VC-2	2 300	108	100	—	—	—	10.0	2 420	89	0.89
VC-3	2 300	115	100	—	—	—	5.6	2 410	96	0.96
VC-4	2 300	100	70	30	—	—	13.9	2 420	81	0.81
VC-5	2 300	108	70	30	—	—	10.1	2 410	89	0.89
VC-6	2 300	115	70	30	—	—	5.4	2 410	96	0.96
VC-7	2 300	80	100	—	JM-II	0.8	15.1	2 420	61	0.61
VC-8	2 300	86	100	—	JM-II	0.8	10.7	2 430	67	0.67
VC-9	2 300	92	100	—	JM-II	0.8	6.0	2 420	73	0.73
VC-10	2 300	80	70	30	JM-II	0.8	14.8	2 420	61	0.61
VC-11	2 300	86	70	30	JM-II	0.8	10.3	2 420	67	0.67
VC-12	2 300	92	70	30	JM-II	0.8	6.5	2 410	73	0.73

about 73 kg/m³, the water-cement ratio is 0.73.

13.4.3.5 Test of the Relation Between CSG Water-cement Ratio and Strength

The relation between water-cement ratio and strength under different conditions are tested according to the basic parameters obtained through the above compaction test and VC value test of CSG materials. NSG in the riverbed is still used in the test. Sand and gravels without the oversize aggregates that the grain size is more than 250 mm are taken as the aggregates in the mix proportion test. The cement is Shilin P.O 42.5 ordinary Portland cement produced by Sanjiang Cement Co., Ltd. The fly ash is Grade II fly ash produced by Kunming No.2 Power Plant. The water content of each combination is the optimum or suitable water content determined under its specific conditions. Cementing material consumption is 90 kg/m³, 100 kg/m³ and 110 kg/m³ respectively. See Table 13.15 below for test parameters.

Table 13.15   Test parameters of relationship between CSG water-cement ratio and compressive strength

Test No.	Sand-gravel material (kg/m³)	Water (kg/m³)	Cement (kg/m³)	Fly ash (kg/m³)	Additive Variety	Additive Mixing amount (%)	Effective water content (kg/m³)	Calculated water-cement ratio
CSG-1	2 300	85	90	—	—	—	66	0.73
CSG-2	2 300	85	100	—	—	—	66	0.66
CSG-3	2 300	85	110	—	—	—	66	0.60
CSG-4	2 300	87	63	27	—	—	68	0.76
CSG-5	2 300	87	70	30	—	—	68	0.68
CSG-6	2 300	87	77	33	—	—	68	0.62
CSG-7	2 300	115	90	—	—	—	96	1.07
CSG-8	2 300	115	100	—	—	—	96	0.96

Continued Table 13.15

Test No.	Sand-gravel material (kg/m³)	Water (kg/m³)	Cement (kg/m³)	Fly ash (kg/m³)	Additive Variety	Additive Mixing amount (%)	Effective water content (kg/m³)	Calculated water-cement ratio
CSG-9	2 300	115	110	—	—	—	96	0.87
CSG-10	2 300	115	63	27	—	—	96	1.07
CSG-11	2 300	115	70	30	—	—	96	0.96
CSG-12	2 300	115	77	33	—	—	96	0.87
CSG-13	2 300	92	90	—	JM-II	0.8	73	0.81
CSG-14	2 300	92	100	—	JM-II	0.8	73	0.73
CSG-15	2 300	92	110	—	JM-II	0.8	73	0.66
CSG-16	2 300	92	63	27	JM-II	0.8	73	0.81
CSG-17	2 300	92	70	30	JM-II	0.8	73	0.73
CSG-18	2 300	92	77	33	JM-II	0.8	73	0.66

During the test, CSG concrete specimens are molded and cured according to the molding method of RCC specimens and relevant regulations, and used for the strength test after reaching the age. The test results are shown in Table 13.16.

Table 13.16 Test results of relationship between CSG water-cement ratio and compressive strength

Test No.	Sand-gravel material (kg/m³)	Water (kg/m³)	Cement (kg/m³)	Fly ash (kg/m³)	Additive Variety	Additive Mixing amount (%)	VC value (s)	Water-cement ratio	Compressive strength (MPa) 14 d	Compressive strength (MPa) 28 d
CSG-1	2 300	85	90	—	—	—	—	0.73	7.2	9.1
CSG-2	2 300	85	100	—	—	—	—	0.66	8.5	12.4
CSG-3	2 300	85	110	—	—	—	—	0.60	9.9	14.8
CSG-4	2 300	87	63	27	—	—	—	0.76	5.2	8.0
CSG-5	2 300	87	70	30	—	—	—	0.68	6.3	10.2
CSG-6	2 300	87	77	33	—	—	—	0.62	7.8	12.3
CSG-7	2 300	115	90	—	—	—	7.5	1.07	6.0	8.4
CSG-8	2 300	115	100	—	—	—	6.4	0.96	7.3	10.5
CSG-9	2 300	115	110	—	—	—	7.7	0.87	8.8	12.1
CSG-10	2 300	115	63	27	—	—	6.9	1.07	5.6	7.2
CSG-11	2 300	115	70	30	—	—	6.2	0.96	6.4	9.0
CSG-12	2 300	115	77	33	—	—	7.6	0.87	8.1	11.1
CSG-13	2 300	92	90	—	JM-II	0.8	8.1	0.81	7.0	11.8
CSG-14	2 300	92	100	—	JM-II	0.8	6.8	0.73	7.9	13.5
CSG-15	2 300	92	110	—	JM-II	0.8	7.3	0.66	10.1	15.6
CSG-16	2 300	92	63	27	JM-II	0.8	6.7	0.81	6.5	10.0
CSG-17	2 300	92	70	30	JM-II	0.8	6.4	0.73	8.1	12.4
CSG-18	2 300	92	77	33	JM-II	0.8	7.0	0.66	9.7	14.3

13.4.4 Performance Test of CSG Materials

13.4.4.1 CSG Initial Mix Proportion

According to the above test results and design requirements, 6 different combinations of raw materials and the preliminary mix proportion of CSG materials under different construction conditions are determined. See Table 13.17 for the determined preliminary mix proportion of CSG materials. The performance proof test of CSG materials is carried out in order to further verify and test whether each performance index of the preliminary mix proportion of CSG materials can meet the design requirements.

Table 13.17　　　　　　　　　　　　　　SG initial mix proportion

Test No.	Sand-gravel material (kg/m³)	Water (kg/m³)	Cement (kg/m³)	Fly ash (kg/m³)	Additive Variety	Additive Mixing amount (%)	Unit weight (kg/m³)	Validation water content (kg/m³)	Actual water-cement ratio
CSG-19	2 220	90	100	—	—	—	Dry volumetric weight: 2 410; wetvolumetric weight: 2 490	72	0.72
CSG-20	2 290	90	77	33	—	—	Dry volumetric weight: 2 400; wet volumetric weight: 2 490	71	0.65
CSG-21	2 217	92	70	30	JM-II	0.8	2 410	74	0.74
CSG-22	2 227	92	90	—	JM-II	0.8	2 410	74	0.82
CSG-23	2 195	115	100	—	—	—	2 410	97	0.97
CSG-24	2 185	115	77	33	—	—	2 410	97	0.88

13.4.4.2 Construction Performance Test of CSG Materials

VC values and volumetric weight are mainly tested in the construction performance test of CSG materials, and the setting time of materials mixed as per CSG-23 mix proportion is tested. The test results show that initial setting time and final setting time indoors are 9:30 a.m. and 22:00 p.m. respectively; initial setting time and final setting time outdoors are 7:30 a.m. and 14:20 p.m. respectively. The mechanical performance is tested through 14 d and 28 d compressive strength test. The test results are shown in Table 13.18. It can be seen that each performance of CSG materials mixed as per the mix proportion can meet construction needs and strength grade requirements in the design.

13.4.4.3 Strength Comparison Test of Specimens with Different Sizes

The maximum grain size of CSG in Gongguoqiao cofferdam is 250 mm. In order to further understand the effect of the aggregates with different grain sizes and specimen sizes on the strength of CSG materials, CSG materials are molded into 300 mm × 300 mm × 300 mm cube specimens after the aggregates that the grain size is more than 80 mm and 40 mm are respectively screened out, and the strength is compared with that of 150 mm × 150 mm × 150 mm cube specimens. See Table 13.19 for test and comparison.

Table 13.18 Construction performance and strength test results of CSG materials mixed as per the mix proportion

Test No.	Sand-gravel material (kg/m³)	Water (kg/m³)	Cement (kg/m³)	Fly ash (kg/m³)	Additive Variety	Additive Mixing amount (%)	VC value (s)	Wet volumetric weight(kg/m³)/ percentage of water content (%)	Dry unit weight (kg/m³)	Compressive strength (MPa) 14 d	Compressive strength (MPa) 28 d
CSG-19	2 220	90	100	—	—	—	—	2 490/3.60	2 400	8.8	12.6
CSG-20	2 290	90	77	33	—	—	—	2 480/3.70	2 390	6.6	11.5
CSG-21	2 217	92	70	30	JM-II	0.8	6.1	2 410/—	—	7.1	10.6
CSG-22	2 227	92	90	—	JM-II	0.8	8.1	2 420/—	—	7.3	10.8
CSG-23	2 195	115	100	—	—	—	7.5	2 410/—	—	6.9	11.2
CSG-24	2 185	115	77	33	—	—	5.9	2 400/—	—	7.2	12.3

Table 13.19 Strength comparison test results of CSG specimens with different sizes made of the aggregates with different grain sizes

Specimen specification	150 mm × 150 mm × 150 mm Aggregate size ≤ 40 mm		300 mm × 300 mm × 300 mm Aggregate size ≤ 40 mm		300 mm × 300 mm × 300 mm Aggregate size ≤ 80 mm	
Test item	Strength of specimen of 14 d (MPa)	Strength of specimen of 28 d (MPa)	Strength of specimen of 28 d (MPa)	Strength ratio of 150 mm × 150 mm × 150 mm specimen of 28 d	Strength of specimen of 28 d (MPa)	Strength ratio of 150 mm × 150 mm × 150 mm specimen of 28 d
CSG-22	7.1	10.8	9.7	0.90	8.7	0.81

According to the comparison of strength test results of (CSG) specimens with different sizes made of the aggregates with different grain sizes in Table 13.19, the larger the specimen size or the grain size of gravel material, the lower the strength of CSG specimen. When the grain size of aggregate in the CSG materials is not more than 40 mm, and 300 mm × 300 mm × 300 mm specimens are used in the test, the strength of specimens is about 90% of that of standard specimens; the grain size of aggregate in the CSG materials is not more than 80 mm, and 300 mm × 300 mm × 300 mm specimens are used in the test, the strength of specimens is about 81% of that of standard specimens.

13.4.4.4 Shear Strength (Calculated with c and φ)

In order to understand the shear resistance of CSG materials, 150 mm × 150 mm × 150 mm cube specimens are molded as per CSG-22 mix proportion for shear resistance test. See Table 13.20 for test results. The results show that f' reaches 1.240; c' reaches 2.215 MPa; the shear strength meets the design requirements.

Table 13.20 CSG shear strength test results

Test No.	Sand-gravel material (kg/m³)	Water (kg/m³)	Cement (kg/m³)	Additive Variety	Additive Mixing amount (%)	Peak strength f'	Peak strength c' (MPa)	Residual strength f	Residual strength C (MPa)
CSG-22	2 227	92	90	JM-II	0.8	1.240	2.215	0.964	0.160

13.4.4.5 CSG Impermeability

According to the design requirements, the impermeability test of the design CSG is performed. See Table 13.21 for test results. The results show that the impermeability of CSG mixed as per six design mix proportions meets the design requirements.

Table 13.21　　　　　　　　　　CSG impermeability test results

Test No.	Sand-gravel material (kg/m³)	Water (kg/m³)	Cement (kg/m³)	Fly ash (kg/m³)	Additive Variety	Additive Mixing amount (%)	Pressure (MPa)	Appearance description	Grade of impermeability
CSG-19	2 220	90	100	—	—	—	0.5	There is no water seepage on the surface of all specimens	≥W4
CSG-20	2 290	90	77	33	—	—	0.5	There is no water seepage on the surface of all specimens	≥W4
CSG-21	2 217	92	70	30	JM-II	0.8	0.5	There is no water seepage on the surface of all specimens	≥W4
CSG-22	2 227	92	90	—	JM-II	0.8	0.5	There is no water seepage on the surface of all specimens	≥W4
CSG-23	2 195	115	100	—	—	—	0.5	There is no water seepage on the surface of all specimens	≥W4
CSG-24	2 185	115	77	33	—	—	0.5	There is no water seepage on the surface of all specimens	≥W4

13.4.4.6 CSG Construction Mixing Proportion

According to the design requirements, CSG mix proportion design for the upstream cofferdam of Gongguoqiao Dam is carried out, and CSG mix proportion in the construction meeting the design and construction requirements is proposed, as shown in Table 13.22.

13.4.5　Field CSG Rolling Test

13.4.5.1　Compacting Process Test Contents

Natural aggregates from the riverbed are used in the rolling test of CSG used in the upstream cofferdam. The maximum grain size of aggregates is 250 mm, and the spreading thickness is 80 cm, 50 cm and 40 cm respectively. According to the design requirements and submitted construction mix proportions, the setting time, uniformity, VC value loss, rolling times, compaction thickness and compactness of CSG mixtures are tested respectively, and CSG materials are sampled to test their compressive strength.

　　Construction parameters, vibrating roller traveling speed, optimum compaction times, spreading thickness and allowable interlayer interval time are determined through CSG rolling test, so as to provide the firsthand construction technical parameters.

13.4.5.2　Natural Aggregate Test and Inspection

The aggregate is natural aggregate in the riverbed. The quality, gradation and sand content of aggregates are

Table 13.22 CSG Mix proportion in construction of upstream cofferdam of gongguoqiao hjydropower station

S/N	Design strength and impermeability grade	Water-cement ratio	Fly ash (%)	Sand ratio (%)	JM-II (%)	VC value (s)	Material amount (kg/m³)				Unit weight (kg/m³)
							Water content	Cement	Fly ash	Sand-gravel material	
1	$C_{28}7.5W4$	0.72	—	20	—	Control of the maximum dry volumetric weight	72	100	—	2 238	Dry 2 410 Wet 2 490
2		0.65	30	20	—	Control of the maximum dry volumetric weight	71	77	33	2 309	Dry 2 400 Wet 2 490
3		0.74	30	20	0.8	5–10	74	70	30	2 235	2 410
4		0.82	—	20	0.8	5–10	74	90	—	2 245	2 410
5		0.97	—	20	—	5–10	97	100	—	2 213	2 410
6		0.88	30	20	—	5–10	97	77	33	2 203	2 410
M-1		0.55	30	—	0.5	—	636	810	347		As per the interlayer bonding grout thickness ≤3 mm
M-2		0.60	—	—	0.5	—	656	1 094	—		

Note: 1. The NSG in the riverbed with the sand and gravel larger than 250 mm removed, with an average moisture absorption of 0.82%; Grade P.O 42.5 cement and Grade II fly ash are adopted.

2. Mix proportions 1# and 2# are suitable for the face rockfill dam construction method, and mix proportions 3#–6# are suitable for the RCC construction method.

tested before mixing. See Table 13.23 and Table 13.24 for test results.

Table 13.23 Test results of grain gradation of natural sand and gravel aggregate excavated from riverbed

Aggregate size (mm)	Sample mass (kg)	Mass of sieve residue (kg)	Screening screen residue performance (%)	Gradation distribution of aggregate >250 mm (%)	Distribution ratio of aggregate >250 mm		Accumulated percentage of residue on sieve (%)
					Grain size range (mm)	Distribution ratio (%)	
>250	1 292	252	20				20
150–250		136	11	13	40–250	41	31
80–150		211	16	20			47
40–80		80	6	8			53
20–40		90	7	9	5–40	29	60
5–20		214	16	20			76
<5		309	24	30	<5	30	100

Chapter 13 RCC Cofferdam Construction and CSG Damming Technology

Table 13.24   Test results of NSG aggregate quality indexes

Aggregate size (mm)	Saturated dry apparentdensity (kg/m³)	Apparent density (kg/m³)	Close packing density (kg/m³)	Moisture absorption of drysaturated surface (%)	Needle-plate like (%)	Crushing index (%)	Clay content (%)	Voidage (%)
5-20	2 650	2 670	1 960	1.01	7	6.4	0.5	26
20-40	2 670	2 690	1 810	0.88	14	—	0.3	33
40-80	2 680	2 690	1 710	0.32	12	—	0.2	36
40-250	2 690	2 700			—	—	—	
DL/T 5144—2001	≥2 550	—		≤2.5	≤15	≤13	D20, D40≤1 D80, D150≤0.5	—

13.4.5.3 Feeding Sequence

In the rolling process test on March 4, 2009, the backhoe is used for mixing according to the actual situation of site construction, and the proposed feeding sequence is: NSG materials → additives → water. After on-site mixing, the appearance and uniformity of CSG mixtures mixed by as per the feeding sequence are tested, and the rollability is observed on site. The test results show that the mixtures of CSG RCC mixed as per the feeding sequence has the same color; the aggregates can be mixed with grout; the performance of the mixtures is better.

13.4.5.4 Mixing Method

Before mixing, the materials are stacked in the cone form. There are small concreted pits around the pile body, which can effectively prevent the loss of cement grout after water addition. The mixing method is that a backhoe is used to mix the materials for twice under the condition of no water first, and then 4 times after water addition. After determining the feeding sequence, the uniformity comparison test of CSG mixtures is performed under the conditions of the same mix proportion and different mixing times. The test results show that the uniformity of CSG RCC mixed for twice under the condition of no water first and 4 times after water addition is better than that of CSG RCC mixed for twice without water first, and then 3 times after water addition.

Through the on-site mixing test, the mixing method that a backhoe is used to mix the materials for twice under the condition of no water first, and then 4 times after water addition is determined.

13.4.5.5 CSG Mixture Test

I Loss of *VC* value

During the CSG RCC rolling test, the *VC* value is controlled at 5-10 s, and the loss of *VC* value of concrete is tested according to field transportation time, weather and temperature. When it is sunny, the average time of CSG RCC transportation by backhoes and loaders from mixing site to rolling site is about 30 min in the rolling test. When the highest temperature is 26 ℃ and the lowest temperature is 6 ℃, the average loss of *VC* value when placing concrete is about 1.8 s. See Table 13.25 for field test results.

The temperature, sunshine, wind speed, and others during construction have great effects on the *VC* value of CSG. The *VC* value must be dynamically controlled according to project location, temperature, raw materials and other actual situations. The setting time is tested indoors and outdoors respectively. Outdoor setting samples are covered with plastic cloth and tested during placing surface construction, which can guide the interval time of CSG RCC placing surface construction to determine the best time for interlayer bonding.

Table 13.25　　Statistics of loss of VC value in CSG field rolling test

S/N	Mixing machinary	Temperature (℃)	Initial VC (s)	After spreading	Spreading time of placing surface	Loss of VC value (s)	Time of setting(h: min)			
							Indoor		Outdoor	
							Initial setting	Final setting	Initial setting	Final setting
1	Backhoe	9.5	2.7	3.6	30	0.9	—	—	—	—
2	Backhoe	7.0	3.1	3.6	30	0.5	—	—	—	—
3	Backhoe	6.0	2.4	3.9	30	1.5	—	—	—	—
4	Backhoe	14.8	2.6	5.4	30	2.8	7:30	22:00	5:00	14:00
5	Backhoe	18.0	3.2	4.8	30	1.6				
6	Backhoe	26.0	3.4	6.8	30	3.4	8:20	20:30	6:10	14:40
7	Backhoe	24.6	4.5	6.4	30	1.9				

II　Uniformity

The uniformity is an important link of mixtures, which directly affects the rollability, interlayer bonding, compactness and imperviousness effect of mixtures. The key point of field control is the VC value and initial setting time of mixtures. The VC value control is the key to the rollability and interlayer bonding of RCC. The VC value at the outlet shall be adjusted in time according to the change of temperature. See Table 13.26 for the table of each material consumption on the basis of CSG mix proportion.

Table 13.26　　Quantity of material consumption on the basis of CSG mix proportion

Strength grade	Water-cement ratio	Sand ratio	VC value (s)	Material amount (kg/m³)			Unit weight (kg/m³)
				Water content	Cement	Sand-gravel material	
C7.5W4	0.72	20	5-10	72	100	2238	Dry volumetric weight: 2 410; wet volumetric weight: 2 490

150 kg samples are taken at both ends of material pile mixed by backhoes respectively in the uniformity test of CSG mixtures to perform strength test and mortar density deviation test, as shown in Table 13.27.

Table 13.27　　Uniformity test of CSG mixtures

S/N	Sampling location	VC value (s)	Concrete compressive strength (MPa)			Mortar density deviation (%)
			14 d		28 d	
			300 mm × 300 mm × 300 mm	150 mm × 150 mm × 150 mm		
1	Before material piling	3.2	2.6	4.6	—	1.00
2			2.2		—	
3	After material piling	5.6	3.9	5.2	—	1.02
4			4.3		—	

When moulding a 150 mm × 150 mm specimen, the aggregates with grain size > 40 mm shall be removed. When moulding a 300 mm × 300 mm specimen, the aggregates with grain size > 150 mm shall be

Chapter 13 RCC Cofferdam Construction and CSG Damming Technology

removed, but the aggregates with grain size < 100 mm are included.

It can be seen from test results that the mixtures are not very uniform. The strength of concrete after material piling is higher than that of material piling. There is also a certain deviation of *VC* value. During actual construction, the mixtures mixed by the mixing method that the materials are mixed for twice under the condition of no water at first, and then mixed for 4 times after water addition is still not very uniform. No matter what mixing method is used, it is important to mix the materials in the middleof the pile.

13.4.5.6 Rolling Times and Compactness

Ⅰ Rolling times (with thickness of 80 cm)

According to the design requirements, the traveling speed is tested in the field rolling laboratory when the speed of the vibrating roller is 1.0 km/h, 1.1 km/h, 1.2 km/h, 1.3 km/h, 1.4 km/h and 1.5 km/h respectively, and the combination of rolling times, such as rolling without vibration twice + rolling with vibration for twice/4 times/6 times/8 times is tested. During the test, all *VC* values of field mixtures are within 3 – 7 s. The test results are shown in Table 13.28.

Table 13.28 Rolling times and compactness

S/N	Rolling speed (km/h)	VC value (s)	Compactness(%) after rolling without vibration twice + rolling with vibration for twice	Compactness(%) after rolling without vibration twice + rolling with vibration for 4 times	Compactness(%) after rolling without vibration twice + rolling with vibration for 6 times	Compactness(%) after rolling without vibration twice + rolling with vibration for 8 times	Bleeding
1	1.0	4.6	91.7	93.7	—	99.6	Fully bleeding
2	1.1		90.3	96.4	—	99.0	Most bleeding
3	1.2	3.3	93.4	96.8	99.8	—	Fully bleeding
4	1.3		93.8	96.2	99.3	—	Fully bleeding
5	1.4	6.7	91.2	95.8	99.0	99.8	Fully bleeding
6	1.5		92.2	94.9	97.1	99.6	Most bleeding
Paving thickness			80 cm				

1. In the case that the traveling speed of the vibrating roller is 1.0 km/h: the combination of rolling without vibration twice + rolling with vibration twice is tested, showing that the average compactness can reach 93% of the design requirements; the bleeding on the surface is insufficient; the combination of rolling without vibration for twice + rolling with vibration for 4 times is tested. The test results show that the average compactness can meet the design requirements, and the bleeding on the surface is sufficient.

2. In the case that the traveling speed of the vibrating roller is 1.2 km/h: the combination of rolling without vibration for twice + rolling with vibration for twice is tested, the results of which show that the average compactness cannot meet the design requirements, and the bleeding on the surface is insufficient; the combination of rolling without vibration for twice + rolling with vibration for 4 times is tested, the results of which show that the average compactness can meet the design requirements, and the bleeding on most of the surface occurs; the combination of rolling without vibration for twice + rolling with vibration for 6 times is tested, the results of which show that the average compactness can reach the maximum, and the bleeding on the surface is insufficient.

3. In the case that the traveling speed of the vibrating roller is 1.5 km/h: the combination of rolling without vibration for twice + rolling with vibration for twice is tested, the results of which show that the average

compactness cannot meet the design requirements, and the bleeding on the surface is insufficient; the combination of rolling without vibration for twice + rolling with vibration for 4 times is tested, the results of which show that the average compactness cannot meet the design requirements, and the bleeding on most of the surface occurs; the combination of rolling without vibration for twice + rolling with vibration for 6 times is tested, the results of which show that the average compactness can meet the design requirements, and the bleeding on most of the surface occurs; the combination of rolling without vibration for twice + rolling with vibration for 8 times is tested, the results of which show that the average compactness can reach the maximum, and the bleeding on the surface is insufficient.

According to the above data, the bleeding on the surface of RCC shall be sufficient on the basis of meeting RCC relative compactness in order to ensure the interlayer bonding quality. It is suggested that the traveling speed of the vibrating roller is 1.0 km/h, and the optimum combination of rolling without vibration for twice + rolling with vibration for 6 times. When the VC value is larger, the rolling times shall be increased.

Ⅱ Compactness test

The CSG is spreading at the upstream cofferdam into strips A, B and C, which are 80 cm, 50 cm and 40 cm respectively, for the rolling test. Based on the rolling times and compactness. The spreading thickness is 80 cm and the rolling is 6 – 8 times in the test. Since the spreading thickness is 80 cm, but the test depth of nuclear density-moisture gauge is 30 cm, the compactness at the depth of 30 cm shall be tested first on the strip with the spreading thickness of 80 cm. After excavating 30 cm downwards, the compactness at the depth of 60 cm is tested by nuclear density-moisture gauge. See Table 13.29 for compactness test results.

Table 13.29 Compactness test results

S/N	Rolling speed	VC value (s)	The test depth is 30 cm	The test depth is 60 cm	Bleeding
			Compactness(%) after rolling without vibration twice + rolling with vibration for 8 times	Compactness(%) after rolling without vibration twice + rolling with vibration for 8 times	
1	1.0 km/h	3.6	99.7	97.1	Fully bleeding
2	1.3 km/h	4.7	98.0	97.1	Most bleeding
3	1.5 km/h	5.3	99.2	96.7	Fully bleeding

The test results show that the compactness at the depth of 60 cm cannot reflect the actual compactness of compaction after the surface has been disturbed and is unrepresentative.

13.4.6 Conclusion

1. The CSG damming technology is a new damming technology. Its design concept and construction method combine those of concrete face rockfill dam and RCC dam. It takes advantage of local NSG mixtures with a small amount of cementing materials to form a hard dam with certain strength. This damming technology has the advantages of rapid construction, safety, environmental protection and economy.
2. Shilin P.O 42.5 ordinary Portland cement, Grade Ⅱ fly ash, SBTJM-Ⅱ high-range retarding water reducing agent produced by Jiangsu Bote New Materials Co., Ltd. and natural materials in the riverbed of Lancang River are used in the test.
3. Through the rolling test of CSG materials, we can deeply understand the CSG dam construction technology and provide technical support for CSG damming technology application at the upstream cofferdam.
4. The CSG damming technology was successfully applied at the upstream cofferdam of Gongguoqiao Dam in 2009. The cofferdam overflow during flood showed that the CSG cofferdam meets the design requirements.

Appendix A

National Method Hydraulic Concrete Mix Proportion Test Method

Construction method No. : SDJTGF 012—2007
Sinohydro Engineering Bureau 4 Co. , Ltd.
Tian Yugong, Gao Jusheng, Hu Hongxia, Wang Huan and Zheng Kai

A.1 Foreword

At present, among dams under construction in China, the water retaining structures of high dams and large reservoirs are mainly made of concrete. The quality of hydraulic concrete is directly related to the construction quality, safe operation and service life of dams. Due to the complexity, persistence, importance, etc. , of working conditions of hydraulic concrete, its design indexes or mix proportion test methods are quite different from ordinary concrete. Design indexes of hydraulic concrete adopt long age (90 d or 180 d), with high requirements for impermeability, frost resistance, crack resistance and temperature control. The mix proportion design adopts large aggregate gradation, low cementing material amount, high admixtures and additives. Concrete mixture adopts small slump (VC value), good workability, appropriate air content, and meets the setting time required by construction in different seasons and climatic conditions. In order to meet the requirements of high quality, high performance and cost-effectiveness of concrete, the concrete mix proportion test is particularly important.

In the development and construction of water conservancy and hydropower projects, Sinohydro Engineering Bureau 4 Co. , Ltd. , as the main force of high concrete dam construction, has undertaken more than 30 concrete dams built and under construction since the construction of Liujiaxia Hydropower Station in the 1950s. They cover Yangtze River Three Gorges Dam, RCC main dam works of Baise Multipurpose Dam Project, Yellow River Xiaolangdi, Laxiwa, Lijiaxia, Longyangxia, Liujiaxia, Yanguoxia, Wanjiazhai, Bapanxia, Shapotou, Lancang River Xiaowan, Jinsha River Jin'anqiao Hydropower Stations, Aqueduct Bridge for Yellow River and Caohe River in the Middle Route of South-to-North Water Diversion Project, Shuikou (Fujian), Longshou (Gansu), Ningxia, Jiangkou (Chongqing), Linhekou (Shaanxi), Zhougongzhai (Zhejiang), Guangguang (Guizhou) and Tekeze (Ethiopia) Hydropower Projects. The concrete mix proportion is the core technology of high concrete dam, and the hydraulic concrete mix proportion test is closely around the core technology. In the long-term practice of the above projects, Sinohydro Engineering Bureau 4 Co. , Ltd. has gradually formed an advanced, mature, feasible and distinctive test method in the process of hydraulic concrete mix proportion test and scientific research project implementation. This construction method is summarized based on relatively typical hydraulic concrete mix proportion tests, and procedures and methods of scientific research projects.

A.2 Characteristics of Construction Method

1. This construction method is formed in the process of long-term and large number of water conservancy and hydropower project site tests. The method is obviously different from the test code for hydraulic concretes, which is supplement and improvement of test procedures, with distinct practicability and operability.
2. The method strictly follows the requirements of hydraulic concrete design indexes to scientifically standardize procedures of the hydraulic concrete mix proportion test closely combined with the regional environment, climate conditions, construction conditions and raw material characteristics of projects.
3. The method focuses on the performance of concrete mixtures, and the study on the relations between the slump (VC value), air content, setting time and apparent density and other performances of fresh concrete and the time, period and temperature change, so as to provide a scientific basis for concreting.
4. The process flow of concrete mix proportion test is established so that the concrete mix proportion test is scientific, standard, reasonable and orderly.
5. Test data is processed by advanced computer programs to ensure the timely, scientific and accurate analysis of test results.

A.3 Applicable Scope

It is applicable to the mix proportion test of mass normal concrete and RCC of hydraulic structures in different regions under different climate and construction conditions.

A.4 Technological Principle

1. According to the concrete mix proportion design index, formulate scientific and reasonable technical route and mix proportion test plans, plan and arrange laboratories, prepare representative and sufficient raw materials, and carry out the mix proportion test in strict accordance with the requirements of regulations, specifications and the construction method.
2. The construction method focuses on the performance of the mixtures based on the hydraulic concrete mix proportion, and studying the relations between the slump (VC value), air content, setting time and apparent density of fresh concrete and the time, period and temperature change through key technologies to find out the inherent law of these factors.
3. The mix proportion test is carried out according to the construction method to ensure the accuracy and reliability of test results, and submit an economical and reasonable construction mix proportion meeting the design and construction requirements.

A.5 Construction Process Flow and Key Points of Work

A.5.1 Construction Process Flow

See the process flow Figure A.1 below.

Appendix A

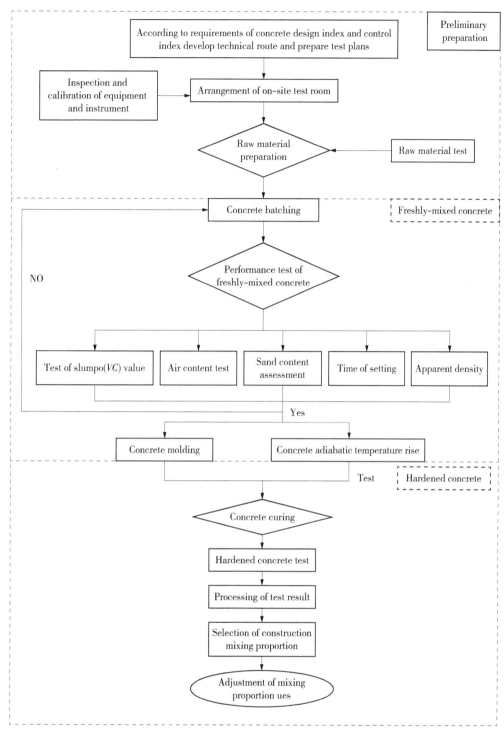

Figure A.1 *Mix proportion design flow*

A.5.2 Operation Key Points

A.5.2.1 Preparation of Test Plan

According to the requirements of design index and quality control index of concrete, different climatic and construction conditions in different project areas and raw material characteristics, formulate a reasonable technical route and prepare a scientific "hydraulic concrete mix proportion test" plan.

A.5.2.2 On-site Laboratory

The hydraulic concrete mix proportion test is performed on the construction site, so it is necessary to arrange the field laboratory at first to ensure that the concrete mixture mix proportion test can be conducted smoothly. The field laboratory shall be arranged scientifically and reasonably according to the test items and existing conditions. In general, 12 – 18 test workshops shall be arranged. The mixing workshop is the most important workshop for hydraulic concrete mix proportion test, so high attention should be paid to the arrangement of mixing workshop. The area of test mixing workshop is generally 50 – 80 m², where the mixer, mixing steel plate, shaking table, pool, silo, workbench and others can be arranged.

1. Mixer: due to the large particle size of mass hydraulic concrete, the mixing capacity of mixer must be sufficient, and 100 – 150 L gravity mixers and 60 – 100 L forced mixers are generally used.
2. Mixing steel plate: the dimension of steel plate shall be length × width × thickness = (2 000 – 2 500 mm) × (1 500 – 2 000 mm) × (8 – 10 mm). The steel plate is generally aligned vertically with the mixer outlet in the longitudinal direction and placed horizontally, which is also 50 mm lower than the ground. In addition, ∠50 mm angle iron may be welded around the steel plate. This way is convenient for the continuous concrete mixing test. Meanwhile, a drainage collecting precipitation tank shall be arranged on the side of the steel plate.
3. Vibrating table: the dimensions of shaking table top is generally 1 000 mm ± 10 mm long and 1 000 mm ± 10 mm wide; the surface is smooth and clean; the frequency is 50 Hz ± 3 Hz, and the amplitude is 0.5 mm ± 0.02 mm; the shaking table shall be installed in a position where other tests and operations cannot be affected; the table top shall be level.
4. Silo 4 – 5 silos with a height of 90 – 110 cm respectively are generally arranged against the wall at the end of mixing workshop to stack saturated and surface dry sand and coarse aggregates separately, and the sand silo shall be large enough. The upper part of the silo can be prefabricated to build a workbench to place materials such as cement and fly ash as well as tools.
5. Mechanical room: the area of mechanical test workshop shall ensure that the arrangement, installation, maintenance, overhaul of various material testers and the normal operation of test personnel are not affected. It is generally 40 – 60 m². According to the mechanical test contents, 100 kN and 1 000 kN universal material testers, 2 000 kN pressure testers and related accessories are generally configured. 5 000 – 10 000 kN pressure testers are configured for the mechanical property test of fully graded concrete.
6. Curing room: the curing room shall be arranged in strict accordance with the requirements of temperature and humidity. The area of curing room is determined according to the factors such as concrete quantity, sampling frequency, curing age and construction peak period, which is generally 40 – 80 m². Its roof should be sealed with precast concrete.

A.5.2.3 Preparation of Raw Materials

The preparation of raw materials for the concrete mix proportion test is very important to ensure the consistency, accuracy, continuity of test results for result analysis. The premise of high-quality mix proportion test is that the sufficient specimens taking from the same batch of representative raw materials (cement, mixtures, aggregates, admixtures and others) to be actually used in the project for the hydraulic concrete mix proportion test must be prepared according to the material consumption plan to avoid secondary sampling as far as possible, thus preventing the difference of test results due to the fluctuation of raw materials. All materials shall be tested in advance according to the plans to understand the quality and performance of raw materials required for the mix proportion test.

1. Cementing materials: cement, mixtures and other cementing materials shall be kept from moisture, and generally sealed and packaged with moisture-proof materials such as plastic film. During test mixing, cementing materials shall be unpacked and put into the plastic bucket with cover, and special instruments shall be used for filling materials. The cover shall be closed in time after filling each time, so that cementing

materials keep in their original state.

2. Aggregate: aggregates are required to be stacked in the indoor silo one day in advance, the saturated and surface dry state should be met, and wet sacks shall be covered on the surface of aggregates to keep them wet. The aggregates stored in the indoor silo shall be mixed evenly before mixing every day, and the water content of aggregates shall be tested to provide the basis for the mix proportion calculation.

3. Additive: the sufficient admixture solutions shall be prepared one day in advance. The concentration of water reducing agent is generally 10% – 20%, while the concentration of air entraining agent is 1% – 2%. Meanwhile, the difficult degree of admixture solution preparation and its precipitation situation shall be observed and evaluated to provide the basis for the preparation and control of admixture solutions in the mixing plants.

A.5.2.4 Concrete Batching

The concrete mixing test is the key point of the mix proportion test, and the temperature in the mixing workshop is kept at 15 – 25 ℃. Before concrete mixing, mortar or lower graded concrete with similar mix proportion is generally applied through mixing by mixers and by mixing steel plates for outgoing concrete. The first tank of fresh concrete is generally only used for preliminary evaluation rather than formal formation. The reason why it should not mix hydraulic concrete manually is that small mixing amount, large aggregate particle size, admixture mechanism and boundary condition effects are easy to lead to a big difference between manual mixing results and mechanical mixing results.

1. Mixing capacity: considering the boundary effect of mixing conditions, the minimum mixing amount of concrete should not be less than 1/3 of the mixer capacity to ensure the uniformity and stability of mixtures.

2. Feeding sequence: it shall be determined through the test. The feeding sequence of a gravity mixer is generally coarse aggregate, cementing material, water and admixture solution and fine aggregate; the feeding sequence of a forced mixers is generally fine aggregate, cementing material, water and admixture solution and coarse aggregate; small amount of water shall be taken out of the calculated water for rinsing admixture containers, and then admixture solutions shall be poured into the remaining water.

3. Mixture discharging: after discharging the mixed concrete at the specified time, the mortar in the tank shall be scraped by a trowel as far as possible, and then the mixer shall be restored to its original position and timely covered with wet sacks or equipped with a cover to prevent the mixer from drying for continuous mixing. The appearance and uniformity of concrete are observed and evaluated by mixing scraped mortar and outgoing fresh concrete for three times. The fresh concrete for formation shall be timely covered with wet sacks to avoid the larger slump loss affecting test results.

A.5.2.5 Workability Test

A large number of test results show that the gradual loss of slump and air content of fresh concrete are inevitable due to the water evaporation on the surface of fresh concrete and cement hydration and other reasons. Therefore, it is the key of the mix proportion test to conduct a large number of repeated test mixing as per the designed concrete mix proportion, so as to master the stability and regularity of concrete mixtures. During the test, parts of instruments and tools contacting with fresh concrete shall be treated by wetting or applied with grout in advance.

1. Wet sieving: when testing the performance of outgoing fresh concrete, the aggregates that the particle size is larger than 40 mm shall be removed by the wet sieve method, and the square hole sieve shall be wet by a sprayer or a wet mop before sieving. After sieving, mixtures are mixed by two people with small square spades for three times.

2. Temperature test: the thermometer is inserted into the outgoing concrete 50 – 100 mm deep, and pulled out after the temperature test. The room temperature is recorded simultaneously.

3. Slump (*VC* value): two people conduct the slump test in parallel on the steel plate simultaneously to reduce human errors; the RCC *VC* value is generally tested twice. The gradual loss of slump (*VC* value) is required to provide the basis for the construction.

4. Air content: for hydraulic concrete with high requirements for frost resistance grade, the air content is tested by a precision air content meter. During charging, the tools cannot collide with the mouth of air content meter. The air valve of air content meter shall be protected and cleaned in time after the test. The relations between air content gradual loss and slump (*VC* value) gradual loss shall also be tested to provide the basis for the quality control of mixing plants, concreting and concrete durability.

5. Sand content assessment: the sand content has a great impact on the performance of concrete, which is generally evaluated by three methods as follows: ①Plaster the surface of concrete mixtures with a trowel. ②Test the concrete bleeding inside the test formwork during shaking table compaction. ③Observe the concrete bleeding on the placing surface when the vibrator is working.

6. Apparent density test: the original graded concrete is used in the apparent density test, which is charged by the quartering method and tested by the shaking table. Normal concrete shall be charged at one time, and RCC shall be charged in layers, which shall be subject to concrete compaction and bleeding.

7. Time of setting: fresh concrete mixtures are sieved by a 5 mm wet sieve, and the mortar is poured into the test formwork to test its setting time. The normal concrete shall be tested frequently near initial setting and final setting, and RCC shall be tested at equal intervals (every $1-2$ h). Test data should be calculated and plotted with a computer.

A.5.2.6 Concrete Molding

1. Particle size for concrete formation: when the standard test formwork is used for concrete formation, the maximum particle size of aggregates in the concrete mixtures shall not exceed 1/3 of the minimum section size of the test formwork. The concrete for formation strength test, elastic modulus test, impermeability test, shear resistance test, bending resistance test and other tests is sieved with a 40 mm wet sieve; the fresh concrete for formation ultimate tensile test, frost resistance test, dry shrinkage test, wet expansion test and other tests is sieved with a 30 mm wet sieve; after sieving, the mixtures must be mixed evenly. The dimensions of test formwork for the fully graded concrete test are 450 mm × 450 mm × 450 mm and ϕ450 mm × 900 mm.

2. Test formwork charging: before formation, the inner walls of test formwork shall be evenly brushed with oil. The oil should not soak the paper as far as possible; during formation, the same model of test molds shall be placed neatly beside concrete mixtures, and the aggregates are evenly charged into the test formwork by a small square spade in the positive and negative diagonal directions to avoid aggregate concentration. During RCC and fully graded concrete formation, the aggregates shall be charged into the test molds in layers.

3. Test formwork tamping: during concrete formation, it is required to use a shaking table to tamp mechanically the concrete. Since there is a large deviation of test results of concrete molded by themanual tamping method, this method should not be used. During tamping, the smooth surface of spatula can be used to tamp along the inner walls of test formwork for several times to remove air bubbles and gap and make the grout cover the whole aggregate surface. RCC vibration formation shall be subject to bleeding. The flexible shaft vibrator should be used for fully graded concrete formation.

4. Plastering and numbering: after formation, specimen positions shall be marked, plastered and numbered in time. The number is generally written in three lines, which are test number, age and test date respectively.

5. Specimen de-formwork: after the specimens are numbered, they should be timely put into the curing room for curing, or covered with films and wet sacks for moisture preservation. The de-formwork time depends on concrete strength level, fly ash content, setting time and climatic conditions. After de-formwork, the

specimens shall be timely put into the curing room for curing.

A.5.2.7 Concrete Curing

It is very important and necessary to cure the concrete to ensure the performance of hardened concrete and the accuracy of test results.

1. Conditions of curing room: the temperature in the curing room must reach 20 ℃ ± 3 ℃, and the humidity must exceed 95%. In addition, the curing room shall be equipped with an automatic constant temperature and humidity controller, spray facilities and an air conditioner and other equipment.

2. Safety of curing room: the curing room shall be equipped with a 36 V low-voltage safelight and automatic power cut-off devices or eye-catching warning signs.

3. Curing frame: the curing frames are generally made of ∠50 mm angle steel (or ϕ32 mm steel bars) and ϕ10 - ϕ14 mm steel bars. The length × width × height of curing frame = (1 500 - 2 000 mm) × (500 - 600 mm) × (1 400 - 1 600 mm). A curing frame is divided into 5 - 6 layers, and the height of each layer should be 250 - 300 mm.

4. Placement of specimens: the spacing between concrete specimens is 10 - 20 mm. The specimens are placed on the curing frames of the specified month according to the test date, so as to facilitate the test and inspection.

A.5.2.8 Hardened Concrete Test

The hardened concrete test must conform to the regulations and specifications. During the test, after the concrete specimens are taken out from the curing room, it is necessary to pay attention to the moisture preservation and conduct the test in time.

1. Physical mechanics test: 1 000 - 2 000 kN testers are generally used for strength test, elastic modulus test, shear resistance test, bending resistance test and other tests, while 100 - 300 kN testers are used for ultimate tensile test. 5 000 - 10 000 kN testers are generally used according to the strength level for the fully graded concrete test.

2. Anti-freezing test: air-cooled rapid freeze-thaw machine controlled automatically by microcomputer should be adopted. Before the test, frost resistant specimens shall be soaked in the water at the standard temperature of curing room for at least 4 d; During the test, the moisture on the surface of specimens is wiped off at first, and then the initial mass and natural vibration frequency of specimens are tested. Reference values must be tested accurately.

3. Anti-permeability test: the concrete impermeability tester is used for the test. After the specimens reach the age, they are taken out from the curing room. When the surfaces are dried, the laitance on the side of the cone is removed with a wire brush, and then the dust is brushed off with a brush. The side of specimen is sealed with cement and grease putty. Its ratio is: cement: grease = 3: 1 - 4: 1. Prepared sealing materials are evenly scraped with a triangular spatula on the side of each specimen, and the thickness is 1 - 2 mm. Then, the specimens are put into an impermeability test formwork and pressed into the formworks mold with a 100 - 200 kN force on the tester. After the test, the specimen shall be timely removed from the tester, split and marked, and the seepage height shall be tested.

4. Dry shrinkage test: install an air conditioner for the dry shrinkage chamber to ensure constant temperature and dry conditions. Apply the horizontal length measuring instrument in the test, and provide a glass observation window on the door to prevent accidents during the test.

5. Data processing: test data should be processed with a computer, and compile the corresponding calculation processing program. Record the test as per the requirements of national metrological certification or laboratory accreditation.

A.5.2.9 Adiabatic Temperature Rise Test

The test is to measure the temperature change and the maximum temperature rise of concrete cementing materials (including cement and mixture) during hydration process under the adiabatic conditions, so as to provide the basis for the calculation of concrete temperature stress. An adiabatic temperature rise tester is adopted for the adiabatic temperature rise test, which is placed in a clean adiabatic temperature rise chamber without corrosive air at 20 ℃ ±5 ℃. As the specimen for adiabatic temperature rise test is large and heavy, and it is difficult to manually load and unload, the beam and chain block are generally used as lifting facilities in the adiabatic temperature rise chamber.

Original concrete gradation is adopted in the adiabatic temperature rise test. The concrete raw materials should be placed in the chamber at 20 ℃ ±5 ℃ 24 h before the test to make the temperature consistent with the room temperature; during the test, the mixing test must be carried out in strict accordance with the provided concrete mix proportion, and the adiabatic temperature rise test can only be carried out after the mixture meets the workability requirements. Evenly coat the inner wall of the container for specimen preparation with a layer of grease or other release agent to facilitate de-formwork. During formation, put the mixed original graded concrete mixture into the container in two layers, and tamp and compact each layer with a tamper. Install a copper temperature tube or glass tube in the center of the specimen, with a small amount of transformer oil in it, which is inserted into the central thermometer, seal the nozzle of the temperature tube with cotton yarn or plasticine to prevent concrete or mortar from falling into the tube, and then cover and seal the upper cover of the container. Put the filled concrete adiabatic temperature rise specimen together with the container into the adiabatic chamber with a chain block, start the instrument and start the test until the specified cement age, and record the test. Complete the mixing and formation of concrete within 30 minutes ready for temperature measurement.

After the test, open the sealing cover of the adiabatic chamber, take out the central thermometer, lift the concrete adiabatic temperature rise specimen together with the container out of the adiabatic chamber with a block chain, and carefully de-formwork it to prevent the container from being damaged during de-formwork.

A.5.2.10 Selection of Construction Mixing Proportion

Determine the scientific and reasonable concrete construction mix proportion according to the concrete design indexes, construction requirements and field review test results and through technical and economic analysis and comparison.

A.5.2.11 Adjustment of Construction Mix Proportion

The concrete mix proportion should be adjusted in time according to the change of construction site conditions and the fluctuation of raw materials. However, key parameters such as water-cement ratio, unit water consumption and fly ash content are generally not allowed to be adjusted; generally, according to the change of sand fineness modulus, super-inferior size ratio of coarse aggregate, temperature and air content, the sand ratio, gradation, admixture content and so on are adjusted according to the relationship law of mix proportion parameters.

A.5.3 Labor Organization

A.5.3.1 Organization

According to the characteristics of the hydraulic concrete mix proportion test procedures and methods, the laboratory organization adopted is shown in Figure A.2.

A.5.3.2 Organization of Manpower

A total of 16 people are involved in the concrete mix proportion test, of which 12 are concrete mixing personnel for alternate operation. See Table A.1 for details.

Appendix A

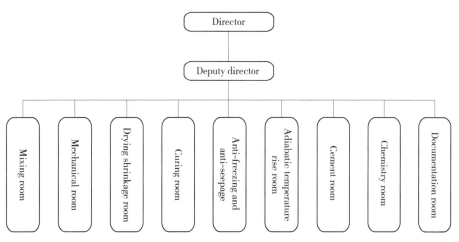

Figure A.2 *Laboratory organization chart*

Table A.1 **Test personnel organization schedule**

Item		Number of people	Scope of work
Test plan preparation		2	According to the requirements of concrete design index and control index, prepare a detailed and specific concrete mix proportion test plan
Organizational implementation		2	Reasonably arrange the resources such as people, materials and equipment for concrete mix proportion test
Documentation		2	Organize and analyze the test data
Concrete batching	Batching sheet	2	Calculate and verify the ingredient list
Concrete batching	Measure and feed the raw materials	6	Weigh the cementing materials, aggregates, admixtures, water and other materials according to the ingredient list, and feed the materials into the mixer according to the specified order
Concrete batching	Mixture performance test	4	Test the slump (VC value), air content, setting time and apparent density
Concrete formation and curing		4	Formation, plastering, numbering, de-formwork, curing, etc.
Thermal test		2	Adiabatic temperature rise test and other tests
Hardened concrete	Physical mechanics test	3	Compression, splitting tensile strength (shear) tests
Hardened concrete	Durability test	3	Frost resistance and impermeability tests
Hardened concrete	Deformation test	4	Ultimate tensile, elastic modulus, and dry shrinkage tests.

A.6 Material and Equipment

A.6.1 Material

See Table A.2 for main materials required for hydraulic concrete mix proportion test.

Table A.2 Main materials required for hydraulic concrete mix proportion rest

S/N	Description		Variety	Principle	Function
1	Cement		Moderate heat Portland cement, low heat Portland cement, low heat slag Portland cement, Portland cement, ordinary Portland cement, sulfate resistance Portland cement, etc.	According to the project site, technical requirements and environmental conditions	Meet the performance requirements of concrete mix proportion, reduce concrete heating and resist environmental erosion
2	Admixture		Fly ash, volcanic ash, slag micropowder, silicon powder, granulated furnace phosphorus slag, magnesium oxide, etc.	According to project technical requirements, mixture quality and resource conditions and demonstration by tests	Improve the performance and quality of concrete, reduce the hydration heat of concrete, inhibit the reaction of alkali aggregate, save cement and reduce cost
3	Aggregate	Fine aggregate	Natural sand and manufactured sand	Hard, clean and well graded	Ensure workability and service performance of concrete, and improve compactness of concrete, etc.
4		Coarse aggregate	Pebble and gravel	High-quality and economical, with local materials applied	Ensure the quality of concrete, determine the strength and durability of concrete, etc.
5	Additive		High-range water reducing agent, retarding high-range water reducing agent, high-temperature retarder and air entraining agent	According to the requirements of concrete design index	Improve concrete workability, save materials, adjust construction performance, improve strength and durability, etc.
6	Water		Potable water	Conform to national standards	Improve the hydration cementation and fluidity of concrete

A.6.2 Machines, Tools and Equipment

See Table A.3 for main instruments and equipment required for hydraulic concrete mix proportion test.

Table A.3 Main instruments and equipment required for hydraulic concrete mix proportion test

S/N	Equipment description	Specification/Model	Unit	Quantity	Purpose
1	Mortar mixer	JJ-5	Set	1	Physical tests of cement, mixtures, admixtures, and aggregates
2	Mortar jolting table	ZS-15	Set	1	
3	High temperature furnace	SX2-4-1300	Set	1	
4	Cement mortar fluidity tester	NLD-2	Set	1	
5	Measuring instrument of paste standard consistency and setting time	(ISO)	Set	1	
6	Tester for determining expansion of Le Chatelier needles	LD-50	Set	1	
7	Cement specific surface area tester	DBT-127	Set	1	

Continued Table A.3

S/N	Equipment description	Specification/Model	Unit	Quantity	Purpose
8	Standard cement curing box	40 ultrasonic humidification	Set	1	Physical tests of cement, mixtures, admixtures, and aggregates
9	Electrothermal blowing dry box	101-2	Set	1	
10	LE cement boiling tank	FZ-31	Set	1	
11	Cement negative pressure sieve analyzer	FYS-150B	Set	1	
12	Electric cement bending resistance machine	5000	Set	1	
13	Cement pressure test machine	YAW-300B	Set	1	
14	Cement mortar mixing machine	160B	Set	1	
15	Thermostat water bath	HHS-6	Set	1	Chemical tests of cement, mixtures, additives and aggregates
16	Flame photometer	6400A	Set	1	
17	Analytical balance	1/10 000 and 1/1 000	Set	2	
18	Acidometer	PHS-3C	Set	1	
19	Viscometer	NDJ-1	Set	1	
20	Electric setting centrifuge	LDZ-4-8	Set	1	
21	Spectrophotometer wavelength	7230G	Set	1	
22	Gravity mixer	100 – 150	Set	1	Cement mixture performance test
23	Forced concrete mixer	60 – 100	Set	1	
24	Mixing steel plate		Pcs.	1	
25	Aggregate sieve	The aperture of square hole is 40 mm and 30 mm	Set	1	
26	Apparent density apparatus	20 – 80 L (Wall thickness 3 mm)	Set	1	
27	Slump cone	—	Nr.	5	
28	Webe meter	HGC-1	Set	2	
29	Concrete air entrainment meter	H-2783	Set	2	
30	Concrete vibrating stand	1 m^2	Set	1	
31	Concrete penetration resistance tester	HT-80	Set	1	
32	Constant temperature and humidity automatic controller	Full-automatic	Set	1	
33	Compression tester	YE-2000	Set	1	Hardened concrete test
34	Universal testing machine	WE-1000B	Set	1	
35	Universal testing machine	WE100 – 300	Set	1	
36	Ultimate tensile strength tester	YJ-26	Set	1	
37	Elastic modulus tester	—	Set	1	
38	Rapid concrete freezing and thawing device	Air-cooled CDR-2	Set	1	
39	Dynamic modulus tester (frost resistance test)	QL-101	Set	1	
40	Impermeability tester	Model HP-40	Set	1	
41	Horizontal concrete length measuring instrument	SP-540	Set	1	
42	Adiabatic temperature rise tester	—	Set	1	Thermal test

A.7 Quality Control

A.7.1 Quality Standards

In addition to strictly implementing the *Test Code for Hydraulic Concrete* DL/T 5150—2001, *Specifications for Hydraulic Concrete Construction* DL/T 5144—2001, *Test Code for Hydraulic Roller Compacted Concrete* SL 48—1994 and *Construction Specification for Hydraulic Roller Compacted Concrete* DL/T 5112—2000 and the national industry standards and existing laws and regulations related to the construction method, the following quality standards shall be noted in combination with the characteristics of the construction method:
1. Technical requirements of actual engineering design;
2. *Laboratory Quality Management System Document* and *Concrete Mix Proportion Test Plan* prepared by the laboratory which has been approved.

A.7.2 Quality Assurance Measures

1. The laboratory should have metrological certification (or national laboratory accreditation qualification), implement the policy of "total quality management, and quality first", implement strict scientific management, effectively control all factors affecting the testing quality, and ensure the authenticity, accuracy and completeness of the testing data and results.
2. Inspection and calibration: Instruments and equipment used for concrete measurement in the laboratory can only be used in various tests after being acceptable in calibration, so as to ensure the accuracy of value in transfer.
3. The measurement and test personnel of the laboratory shall obtain the test qualification certificate after examination, regularly participate in learning and training, and constantly improve technical and work capabilities.
4. The laboratory regularly organizes contrast tests with the authoritative third-party test institution recognized by the state to verify the standard of environmental facilities, the accuracy of instruments and equipment, the standardization of inspection process, and the work level of test personnel, so as to improve the test and measurement capacities.
5. In order to ensure the smooth implementation of concrete mix proportion test, sufficient and reasonable personnel shall be provided for the test, and the capabilities of all personnel who use special equipment, measure and evaluate results and sign the test report shall be guaranteed.

A.8 Safety Measures

1. Strictly comply with and execute the requirements of the national, local and industrial laws, regulations and standards in safety and laboratory safety system for testing, establish a sound occupation, health and safety management system, and prepare work safety regulations. Test personnel must be trained in safety, be clear about the work safety regulations of the instruments used, and master the common sense and knowledge in safe use of electricity, water, etc..
2. Before and after the test, all electrical control switches, electrical appliances and line connections used shall be checked, and the switch shall also be controlled in place.
3. Before and after the test, the state of the test instruments shall be checked, and the test personnel must be familiar with the work procedures.

4. The curing room shall be equipped with low-voltage lighting system, and provided with the switches automatically controlling power supply or the obvious signs during access.
5. All kinds of drugs used in chemical test shall be stored by cabinets, strictly managed and registered before and after use; special medicals shall be controlled strictly according to the test process required, and marked clearly.
6. In all kinds of tests, the procedures must be strictly followed to prevent personal injury accidents.
7. In case of any safety accident during the test, the test personnel shall take emergency measures promptly according to the safety knowledge and professional knowledge mastered, and handle the accidents in accordance with the control procedures of relevant accidents, non-conformance items and preventive measures.

A.9 Environmental Protection Measures

1. Strictly comply with and execute the national, local and industrial laws, regulations and standards in environmental protection, establish a sound environmental management system, and prepare the regulations of relevant procedure documents in environmental protection. Test personnel must be trained in environmental protection knowledge, be clear about the hazards of the instruments, equipment and materials used to environment, control the hazard extent caused by the test to environment, and implement the environmental protection indexes specified in the national, regional and industrial regulations related to environmental protection performed.
2. The laboratory shall be kept clean, orderly, quiet and hygienic, and the residues, sundries and hazardous substances in the test shall be strictly controlled and managed to meet the requirements of environmental protection.
3. In case of mutual interference or influence between adjacent areas during the test, effective isolation measures shall be taken to prevent cross contamination. If specified requirements are still not met after isolation, the test and measurement shall be suspended until conditions are met.
4. Test personnel shall check and record environmental monitoring parameters at the beginning of, during and after the inspection, so as to avoid the adverse impact on the inspection results caused by the deviation of environmental conditions.

A.10 Benefits Analysis

Sinohydro Engineering Bureau 4 Co., Ltd. has always adhered to the concept of practice-summary-improvement-innovation in the long-term continuous tests of hydraulic concrete mix proportion, and has formed a set of hydraulic concrete mix proportion test method with mature theory, advanced technology, simple work, accurate data and reasonable cost-effectiveness. Hydraulic concrete mix proportion test method plays a strong supporting role in improving the core technology of concrete dam.

In particular, the Yangtze River Three Gorges Dam Project, the RCC main dam works of Baise Multipurpose Dam Project in Guangxi, the high arch dam of the Yellow River Laxiwa Hydropower Station Project, and the high arch dam of the Lancang River Xiaowan Hydropower Station Project and other projects constructed by Sinohydro Engineering Bureau 4 Co., Ltd. are distinctly characterized by advanced technology, strong practicability, remarkable economic and social benefits and others by means of the concrete mix proportion determined in the test by the method, reaching the domestic leading and international advanced level.

Roller Compacted Concrete Rapid Damming Technology

1. For Phase II powerhouse dam section on the left bank and Phase III powerhouse dam sections on the right bank in Yangtze River Three Gorges Dam Project, the test center of the Sinohydro Engineering Bureau 4 Co., Ltd. carried out the concrete mix proportion test of the dam from September 1997 to May 2006, and submitted an advanced technical and economic index (mix proportion in concrete construction of dam). Such mix proportion has been successfully applied to the Yangtze River Three Gorges Dam Project. Compared with the bid winning mix proportion expense, only the mix proportion in construction of powerhouse dam sections on the left bank can save investment of about CNY 36 million.

2. For RCC main dam works of Baise Multipurpose Dam Project in Guangxi, a topic on use of rock powder of manufactured sand processed by diabase in RCC was studied from January 2002 to December 2006 for the adverse conditions such as adoption of Diabase aggregate for Baise RCC and high content of rock powder of the manufactured sand, and a large number of scientific and research achievements on ECC damming technology were obtained, which ensured the construction of RCC main dam in Baise and acquired remarkable social benefits.

3. For Yellow River Laxiwa Hydropower Station Project, its concrete mix proportion test was started in April 2003 for the special grade durability and F300 frost resistance grade of concrete in cold plateau regions; through technical innovation and thorough test research, the unit water consumption, cementing material consumption, durability, crack resistance and others of the concrete reached the domestic leading and international advanced level. According to the construction and mass control results, the construction mix proportion submitted was very consistent with the site application results, creating substantial technical and economic benefits.

Appendix B

National Construction Method GEVR Construction Method in RCC Dam Construction

Construction method No. : YJGF 079—2006

Sinohydro Bureau 11 China Gezhouba Group Co. , Ltd.

Sinohydro Bureau 8 Sinohydro Engineering Bureau 4 Co. , Ltd.

Ge Jianzhong, Ren Xuewen, Lu Dawen, Huang Wei, Tian Yugong, Li Zhepeng

B.1 Foreword

RCC dam construction technology has the characteristics of high degree of mechanization, rapid and economic construction. Since the technology was introduced into China in the 1980s, the development is very rapid and the construction technology is improving day by day. Through more than 20 years of innovation and development in design, materials, construction technology, etc. , the technical mode with Chinese characteristics has been formed. Among them, GEVR is a typical innovative technology and process and is a grout-enriched RCC by spreading grout during paving, and then tamped and compacted by vibration method. The technology has successfully solved the problems of RCC at the areas around formwork, abutment bank slope, corridor and holes where vibration and rolling are difficult and solved also the construction interference problem. Since the early 1990s, the GEVR technology has been adopted in Rongdi, Puding, Shimantan, Shapai, Baise, Zhaolaihe and Guangzhao hydropower stations. The application scope has been constantly expanded and developed. At present, this technology has been widely used in all RCC dam projects. In October 1997, the GEVR construction technology of Shimantan project passed the technical appraisal organized by Henan Province and reached the domestic leading level. Puding RCC dam technology won the state "The Eighth Five Year Plan" science and technology tackling major scientific and technological achievements award and the first prize of national science and technology progress award. This construction method is summarized and formed on the basis of the construction practices of many GEVR projects mentioned above and can be improved and applied continuously.

B.2 Characteristics of Construction Method

In this method, GEVR is used instead of normal concrete. It solves the construction problem of RCC dam where the vibration roller cannot roll. It is not only convenient for transportation and paving construction, in addition, it is beneficial to combine with adjacent concrete, reducing the construction interference.

1. The construction method is simple. In this method, a certain amount of mortar (cement + fly ash + water +

admixture) is added into the paved RCC and the normal concrete vibration method is used to complete the concrete construction. The concrete type in the dam is reduced and the construction interference of different kinds of concrete is greatly simplified. The construction method is simple.

2. Construction interference is reduced and construction efficiency is high. The RCC can be modified by this technology. The GEVR is used for the areas around the RCC formwork, the bank slope of the dam abutment, the corridor and the holes which cannot be vibrated and interfered by the construction. It can ensure that the paving thickness of RCC with grout spreading is the same as that of leveling surface. The construction interference and manual work amount are greatly reduced and the construction efficiency is improved.

3. Good construction quality and low investment. GEVR solves the problem of poor bonding between different kinds of concrete. Its performance can meet the requirements of normal concrete with the same design grade and can effectively improve the bonding ability and anti-permeability, uniformity of RCC layers. At the same time, due to the reduction of cement consumption, it is beneficial to reduce the temperature rise of concrete hydration heat and save project cost.

B.3   Applicable Scope

Around RCC formwork, corridors, structural holes, steel bars, water stops and embedded parts, etc. ; the GEVR can be used in the dam abutment bank slope, cushion concrete and other areas that cannot be rolled by vibration roller.

B.4   Technological Principle

The GEVR is produced by adding a certain amount of grout into RCC mixture (cement + fly ash + water + admixture) to modify its work ability. It is a construction technology that the GEVR can be tamped and compacted by normal concrete operation method. The grout shall be spread at the bottom or middle of the newly paved RCC, then, the normal concrete operation method is adopted to tamp and compact RCC. This construction method can speed up the construction progress, ensure the quality of the project and can obtain better economic benefits. The paving thickness of RCC paved with grout can be the same as that of the placing surface, so as to reduce the amount of manual work, improve work efficiency and reduce project cost. The amount of grouting shall be determined by test according to specific requirements.

B.5   Construction Technology and Operation Points

B.5.1   Process Flow

RCC paving→leveling→rolling→trenching→grouting→trench filling→vibration→re-rolling.

Concrete after grouting and vibration must meet all physical performance indexes and durability indexes required by the design. At the same time, it shall meet the current relevant provisions of *Code for Construction of Hydraulic RCC*.

B.5.2   Mix Proportion Design of Grout

1. The materials used for mixing grout shall be consistent with the raw materials of RCC. No other materials shall be selected.
2. The important parameters such as the concentration of grout, water binder ratio and the amount of mortar

shall be determined by tests. The material used for grout shall be consistent with RCC mixture. The amount of mortar is related to the *VC* value of RCC and shall be fully considered in the test (Generally, when the *VC* value is 5 – 10 s, the volume ratio of mortar is 5% – 7%. The work ability of the concrete after grouting can reach 1 – 4 cm and the water binder ratio shall not be greater than that of the same kind of RCC, meeting the design requirements of the physical performance indicators and durability indicators.

3. In order to control the water binder ratio of grout on site, the grout concentration control method is adopted. The specific gravity method shall be used to control the grout concentration on site.

B.5.3 Construction Technology Requirements

B.5.3.1 Grout

1. Grout mixing station must be set up for mortar mixing and the net mixing time shall not be less than 120 s. Manual mixing is strictly prohibited.
2. The grout shall be prepared in strict accordance with the mix proportion approved. The precipitated mortar or the grout more than 8 h shall be removed and new grout shall be prepared again.
3. When the specific weight of the fresh grout fails to meet the requirements, no more water and fly ash shall be added and cement shall be added until the specific gravity meets the requirements.

B.5.3.2 Technology and Requirements of Grouting and Vibration

1. The GEVR shall be poured layer by layer. In general, the bottom or middle grouting method shall be adopted; the paving thickness of GEVR shall be the same as that of leveling thickness.
2. The process of grouting and vibrating is as follows: Leveling→rolling→trenching→grouting→trench filling →vibration→rerolling. As shown in Figure B.1.

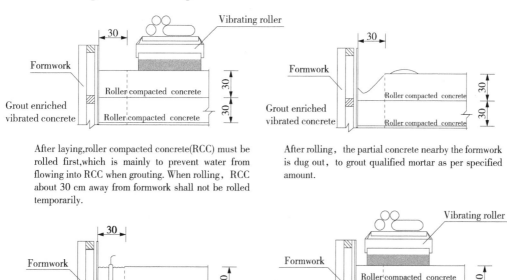

Figure B.1 Key points of GEVR construction technology

3. When rolling, it is not allowed to roll at the grouting and vibrating place. Other areas shall be rolled according to the required compactness. The purpose is to prevent the mortar from spreading out of the mortar vibrating area, ensure the effect of grouting and vibration.

4. The amount of grouting shall be determined by test, so as to control strictly the dosage according to the regulations. Special grouting equipment able to meter should be used, and the amount of mortar shall be controlled by calculation [such as: the vibration width of upstream dam surface is 30 cm and the thickness of rolling layer is 30 cm; mortar volume (the amount of mortar is 6%) $3 \times 10 \times 3 \times 0.06 = 5.4$ L/m]. Depending on the change of *VC* value of RCC, the amount of mortar can be adjusted properly.
5. High frequency vibrator shall be used for vibration of GEVR and the vibration time shall be 1 – 2 times longer than that of normal concrete. Making sure the grout turns upward to the surface. Meanwhile, the vibrator shall be inserted 5 cm into the lower concrete to ensure uniformity and the combination of upper and lower layers.
6. At the joint of GEVR and RCC, the vibration roller is used to re-roll the joint. The range of re-rolling shall be more than 20 cm into the GEVR. The number of rolling passes is one-half of that of leveling surface.
7. The area around water stop is one of the key areas of GEVR construction. The construction shall be carried out in strict accordance with the design requirements and the measures shall be taken to support and protect water stop, ensuring vibratingd densely.

B.6 Mechanical Equipment Configuration

See Table B.1 for the machines and equipment used in this method.

Table B.1   Table of machinery and equipment allocation

S/N	Equipment description	Model and specifications	Unit	Quantity
1	Net grout mixer	ZJ-250 or 100	Set	1
2	Vibrating bar	Model F100 or 125	Set	2
3	Frequency converter	Matching with vibrator rod	Set	1

B.7 Quality Control

1. Operators of the construction team shall be provided with technical training in advance, and the technicians on site shall conduct whole-process supervision on grouting concentration and volume and other operations.
2. The technicians on site shall carry out tests on the mortar concentration every 2 h, while the quality inspectors shall be responsible for supervising the grouting and vibration process, and for at least one spot inspection on the specific gravity of mortar per shift.
3. The laboratory shall conduct tests on the GEVR. A least one group of samples shall be taken from the concrete which has been grouted and vibrated for every 3 m rising for purpose of evaluation on physical property indexes of the concrete.
4. Physical objects on the site shall be sampled for mathematical statistics for purpose analysis onconstruction level and concrete acceptance.
5. During construction in winter, in addition to those provisions on concreting in winter in the project, special attention shall be paid to the anti-freezing of GEVR, which shall be covered with the insulation materials upon completion of the vibration. It is strictly prohibited to use straw bags and materials which may contaminate the concrete as the insulating material.

B.8 Safety Measures

1. Enforce the Labor Law of the *People's Republic of China* and relevant safety laws and regulations.
2. Establish and improve the safety organization and appoint appropriate full-time and part-time safety officers.
3. Lifting facilities shall be secure and under sound conditions, and all machines and tools throughout the site shall be free of unsafe condition. Lifting operators shall have been properly certified before working on the job.
4. Develop safe operation instructions and carry out operations in accordance with such instructions.

B.9 Environmental Protection Measures

In the process of construction, observe the applicable national and ministerial laws, regulations and rules regarding environmental protection, and provide environmental protection in the construction area.

B.10 Technical and Economic Analysis

In perimeter of the RCC formworks, around the corridors and openings, and around locations including the steel bars, water stops and embedded parts; GEVR is adopted for the dam abutment cushion according to the design requirements. Compared with the construction of GEVR and normal concrete, the efficiency of GEVR construction is significantly improved, and the construction quality is well guaranteed, with smooth surface and compacted inside. There is also significant economic benefit.

B.11 Project Examples

1. Shimantan Reservoir Reconstruction Project, which is a RCC (RCC) gravity dam, adopts this construction approach. The project requires 360 000 m³ of RCC. Instead of normal concrete, about 16 000 m³ of GEVR is placed in perimeter of the RCC formworks, around the corridors and openings, and around locations including the steel bars, water stops and embedded parts. The GEVR is placed to 30 cm wide, the roller compacted layer is 30 cm thick, and with grouting volume of 5%, 5 L of grout is grouted per meter. After grouting, the concrete slump varies ranges from 1 – 3 cm; The statistics of field sampling tests on GEVR shows that all physical property indexes comply with the design requirements, and the project investment is saved by CNY 2 million. The project has been recognized as one of the top-quality projects in Henan Province.
2. Zhaolaihe Hydropower Project is a high RCC double-curvature thin arch dam which requires 18 000 m³ of quantities of RCC. In perimeter of the RCC formworks, around the corridors and openings, and around locations including the steel bars, water stops and embedded parts, GEVR to quantities about 10 000 m³ is constructed; The GEVR is placed to 50 cm wide, the roller compacted layer is 30 cm thick, and with grouting volume of 5%, 7.5 L of mortar is grouted per meter. After grouting, the concrete slump varies ranges from 2 – 4 cm. The statistics of field sampling tests on GEVR shows that all physical property indexes comply with the design requirements, and the project has achieved significant economic benefits, saving project investment by more than CNY 1 million.
3. In Puding Hydropower Station, the RCC arch dam is 75 m high, and the dam body requires 149 900 m³ of concrete in total, including 127 000 m³ of RCC. The project, which was commenced in 1991 and completed

in May 1993, adopts the whole-block placement & thin lift roller compaction and continuous lift construction process. It is the first dam in China which relies on RCC for imperviousness. The GEVR construction technology is substantially used in the project. With indoor and outdoor tests, the relationship between the grouting volume of GEVR and the vibration of bleeding, the homogeneity and quality dispersion of GEVR are studied. The GEVR in the project has reached high technical indexes in the unit weight, compressive strength and dispersion, and it is with good uniformity, and coefficient of variation (CV) < 0.081. *Study on Whole-block Placement Continuous Construction Process of Roller Compacted Concrete* and *Study on Quality Control of Roller Compacted Concrete on Site* have been listed as the sub-topics of national scientific and technological key project during "the Eighth Five-year Plan Period", and its research on RCC arch dam construction technology has won the national Scientific and Technological Achievement Award during "the Eighth Five-year Plan Period" and the first prize of national science and technology progress award.

During the construction of Shapai RCC Arch Dam in 1999, with indoor and field tests and researches, several technical problems with mortar design, mortar volume control, construction technology and quality control, etc. are solved, and the application of GEVR is expanded.

4. As a complete RCC gravity dam, the volume of RCC Baise dam was 2.104 million m^3. During the construction of Baise dam, GEVR had been widely used and the total volume of GEVR was about 163 300 m^3. The GEVR was used around the normal concrete cushion of the whole dam bank slope and some parts of the bank slope dam foundation, upstream and downstream formwork edges, expansion joints, burying places where upstream and downstream water stops were embedded, corridor, elevator shaft periphery, observation cable periphery, joint reinforcement mesh part and parts that cannot be rolled by vibration roller. The concrete completed is smooth on surface and compacted inside, and free of crack or other defects. The core drilling and sampling show that the core sample is dense and smooth, and it is impossible to identify the junctions. The mechanical performance indexes of modified concrete can fully comply with the design requirements.

Appendix C

Example of Construction Methods: Construction Method of RCC for Dam of a Certain Project

C.1 General Principles

1. To construct the RCC main dam project of a certain hydropower station to a high-quality project, the RCC dam construction has to be standardized, the *Construction Methods of RCC for A Certain Dam* is hereby formulated. By closely combining with the development of RCC construction in hydropower and water conservancy industry in recent years, actual condition of adopted new technologies, new materials, new process and new equipment, in reference to relevant project construction methods and standards, this construction method has carried out necessary supplement and improvement to "Construction Methods of RCC Dam".

2. The main basis for this construction method is *Contract for Construction of Civil and Metal Structure Installation Project of A Certain Hydropower Station Dam* Volume Ⅱ *Technical Terms*, *Construction Specification for Hydraulic RCC* DL/T 5112, *Specification for Hydraulic Concrete Construction* DL/T 5144 and *Technical Requirement of Concrete Construction for Permanent Hydraulic Structure of A Certain Hydropower Station*. The parts not covered by this construction method shall be implemented according to the current national and industrial standards.

3. The concrete tests are carried out in accordance with *Test Code for Hydraulic Concrete* DL/T 5150, *Test Code for Hydraulic Concrete* SL 352, *Rock Test Regulations for Water Conservancy* SL 264 and *Hydropower Projects* and *Field Rolling Test Requirements of Dam RCC* by Design and Research Institute.

4. All personnel participating the RCC construction and management shall follow this construction method. Those who violate the provisions of this construction method and cause influence and consequence on the quality, schedule and safety of the project, and those who have made outstanding achievements in strictly implementing the construction method will be rewarded and punished according to *Rewards and Punishment Measures for RCC Dam Project Quality of a Certain Hydropower Station*.

5. This method is prepared based on the summary of the experience from RCC construction of Shapai, Lanhekou, Baise, Longtan, Guangzhao, Jinanqiao, Gelantan and other Projects, by reference to the construction method of dam RCC of above projects, and by combining with the specific condition of dam RCC construction of this project.

C.2 Normative References

The following standards are not dated, and the latest version of the standards cited is used and is applicable to this work.

GB 175 *Common Portland Cement*

GB 200 *Moderate Heat Portland Cement, Low Heat Portland Cement and Low Heat Portland Slag Cement*

GB/T 1596 *Fly Ash Used for Cement and Concrete*

GB 2938 *Low Heat Expansive Cement*

GB/T 18046 *Ground Granulated Blast Furnace Slag Used for Cement and Concrete*

DL/T 5055 *Technical Specification of Fly Ash for Use in Hydraulic Concrete*

DL/T 5100 *Technical Code for Hydraulic Concrete Admixture*

DL/T 5150 *Test Code for Hydraulic Concrete*

SL 352 *Test Code for Hydraulic Concrete*

DL/T 5151 *Test Code for Aggregates of Hydraulic Concrete*

DL/T 5152 *Test Code of Water Analysis for Hydraulic Concrete*

DL/T 5330 *Code for Mix Design of Hydraulic Concrete*

DL/T 5112 *Construction Specification for Hydraulic Roller Compacted Concrete*

DL/T 5144 *Specifications for Hydraulic Concrete Construction*

SL 314 *Design Specification for Roller Compacted Concrete Dams*

SL 319 *Design Specification for Concrete Gravity Dams*

DL 5108 *Design Specification for Concrete Gravity Dams*

SL 275 *Field Test Methods for Nuclear Density-moisture Gauges*

DL/T 5169 *Specification for Construction of Hydraulic Concrete Reinforcement*

DL/T 5178 *Technical Specification for Concrete dam Safety Monitoring*

DL/T 5148 *Technical Specification for Cement Grouting Construction of Hydraulic Structures*

DL/T 5110 *Construction Specification for Formwork of Hydropower and Water Conservancy Engineering*

DL/T 5113.1 *Quality Degree Evaluate Standard of Unit Engineering for Hydropower and Water Civil Engineering*

DL/T 5113.8 *Quality Degree Evaluate Standard of Unit Engineering for Hydropower and Water Conservancy Construction Engineering (Part 8) Hydraulic RCC Engineering*

DL/T 5123 *Specification for Engineering Acceptance of Hydropower Station Capital Construction*

C.3 Terminology

C.3.1 Extender

Admixture is the mineral substances such as fly ash, etc., that is used to mix when mixing the RCC, mortar and grout.

C.3.2 Rock Powder Content

The mass percentage of the content of particle with a size smaller than 0.16 mm in fine aggregate rock powder. The content of rock powder should be controlled from 16% and 22%.

C.3.3 Strength Grade

The cube is divided into many grades according to its compressive strength under standard conditions, known as concrete strength grade.

C.3.4 Amount of Cementing Materials

The sum of cement and admixtures per cubic meter of concrete.

C.3.5 Cementing Grout Consumption

The sum of content of cement + admixtures + 0.08 mm rock powder in a cubic meter of concrete, is the main factor that affects the P_V value of grout to sand ratio.

C.3.6 Water-cement Ratio

The ratio of water consumption to cementing material consumption in a cubic meter of concrete.

C.3.7 Batching Time

The time from all materials are added to start of discharge.

C.3.8 Concrete Transport Time

The time from complete discharge from the mixer outlet to the concrete is discharged into block.

C.3.9 Rolling Layer Thickness

The concrete thickness before rolling of each rolling operation.

C.3.10 Compaction Thickness

The thickness when each rolling operation layer meets the compactness or volumetric weight required by design after rolling.

C.3.11 Relative Compaction

The ratio of apparent density measured on the construction placement surface to the average reference apparent density obtained from the indoor test of the roller compacted concrete. The relative compactness is more than 97% inside the dam, and more than 98% in external imperviousness area.

C.3.12 Reference Apparent Density

The average of apparent density of roller compacted concrete with selected mix proportion obtained from the indoor test.

C.3.13 GEVR

The concrete that is compacted by vibration after certain proportion of grout (cement + admixture + additive) is mixed in the spread roller compacted concrete.

C.3.14 Bedding Mixture

The grout, mortar or small graded grout enriched concrete that are spread on the placement layer or bed rock surface, and adapted to the roller compacted concrete.

C.3.15 *VC* Value

That is the operation temperature of the roller compacted concrete admixture, which refers to the time required for the roller compacted concrete admixture to start vibration to surface bleeding, in the unit of measurement of second.

C.3.16 α Value

The ratio of grout volume to mortar voids volume.

C.3.17 β Value

The ratio of mortar volume to aggregate voids volume.

C.3.18 P_V Value

That is the ratio of cementing grout volume to mortar volume, which is called as "grout to sand ratio" for short. The grout to sand ratio shall not be less than 0.42 P_V, the main factor that affects the P_V value is the content of rock powder.

C.3.19 Intermittent Time between Layers

The time duration starting from adding water into the lower layer of concrete mixture to the completion of upper layer concrete rolling.

C.3.20 Permissible Time Interval between Placing Layers

Maximum intermittent time between layers that can meet the required interlayer bonding quality by direct placing upper layer of concrete without any layer surface treatment.

C.3.21 Cold Joint

The roller compacted concrete surface with intermittent time between layers more than permissible time interval between placing layers.

C.3.22 Construction Joint

The joint that is set according to the construction requirement.

C.3.23 Hair Surface

The rough concrete surface without milk skin formed after treatment.

C.3.24 Temperature of Machine Outlet

The temperature at a depth of 10 cm under the concrete surface when the concrete mixing is completed and discharged to transporting equipment.

C.3.25 Placing Temperature

The temperature at a depth of 10 cm under the concrete pile surface after the concrete is transported to the placement surface, discharge and spread.

C.3.26 Placement Temperature

The temperature at a depth of 10 cm from the concrete surface after the roller compacted concrete is leveled and rolled before upper layer of concrete is covered.

C.3.27 Placement Surface Environment Temperature

The temperature at 1 m above the concrete placement surface.

C.3.28 Sudden Temperature Drop

Daily average temperature drops by more than 6 ℃ continuously within 24 h.

C.3.29 Cold Wave

The sudden temperature drop of daily average temperature of 5 ℃ and less.

C.4 Concrete Compaction Process Flow Chart

See Figure C.1 for the process flowchart of roller compacted concrete construction.

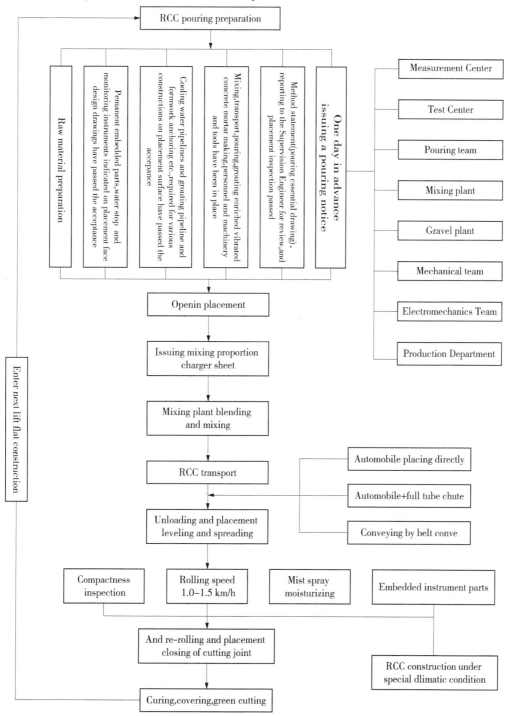

Figure C.1 *Concrete compaction process flow chart*

C.5 Control and Management of Raw Materials

C.5.1 Composition of Roller Compacted Concrete Materials

The roller compacted concrete is composed of cement, admixture (fly ash), water, coarse and fine aggregates (sand, stone), additive, rock powder, etc. The quality of raw materials for roller compacted concrete shall comply with the requirements specified in currentnational standards and this construction method. The control of the raw materials is carried out jointly by the Quality Department, Test Center and relevant operation departments. All cement, fly ash, admixture and steel used in the main construction shall be subject to the relevant provisions of the contract and the specification, the sampling re-inspection, the sampling items and frequency shall be subject to Table C.1.

C.5.2 Cement, Fly Ash

Receiving and storage of cement and fly ash. The Test Center is responsible for the inspection of cement and fly ash according to the relevant items in above Table C.1, the purchasing staff is responsible for the receiving work of the cement, the warehouse keeper is responsible for the storage work of the cement. The cement or fly ash of different manufacturers and lots shall be stored in tanks for separation, the storage capacity shall meet the requirement of placing strength of the roller compacted concrete.

Table C.1   Roller compacted concrete inspection item sampling frequency table

Description	Test items	Sampling location	Control index	Number of times of sampling	Test purpose
Cement	Specific surface area, fineness, stability, standard consistency, setting time, strength	Warehouse	National standards	Each lot, 400 t or 1 time as required	Verify incoming cement quality
Admixture	Fineness, water content ratio, loss on ignition, water content	Warehouse	Meet requirements of specification and design	1 time for a lot of 200 t continuous feeding	Verify incoming fly ash quality
	SO_3 content	Warehouse	Meet requirements of specification and design	Once a quarter	Verify incoming fly ash quality
Additive	Concrete water reducing rate, bleeding rate, air content, setting time	Warehouse	It meets the requirements of the specification	1 time for each lot	Verify incoming additive quality

Continued Table C. 1

Description	Test items	Sampling location	Control index	Number of times of sampling	Test purpose
Fine aggregate (sand)	FM, rock powder content, water content	Final product warehouse of sand factory or on discharge belt	FM: Note: 2.2 – 2.9, Rock powder content 16% – 20%	FM: Note: 1 time/shift Rock powder content: 1 time/shift Moisture content: 2 h/time	Production (product) quality control
	Saturated surface dry density, saturated surface dry water absorption, volumetric weight (loose, compaction by vibration), percent of void, organic matter, mica content, sulphide and sulfate content	Final product warehouse of sand factory or on discharge belt	It meets the requirements of the specification	Once/month	
Coarse aggregate (large, middle, small stone)	Super-inferior size ratio	Final product warehouse of sand factory or on discharge belt	Excess < 5%, shortage < 10%, flat and elongation < 15%	1 time/shift	Production (product) quality inspection
	Small stone water content	Final product warehouse of sand factory or on discharge belt	Allowable deviation less than 0.2%	1 time/shift	
	Silt content, content of clay lump, crush index, flat and elongation, saturated surface dry density, saturated surface dry water absorption, volumetric weight, percent of void, grain gradation	Final product warehouse of sand factory or on discharge belt	It meets the requirements of the specification	As the case may be	
Steel	Mechanics	Warehouse	Meet requirements of specification and design	1 time for each lot of 60 t	Verify incoming steels quality

Note: 1. The inspection frequency of water for test as appropriate.

2. The contents of large, middle, small stone, sand, etc., when necessary.

3. The sampling method complies with the specification, in case the one sampling is not acceptable, the number of sampling shall be doubled, if the design requirement still is not met, this product is listed as nonconforming product or used by degrade.

C.5.3 Water

The water for mixing and curing shall meet the requirements in *Specification for Hydraulic Concrete Construction* DL/T 5144—2001.

C.5.4 Additive

The preparation of the additive shall be done according to the charger sheet issued by the Test Center, the additive solution is required to be accurate in measurement, uniform in mixing, which shall be inspected and tested by the Test Center.

C.5.5 Aggregate

The Test Center is responsible for the inspection of the manufactured aggregates according to specified items and frequency, and report the test result to Quality Department or supervision engineer at every weekend.

C.6 Selection of Roller Compacted Concrete Mix Ratio and Issuing of Charger Sheet

C.6.1 Selection of Mix Proportion

1. The mix proportion and parameter selection of the roller compacted concrete, bed course concrete, contact mortar and cement grout shall be determined according to the approved mix proportion design test report and relevant test results.
2. The selection of roller compacted concrete construction mix proportion and the adjustment of important mix proportion parameters shall be reported by the Test Center, which shall be examined and approved by the chief engineer and submitted to the supervision engineer for approval.

C.6.2 Issue of Concrete Charger Sheet

1. The roller compacted concrete placement notice shall be filled by the placement team and sent to the Test Center 2 – 6 h in advance, the Test Center shall issue the concrete charger sheet 30 minutes before starting the machine.
2. Before sending the construction batching sheet, the Test Center shall conduct inspection and sampling inspection to the raw materials to be used.
3. The Test Center shall be responsible for the construction batches sheet that has been sent. The charger sheet can be distributed only after it is checked by the checker (no mistake is found) and signed by the supervision engineer.

C.7 Inspection and Acceptance before Placement Construction of Roller Compacted Concrete

C.7.1 Inspection of Preparatory Work

1. The Production Headquarter is responsible for inspecting various items of preparatory work before placing

the roller compacted concrete, such as the conditions of mechanical equipment, staffing, placement access, communication facilities, the inspection of block lighting as well as water supply and drainage condition, the Quality Control Department shall be responsible for inspecting the state of the detecting instrument.

2. When direct placing from the truck transportation is used, the facilities at the place where the truck tires are flushed shall meet the technical requirements. There shall be sufficient dehydration distance from the dam placing entrance, the access to the placing shall be paved with gravel and flushed clean without contamination.

3. The inspection work of the construction equipment is the responsibility of the unit using the equipment. The placement team is responsible for the truck tire washing facilities, the Quality Control Department is responsible for the inspection; the Production Headquarter is responsible for the access to placing, the placement team helps with the arrangement of the truck washing table.

C.7.2 Inspection and Acceptance of Unit Engineering in Block

The inspection and acceptance of the quality of unit engineering in block shall adhere to "Three inspection system" (construction group (team) caries out the first inspection, the placement team carries out the second inspection, the Quality Control Department carries out the third inspection), see the acceptance organization flow table below. Prior to the acceptance, the survey crew shall complete the checking work of the formwork, embedded parts, etc., each construction team shall carefully complete the first inspection and fill the form, the quality inspector shall strengthen the inspection of the construction process, and carry out quality evaluation after the final inspection of the unit engineering is acceptable. The acceptance organization flow to see Figure C.2.

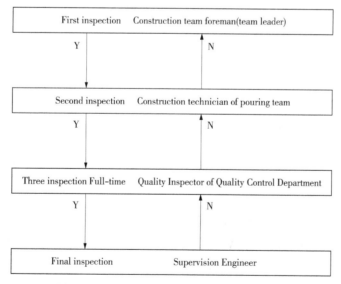

Figure C.2 *The acceptance organization flow*

C.7.3 Inspection Items for Foundation or Concrete Construction Joint Treatment

1. Foundation base
2. Surface water and groundwater
3. Rock cleaning
4. Surface roughening
5. Placement surface cleaning
6. Placement surface ponding
7. Mortar delivery pipe erection and connection
8. Rain-proof, heat insulation facilities are available

9. Vibrating and leveling tools allocation
10. Measuring, inspection, re-check data

C.7.4 Inspection Items of Formworks

1. Formworks and supports material quality
2. Stability and rigidity of formwork and supporting structure
3. Height difference between neighboring tow panels of formwork
4. Local unevenness
5. Surface cement mortar binding
6. Application of form release agent on surface
7. Joint gap
8. Deviation of formwork-erection line from design line
9. Reserved hole, cavity dimension and position deviation
10. Measuring, inspection, re-check data

C.7.5 Inspection Items of Reinforcement

1. Batch No., steel grade, specification
2. Reinforcement surface treatment
3. Local deviation of protective layer thickness
4. Local deviation of main reinforcement spacing
5. Local deviation of stirrup spacing
6. Local deviation of distributing reinforcement
7. Rigidity and stability after installation
8. Weld surface
9. Weld length
10. Weld heighte
11. Welding test resulte
12. Mechanical connection mode
13. Mechanical connection test result

C.7.6 Inspection Items of Waterstop, Expansion and Contraction Joint

1. Geometric dimension of metal waterstop strip and plastics waterstop strip
2. Overlapping length of metal waterstop strip and plastics waterstop strip
3. Installation deviation
4. Foundation inserting part
5. Asphalt applying
6. Bituminous felt affixing
7. Layout prefabricated asphaltic felt
8. Asphalt well, column installation
9. Asphalt heating panel

C.7.7 Inspection Items of Embedded Parts

1. Specification of embedded parts
2. Surface of embedded parts
3. Position of embedded parts
4. Reliability of embedded parts installation
5. Connection of embedded pipes

C.7.8 Installation of Precast Concretes

1. Overall dimensions and intensity of concrete precast parts shall be in compliance with design requirement.
2. Type and installation position of the precast concrete shall meet the design requirement.

3. When installing the precast concretes, the connection at the location where its bottom contacts with the member shall meet the design requirement.
4. The fabrication of precast concretes for placing main construction shall be constructed according to the mix proportion issued by the Test Center, and inspected and sample inspected by the Test Center, the acceptance shall be carried out before delivery, only the eligible product can be used after delivery.

C.7.9 Grouting System

1. The materials, specification and dimension of the embedded parts in the grouting system shall meet the design requirement.
2. The embedding position shall be accurate and fixed, and connected reliably.
3. The embedded pipeline shall be unblocked.

C.7.10 Issuing of Acceptance Conformity Certificate and Inspection in Construction

1. The full-time quality inspector shall ask the supervision engineer for acceptance only after the inspection of all items in Section 2 is eligible, after the acceptance is passed, the supervision engineer shall issue the "approval to placing" conformity certificate.
2. The conformity certificate is in triplicate, one copy is kept by the supervision engineer, two copies are submitted to the Quality Control Department for the record of completion acceptance.
3. If the placing conformity certificate is not issued, the placing of concrete is strictly forbidden, otherwise, it will be dealt with for serious violation of regulations.
4. The dam placing, formworks, requirement, grouting pipeline, pre-fabricated members, etc., shall be inspected in sections, those that continuous construction are required in roller compacted concrete construction, their quality requirements and inspection and acceptance procedure are same as that of Section 2 in this Chapter.
5. All acceptance items mentioned in Section 2 of this Chapter shall be put on duty and carefully protected during roller compacted concrete construction. Whenever any abnormality is observed, carry out careful inspection and treatment, immediately report it to the placement surface commander if the damage is serious, the placement surface commander shall inform the operation team to take prompt measures.
6. In the roller compacted concrete construction, the full time quality control personnel for each shift on the placement surface includes 1 full time quality inspector from Quality Inspection Department, 3-4 test man from Test Center and 1 placement surface commander, each quality control person shall coordinate with each other, reflect the problems occurred in the construction to the placement surface commander as quickly as possible, which shall be handled solely by the warehouse placement surface commander. When the placement surface supervision engineer or full time quality inspector observes any problem, the placement surface commander shall follow the opinion from the placement surface supervision engineer or full time quality inspector, if there is any different opinion, report it to the superior leadership.

C.8 Concrete Mixing and Management

C.8.1 Management of Mixing

1. The mixing plant shall be fully responsible for the mixing production and quality of the roller compacted concrete. The Test Center shall be responsible for fully monitoring the concrete mixing quality, carry out

sampling inspection and forming according to the stipulation in 11.3.1.

2. In concrete mixing production, each mixing plant shall have one chief on duty, one quality controller and one mixing foreman. Enough personnel shall be allocated on posts such as batching operation, mixing layer monitoring, discharge monitoring, concrete sign issuing, etc., so as to guarantee the mixing production can carry out normally. The Test Center shall have 6 quality control personnel in the mixing plant as a minimum for each shift (3 persons for each of left and right banks).

3. To assure the continuous production of the roller compacted concrete, the operator on duty in mixing plant test center have to stick to their posts, be carefully responsible and fill the quality control original record, and do a good job on site shift change work.

4. The mixing plant and Test Center shall cooperate closely, work together to ensure good quality, deal with the quality problems occurred in concrete mixing and production through consultation in time, in case of any disagreement, the handling opinion of the Test Center shall be taken as final.

C.8.2 Concrete Batching

1. The mixing plant is responsible for the accuracy inspection of the weighing equipment in the mixing plant, the Quality Control Department, Engineering Technology Department and Equipment Department are responsible for joint inspection and acceptance. The weighter that weighs the constituent materials of the roller compacted concrete shall have its accuracy inspected before the operation starts, the accuracy of the weighing equipment shall meet relevant provisions, the machine can be turned on only after it is confirmed to be normal.

2. Before turning on the machine in each shift (including the change of charger sheet), carry out calibration according to the charger sheet issued by the Test Center, the mixer can be turned on only after no error is found in the checking by the quality control personnel of the Test Center. The right to adjust the water consumption belongs to the quality controller of the Test Center, without the consent of the on-duty quality controller, no one may change the water consumption at will.

3. Evaluation criteria the material weighing error shall not exceed the range mentioned below (amounts by weight):

    Water, cement, fly ash, additive: ±1% Coarse and fine aggregate ±2%. When the bigger range of fluctuation frequently occurs and the quality is not assured, the operator shall report timely and find the causes, when necessary, shut down the mixer temporarily, immediately carry out inspection, eliminate the fault, the machine can be turned on after checking.

4. The materials for roller compacted concrete shall be mixed sufficiently and uniformly, and meet the requirement of working temperature in construction. Its batch charging sequence is sand → cementing material → water → additive → small stone → middle stone → big stone. The mixing time is 80 s for III gradation and 70 s for II gradation.

5. During the mixing of concrete, the quality controller of mixing plant in the Test Center shall strengthen the patrol inspection about the concrete quality at the mixer outlet, whenever any abnormality is observed, find out the cause and deal with it in time, it is strictly forbidden for the out-of-spec concrete to be placed. It shall be disposed as waste if any of following condition is met: a. Raw materials without sufficiently mixed; b. VC value is bigger than 12 s; c. The density requirement cannot be met due to very poor concrete mixture uniformity.

6. The operator on duty in mixing plant shall observe the binding condition of grout on the mixer blade during the mixing, if the binding is very serious, it shall be cleaned in time. Before shift handover, the ticking substance in the mixer shall be removed.

7. The service shall be carried out timely when water leakage, liquid leakage, cement leakage and electronic weighter shifting occur during the mixing, shut the mixer down if the concrete quality is seriously affected.
8. The production staff and the quality controller in the mixing plant shall carry out shift handover on the post at site, no production interruption is allowed due to shift handover.
9. For the control of VC value at the outlet of the mixing plant, dynamic control shall becarried out according to the climate and the value of loss in transit within the design range of the mix proportion, if it exceeds the adjustment range of the mix proportion design, adjust the water consumption under the condition that the $W/C+F$ is kept unchanged, the adjustment of the placement surface VC value is determined by the commander, and the personnel of the placement surface test center inform the mixing plant of execution.

C.9 Concrete Transport

C.9.1 Transport by Dump Trucks

1. The driver is responsible for the relevant work during the dump truck transportation, before and after each block placement, rinse the carriage to keep it clean, the roller compacted concrete shall be covered with awning during the transportation in high temperature season as required, the quality inspector and placement surface commander are responsible for inspecting the condition.
2. When the dump truck is used to transport the concrete, the road for vehicle travels shall be flat.
3. Before placing of the block, the Production Headquarter is responsible for the construction of access to the placement and inspecting the condition of other roads, the verification shall be arranged timely when any problem is found. The flushing workers shall be responsible for rinsing the tires clean by truck washing table or high-pressure water gun before placing from the dump truck and prevent any water from being brought into the placement surface. The quality inspector shall be responsible for inspecting the truck washing condition.
4. When the concrete is loaded to the truck, the driver shall obey the command of the discharging personnel. When discharging from aggregate bin to the truck, multi points discharge shall be adhered to, after it is fully loaded, the sign shall be hung on the cab, indicating that the loaded concrete type from the mixing plant, any truck without the sign is not allow to drive away from the mixing plant and enter the placing block. A sign issuing post shall be set in the mixing plant, the person in charge of the sign issuing shall issue signs to each transportation truck, the person in charge of the sign issuing is the placement team member.
5. The drivers shall be responsible for keeping the truck transporting concrete on the placement surface to be clean, strengthening the maintenance and service, maintaining good vehicle condition, free from oil or water leakage.
6. The dump truck is used to transport the roller compacted concrete for the placement surface, the dump truck shall be rear discharging, the dump truck shall travel smoothly on the placement surface, and the traveling speed shall be strictly controlled. No matter the vehicle is empty or loaded, its traveling speed must be controlled within 10 km/h, the drive route should avoid the areas with mortar or cement paste spread, any operation that may damage the quality of roller compacted concrete shall be avoided, such as sudden braking, sharp turning, etc. . For those who do not listen to the advice, the full-time quality inspector and placement surface commander have the right to punish them in accordance with the relevant provisions, and those who do not change repeatedly will be expelled from the site.

C.9.2 Operation Management of Full Pipe Chute and Negative Pressure Chute

1. The installation of full pipe chute and negative pressure chute shall meet the design requirement so as to assure the vacuum degree of the chute.
2. Inspect the lining plate and compound soft skin at regular interval (daily), repair any damage when it is found or replace it if necessary. The chute operator on duty is responsible for the regular inspection, the quality inspector is responsible for inspecting each block once before placing.
3. The chute discharge controller shall strictly control the discharge to the chute aggregate bin, the discharge shall be carried out according to the size of the truck capacity, any operation at will is not allowed.
4. Before final setting of the roller compacted concrete after the completion of the placing, if it becomes to carry out maintenance of the chute by flushing, a water tank is placed at its outlet section for receiving water to prevent the wash water from entering the block.

C.9.3 Horizontal Belt Conveyor Transportation

1. The installation of horizontal belt conveyor shall meet the design requirement to assure the conveying angle of the horizontal belt conveyor.
2. Inspect the carrier rollers and the belt at regular interval (daily), repair any damage when it is found or replace it if necessary. The belt conveyor operating personnel onduty is responsible for the inspection of each shift, the quality inspector is responsible for inspecting each block once before placing.
3. The belt conveyor discharge controller shall strictly control the discharge to the belt conveyor, any operation without authority is not allowed.
4. After the completion of the placing, the mortar on the belt conveyor shall be completely removed by using a scraper, and necessary maintenance to other members shall be carried out.

C.10　Construction Management inside Block

C.10.1　Placement Surface Management

1. The placement team is responsible for the placement construction of the RCC, one placement surface commander is set for each shift, two command conductors, the placement surface commander will fully arrange, organize, command and coordinate the rolling construction, and be responsible for the quality, progress and safety, the placement surface commander accepts the technical guidance of position engineer and the quality supervision of full-time quality inspector. When the commander encounters any technical problem beyond his ability, timely reflects it to the position engineer so as to have it solved as quickly as possible.
2. The placement surface commander command conductor, quality inspector, Test Center site inspector shall wear a work permit on duty. The quality controller and Test Center site inspector shall carry out inspection and sampling inspection to the construction quality, and fill the record as required.
3. Except the commander and the placement surface conductor, others shall not direct production in placement surface. Project Department Leader, position engineer, full-time quality inspector, etc. find problem and make handling decision, which shall be implemented by the commander, the placement surface commander shall carry out according to the suggestion, if there are different opinions, they shall report to the superior leadership. If the Test Center site inspector observes any problem, timely reports it to the full-time inspector

or placement surface commander, helps to find out the causes and make a detailed record.
4. All personnel involved in RCC construction shall follow the on-site shift handover system, stick to their job, make the construction record as specified, when it becomes necessary to leave the post temporarily, the consent of the commander shall be obtained.
5. To keep the placement surface clean, all personnel are prohibited from throwing any foreign matters onto the placement surface (such as cigarette ends, matchstick, shredded paper, etc.).
6. The management of placement surface shall be carried out by strictly in compliance with the flow in Figure C. 3.

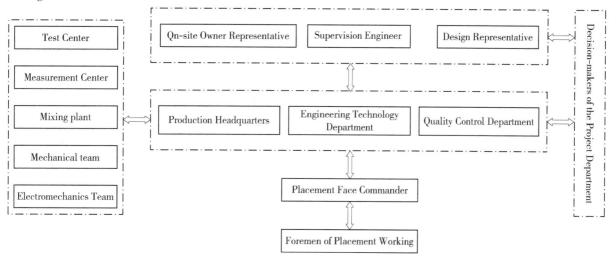

Figure C. 3 *Flow for message passing at RCC construction site*

C. 10. 2 Placement Surface Equipment Management

C. 10. 2. 1 Equipment Moves in the Block

1. Before the equipment moves in, it shall be fully inspected and maintained so as to enable the equipment in good working condition. The inspection of the equipment is the responsibility of the operator, detailed record is required to take and the inspection of the Equipment Department shall be accepted.
2. The equipment shall be fully washed before moving in the block, the sludge inside and outside of the carriage, tires, chassis, fender and frame of the truck shall be cleaned by washing, then it can enter the placement area after it is subject to drying after washing. The equipment cleaning condition is inspected irregularly by the quality inspector.

C. 10. 2. 2 Equipment Operation

1. The operation of the equipment shall be carried out according to the operating procedure. The equipment shall be operated by specially-assigned person with certificate, the operator shall take good care of the equipment and not let others use it at will, otherwise will be treated according to the severity and repelled from the site.
2. When the truck travels on the RCC placement surface, the driver shall be responsible for avoiding the operation that impairs the quality of concrete such as sudden braking, sharp turn and so on, the discharging from the truck, the command from the conductor shall be followed, the conductor shall carry out instruction through the modes of flag and whistle.
3. The construction equipment shall utilize the access to the construction placement to leave for fueling, if the fueling is done on the placement surface, the measures such as carpet laying, etc. , shall be taken to protect the placement from being contaminated, the quality inspector shall be responsible for supervision and

inspection.

C.10.2.3 Equipment Parking

1. The parking of equipment on the placement surface is arranged by the commander, the equipment shall be parked in good order, the operators have to obey the command unconditionally, the equipment that is not in use shall be withdrawn from placement surface.
2. All equipment, inspection instruments and tools on the construction placement surface shall be parked at the location designated by the site commander when they are not working temporarily or the position where has no influence on the construction.

C.10.2.4 Equipment Maintenance

1. The equipment shall be serviced and maintained by the operator at regular interval, the detailed maintenance records are required, any equipment failure shall be reported to placement surface commander and Equipment Material Department in time.
2. The equipment to be serviced shall run out of the placement surface from the access for concrete placing, or it is hoisted from the placement surface, if the repair and service has to be carried out on the placement surface, the carpet shall be laid to protect the placement surface from being contaminated.

C.10.3 Management of Construction Personnel on Placement Surface

C.10.3.1 The Qualification Examination of Personnel Allowed to Enter the Placement Surface

1. No visit on the placement surface is allowed without the permission from the Project Department, only the workers involved in placement surface construction (including employer's representative, supervision engineer, design representative) can enter the placement surface. The persons entering the placement surface has to wear the placement surface entrance permit (or work permit), non-construction personnel who has to enter the placement must have the approval of the General Affairs Department.
2. Anyone entering the RCC placement surface has to clean the mud on the shoes, throwing any foreign matters onto the placement surface (such as cigarette ends, matchstick, shredded paper, etc.) is forbidden, the warehouse commander or full-time quality inspector shall have the right to impose a corresponding fine on those who do not listen to the advice, regardless of their position.
3. The walking route or stop position of the non-operating personnel entering the placement surface shall not have any influence on normal construction.

C.10.3.2 The Use of Placement Surface Entrance Permit

1. The placement surface entrance permit is uniformly handled and issued with numbering by the General Affairs Department.
2. The placement surface construction personnel has to work with certificate, no one without a certificate is allowed to enter.
3. The placement surface entrance permit (work permit) shall be used by special persons and shall not be given to others.
4. Personnel on duty of the concrete pouring unit are responsible for verifying the permit before entering at site.

C.10.3.3 Regular Inspection of Placement Surface Personnel

1. The site guard on duty has to inspect the persons on the placement surface, any non-construction personnel shall be cleared from the placement surface whenever he/she is found.
2. The placement surface commander can inspect the workers on the placement surface at any time to see if any watch keeper at various construction position leaves without permission.

C. 10. 3. 4　Training and Education of Construction Personnel

1. Construction personnel must be trained and qualified to participate in the construction, special work has to work with certificate.
2. The constructor technicians shall be trained at regular interval so as to improve the technical level.
3. The training job is organized and implemented by the General Affairs Department.

C. 10. 4　Discharging

C. 10. 4. 1　General Principles

The large block face thin-lift continuous placement is used for RCC, the rolling layer thickness is 30 cm, the spreading thickness is from 34 cm to 36 cm, if any change becomes necessary, it shall be tested at site and approved by the supervision engineer.

C. 10. 4. 2　Discharging

1. When discharging directly from the dump truck for placing, to minimize the separation of the aggregate, the double point overlapping discharge should be used, the dump truck discharges the concrete on the step at the spreading front edge of the pavement layer, then the concrete is pushed down from the step by the leveling machine and the displacement type leveling is carried out. The discharge shall be uniform possibly, when separated aggregate occurs at the side of the pile, it shall be spread evenly on the unrolled concrete surface by workers or other machines.
2. The placement surface conductor shall, according to strip by strip spreading sequence of the requirement on the placement instruction drawing, direct the route and unloading location of dump truck, the driver shall obey the orders.
3. When bucket is used for placing, the instruction is given by the bucket commander, the free discharging height should not be more than 1. 5 m.
4. The distance from the edge of the dump heap to the formwork shall be not less than 1. 2 m.
5. The conductor shall strictly control the boundary between Ⅲ and Ⅱ gradation concrete, the boundary is marked by a red flag every 20 m, any negative error for Ⅲ gradation concrete after rolling is not allowed nor is greater than 30 cm, this is the responsibility of the quality inspector.
6. the mixing plant control room operator, quality controller and the test personnel shall make joint checks at the mixer outlet, placing of any nonconforming RCC is strictly forbidden, once the nonconforming concrete enters the block, it shall be handled by the placement surface commander.

C. 10. 5　Leveling

1. The surveyor is responsible for line-drawing and setting out on the surrounding formworks every 20 m, mark the pile number, elevation and leveling control line, which is used for control the spreading thickness, etc. ; one point every 20 m shall be set out for the boundary of between different concretes in Ⅱ and Quasi -Ⅲ gradation areas, the placement team shall insert red flag according to the setting-out points for identification.
2. The leveling machine is used for placement leveling, the damage of the rolled concrete by tracks while traveling is not allowed. The manual operation is used for assisting the discharging and placement leveling operation at the boundaries and other specified or approved location, the manual operation is instructed by the placement surface commander, and the full-time quality inspector is responsible for overseeing.
3. The placement leveling operation is the responsibility of the leveling machine operator, the leveling direction of the RCC shall meet the requirement of placement instruction drawing, the spreading shall be uniform, and leveling once for each rolling layer. The quality inspector shall carry out inspection by pulling line according to the leveling line drawn nearby. The thickness of each leveling layer shall be 34 – 36 cm more or less, if

any location is found through inspection that the specified value is exceeded through inspection, carry out leveling again, the location which is locally uneven shall be spread out manually.

4. The concrete discharged from the truck shall be leveled immediately so as to enable the concrete is discharged on the step at the leading edge of paving layer spreading, and meet the requirement that the mixture is charged and mixed till completion of rolling on the placement surface within 2 h. The leveled concrete surface shall be flat and pit-free.

5. The aggregate concentrated on both sides a slope toe during the leveling shall be scattered on the strip manually. The implementation of such job is to be arranged by commander on duty of the placement surface, the quality inspector is responsible for inspection and supervision.

6. If any local aggregate concentration with the layer surface after leveling, it can be treated by spreading fine aggregate manually.

C. 10. 6 Rolling

1. Rolling equipment: the selection of the type of vibration roller shall take into consideration of the rolling efficiency, exciting force, drum size, vibration frequency, amplitude, traveling speed, maintenance requirement and reliability of operation. For concrete rolling on special location, small vibration roller or heterogeneous concrete placing can be used.

2. Various types of rolling engineering proposed to use in plan shall be controlled according to various rolling parameters determined by rolling test approved by the supervision engineer and that the design requirements of the concrete are met.

3. The roller operator is responsible for the roller operation, after each paving layer is spread, carries out rolling according to the specified number of times of rolling, if Hummer double drum vibration roller is used, the number of times of rolling is "2 + 8 + 2", that is, 2 times of rolling without vibration, 8 times of rolling with vibration and 2 times of rolling without vibration. The roller operator has to count the number of times of rolling during the rolling on each strip, no change without authorization is allowed. The placement surface commander and full-time quality inspector can make a decision whether the number of times of rolling is increased or not according to the surface bleeding condition and test result of nuclear density gage. The full-time quality inspector is responsible for the random inspection of the rolling operation. The rolling direction shall meet the placement instruction drawing, the overlap width between the rolling operation strips is 20 cm, the overlap width at the end should be 100 cm.

4. The placement surface inspector of the Test Center is responsible for the rolling test results, after completion of each rolling operation, inspect the compactness and volumetric weight according to grid points, the nuclear density gage is arranged in a grid of $100 - 200$ m^2 and not less than 3 points for each rolling surface in each placement unit, the sum of compactness and volumetric weight of the placement surface RCC shall meet the requirements of evaluation terms in 8. 8. 1 or the design notification document. When the inspection result is not pass, carry out re-inspection immediately. When it is confirmed that the result is lower than the specified index, immediately report to the placement surface commander and help him to find out the cause, and take the treatment measures such as make-up rolling, the roller operator shall unconditionally obey the make-up rolling required by the placement surface commander.

5. The roller operator is responsible for controlling the traveling speed of the roller operator within the range from 1. 0 – 1. 5 km/h. The full-time quality inspector shall inspect whether the traveling speed of the vibration roller meets the requirement or not, at least 2 times in each shift is required. If the over-speed is found, point it out to the roller operator immediately and ask for correction.

6. The placement surface commander is responsible for controlling the intermission time of the rolling layer: the

permissible intermission time between RCC layers for continuous lift placing (That refers to the time duration starting from mixing and adding water of the lower layer concrete mixture to the completion of rolling of the RCC) shall be controlled within 6 – 8 h, the setting time of concrete is not allowed to exceed as maximum, if the concrete intermission time is exceeded, pave 1.5 cm thick cement mortar or 5 mm cement fly ash grout. As a result, the time from mixing of the RCC mixture to completion of rolling shall not be more than 2 h.

7. The decision for water compensation rolling is made by supervision engineer and quality inspector, which shall be arranged to implement by the placement surface commander, when no bleeding after long time pressing occurs if there is influence by the factors such as air temperature, wind power and so on, the VC value is too high due to evaporation of water from the rolling layer surface, the water compensation rolling can be carried out by using the water tank equipped with the roller to spray the rolling drum surface at the second times of rolling with vibration, for the degree of water compensation, the moisture level after rolling prevails, excessive water compensation is not allowed. The state of flowing of layer surface cement mortar with the vibration roller is not allowed when vibration roller is rolling.

8. When the compactness is less than the design requirement, immediately inform the roller operator to carry out rolling according to instruction, if the requirement is still not met after makeup rolling, it shall be treated by excavation. During the rolling, the placement surface quality controller shall keep good construction records, the quality controller shall make quality control record.

9. The direct approach rolling is used around the formworks and bed rock, the 50 cm to 100 cm zone close to it or complicated structure around where the rolling is impossible is directly placed with GEVR.

10. If the rolling is not compacted at locations where concrete cracks and surface aggregate concentrates when rolling, the quality inspector shall ask the commander to excavate manually, pave it again and roll it till the design requirement is met.

11. The VC value on the placement surface is controlled between 3 – 12 s in principle, the site Test Center shall properly adjust the VC value at the mixer outlet according to climate conditions such as air temperature, day or night, rain or shine, humidity, etc., criteria for rolling is that concrete layer surface reaches full bleeding after completion of the rolling, slight elastic feeling when people walk on it, and there is no aggregate concentration on the surface.

12. The surface whitening caused by too long stacking time of RCC shall be treated by grout paving or mist spraying.

C.10.7 Construction of GEVR

1. The foundation normal concrete of the riverbed area should have an intermission time of 3 – 7 d after completion of the placing, then RCC can be spread on it, however, the long term intermission shall be avoided, the longest intermission time should not be more than 28 d, its surface layer shall be treated as construction joint. The foundation bedding GEVR close to the bank slope shall be cast before the RCC of main body, in the RCC construction, an GEVR shall be vibrated by adding grout within an area 50 cm foundation bedding GEVR close to the bank slope.

2. The GEVR is the concrete that is to pave cement fly ash grout before and after spreading the RCC mixture to make it be abnormal, which is compacted by the operation of using normal concrete vibration method, and meet design requirement. The preparation of cement fly ash grout used for GEVR is determined through tests, the grouting mode adopts spreading by using manual hand-held bucket, it is required that the length and width for spreading each bucket of grout is fixed, the spreading shall be uniform. The quantity of grout spreading is 5% of that of the concrete, the grouting method is used (that is grouting after hole-inserting),

the ϕ 100 mm high frequency vibrator or ϕ 70 mm flexible shaft vibrator can be used for vibration.

3. The GEVR is mainly used for surrounding area such as bedrock surface of two banks, dam upstream and downstream face formworks surface, expansion joint, upstream and downstream waterstop material laid places, gallery and elevator shaft as well as the places where the rolling by vibrations roller cannot reach, it can also be used at the location where the normal concrete and RCC transits. The GEVR and RCC can be cast synchronously or crossly, the vibration or rolling shall be completed within the time specified for these two kinds of concrete.

4. The placement surface commander is responsible for arrange the time of GEVR construction, the construction mode of RCC before GEVR should be used. At the place where the GEVR transits to the RCC, the vibrator is used to carry out vibration toward the RCC, after the overlapped place is rolled, the high frequency vibrator is used to insert vertically and vibrate so as to allow them is blended and compacted. If the mode of GEVR before RCC, the vibration roller shall complete the rolling by riding on the joint.

5. For construction of GEVR at the upstream face 50 cm GEVR area, bank slope, gallery surrounding and downstream face formworks nearby, artificial pore-forming by ϕ 20 mm hole making device on the spread RCC, the holes are formed in plum-shaped arrangement, the hole interval is about 30 cm, and the hole depth is 20 cm. Then, the middle and top grouting mode is used at fixed quantity and fixed number of holes by manual hand-held bucket, the grouting quantity is controlled at 5% of that for concrete, the vibration is carried out 15 minutes after grouting, the vibration time is 30 s and more. For the convex part where the GEVR overlaps with the RCC, the large vibration roller is used to roll flat the overlap location in principle, if the use of large roller is really impossible, the small vibration roller can be used for rolling.

6. For the location where the abnormal concentration occurs such as the bank slope platform foundation surface, etc., the GEVR can be directly mixed in mixing plant, after two layers (greater than 60 cm) of GEVR are placed for the platform, the RCC construction is used from the third layer on.

7. For the contact joint of heterogeneous concrete, the large vibration roller can be used to carry out the rolling by riding on the joint.

C.10.8 Placement Surface Quality Management

C.10.8.1 Quality Control

The Quality Control Department is responsible for the quality control jobs such as quality inspection, supervision, evaluation, etc., the on-site laboratory is responsible for the quality control according to Table C.2.

Table C.2 Dam RCC placement surface construction quality inspection and frequency table

Item		Control index	Test (inspection) method	Times of test (inspection)	Purpose of test (inspection)
Test Items	Compacted volumetric weight	Design value	Test by nuclear density-moisture gage	At least 1 measuring point for 100 to 200 m^2, not less than 3 points for each layer in each unit	Control construction quality
	VC value	3 – 5 s	Sampling test	Not less than twice per shift in each unit	Control construction quality
	Compressive strength	Design value	Forming	Sampling of each unit is equivalent to 5% – 10% of sampling quantity at mixer outlet	Control construction quality

1. Evaluation criteria: Compacted volumetric weight 85% measured value of compacted volumetric weight for each layer of RCC is not less than design value.

    Relative compactness: the measured value of relative compactness shall reach 98.0%.

    Depending upon the progress of the project, design notice shall prevail in case of any other notice from the designer.

2. Measures: During the construction, the Test Center shall dynamically control the *VC* value at the mixer outlet of mixing plant according to the loss range of *VC* value under conditions of different weather, different air temperature and humidity, adjust it timely according to the construction site condition, the Quality Control Department shall strictly control 2 h of rolling time limit for RCC and the time interval between two rolling layers. When the relative compactness of the RCC fails to meet the specified requirement, inform the roller operator for makeup rolling, if the requirement is still not met after rolling, immediately notify the engineer, find out and analyze the cause, the construction can be continued only after the problem is resolved.

3. Division of units: The left bank non-overflow dam section is a unit, right bank non-overflow dam section is a unit, powerhouse dam section is a unit, the overflow dam section is a unit, the left and right bank sand-washing port dam section work is a unit.

C.10.8.2 Water Stop

1. The requirement of the material for expansion joint upstream and downstream water stop strip shall comply with relevant provisions in *Specification for Hydraulic Concrete Construction* DL/T 5144—2001.

2. The placement team is responsible for the construction of water stop materials, there shall be measuring and setting out data at the position (provided by Survey Center), the setting out is required to be accurate. During the GEVR construction at the place where the water stop material is buried, proper measures shall be taken to support (such as fixing by reinforcing steel bracket) and protect these water stop materials, the vibration shall be carefully carried out, if there is any damage to the water stop material, it shall be repaired, the large aggregate at this place shall be removed manually so as to avoid any water seepage channel from producing. The quality inspector shall take the construction of water stop material as the important quality control item and carry out inspection and supervision.

C.10.8.3 Embedded Parts

1. The person in charge of embedding is responsible for the jobs such as embedding, inspection and record, etc., of embedded parts. When the concrete construction is carried out at places where the pipeline, observation instrument and other embedded parts are buried, the embedded parts shall be protected properly and the construction is carried out strictly according to the design requirements.

2. The laying of observation instrument and electric cable in the RCC shall adopt cutting method, that is after the previous layer of concrete is compacted, cut the slots to install the instrument according to the position of instrument and leading lines, fill the concrete back and carry out vibration manually after the inspection is passed, then the next layer can be spread and rolled.

3. For the instrument that does not have direction requirement such as thermometer and so on, the cutting is okay if the depth is sufficient to cover the instrument and cable. For the instrument with direction requirement, lay it as deeply as possible, and a layer of mortar is paved at the bottom of the slot, at least 60 cm depth of GEVR is backfilled on the top. The backfill job shall be completed before initial setting time of the concrete. The aggregate that is bigger than 40 mm shall be removed from the backfill materials, the small plate vibrator is used to vibrate carefully.

4. A margin of 0.5 – 1.0 m outgoing cable shall be reserved in the vicinity of the buried point; the cable that runs vertically or diagonally upwards shall be laid horizontally into the corridor and then extended upward (or outward).

5. After the instrument and cable are buried, the detailed construction process shall be recorded, draw the actual embedding diagram timely and submit it to the engineer. Specially-assigned person shall be assigned by the construction unit in charge of the observation project is responsible for backfilling of concrete for installation and burying of the instrument, which shall be properly protected, if any abnormal change or damage is found, the contractor shall report to the supervision engineer timely and take remedial measures in a timely manner, the observation project construction shall not affect the construction of RCC.

C.10.8.4 Treatment of Construction Joints

1. The entire RCC block must be cast in such a way that it condenses into a single block without interlayer weakness or seepage channels.
2. Cold joint and construction joint on the joint surface must be treated, the construction can be continued only after the treatment is acceptable.
3. In case of the treatment of joint surface, the methods such as high pressure water flushing and so on can be used to remove the floating sand and loosened aggregate on the concrete surface (criterion is that sand grains and small stones are exposed). After the treatment is acceptable, a layer of 1.5 cm mortar layer is scraped and spread first (strength classes of the mortar is one class higher than that of the RCC), then immediately spread the RCC on it, and complete the rolling before the initial setting of the mortar.
4. The concrete flushing roughened time can be determined based on factors such as season, concrete strength, equipment performance and so on and by site tests, it is not allowed to roughen in advance.
5. The paving layer surface of the RCC shall reach basically same elevation when completing the placement, if any interruption is caused by change in construction plan, rainfall or other causes, the spread concrete shall be rolled timely, a slope of 1:4 should be rolled on the concrete surface where the spreading stops.
6. The placement surface commander is responsible for keeping the joint surface to be clean and wet during the placement, any contaminated, dry and ponding zone is not allowed. Minimize the secondary pollution, the mortar should be spread strip by strip in different stages. The contaminated joint surface shall be flushed clean by high-pressure pump before spreading the mortar.

C.10.8.5 Joint Forming

1. The placement surface commander is responsible for arranging the joint cutting time, the cutting by joint cutting machine adopts "cutting after rolling", the joint cutting depth is 2/3 of the spreading thickness at that layer, the joint forming area shall not less than 60% of the design surface, the joint is filled by using colored strip.
2. Joint forming can be done by selecting forming time, controlling the joint space, direction and slope according to the requirements approved by the engineer so as to assure the forming quality.

C.10.8.6 Layer Surface Treatment

1. The placement surface commander is responsible for the layer surface treatment job: except that the RCC layer surface of upstream impervious body is treated according to Item 3 in this section, the layer surface exceeding the initial setting time shall be paved with 1.5 cm cement mortar or 5 mm cement fly ash grout, then continue to spread the upper layer concrete.
2. The concrete layer surface that exceeding the final setting time is cold joint, there are two modes of cold joint surface treatment. If the interval time is 24 h and less, treat by paving mortar; If the interval time is more 24 h, treat as construction joint surface.
3. The special mortar spreading worker is responsible for implementation of the mortar spreading. Within the scope of the upstream II gradation impervious area, each rolling layer shall be paved with 5 mm thick cement fly ash grout. The quality inspector of placement surface test center shall inspect the specific gravity of the prepared grout, and provide the data for specific gravity of the grout at least once a shift.

4. After the heterogeneous concrete is initially set, the upper layer can be cast only after removing surface skin and paving mortar or grout.

C.10.8.7 Drain Hole Construction

The specification of the dam body drain hole is $\phi 200$ mm, the construction adopts embedded blind pipe to form holes.

C.10.9 Grout Paving

1. The grout is prepared by cement + fly ash + water reducing agent, the cement fly ash grout grouting construction is mainly used for the treatment of upstream II gradation impervious area, various kinds of GEVR, interlayer and slope toe, the grouting construction of various kinds of GEVR. The preparation of the cement-fly ash grout shall be done according to the charger sheet issued by the Test Center, the charge mixture dose is required to be accurate in measurement and mixed uniformly. The Test Center shall supervise the quality of prepared grout.
2. The whole laying process of cement-fly ash grout is arranged by the placement surface commander, who shall inform the person on duty of paste making team to carry out preparatory work of mortar making 1 h before spreading operation, it shall be assured that the operation can start whenever is required.
3. Before paving the cement-fly ash grout, the placement surface commander is responsible for monitoring that, the spreading area is clean and ponding free, to avoid the cement-fly ash grout from occurring precipitation problem.
4. It should not be too early to spread cement fly ash grout in the II gradation impervious area, it shall be done in sections before discharging for this strip, it is not allowed to keep uncovered with concrete for a long time after spreading the cement paste.

C.10.10 Inclined Layer Paving and Rolling

1. The placement team shall draw detailed placement instruction drawing for each block according to the construction measures, design documents and drawings, and submit to Project Technical Department for approval. The placement surface commander, conductor, quality inspector, etc. shall get familiar with the main points of placement before pouring, and organize the implementation according to the requirement of the placement instruction drawing.
2. The placement team technician is responsible for marking out and setting out every 20 m, marking the pile number, elevation and leveling control line on the surrounding formworks according to the requirement of the placement instruction drawing, which is used to control the thickness of the slope spreading layer.
3. Treatment of slope toe of inclined layer rolling.
   a. Spread mortar for a width of 3 m at the slope toe of the rolling layer set out according to 1:10 to 1:15 gradient (the slope toe width can be adjusted when the gradient changes). The spreading length of the mortar corresponds to the rolling strip width.
   b. The vibration roller will roll the slope toe at the same time, the number of times of rolling is carried out as per mode of "2 + 8 + 2".
   c. The slope toe RCC outside the slope toe setting out shall be excavated before next layer starts, the slope toe cutting height is that when it is cut to mortar, the initially set concrete will be disposed as waste.
4. Discharging Requirements is same as those in Section 4 of Chapter 8.
5. Leveling Incline from downstream to upstream or from one bank to the other, the slope ratio is 1:10 and more, other requirements are same as those in Section 5 of Chapter 8.
6. Other requirements of rolling are same as those in Section 6 of Chapter 8.

7. To reduce the secondary pollution on the placement surface, the old concrete strip at the slope toe of the inclined layer shall be flushed clean by high-pressure water of high-pressure pump before paving mortar, the mortar is spread in sections, the ponding should be absorbed clean by using sponge, etc. before mortar is spread.

C.11 Construction under Special Climatic Conditions

C.11.1 Rainy Day Construction

1. The rainproof materials should be well prepared, rainproof materials should be slightly larger than the placement surface area and should be placed on site. The precipitation measuring job shall be strengthened for the construction in rain day, the full-time quality inspector is responsible for measuring the precipitation. When the precipitation is close to 3 mm/h, report the measured results to the headquarter and placement surface commander once every 60 minutes.

2. When the precipitation is more than 3 mm/h, no placement shall be done; if an intensity of the 3 mm/h is encountered during the placement, stop the mixing, and the placed concretes shall be spread and rolled as quickly as possible and covered properly. It is required to use plastic sheeting to cover the entire fresh concrete surface, the coverage of the plastic sheeting shall adopt the overlap method, the overlap width shall not be less than 30 cm, which can prevent the rain water from flowing into the concrete surface from the overlap position. The rain water is intensively drained to outside of the dam, the manual treatment shall be used for the individual puddles that cannot be discharged automatically.

3. After the construction suspension order is issued, all personnel related with RCC construction shall stick to their posts and be ready to return to work at any time. The placement surface commander shall first issue the construction suspension order to the mixing plant, and inform the Production Headquarter, Quality Control Department and Project Technical Department.

4. When the rain stops or the precipitation is less than 3 mm/h, if the duration is 30 min or more and the concrete on the placement surface is not initially set, the construction can be restored. The restore of construction after raining can be done only after the treatment is completed and approved by the supervision engineer, and following jobs shall be done:

    a. Properly increase the VC value at the outlet of the mixer, the commander will inform the Test Center to carry out adjustment according to the condition of construction in blocks;

    b. The driver of transportation equipment is responsible for completely removing the ponding in the concrete transportation equipment that is parked in the open air;

    c. The site test center is responsible for adjusting the VC value at the outlet of the mixer;

    d. The quality inspector carries out careful inspection, the concrete that has seriously soaked by rain water shall be excavated.

    e. The placement surface commander organizes the removal of the ponding, first is the ponding within the range of discharging and leveling;

    f. The locations where there is exposed sand and gravel on the concrete surface subjected to rain water flushing shall be treated by paving the cement mortar.

C. 11. 2 Construction under High Temperature and Low Temperature Conditions

1. Placement temperature the casting temperature of the RCC is defined as the temperature measured 10 cm under the concrete surface after the concrete is compacted, see the table below for allowable casting temperature of dam body at various temperature control zones from January to December.
2. The average temperature for the construction of RCC should be between 3 – 25 ℃, when the daily average temperature is more than 25 ℃, the intermittent time between layer shall be reduced significantly, the measures such as protection against high temperature and sun, adjusting partial small climate of the placement surface, etc. shall be taken to prevent the surface moisture of the concrete from quickly evaporating and dissipating during transportation, spreading and rolling.

    From the end of February to the end of October each year, the corresponding temperature measures are taken for the production of the concrete according to the dam body concrete temperature control requirements, such as the measures of aggregate pre-cooling, mixing with clod water and ice, sunblock equipped with truck, high stacking of finished aggregate, VC value adjustment, use of high efficient retarding agent, spraying on placement surface.
3. When the daily average temperature is higher, it is proposed to carry out RCC construction quickly when the air temperature is low in the morning and evening (avoiding the high temperature period at noon), and set up sunblock for concrete transportation trucks.
4. When the temperature is 20 ℃ or below, the intermittent time between layers is 8 h and more, the cement-fly ash grout is spread between layers for treatment; when the temperature is more than 20 ℃, intermittent time between layers is 6 h and more, the cement-fly ash grout is spread between layers for treatment; when the intermittent time is exceeded and within 24 h, the cement mortar is spread between layers for treatment.
5. When the daily average temperature is less than 3 ℃ or minimum air temperature is less than – 3 ℃, the low temperature measures shall be taken.

C. 12 Quality Control Management

C. 12. 1 Raw Material

1. The inspection of the raw materials is carried out by the Test Center, see the raw material indoor quality control instruction drawing below for the quality and control of cement, water, fly ash, steel and additive. The inspection items and sampling frequency of the raw materials (cement, fly ash, additive, steel, aggregate) comply with " Table for RCC Raw Material Inspection Item and Sampling Trequency of a Certain Hydropower Station RCC Ham Project".
2. Evaluation criteria. Coarse aggregate oversize and sub-size content: in oversize and sub-size screen inspection, the control criteria is: oversize = 0, sub – size < 2%, fineness modulus change ≤ ±0. 2. Otherwise, adjust the mix proportion of RCC. Fine aggregate water content < 6%, water cut change ≤ ±0. 5%, aggregate surface water content < ±0. 2%. Otherwise, adjust the water consumption of RCC.
3. The sand and gravel inspection is implemented by the Test Center according to "Instruction drawing of sand and gravel yard quality control". When the aggregate gradation inspection results exceed the requirements, immediately inform sand and gravel factory and report to the engineer, the sampling inspection is carried out again. If the specimen still exceeds the specified limit, immediately report to the engineer again, this production process is considered as out of control, effective measures shall be taken to carry out adjustment.

C.12.2 RCC Mixing

C.12.2.1 RCC Mixing

1. The sampling inspection is carried out by the site test center, the concrete is mixed by the concrete mixing plant, when the concrete mixing is running, all constituent materials including cement, fly ash, various sizes of aggregates, water, additive and so on need to be controlled frequently, the aggregate weight and the water addition quantity for adjusting the aggregate humidity need to be adjusted. The daily report has to be prepared, which shall indicates that the categories and source of all cement and fly ash used today, grouping of all aggregate particles, various mix proportion required per cubic meter, natural water content of each kind of aggregate, and the weight and water consumption per cubic meter of aggregate in various design mix proportion during the running of the mixing plant.

2. Weigh. the weigher that weighs the constituent materials of the RCC shall have its accuracy calibrated before the operation starts, the accuracy of the weighing equipment shall meet relevant provisions, recheck the calibration at a regular interval of one time a month during the construction, if any change in RCC performance due to weigh error is observed in the construction, carry out recheck and calibration. The calibration of the weigher is carried out by the mixing factory.

C.12.2.2 Quality Test of RCC

The Test Center is responsible for the implementation job of RCC quality test. And, carry out quality control according to Table 3 "RCC mixing plant production quality control test items and sampling frequency table" below.

1. Evaluation criteria The VC value fluctuation range of the RCC should becontrolled within ±5 s.
2. Measure When the VC value of the RCC exceeds the specified range, adjust the water consumption of the RCC and keep the water-cement ratio unchanged, and record the results of this adjustment.

C.12.3 Inspection at Mixer Outlet

It is the responsibility of the Test Center, the inspection at mixer outlet includes VC value at mixer outlet, test of test pieces from concrete at mixer outlet, etc.. Its test frequency and sampling group quantity is carried out according to Table C.3 main Dam RCC mixing plant production quality control test Items and sampling frequency table. At the end of each month, the statistics of above test results shall be carried out according to unit engineering to form a summary table of specimen results, and the test results shall correspond to the pouring blocks.

Table C.3   RCC mixing plant production quality control test items and sampling frequency table

Description	Test Items	Sampling location	Number of times of sampling	Control index	Test purpose
Additive preparation	Specific gravity, concentration	Preparation workshop or mixing plant	1 – 2 times per shift	In compliance with preparation requirement	Control the preparation concentration:
Fine aggregate (sand)	FM, Rock powder content	Batching and mixing plant	1 times per shift	2.2 – 2.9, Rock powder 16% – 22%	Control the range of change, whenever required, find out the causes and adjust the mix proportion
	Moisture content	Batching and mixing plant	2 h/time	Gradient of water content within 4 hours does not exceed 1%	Control W/C

Continued Table C.3

Description	Test Items	Sampling location	Number of times of sampling	Control index	Test purpose
Coarse aggregate (stone)	Super-inferior size ratio	Batching and mixing plant	1 times per shift	Round hole screen: Oversize <5%, sub−size <10%	Control the range of change, whenever required, find out the causes and adjust the mix proportion
	Clay content	Batching and mixing plant			
	Small stone water content	Batching and mixing plant	2−4 times per shift	Gradient of water content within 4 h does not exceed 1%	Control W/C
Concrete mixture	VC value	Material at mixer outlet	2 h/time	Meet the value specified in charger sheet	Control batch charging error
	Air content	Material at mixer outlet	1 times per shift	Meet the mix proportion design value	Control the additive batch charging error
Concrete mixture	Compressive strength forming	Material at mixer outlet	28 d age 500 m^3 forming 1 group, design age 1 000 m^3 forming 1 group	Design index	Quality evaluation after hardening of concrete
	Concrete performance forming	Material at mixer outlet	Otherwise provisions of text	Design index	Performance evaluation after hardening of concrete
Temperature test	Water temperature	Batching and mixing plant	Once per 2 h		Temperature control requirement
	Temperature	Batching and mixing plant	Once per 2 h		Temperature control requirement
	Concrete temperature	Batching and mixing plant	Once per 2 h		Temperature control requirement

Note: 1. Once a month weigh error calibration.

2. Sampling method is in compliance with specification requirements.

C.12.4 Placement Surface Construction Quality Test

1. In the construction of RCC, the quality inspector, personnel on duty in the Test Center shall carry out the specified inspection items and test items listed in Table C.4, and keep a record.
2. In the construction of RCC, the constructors shall carry out careful construction and assure the quality according to the criteria specified in relevant specifications and this construction method.
3. The quality inspector and test personnel shall inspect and control the construction quality, if any problem is observed, have it solved through consultation with the commander in time, if the opinion is different, it shall

be executed according to thehandling suggestion of quality inspector and test personnel, the placement surface commander can reserve his opinion or report it to the Chief Engineer for adjudication.

4. During the construction of the RCC, the major quality problem must be dealt with in time without leftover, otherwise, they will be investigated for responsibility. If the fault is caused by the failure of the construction personnel without carrying out the opinions of the inspection and test quality control personnel, the construction personnel shall take full responsibility; if the fault is caused by the inspection missed by the quality inspector and test personnel, the construction personnel shall bear the responsibility for construction and the quality inspector and test personnel shall bear the responsibility for inspection.

5. This construction method authorizes the placement surface commander to have the direct administrative penalty power to the RCC construction operators, the penalties include criticism, fines and disqualification from construction (exit). The Project Department will give full support to the placement surface commander's work and punishment decisions. If the quality accident caused by the placement surface commander's command error or refusal to implement the quality inspector's rectification opinions, the quality inspector (full-time quality inspector of the Quality Department, Test Center placement surface inspector) may report to the leader of the Project Department for punishment, if any serious quality accident is caused, it shall be handled according to dereliction of duty and transferred from the position of commander of placement surface. The Project Department adopts the department director on-duty inspection system to the RCC construction, inspects the quality inspection personnel, the placement surface commander, and the personnel on duty of each working operation, and punishes the delinquent.

Table C.4   Placement surface construction quality inspection, test item division table

No.	Inspection item	Quality standards	Inspector
I	1. Truck flushing	No brought-in muddy water	Quality inspector, placement surface commander
	2. Placement surface clean	Free from sundries or grease dirt	
	3. Ponding, extraneous water	Ponding-free	
	4. Mortar, cement paste spreading	Uniform, no missed zone	
	5. Intermittent time between layers	Lower layer concrete not initially set	
II	6. Aggregate separation treatment	Dispersion treatment	Quality inspector, placement surface commander
	7. Leveling thickness	Height difference ±5 cm	
III.	8. Rolling layer surface	Flat, slight bleeding	Test center inspector
	9. Compacted volumetric weight	98%	
IV	10. Compaction times	As required	Quality inspector
	11. VC value	3 – 12 s	Test center inspector
V	12. Disposal of waste material	To be disposed as required	Quality inspector
	13. Heterogeneous concrete bonding	Conform to requirements	
VI	14. Rainy day construction	Measures in compliance with requirement	Position engineer
	15. Construction under high temperature and sunshine	Rain and sun protection measures	Quality inspector

C.12.5 Concrete Layer Surface Quality Defect Treatment

1. In case of bleeding on the placement surface, the individual bleeding locations occurred on the placement surface shall be collected and removed by manual digging, the place where the bleeding is more serious

shall be vibrated by high frequency vibrator along the bleeding channels.
2. For any poor layer surface bonding, poor bonding of heterogeneous concrete, poor bonding of concrete with bed rock, etc., it shall be rolled by vibration roller or vibrated by high frequency vibrator after paste is added at the junction.
3. The Quality Control Department shall inspect and evaluate the quality condition after the concrete surface quality defect is solved.
4. The full-time quality inspector shall carry out visual inspection to the concrete afterformwork removal and make a detailed record, ask the formwork construction team to eliminate the defect after it is observed in one day, the major defect shall be removed within one week. Whether the defect treatment meets the requirement or not shall be confirmed by the supervision engineer.

C.13 Management of Placement Surface after Rolling and Placement Finishing

C.13.1 Placement Cleanup

1. After the placement of RCC finishes, the placement surface commander is responsible for arranging the cleaning up of the placement surface to keep the placement surface in clean state.
2. The placement surface construction machinery and equipment is placed neatly and orderly, the materials and so on are stacked in order, enabling civilized construction of dam surface.

C.13.2 Curing

1. The curing is carried out by special curing worker during the construction, the measures such as mist spray, watering, plastic sheeting or jute bag covering, etc. are taken to keep the placement surface of the roller compacted concrete always wet.
2. The curing worker is responsible for starting the curing work immediately after the roller compacted concrete is finally rolled during the construction interval. For the horizontal construction layer surface, the curing work will continue until it starts to pave the upper layer of roller compacted concrete. For the permanent exposed surface, the surface should be cured for 28 d and more. The specially-assigned person is assigned to be responsible for the curing work of roller compacted concrete, the curing record shall also be made.
3. If the temperature difference of the internal and outside is more than 12 ℃ resulted from sudden temperature drop in winter construction, the temperature drop of the concrete surface is less than 2 ℃ or the air temperature drops to below 0 ℃ within 4 h, the RCC surface has to be covered by thermal insulation materials until the surface temperature of the concrete returns to more than 2 ℃.
4. The curing worker is responsible for protecting the concrete before acceptance of the roller compacted concrete to prevent any damage, special care shall be taken to the surface of the concrete to prevent any crack from producing due to sudden temperature drop.
5. In dry, hot or windy weather, the water evaporation is high, spray mist on the placement surface by high pressure water gun to keep the placement surface wet, in the interval of construction, the layer surface before final setting shall be covered by jute bags and spray water by high pressure water rotary nozzle for curing.

Main References

[1] SL 314—2004. *Design Specifications for RCC Dams*[S].
[2] SL 319—2005. *Design Specification for Concrete Gravity Dams*[S].
[3] DL 5108—1999. *Design Specification for Concrete Gravity Dams*[S].
[4] DL 5077—1997. *Specification for Load Design of Hydraulic Structures*[S].
[5] DL/T 5144—2001. *Specifications for Hydraulic Concrete Construction*[S].
[6] DL/T 5122—2000. *Hydraulic Specification for Hydraulic RCC*[S].
[7] DL/T 5112—2009. *Hydraulic Specification for Hydraulic RCC*[S].
[8] GB 175—2007. *Common Portland cement*[S].
[9] GB 200—2003. *Moderate heat Portland cement Low heat Portland cement Low heat Portland slag cement*[S].
[10] DL/T 5055—2007. *Technical Standard of Flyash Concrete for Hydraulic Structures*[S].
[11] DL/T 5100—1999. *Technical Code for Hydraulic Concrete Admixtures*[S].
[12] DL/T 5151—2001. *Test Code for Aggregates of Hydraulic Concrete*[S].
[13] DL/T 5150—2001. *Test Code for Aggregates of Hydraulic Concrete*[S].
[14] SL 352—2006. *Test Code for Hydraulic Concrete*[S].
[15] DL 5108—1999. *Design Specification for Concrete Gravity Dams*[S].
[16] DL/T 5112—2009. *Construction Specification for Hydraulic RCC*[S].
[17] DL/T 5144—2001. *Specifications for Hydraulic Concrete Construction*[S].
[18] SL 264—2001. *Specifications for Rock Tests in Water Conservancy and Hydroelectric Engineering*[S].
[19] SL 31—2003. *Specifications for Drilling Pump-in Tests in Water Conservancy and Hydroelectric Engineering*[S].
[20] Zhu Bofang. The History of Ending "All Dams Suffer Cracks" by *Comprehensive Temperature Control and Long-Term Thermal Insulation*[C]. Beijing:
[21] Lu Min'an. Study on Temperature Creep Stress of Baise RCC Gravity Dam[J]. *Guangxi Water Resources & Hydropower Engineering*, 2004(B05).
[22] Tian Yugong, Discussion on the Influence of Temperature, *VC* Value and Admixture on the Setting Time of RCC[J]. Guangxi Water Resources & Hydropower Engineering, 2004 Supplement.
[23] Tian Yugong and Gao Jusheng. Experimental study on fiber mixed RCC in impermeable zone[J]. *Guangxi Water Resources & Hydropower Engineering*, 2004 Supplement.
[24] Tian Yugong. Analysis on Characteristics of RCC Mix Proportion Design Test in China[J]. *China Water*, 2007(21).
[25] Tian Yugong. Discussion and analysis of RCC VC. Beijing. *Hydropower*, edition 2, 2007.
[26] Tian Yugong. Property Research and Application of RCC admixture. 20 *years of RCC dam in China-Big step from Kengkou dam to Longtan dam*[J]. Beijing: China WaterPower Press, 2006.
[27] Tian Yugong. The role of rock powder in RCC [J]. *Qinghai Hydropower*, 2006 (1).
[28] Yang Songling, Tian Yugong. Research and utilization of diabase manufacturerd sand and rock powder[J]. *People's Yangtze River*, 2007, 38(1).
[29] Tian Yugong. Evaluation and Analysis of Ten Innovative Technologies for RCC Dam Construction. [J]. *Water Conservancy and Hydropower Construction*, 2008(4).
[30] Li Linghong, Tian Yugong, Jin Anqiao. Study on characteristics of RCC with basalt aggregate for hydropower station. [J]. *Water Conservancy and Hydropower Technology*, 2009(5).
[31] China Three Gorges Projects Development Co., Ltd. *Implementation Guidance Materials for Specifications for Hydraulic Concrete Construction*[M]. Beijing: China Electric Power Press, 2003.
[32] RCC Damming-Design and Construction. *RCC Promotion Leading Group of the Ministry of Energy of the People's Republic of China and the Ministry of Water Resources of the People's Republic of China*[M]. Beijing: Publishing House of Electronics

Industry, 1990.

[33] Sun Gongyao, Wang Sanyi, *Feng Shurong. High RCC Gravity Dams*[M]. Beijing: China Electric Power Press, 2004.

[34] Jia Jinsheng, Chen Gaixin, Ma Fengling, et al. *Development Level and Project Cases of RCC Dams*[M]. Beijing: China Water & Power Press, 2006.

[35] Gezhouba Joint Venture, No. 8 Bureau and No. 7 Bureau, Longtan Hydropower Station, Guangxi. *Construction and Management of RCC Gravity Dam of Longtan Hydropower Station*[M]. Beijing: China Water & Power Press, 2007.